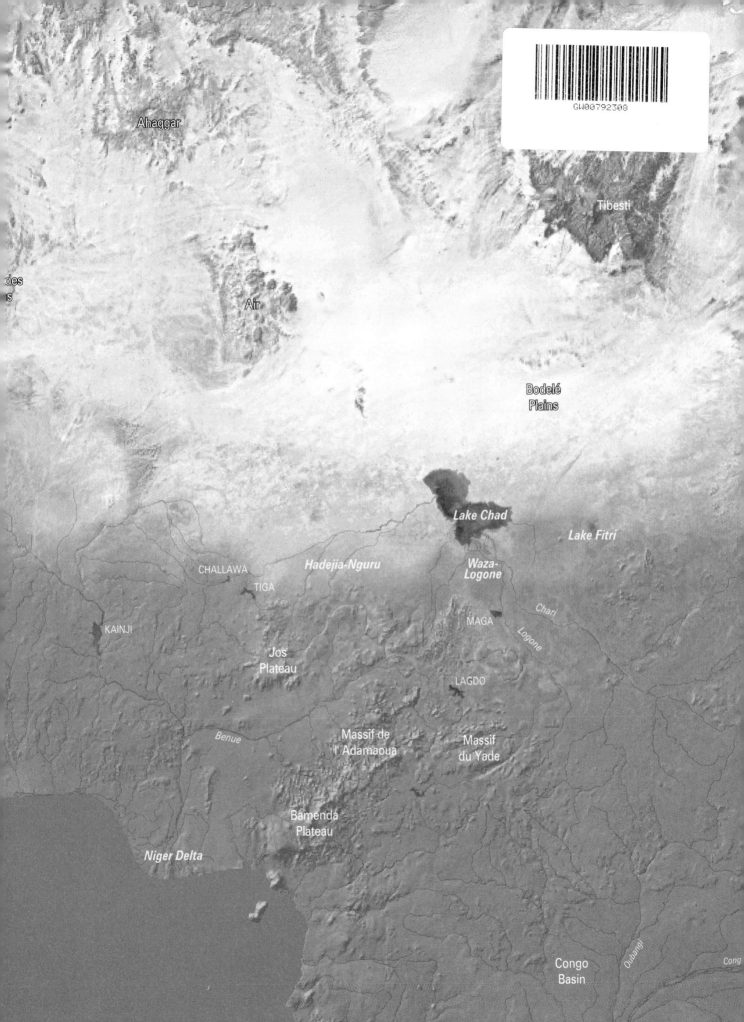

Living on the edge

Authors Leo Zwarts · Rob G. Bijlsma · Jan van der Kamp · Eddy Wymenga
Contributions Peter van Horssen · Ben J. Koks · Pierfrancesco Micheloni
· Wim C. Mullié · Otto Overdijk · Martin J.M. Poot · Paul Scholte · Vincent Schricke
· Christiane Trierweiler · Patrick Triplet · Jan van der Winden
Watercolour paintings and pen drawings Jos Zwarts
Diagrams and maps Dick Visser
Editorial support Mike Blair · Marie-Anne Martin
Graphic layout and design x-hoogte · Tilburg, with contribution of *Ju dith it*
Printer DZS Grafik d.o.o.

Publication of KNNV Publishing
© 2009 KNNV Publishing · Zeist · The Netherlands
2nd edition (reprint with minor corrections) · 2010
ISBN 978 90 5011 280 2
NUR: 942
www.knnvpublishing.nl

Discovering nature and getting close to it
KNNV Publishing specialises in unique works on nature and landscape: easily accessible field guides, conservation manuals, distribution atlases and much more. KNNV Publishing also produces chronicles on nature conservation, beautifully illustrated books on natural, cultural and landscape history, travel guides, children's books and, last but not least, the journal 'Natura'. All these works help to make valuable knowledge gathered by scientists and amateur researchers available to a broad public. By producing these works, KNNV Publishing contributes to nature conservation and to the enjoyment of nature in The Netherlands.

No part of this book may be reproduced in any form by print, photocopy, microfilm or any other means without the written permission from the publisher.

to be cited as:
Zwarts L., Bijlsma R.G., van der Kamp J. & Wymenga E. 2009. Living on the edge: Wetlands and birds in a changing Sahel. KNNV Publishing, Zeist, The Netherlands
Chapter 14: Mullié W.C. 2009. Birds, locusts and grasshoppers. In: Zwarts L., Bijlsma R.G., van der Kamp J. & Wymenga E. 2009. Living on the edge: Wetlands and birds in a changing Sahel. KNNV Publishing, Zeist, The Netherlands
Chapter 26: Trierweiler C. & Koks B.J. 2009. Montagu's Harrier *Circus pygargus*. In: Zwarts L., Bijlsma R.G., van der Kamp J. & Wymenga E. 2009. Living on the edge: Wetlands and birds in a changing Sahel. KNNV Publishing, Zeist, The Netherlands.

This publication was financially supported by
Altenburg & Wymenga ecological consultants · Feanwâlden
DLG Government Service for Land and Water Management · Utrecht
Dutch International Policy Programme on Biodiversity (BBI) · Den Haag
Rijkswaterstaat, Centre for Water Management · Lelystad
Vogelbescherming Nederland · Zeist

Technical and scientific support
Altenburg & Wymenga ecological consultants · Feanwâlden
Rijkswaterstaat, Centre for Water Management · Lelystad
Wetlands International - West Africa Office · Dakar · Sévaré · Bissau.

Living on the edge

Wetlands and birds in a changing Sahel

Leo Zwarts   Rob G. Bijlsma   Jan van der Kamp   Eddy Wymenga

KNNV Publishing

Contents

1. Introduction — 6

Sahel
2. Rainfall — 14
3. Rivers — 26
4. Vegetation — 36
5. Land use — 46

Wetlands
6. Inner Niger Delta — 60
7. Senegal Delta — 104
8. Hadejia-Nguru floodplains — 138
9. Lake Chad Basin — 144
10. The Sudd — 158
11. Rice fields — 170
12. Wetlands in the Sahel: a summary — 180

Birds
13. The Sahel as wintering area for Eurasian bird species — 188
14. Birds, locusts and grasshoppers **by Wim C. Mullié** — 202
15. Grey Heron *Ardea cinerea* — 224
16. Purple Heron *Ardea purpurea* — 228
17. Black-crowned Night Heron *Nycticorax nycticorax* — 234
18. Squacco Heron *Ardeola ralloides* — 240
19. Little Egret *Egretta garzetta* — 246
20. White Stork *Ciconia ciconia* — 252

21.	Glossy Ibis *Plegadis falcinellus*	266
22.	Northern Pintail *Anas acuta*	272
23.	Garganey *Anas querquedula*	278
24.	Osprey *Pandion haliaetus*	292
25.	Eurasian Marsh Harrier *Circus aeruginosus*	304
26.	Montagu's Harrier *Circus pygargus* **by C. Trierweiler & B.J. Koks**	312
27.	Black-tailed Godwit *Limosa limosa*	328
28.	Wood Sandpiper *Tringa glareola*	344
29.	Ruff *Philomachus pugnax*	350
30.	Gull-billed Tern *Gelochelidon nilotica*	364
31.	Caspian Tern *Sterna caspia*	370
32.	Eurasian Turtle Dove *Steptopelia turtur*	378
33.	Eurasian Wryneck *Jynx torquilla*	390
34.	Common Sand Martin *Riparia riparia*	396
35.	Barn Swallow *Hirundo rustica*	406
36.	Yellow Wagtail *Motacilla flava*	418
37.	Common Redstart *Phoenicurus phoenicurus*	426
38.	Sedge Warbler *Acrocephalus schoenobaenus*	436
39.	European Reed Warbler *Acrocephalus scirpaceus*	444
40.	Lesser Whitethroat *Sylvia curruca*	450
41.	Common Whitethroat *Sylvia communis*	456
42.	Crossing the desert	464
43.	Carry-over effects of Sahel drought on reproduction	472
44.	Impact of the Sahel on Eurasian bird population trends	480
	References	504
	Index	545

Introduction

Our pirogue steadily weaves its way through the Inner Niger Delta. It is late November 2008. We enjoy the croaking song of Great Reed Warblers. The flood has been high this year, promising waterbirds aplenty. The totals of our count stand at 380 000 Cattle Egrets (more than ever before in the past 20 years), 55 000 African Cormorants (fully recovered from a still puzzling dip in the population, when thousands were dying on the breeding colonies in 2004) and 6000 Glossy Ibis (halting the long-term decline?). What better setting than this impressive floodplain in Mali in which to proof-check a book intended to address the huge subject of Eurasian breeding birds and their relationship with the Sahel. Mixed feelings are the order of the day: reading brings relief (it's done), but raises nagging unresolved doubts as new subjects present themselves – should some be incorporated and should we defer the book's completion until exciting new lines of research have been followed? Mostly, relief predominates. How much data collected over the years have been lost, simply because they remain in notebooks, unpublished? We have been fortunate in that "we stood on the shoulders of giants" - those who did publish - first and foremost of whom was Reginald Ernest Moreau. He published many papers on African ecology, including two books: *The Bird Faunas of Africa and its Islands* (1966) and *The Palaearctic-African Bird Migration Systems* (1972). The latter was completed on his deathbed.

Of the more than 500 bird species breeding in Europe, some 1300 – 2600 million pairs altogether, about a quarter migrate south to sub-Saharan Africa. These migrants spend the greater part of their life in Africa and return north to breed. Take, for example, European Honey Buzzards: being present on the breeding grounds between mid-May and mid-September, or 120 days, they need 105-110 days to complete a full breeding cycle, a narrow 'window', with a maximum of two weeks' adjustment possible. This is such a tight schedule that many refrain from breeding unless their arrival on the breeding grounds is timely and conditions favourable. On an annual basis, two-thirds of their time is spent on migration and in Africa, this proportion being even greater when taking into account the juvenile and non-breeding phases. Juveniles spend the first and often second years of their lives in Africa; adult birds may skip one or more breeding seasons and remain in Africa during the northern summer. For such long-lived birds, it is easy to understand that, for survival, Africa truly matters. However, the same applies to short-lived passerines, many of which may make but a single trip to Africa, only to die before they have a chance to fly back to the breeding grounds and pass on their genes. The survivors have to carry home the burdens of their African stay, effectively extending Africa's influence into their breeding performance. For a Spotted Flycatcher, the "endearing ultimate *persona*" in Moreau's 1972 treatise, to raise two broods in a single season after having completed a 7000 km flight, is surely an extraordinary feat.

The Palearctic-African bird migration system draws birds from the geographical range between 10°W (Ireland) and 164°E (Kolyma Basin), or even further east. The long-distance migrants in this vast region, apart from those wintering on the Indian subcontinent or in SE Asia, pour into sub-Saharan Africa, not just scattered across the continent, but in the main amassing in the northern savannas and avoiding the equatorial woodlands. The savannas and woody savannas of the Sahel and Sudan-Guinea zone (or what remains of it) between 8° and 18°N run the entire width of Africa from Senegambia and Mauritania in the west to Sudan, Ethiopia and Somalia in the east. The northern savannas are dry during the northern winter, the last rains normally falling in September or October as the Intertropical Convergence Zone retreats. From then on, conditions become drier by the day, and only after most of the surviving Palearctic birds have vacated the region in spring do the first new rains arrive. This counterintuitive finding, that the highest densities of Palearctic birds are to be found in the most desiccated vegetated region of the northern tropics where there is little prospect of rainfall during their stay, gave rise to Moreau's Paradox: "How can enormous numbers of birds and indeed the majority of all migrant passerines and non-passerines (other than raptors and water-birds) support themselves and put on migratory fat they need for the spring journey, on a food supply that *prima facie* they and the native birds have been eating down with little or no replenishment for some six months?" (Moreau 1972). Gérard Morel (1973), who was intimately familiar with Sahelian conditions in the lower Senegal Valley, answered that question: the paradox is an apparent paradox. What to the observer appears superficially as poor and desiccated habitat, is actually a complex *Acacia-* and *Balanites*-dominated ecosystem, where seasonal variations in leafing, flowering and fruiting of trees and shrubs provide food and shelter for insects and birds throughout the northern winter. This beneficence occurs even when the once-a-year production of grass, seeds and invertebrates dwindles after the rains.

It may come as a surprise that in a region so close to the Sahara, there are four huge Sahelian wetlands, *i.e.* the Senegal Delta, the Inner Niger Delta, Lake Chad and the Sudd. In his 1972 review, Moreau hinted that these wetlands played a pivotal role in the lives of millions of Palearctic birds, but he could substantiate that only for the Senegal Delta. We hope our book will reveal the Inner Niger Delta in its full glory as the recipient of millions of Palearctic migrants. Lake Chad and the Waza Logone have also received much attention in the past decade or so, but our knowledge for that region is still incomplete. In Moreau's time, the Sudd was a complete void in terms of knowledge, and little has changed.

In the early genesis of the book, we felt that our data needed a broader perspective than just bird studies in the Sahel. Coupling the vicissitudes of Eurasian birds with sub-Saharan conditions demanded the inclusion of ring recovery data, especially because we could outline migratory connectivity (the tendency with which individuals of a given breeding area migrate to the same wintering area) and 'winter' mortality. By 'winter', we mean the northern hemisphere winter season, very much a European-centred view, although many migrants spend much of the non-breeding season in the southern hemisphere summer. We used a selection of ring recoveries made available to us by EURING, the co-ordinating organisation for European bird-ringing schemes (hosted by the British Trust for Ornithology), our selection embracing recoveries in Africa below 20°N and species for which sufficient material was available.

Our main focus has been the ecology of Palearctic birds in the Sahel. Life and death here are closely linked to rainfall and river discharges. Good rains spell high river flows, large floods in the inundation zones, high productivity, green vegetation and an abundance of insects and seeds, providing the wherewithal for survival in a harsh environment. However, rainfall is unpredictable in the region. Periods of abundant rains alternate with droughts, whose frequency and duration have increased since 1969. For decades now, rainfall in the Sahel has been below the long-term average of the 20th century. The resounding adverse impact of this apparent regime shift on birds has been reinforced by a suite of man-made changes in the Sahel. We have endeavoured to clarify the extent to which Sahelian conditions affect Palearctic bird populations and to relate that to the generally negative changes in conditions on the breeding grounds. Our endeavour is centred within the Sahel, but we felt free to expand our interest into Africa as a whole when relevance demanded. To mention only the Sahel in the context of migrant birds' life-cycles without bringing the Sudan and Guinea zones and the Sahara into the argument would be to omit the existence of an immensely intricate and interrelated ecosystem. Birds are wonderful creatures, governed by a mixture of evolutionary legacies and local conditions, yet many are highly adaptable. We feel priviliged to have observed 'our' birds in their African homes. Watching birds is a pleasure in itself, but doing so in Africa brings joy beyond measure.

Acknowledgements

This book is a synthesis of the multifarious data collected over decades of hard-core fieldwork from Asia, Europe and Africa. Thousands of people have been involved. Each data point in the graphs represents untold hours in the field and behind the desk. Each graph is a stylised commemoration of hardship and joy. Many of the people involved appear in the list of references, others are mentioned in the text, but the majority belong to the innumerable unrecorded helpers and volunteers out there. We have no doubt that they, as we did, applied the Hamerstrom Rule of Thirds: spend one

We regard Reg Moreau as the forefather of African bird ecology. Lacking scientific training, Moreau nevertheless made hugely important contributions to several fields of ornithology, all from a relatively late start at 23 years of age when he went, partly for health reasons, to Cairo. In his words, he was, "about as ignorant and unformed as I had been half-a-dozen years before", but this state changed rapidly. In 1966, David Lack suggested that Moreau should compile an autobiographical sketch (Moreau et al. 1970). In it, Moreau downplays his own role in African ornithology by suggesting that 'I just happened to be in the right place at the right time', not once but several times. For an "Opsimath as I am proud to reckon myself", his achievements are remarkable. Although some of his ideas have since been refuted, others have stood the test of time. Much research, including our own, over the intervening decades pays tribute to the ideas aired in *The Palaearctic-African Bird Migration Systems*. "No bum swan-song", indeed. (Photo by courtesy of: The Alexander Library, Oxford University, Library Services).

third of your time on mostly worthless red tape, one third on what is wanted of you, and one third on what you please (Hamerstrom 1994). This is not a formula of the lazy, but rather one that applies to those of dedicated application and meticulous attention to detail; relaxation is necessary for sanity! During our quest we did indeed encounter an amazing number of people who had that splendid character. We are indebted to them all.

The original plan for this book was envisaged in Mali in April 2002. As is usually the case, the best ideas are born when unburdened by everyday routine. As we were discussing the prospects of survival of European birds in Africa and the knock-on effect on breeding bird populations, we came to the realisation, in agreement with Vincent van den Berk, that a review was badly needed. It took much longer for him to convince us that we should undertake the task! Much as a skilled angler pits his wits against the fish, he put down the groundbait first: think of the opportunity this presents of continuing your field work across West Africa for another couple of years! The hook of course was the proposal that we should write the book concomitantly.

The result is a book that has built on the work of others, and comprises the field studies we have undertaken mostly in West Africa but also elsewhere on the continent, between 1982 and early 2009. These studies began as low-budget affairs, partly financed by the participants themselves, but have since developed into larger-scale projects encompassing Mali, Mauritania, Senegal, The Gambia and Guinea-Bissau. Paid, underpaid or unpaid, the passion remained the same.

The help of Wetlands International in Senegal, Mali, Bissau and The Netherlands, DNCN (Mali), RWS-RIZA (The Netherlands, till 2004), Alterra (The Netherlands, till 2002) and Altenburg & Wymenga ecological consultants (The Netherlands) paved much of our way in West Africa. In Mali, the help and company of Bakary Kone, Bouba Fofana, Mori Diallo and Sine Konta (Wetlands International, Se-

Male Montagu's Harrier (above), Little Egrets near fish nets (below)

African Swallow-tailed Kite. Its graceful flight is a wonder to behold.

vare) was much appreciated; their knowledge of the people, plants and wildlife along the Niger and in the Inundation Zone put us in good stead during our surveys. Willem van Manen was a great companion during the bird density counts and other escapades, during which we indulged in our shared passion for raptors. We made the most of his two field stints by using his tree climbing prowess to check nests of Yellow-billed Kites in the Inner Niger Delta (Bijlsma et al. 2005). Two other appreciated participants in Mali involved in density counts were Allix Brenninkmeijer and Marcel Kersten.

In Senegal, Seydina Issa Sylla of the Dakar office of Wetlands International and his team offered a welcome base and logistical help. We are greatly indebted to Mme Dagou Diop, who introduced us to the Senegal Delta and was instrumental in bringing together existing information related to the delta. We would like to thank the *Organisation pour la Mise en Valeur du fleuve Sénégal* for making hydrological information available, the *Centre de Suivi Écologique* (CSE; Dethié Soumaré Ndiaye) for map information, the *Direction des Parc Nationaux* (in particular Jacques Peeters) for information concerning the ecology of the Senegal Delta and finally the *Société Nationale d'Aménagement et d'Exploitation des terres du Delta du fleuve Sénégal et des Vallées du Fleuve Sénégal et de la Famille* (SAED) for their cooperation, in particular Abdou Dia, Adama Fily Bouso and Landing Mandé, who assisted in spatial data processing. Mme Dagou Diop also introduced us to the management teams and inhabitants of the Djoudj National Park and the Diawling National Park. We vividly recall the animated discussions regarding the delta's socio-economic developments. For their open-mindedness and information on the national parks' management we acknowledge: Issa Sidibe, the director (until 2007) of the Djoudj NP, Idrissa Ndiaye (consultant Djoudj NP, Senegal); Moctar ould Daddah and Zein el Abidine ould Sidaty (both Diawling NP, Mauritania). Cheikh Hammalah Diagana, Abdoulaye Diop, Ibrahima Diop, Abdoulaye Faye, Idrissa Ndiaye, Hassane Ndiou, Ousman Sane, Mamadou So and Fousseini Traore also assisted in the fieldwork in the Senegal Delta and in the Casamance. The amicable and cooperative support of Joãozinho Sá (Wetlands International) made working in Guinea-Bissau a pleasure. Hamilton Monteiro cheerfully tramped the rice fields with us in Guinea-Bissau, as did Kawsu Jammeh in The Gambia and in the Casamance. Their friendly quizzing of the local populace greatly facilitated our understanding of the system of the *bolanhas*.

We were fortunate to have dealt with Chris du Feu as the co-ordinator of the EURING database, who complied promptly and with great kindness with our requests. So far, this source of information has been only lightly tapped, and notwithstanding the advent of spectacular new methods of studying migration patterns involving radio transmitters, satellite telemetry and stable isotope ratio techniques, it begs for much more elaborate analyses than our basic attempts.

Tapping a very different source of data, *i.e.* the ornithological literature of Russia and other countries of the former Soviet Union, the hidden treasures revealed demand a special word of thanks to Jevgeni Shergalin, who translated so many relevant documents. We have also been fortunate to have experienced the benefits from the

Inner Niger Delta at a high flood. (Kakagnan, November 2008).

first era in which digital photography became commonplace, and so we have been able to select from a wide choice of material collected throughout the Afro-Palearctic region. This book contains a selection of photographs; each with a story, rather than plates chosen just for their beauty (although combining the two is possible, as we show). The unselfish support of photographers, even to the extent of delving in decades-old archives, digitising old slides and browsing through dust-covered notebooks to unearth relevant data, gave us the opportunity to enliven and expand the text beyond mere science. For those without first-hand experience of Africa, it may perhaps be hard to fully comprehend the sweltering heat and lack of shade in the Sahara, or to appreciate the changes wrought on the vegetation by the rains that can turn a withered and dust-laden Sahel into a green 'paradise' in little more than a few days. A single glance at the photographs taken by Wilfried Haas and Gray Tappan suffices to make that point. Our selection of pictures should help bring to life in your mind the African and Eurasian environments and their inhabitants. We are enormously grateful to all who supplied material to that end. The Alexander Library Service of the Oxford University kindly gave permission to reproduce a picture of Reg Moreau.

The water-colours and other drawings by Jos Zwarts are an evocation of the birds in their natural setting, Sahelian or otherwise. Intended to be equally pleasing to the eye are Dick Visser's graphics that condense huge amounts of information in a clear and straightforward manner. We are truly honoured to have such a lavish collection of their art throughout the book. Fred Hustings produced an adaptation of his famous (in The Netherlands) cartoon of the real monitoring spirit, to align with the essence of our book. He also read the book from cover to cover, and we profited from his comments and attention to detail.

The twists, turns and pertinence of the many tales in this book, including the asides and addenda in endnotes and figure and photo captions, have been spun together, to our delight, from the expertise, data (both published and unpublished), personal libraries, comments and encouragement of (in alphabetical order) Joost Backx, Albert Beintema, Mike Blair, Anne-Marie Blomert, Gerard Boere, Daan Bos, Christiaan Both, Bennie van den Brink, Joost Brouwer, Nigel Collar, Floris Deodatus, Gerard van Dijk, Hans Drost, Michael Dvorak, Sara Eelman, Martin Flade, Bart Fokkens, Gerrit Gerritsen, Kees Goudswaard, Matthieu Guillemain, Manfred Hölker, Hermann Hötker, Jos Hooijmeijer, Fred Hustings, Joop Jukema, Guido Keijl, Roos Kentie, Kees Koffijberg, Pertti Koskimies, Jim van Laar, Karl-Heinz Loske, Chris Magin, Willem van Manen, Sven Mathiasson, Peter Meininger, Johannes Melter, Pierfrancesco Micheloni, Theunis Piersma, Steven Piper, Maria Quist, Jiri Reif, Magda Remisiewicz, Riche Rowe, Geoske Sanders, Vincent Schricke, Guus Schutjes, H. Stel, Torsten Stjernberg, Ekko Smith, Roine Strandberg, Tibor Szép, Dirk Tanger, Simon Thomsett, Ludwik Tomiałojć, Patrick Triplet, Bertrand Trolliet, Yvonne Verkuil, Jan Visser, Jan Wanink, Jan Wijmenga, Mike Wilson and Dedjer Wymenga; their support meant much to us.

Several institutions and organisations kindly allowed us the use of their data: Finnish Ringing Centre, Helsinki (Jari Valkama),

Sahel landscape NW of Timbuktu at 16°N (175 mm rain per year; below), north of Niamey at 14°N (350 mm rain per year; left) and Sudan landscape at 12°N in the W National Park (700 mm rain per year; right). (January 2009).

Gannet Flemming Inc., (Frank J. Swit), SOVON Vogelonderzoek Nederland, Beek-Ubbergen (Chris van Turnhout), Swedish Bird Ringing Centre, Stockholm (Roland Staav), Tour du Valat, Le Sambuc (Patrick Gillais), Vogeltrekstation, Heteren (Hans Schekkerman), Vogelwarte Helgoland, Wilhelmshaven (Olaf Geiter), Vogelwarte Radolfzell, Radolfzell (Wolfgang Fiedler) and Wetlands International, Wageningen (Ward Hagemeijer).

Throughout the years we enjoyed and benefited from the logistic, administrative and scientific support of the staff of Altenburg & Wymenga ecological consultants, notably Hieke van den Akker, Wibe Altenburg, Daan Bos, Leo Bruinzeel, Lucien Davids, Harmanna Groothof and Franske Hoekema. We should like to give especial thanks to the specialists who contributed texts, adding themes and discussions to an already widely branching tree of knowledge: Peter van Horssen, Ben Koks, Pierfrancesco Micheloni, Wim Mullié, Otto Overdijk, Martin Poot, Paul Scholte, Vincent Schricke, Christiane Trierweiler, Patrick Triplet and Jan van der Winden.

Story telling in Africa is an oral tradition, embedded in music, chants and poetry. We, on the other hand, have been taught to place great value on written sources. To successfully convey the message, a language is needed, honed to perfection to suit the intended audience. We are immensely indebted to Mike Blair, who volunteered to edit our English, but who, at the same time, took us back to the school-benches, pointed out idiosyncrasies, inconsistencies and plain errors, never tired of repeating lessons, provided extra information yet still kept his humour – a humbling and gratifying experience. Similarly, Marie-Anne Martin expertly fine-tuned the final versions, also accompanied by comments and asides covering life in general and the book in particular. This may sound like a platitude, but how very true it is nonetheless: any mistakes left are entirely our own.

Financial support was given by Vogelbescherming Nederland, Vereniging Natuurmonumenten, the Dutch Ministry of Foreign Affairs (Directorate for International Cooperation), the Dutch International Policy Programme on Biodiversity (BBI), the DLG Government Service for Land and Water Management, the Dutch Partners for Water Program, Altenburg & Wymenga ecological consultants and, in particular, by the Dutch Ministry of Agriculture, Nature and Food Quality (Department of Nature) and the Dutch Ministry of Transport, Public Works and Water Management (RIZA; since 2007 Rijkswaterstaat Centre for Water Management). Migratory birds constitute 23% of the Dutch avifauna, and the support was motivated by the realisation that science-based protection needs information on the birds' whereabouts and ecology in the non-breeding areas. Without their generous input, this book would neither have been written, nor published.

Photography credits (l=left, r=right, a=above, c=centre, b=below)

Wibe Altenburg • 69a, 432-3
Corstiaan Beeke/Buiten-Beeld • 273l
Albert Beintema • 32b, 70a, 71, 106
Anne-Marie Blomert • 295
Bennie van den Brink • 143, 165, 166, 167al, 184, 407, 410, 413, 415, 417, 473
Frank Cattoir/Buiten-Beeld • 236
Floris Deodatus • 14, 27, 39r, 40a, 46, 55bl, 58r, 59, 150r, 256
Nicolas Gaidet • 286, 287
Hans Gebuis/Buiten-Beeld • 267
Gerrit Gerritsen • 32al, 151b, 354r, 392r
Wim de Groot/Buiten-beeld • 158
Wilfried Haas • 258, 259, 380-1, 467, 468-9
Johan Hanko/Buiten-beeld • 255a
Fred Hustings • 387
Hans Hut • 18, 44, 45, 51, 53ar, 89, 104, 127, 129, 130l, 131, 136, 138, 157, 197, 198 (Kittlitz), 206, 207, 208, 209, 213, 214, 215, 219, 253b, 255b, 257, 271, 293, 298, 308, 315al, 316-7, 319l, 321, 322, 325, 349, 400, 421, 434b, 448, 453, 458, 471r, 484, 488l, 491
Jan van de Kam • 70b, 87, 90, 93al+ar, 103, 114, 141t, 153a, 167ar, 168, 178a+c, 183r, 185, 191l, 197br, 198, 199 (except Kittlitz+eagle), 226, 227l, 248, 249, 251, 260, 263, 268, 269, 276, 279, 282, 290, 307r, 331, 342, 346, 347, 348, 351a, 353, 363, 365, 367a, 371l, 374, 392l, 394, 403, 416, 440, 476, 501
Jan van der Kamp • 105, 108bl, 130r, 174, 197b, 299
Johannes Klapwijk/Buiten-beeld • 227r
Benny Klazenga • 180, 273r, 309, 336, 351l+r, 424, 425, 431, 442, 477, 478
Alexander Koenders/Buiten-beeld • 384
Dries Kuijper • 118al
Ben Koks • 9a, 10, 12a, 13a, 38r, 39l, 55a, 210, 307l, 314a, 315r+b, 318a, 319r, 320a, 488r
Alexander Kozulin • 361
Michel Lecoq • 204l
Willem van Manen • 77ar
Wim Mullié • 16r, 17r, 36, 40c, 53al, 110-1, 116-7, 124, 125, 126, 140, 167c+cr+b, 172br, 203, 204r, 211, 218, 220-1, 238, 241, 242, 243, 297, 320b, 371r, 376, 465, 471l, 479
Paul Scholte • 32ar, 58l+c, 144, 145, 150l, 151a, 152, 156
Guus Schutjes • 28l, 200, 201
Leen Smits • 318b
Onno Steendam • 310
Dirk Tanger • 475
Gray Tappan • 20, 28r, 30b, 42
Christiane Trierweiler • 314bl+br
Patrick Triplet • 123l
Larry Vaughan • 217
Frank Willems • 253a
Jan Wijmenga • 40b, 75, 77al, 101a, 102, 153c+b, 191r, 198 (eagle), 330b, 358
Eddy Wymenga • 108a+bc, 109, 118rb, 119, 120, 123r, 132, 135b, 435r
Sergey Zuyonak /APB-BirdLife Belarus • 135a
Leo Zwarts • 6, 9b, 11, 12-13b, 16l, 17l, 24, 26, 30a, 34, 35, 38l, 43, 47, 48, 50, 53bl+br, 55bc+br, 56, 60, 63, 64, 65, 66, 67, 69b, 72, 73, 74, 76, 77b, 79, 81, 84, 92, 93b, 99, 101b, 114b, 148, 162, 170, 172a+l, 173, 175, 177, 178b, 183l, 222, 225, 231, 244, 245, 250, 274-5, 285, 301, 202, 303, 306, 330a, 333, 354l, 359, 367b, 382-3, 388, 398, 401, 404-5, 420, 422, 428, 434al+ar, 435l, 438-9, 446, 455, 459, 460-1, 463, 481, 490, 493, 496, 498-9, 503

Rainfall

When travelling from southern Mauritania to Guinea-Bissau, a distance of just 500 km, you find that the scenery changes from dry desert daubed in reds and ochres into a lush, green world of bushes and trees. This step-by-step transition makes a deep impression. Bare sand dunes give way to dry, almost treeless grassland that in turn evolves into sparsely forested plains. Further south, tree density increases and eventually changes into dense gallery forest, which in the not too distant past consisted of tropical rainforest. For each degree of latitude south (*i.e.* 110 km) between Mauritania and Guinea, rainfall increases by about 45%, the amount effectively doubling every 160 km. Apart from the world's mountainous areas, it is difficult to find such a degree of change elsewhere. The sand dunes in Mauritania are part of the southern edge of the world's largest desert, the Sahara (8.5 million km^2, comprising 25% of the African continent). The transition zone south of the Sahara is known as the Sahel, a huge area of 3 million km^2, a strip some 500 km wide, stretching 5500 km from the Atlantic coast to the Red Sea.

Introduction

The steep increase in rainfall over a latitudinal range of only 500 km, from less than 100 mm annually in the north to more than 2000 mm in the south (Fig. 1), is due to the Intertropical Convergence Zone (ITCZ), a belt of low pressure hugging the equator. Ascending warm and moist air immediately north and south of the equator is sucked into the ITCZ and transported at altitudes of 10-15 km further north and south of the equator. To compensate for the rising air in the convergence zone, the northern flow descends in the desert zone, normally centred between 20° and 30°N. Descending air heats up as the pressure increases, becoming undersaturated with water vapour and leading to the typical clear skies and general aridity of the Sahara. This air circulation system, known as the Hadley cell, is affected by the rotation of the earth, which ensures that the prevailing wind over the Sahara, the *Harmattan*, always blows from the northeast and not from the north. The *Harmattan*, a well-known phenomenon in West Africa, brings dry, dusty air to the Sahel and further south. When, during the northern summer, the sun is overhead in the Sahara, a low pressure belt forms over the Sahel, bringing clouds, rain, frequent thunderstorms and a monsoon from the SE.

The ITCZ moves seasonally, according to the position of the sun, between the Tropic of Cancer (21 June) and the Tropic of Capricorn (21 December).[1] The northern limit of expected regular rainfall in the Sahel is defined by the position of the ITCZ (with rain-bearing clouds usually lying 150-200 km south of it). The ITCZ, being the dominant rain-making mechanism over Africa, explains why northern and southern Africa are dry, and why central Africa is wet (Fig. 2). It also explains why rain occurs in the Sahel during the northern summer, and in southern Africa six months later (Fig. 4).

Although the tropical rain belt movement annually cycles across the equator and back again, there are year-on-year differences. These differences cause large variations in local rainfall (Fig. 1). The Sahel not only experiences unparallelled rapid transitions from dry to wet, but also has suffered a historically unprecedented decline in rainfall since about 1960. According to Hulme (1996), it is the most dramatic example of climate variability ever measured. Much research has been done to understand what lies at the heart of this decline.

The most obvious hypothesis postulated that the reduced rainfall in the Sahel was due to a southward shift of the entire ITCZ. Indeed, the ITCZ remained further to the south in 1972, the first extremely dry Sahel year of the Great Drought between 1972 and 1992 (Fontaine & Janicot 1996). However, when long series of dry and wet years are compared, a southward shift of the ITCZ could not explain the ongoing decline of rainfall in the Sahel – there was no commensurate increase in rainfall south of the equator. Instead, it appeared that the action of the Hadley cell had become less intense (Nicholson 1981a). Further research confirmed that the atmospheric circulation in the Sahel simply had varied between dry and wet years (Nicholson & Grist 2003). This meant that a new hypothesis was needed to explain the changing atmospheric circulation. Since 1977 it has been understood that rainfall in Africa largely depends on the sea surface temperature (SST) of the oceans, although the relationship is complex (Hastenrath & Lamb 1977) and remains so 30 years later (Balas *et al.* 2007).[2] Drought in the Sahel occurs during years of relatively warm equatorial oceans and relatively low temperatures in the subtropical oceans. Given this complexity, it is not surprising that it took more than 25 years of intensive research to begin to understand the relationship between Sahel rainfall and the temperature sequences in oceans on both sides of the continent.

The first studies on the relationship between seawater temperature and Sahelian rainfall could only partly explain the recent drought in West Africa. An alternative explanation for changes in rainfall patterns was the rapid changes in land usage and vegetation cover in the Sahel and adjacent vegetation zones. For example, in West Africa, almost 90% of the original moist forest has disappeared, and what remains is heavily fragmented and degraded

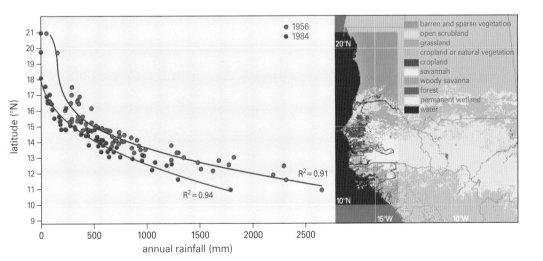

Fig. 1 Annual rainfall at 67 sites in Mauritania, Senegal, Gambia, Guinea-Bissau and Guinea in an extremely wet (1956, blue) and an extremely dry year (1984, red) at different latitudes within the rectangle (10-21°N and 13.5°-17.5°W) shown on the map. The vegetation map is based on Fig. 22.

The belt between the Sahara (here exemplified by southern Mauritania) and the gallery forests in southern Senegal (September 2007), a distance of only 500 km, covers a surprising suite of habitats, including the Sahel.

(Hennig 2006; Chapter 5). It is not certain, however, that this has contributed to the decline in the amount of rainfall. Studies on the other side of the Atlantic Ocean revealed that large-scale deforestation in the Amazon basin has not been followed by a decline in rainfall, but rather by an increase locally (Chu et al. 1994, Negri et al. 2004).

Another explanation put forward was that climate change had apparently caused an advance of the Sahara desert into the Sahel (Nicholson et al. 1998). The United Nations Convention for Combat Desertification (UNCCD) suggested this process had been accelerated by 'desertification', i.e. 'land degradation in arid, semi-arid, and dry sub-humid areas resulting from various factors, including climatic variations and human activities'. Charney (1975) hypothesised that this loss of vegetation caused a change in the energy flux between surface and atmosphere. Later research showed that the change of climate was unlikely to be *triggered* by desertification itself (Nicholson 2000). This does not imply, however, that surface processes do

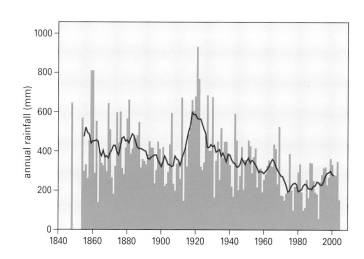

Fig. 2 Average annual rainfall (mm) in Africa calculated over the years 1995-2005. Data, taken from http://www.fews.net, were collected by CPC-NOAA.

Fig. 3 Annual rainfall in Saint Louis, Senegal, from 1848 to 2005. The line gives the running mean over 9 years.

Fig. 4 Shift of the rain zone between northern and southern Africa during the course of the year. The figures show the rainfall in 2005, selecting for each month the rainfall during the middle ten days. The colours (white→brown→ light green→dark green→light blue→dark blue→light red→dark red) indicate the rain intensity. Data, taken from http://www.fews.net, were collected by CPC-NOAA.

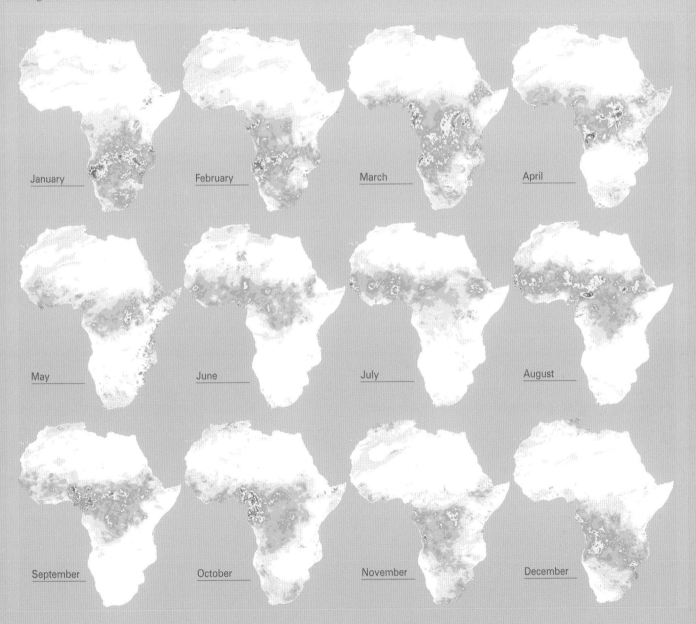

Rainfall in West Africa 17

Baobab trees belong to the Sahelian landscape; mid-Senegal, January 2008.

not play a role in *maintaining* the drought. Wang *et al.* (2004) and Giannini *et al.* (2005) actually concluded that vegetation dynamics have enhanced the severity of the drought. The regional climate may also have been affected by dust in the atmosphere (Tegen & Fung 1995). During the drought in the 1970s and 1980s, the number of dust storms in the Sahel increased sharply (Prospero *et al.* 2002). Because recent climatological models are better equipped to capture the variation in the Sahel rainfall in conjunction with ocean temperatures, the significance attributed to regional effects has diminished.

What, then, determines the variation in surface temperature of the ocean? If sea temperature is so important, what has changed in the oceans to explain enduring droughts in the Sahel? Furthermore, if the Sahel drought has a bearing on greenhouse gases and global warming, what will the future bring? We will return to these questions at the end of this chapter, but first, we describe the variations in rainfall in the Sahel and analyse whether regional differences in the annual rainfall within the Sahel zone can be discerned.

Data

The longest series of daily measurements of rainfall in the Sahel is available from Senegal for Saint Louis, whose annual rainfall, registered since 1848, fluctuated more recently between 937 mm in 1921 and 59 mm in 1992 (Fig. 3). This huge variation nevertheless is consistent with a downward trend over the last 150 years. The rainy period in the northern Sahel is limited to a few months, typically punctuated by local downpours and tropical thunderstorms causing such local variation in daily rainfall that the tallies of adjacent rainfall stations can show remarkable differences over the entire rainy season. This local variation is nicely quantified by Taupin (1997, 2003), using 25 rain gauges in one km^2 and 98 gauges over 16 000 km^2, respectively.

The existence of such large variations requires many rainfall stations to record adequately the annual variation of rainfall over the entire Sahel. In the 19th century, there were but a few operational rainfall stations, initially only along the Atlantic coast (*e.g.* Lungi, Sierra Leone, since 1875 and Banjul, The Gambia, since 1884), but later also inland (*e.g.* Timbuktu, Mali, since 1897). From about 1920, rainfall was measured over the entire Sahel, the number of rainfall stations reaching a peak between about 1950 and 1990, but decreasing since then.

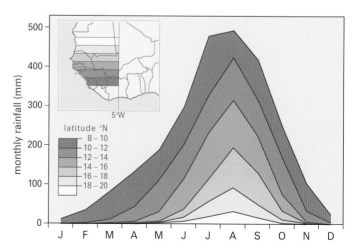

Fig. 5 The seasonal monthly rainfall in the western Sahel between the Atlantic coast and 5°W, given separately for six latitudes.

We selected weather stations from Mauritania, Senegal, Guinea-Bissau, Guinea, Sierra Leone, Mali, Burkina Faso, Niger and Chad, and also from the northern regions of Ivory Coast, Benin, Ghana, Nigeria, Cameroon and the Central African Republic. The data used here were taken mainly from the data set produced by FAOCLIM.[3]

A serious problem with using rainfall data from the Sahel is the large, and increasing, number of missing values. The simplest solution is to restrict the analysis to the few months covering the rainy season. In fact, some papers do use rainfall only from August or from July, August and September. August contributes, on average, 42% to the total annual rainfall in the Saharan margin (18-20°N), but represents only 31% in the southernmost Sahel zone (12-14°N) and a mere 20% further south (8-10°N) (Fig. 5; see also Nicholson 2005 for comparable data for the Sahel west of 15°E). When July, August and September are combined, we capture 85% of the annual rainfall in the northern Sahel (14-20°N), decreasing to 76% in the southern Sahel and 55% in the 8-10°N zone. We decided to include data only when there were no missing values in the wet part of the year, *i.e.* from May until October. Missing values in the six dry months do not significantly affect the summed annual precipitation.

Missing data in series of average annual rainfall tallies, based on a set of different stations, corrupt the results in analyses. This problem is usually resolved by standardising the data for all rainfall stations. First, the average yearly rainfall is calculated. Subsequently, the rainfall for each year is converted into the difference relative to the long-term average, divided by the standard deviation – this is known as the 'anomaly' or 'rain index'.

Satellite technology allows an alternative method of monitoring rainfall (Fig. 6). The US Climate Prediction Centre (CPC-NOAA) combines satellite estimates of rainfall, as provided by NASA (the US National Aeronautics and Space Administration), with ground truth data from rainfall gauges (Xie & Arkin 1995). Daily estimates have been combined per ten-day-period from July 1995 onwards; all images are at http://www.fews.net. Fig. 6 compares the rainfall for two years. In a wet year, such as 1999, the Sudan zone to the south of the Sahel received much more rain than in a dry year (1996) and the rainbelt reached further north (Fig. 6). The images of the ten-day periods were converted to grid cells in order to calculate rainfall between 1995 and 2005 split up into 2°-wide latitudinal zones (8-10°N through 18-20°N). From these data, the average amount of rainfall in 1999 appeared to be 38% higher than in 1996, for instance (Fig. 6).

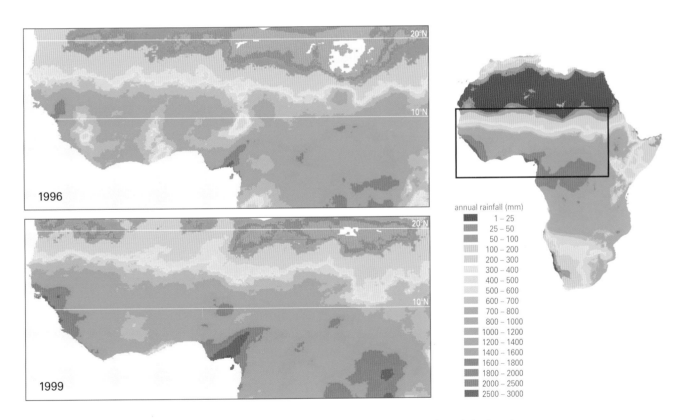

Fig. 6 The annual rainfall in West Africa during a dry year (1996) compared to a wet year (1999); the long-term average is shown right (from Fig. 2). Data, taken from http://www.fews.net, were collected by CPC-NOAA.

Aspects of dry and wet season landscapes, going from northern to southern Senegal: *Acacia* savanna (with communal nests of White-billed Buffalo Weavers; 1st row), scattered *Balanites aegyptiaca* (small evergreen trees) on sandy soil (2nd row), savanna with trees and shrubs on rocky soil (3rd row) and woodland in the Casamance (4th row).

Annual variation

An important question to address is whether year-to-year variation in rainfall differs across the Sahel. This is particularly relevant in the context of Palearctic migrants wintering in discrete sections of the Sahel. To remain valid, such an analysis must be derived from data from a large number of rainfall stations, in order to take into account the substantial local variation. Between 1960 and 1993, rainfall in the Sahel was rather well synchronised between countries, with, on average, good rainfall in 1960-69 and poor rainfall in 1973, 1984, 1987 and 1990 (Fig. 7). When annual rainfall tallies are compared between different countries, high correlations were found between neighbouring countries along the east-west gradient: Mauritania – Mali (R = +0.81), Mali – Niger (R = +0.80), Niger – Chad (R = +0.78) and Chad – Sudan (R = +0.79). The correlations are also high along the north-south gradient: Mauritania – Senegal (R = +0.79), Senegal – The Gambia (R = +0.94), Senegal – Guinea-Bissau (R = +0.85). Correlations decrease to about R = +0.70 when non-contiguous countries are compared. The cluster analysis shows that the mutual linkage between the Sahel countries is high and – to a lesser degree – that between the countries along the Gulf of Guinea. We may conclude that rainfall in Sahel countries is well-synchronised (east-west), but that synchronisation is smaller between the Sahel, Sudan and Guinea zones (north-south).

To analyse the annual variation in rainfall further within the Sahel zone, the gauge data up to 1997 were combined with the satellite data (1998-2003) and split up into four 2°-wide latitudinal zones from 12-14°N through 18-20°N (Fig. 8A). The annual rainfall fluctuated synchronously for the latitude bands 12-14°, 14-16° and 16-18°N, but much less so for the desert latitude band.[4] This is even more apparent from Fig. 8B, which shows the 'anomalies', the standardised rainfall in the different latitudinal zones. Thus, for the actual Sahel zone (12-18°N), the year-to-year variations in rainfall for the different latitudes are quite alike. As is already evident from the comparison of the rain gauge measurements in the various Sahel countries (Fig. 7), rainfall in the western and eastern Sahel is synchronised too.

In this analysis, we want to ascertain whether fluctuations in rainfall in the Sahel can be described under a single rain index, and,

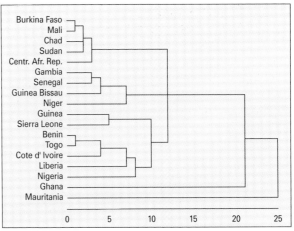

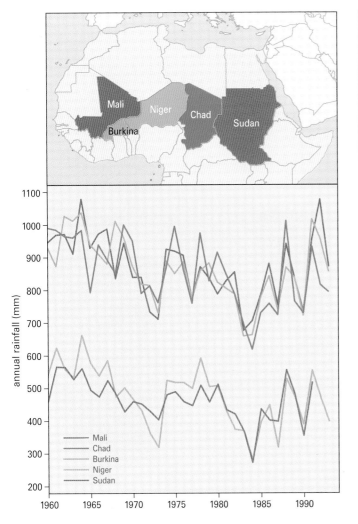

Fig. 7 Annual rainfall in five Sahel countries from 1960 to 1993. The averages are calculated over a large number of stations (Mali: n=177, Chad: n=140, Burkina Faso: n=156, Niger: n=205, Sudan: n=51). A hierarchical cluster analysis was carried out over the same period for 16 countries. The dendrogram shows that the average linkage between countries in the Sahel, Sudan and Guinea zone is large in neighbouring countries, especially within the Sahel zone.

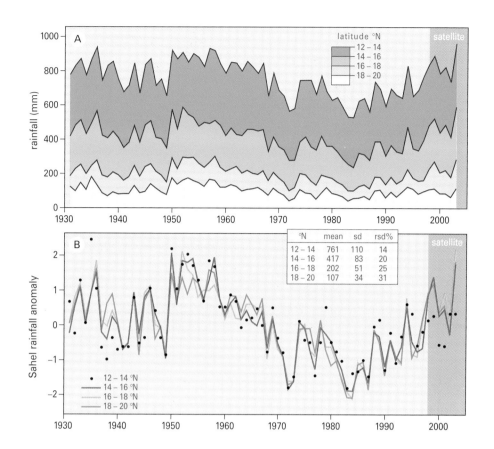

Fig. 8 Annual rainfall in the Sahel (west of 15°E) between 1931 and 2005, given separately per latitude band (four, thus: 12-14°N, ... 18-20°N). (A) rainfall in mm, (B) same data, but standardised (annual rainfall minus average rainfall in the 20th century divided by the standard deviation). All data are taken from Nicholson (2005), comprising gauge data up to 1997 and satellite data since then.

if not, whether different indices are required to describe rainfall patterns in the eastern and western part of the Sahel, for different latitudes, or for both sets of circumstances. Although Figs. 7 and 8 show subtle differences (see also Nicholson 1986, 2005, Nicholson & Palao 1993, Balas et al. 2007), we conclude that rainfall in the Sahel zone proper can be fitted adequately into a single index.

Fig. 8 is based on gauge data for 1930 to 1997, and on satellite data for 1998-2003. Although the satellite data, available since 1995, have been calibrated with field stations, one may doubt whether the rainfall recovery around 2000 was as large as suggested in Fig. 8. The satellite measurements suggest an exceptionally high rainfall in 1999 and 2003, but available gauge data do show lower amounts (Dai et al. 2004).[5] We used river discharge data (Chapter 3) to check which set of data was closer to the truth, from which it became apparent that the 1997-2005 rainfall in Fig. 8 had been overestimated. Hence, we use the gauge data to derive the annual variation in rainfall in the Sahel (Fig. 9).

During the 20th century, three periods of drought can be discerned: the first two, in 1900-15 and 1940-49 respectively, were followed by periods of improved rainfall. Again 30 years later, another drought occurred, but instead of the expected recovery in rainfall, there was a further decline in rainfall until 1984 (Fig. 9). This last period is known as the Great Drought in Africa – *La Grande Sécheresse* (1972-1992). Since then, rainfall has gradually improved somewhat.[6]

The last drought raises some major questions. Is the Great Drought unique or have similar episodes occurred in previous centuries? Has the recent drought been reinforced by desertification, with consequent local land degradation? Does rainfall in northern

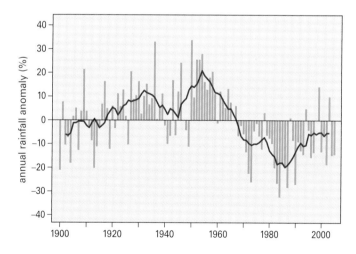

Fig. 9 Annual rainfall in the Sahel (1900-2005), expressed as percentage departure from the average calculated over the whole of the 20th century. The smooth curve gives the 9-year running mean.

and southern Africa fluctuate synchronously with the Sahel rainfall? Does global climate change play a role and, if so, what will happen in the future should the greenhouse effect continue to increase?

The answer to the first question is straightforward: the recent Great Drought is not unique. The dip in the 1980s looks prominent because the peak in the late 1950s was exceptionally high. The reconstructed climate history of Africa, based upon meteorological, hydrological and historical sources (Nicholson 1981b, 1982, 2001), showed several such climate epochs over the last centuries. The Sahelian kingdoms of Ghana, Mali and Songhay developed, thrived and declined in conjunction with prolonged periods of abundant rainfall or drought (McCann 1999). The period from 800 to 1300 is thought to have been relatively wet, but was followed by a drier period from 1300-1450, a wetter period until 1800 and a gradual decline since then (Fig. 10).

The answer to the second question was given earlier: reduced rainfall has caused temporal desertification, but effects the other way around appear to be limited: *i.e.* degradation in itself has little effect on climate.

The third question, whether rainfall in northern and southern Africa fluctuates synchronously with the Sahel rainfall, is studied by comparing three regions: the Maghreb, the Guinea zone (east of the Dahomey gap between 4° and 8°N) and South Africa (Fig. 11). The annual rainfall in the selected zones is positively, but weakly, correlated. The long-term trend over the 20th century is – just as in the Sahel - downwards in northern and southern Africa. The information given in Fig. 11 is particularly relevant for bird species which cross North Africa and the Sahel to winter in the Guinea zone and in South Africa, such as the Barn Swallow.

The fourth question, regarding climate change, is worthy of a separate paragraph.

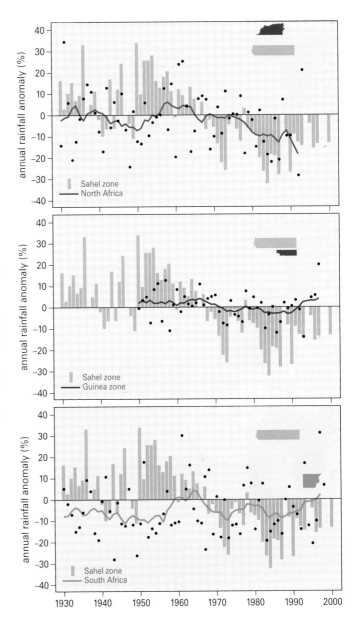

Fig. 11 Annual rainfall in North Africa (Morocco, Algeria, Tunisia; 20 rainfall stations), the Guinea zone between Nigeria and the Central African Republic (45 stations) and Southeast Africa compared to the Sahel rainfall (Fig. 9). From: Hulme *et al.* (2001) for SE Africa, and FAOCLIM-database. The dots give the annual rainfall anomaly, the smooth curve the 9-year running mean. Too few stations of the selected Guinea zone were available before 1950 to calculate a reliable rain anomaly.

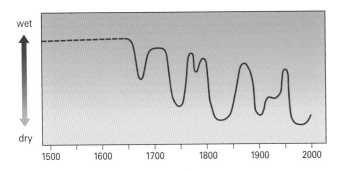

Fig. 10 Reconstruction of the climate in the Sahel, based on rainfall data since 1850 and historical research from the three preceding centuries (*e.g.* river discharges, harvests). From: Nicholson (1982).

Climate change

Warming is – usually – a global phenomenon, and the northern half of Africa does not escape this trend (Fig. 12). The six warmest years in northern Africa since 1860 were registered after 1998. The rise in temperature since 1970 has been even faster in the Sahara-Sahel zone than worldwide, with a 0.2°C rise per decade in the 1980s. This rate increased to 0.6°C per decade at the end of the 20th century.

Global Circulation Models (GCMs) predict a further warming of Africa in the 21st century, varying between 0.2 and 0.5°C per decade (Hulme *et al.* 2001). The warming is expected to be even greater in the Sahel. Consequently, the temperature may rise another 2-7°C in the next 80 years – a daunting prospect!

Global Circulation Models also provide predictions about rainfall. Given the important role that ocean surface seawater temperatures exert on rainfall in Africa, it is to be expected that a continuing warming of the tropical oceans would lead to a further reduction in rainfall. However, global warming may also have an impact on the temperature gradient in tropical and subtropical oceans, and this would complicate predictions of African rainfall. After comparing four climate change scenarios and seven global climate models, Hulme *et al.* (2001) concluded that annual rainfall in the western Sahel would possibly remain at the same level, but that a decrease of 10-20%, or even 40%, would be more likely.

One of the problems with Global Circulation Models is that, when applied to the Sahel, they are not able to capture the Great Drought. Recently, however, Held *et al.* (2005) presented a model that appears to simulate 20th century rainfall reliably for the Sahel. Their model includes the effects of greenhouse gases, carbonaceous and volcanic aerosols, ozone, solar irradiance and land use. They conclude that the decline in the Sahel rainfall in the second half of the 20th century is the result of anthropogenic forcing, partly due to increased aerosol loading and partly to an increase of greenhouse gases. They predict that rainfall until 2020-40 will remain at about the same low level as the last twenty years of the 20th century, but will then gradually decrease by about 20% in the next 50-100 years.

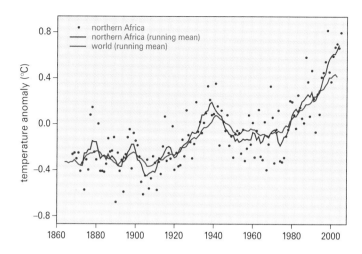

Fig. 12 The average temperature given as departures from the 1961-1990 average in northern Africa (0-40°N, 20°W-60°E) and worldwide (source: www.met-office.gov.uk/research/hadleycentre/CR_data/Monthly/HadCRUG.txt). The trends show the running mean over 9 years.

Rice planting women continue working despite an approaching downpour, Casamance (Senegal), September 2007.

Endnotes

1 The actual daily position of the ITCZ can be followed online via the Famine Early Warning System Network (http://www.FEWS.net) website. The satellite data are collected by the American Climate Prediction Center (CPC) of the National Oceanic & Atmospheric Administration (NOAA). The FEWS-site also gives per year the ten-days' rainfall since July 1995. Figure 3 illustrates the seasonal shift of the rain belt between January and December 2005, as described in the text.

2 Seawater temperature in the nearby eastern Atlantic Ocean affects rainfall in the Sahel. There is less rain in the Sahel when the equatorial Atlantic is relatively warm (Folland et al. 1986, Palmer 1986, Rowell et al. 1995, Chang et al. 1997, Janicot et al. 1998, Thiaw et al. 1999, Zheng et al. 1999, Vizy & Cook 2001, Giannini et al. 2003, Messager et al. 2004, Giannini et al. 2005). In contrast, the variation in the temperature of the Indian and Pacific Oceans has a long-term effect on rainfall in the Sahel, where again warm temperatures in both oceans tend to promote less rain in the Sahel. The effect of the Indian Ocean on rainfall in the Sahel was shown by Latif et al. (1998), Bader & Latif (2003), Giannini et al. (2003) and Lu et al. (2005). A similar effect of the Pacific Ocean (i.e. the El Niño Southern Oscillation (ENSO)) was demonstrated by Palmer et al. (1992), Myneni et al.(1996), Latif et al. (1998) and Janicot et al. (1996, 1998, 2001). To make it even more complex, rainfall in the Sahel does not depend simply on seawater temperature in the tropical zone of the three oceans, but on the difference in temperatures between the tropical and subtropical zone.

3 Historical rainfall data are published in the annals of the national meteorological organisations, and for Francophone Africa in the past also in the annals of ORSTOM. These data were summarised for 10-year periods for a number of key stations by the Smithsonian Institution (1920-1940) and the Department of Commerce (UAS) (1941-1960). Since 1961, monthly data have been published by the Department of Commerce in the "Monthly Climatic Data for the world". The latter data can also be downloaded from: www7.ncdc.NOAA.gov/IPS/NCDWPubs. Several institutes and organisations have digitalised the historical meteorological data. A large dataset was collected by the Climate Research Unit (CRU), University of Norwich (UK). These data are available from http://www.CRU.uea.ac.UK/. Another source is the Global Historical Climate Network (GHCN; ncdc.NOAA.gov/pub/data/ghcn/v2), jointly produced by the National Climate Data Center (NCDC) of the Arizona State University (uas) and Oak Ridge National Laboratory (ORNL). The National Disaster Management Centre (NDMC) of the University of Witwatersrand (South Africa) and the ORNL produced a file with monthly African precipitation data (sandmc.pwv.gov.za/safari2000/nasa/rain/cntrlist.asp), which had been extracted from the GHCN. The Agrometeorology Group of the FAO produced a database, called "FAOCLIM", containing data from 28 000 weather stations. Its interface is more user-friendly than on the sites mentioned above. The data are produced on CD, but can also be downloaded from the ftp-site of the fao (ftp://ext-ftp.fao.org; see http://www.fao.org/sd/2001/en1102_en.htm). Recent data were obtained from the Direction Nationale de la Météorologie in Mali and in Senegal. We also added long rainfall series from local organisations in Mali (ier, on, ors, orm). To include data from other Sahel countries, the up-to-date 'Monthly Climatic Data for the world' was consulted. Precipitation data for the past few decades may also be taken from the Climate Prediction Centre (cpc) and Global Precipitation Climatology Centre (GPCC) of NOAA, but we refrained from doing so, since these data appeared to deviate from the other sources (as already noted by Beck et al. 2004, Nicholson 2005).

4 The annual rainfall for the latitude bands 12-14°, 14-16° and 16-18°N were highly correlated (R= 0.93 - 0.96), but the correlation between the band 18-20°N and the bands further south was lower: (16-18°N: R = +0.85, 14-16°N: +0.76 and 12-14°N: +0.78).

5 The anomaly produced from derived data from satellites (1997-2003; Fig. 8) in all years represents 1 SD higher than the gauge data for the same years (Fig. 9); the correlation between both series is high ($R^2=0.90$).

6 Many papers give the anomalies relative to the 1961-1990 average. Since the average rainfall in the Sahel, calculated over the entire 20th century is 10% higher than for the period 1961-1990, one has to move the y-axes 10% downwards to read the data in Fig. 9 as a departure from the 1961-1990 average. The annual rainfall anomaly has been calculated by several authors. The data shown in Fig. 9 correlate well with, for instance, the Sahel anomaly given by Dai et al. (2004) ($R^2=0.944$) and also with the average anomaly for 14-20°N (Nicholson 2005, selection made for the years until 1996; $R^2=.950$). The correlation is poor ($R^2=0.545$), however, with another Sahel rainfall index, based on fewer gauge stations in the set (http://jisao.washington.edu/data_sets/sahel/). Unfortunately this latter index is used regularly in ornithological literature.

Rivers

La Grande Sécheresse – the disastrous Great Drought in the 1970s and 1980s – was a major catastrophe for the people in the Sahel. Rainfall was poor, but the decline of the river flow was even greater. Many people in Mali were convinced that the recently built Selingue dam was the cause of the low discharge from the Niger River. Environmentalists used this argument in international debates about dams. Hydrologists, on the other hand, argued that it was impossible for relatively small reservoirs to have such a large impact downstream. This idea was undoubtedly correct under normal conditions, but not necessarily so during extreme droughts. Irrespective of the debate, the discharge of rivers in the Sahel is known to fluctuate significantly. The reasons for these fluctuations are both natural and man-made. This chapter quantifies the natural variations in river discharges, and the impact of man-made structures thereon.

Introduction

The coastal zone along the Gulf of Guinea belongs to one of the wettest areas in the world, with an annual precipitation exceeding 3000 mm. Most of this water drains almost directly into the ocean, but geography forces some rivers to make a lengthy detour inland. The Senegal and Niger Rivers, both originating in Guinea-Conakry, run initially north eastwards. The Senegal then bends to the west, whereas the Niger makes a huge loop to the east, crossing the southern edge of the Sahara, before turning south towards Nigeria. Further to the east, the Chari and the Logone Rivers drain into Lake Chad. The catchment areas of the Senegal River, Niger River and the Lake Chad Basin cover almost the entire western and central Sahel (Fig. 13) and are of paramount importance to people and birds alike.

The Niger River is 4184 km long and its drainage basin covers an estimated 2.2 million km^2. The Senegal River is 1800 km long, draining an area of 0.44 million km^2. The catchment of the Chari (1200 km) and the Logone Rivers (965 km) comprises the southern part of the Lake Chad Basin (2.4 million km^2). Many separate and mostly small rivers drain the land between the Atlantic Ocean and the basins of the Senegal and Niger Rivers, but some of the larger rivers, such as the Gambia, are significant. The latter is 744 km long and drains central Senegal and the nation of The Gambia (0.18 million km^2). The White Volta River (1800 km) originates in the Mossi Highlands and drains the eastern part of Burkina Faso and Ghana, in total an area of 0.37 million km^2.

West-African rivers are constricted by hundreds of dams, many holding tiny reservoirs but some much larger (Fig. 13); notably the Akosombo storage reservoir in Ghana, which was completed in early 1966, is larger (8482 km^2) than all other West-African reservoirs put together. Other large reservoirs are found in Ivory Coast (the Kosso, 1780 km^2) and in Nigeria (Kainji, 1250 km^2). In the Sahel itself, large reservoirs are absent except for the Manantali (477 km^2) on the Upper Senegal. Although the other dams in the Sahel zone are much smaller, their impact on the river flow, as we show in this chapter, is relatively large.

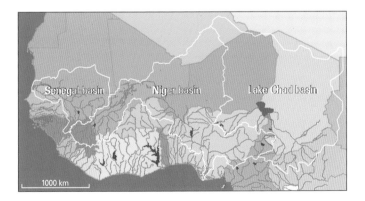

Fig. 13 The three major drainage basins in West Africa: Senegal, Niger and Lake Chad, and (shown in purple) the 12 largest water reservoirs.

Downpours during the rainy season bring a lot of water in a short period; dry or near-dry rivers may become flooded in a matter of hours; Burkina Faso, August 1985.

The rivers in West Africa demonstrate huge between-season variations in discharge. For instance, the average discharge of the Niger at Koulikoro (SW Mali) in September is eighty times larger than in April. This is caused by the short but intense rainy season, reaching a peak in August. It takes some time before all this surface water finds its way via shallow gradients into the Upper Niger, which normally reaches its maximum height in September or October. Sahel rivers, draining lowland catchments, have a characteristically low flow rate. The length of the Niger means that it takes 6 months before the rainfall in Guinea reaches the ocean.

The rivers and their floodplains function as lifelines, not only for the local people, but also for waterbirds and other wildlife. The floodplains of the Sahel are green oases amidst vast dry lands (Chapter 6-10). Variations in annual rainfall cause large variations in river flow and also in the extent of flooding between years. This chapter analyses yearly variations in river discharges and investigates to what degree these are related to rainfall fluctuations. As we will show, the river flow not only depends on the rainfall in preceding months, but also on rainfall in the preceding years. This system is complicated by irrigation schemes and water reservoirs that, by extracting and diverting water from the rivers, interfere with the natural timing of the flow.

Senegal River Basin

The main branch of the Senegal River is the Bafing, originating in the Fouta-Djalon hills in central Guinea; nearly half of the flow of the Senegal River depends on the Bafing. The Bakoye starts on the Mandinga Plateau in West Mali and contributes nearly a quarter of the entire river flow. Like the Bafing, the Falémé rises on the Fouta-

The Diama dam at the mouth of the Senegal River, December 2008.

The Senegal River during high water, partly tamed with dikes and canals.

Djalon hills, but takes a more westerly course; it delivers 10% of the entire flow to the Senegal River. Other tributaries are mostly much smaller (Fig. 14). The outer boundary of the Senegal Basin extends far into the Sahara desert. The watershed is determined by the elevation pattern, but what little rain falls in the northern half of the basin scarcely adds to the river discharge.

To relate the annual river discharge to rainfall, we selected 10 rainfall stations from the humid part of the Senegal Basin, for which measurements were available from around 1920 (red dots in Fig. 14). These measurements were averaged and related to the peak river discharge in September at Bakel (blue dot in Fig. 14). The river flow has been measured in Bakel since 1903.

Since 1922, periods of good rains and high river discharges (prior to 1930 and 1950-1965) alternated with periods of poor rainfall and low river flows (around 1940 and 1970-1990). Rainfall is less variable than river flow (Fig. 15). Rainfall on the Upper Senegal usually varies between 1100 and 1500 mm per year, but the river flow in September fluctuates between 1600 and 4000 m^3/s. Since 1987, the river flow of the Senegal at Bakel has been partly determined by water releases from the Manantali dam. River flow in September has been at a relatively

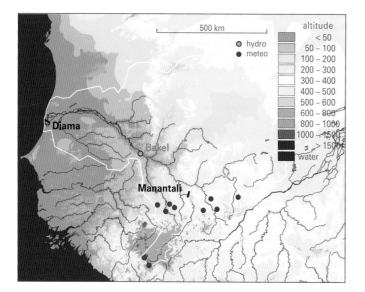

Fig. 14 The Senegal Basin, showing two dams (Diama near the river mouth and Manantali reservoir on the Upper Senegal), the main hydrological station (Bakel) and ten long-term meteorological stations on the Upper Senegal.

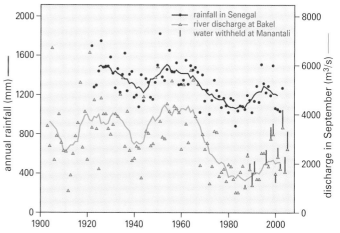

Fig. 15 Annual rainfall in the catchment area of the Upper Senegal, averaged over 10 rainfall stations (dark blue symbols, left-hand y-axis), and the river discharge in September at Bakel (light blue symbols, right-hand y-axis). The trends show the running means over nine years. Red bars indicate the effect of the Manantali reservoir on the river discharge. See Fig. 14 for the location of Manantali, Bakel and the rainfall stations.

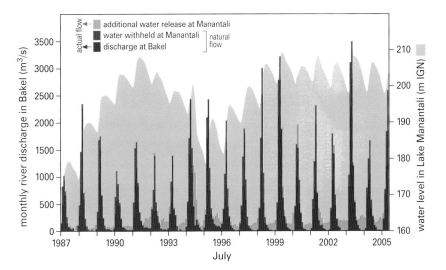

Fig. 16 Actual and reconstructed natural monthly flow (m³/s) of the Senegal River at Bakel since July 1987, as determined by the flow into Lake Manantali, additional releases and water retained at the Manantali dam – this retention occurs during the period of peak flow in most years. The grey shading shows the variation in monthly water level in the reservoir (m average sea level (IGN); right axis). Data from OMVS.

low level since then, despite increased rainfall on the Upper Senegal.

Lake Manantali came into existence in 1987 when a 70 m-high concrete dam across the Bafing valley was completed, but it took five years to fill the lake. Subsequently, the water level was lowered for four years, after which it took another four years to refill the lake. Since 1998, the lake has been full, and its water level has been kept between 200 and 210 m above sea level (Fig. 16). The Manantali Lake stores up to 11.3 km³ of water. The average annual river discharge at Bakel since 1928 (for all tributaries) has been 669 m³/s, equivalent to an annual flow of 21.1 km³. Given an evaporation rate at Manantali Lake of 7 mm per day, the total annual evaporation can be estimated at 1.2 km³, some 6% of the average total river flow. Hydropower plants also change the seasonal variation in the river discharge, as at Manantali. Much of the water accumulated during the flood season is released in the dry period. How much is collected and released is known precisely, because the monthly river flow into the lake has been registered, as has the amount of water released at the dam. These records permit the reconstruction of the natural, unmanaged river discharge at Bakel as if the upstream Manantali reservoir was not there (Fig. 16). In the unmanaged situation, the river flow would be close to zero in the dry period, but due to the water releases in Manantali, it is now hardly ever below 100 m³/s. In fact, the water taken from the river during the peak flow (minus the evaporated 1.2 km³) is released to the river in the dry season. With the exception of the year 2000, about 500 m³/s is normally retained in September, including the dry years of 1990 and 2004 when the September flow, without the Manantali dam, would have been just over 1000 and 1500 m³/s, respectively.

The reconstructed river flow discharge without the Manantali dam shows that the flow of the Senegal has recovered to pre-drought levels since 1990, although it has not yet reached the high levels of 1930 and 1950-1960 (red bars in Fig. 15), when rainfall was higher still. Chapter 7 will describe the consequences of the Manantali dam downstream.

Niger River Basin

The Niger and its Niandan, Milo and Sankarani tributaries rise in the Guinea Highlands (Fig. 17). The most northerly branch, the Tinkisso, originates in the Fouta-Djalon. The main tributary of the Niger is the Bani, which drains southernmost Mali and the northeastern corner of Ivory Coast. After the Bani flows into the Niger near Mopti, at the southern edge of the Inner Niger Delta, there is no further run-off from eastern Mali and Niger. Consequently, evaporation gradually diminishes the river flow. The discharge of the Niger trebles during the last 500 km before it empties into the Gulf of Guinea. Much water is discharged by the Benue, a large river draining the extensive rainy area of eastern Nigeria.

The total catchment area of the Bani (129 000 km²) is nearly as large as the rest of the Upper Niger Basin (147 000 km²). Yet the discharge of the Bani is less than half that of the Niger, because the Bani sub-basin receives less rainfall than the other sub-basins of the Upper Niger. We investigate separately the relationship between rainfall on the Upper Niger and the river discharge, and that for the Bani and the rest of the Upper Niger. We relate the data from the 28 long-term rainfall stations (red diamonds Fig. 17) on the Upper Bani to discharge at Douna (blue diamond in Fig. 17). Only 15 long-term stations (red dots in Fig. 17) are available for the Upper Niger. To increase sample size, two stations just across the border of the basin were included in the calculation of the average annual rainfall. These data are compared with the discharge at Koulikoro (blue dot in Fig. 17).

The annual rainfall in the Bani catchment area usually varies between 1000 and 1200 mm (Fig. 18). Wet and dry periods, as reflected in water flow, closely agree with those registered for the Senegal River (Fig. 15). The Bani's flow in September fell back from 3000 m³/s to only 250 m³/s during the drought in the early 1980s, but subsequently gradually increased again. The long-term effect of a series of dry years on the flow of the Bani is evident. The Bani was a river

The Niger River plays an important role in the transportation of goods and people, particularly in the wet season; harbour of Mopti, September 2006.

with a fully natural flow until 2006 when the Talo dam became operational.

Since 1922, the average annual rainfall on the Upper Niger has varied between 1300 and 1600 mm (Fig. 19). The trend resembles those shown for the Senegal (Fig. 15) and Bani (Fig. 18), but where the Senegal and the Bani lost 80% of their flow during the Great Drought in the 1980s, the decrease in the Niger was slightly less than 50%. Since 1982, the flow of the Upper Niger has not been fully natural due to the construction of the Selingue reservoir in the Sankarani. The Selingue reservoir covers 450 km^2 when full (2.1 km^3). The monthly inflow and outflow have been registered since its inception. The annual loss amounts to 0.83 km^3, of which 0.57 km^3 is lost to evaporation; the rest becomes groundwater (Zwarts *et al.* 2005a). More important than water loss is the change in seasonal water flow. The flow into the reservoir (representing natural flow) is reduced by, on average, 61% in August and by 36% in September due to the filling of the reservoir. In contrast, when water is released from the dam in the dry season, the outflow between February and

The Niger River near Gao (eastern Mali), an example of an untamed section of the river.

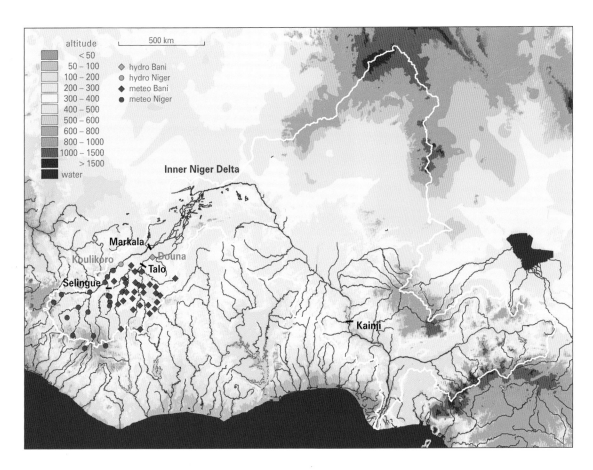

Fig. 17 The Niger Basin, showing four dams (Selingue and Markala on the Upper Niger, Talo on the Bani and Kainji on the Lower Niger), the hydrological stations (Koulikoro and Douna) and 45 long-term meteorological stations on the Upper Niger.

April is about three times the amount of the natural flow. During the first years of its existence, the effect of the Selingue dam was limited (red bars in Fig. 19), because the lake was only partly emptied in the course of the year and not fully refilled.

Another dam in the Upper Niger, the Markala dam, is a low weir across the river, used to raise the water level sufficiently to irrigate the area managed by *Office du Niger*. The irrigation zone is located in the *Delta mort*, the dead Delta, an ancient branch of the Niger.

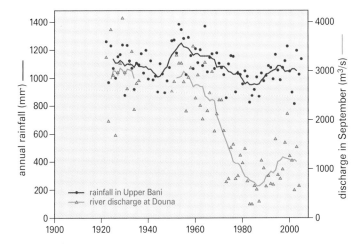

Fig. 18 Annual rainfall on the Upper Bani averaged for 28 rainfall stations in the catchment area (dark blue symbols, left y-axis) and the river discharge in September at Douna (light blue symbols, right y-axis). The trends show 9-year running means. See Fig. 17 for the location of Douna and the rainfall stations.

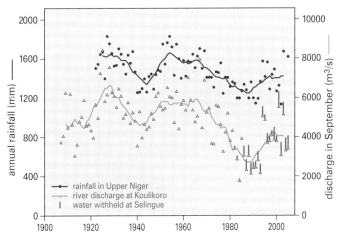

Fig. 19 Annual rainfall on the Upper Niger (dark blue symbols, left y-axis) averaged for 17 rainfall stations in the catchment area, and the river discharge in September in Koulikoro (light blue symbols, right y-axis). The trends show 9-year running means. The red bars show the effect of the Selingue reservoir on the river discharge in September. See Fig. 17 for the location of Selingue, Koulikoro and the rainfall stations.

Rivers

Sandbanks emerge in the Logone River as water levels fall; January 1999. The Chari-Logone system is the main tributary to Lake Chad.

The Logone River rises from the Adamawa mountains.

The Markala dam has been operational since 1947, but it took many years before the irrigation scheme was developed. In 2000, 740 km² of land was under irrigation and a further expansion of between 140 and 400 km² is planned (Keita et al. 2002, Wymenga et al. 2005). The water intake, as registered by *Office du Niger*, amounts to 2.69 km³ per year, equivalent to 86 m³/s. Despite the gradual extension of the irrigated zone, water extraction remained at the same level between 1988 and 2006. Over the same period, the annual river discharge at Koulikoro varied between 624 and 1258 m³/s. As a consequence, water use by *Office du Niger* is not more than 6% of a high river flow (1995), but this may increase to 16% when the flow is low (1989). The monthly water use by *Office du Niger* is 60 m³/s in January, gradually increasing to 130 m³/s in October, decreasing to 90 m³/s in November and 50 m³/s in December. The available water supply varies between 100 m³/s in March and 3200 m³/s in September. Hence 60% of the flow is tapped in March, as opposed to only a few percent in September. The downstream consequences of the water use by *Office du Niger* will be discussed in Chapter 6.

Lake Chad Basin

The northern half of the Lake Chad Basin is dry. The lake itself is fed by the run-off from the southern quarter of the Basin. Three river systems may be distinguished. The Komadugu/Hadejia/Yobe river complex rises from the eastern Jos Plateau and flows east to enter Lake Chad in its northwestern corner. The annual flow into Lake Chad, measured at Diffa, is limited compared with the flow of the rivers measured upstream, varying between 4.5 m³/s in 1984 and 20.9 m³/s in 1986 (source: Global Run-off Data Centre, GRDC; data from www.grdc.bafg.de). For example, the annual flow of the Hadejia River near the town of Hadejia, 300 km upstream of Diffa, varies between 200 and 900 m³/s (Goes 1997). The decline in river flow can be attributed to seepage and to evaporation, mainly in the Hadejia-Nguru floodplains (Goes 1999), but, in recent decades, the construction of dams has also played a part. The Tiga dam (1974), the Challawa Gorge dam (1992) and the Hadejia dam (1992) withheld 1.3, 1.0 and 1.2 km³ of water respectively, while the Kafin Zaki dam (2.7 km³) on the Jama'are River is still under construction (Fortnam & Oguntola 2004). The total volume of the Tiga and Challawa Gorge reservoirs amounted to 8% of the total discharge of the Hadejia River in a year

Markala dam on the Niger River near Segou, November 1999. This dam has been operational since 1947 to irrigate the rice fields managed by Office du Niger in the Delta mort, an ancient branch of the Niger.

of high discharge (2.4/28.9 km^3), but this increased to 38% in a low-flow year (2.4/6.3km^3). These dams have an overriding impact on local river discharge. However, the effect on the flow into Lake Chad seems to be limited. According to Oyebande (2001; cited by Fortnam & Oguntola 2004), only 10% of the total run-off from the Komadugu/Hadejia/Yobe system reached Lake Chad before the construction of the dams. In other words, the dams created storage reservoirs at the expense of the floodplains downstream (see also Chapter 8), but did not affect the flow in the lower reaches of the rivers.

A much smaller river system is formed by the Ngadda and Yedseram river complex. Both flow from the Mandara mountains to the north, but empty into local floodplains before reaching Lake Chad. This leaves only one river system to feed Lake Chad: the Chari-Logone system. The Logone rises in the Adamawa mountains and joins the Chari at N'Djamena. The Chari itself drains the Mongo hills on the border of Chad and the Central African Republic. To investigate the relationship between rainfall and discharge, we compared the combined river flow of the Chari and the Logone at N'Djamena with the annual rainfall averaged over 24 meteorological stations in the catchment area of both rivers (Fig. 20). The annual variation in rainfall is relatively small, compared with rainfall on the Upper Senegal (Fig. 15) and Upper Niger (Figs. 18 and 19). In the wet years, around 1930 and 1950, annual rainfall amounted to 1300 mm. This decreased to 1200 mm in the dry 1940s and to 1100 mm during the Great Drought in the 1980s. Although annual rainfall does not vary much, it has a large impact on the discharge. During the Great Drought, the discharge decreased from 1400 to 500 m^3/s. When rainfall gradually increased again after 1985, the discharge increased to 900 m^3/s, but the river flow is being increasingly reduced by upstream irrigation.

Olivry *et al.* (1996) analysed the river discharge in the Lake Chad Basin for the period 1953-1977 and concluded that annually about 1.9 km^3 (or 60 m^3/s) was extracted for irrigation, equivalent to 5% of the average total water flow. Vuillaume (1981) arrived at an irrigation loss of 2.5 km^3 (or 79 m^3/s) for 1965-1977. This further increased to 10 km^3 (292 m^3/s) in 1990-1991 (Coe & Foley 2001), with a marked impact on the discharge in N'Djamena (red line in Fig. 21). Because statistics on annual water use are lacking, the red line is based on the assumption that water use was 0 in 1953, 79 m^3/s in 1980 and 292 m^3/s in 1991, with interpolated values for the years in between. We found no data on further extension of the irrigation schemes, and therefore assume that irrigation between 1991 and 2006 remained at the same level. Chapter 9 will describe the consequences downstream.

Hydrological versus meteorological drought

The comparison between rainfall and the (reconstructed) discharge in the various basins reveals a large variation in river flow and a much smaller variation in rainfall.[1] This larger variation in river discharge is an important ecological asset of the Sahel. The explanation hinges

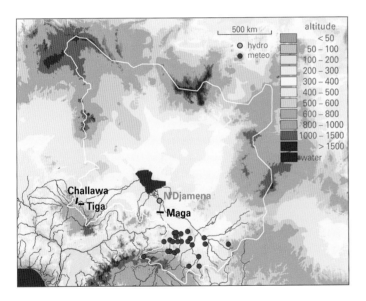

Fig. 20 Lake Chad Basin, showing three storage reservoirs (Lake Maga along the Logone River; Lake Tiga and Lake Challawa on the tributaries of the Hadejia River), the hydrological station of N'Djamena and 24 long-term meteorological stations in the catchment area of the Chari and Logone.

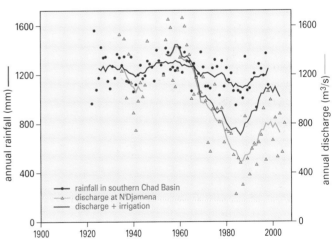

Fig. 21 Annual rainfall in the catchment area of the Chari (dark blue symbols, left-hand y-axis) averaged over 24 rainfall stations, and the average annual river discharge in N'Djamena (light blue symbols, right-hand y-axis). The trends show 9-year running means. The red line indicates the reconstructed river discharge had there been no irrigation. See Fig. 20 for the location of N'Djamena and the 24 rainfall stations.

The Niger is a typical dryland river with a large seasonal variation in water level; downstream of Mopti, 26 September 2006 and 8 February 2007.

on the cumulative effect of rainfall on river discharges: dry years lower the discharge, but it takes a number of wet years subsequently to attain a high discharge (see Figs. 15, 18, 19 and 21). In other words, river discharges not only relate to rainfall in the preceding wet season, but, to a large extent, to earlier wet seasons as well. [2]

Olivry (1987), studying the Senegal River, was the first to note the decrease in discharge after a series of dry years. He hypothesised that the discharge was insufficient to keep the groundwater table at a certain level; in turn a low groundwater table increased seepage of surface water. In the catchment area of the Bani, Mahé et al. (2000) did indeed find that a low discharge goes with a low groundwater level. Similarly, for the Hadejia-Nguru floodplains, Goes (1999) showed that regular flooding is necessary to recharge groundwater resources. According to Ngatcha et al. (2005), the groundwater level in the Logone floodplains decreased by 3-10 m after a series of dry years. Groundwater deficits apparently differ between river systems, being more pronounced for the Bani and the Chari/Logone than for the Senegal and the Niger. The most likely explanation is that the exchange of water between river and groundwater varies in conjunction with soil type, reduced exchange in rocky areas and increased exchange in sandy soils.

Over a period of years, the cumulative effect of rainfall on river discharge has been christened 'phreatic drought' (Bricquet et al. 1997), to indicate specific stream drought as opposed to 'meteorological drought'. Since aquifers have a response time of several years following changes in climate, phreatic drought will always show a delayed effect compared with meteorological drought. Any restoration of the hydrological regime should last as long as it took to bring it to its current state of degradation (Bricquet et al. 1997). Hence, meteorological drought produces a prolonged and delayed hydrological drought. But the effect also works the other way around: hydrological drought affects meteorological drought because evaporation in-

creases when there is more surface water and higher soil moisture content (Nicholson 2001). This phenomenon partly explains why wet and dry years in semi-arid areas may persist for several years in a row.

The flow of West-African rivers will decrease in the future, owing to the expected reduction in rainfall (Chapter 2). But even if rainfall does not decrease, a reduced river flow is to be expected, given an increased evaporation rate in conjuction with the rise in temperature. A small reduction in rainfall causes a substantial drop in river flow, most evident in an arid land river such as the Bani. This effect was confirmed by De Wit & Stankiewicz (2006) who compared rainfall with drainage of African rivers. They found that with a 10% decrease in rainfall, drainage would drop by 17% in regions where annual rainfall was 1000 mm, but the impact was much greater (50%) in regions with only 500 mm of rainfall.

Endnotes

1 The discrepancy between discharge and rainfall, as measured by the relative standard deviation (SD as percentage of the mean, or RSD), amounts to a factor 2 in favour of discharge in the Chari/Logone and the Niger, a factor 3 for the Senegal and as much as a factor 4.4 for the Bani (Table 1).

Table 1 Rainfall and reconstructed 'natural' discharge (excluding the effects of storage reservoirs and irrigation): average (x), standard deviation (SD) and standard deviation as percentage relative to the mean (RSD), calculated over the same period (1922-2005). Data extracted from Figs. 15, 18, 19 and 21. Note that the discharge refers to the peak discharge in September, but to the annual average for the Chari/Logone.

country	rain, mm			natural discharge, m³/s		
	x	SD	RSD,%	x	SD	RSD,%
Senegal	1298	193	14.9	2977	1311	44.0
Bani	1081	125	11.6	2007	1027	51.2
Niger	1467	170	11.6	5001	1292	25.8
Chari/Logone	1220	126	10.3	1039	230	22.1

2 A multiple regression analysis shows that – at least for the Bani and the Chari/Logone – discharges depend on the rainfall of as many as four preceding years. For the Senegal and the Niger, this effect was not significant, however.

Vegetation With contributions of Paul Scholte

The vegetation in the Sahel is primarily governed by rainfall. The 50 mm rainfall isocline is often considered as the transition between Sahara and Sahel, as it delineates the minimum precipitation required for grass to grow, which in turn determines the northern limit of grazing grounds for cattle and wildlife. Another important bottleneck is set by the 400 mm isocline, below which rain-fed agriculture is seldom possible. In the Sahel, drought-resistant cereals (e.g. sorghum and millet) are grown where the annual rainfall range is 400-800 mm. Nicholson (2005) calculated the position of the three key boundaries (50, 100 and 400 mm) in the western Sahel for the period 1998-2003. The position of the 50 mm isocline varied between 19.3 and 27.0°N, a 400 km-wide belt. Such variation was smaller for the 100 mm isocline, fluctuating between 18.3 and 21.1°N, a belt some 150 km wide. The 400 mm isocline showed very little variation in these six years, shifting only between 15.1 and 15.8°N (less than 40 km). This means that the area annually available for grazing varies more in extent than that used as cropland. Although a dry year has an immediate effect on annual vegetation, the effect on perennial vegetation is less direct. The scattered old acacia trees along the southern edge of the Sahara are living proof of a wetter climate in the past, although the latest long-lasting drought claimed many such trees.

Introduction

Five different eco-climatic zones are usually distinguished in West Africa: the Sahara desert, the Saharo-Sahelian zone, the Sahelian zone, the Sudano-Sahelian zone and the Guinea zone. The Saharo-Sahelian zone comprises essentially barren soil that holds a cover of perennial grasses in depressions and wadis. Rain permitting, the sparse vegetation is grazed by cattle, sheep and goats until the burning sun withers the grass. The Sahel can be characterised as savanna, dominated by annual grasses, scattered spiny shrubs and acacia trees. This land is used by resident and nomadic herders. Depressions allow some cultivation of millet, but only as a risky enterprise. The Sudano-Sahelian zone is also covered by annual grasses, but acacias are replaced in part by other tree species. Farmers grow not only millet, but also sorghum. The Sudan zone appears somewhat park-like – its biomass is greater, comprising taller, perennial grasses, shrubs and trees. The growing season extends over four months, one month longer than in the Sudano-Sahelian zone. A greater variety of crops is grown (e.g. maize, peanuts and rice). The zoning of the major vegetation types relates closely to rainfall (Fig. 22).

There is some disagreement in the literature about how to differentiate these five zones. For instance, the Sahel proper has been variously defined as the zone limited by the 100-700 mm, 150-600 mm, 200-400 mm, 250-500 mm and 300-700 mm isoclines. This is not surprising because the transition from dry Sahara to wet Guinea zone is gradual. This introduces an unavoidable arbitrary element into definitions of zones, which also vary between years, depending on rainfall.

As well as rain, fire is another major ecological factor in the region, eliminating fire-sensitive species from the savanna and favouring grassy vegetations (van Langevelde et al. 2003). Farmers and herders both use fire as a management tool, benefiting from the direct effect of the additional release of nutrients, thus stimulating plant growth, but the long-term effect may be negative due to over-use reducing soil fertility. [1]

Vegetation index

The primary production of the vegetation in the Sahel has been studied since the 1980s by remote sensing techniques, which provide red and near-infrared spectral data (Tucker 1979, Tucker et al. 1985). The advantages of satellite-produced information are that large areas can be measured directly and that green-leaf biomass integrates many ecological variables (Skidmore et al. 2003, Kerr & Ostrovski 2003, Turner et al. 2003, Seto et al. 2004, Pettorelli et al. 2005, Xiao & Moody 2005). This kind of data is now available for a long series of years. [2]

NDVI-images vividly show the seasonal variation in the 'greenness' of the Sahel. During the peak of the rainy season (mid August), this differs dramatically from that seen two and four months later (Fig. 23). Annual differences may also be striking. In the extremely dry year 1984, the southern region of West Africa started off greener in August than it did in the relatively wet year 1999. Each year, the vegetation successively dies back until the next rainy season, but this process accelerates during the dry season, when the deficiency in precipitation from the preceding rainy season is greater than average (compare October and December for 1984 and 1999; Fig. 23).

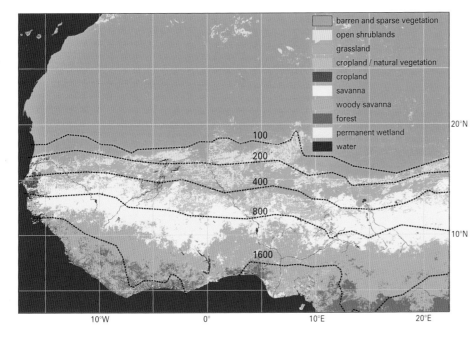

Fig. 22 Simplified vegetation map of western Africa, with 100, 200, 400, 800 and 1600 mm isoclines. The vegetation map is based on the global land-cover map, produced by the International Geosphere-Biosphere Programme-Data and Information System (IGEP-DIS) from remote sensing data covering 1992-93 (http://edcsns17.cr.usgs.gov/glcc/glcc.html). Isohyets taken from Fig. 2.

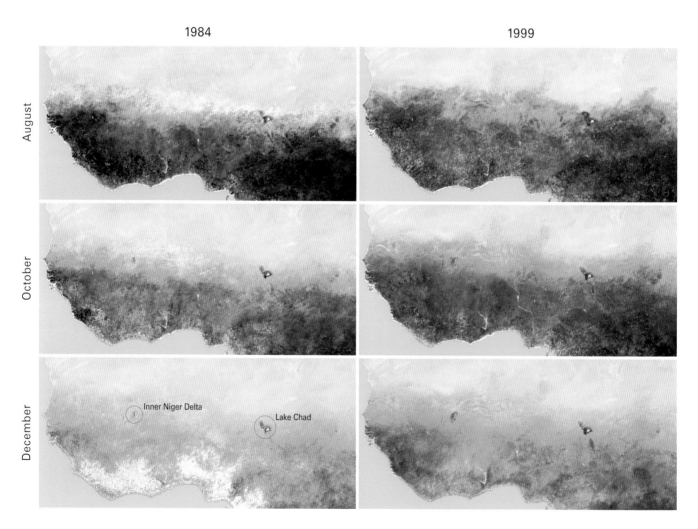

Fig. 23 The green vegetation in West Africa in August, October and December of a very dry year (1984; left) and a relatively wet year (1999; right). The greenness is based on NDVI-values, as measured by the NOAA-AVHRR. Images downloaded from http://www.fews.net. The vegetation-index has not been measured in the Sahara proper; in all images the desert is depicted with the same sandy colour.

The Sahel may seem untouched by man from a distance, but a closer look reveals that the landscape is shaped by grazing (sheep: Mali, cow: Niger), farming (Mali) and burning (Burkina Faso).

Seasonal and annual variation in the NDVI

The Advanced Very High Resolution Radiometer series [3] can be used to study the seasonal and annual variation in greenness. In our analysis, we divided the Sahel into a western (20-0°W) and an eastern sector (0-20°E), both sectors covering seven latitudinal zones (8-10, 10-12, 20-22°N). For each of the 14 zones, we calculated average NDVI values for available decadal data. The seasonal variation in the NDVI for the western sector (10-12°N) is large, *i.e.* above 0.5 during the peak of the rainy season (indicating a dense green vegetation), and below 0.2 at the end of the dry season (indicating a low coverage of green vegetation). Annual maxima and minima also show wide year-to-year variations in the NDVI-values (Fig. 24).

In the zones of 12-14°N and 14-16°N the peak in greenness is reached in early August, and slightly later (mid-August) in the zones of 8-10°N and 10-12°N. The rains arrive earlier in the south (Fig. 5), which explains why the NDVI starts to increase 1-3 months earlier than in the north. NDVI-values in the northern regions decrease until May.

Peak NDVI-levels differ little between the 8-10°N and 10-12°N belts, primarily because green biomass can be measured only up to a certain saturation level. The seasonal pattern in NDVI for the 12-14°N belt in the western sector of the Sahel resembles that for 10-12°N in the eastern sector. Another latitudinal shift in NDVI within West Africa is visible between the western 14-16°N belt and the eastern 12-14°N belt (Fig. 27). These shifts parallel the vegetation zones that do not – as shown in Fig. 2 - align strictly with latitude. Similarly, the isoclines show that the Sahara desert encroaches further southwards in the eastern section of West Africa (Figs. 2, 22).

Since the start of the NDVI-measurements in 1981, the Sahel has been less green up to 1994, with particularly poor results in 1984 and 1993 (Fig. 26). From 1993 onwards, greening has improved substantially, a trend also visible in the rainfall anomaly (Fig. 26).

Rain has an immediate and significant effect on the vegetation

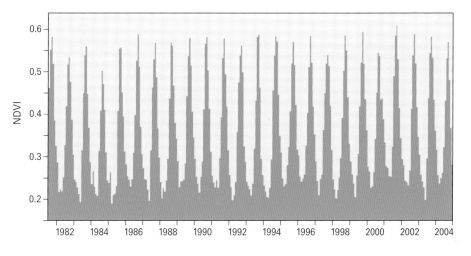

Fig. 24 Decadal variation of the NDVI-value in the regions covered by 18-0°W and 12-14°N between July 1981 - December 2004 (846 NDVI-measurements referring to data from 36 decades per year). From: NOAA.

Box 1
Desertification

Do droughts create deserts? Or do deserts create droughts? The extension of the Sahara between 1970 and 1990 (Nicholson *et al.* 1998) was generally believed to be caused by human-induced irreversible degradation. The reasoning was simple. Humans turn semi-arid drylands into deserts via deforestation, overstocking of pasture land and depletion of cropland. This non-sustainable use of land causes a decrease of vegetation cover, soil moisture and ground water table, as well as an increase of wind and water erosion and a loss of nutrients. The Sahel drought in the early 1970s and the apparent marching south of the Sahara Desert were the triggers to convene the United Nations Conference on Desertification (UNCOD) in Nairobi in 1977. The follow-up was entrusted to the United Nations Environmental Programme (UNEP). The United Nations Conference on Environment and Development (UNCED), held in Rio de Janeiro in 1992, concluded that desertification affects 70% of the world's dryland (UNEP 1993). It was decided to establish the United Nations Convention to Combat Desertification, the UNCCD. Desertification was defined at the Rio conference as "land degradation in arid, semi-arid and dry sub-humid areas resulting from various factors, including climatic variations and human activities". By accepting climatic variation as one of the causes of desertification, the concept has lost its original meaning as man-induced, irreversible land degradation (Mortimore & Turner 2005). The original definition was based on the misconception that arid environments are in equilibrium. Moreover, the claim of desertification in the Sahel was derived from observations and some empirical data collected during a period of declining rainfall.

Based on remote sensing data and field work in Sudan, it could be shown that changes took place in response to drought, but that there was no reason to assume a longer-term human impact (Helldén 1991). Rasmussen *et al.* (2001), working in fossil sand dunes in northern Burkina Faso, combined aerial photographs with recent high-resolution satellite data, field work and interviews with local people. The title of their paper 'desertification in reverse' already draws their conclusion: the apparent desertification between the early 1970s and mid-1980s has been reversed in the 1990s after the rains returned. Tucker *et al.* (1991), Tucker & Nicholson (1998), Nicholson *et al.* (1999) and Prince *et al.* (1998) reached a similar conclusion for the Sahel as a whole. What has been interpreted as human-induced desertification appears to be the natural response of a semi-arid environment to climatic variation. This does not imply that humans do not influence the Sahelian environment. Irrefutably, there has been deforestation; the hypothesized soil nutrient depletion will be covered in Chapter 5. However, we will first consider the effect of soil erosion.

Clear proof of the irreversibility of desertification seems to be the so called 'Sahara plume', dust blown from the Sahara and the Sahel. Satellite images provide spectacular examples of Sahara

The dry season in the Sahel is characterised by dust, obscuring the view and coating everything with a film of fine sand particles. Sandstorm and whirlwind in Burkina Faso, August 1985 and in Senegal (with Baobab). Even when the wind drops after gusty days, dust remains in the air for a considerable period of time. Blue skies are a rare commodity in the dry season; instead a yellow mist predominates in dry years (Great White Pelicans in the Inner Niger Delta).

plumes, as shown here for 11 February 2001 (Fig. 25). Seven studies estimated the dust produced in the Sahara + Sahel at a minimum of 130 million tonnes per year and a maximum of 760 million tonnes (Goudie & Middleton 2001). Total dust emission to the atmosphere varied, according to five of these studies, between 500 and 3000 tonnes per year. Two questions are relevant. What is the contribution of the Sahara proper, and of the Sahel to the Sahara plume? And if wind erosion also removes part of the topsoil in the Sahel, to what degree is this due to human land use?

Visibility measurements are an obvious, albeit indirect, indication that the Sahel produces a significant part of the Sahara plume. Visibility is a standard measure of many meteorological field stations, including those in the Sahel. There were only relatively few 'dust days' per year until about 1970, but many more during the droughts. In Gao (eastern Mali), the number of dust days increased from nearly none in the early 1950s to almost daily in the mid-1980s. The number of dust days was highly negatively correlated with rainfall in the three foregoing years (N'Tchayi et al. 1994, 1997).

Darwin had already noted in 1833, while he was on the *Beagle* near Cape Verde, that the hazy atmosphere must be due to very fine dust blown by the wind from Africa. Visibility has been studied and measured in the Caribbean in recent decades; at Barbados a significant linear relationship was found between Sahelian rainfall and local visibility (Nicholson et al. 1998). Since annual rainfall in the Sahara desert is always close to zero, it is unlikely that all dust would originate from the Sahara. N'Tchayi et al. (1994, 1997) suggested that normally most dust originates from the Sahara, but that during the Great Drought more than half of the dust came from the Sahel. Tegen & Fung (1995) arrived at the same conclusion and estimated that 20-50% of the global dust load was derived from 'disturbed soil'. These data have been challenged in recent years.

Brooks & Legrand (2000), using precise infra-red channel measurements from the METEOSAT satellite, described in detail where and when dust was produced. They concluded that nearly all dust is being produced in the Sahara desert, particularly in the western Sahara between 20 and 25°N (Mauritania, Mali, Algeria) and further east between 13 and 25°N. Despite a high correlation between dust production and rainfall in the preceding 1-2 years in the northern Sahel zone (15-17°N), they found no evidence that dust came from the Sahel.

To detect aerosols, Goudie & Middleton (2001) and Prospero et al. (2002) used UV measurements from the Total Ozone Mapping Spectrometer (TOMS) on board another satellite. A comparison of dust production worldwide in 1980-92 showed that the Sahara produced more dust than all deserts and dry lands in the world together, that three of the four largest dust producers in the world are found in the Sahara, and that the Bodélé area north of Lake Chad is by far the biggest dust producer in the world (Prospero et al. 2002). The extremely dry Bodélé depression produces so much dust because its soil consists of clay; about 10 000 years ago water filled this drainage basin of Lake Chad almost to capacity. The strong *Harmattan* wind completes the optimal scenario for dust production in this inhospitable region. Goudie & Middleton (2001) and Prospero et al. (2002) concur with Brooks & Legrand's (2000) conclusion that nearly all dust is being produced in uninhabited areas with less than 200 mm rain per year. Although fertile ground has been blown away from the Sahel during the Great Drought, the suggestion that the 'Sahara plume' in dry years was mainly due to dust produced in the Sahel has been convincingly shown to be false.

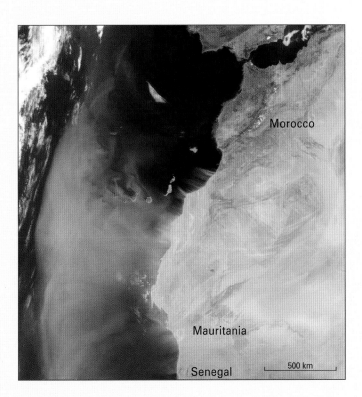

Fig. 25 Dust from the western Sahara is blown over the Atlantic Ocean, 11 February 2001. From: http://veimages.gsfc.nasa.gov.

Paired photo from the Sahel-Sudan zone in central Senegal in 1984 and 1997 (top) showing extension of cropland (groundnut and millet), but also more trees; *Acacia* woodland near Dagana (North Senegal) on 14 october 1983 and 31 january 1994, showing loss of trees from cutting (bottom);

satellite images of the surroundings of Dara (40 km NE of Dakar, Senegal) showing denser tree and scrub cover in 1965 and open tree and scrub cover in 2005. Grazing in the Sahel extends beyond mere grasses and herbs. In the dry season, the diet of goats in the Sahel includes leaves of trees and bushes. Goats normally can reach up to a height of 120 cm (left), although they can sometimes be seen climbing a tree), dexterously clipping the leaves in between the long spines of Acacia seyal (middle). Due to the permanent grazing pressure, palatable branches and leaves below the 120 cm line are systematically sheared (*Salvadora persica tree*; right). (Northern Mali, February 2009).

Vegetation

Bush fires in the dry season may turn into big fires.

not decrease too much. This explains why the seasonal variation in NDVI is much smaller for floodplains than for the surrounding dry land, and also why the Inner Niger Delta and Lake Chad remain visible as green 'oases' in the October and November images (Fig. 23). For the same reason, the NDVI in open grassland varies more than ecosystems dominated by trees (Chamberlin et al. 2007).

Analyses of NDVI-data in the Sahel, usually based on satellite images with a higher spatial resolution, could show long-term effects of land use change, deforestation, overgrazing and land degradation (Thiam 2003, Li et al. 2005). The NDVI-index is a powerful tool to describe the variation in green biomass: the data summarised in Fig. 26 will therefore be used in later chapters as a green vegetation index.

in the Sahel (Hess et al. 1996, Olsson et al. 2005, Anyamba & Tucker 2005). Statistical analysis shows that rainfall in the preceding year also has an impact on vegetation [4] (Camberlin et al. 2007). On a longer timescale, delayed effects of climate change on vegetation are even more marked (Breman & Cissé 1977). Olsson et al. (2005) argued that the recent greening of the Sahel cannot be fully explained by increased rainfall, and put forward other contributing factors such as improved land management, higher agricultural productivity and - regionally important - neglect of cropland due to political unrest and armed conflicts. Herrmann et al. (2005) arrived at the same conclusion.

Other factors, beside rainfall, determine the variation in NDVI. The seasonal course of the NDVI in the Sahel has been shown to differ, for instance, between cropland and forest (Li et al. 2005). It also differs for sandy and clay soils (Kumar et al. 2002), related to the variation in soil moisture. The vegetation may remain green longer in the dry season, despite the burning sun, if soil moisture does

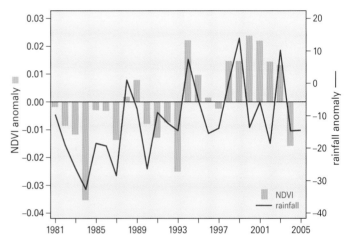

Fig. 26 Annual variation in the NDVI (anomaly in the period September-February; data from Figs. 24 and 27; left axis) and rainfall (anomaly; data from Fig. 9; right axis) in the Sahel. From: NOAA.

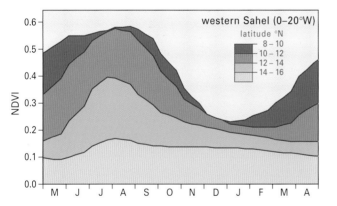
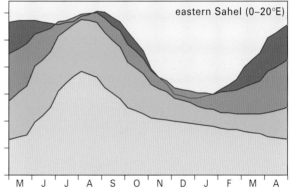

Fig. 27 Seasonal variation of the vegetation-index (average NDVI for all decades between May and April for 24 years (1981 - 2004)), given separately for the western and eastern Sahel and four different latitudinal zones. From: NOAA.

An early sunset due to dust in the atmosphere during the dry season.

Floodplains

The major African floodplains are associated with rivers that have strong seasonal differences in volume (Denny 1993). These include the Zambezi, Nile and Niger Rivers, as well as the rivers flowing into Lake Chad and their tributaries. Spilling of the river water over the levees onto the associated floodplains can take place from once to several times a year. The regularity of flooding and its depth and duration obviously influence what plant and animal species are present (Denny 1993).

Maximum flood depth also determines above-ground biomass production of perennial grass communities in African seasonally flooded grasslands (Scholte 2007). Under deeply inundated circumstances, *i.e.* more than 2 m, above-ground standing herbaceous biomass may reach 30 ton dry mass per ha (Hiernaux & Diarra 1983), up to ten times as high as in surrounding dryland areas (Le Houérou 1989). The forage quality on floodplains, when characterised by its protein-content, is generally negatively correlated with above-ground biomass. At the end of the flooding season, these floodplains are covered with a large quantity of grasses of below-maintenance quality (Breman & de Wit 1983; Hiernaux & Diarra 1983; Howell *et al* 1988). The main grazing asset of the seasonally flooded grasslands is regrowth, which is of much higher quality. This regrowth is triggered by burning and grazing, and gradually becomes available during the dry season (Hiernaux & Diarra 1983; Howell *et al.* 1988; P. Scholte unpubl.). On the Inner Niger Delta floodplains, regrowth biomass was found to be a linear function of previous above-ground biomass (Breman & de Ridder 1991), and thus indirectly a function of maximum depth of the preceding flood.

Endnotes

1. The MODIS spectroradiometer on board of the TERRA spacecraft may register thousands of fires throughout the Sahel simultaneously, varying in extent and number during the year; see http://www.terra.nasa.gov.
2. One of the measures frequently used is the Normalised Difference Vegetation Index (NDVI), which is defined by the formula: NDVI = (NIR − VIS) / (NIR + VIS), where NIR stands for solar radiation in the near-infrared and VIS for solar radiation in the visible wavelengths. The NDVI function is indicative of the abundance and activity of leaf chlorophyll pigments and thus a proxy for photosynthetic activity of terrestrial vegetation. When vegetation occurs the NDVI is usually larger than 0.1. Dense vegetation exceeds the value of 0.5. The NDVI proved to be a goldmine for researchers. The Advanced Very High Resolution Radiometer (AVHRR) aboard the NOAA meteorological satellites, used in measuring spectral data, has been operational since July 1981. The data have a spatial resolution of 4 km, but are distributed at a resolution of 8 km. The latter results in a coverage of the Saharo-Sudan transition zone by 360 rows times 750 columns, thus 270 000 'pixels'. The data (maximal values per decade) are freely available and can be downloaded from http://www.fews.net. The NDVI data set of the Sahel has been the topic of several studies (e.g. Tucker *et al.* 2005).
3. To analyse the annual variation in NDVI, we calculated the deviation between the average decadal values, as shown in Fig. 27 and the measurements of the different years. These deviations were averaged for the following September through February, *i.e.* covering 24 decadal measurements. Since the average anomaly values for the different zones did not appear to differ between the years, they were again averaged.
4. The vegetation anomaly in year I (V_i) may be described as a function of the rainfall anomaly (R_i) in the same year: $V_i = 0.00956 + 0.00082R_i$. The relationship between rainfall and vegetation cover, although significant (p<0.001) is far from perfect: $R^2 = 0.41$. The fit improves significantly ($R^2 = 0.52$) if the rainfall during the year before (R_{i-1}) is added to the equation: $V_i = 0.01384 + 0.00075R_i + 0.00042R_{i-1}$.

Land use
With contributions of Paul Scholte

Africa is a continent where ecological problems are larger than life. This holds in particular for the Sahel, which has been badly hit by simultaneously occurring disasters: loss of vegetation cover and soil moisture, desertification, exhaustion of cropland, land degradation, soil erosion, deforestation... The list seems endless, and these problems become even more pressing in the light of Africa's steep population growth. The questions surrounding human-induced climate change and desertification have already been discussed in chapters 2 and 4. The idea that the Great Drought in the 1970s and 1980s was linked to land abuse and vegetation loss in the Sahel was universally accepted in the 1980s and 1990s. This hypothesis lost much of its attraction when it became clear that rainfall in the Sahel largely depends on variations in the surface temperature of the tropical and subtropical oceans. Most researchers now believe that land use has not triggered, but rather reinforced, the decline in rainfall, thus hampering recovery to pre-drought levels. In this chapter, we investigate the changes in land use and the concomitant changes in land cover in the Sahel, particularly in the face of the growth of the human population in the Sahel and the expanding demands on food, water and energy.

The food demand of an expanding human population

In the early 2000s, the human population in most African countries increased by about 3% per year. According to the FAO (FAOSTAT 2005), the population increase is highest in Niger (3.15%) and The Gambia (3.41%), but still reaches about 2.5% in the other Sahel countries. At a rate of increase of 2.5%, the population doubles within 28 years, but at 3%, doubling takes only 23.5 years. From fewer than 20 million inhabitants in the mid-20th century, the population in eight Sahelian countries has expanded to over 60 million at the start of the 21st century. Contrary to most other countries with a high population growth, this increase is projected to increase further in the 2010s (e.g. Niger 3.5% according to the Human Development Report 2007-2008 of the UNDP. The Sahelian population size may reach 130 million in 2030 (Fig. 28).

Population growth differs markedly between urban and rural populations. The figures for Mauritania are most striking in this respect: between 1961 and 2003 many people moved to the cities of Nouadhibou and Nouakchott, which expanded from 65 000 to 1.8 million inhabitants; the total population increased from 1 to 2.9 million, but the rural population by only 10% to 1.1 million. A similar pattern applies to the other Sahel countries. In 1970, at most 20% of the people in West Africa were city dwellers. In the early 2000s, half of the population in West Africa already lived in cities. Because the growth rate in cities is twice that in the countryside, city dwellers will vastly outnumber the rural population by 2020.

The change in land use would probably have been different without urbanisation. The environmental effects of urbanisation are manifold. First, people in Sahelian cities use charcoal instead of firewood to prepare their meals and, although charcoal stoves are more efficient than woodstoves, this does not compensate for the

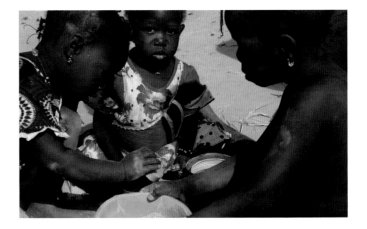

The human population in the Sahel increases by 3% per year.

75% loss of wood energy during charcoal production (Ribot 1993). As a consequence, city dwellers in Senegal in the late 1980s (25% of the population) consumed well over half of the total primary wood energy. Secondly, the growing urban population needs ever more drinking water. Dakar (2.3 million inhabitants in 2004) derives its drinking water from Lake Guiers, a large reservoir created along the Senegal River. And Kano, in northern Nigeria, could not have expanded to a city of 1.2 million people without the dams built across nearby rivers. Thirdly, the per capita energy demand, although still low in the Sahel, is higher in cities than in the countryside. The main reason for creating the Selingue reservoir in the Upper Niger was to produce electricity for Bamako, the capital of Mali (1.7 million inhabitants in 2007).

Urbanisation also has indirect effects on land use. In West Africa, many city dwellers eat bread and rice, rural people mostly millet and sorghum. Consequently, urbanisation affects the crops being grown. For example, in Mauritania, the production of millet and sorghum actually decreased between the early 1960s and early 2000s by 30% (FAOSTAT 2005). To cover the shift to eating baguettes, wheat imports increased in Mauritania from 11 000 tonnes in the early 1960s to 300 000 tonnes in the early 2000s. At the same time rice consumption rocketed. Because rice was not grown in Mauritania in the 1960s, 10 000 tonnes had to be imported annually to cover the demand. Forty years later, 40 000-50 000 tonnes were locally produced to supplement the import of another 40 000 tonnes. Millet, and to a lesser degree sorghum, can be grown in semi-arid areas, but rice plants need wet roots. The shift from millet to rice production in the Sahel required the cultivation of natural floodplains and the creation of irrigated areas.

The Mauritanian data in the FAO *food balance sheets* show that its trebled population was fed mostly with imported cereals and imported milk. Similarly, in Senegal food consumption increasingly is covered by imports (Fig. 29). Here, the shift from millet to rice consumption was covered partly by an increase in rice production

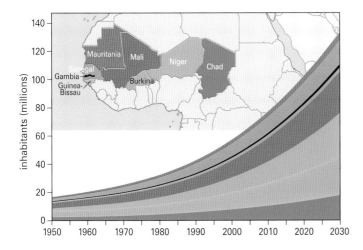

Fig. 28 Population growth in eight West-African Sahelian countries. Source: FOASTAT (2005).

The staple food of the rural population consists of sorghum and millet, crops that are adapted to dry environments. City dwellers, on the other hand, prefer rice and bread, necessitating increasing imports and the construction of irrigation schemes.

to Mauritania and Senegal, the other Sahelian countries have enlarged their national food production to a greater extent, as shown for Mali (Fig. 29), and imports are limited mainly to wheat and rice. Extremes in annual food production (poor years were 1971-1973, 1982-1987 and 1990; good years were 1978, 1988, 1994, 1998, 1999 and 2003) are related to local rainfall (Figs. 7 and 9). The early 1970s and 1980s were years of starvation, and much food was imported to prevent a still larger disaster. Nevertheless, even in these drought-stricken years, local food production was still much higher than the imported totals.

Dry and wet years have a direct bearing on cereal production. The effect on livestock is more delayed, because it takes time to replace the depleted herds of cows, goats and sheep. Nevertheless, the general pattern has been one of marked increase in livestock since the 1960s. For the five western Sahel countries combined, meat production increased from 250 000 to over 800 000 tonnes between 1961 and 2003 (c. 3.5×), just keeping ahead of the growth of the human population (Fig. 28). In this 42-year period, milk production doubled from 750 000 to 1 500 000 tonnes, but when the import of 300 000-400 000 tonnes of milk since the 1980s is taken into account, the per capita milk consumption remained stable.

From these data we conclude that, with the exception of Senegal

from 50 000 tonnes in 1960 to 150 000 tonnes in 2000; the import of rice rose more steeply from 100 000 to 800 000 tonnes. In contrast

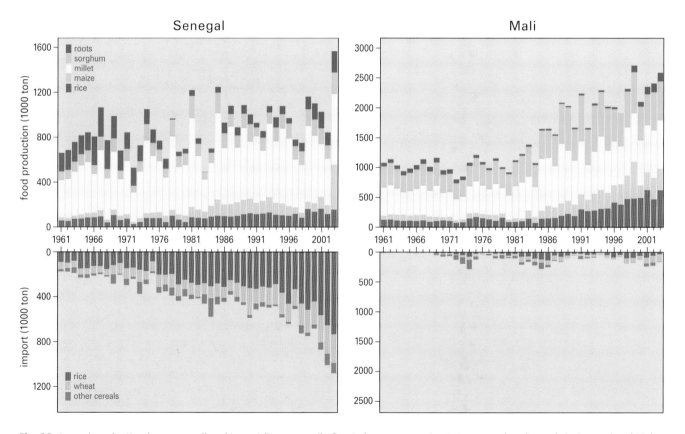

Fig. 29 Annual production (upper panel) and import (lower panel) of roots (cassava, sweet potatoes, yams) and cereals in Senegal and Mali. From: *Annual food balance sheets* of the FAOSTAT.

and Mauritania, the increase in agricultural production sufficed to cover the food demands of the growing population. How could this be, given the plethora of studies on desertification, land degradation and declining soil fertility in the Sahel?

Cropland

The Sahel has poor soils and an unfavourable climate for agriculture (Breman *et al.* 2001). Nevertheless, the area covered with millet in the eight western Sahel countries doubled from 4.5 million ha in the early 1960s to 9 million ha 40 years later. In the same period, sorghum increased from 3 to 5.5 million ha and rice from 0.4 to 0.8 million ha (Fig. 30). The increase of arable land differs per country (Table 2), with a decline in Mauritania, small expansions in Senegal, Guinea-Bissau and Chad and large increases in The Gambia, Mali and Niger. The area of productive arable land covers only a small fraction of the northern countries (Mauritania, Mali, Niger), but takes a larger fraction of the southern countries (The Gambia, Guinea-Bissau, Burkina Faso). Part of the arable land is not cultivated but laid fallow and, especially in dry years, a harvest is not always forthcoming from the cultivated land. According to FAO-statistics, the total area of arable land in most countries and most years is about twice as large as the harvested area. Land use statistics collected by the FAO pertain to very global data, but there are several detailed studies which have quantified the increasing surface of cropland.[1]

Table 2 Total surface area of seven countries and the area used for cereal production, averaged for 1961-1963 and 2001-2003 (million ha). The last column gives the average annual change between 1961-1963 and 2001-2003 (same data as Fig. 29, but including the surface area of maize, fonio and wheat).

Country	Total surface	Harvested 1961-1963	Harvested 2001-2003	Change,%
Mauritania	102.52	0.25	0.12	-1.3
Senegal	19.25	1.01	1.16	0.4
The Gambia	1.00	0.08	0.18	3.1
Guinea-Bissau	3.16	0.10	0.13	0.8
Mali	122.02	1.60	3.28	2.6
Niger	126.67	2.27	7.17	5.4
Burkina Faso	27.36	1.86	3.25	1.9
Chad	126.00	1.27	1.89	1.2

A second way of improving food production would be to increase the yield per ha. With similar amounts of rainfall, the yield of millet after 1984 is 5% higher compared to the period before (Fig. 31). Farmers apparently increased their yield slightly, a conclusion supported by detailed studies (*e.g.* Mortimore & Harris 2005, Rey *et al.*

2005). Until the mid-20th century, farmers used a system of shifting cultivation in which fallow land was pivotal to avoid depletion of soil nutrients. The population growth caused a gradual replacement of this system by permanent cultivation. Despite declining soil fertility (Penning de Vries & Djitèye 1982, Breman & de Wit 1983), this was achieved without artificial fertilizers (too expensive to be used on a large scale). This unavoidably led to nutrient mining, *i.e.* the lowering of organic matter in the soil and loss of nutrients, as con-

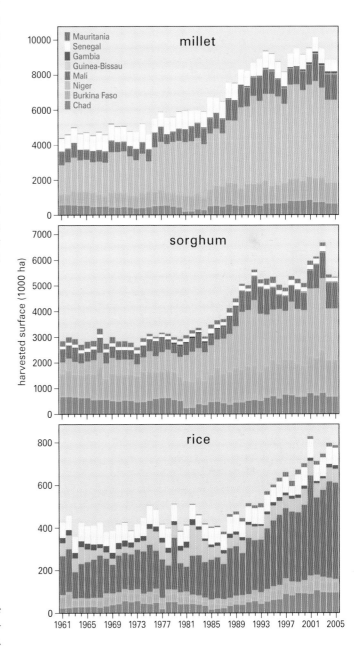

Fig. 30 Surface area of arable land where millet, sorghum or rice was produced in 1961-2005. From: FAOSTAT.

Millet is grown on 1 million ha in West Africa.

firmed in later studies (*e.g.* Stoorvogel *et al.* 1993). Despite the negative "nutrient budget", the yield per ha increased over the years (Ben Mohammed *et al.* 2002; Fig. 31). Ridder *et al.* (2004) reviewed earlier studies and concluded that fertility decline of communal land far from the villages was not as large as predicted. The land close to the villages became more fertile by increasing the usage of organic material. At least locally, land degradation could thus be avoided by integrating livestock with cropping (Rey *et al.* 2005, Mortimore & Turner 2005).

Livestock

The African livestock population in 2004 was estimated by the FAO at 254 million sheep, 239 million cattle and 232 million goats. Some 60% of the sheep and cattle, and up to 66% of the goats, are found in the 21 countries situated between the Sahara and the tropical rainforest, roughly the zone between 5° and 20°N (Fig. 32). In the transient zone from dry Sahara to rainforest, the distribution of cattle, sheep and goats is generally limited to the zone where annual rainfall ranges between 50 and 1000 mm. With less than 50 mm of rain annually, vegetation is too scarce to support livestock other than camels. In regions with more than 1000 mm of rain, the tsetse fly abounds and sleeping sickness prevents herding the large humped Zebu cattle. The humpless, longhorn *Ngama* cattle are resistant to sleeping sickness but less adapted to the Sahel. Consequently, 80% of the livestock in the transient zone is found in the Sahel, and 20% in the Sudan zone (Le Houérou 1980).

Grazing conditions vary seasonally in the area limited by rainfall (50-1000 mm). At the beginning of the rainy season, herds start moving northwards to reach the northernmost grazing grounds about two months later; rainfall in the northern Sahel is limited to showers in June – September. The grasslands in the northern Sahel provide high-quality grazing, but as soon as the green biomass withers at the start of the dry season, herders move southwards again. During the dry season, livestock feed on left-overs of the harvest, grasses, and leaves of scrub and trees. Many herders move with their animals into the floodplains which provide high quality forage (Box 2). This seasonal land use system, known as transhumance, makes optimal use of temporal and spatial variations in the quality of grazing grounds.

There have always been more farmers than herders in the Sahel, but the total surface area of the grazing grounds is much larger than that occupied by croplands. In Mauritania and Chad, for instance, the arable land only covers respectively 1.3 and 7.5% of the total land area, but in both countries 30% of the land is used as grazing grounds (source: FAO). In the northern Sahel, where rainfall is insufficient for cropping, there is no competition between herders and farmers. However, wherever precipitation exceeds 400 mm per year, millet can be grown, and farmers will do anything to prevent herds from grazing on their cropland before it has been harvested. Direct competition is absent outside the rainy and harvest seasons, because grazing on crop residues offers forage to the herds and manure to the farmers (Turner 2004). The situation is more prob-

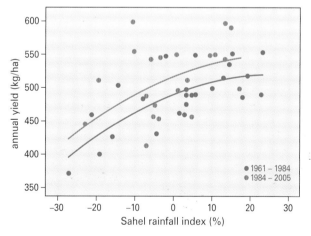

Fig. 31 Annual yield of millet (kg/ha), before and after 1984, combined for eight western Sahel countries, as a function of the annual rainfall in the Sahel and expressed as percent deviation from the average in the 20th century (Fig. 9). The curvilinear lines are based on second degree polynomials. From: FAOSTAT.

Fig. 32 Density of cattle and sheep per km² in Africa in 2004. The area indicated as unsuitable includes deserts, rainforests and protected areas (Fig. 35). From: *Global Livestock Production and Health Atlas*, produced by FAO-AGA, in collaboration with ERGO and the TALA research group, University of Oxford, UK (http://www.fao.org/ag/aga/glipha). Land cover is shown for 2000; from http://www-gem.jrc.it/glc2000).

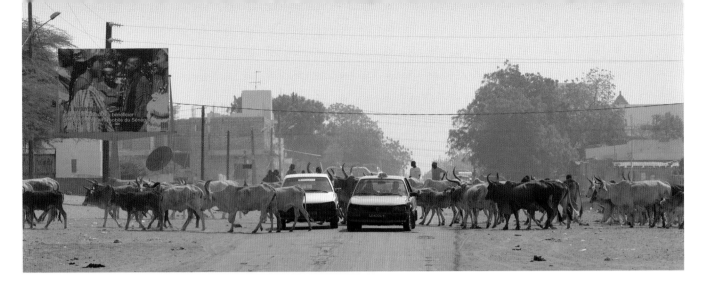

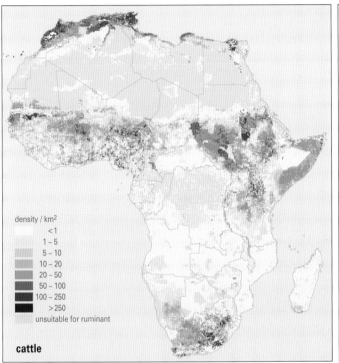

density / km²
- <1
- 1–5
- 5–10
- 10–20
- 20–50
- 50–100
- 100–250
- >250
- unsuitable for ruminant

cattle

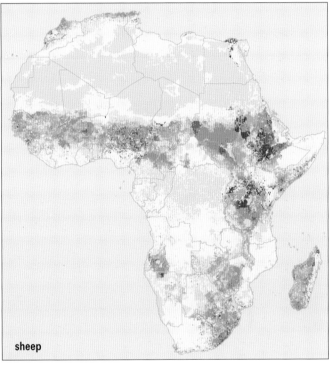

sheep

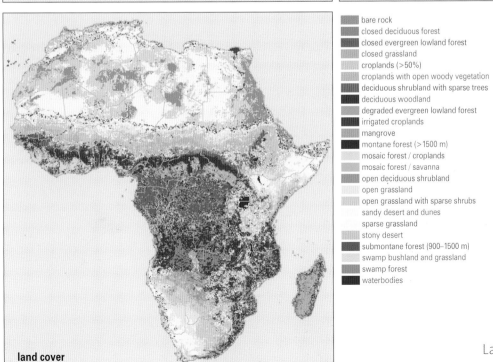

land cover

- bare rock
- closed deciduous forest
- closed evergreen lowland forest
- closed grassland
- croplands (>50%)
- croplands with open woody vegetation
- deciduous shrubland with sparse trees
- deciduous woodland
- degraded evergreen lowland forest
- irrigated croplands
- mangrove
- montane forest (>1500 m)
- mosaic forest / croplands
- mosaic forest / savanna
- open deciduous shrubland
- open grassland
- open grassland with sparse shrubs
- sandy desert and dunes
- sparse grassland
- stony desert
- submontane forest (900–1500 m)
- swamp bushland and grassland
- swamp forest
- waterbodies

Land use

Box 2
Sahelian floodplains: important for livestock and wildlife

The almost featureless floodplains of the Inner Niger Delta, the Sudd and the Lake Chad wetlands are, once the floods have receded, home to millions of cattle. The same can be said of the floodplains of the Zambezi and other major rivers in Africa, where lethal diseases do not preclude grazing of cattle. Grazing the floodplains involves a delicate and complicated system of rights of passage, water access and management fine-tuned over centuries but now threatened by recent changes. By and large the herders use the floodplains when they can, i.e. after flood recession, when grasses and herbs of sufficient quality are available. Water availability generally does not pose problems. Good quality regrowth is triggered by setting fire to the old, poor quality growth (Chapter 4).

Sahelian floodplains provide forage for up to ten times as many cattle compared to surrounding drylands. During the peak of the dry season, a crucial time for survival, cattle densities in the floodplains are up to 20-30% higher than in adjacent drylands. Contrary to common belief the individual condition of cattle is similar to that of cattle grazing elsewhere; rather, it is the sheer number of livestock in the floodplains that stands out. Another misconception relates to the idea that floodplains form a buffer in times of drought: they do not. Nonetheless, in the absence of floodplains, the carrying capacity of the Sahel would be disproportionally smaller. Intensified land use, such as rice cultivation competing for the same land and water resources, hits the livestock system at its most sensitive spot, and reduces the entire region's capacity to hold livestock.

The Sahelian floodplains are equally important for wildlife. The collapse of antelope populations, such as Topi (Box 4), strongly suggests that floodplains have become inhospitable, notably so during the dry season. Where floodplains and swamps are not yet grazed with cattle, as for instance in the Bengweulu Basin in Zambia, antelopes still abound; after the flood has receded some 30 000 Black Lechwe *Kobus leche smithemani* leave the woodlands and roam the grasslands unobstructed by cattle or man (and thanks to the tsetse fly). Floodplains are equally important to birds, both African and Palearctic (Chapters 6 onwards). From: Scholte & Brouwer (2008).

lematic in floodplains and other wetlands, where crops compete with dry season grazing (Box 2).

The trend among herders to settle (Turner & Hiernaux 2008), and the steep increase in arable land since the 1960s (to the detriment of grazing grounds; Fig. 30), did not prevent livestock numbers from increasing. In fact, FAO statistics for the eight western Sahel countries show a doubling of cow numbers between 1961 and 2003, and a still larger increase for goats and sheep (Fig. 33). The dry years 1973 and 1984 killed off large numbers of cattle, sheep and goats, but after 1984 numbers have doubled in just two decades.

The annual increase of livestock since 1960 differed per country (Fig. 33); cattle populations in Mauritania and Niger even decreased. Sheep and goats increased more than cattle, possibly due to an increase of village-based (compared to nomadic) livestock. Sheep and goat numbers increased least in the two countries where cattle numbers decreased, i.e. Mauritania and Niger. The steeply growing number of livestock in Burkina Faso, already in evidence in the 1950s (Le Houérou 1980), and the slow increase or even decline in Mauritania, Niger and Chad, suggest that the increase in livestock numbers is biased towards the southern Sahel, possibly a side-effect of reduced rainfall since 1970 having a larger negative impact on semi-desert than on the Sudan zone.

There are several explanations why livestock in the Sahel has increased during the second half of the 20th century. The first is rainfall. The extremely dry years 1973 and 1984 caused mass mortality among cattle, sheep and goats. The recovery of these losses since 1984 was facilitated by a gradual improvement of rainfall.

Secondly, in the second half of the 20th century thousands of boreholes and wells were created across the Sahel, allowing access to areas without natural watering points. In the past, grazing grounds were restricted to within 15 km of rivers, floodplains or temporary ponds (Lind et al. 2003). Thus, transhumance was not only governed by spatial and temporal variations in the quality of grazing grounds, but also by the availability of watering places. Man-made watering points allowed herders to graze their livestock in areas and during times when they would be otherwise inaccessible. However, although these watering points created security and an extension of grazing opportunities in the northern range, they also initiated social strife (Théboud & Batterbury 2001).

Thirdly, tsetse eradication has opened much of the southern grazing range. In the area in central Africa infested with tsetse fly, covering some 10 million km^2 and known as "the Green Desert", neither cattle nor horses can survive (Fig. 32). Pesticides such as DDT, and, more recently, sterile insect technology, have eradicated the tsetse fly in many places, albeit often only temporarily. If such methods were successfully applied on a large scale, it would facilitate vast areas for use by livestock and people, wiping out the remaining wildlife concentrations in West-Central Africa.

Fourthly, livestock husbandry and cropping have become increasingly integrated. During the dry season, cattle are nowadays fed with supplementary fodder, such as groundnut, cotton cake and

Herders move with their cattle, goats and sheep between the northern Sahel (July-September) and the Sudan zone (November-April).

cowpea haulms, sometimes imported from far away (Mortimore & Turner 2005). Herders in the Inner Niger Delta grow a floating grass species, *Echinochloa stagnina*, locally known as *bourgou*, to be used as fodder during the dry season (see Chapter 6). As a consequence, cattle numbers may exceed the maximum as determined by the carrying capacity of grazing land alone (Mortimore & Turner 2005). The dichotomy between nomadic herders and resident farmers has also become less clear-cut. Increasingly farmers have some cattle that graze on village land. Herders have also diversified into owning some arable land. Many city-dwellers now hire herders to graze their livestock (Rey *et al.* 2005, Turner & Hiernaux 2008). Although these changes have enhanced the efficiency of the system, they have also caused tensions, for instance when centuries' old agreements to graze nomadic herds on farms were cancelled to allow grazing on crop residues by resident cattle exclusively (Théboud & Batterbury 2001, Turner 2004, Moritz 2002).

Forest

The expansion of cropland and the increase in livestock were preceded by clearance and degradation of woodland, where cutting and felling exceeded regeneration. It is widely accepted that the Sahel and Sudan zones have been subjected to deforestation. However, Fairhead & Leach (1995) found no 'savannisation' near Kissidougou (Guinea) and concluded from this that 'eco-pessimistic' sto-

Local authorities and non-governmental organisations advocate planting of Eucalyptus near villages to reduce the cutting of natural woodland; Inner Niger Delta, November 2008 (left) and February 2007. In terms of food supply for Palearctic birds Eucalyptus is inferior to indigenous trees.

ries about deforestation were exaggerated or even untrue (Fairhead & Leach 2000).

Detailed quantitative data on land cover change and deforestation are available for Senegal (Tappan *et al.* 2004, Wood *et al.* 2004). Tappan *et al.* (2004) showed that changes in land use and rates of deforestation in Senegal differ regionally. Forests (defined as areas with over 80% canopy closure) covered 4.4% of the country in 1965 and 2.6% in 2000, a loss of 41% in 35 years. For Senegal's wooded savannas and forest combined, the cover decreased from 78.1% in 1965 to 72.2% in 2000. The loss of woodland is mainly caused by expansion of cropland. The loss of wooded cover is, however, much larger than is reflected in the above figures. For instance, the agricultural parkland in the Saloum area had a tree cover of 40-70% in 1943, but this had declined to 10-20% by 2000. Moreover, in northern Senegal many trees died during the Great Drought (Gonzalez 2001, Tappan *et al.* 2004), while throughout Senegal woodlands have been thinned for charcoal production. The annual legal charcoal production has increased from some thousand tonnes in the 1930s, to 15 000 tonnes in the 1950s and 100 000 tonnes since 1970 (Ribot 1993). According to the same author, in the 1990s the official quota covered only 50-67% of the actual urban use, so charcoal production must have increased again in the 1990s. In 1950, charcoal came from forests within a radius of 70-200 km around Dakar, but in 1990 this source area had increased to within 300-450 km from Dakar. In fact, in the 1990s charcoal producers already reached as far as the borders of the country. It is obvious that the deforestation of Senegal is rapid, but measurements show that the rate is less than stated in several political documents and semi-popular papers. For instance, Tappan *et al.* (2004) found that the absolute decline of Senegal's woodland amounted to 333 km^2/year, which is less than the estimate of 520 km^2/year in FAO statistics. Ribot (1999) concluded that the actual urban demand for firewood, although very high (Ribot 1993), does not support crisis scenarios, which in the 1980s predicted shortage of firewood for the 1990s or early 2000s. The impact of collecting wood for fuel has been overrated as a deforestation factor, because natural regeneration was underestimated and rural people often collect wood from dead trees (Benjaminsen 1993, Thiam 2003).

Other empirical studies also found less deforestation than generally anticipated. Most studies show a decline in woodland cover from agricultural encroachment and collection of wood for fuel, but the data often refer to local studies and are difficult to summarise in simple statistics. Moreover, the transition from woodland to agricultural land often goes along a gradual pathway, from forest through open forest, agricultural parkland and agricultural land with scrub to pure agricultural land. This impedes straightforward quantification of loss of wooded cover and complicates comparisons between different studies. Mortimore & Turner (2005) and Rey *et al.* (2005) show that the degree of deforestation has not been the same throughout the Sahel, because farmers have planted trees along their fields or around their houses, and valuable trees have been protected during clearance of woodlands. In northern Ghana and southern Burkina Faso, Wardell *et al.* (2003) found that between 1986 and 2001 25% of the gallery forests had been changed into savanna woodland and 45% into agricultural parkland. At the same time, a small part of the agricultural parkland of 1986 had transformed back into woodland because the clearance of the Sudan woodland had facilitated tsetse flies and the spread of sleeping sickness, forcing people to retreat and promoting natural reforestation.

Protected areas

IUCN listed 26 832 protected areas in the world, covering in total 7539 million ha. Of these areas, 974 are situated in Africa and cover 175 million ha. Within Africa, more than 60% of the protected areas lie in only seven countries, of which two have more than 10% of the African total: Zambia 13% (22.8 million ha) and Ethiopia 10% (17.6 million ha). Protected areas are mostly found in eastern and southern Africa (Fig. 35A). The total surface of the protected areas in the western and central Sahel countries amounts to 24.9 million ha, or 14% of the entire surface of the protected areas in Africa.

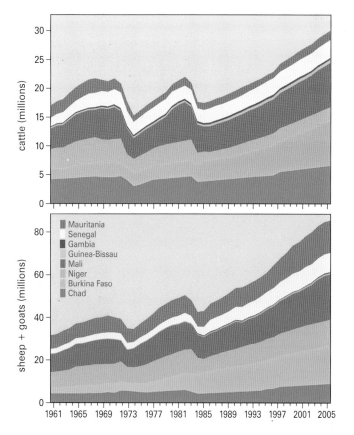

Fig. 33 Number of cattle and sheep+goats in western and central Sahel countries in 1961-2005. From: FAOSTAT database.

In total, 6.8% of Africa has the status of protected area. Countries with a relatively high coverage of protected areas again mostly lie in eastern and southern Africa, except for Equatorial Guinea (21%), the Central African Republic (12%), Senegal (11%) and Chad (9%). In other Sahel countries, the percentage coverage is low, at least compared to the rest of Africa,: 3.0% in Mali, 1.5% in The Gambia, 1.3% in Sudan and 1.1% in Mauritania. Since most of these protected areas are found in the Sahara desert (Fig. 35), the Sahel zone proper is almost entirely unprotected.

Not all protected areas have the same status. IUCN distinguishes eight categories (Fig. 35A). Scholte et al. (2004) reviewed the management effectiveness of the 14 main protected areas in the Lake Chad Basin totalling 14 million ha. Four (representing 71% of the total surface) exist on paper only, four have a low to moderate effectiveness (21%), and another four, including Waza National Park in Cameroon have a moderate effectiveness (4%). The protection can be considered effective in only two areas (4%), *i.e.* Lake Fitri (with a strong authority based on tradition) and Zakouma National Park. Conservation in western and central Africa depends more on the absence of people (Fig. 35B) and livestock (Fig. 32) than on declared status. The average population density in Africa is 27.5 inhabitants per km^2, with highest densities in the Nile delta, the Maghreb, western Africa, on the Ethiopian highlands and around Lake Victoria (Fig. 35B). Within the Sahel, the western part is more densely populated than the central and eastern parts (Fig 35B), possibly the main reason why the eastern Sahel still harbours large herbivores that have now disappeared from the western Sahel (Box 4).

Neem *Azadirachta indica* is native to SE Asia and known for its antimicrobial properties. It provides effective ingredients for medicines, cosmetics and insect repellents. This tree species has been planted on a large scale near human settlements in West Africa. Insect densities are very low in Neem, except for ants. (Niger (above), Burkina Faso (below left), Mali (below centre and below right)).

Land use

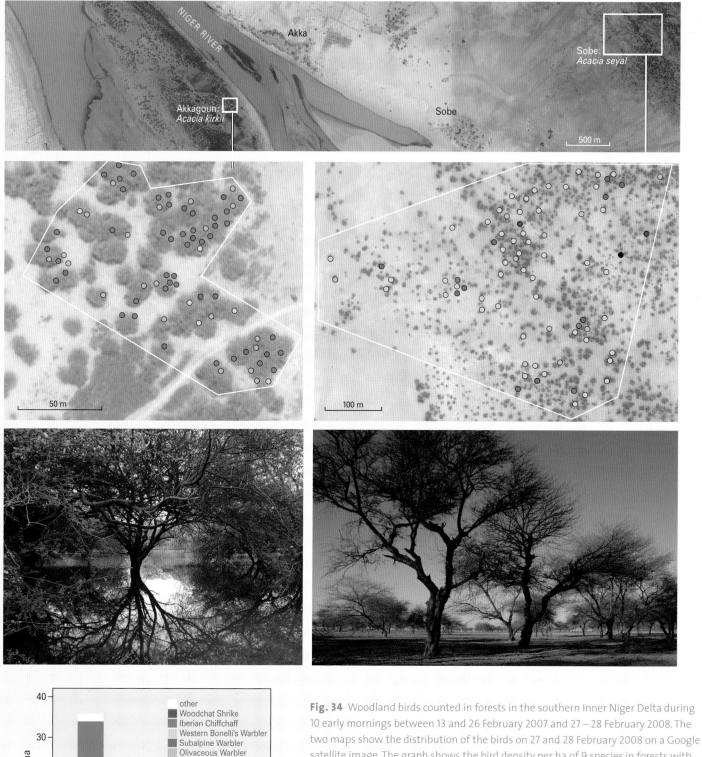

Fig. 34 Woodland birds counted in forests in the southern Inner Niger Delta during 10 early mornings between 13 and 26 February 2007 and 27 – 28 February 2008. The two maps show the distribution of the birds on 27 and 28 February 2008 on a Google satellite image. The graph shows the bird density per ha of 9 species in forests with *Acacia seyal* (4 plots; 22.1 ha), *Acacia kirkii* and *Ziziphus mucronata* (3.72 ha; 5 plots in the flooded forest of Akkagoun and Dentaka), Neem *Azadirachta indica* (3 plots; 4.48 ha;) and *Eucalyptus* (1 plot; 1.08 ha). Grasshopper Warbler, Sedge Warbler, Melodious Warbler, Common Whitethroat and Willow Warbler are lumped as 'other species'. Note the close canopy of *A. kirkii* compared to the open *A. seyal* woodland (where most woodland bird species are concentrated where the canopy is relatively dense).

Box 3

Palearctic birds in exotic and indigenous trees

Deforestation, as described in this chapter and many other studies, in terms of loss of wooded cover ignores the shift during the last decades from natural woodland into plantations of exotic species such as *Eucalyptus camaldulensis* and, especially, Neem *Azadirachta indica*. For example, the surface area of forest plantations in Senegal has increased from 2000 km^2 in 1990 to 3600 km^2 in 2005, equivalent to 2.2 and 4.4% of the total forest surface. In the same 15 years the natural forests lost 1600 km^2 of their area (FAO 2006). The decline of natural forests in the Sahel is thus larger than is obvious from global remote sensing studies. For the same reason, the greening of the Sahel in the 1990s partly refers to plantations of exotic tree species, which are indeed valuable to the local people, but have poor ecological value compared to indigenous tree species. *Eucalyptus* and Neem are popular since they grow fast, even in the dry Sahel climate, while Neem produces excellent, termite-resistant timber. Moreover, they appear insensitive to plague insects. Actually, Neem and *Eucalyptus* hardly hold any insects, caterpillars or larvae. Leaves and crushed kernels of Neem are even used as a natural pesticide.

There are two studies on bird densities in exotic and indigenous forests in Africa. Stoate (1997) measured the bird density in Neem plantations and on farmland with indigenous acacia trees and *Pilostigma reticulatum* shrubs in northern Nigeria. The natural wooded vegetation held densities of 1.3 Whitethroats and 0.4 Subalpine Warblers per ha, but not a single warbler was observed in Neem stands. The author also quantified insect density and collected a large variety of such potential prey as ants, plant hoppers, plant bugs, caterpillars, beetles, spiders and flies in indigenous trees and shrubs, but only ants in Neem. John & Kabigumila (2007) found that, compared to natural forests, only a few bird species bred in eucalyptus plantations, and only in low densities.

We investigated bird density in exotic and indigenous forests in the southern Inner Niger Delta in February 2007 and 2008. Ten forests were visited in the early morning, during which we observed 280 individuals of 14 Palearctic bird species within the plots (total surface 30.5 ha). The highest bird density was observed in flood forests with 35.5 birds per ha, of which Chiffchaff (14.8/ha) and Olivaceous Warbler (9.4/ha) were the most common. The density averaged 5.3 bird/ha, mainly Bonelli's Warblers (3/ha), in dry forest with *Acacia seyal*. The observed densities of the different species are quite similar to data collected in Sahelian forests in Senegal (Morel & Roux 1966b, 1973, Morel & Morel 1992) and North Nigeria (Jones et al. 1966, Cresswell et al. 2007). In natural forests we found many insects in the canopy, but there was no insect or animal life in the canopy of Neem, which explains its poor diversity of passerines. Neem berries were eaten by the resident Common Bulbul, whereas the only migratory bird species seen in these trees was the Olivaceous Warbler (2.5/ha). Given the observations of Stoate (1997), these birds may have fed on ants, which were in fact mentioned by Cramp (1992) as prey of Olivaceous Warblers. These data, although preliminary, clearly show the ornithological significance of local acacia forests compared to exotic trees.

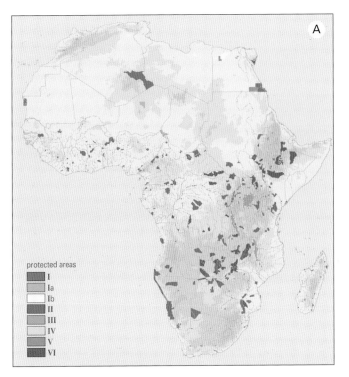

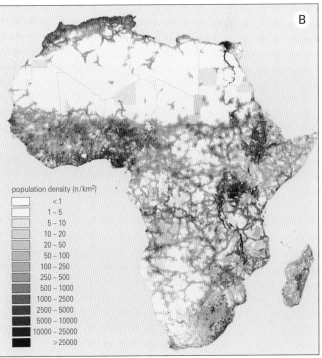

Fig. 35 (A) Protected areas in Africa. Eight categories are distinguished: I=nature reserve, Ia=scientific reserve, Ib=wilderness, II: national park, III: scenic reserve, IV: national reserve, V: area of outstanding beauty, VI: flora and fauna protection area; from WDPA Consortium (2005); (B) density of the human population in 2000; from Nelson (2004).

Box 4

The Sahel as former grazing ground for millions of wild herbivores

Sahelian grasslands are now grazed by 150 million cattle and 300 million sheep and goats, where formerly huge numbers of antelopes and other grazers roamed. Showing seasonal movements just like present-day livestock, wild herbivores facilitated and maintained grasslands by grazing. Early European travellers, as cited by Spinage (1968), were unequivocal in their observations. Barth, who traversed the Sahara in 1850, noted for the region of Aïr ..."bare and desolate as the country appears, it is covered (..) with large herds of wild oxen (*Antelope bubalis*)"; in 1871 Nachtigal saw Addax Antelopes *Addax nasomaculatus* north of Lake Chad ... "the numbers of these animals were almost unbelievable. They were to be seen in every direction, singly, or in small groups, or in herds of hundreds"; and the Topi *Damaliscus korrigum* was still so common in the western Sahel in 1935 that, "if the Topi could be counted it would no doubt be shown to have the greatest population of any African antelope" (Spinage 1968).

The Red-fronted Gazelle *Gazella rufifrons* is a Sudano-Sahelian species, whereas the Dorcas *Gazella dorcas* and Dama Gazelle *Gazella dama* are species typical of the Sahel (Scholte & Hasim 2008). Addax, and especially the Slender-horned Gazelle *Gazella leptoceros*, are inhabitants of almost pure desert. Once they were abundant and widespread, but they have been virtually exterminated, as happened with Oryx *Oryx dammah*, during the last two decades. The elimination of wildlife in the western Sahel and Sudan zone was caused by modern hunting methods (rifles and four-wheel drives), disturbance, and competition with livestock for grazing grounds and water.

Kobus kob, a typical floodplain antelope, were once common in the Inner Niger Delta, but they have been eradicated. Also giraffe and crocodile no longer occur in the area, and of all large African mammal species, only some dozen hippos remain (Wymenga *et al.* 2005). In the 1920s elephants were still common in the Inner Delta and widespread elsewhere in southern Mali. There were four small, discrete elephant populations left in Mali in 1983, of which one remained after the Great Drought (Blanc *et al.* 2003). This population, decreasing from 550 in the 1980s to 322 in 2002 (Blake *et al.* 2003), is found in the Gourma, east of the Inner Niger Delta. The migration patterns of the last of these desert elephants have been studied in detail using radio-tracking (Blake *et al.* 2002). In the course of the year the elephants follow a circular route within an area of 30 000 km^2. Given the recent developments, including the construction of a new road, the future of these desert elephants looks bleak.

Northern Cameroon is the only area in the West-Central Sahel where large numbers of antelopes can still be seen: the Waza National Park harboured 25 000 kobs in the 1960s and 5000 in the 1990s (Scholte *et al.* 2007). Waza National Park, with its giraffes, elephants and six antelope species still gives an idea how the other Sahelian floodplains must have looked. One has to go further east to encounter more wildlife (Zakouma in Chad and Gounda-Manovo in the Central African Republic).

Even larger numbers of antelopes still migrate seasonally in and around the Sudd floodplain in southern Sudan (Chapter 10). In the early 1980s, the Boma National park, close to the Ethiopian

Large animals have disappeared, or become scarce, in much of West Africa, the outcome of a century of competition with man. The distribution of Giraffe (Central African Republic), Lion (Cameroon), Roan Antilope (Niger) and Elephant (Burkina Faso) in the western Sahel is largely restricted to parks.

border, harboured over 800 000 White-eared Kobs *Kobus kob leucotis* moving between floodplains in the dry season and savanna in the wet season (Fryxell 1987). This marsh antelope declined to 210 000 in 2001, according to a count by the New Sudan Wildlife Conservation Organization. Since the widespread introduction of weapons was the expected outcome of the civil war between 1983 and 2003, it was surprising that such large numbers of kobs were still present. Numbers appeared even larger during an aerial count in January 2007 by J.M. Fay (news.nationalgeographic.com). In the Boma and Southern National Park, and the Jonglei area between them, including the Sudd, his team counted 800 000-1 200 000 White-eared Kob. The area is also the grazing ground of other antelopes, including 160 000 Tiang *Damaliscus korrigum korrigum*, 250 000 Mongalla Gazelle *Gazella thomsonii albonotata*, 13 000 Reedbuck *Redunca redunca* and 4000 Nile Lechwe *Kobus megaceros*. The 2007 count shows that seasonally migrating animals which spend a part of the year in marshland could, at least temporarily, escape poachers and rebels. However, more sedentary animals, especially the species grazing outside the marshes, were severely hit. In 1976, 134 000 elephants were counted in southern Sudan, but numbers had dwindled to 22 000-45 000 in 1991 (Blanc *et al.* 2003) and only 8000 in 2007. Of the 60 000 buffalos in the Southern National Park (SW Sudan) and 30 000 zebras in SE Sudan present before the civil war started, none remained by 2007. Despite the decline of some large herbivores in southern Sudan, numbers are still exceptionally high compared with the rest of the Sahel and Sudan zone.

Concluding remarks

Empirical studies show that nutrient mining, land erosion and deforestation proceed less rapidly than formerly thought. The bad news remains that, in general, land degradation in one form or another continues. Bird species depending on natural woodland or grazing land face extensive habitat loss. This is also the prospect faced by birds feeding on fallow land, since the system of shifting cultivation is gradually replaced by permanent cultivation. Range expansion of cattle, sheep and goats, formerly limited to the surroundings of natural watering sites, increased competition for local resources, as evidenced by the eradication of wild herbivores in the western Sahel. For birds, reforestation programmes cannot replace the loss of natural forests as non-native trees harbour far fewer insects than indigenous trees.

Endnotes

1 Chomitz & Griffiths (1997) found an annual increase of 13% of the agricultural land in the Chari-Baguirmi region between 1983 and 1995 (SW Chad). The annual rate of change in agricultural land in Fouta Djalon (Guinea) amounted to 5.31% between 1953 and 1989 (Gilruth & Hutchinson (1990), 4.87% between 1955 and 1994 in northern Burkina Faso (Lindqvist & Tengberg 1993), 2.59% between 1945 and 1995 in NE Burkina Faso (Reenberg *et al.* 1998), 2% between 1956 and 1996 in SW Niger (Moussa 1999) and 3.2% between 1973 and 1999 in south-central Senegal (Wood *et al.* 2004). A long series is also available for Maradi (Niger) where between 1957 and 1975 the cropland increased by 3%/year (Raynaut *et al.* 1988), slowing down to 1% between 1975 and 1996 (Mortimore & Turner 2005). The regional differences are thus large. One explanation is that suitable land still avaiable for cultivation varies locally. For instance, in Diourbel (Senegal) 82% of the land was already cleared for cropland in 1954, so the possibilities for expansion were limited. The cover by cropland increased to 87% in 1987 and 93% in 1999 (Mortimore & Turner 2005). In contrast, in the nearby 'empty' Ferlo region, the arable land increased from 1% in 1965, to 13% in 1984 and 16% in 1999 (Tappan *et al.* 2004). For Senegal as a whole, the proportion of arable land increased from 17% in 1965, to 19.8% in 1985 and 21.4% in 2000 (Tappan *et al.* 2004).

Inner Niger Delta

Floodplains everywhere are disappearing like snow in the sun. Dams, irrigation schemes and changes in land use wreak havoc. Even as large a floodplain as the Inner Niger Delta, at first sight pristine, is threatened by ever more dams in the Niger River upstream and downstream of the Inner Niger Delta. Dams interfere with the flood pulse of a delicate ecosystem. The evocative term "flood pulse" was coined by Junk *et al.* (1989), to describe the seasonal flooding of the aquatic-terrestrial transition zone along rivers. The flood pulse is like a wave rolling over a saucer, magically transforming arid land into a huge but temporary wetland, a wonder to behold and a source of legends and myths over the centuries. In the Inner Niger Delta, flood height may reach six metres, slowly engulfing an area of 400 by 100 km. Fortuitously, the seasonal rise and retreat of the flood in this area has been measured daily at several hydrological stations over many decades, producing a time series of great value, as we will show in this book. Usually, lower-level floods cover floodplains for four months only (October-February), but high floods inundate them for twice as long (September-April). Floodplains are highly dynamic wetlands that attract large numbers of waterbirds, and people as well. The Inner Niger Delta is no exception.

Introduction

The Inner Niger Delta is one of the largest floodplains in Africa. The topographical maps of the Institut Géographique National (IGN) reveal that the inundation area measures 36 470 km^2, including 5340 km^2 of levees, dunes and other islands within that area. They also show that water coverage declines from 31 130 km^2 in wet periods to 3840 km^2 in dry periods (Fig. 36). The entire floodplain area is included in the 41 195 km^2 designated as a Ramsar Wetland Site of International Importance in January 2004. This chapter deals with the ornithological significance of the area, but first explores hydrological and ecological information to provide a background for understanding the ecosystem.

We will first describe the relationship between river flow and flooding, using satellite images to quantify the surface of the inundation zone at different water levels. A simple equation accurately predicts the size of the floodplains as a function of river flow. Because water extraction through dams and irrigation is known, it is also possible to indicate the impact these infrastructures have on the flooding cycle. Next we deal with the ecological consequences of variations in flooding. Vegetation types are clearly associated with flood duration and height. We used a digital flooding model to calculate the surface area of sets of optimal water depths for the various vegetation types. At a lower flood level, plants have to colonise lower-lying zones which will provide optimal growing conditions. Thus, reduced flooding affects the coverage of different vegetation types. At least, that is to be expected in a fully natural system. However, the Inner Delta is inhabited by one million people, who (as described in a separate Section) largely depend on its natural resources. They have reshaped the floodplains into a semi-natural landscape. In the remaining Sections we analyse bird counts to outline the significance of the area for a number of bird species. We used aerial counts of the entire Inner Niger Delta, ground surveys of the central lakes in the Inner Niger Delta and counts of birds flying to nocturnal roosts. We plotted density counts in different habitat types on a vegetation map and correlated the result with the digital flooding model to estimate the total number of birds present in the Inner Niger Delta.

River flow and flooding of the Inner Niger Delta

The Inner Niger Delta is awe-inspiring in size and dynamics. Starting in July, the water rises about 4 m in 100 days. In years of high river discharge, peak level may be 6 m higher than a few months previously (Fig. 37). Large between-year differences in flooding make the system even more dynamic.

Topographical maps show the floodplain as if it were flooded at a maximum level (Fig. 36). However, the area actually inundated varies considerably between years. Zwarts & Grigoras (2005) used satellite images to produce a continuum of 24 water maps of the Inner Niger Delta, covering the range of water levels between -2 and +511 cm, as measured on the gauge at Akka in the central lakes. The maps allow the determination of the relationship between water level and the area inundated. Water maps were combined to construct a digital flooding model. This was done separately for rising and reced-

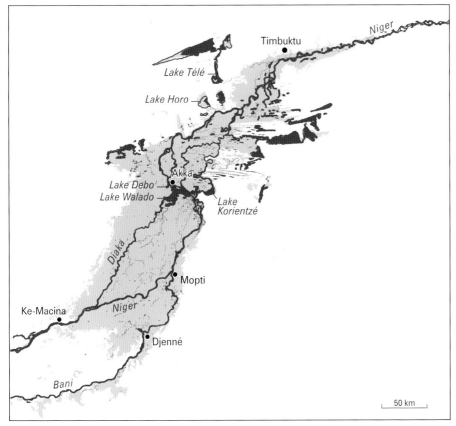

Fig. 36 The floodplains (light blue) and permanent water bodies (dark blue) of the Inner Niger Delta, as indicated on the topographical maps of the Institut Géographique National. The maps are from 1956, and based on aerial photographs and field work in the early 1950s, a period with very high floods.

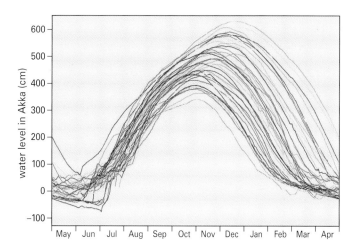

Fig. 37 Annual variation in daily water level at Akka, from 1 May to 30 April, in 1956-2007. From: Direction Nationale de l'Hydraulique (DNH).

ing water. For receding water, we had to make two flooding models: one for years when the peak flood level had been high (inundating a large area) and one for years of low flood (when only the lowest-lying floodplains connected to the river became inundated); see Fig. 38 for both extremes.

Without first-hand experience of the large annual variations in flooding, it is difficult to appreciate what this cycle signifies for those depending on its regime. Fig. 39 is merely a graphical attempt to show the difference between a dry season (1984/85) and a relatively wet season (1999/2000). In the dry year, just one third of the southern Delta became inundated; the northern Delta was not even reached by the flood. In 1999 though, the southern Delta was fully flooded, as well as a large part of the northern Delta, including several of the lakes just north of the Delta.

For migrant birds arriving in the Inner Niger Delta in August, flooding conditions at that time do not overly differ between years, because the flood has not yet reached all of the area (Fig. 37). Later in

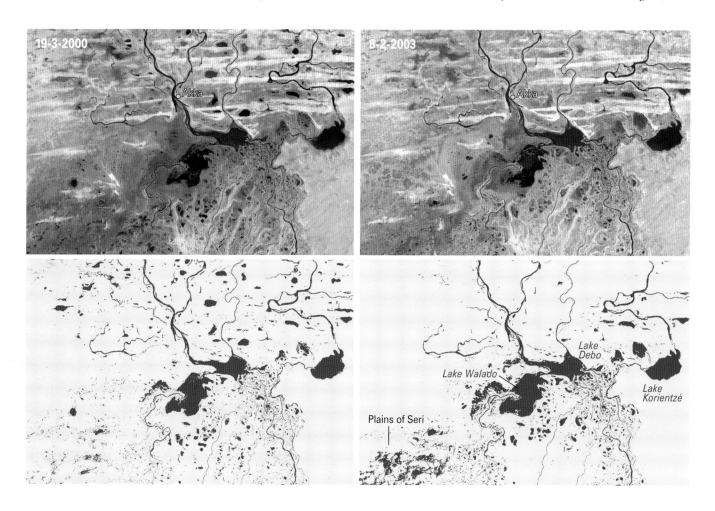

Fig. 38 True colour composite (top) and water map derived from the same Landsat scene (bottom) of the central part of the Inner Niger Delta (68x103 km) during receding water with exactly the same water level in Akka (86 cm), shown for a year when peak flood level had been high (1999/2000: 511 cm; left) and low (2002/2003: 411 cm; right). Note the many isolated, temporary lakes on the lefthand image, lacking on the righthand one (where the flood never reached these sites). On the other hand, since the righthand image was taken 5.5 weeks earlier than the lefthand one, some low-lying plains were still flooded (e.g. the Plains of Seri).

Fig. 39 The graphs show daily measurements of the water level at Akka between June and May in 1984/85 (the lowest flood ever measured) and in 1999/2000 (one of the highest floods since 1970, but of normal height when compared with pre-1973 floods). The surface areas flooded in 1984/85 and 1999/2000, shaded in the same graph, are derived from the water level data, using the digital elevation models of Zwarts & Grigoras (2005). The maps show the flooded area when the water level at Akka reached its peak, and for 1 March.

When the flood recedes, herders and their cows have to cross the Niger River to reach the emerging floodplains.

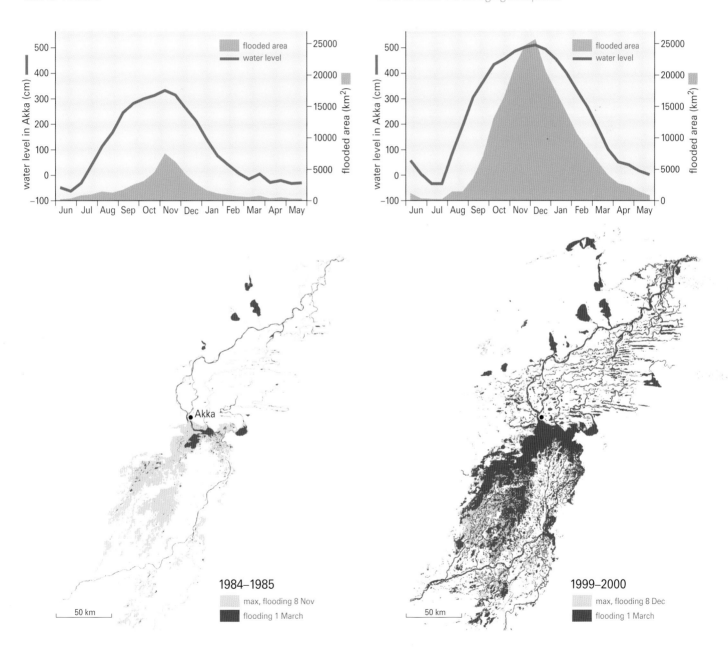

Inner Niger Delta

Millions of cattle graze on the floodplains of the Inner Niger Delta between November and May.

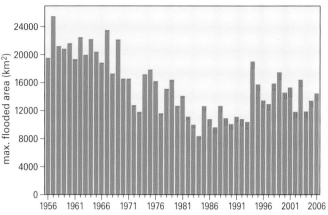

Fig. 41 Peak flooding of the Inner Niger Delta between 1956 and 2006. Blue bars show the actual flooding. Red bars indicate the additional flooding had there been no dams in the Upper Niger. From: Zwarts & Grigoras (2005).

the season, however, very large differences may occur. For example, between December and February 1999/2000 the size of the flooded area declined from 17 400 to 6200 km², but in 1984/85 the flooded area declined from 1680 to 480 km² during the same months. During flood recession, the water level in the river system drops at 2-3 cm per day, but in lakes isolated from the river system only 1 cm evaporates daily. When the water level in Akka has receded from 300 to 100 cm, half of the floodplain remains connected to the river system after a high flood (such as in 1999), but the low flood level in 1984 was insufficient to reach many lakes and depressions, which were consequently deprived of their seasonal water boost. Many of these lakes are liable to dry out completely if unreplenished over a number of years. As we will show later, this has far-reaching consequences for migratory waterbirds.

The flooding of the Inner Niger Delta is determined by the inflow from the Niger and Bani Rivers, which in turn relate to the rainfall experienced 600-900 km SW of the area. Local rainfall is too limited to have an effect on flood height. Thus, the maximum water level in Akka, usually reached in November, can be predicted reliably from the average river flow in August, September and October combined for these two rivers (Fig. 40, left axis). A high river discharge not only produces a high flood, but also floods a more extensive area (Fig. 40; right axis). Since the mid-1950s, the average flow in August-October for Bani and Niger has varied between 1850 and 7200 m³/s, equivalent to a total seasonal flow of 14.7 and 57.2 km³ respectively. In

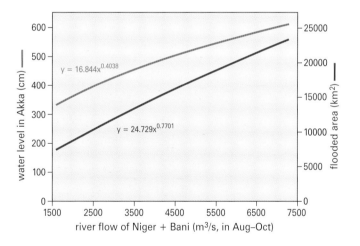

Fig. 40 The highest water level in Akka (cm; left axis) reached within a year as a function of the combined river discharge of the Niger (Ké-Macina) and the Bani (Douna), averaged for August-October that year. The blue line gives the same relationship for the flooded surface (km²). From: Zwarts & Grigoras (2005).

Box 5
Impact of dams in the Niger River on the Inner Niger Delta

Several dams have been constructed on the Niger River since the mid-20th century, and more are planned.

The Selingue reservoir has been operational since 1982. It withholds more than 2 km³ water from the water flow, mainly in August and to a lesser degree in September, and releases 1.2 km³ during the dry season (Zwarts et al. 2005a). Such management has an obvious effect on the flooding of the Inner Niger Delta. Without Selingue, the water level in September-December would be 15-20 cm higher, implying a dam-caused reduction of the flooded area by 600 km².

The Markala dam, at the very entrance of the Niger River into the Inner Niger Delta, was built in 1934-1947. The annual water intake is reasonably constant at 2.69 km³, used for irrigating 740 km² of land. The monthly water intake varies between 60 m³/s in January and 130 m³/s in September, but the river flow of the Niger shows much wider fluctuations, which can be as much as 80 times greater in September than in April. The water intake amounts to only a tiny percentage of the river flow in August-October, but increases to 50-60% in February-June. Indeed, the additional flow from Lake Selingue in the dry months is fully consumed by irrigation near the Markala dam and does not reach the Inner Niger Delta. During high water flows though, the impact of irrigation is limited: a reduction of flood level by 5-10 cm and a loss of 300 km² of effective floodplain size. The impact is somewhat larger in December-February.

The Talo dam, built in 2005 on the Bani River, facilitates irrigation. Although the reservoir is small (maximum surface 50 km², maximum volume 0.18 km³), the effect on the lower Bani may be considerable, especially in dry years, when the river discharge does not reach a peak higher than 200 m/s or 1.55 km³ per month (Fig. 18). The actual impact

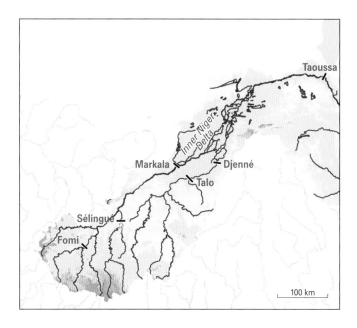

Fig. 42 Location of existing (Selingue, Markala and Talo) and planned (Fomi, Djenne and Taoussa) dams in the Upper Niger Basin.

downstream will depend mainly on the amount of water used for irrigation.

Two other dams are still under study, the Djenne dam on the Lower Bani and the Fomi dam on the Upper Niger. The Djenne dam is intended to counteract water losses caused by the Talo dam. As a consequence, even less water will be available downstream of Djenne. It is estimated that the effect of the Djenne dam on the Inner Niger Delta would be similar to that of the Markala dam. The Fomi dam will create a reservoir nearly three times as large as the Selingue reservoir, resulting in a reduction of the floodplain of the Inner Niger Delta by another 1400 km².

Apart fom the three existing and two planned dams upstream, another dam is planned downstream of the Inner Niger Delta, the Taoussa dam. Its potential hydrological effects on the Inner Niger Delta are still unknown. The dam will probably not affect maximum flooding, but a higher water level in the dry period, at least in the northern part of the Inner Delta, is a likely scenario. Such a reduction of the seasonal dynamics has several ecological and socio-economic effects that are difficult to evaluate without a detailed study.

To summarise, the flood extent of the Inner Niger Delta has been reduced, on average, by 5% due to the Selingue dam and another 2.5% due to irrigation by Office du Niger. Since the irrigated area of Office du Niger will be expanded in the future, it is to be expected that the impact downstream will increase. Building the Djenne and Fomi dams would increase the total loss of floodplains by upstream dams in the Inner Niger Delta to about 15-20%, or 2500-3000 km². From: Zwarts & Grigoras (2005).

More than half a million people in the Inner Niger Delta depend for their living on fishing and cattle breeding.

1984, the water level at Akka did not exceed 336 cm and the flooded area was limited to a mere 7800 km². In contrast, in 1957 and 1964, the water level at Akka reached the very high level of about 600 cm, leading to a flooded area of 22 000 km². It should be noted that this is still substantially smaller than the total floodplain of 31 000 km² as shown on the IGN maps (Fig. 36). This apparent discrepancy is caused by the shallow northward slope of the floodplain that delays flooding in the north by three months; by that time the southern floodplain has already been drained of water. Because our remote sensing analysis is based on actual water coverage, the area flooded at any one time is always less than the total area flooded in the course of a year.

Maximum flood height shows wide between-year variations (Fig. 37). Because flooding is closely related to the river flow entering the Inner Niger Delta, the variations reflect the river discharge of the Bani (Fig. 18) and the Niger (Fig. 19). The 60% decline in flooding between the 1960s and 1980s was due mainly to a decrease of the river discharge of the Bani (down 80%) and the Niger (down 55%).

People cut *bourgou* to feed the cattle; this floating grass is locally planted to provide fodder for cattle.

The impact of dams on flooding is smaller (red bars in Fig. 41) but significant and increasing (see Chapter 3 and Box 5).

Vegetation zones

The Inner Niger Delta is vegetated with plant and tree species which are adapted to steep fluctuations in water level, seasonal flooding and long dry periods. Wild rice *Oryza barthii* and *O. longistaminata*, for example, produces long stems and occupies the zone where the water column reaches up to 2 metres. Another grass species, *Echinochloa stagnina*, locally known as *bourgou*, has stems up to 6 metres and grows where the water depth is 4 m on average. During flooding, wild rice, *bourgou*, and also *Vossia cuspidata* (known in eastern Africa as Hippo Grass, but as *didere* in the Delta), form huge floating meadows. *Bourgou* has a high nutritional value and is therefore also planted by local people to be used as fodder for cattle during the dry period. Since the production of *bourgou* increases with water depth, the people plant *bourgou* in deeper water than wild *bourgou* would normally occupy. Planted *bourgou* therefore runs the risk of drowning during particularly high floods, but during normal floods these anchored plantations represent an addition to the natural floating meadows in the lower floodplains (Zwarts *et al.* 2005b).

People are cultivating an increasing proportion of the floodplains to grow rice. Cultivated rice *Oryza glaberrima* requires the same water depth zone as wild rice and flood forests, and so extension of cultivated rice fields occurs at the expense of natural habitats. For similar reasons, forests, except for tiny fragments, have now been removed.

The highest floodplains are covered by a tall grass species, Black Vetivergrass *Vetiveria nigritana,* and locally by *Acacia seyal* forests (Red Acacia or Shittim Tree). The lowest floodplains often become green as soon as a dense vegetation of grasses and Guinea Rush *Cyperus articulatus* emerges after the flood has passed. However,

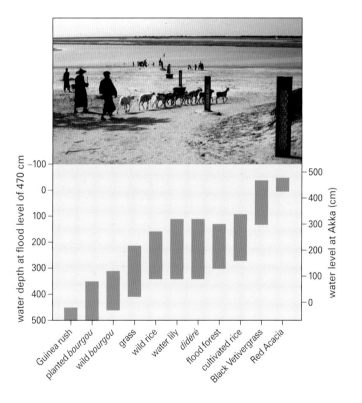

Fig. 43 Zones of 11 vegetation types relative to maximum flood level in the Inner Niger Delta. Sites with more or less homogeneous vegetation were combined with the digital flooding model to determine the frequency distribution of the elevation per vegetation type. The bars show the 10-90% range of elevation. Data were collected in the late 1990s when the average peak water level in Akka (see photograph) amounted to 470 cm. From: Zwarts *et al.* (2005b).

View from the Rock of Soroba across the floating meadows so characteristic of the Inner Niger Delta: *bourgou* (dark green) grows in deeper water (4 m) than wild rice (light green, about 2 m). January 2003

these green floodplains are short-lived and quickly transform into dry dusty steppe with hardly any vegetation, a combined effect of the withering sun and intensive grazing by cattle, sheep and goats. Twenty percent of the 20 million goats and sheep and 40% of the five million cows in Mali are concentrated in the Inner Delta and its surroundings during the dry period.

The distribution of vegetation zones can be described accurately as a function of water depth (Hiernaux & Diarra 1983, Scholte 2007; Fig. 43), but as the flood level undergoes considerable annual variation, three questions arise: (1) do plant species colonise different zones with a change in flood level, and if so, does this occur immediately or is it delayed, (2) is a change in distribution mirrored by a change in the surface area covered by water, and (3) what is the human impact on these natural changes?

There was indeed a shift in the distribution of *bourgou*, *didere* and rice following changes in flood level. For instance, the low-lying Lake Walado has always been a lake where floating vegetation was restricted to the border zone. The lake was colonised by *bourgou* in 1985 and 1986, after the flood level had been low for a number of years (Zwarts & Diallo 2002). In the same period that *bourgou* settled in Lake Walado, elsewhere much larger *bourgou* fields were replaced by *didere*. During the 1990s and early 2000s, we plotted *bourgou* fields for which we calculated maximum water depth using the gauge measurements at Akka and the digital flooding model. These data clearly show that *bourgou* usually grows where maximum water depth fluctuates between 4 and 5 m; *bourgou* showed the expected colonisation in response to a change in water level, albeit with a delay of about two years.

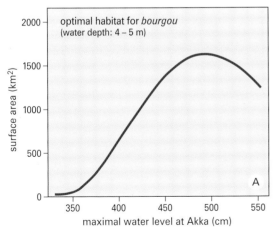
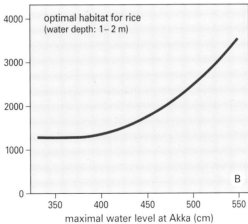

Fig. 44 Surface (km²) of the floodplains in the Inner Niger Delta where we expect (A) *bourgou* (water depth 4-5 m) or (B) rice (water depth 1-2 m), as a function of water level at Akka. Optimal *bourgou* habitat disappears at lower floods, while the effect is less pronounced for optimal rice habitat. From: Goosen & Kone (2005), Zwarts & Grigoras (2005), Zwarts & Kone (2005b).

Given a water depth of 4-5 m, the surface area of optimal *bourgou* habitat can be calculated for different flood levels, using the digital flooding model (Fig. 44A). In 1984, when peak water level reached only 336 cm, none of the floodplains in the Inner Niger Delta had a water column in excess of 4 m. *Bourgou* is outcompeted by plant species such as *didere* in suboptimal habitats with less than 4 m of water. Relatively small changes in flood level of the Inner Niger Delta thus have a large impact on plant species restricted to a narrow range of water depths. At a reduction of peak flooding from 420 to 400 cm, caused by the construction of both the existing two dams, the surface area of the floodplain was reduced by 12% (from 12 600 to 11 200 km²), but the extent of optimal *bourgou* habitat decreased by 45% (from 970 to 540 km²). Steeper declines are to be expected when more dams are built and flood level is further reduced. The emphasis on *bourgou* is particularly relevant, given its ecological significance as fodder for cattle, as a nursery for fish and, as we demonstrate below, as the habitat where waterbirds concentrate in high densities.

In contrast to *bourgou*, cultivated rice grows in shallower water 1.0-2.5 m deep (Fig. 44B). The area where water depth varies between 1 and 2 m measures only 800 km² during a low flood (360 cm at Akka), but increases to 4300 km² during a high flood (580 cm). Water depth is not the only criterion for farmers to cultivate an area for rice growing. Rice cultivation is restricted to substrates that are rather claylike, which explains why rice is almost absent from the sandy northern half of the Inner Delta. The clay content of the soil being rather high in the southern section of the Delta, it is there that most rice fields are concentrated. The area of the Inner Niger Delta cultivated for rice has increased from 160 km² in 1920 to about 1600 km² in 1980-2000 (Gallais 1967, Marie 2002, Zwarts & Kone 2005b).

The IGN 1950s topographical maps mentioned before also show the distribution of cultivated rice fields, which makes for telling comparisons with similar maps from 1987 (Marie 2002) and 2003 (Zwarts *et al.* 2005b). In 1952, most rice fields were found on sites flooded and deflooded when the water level was 310-410 cm on the gauge of Akka (average 382 cm). Since the average flood level in the early 1950s amounted to 580 cm, this implies that the rice fields were then covered by 170 to 270 cm of water at most (on average 198 cm). In the mid-1980s, peak water level declined to 360 cm, and rice cultivation was forced down to lower sites, flooded when the water level at Akka was between 230 and 360 cm (on average 303 cm, a

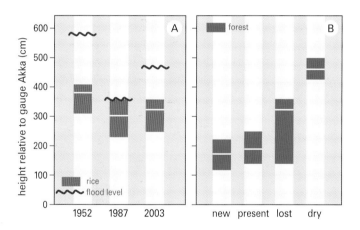

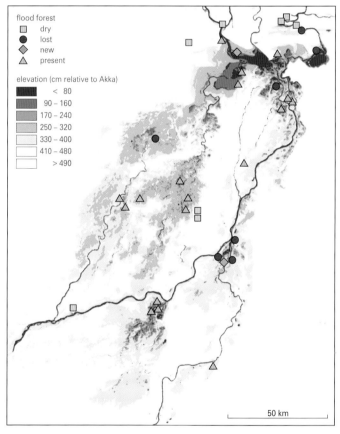

Fig. 45 (A) When peak water level at Akka went down from 580 cm in 1952 to 360 cm in 1987, rice farmers were forced to move to lower-lying floodplains. They abondoned the lowest-lying sites when the floods were higher again in 2003 (470 cm). (B) The occurrence of 35 flood forests in the southern Inner Niger Delta (see Fig. 46). New forests are situated in low-lying parts of the floodplain; forests on the higher floodplains changed into dry forests, owing to the generally lower floods in the late 20th century. Average and 20-80% range indicated by line and bars. From: Beintema *et al.* (2007), Zwarts & Grigoras (2005), Zwarts *et al.* (2005b).

Fig. 46 Present status of 35 flood forests in the southern half of the Inner Niger Delta: new (existing in 2000 but not in the 1980s), present (in the 1980s as well as in 2000), lost (present in the 1980s but lost by 2000), dry (flood forest before 1980s, but now dry land forest). The flooding is shown for different water levels.

Large expanses of *bourgou* are traversed via 1-2 m wide waterways, created by removing the vegetation. The small canals enable easy penetration of dense *bourgou* vegetation with *pirogues*, to put up nets and fish traps. We used the waterways to perform transect counts of birds where the water level was more than waist-deep.

drop of 79 cm from the 1950s). Despite this move, flooding of the new rice fields was poor or even non-existent, amounting to 0-130 cm of water (Fig. 45A). Consequently, rice production in the Inner Niger Delta crashed from 100 000 tons during normal and good floods to only -20 000 tons in the mid-1980s. As floods improved in the late 1990s and early 2000s, the low-lying rice fields were abandoned in favour of fields flooded at a water level of 250-360 cm in Akka (on average 321 cm). The accompanying average water depth during flooding of 149 cm is much better for rice cultivation than the 57 cm in the mid-1980s, but has not yet attained the level of 180 cm in the 1950s.

In the 1980s farmers adjusted their land use to the prevailing low flood levels by creating rice fields in the lower floodplains. *Didere* and wild rice had to be removed, and most of the remaining flood forests were cut. Of the forests still present in the 1980s, seven had disappeared by 2005 and eight had transformed into dry forest (Beintema *et al.* 2007). Only 18 forests survived, mostly in degraded form. Two new forests came into existence (Fig. 46). Altogether, forests now cover not more than 20 km^2, a small fraction of the several hundreds of km^2 before the 1980s and a tiny fraction of that existing in pre-colonial times.

Flood forests that had changed into dry forest were exclusively situated on the highest sites, beyond the reach of most floods. Lost flood forests were restricted to sites that became flooded when the water level at Akka was above 140 and below 360 cm (median 320 cm). The few remaining flood forests were confined to sites flooded when the water level at Akka was above 150 and below 250 cm (median 190 cm) (Beintema *et al.* 2007; digital elevation model). This fits in nicely with our knowledge of newly-created rice fields of the 1980s (Fig. 45A); flood forests survived only when situated in the lowest part of the floodplain (120-220 cm on the gauge of Akka), *i.e.* out of reach of new rice cultivations at 240-300 cm on the gauge of Akka. Similarly, new flood forests are found only in the very low floodplains (120 to 220 cm). When flood levels decline again, these forests are in danger of conversion into rice fields.

People

The national census of 1998 showed that the Inner Niger Delta is inhabited by 1.1 million people, of which 230 000 live in cities like Djenne, Mopti and Timbuktu (Zwarts & Kone 2005a). The remaining

View from the Rock of Gourao over a large *bourgou* meadow, planted to augment fodder for the cattle during the dry period.

Inner Niger Delta

Settlements in the Inner Niger Delta are situated on the levees, like Pora in the southern Delta, November 1999, during a high flood.

870 000 are thinly distributed over an area of about 50 000 km². As detailed in previous censuses (1976, 1987), the rural population has remained stable, which is remarkable in the light of the overall annual population growth of 2.3% in 1976-1998 in Mali. Many people have left the Inner Niger Delta, especially in the northern half where the population has actually decreased by 0.6% per year between 1976 and 1998. Most moved to cities in the region and elsewhere in Mali, or went abroad. The depopulation is a direct result of the prevail-

All someone's belongings are transported by large *pinasses*.

Extensive rice fields on the floodplains in the southern Inner Niger Delta, November 1999.

ing drought and low floods (Fig. 41), which make survival difficult. About 40% of the rural population in the Inner Niger Delta are farmers, 30% herders and 30% fishermen. Altogether these people tax the natural resources heavily, creating the landscape as it now exists.

Rice cultivation   About 1600 km² in the southern half of the Inner Delta (5.1% of the total floodplains) are cultivated by farmers growing rice on the floodplains. Another 680 km² are managed as rice fields by *Opération Riz Mopti* and *Opération Riz Ségou*. Both areas lack active irrigation but employ dikes and sluices to delay flooding, and to manage the water level during deflooding. However, if the flood is not high enough, the areas remain dry. This means that rice production, as elsewhere on the floodplains, depends exclusively on local rain and the river floods; pumping occurs locally, albeit on a small scale.

Farming is not easy in the Inner Delta, and rice farming especially so. The farmers grow a West-African rice variety of *Oryza glaberrima*, known as *riz flottant* (floating rice), well-adapted to grow as the water rises during the flood. Ideally, the seed should germinate before the flood arrives. Farmers have to sow before the first rains, in the hope that the rain will precede the flood, allowing rice to sprout before the flood arrives and before the water starts to rise by several cm a day. Because rice plants can grow up to 3-4 cm per day, they do not drown. The stems may attain lengths of 5 metres, but usually only 2 metres of growth suffice. After a 3-month flood period, the rice is harvested when the floods recede. Much can go wrong in such an unpredictable cycle, and annual rice production therefore varies between 50 000 and 170 000 tons (Zwarts *et al.* 2005b). The few irrigated rice fields in the Inner Niger Delta have a more stable harvest of 40 000-60 000 tons per year. Floodplain rice yield is low (1.0-1.5 ton/ha) when compared with that of irrigated rice fields (5.0-5.5 ton/ha), but the latter incur high costs related to investments and irrigation schemes. Farmers on the floodplains have few if any overheads.

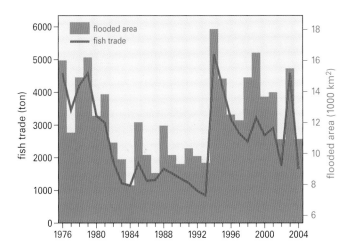

Fig. 47 Annual fish trade in the Inner Niger Delta and the surface maximally flooded in the previous year between 1976 and 2005. From: Zwarts & Diallo (2005, updated for recent years).

Livestock   Two million cattle and four million sheep and goats graze the floodplains of the Inner Niger Delta in the dry period (Goossen & Kone 2005). When floods still cover the plains, the stock grazes nearby, but as soon the water begins to recede, phalanxes of cattle, then of sheep and goats, move into the Inner Niger Delta. The silence of the Delta is now broken by mooing and bleating, and dust blurs the glare of the sun. At low flood levels, most livestock concentrates on

During receding floods in the Inner Niger Delta, nearly all fish are captured in traps of up to 5 m long and 2 m high, strategically placed in rivers and gulleys. Increasingly, many fish are now caught in standing nets with a mesh size of 10 mm encircling large *bourgou* fields. Few fish survive the deflooding period, and fish older than 6-8 months have become rare.

the low-lying parts of the floodplains, where cattle reach a density of 100 per km² and goats and sheep a density of 30 per km² or more. As zebu-cows weigh 250 kg and goats and sheep about 20 kg, the total annual grazing pressure is equivalent to 26 tonnes/km². This is high compared with an average annual grazing pressure of 2-4 tonnes/km² on grassland in the western Sahel (Penning de Vries & Djitèye 1982). A high stocking rate is feasible because floodplains are highly productive (Box 2). Young grasses cover the emerging floodplains and floating, aquatic grass species like *didere* and *bourgou* gradually become available after flood recession.

During the dry months which follow, the vegetation withers and decreases in quality. To improve survival conditions for their stock prior to the next rainy season, pastoralists burn remaining vegetation to stimulate regrowth. Moreover, *bourgou* – planted on a large scale by the herders – is used as fodder in the dry period. *Bourgou* is also cut in the dry season to stimulate sprouting. Heavy grazing on *bourgou* sprouts, however, leads to its eradication, as happened during the Great Drought in the 1980s.

Fishing Annually, fishermen in the Inner Niger Delta catch 60 000-120 000 tonnes of fish. This FAO estimate is based on several untested assumptions, such as a fixed daily consumption by the fishermen themselves, independent of the catch, and an overestimation of the annual increase in the number of fishermen (Zwarts & Diallo 2005). The annual trade is registered: it varied between 10 000 and 50 000 tonnes between 1977 and 2005. The variation is closely associated with the previous year's flood level (Fig. 47). Theoretically, because the number of fishermen increased from 70 000 in 1967 (Gallais 1967) to 225 000 in 1987 (Morand *et al.* 1991) and to 268 000 in 2003 (Zwarts & Diallo 2005), the trade should have increased at the same rate. However, correcting for flood level, there was no increase in trade at all. When 270 000 fishermen are unable to bring more fish to the market than 70 000 fishermen, this strongly suggests that fish capture is constrained by an absolute ceiling in biological production. The same conclusion arises from seasonal catch data. On average, the daily catch per fisherman decreases from 35 kg/day in early February to 7 kg/day at the end of June (Kodio *et al.* 2002). This decline is consistent with depletion of the available fish stocks; at the end of the fishing season, nearly all fish have been removed from the floodplains.

Fish more than one year old have become increasingly scarce in the Inner Delta (Laë 1995). The only way for a fish species to survive here is to reproduce as early in life as possible. Indeed, Bénech & Dansoko (1994) found that the fish species in the Inner Niger Delta have adapted to the extreme predation pressure by reducing their

Inner Niger Delta

age of reproduction. Reproduction for most species is limited to the high water period (Bénech & Dansoko 1994). Therefore, the annual fish stock entirely depends on the spawn and fry produced by the few fish still alive at the end of their first year and by the very few fish over one year old.

The introduction of nylon nets in the 1960s resulted in the near-depletion of fish stocks in the Inner Niger Delta, thus changing the exploitation system significantly (Laë *et al.* 1994). Concomitant with the steady decline in the size of captured fish, the mesh size of nylon nets decreased: before 1975, most nets had a mesh width of 50 mm, but this declined to 41-50 mm between 1976 and 1983, and to 33-41 mm between 1984 and 1989 (Laë *et al.* 1994). This downward trend has continued: in 2007 we measured many nets with a mesh size of only 10 mm.

The fish population depends entirely on the young produced by the fish which manage to survive the previous year's intensive fishing campaign. As a consequence, fish species restricted in their distribution to flooded areas have decreased, whereas species able to reproduce at one year old have become more abundant (Laë 1995). The history of fish exploitation in the Inner Niger Delta is a classic example of overexploitation, but it is also a typical "tragedy of the Commons", where an individual takes a (small) benefit at the expense of the community as a whole. Annual catches have become more variable in volume as most fish are now less than one year old and the total fish population is determined by flood level. Meanwhile, the future prospects for fishermen have become precarious – they may use more nets, but catch fewer and smaller fish.

We can only speculate about the ecological impact of these changes in fish exploitation and fish stocks. Many bird species are fully dependent on the fish they capture on the floodplains. In the past, people captured Nile Perch two metres or more in length. Such giant predators were much too large to be eaten by herons and other fish-eating birds, and this applied to the larger part of the fish biomass. Dependent on fish shape, piscivorous birds can swallow fish of up to 20-25 cm long. At present, fish of this size class and smaller abound in the Inner Niger Delta, and so the general decrease in fish size may appear an advantage to fish-eating birds. Furthermore, several fish-eating bird species profit from the contraptions put up by fishermen, concentrating their feeding activities where fish traps are operated or where fishermen empty their nets. The dwindling fish stocks suggest these advantages to be short-term. In any case, for birds there are also huge disadvantages associated with the present catching methods. Many birds on the floodplains are captured and killed by accident in fykes and nets, and many more become victims of hooklines. The rapidly increasing number of nets and hooklines must constitute an important and in-

Wood (here taken from dead Red Acacia) is transported in the Inner Niger Delta over dozens of km to be used by fishermen to dry and smoke their fish.

Box 6
Stories from long gone days

What did the Inner Niger Delta look like 50 or 100 years ago, and what has changed in its bird life? The reports of the first ornithologists visiting the area, although a pleasure to read, are a bit frustrating as they are purely descriptive. The observations are difficult to compare with the more quantitative data available since the 1970s. Some differences are evident, however. Goliath Heron, Hamerkop, five stork species (Yellow-billed, African Open-bill, Abdim's, Saddle-billed and Marabou) and Hadada Ibis were seen frequently in 1931 by Bates (1933), in 1943-1944 by Guichard (1947), in 1954-1959 by Malzy (1962) and in 1956-1960 by Duhart & Descamps (1963). The Goliath Heron must have vanished from the Inner Niger Delta in the 1960s. The Hadada Ibis, still common in the wooded parts of the Niger and Bani valleys in 1972-1974 (Curry & Sayer 1979), has been rare since the 1980s and does not seem to breed in the area any more. The same can be said of the above-mentioned storks. For example, Abdim's Stork was classified as abundant in the late 1950s (Duhart & Descamps 1963), but has become noticeably scarcer since the 1980s. When Guichard (1947) wrote about the Black Crowned Crane as "not very common and does not occur in huge flocks such as were seen in Bornu and near Lake Chad", nevertheless hundreds or thousands must still have been around. In the late 1950s, Duhart & Descamps (1963) observed some flocks of several hundred. By the year 2000, numbers had dwindled to only fifty cranes, unable to produce fledglings as all chicks were systematically removed and traded by the local people. The African Skimmer qualified as "occurring in large groups" (Malzy 1962) and "being common along the Bani and Niger Rivers, where the species can be seen the entire year" (Duhart & Descamps 1963), but it has disappeared from the Inner Niger Delta. We have seen twice a Skimmer in this region since we started our fieldwork in the mid-1980s.

Regarding the raptor guild, ornithologists visiting the Inner Niger Delta since the 1980s mostly encounter Eurasian Marsh Harriers, and the occasional Montagu's Harrier, Osprey or African Fish Eagle. Several other raptor species, albeit in small numbers, inhabit the dry or wooded habitats. How different were the reports in the past. The African Fish Eagle was mentioned by Duhart & Descamps (1963) as "common", which may have indicated at least 100 pairs in such a large area with ample breeding sites. Around the turn of the 21st century no more than 15-30 pairs were estimated (van der Kamp et al. 2002a).The Egyptian Vulture was considered "common" by Guichard (1947), but has been scarce since the mid-1980s, except for the dry years of 1992-1994 when flocks of up to 50-60 birds appeared in the Debo area. The decline of raptors in the Inner Niger Delta is similar to the overall decline in the Sahel and Sudan zone of West Africa, as quantified by Thiollay (2006a), who saw, for instance, 204 Egyptian Vultures along 3700 km of road transects in 1971-1973, but only 1 in 2004.

Yellow-billed Kite on Rock of Gourao, January 2005.

The Arabian Bustard was "very common" in the Inner Niger Delta in the late 1950s (Duhart & Descamps 1963), but constitutes an exceptional encounter today. In the early 1970s, Thiollay (2006b) observed 152 Arabian and 64 Nubian Bustards along his road transects, but none in 2004. He argues that the decline had already started before the 1970s.

All the species mentioned so far have one thing in common - they are large and thus form an attractive target for the local people. This may in part explain their decline. The older reports do not pay much attention to smaller bird species, but Duhart & Descamps (1963) and Curry & Sayer (1979) made some interesting remarks about the Common Redstart ("very common in broadleaved trees"), Quail ("very frequent" and "throughout the drier areas"), Tawny Pipit ("common, occurring in flocks", "found in most situations"), Turtle Dove ("immense large flocks"), Bluethroat ("small numbers in many different habitats"), and Great Snipe ("very common"). Thirty years later, all these species have become less common or even rare. The Common Redstart has lost a large part of its wooded habitat, and no one would classify its present occurrence in the remaining forests as "very common" (see Box 3). The present occurrence of the once common Great Snipe, of which we estimate the total wintering population at 2000 birds (Table 41), can be described as 'rare and very local'. In our density counts, the Bluethroat was 15 times less common than the Sedge Warbler, but it used to be more common than this species in the late 1960s.

The early reports contain many such indications of profound changes, and none for the better. We will, though, never know for certain just how large the losses have been.

creasing mortality factor in fish-eating birds in the Inner Niger Delta (see for example Caspian Tern, Chapter 31).

Wood   The loss of flooded forests has been described above. In the past, the Inner Niger Delta was surrounded by extensive forests of mainly *Acacia seyal*, inundated briefly at the peak of flooding (Fig. 43), and *A. nilotica* and *A. albida* growing on the higher levees. Relicts of these forests still remain at sacred sites, such as traditional burial grounds, or in thinly-inhabited parts of the Inner Niger Delta. The extensive *A. seyal* forests hugging the northwestern section of the Inner Niger Delta died during the prolonged drought in the 1980s. The dead trees were cut and sold to local fishermen to smoke their fish, but after 30 years, this source was exhausted, and the price of wood started to rise in 2006 and 2007. The pressure on the remaining local forests is likely to increase, because 45% of the total fish capture (estimated at on average 56 000 tons, according to the annual reports of the OPM 1977-2003; Zwarts & Diallo 2005) is smoked above a fire. Since 2 kg of wood is needed to smoke 1 kg

of fish (Dansoko & Kassibo 1989), the annual wood consumption to smoke fish is estimated at 50 000 tons. Wood production from *A. seyal* is 10-35 m^3/ha if managed on a 10-15 year rotation (Hall 1994). In theory, the fish in the Inner Niger Delta can be smoked in a sustainable way if wood is collected from forests encompassing 2000-5000 ha. In practice, forests along rivers and near villages are being overexploited, while the more remote forests are not yet visited frequently.

Much wood is also needed to build and maintain the thousands of boats. Every day fishermen in the Inner Delta use *pirogues*, small slender boats made of planks some 3.0-3.5 cm thick and 40-80 cm wide. The much larger *pinasses*, used for transportation of goods and people, are made of larger planks. To build such boats, large trees are needed. Indeed, the boats in the Inner Niger Delta are made of very large tree species, mainly *Khaya senegalensis*, the mahogany tree, reaching a height of 30 m and a diameter of 1 m. The lifespan of boats, although made of tropical hardwood, is about 12 years (Kassibo & Bruner-Jailly 2003). Given the presence of 25 000 small pirogues (5-8 m long), 1500 large pirogues (10-25 m long) and 75 pinasses (30-50 m long) in the Inner Delta (Kassibo & Bruner-Jailly 2003), it may be estimated that some 3000 large trees have to be cut down annually in order to maintain the existing fleet. Since large trees are now scarce in the surroundings of the Delta, the wood is currently imported from Ivory Coast and Ghana.

Counting birds in the Inner Niger Delta

Various census methods were employed to quantify important aspects of the bird fauna of the Inner Niger Delta (Fig. 48). First, counts of breeding birds were focussed on the larger species, mainly those breeding colonially in the few remaining flood forests. Secondly, counts at roosts in the same forests were used to quantify the number of cormorants, darters, herons, spoonbills and ibises, mainly outside the breeding season. The following Section deals with the aerial counts of wintering ducks and geese. Counts covering the entire floodplains are lacking for most other waterbirds. In-

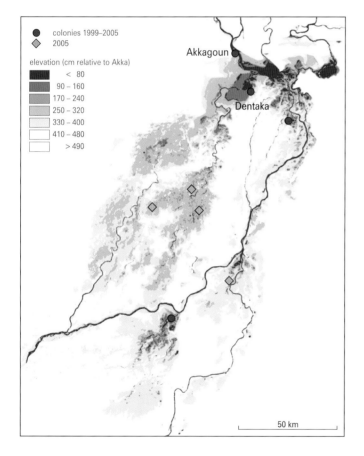

Fig. 49   Flood forests used as breeding colonies in the Inner Niger Delta since 1986. Colonies counted in 1999-2005 and newly discovered colonies in 2005 are separately indicated. Most birds breed in Akkagoun and Dentaka. From: van der Kamp *et al.* (2002c, 2005a, 2005c).

To count birds on the flat floodplains of the Inner Niger Delta, a suite of supplementary methods has been applied, from aerial and ground surveys to roost counts and plot sampling. A wide field of vision is needed to count the birds heading to nocturnal roosts or present by day on the floodplains, and any elevation in the terrain – whether man-made (*pirogue*) or not (termite mound) – is exploited to improve the view. Plot sampling, here in wild rice, was used to obtain quantitative data on widely dispersed birds.

Long-tailed Cormorants in Lake Debo.

stead, we use counts of the central lake area, where large numbers of waders and other waterbirds concentrate during the receding flood. Finally, we describe density counts used to estimate the wintering population of passerines and skulking species. All census methods are, not surprisingly, fraught with interpretation problems in so vast an area, where logistics are a nightmare, and so they should be viewed against this background.

Fig. 48 Timing and methods of bird counts employed in the Inner Niger Delta, covering 62 ground surveys of the central lake area, 22 aerial counts, 7 systematic counts of large colonially breeding waterbirds, 17 counts of herons and other waterbirds on major roosts, and 1617 density counts performed between 15 November and 15 March.

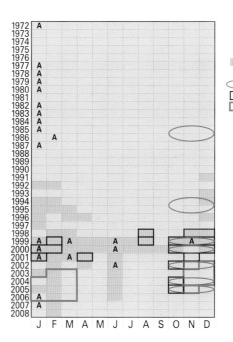

Inner Niger Delta

Breeding birds

Quantitative data of birds breeding on the emerging floodplains are scarce (see below). Large colonial waterbirds, restricted to breeding in forests, have been censused intermittently over the years (Fig. 48). The Inner Niger Delta still harbours large heronries, even though many flood forests have disappeared or been turned into dry forest (Fig. 46). Cormorants, darters, herons, ibises and spoonbills breed mostly in Dentaka forest along the eastern border of Lake Walado and, to a lesser degree, in Akkagoun forest in the northwestern part of Lake Debo. Two smaller colonies are near-annually used in the southern and eastern part of the Inner Delta. Four new colonies were discovered in 2005 (Fig. 49); according to local people the birds here had already been breeding for several years before that.

Censusing colonial birds in the Inner Niger Delta is time-consuming, and a full census is stretched over months. Cattle Egrets start to breed in the second half of June, while cormorants and most other species do so in August-September. When the flood is high, the breeding season is extended until January, but during a low flood the colonies are already deserted by November. Quantitative data are not easy to collect during the flood peak, since the forests are standing in water 2-3 metres deep (and to complicate matters, *Acacia kirkii* and *Ziziphus mucronata* are thorny). Only in the later flooded stage is it possible to wade or walk across the forest to count canopy nests of species which breed when the flood recedes (Grey and Purple Heron, spoonbills, ibises).

Skinner *et al.* (1987b) based their population estimates on counts of birds flying back to the colonies at the end of the day. This was done throughout the breeding season, taking into account the reproductive stage (one or both parents flying back at dusk) in order to arrive at justified estimates. During the 1990s, roost counts were carried out in late December and January and focussed on the flood forests of Dentaka and Akkagoun, which contained important breeding colonies. Since 1999 these counts have been performed in late October and November during systematic boat surveys throughout the southern half of the delta. Both counting periods have in common that bird numbers were quantified by the end, or just after, the breeding season, when practically all birds commute between

Table 3 Estimated number of breeding pairs of colonial waterbirds in the southern half of the Inner Niger Delta, 1986-2006 (the same 4 colonies in 1986/87 to 2005/06, 8 colonies - including 4 newly discovered - in 2005/06*; Fig. 49). Egrets, excluding Cattle Egret, were lumped in the counts of 2005/06. Sources: Skinner *et al.* (1987b); van der Kamp *et al.* (2002a, 2005c).

Year	1986/87	1994-96	1999-2001	2002/03	2004/05	2005/06	2005/06*
Max. water level, cm	388	534	511	411	410	442	442
Long-tailed Cormorant x 1000	17-17.5	16-17	18-20	16-17	17-19	4.8	7.2
African Darter	40-45	15-30	300-350	210-230	130-150	240-250	240-250
Grey Heron	10-15	30-50	1-10	0	0	?	?
Black-headed Heron	10	1-5	<5	<5	<5	<5	<5
Purple Heron	0	2-10	0	0	0	0	0
Black-crowned Night Heron	<10	100-300	1-10	<10	<10	?	<10
Black Heron	200-250	150	130	80	<50	?	45-60
Cattle Egret x 1000	63-65	65-90	50-60	55-60	40-50	65-70	110-115
Squacco Heron	550-650	?	500	500	500		
Little Egret	900-1000	500-1000	500-1000	1000	1500		
Little Egret dark morph	80-110	?	80	80	50	3500	5000
Intermediate Egret	800-875	>200	1700	?	1800		
Great Egret	2800-3100	500-1000	1500-1800	800-1000	800-1000		
African Openbill	30-40	0-1	0	0	0	0	0
African Spoonbill	300-350	50	100-150	?	100-150	?	?
Glossy Ibis	0	150	0	0	0	0	0
Sacred Ibis	30-40	50	200-250	?	100	?	?

Whiskered Terns following a *pirogue*, Lake Debo.

Pied Kingfishers.

feeding grounds and breeding sites. The number of breeding pairs was assessed by dividing the counted totals by three (Delany & Scott 2006). Numbers were rounded up to account for missed birds (those foraging between observation point and flood forest) and for birds flying to the roost before the start of the counts at 16:30 (especially Long-tailed Cormorants). Estimates for Great, Intermediate and Little Egrets were based rather on numbers counted during waterbird counts in the central Inner Niger Delta lakes combined with observations at breeding sites. The division factor (3 = 2 adults plus one juvenile) is a rigid one as hardly any information on breeding success is available. Walsh *et al.* (1989) recorded breeding success during a Great Drought season (1986/87) in the Inner Delta for Cattle Egret (1.87 young/nest), Great Egret (1.25) and Intermediate Egret (1.13). It is likely that the breeding success is substantially higher during years of high floods. For instance, Black-necked Herons breeding on high trees in the southern Delta (>100 nests) showed on average three large young per nest during the high flood of 1994; this would give a division factor of 5 instead of 3. Since we were not able to correct for annual variations in breeding success, the derived number of breeding pairs will constitute an underestimate in dry years and an overestimate in wet years.

The systematic counts show that over 100 000 pairs of waterbirds breed colonially in the flood forests in the southern half of the Inner Niger Delta (Table 3). The Dentaka forest harbours the most impressive colony, with 16 species and about 60 000 breeding pairs. The large Dentaka colony should not be interpreted as a sign that colonial waterbirds are doing well in the Inner Delta. On the contrary, the birds are concentrated there because of a lack of suitable, undisturbed breeding areas elsewhere in the area. It is possible that some colonies still exist in the northern part of the Inner Delta (though recent field data are lacking), but this is unlikely, as the six known flood forests have disappeared or turned into dry forest, reducing breeding opportunities to virtually nil (Beintema *et al.* 2007). At the same time the lower flood levels since 1973 reduced the extent of potential feeding areas (= area inundated) in the northern Delta much more than in the southern part.

The **Long-tailed Cormorant** is a common species in the Inner Niger Delta. Despite large annual variations in flooding, breeding numbers remained more or less stable at 17 000–19 000 pairs for at least 20 years (Table 3), but numbers suddenly collapsed in 2005. For reasons which are still unknown, large numbers of adult and juvenile cormorants died during the second half of 2004. Many thousands of birds, mainly cormorants, were found dead and dying in the colonies, except in those situated along the Niger upstream of Mopti (van der Kamp *et al.* 2005c). Local food shortage may have played a role, as first observations of ´unnaturally´ high numbers of dying birds were reported by fishermen in June-July, at the end of the dry season and before the onset of the breeding season. Locust-spraying later on may have exacerbated the situation, but this seems unlikely as Cattle Egrets, which largely depend on locusts, were not affected.

African Darters have a patchy breeding distribution. Decades ago they bred across the entire Delta. A single breeding site remains in the Dentaka flood forest. Population size in 1999-2007 fluctuated between 70 and 350 pairs, as evident from early morning counts in February, around the end of their breeding cycle. The higher the flood level, the larger was the increase in the population (van der Kamp *et al.* 2005a, 2005c). However, population growth is limited by human exploitation, which was excessive in 2003 and 2007. According to local people, nomadic fishermen took all eggs and young from the nests. Without human disturbance, the Darter would be much more common in the Inner Delta than is presently the case.

Several heron and egret species breed in small numbers in the Inner Niger Delta (Table 3). The breeding populations of Black-headed Heron, Black Heron, dark morph of Little Egret, Black-crowned Night Heron, Squacco Heron, Sacred Ibis and African Spoonbill in the central lakes are counted mostly during the final stage of the flood or even in April-June when the rest of the delta is almost dry.

Black-headed Herons breed in very small numbers in Dentaka (<5 breeding pairs). So far, breeding has been observed mainly during the flood, on high Kapok trees *Ceiba pentandra* standing on dry elevations (*toguérés*), but in May 2005 (Beintema *et al.* 2007) and June 2006 (B. Fofana) 80 nests were seen on *Acacia albida* at the graveyard of Kadial and >35 birds in breeding plumage on Dentaka's - then standing dry - *A. kirkii* trees, respectively.

Purple Heron and **Glossy Ibis**, which had not bred in the Inner Niger Delta for decades, started to do so during the flood of 1994, the first high flood since 1972. At the same time, numbers of breeding **Grey Heron** and **Black-crowned Night Heron** were also exceptionally high (Table 3). All four species are Palearctic migrants which might have profited from the favourable flood to double-brood in 1994 (first in mid-summer in Europe, and again in December in the Inner Niger Delta). An alternative, or additional, explanation considers the first breeding by subadult birds which have not yet returned to Europe, as suggested by a subadult Purple Heron attending one of the nests. Lamarche (1981) also states that in wet years **Grey** and **Purple Heron** started nest-building between December and February. None of the local large, colonial breeding species, twelve in total, showed higher breeding numbers in 1994 (Table 3).

The white egrets are common breeding birds. Population trends differ between species, with **Great Egret** showing a decline, **Intermediate (or Yellow-billed) Egret** an increase and **Little Egret** stable numbers (Table 3). The breeding population of the **Little Egret** (1000 pairs) constitutes a small fraction of the birds present in the Inner Niger Delta, which are, for the most part, migrants from Europe. The commonest colonial waterbird species in the Inner Niger Delta is the **Cattle Egret**. Its numbers have been remarkably stable since the late 1980s, with about 60 000 pairs in the well-known colonies. As some colonies have been overlooked, it is likely that the breeding population may have comprised up to 110 000 pairs (Table 3).

The status of eight other large waterbird species has to be written in a minor key. **Glossy Ibis** may have been a regular local breeding bird under well-flooded conditions such as, for instance, those pertaining during the 1950s and 1960s. Morel & Morel (1961) found Glossy Ibis breeding in the northern Inner Delta in March 1960. Nowadays they may only breed - in small numbers - when flooding at the nesting sites extends well into January-March, as was the case in the 1994-1995 season (Table 3). The only *Ciconia* species breeding in the Inner Niger Delta is **Abdim's Stork**, with nesting sites reported along the central lakes (Korientzé, Akka), in the northern reaches (Goundam, near Timbuktu), and possibly in the *Centre vide*. **Marabou, Yellow-billed** and **Saddle-billed Stork** have disappeared as breeding species. Lamarche (1980) mentions breeding for the last-named species in the western Inner Niger Delta in January through March, but the only records since the early 1990s were of two immature birds in June: one third-year bird in 2001 and one second-year bird in 2003, both in Walado Debo.

Although listed as a breeding species in Table 3, reproduction of **African Openbills** has not been recorded since the mid-1980s. The latest observations, of 2-3 birds near or at their breeding site in the Dentaka flood forest, date from January 1992 and December 1993. Since then African Openbills have been seen only once, during an aerial count in early March 1999 (2 birds, Seri plain). A decline has been noted across the western part of their West-African range (Borrow & Demey 2001).

Pelicans no longer breed in the Inner Niger Delta, at least not in the last decades of the 20th century. Roux (1973) mentioned young **Pink-backed Pelicans** held in captivity by local fishermen and assumed that the species might still be breeding somewhere in the Inner Niger Delta. However, breeding evidence is lacking here. During the mid-1980s J. Skinner and S. Konta (pers. obs.) checked the evidence for presumed breeding of **Great White Pelicans** on the Dogon table mountains, 80 km east of the Delta. They recorded the species (together with Yellow-billed Stork), but were unable to reach the possible breeding site. Nearby villagers told that the species disappeared in the early 1990s due to repeated poaching (van der Kamp *et al.* 2002c).

The **Black Crowned Crane** is presently qualified as near-threatened (Kone *et al.* 2007). In the Inner Niger Delta, the species would certainly deserve the designation (critically) endangered, as only fifty adult birds remained in 2001. Hardly any juveniles have been observed since the early 1990s (van der Kamp *et al.* 2002c). In 1999-2001 only one first-year bird was seen during monthly counts in the central lakes; none were found in aerial surveys, during an overland survey focussed on the species in May, or during earlier counts. The overland survey team, including the regional crane expert who remained convinced of a population comprising several hundreds of birds, was unable to show us more than the 50 birds recorded from the air. An inquiry into its status in Mali in the early 2000s revealed

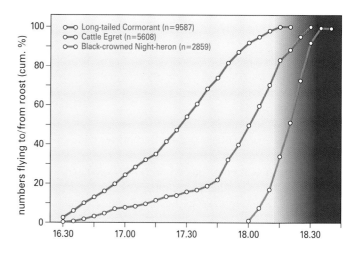

Fig. 50 Cumulative % of Long-tailed Cormorants and Cattle Egrets arriving at, and Black-crowned Night Herons leaving, the Dentaka roost, as they passed Garouye on 22 January 2004; sunset 18:15, civil and nautical twilight 18:32 and 18:58 local time respectively.

Communal feeding of Little Egrets and Spotted Redshanks on emerging floodplains with scattered *Ziziphus* trees; Lake Debo.

fewer than 100 birds in the wild throughout the country (Beilfuss et al. 2007), and another 400 cranes in captivity (Kone et al. 2007). Considering its alarmingly low reproductive success, mainly due to poaching, the resident population in the Inner Niger Delta is bound to disappear.

A small colony of **Whiskered Terns** was found in Lake Debo in August 1991, the first record of breeding in West Africa. Since then, we found evidence for annual reproduction at the central lakes between August and January. Colonies comprised up to several dozen nests; the timing of laying within colonies varied considerably. Fledglings were rarely observed, possibly due to natural and human predation. This makes for a vulnerable population, estimated at some 200-250 breeding pairs in November 1999 (van der Kamp 2002c). Breeding of **Little Tern** has been recorded almost annually in Lake Debo since a small colony (7 breeding pairs) was found in May 1999 (van der Kamp et al. 2002c).

Birds at communal roosts

Thirteen large waterbird species (Long-tailed Cormorant, African Darter, African Spoonbill, eight heron species, Glossy and Sacred Ibis) exploit the Inner Niger Delta, but concentrate in the few remaining flood forests to roost. This behaviour provides excellent opportunity to count them, either early in the morning when they leave the roost, or when they are heading back there in the late afternoon (van der Kamp et al. 2002c). All species roost at night, except the Black-crowned Night Heron, which feeds at night and roosts by day.

Birds heading for, or leaving, the roost can be counted accurately in the two hours preceding darkness or from civil twilight onwards in the early morning. The method is labour-intensive, because roosts attract birds from a 360° radius and several counting stations are often needed to cover all incoming or outgoing flights. For a complete survey, roosts should ideally be counted simultaneously, a prerequisite never met because of logistic problems associated with the expanse of the region. Most birds roost in Dentaka forest on the east side of Lake Walado in the central Inner Niger Delta, but to obtain total numbers, at least a dozen other roosts should be counted as well. At higher flood levels, birds are more evenly distributed across the Delta, and hence over the available roosts. A complete survey of all roosts takes more than a fortnight. Van der Kamp et al. (2002c) performed more or less complete monthly counts of roosting birds between October and February during three seasons (1998/1999 to 2000/2001). The same team managed to repeat this count in October/November 2002, 2004 and 2005 (van der Kamp et al. 2005c).

Numbers counted will often be too low. First, not all birds fly to a roost – Purple Herons, for instance, increasingly remain on the feeding site at night when the flood recedes. Secondly, due to lack of time or logistic problems, it is not always possible to count *all* of a roost's sectors through which birds arrive or depart. Thirdly, birds feeding just outside or even within the colony, and juveniles staying behind (whether or not capable of flying) on the breeding site, cannot be accounted for; counting positions are always situated at some distance from the forests in order to have an overview of movements. Finally, counts start at 16:30, but Long-tailed Cormorants may start earlier (involving small numbers only, and therefore not relevant to the overall picture; Fig. 50). Most species still arrive in numbers well after sunset, and some even well after civil twilight. In particular, Black-crowned Night Herons are difficult to count, as most individuals leave the roost when darkness has greatly reduced visibility. Unless counts are made directly beneath the line of flight, such birds are easily missed (although their loud croaking calls can still be heard). However, a well-positioned observer may still cover departing birds reliably, as checked on 22-23 January 2004 near Garouye: 2859 departing Night Herons were counted, against 2810 returning birds next morning. Maximum numbers given here are therefore approximations of actual numbers present in the Inner

Niger Delta. Van der Kamp *et al.* (2002a) estimated the maximum number of Black-crowned Night Herons in the Inner Niger Delta at 10 000 birds, but our most recent counts arrive at higher numbers: 10 730 birds left at sunset from the Dentaka forest on 30 January 2005, and as many as 21 000 on 7 February 2008.

Up to 50 875 **Long-tailed Cormorants** were counted on the roosts after the breeding season (van der Kamp *et al.* 2002a, van der Kamp *et al.* 2005c). This total presumably represents the entire postbreeding population (about 1.3 times the number of 18 000-20 000 breeding pairs; Table 3). The highest total of **Cattle Egret** (335 377 birds) was recorded in November 2005, equivalent to 1.5 times the estimated breeding population of 110 000 pairs (Table 3).

Not only large waterbirds, but also some passerine species concentrate in impressive nocturnal roosts. In 1991/92 and 1998-2000 **Yellow Wagtails** had huge roosts near Lake Debo. A major roost on *bourgou* on the eastern shore of Lake Debo held up to 200 000-300 000 birds by the end of November, some 2500-5000 already being present in the first half of September. This seasonal trend was more or less paralleled by **Sand Martins** at the same site; starting in September, roosting numbers (100-500) increased to a maximum of over 500 000 birds. Both species had abandoned this roost by late November/early December, in the last weeks before the peak flood arrived in the central lakes. Three surveys in October-November in the southern Inner Niger Delta showed very few Yellow Wagtails and Sand Martins, suggesting that most birds stayed in the northern Inner Niger Delta awaiting the improvement of feeding conditions in the southern delta's drainage. In Mopti numbers built up as soon as numbers on roosts in Debo started to decline (van der Kamp *et al.* 2002a).

Bird counts from the air

The Inner Niger Delta is large, not to say huge. Even when a low- and slow-flying aircraft is used, it takes two trained ornithologists about

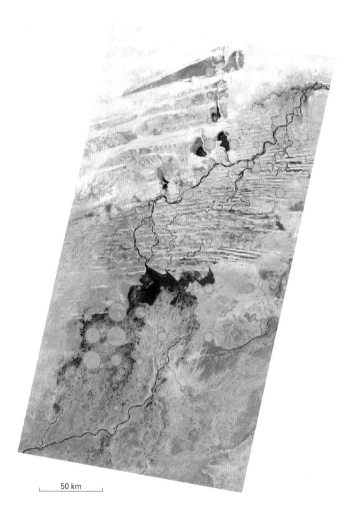

Fig. 51 Distribution of 900 000 Garganeys across the Inner Niger Delta in January 1987; the satellite image is of another date (10 November 1984). Numbers were summed for all 18x18 km squares. From: Skinner *et al.* 1987b.

Table 4 Aerial counts of 13 waterbird species in the Inner Niger Delta in January (February in 1987). Data from the African Waterbird Census database of Wetlands International. Original sources: Roux (1973) for 1972 count, Skinner *et al.* (1989) for ducks in 1977-1983, Trolliet & Girard (2001) for Ruff in 1977-1980, Roux & Jarry (1984) for 1984, Skinner *et al.* (1987b) for 1986 and 1987, Girard *et al.* (2004; 2006) for 1999-2006; the 2007 count is pre-published in African Bird Club Bulletin 14: 223-4; the original source of the 1985 count was not found. The water level on 15 January refers to the gauge of Akka.

Year	1972	1977	1978	1979
Water level 15 January, cm	372	459	227	368
Glossy Ibis	?	?	?	?
White-faced Whistling Duck	19267	16887	73647	40759
Fulvous Whistling Duck	300	2970	23075	1916
African Pygmy Goose	130	10	68	52
Knob-billed Duck	2655	1193	19818	9196
Spur-winged Goose	1513	2476	1686	5545
Egyptian Goose	1729	730	1829	343
Northern Shoveler	10	2350	2207	588
Northern Pintail	26788	100610	384685	65852
Garganey	92427	306465	492917	112867
Ferruginous Duck	37	4000	892	4015
Black-tailed Godwit	20700	37300	19009	55060
Ruff	103705	185750	246850	75640

50 hours to cover the entire area and to obtain a rough estimate of the large, conspicuous bird species, such as ducks and large herons (Girard *et al.* 2004). This method is less suitable for smaller bird species like African Pygmy Goose and small waders (Roux & Jarry 1984, Girard & Thal 1999, 2000, 2001; van der Kamp *et al.* 2002a; Girard *et al.* 2004).

Thanks to the efforts of CRBPO, WWF/IUCN and ONCFS, a long series of aerial counts is available of waterbirds in the Inner Niger Delta during the northern winter. Counted numbers were summed for 171 18x18 km squares (see Fig. 51 for Garganey in January 1987).

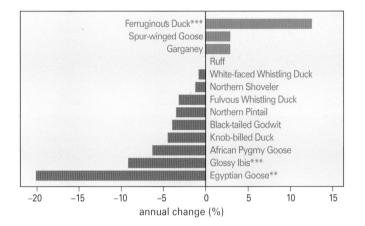

Fig. 52 Long-term change in waterbirds during aerial counts in the Inner Niger Delta in January or February, 1972-2007 (data from Table 4). The annual change is derived from an exponential function of counted number against year. The significance of this function is given: $P<0.01$ = **, $P<0.001$=***.

The 15 mid-winter counts since 1972 (Fig. 48) highlighted the presence of about one million ducks and geese, of which Garganey is invariably the commonest species. The figures show wide annual variations, which may be partly related to methodological problems. Girard *et al.* (2004) stressed how difficult it is to perform a complete survey; even from an aircraft, birds may easily be overlooked or misidentified. This is certainly true in years of extensive flooding when birds are scattered over the entire Delta. However, birds may also be overlooked in extremely dry years, when they are concentrated on the Niger River outside the Inner Niger Delta (Roux & Jarry 1984).

Aerial surveys were performed from a variety of aircraft, either low-winged or high-winged, the latter permitting superior visibility and therefore being used consistently in later years. Flights were made at less than 100 m above ground at an average speed of 150 km/h (Roux 1973, Trolliet & Girard 2001). From this altitude flocks of Anatidae were easily located up to 400-500 m, or even further, away when emergent vegetation was scarce and light conditions favourable. Even so, large flocks of ducks feeding in pools covered with water lily, so common in the centre of the Inner Niger Delta, were difficult to discover. Two observers, one each on either side of the plane, noted birds and numbers using a tape recorder, marking the route on a map and noting the time every 15 minutes. The Delta was fully covered. The uneven distribution of birds does not justify extrapolation, and this also has not been attempted.

When numbers counted are plotted against water level (data given in Table 4), no significant correlation was found for any of the 13 species, contrasting with our expectation that low flood levels would coincide with fewer birds. Several explanations can be put forward to account for this counter-intuitive finding. First of all, methodological problems differ for counting during high and low floods. Secondly, whereas most aerial censuses took place in January, the

1980	1982	1983	1984	1985	1986	1987	1999	2000	2001	2006	2007
362	279	202	129	78	166	142	325	401	331	250	270
?	?	7735	25800	20224	12756	23548	4520	3517	6150	2630	2338
28755	19447	87245	37720	7153	10538	19839	7760	47310	70950	14739	?
2762	2470	72700	31720	13267	13246	1513	88	7733	2795	2039	?
0	62	0	63	50	0	0	0	5	12	23	?
7525	1725	765	1719	321	1939	6904	821	4299	627	1239	?
4193	5518	1063	5580	2094	18519	5579	2450	5760	3220	7804	6450
262	10	195	2145	612	724	297	6	67	0	14	?
554	50	160	200	20	25	35	0	200	195	13907	?
202536	149368	329400	129473	103153	79855	208792	41100	116650	164160	59379	10612
460105	506769	304960	156391	105882	484426	899916	209130	515680	744000	815800	226250
3805	924	1252	2928	6442	3312	5601	7800	13020	14300	13590	15066
4700	21904	13900	38964	41298	45353	40492	10077	3075	40280	10495	5990
110400	179665	91650	61665	30500	114135	175217	147936	135180	188095	80935	98265

Cattle Egret feeding in dry *bourgou* are often found in association with cattle, preying on locusts and grasshoppers disturbed by the thousands of hooves.

biggest impact of low flooding on mortality materialises at a later date, *i.e.* in the last stage of the deflooding, when feeding and resting sites are in short supply. Thirdly, waterbirds may have redistributed in response to changing conditions in the Inner Niger Delta. In dry years, waterbirds move in part from the Inner Niger Delta to the Niger River, downstream of Timbuctu. Aerial counts along the Niger River between Timbuctu and Gao yielded low numbers for all species in relatively wet years (1972, 1977, 1979, 1980, 1999), whereas substantial numbers may occur in dry years, as for instance in 1983, when 29 000 Pintails, 26 000 Garganey, 33 800 Fulvous Whistling Ducks, 3800 Glossy Ibises, 1350 Black-tailed Godwits and 7565 Ruff

Fig. 53 The central lakes (Debo, Walado and Korientze) on four days (true colour composite based on Landsat scenes; 51 x 67 km) with different flood levels (water level at Akka varying between -2 and +511 cm). The census area is outlined.

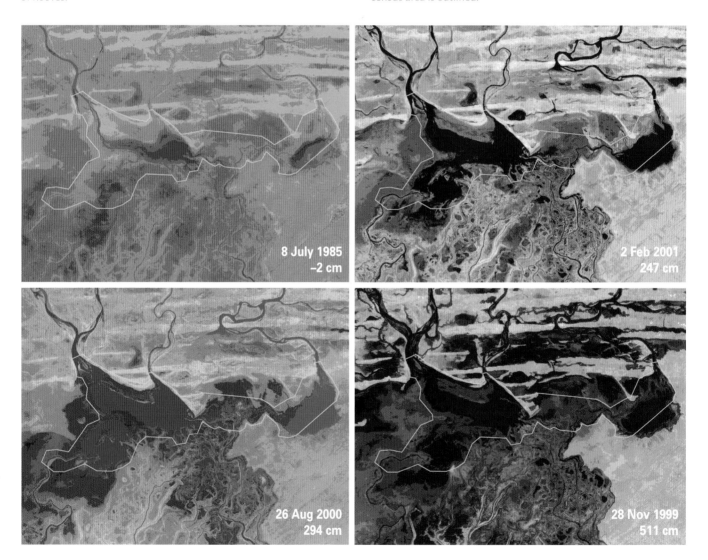

Table 5 Maximum numbers recorded during 62 counts of the central lake area between December 1991 and February 2007. From: van der Kamp et al. (2002a, 2005a, 2005c).

Species	Maximum	Month	Species	Maximum	Month
African Darter	493	June 2002	Pied Avocet	86	Jan 1994
Long-tailed Cormorant	13521	Feb 2005	Great Snipe	140	Mar 2001
Great White Pelican	4300	May 2000	Egyptian Plover	753	Aug 2000
Grey Heron	6218	Mar 2001	Greater Painted Snipe	102	Jun 2002
Black-headed Heron	94	June 2004	Collared Pratincole	20904	Jan 2005
Purple Heron	4606	Mar 2000	Black-headed Lapwing	16	Sep 2000
Black-crowned Night Heron	4620	Apr 2001	Spur-winged Lapwing	5829	Jul 2000
Squacco Heron	1667	Feb 2000	Afican Wattled Lapwing	12	Jul 2001
Western Reef Egret	261	Apr 2000	Kittlitz's Plover	13887	Jun 2004
Black Heron	390	Mar 1999	Common Ringed Plover	6073	Feb 1998
Cattle Egret	5571	June 2003	Kentish Plover	3	Jun 2003
Little Egret	11100	Mar 2001	Little Ringed Plover	36	Feb 2007
Intermediate Egret	1503	Feb 2000	White-fronted Plover	791	Aug 1998
Great Egret	5534	Jan 2000	Grey Plover	7	Jan 2005
African Spoonbill	900	Apr 2001	Black-tailed Godwit	37654	Jan 2005
Eurasian Spoonbill	75	Feb 2005	Eurasian Curlew	372	Aug 1998
Glossy Ibis	15375	Jan 1992	Little Stint	38362	Apr 2001
Sacred Ibis	1160	Apr 2001	Curlew Sandpiper	3754	Mar 2000
White-faced Whistling Duck	6643	Jan 2005	Green Sandpiper	2	Mar 2001
Fulvous Whistling Duck	9500	Dec 1991	Wood Sandpiper	755	Apr 2001
African Pygmy Goose	664	Jan 1995	Common Sandpiper	74	Aug 2000
Knob-billed Duck	9124	Dec 1993	Ruff	90470	Jan 1992
Spur-winged Goose	11481	Jun 2004	Common Redshank	8	Feb 2006
Egyptian Goose	900	Apr 1996	Spotted Redshank	4557	Feb 2004
Northern Pintail	150000	Dec 1993	Common Greenshank	3031	Feb 2007
Garganey	273000	Dec 1993	Marsh Sandpiper	185	Mar 2001
African Fish Eagle	5	Feb94+Jan97	Lesser Black-backed Gull	236	Feb 2002
Osprey	5	Feb 2006	Grey-headed Gull	242	Jan 1994
Yellow-billed Kite	346	Feb 2003	Black-headed Gull	33	Feb 2000
Eurasian Marsh Harrier	458	Feb 2007	Gull-billed Tern	3896	Feb 2007
Pallid Harrier	2	Feb 2005	Caspian Tern	3545	Feb 2002
Montagu's Harrier	26	Jun 2003	Little Tern	346	Mar 2000
Black Crowned Crane	50	Jan 1992	Whiskered Tern	6679	Feb 2005
Senegal Thick-knee	119	Jan 1994	White-winged Tern	5874	Jan 1997
Black-winged Stilt	5536	Feb 2004	Marsh Owl	70	Jun 2002

were recorded. Similar or even higher numbers were observed in two other dry years, 1978 and 1984 (e.g. 4260 Black-tailed Godwits in 1984). In those years of drought, large numbers were also counted further downstream, between Gao and the border between Mali and Niger (e.g. 34 500 Garganeys in 1978, 28 500 Fulvous Whistling Ducks in 1983 and 19 700 Ruff in 1984). However, such numbers are still small in comparison with those staying in the Inner Niger Delta at large.

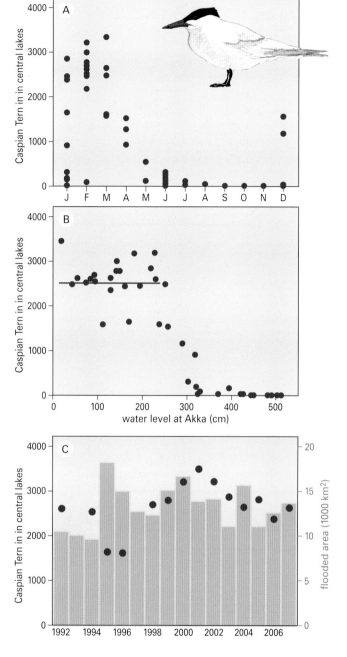

Fig. 54 Numbers of Caspian Terns in the central lakes of the Inner Niger Delta in 1991-2007 (see Fig. 48 for months censused, and Fig. 53 for outline of census area). (A) Monthly totals. (B) Numbers in October – March as a function of water level. (C) Maximum number per year in January – March (water level less than 250 cm; left axis); the bars show the maximum flooding in the preceding months (right axis).

Over a 35-year period, most waterbirds show a negative population change (Fig. 52). Declines may have been larger than indicated, since the earlier counts in 1972 and 1982 were incomplete (Roux & Jarry 1984). The aerial counts (Table 4) show that around 1980, about half a million **Garganey** spent the northern winter in the Inner Delta. Numbers were severely reduced in the dry years 1984 and 1985 to 100 000-150 000, but in 1999-2006, between 200 000 and 900 000 birds were counted. The counts suggest no overall decline for this species in the Delta, constrasting with trends in the breeding areas (see Chapter 23). The 200 000-400 000 **Northern Pintails** counted before the Great Drought were reduced to about 100 000 in 1984, after which the population remained at this much lower level in the following years (see Chapter 22). **Northern Shovelers** showed up in surprising numbers in January 2006. The species, if seen at all during aerial censuses, so far had been counted in negligible numbers in the Inner Niger Delta (Table 4). Recent counts revealed an increasing number of **Ferruginous Duck**, with major concentrations in Lac Horo and Lac Fati in the northern Delta. The species avoids the floodplains and feeds in lakes with stagnant water which still contained water as long as their stay lasted, even during the Great Drought. **Knob-billed Ducks** declined during the Great Drought, without recovering in later years. Up to 74 000-87 000 **White-faced Whistling Ducks** were counted before the Great Drought, and similar numbers were present in 2001. Also **Spur-winged Geese** remained stable; over the last 30 years, aerial surveys showed 3000 to 6000 birds and in 1986 an outlier of 19 000. The **Egyptian Goose**, with some 2000 present until 1984, has decreased to a few dozen since then. Also **African Pygmy Goose** shows a clear decline. The largest number (130 birds) was seen in 1972 during an incomplete survey, with relatively low numbers recorded for all other species. Recent aerial counts revealed only 0-23 birds, but this species is easily overlooked from the air (Roux & Jarry 1984). The **Glossy Ibis** showed an obvious decline (see also Chapter 21). The counted numbers of **Black-tailed Godwit** and **Ruff** varied annually, but showed no clear trend.

Bird counts of the central lakes

Within the Delta, the central lakes (Lake Debo, Walado and Korientze; Fig. 53), an area of 460 km², have been covered by systematic bird counts since 1992 (van der Kamp et al. 2002a, 2002b, 2005a). It takes four to seven days to count all waterbirds in the central lake area. Depending on water level, the counts were made from a boat or whilst walking along the shores (van der Kamp 2002a). No attempt has been made to correct for underestimates in species that are secretive or are otherwise easily overlooked. Moreover, small passerines, such as Yellow Wagtail, have not been included in these counts; for such species, density counts have been employed (see below). The ornithological significance of the central lakes is obvious from Table 5. Maxima are not that different from near-maxima, i.e. they do not refer to outliers. Most Palearctic species are commonest in Feb-

Many fish-eating species assemble in shallow waters during the receding flood, like Long-tailed Cormorant, Black Heron, Little Egret, Black-winged Stilt, Greenshank, Whiskered and White-winged Terns.

ruary-March, whereas local bird species often peak in May-June.

The bird counts of the central lakes will be used in Chapters 15-41, including comparisons with density counts. Here we deal with the question of how to interpret the bird counts in the central lakes, which cover about 2% of the entire floodplain, but at a water level of 0 cm in Akka, constitute 70% of all water available; the remaining water is to be found mainly in the permanent lakes in the far north of the Delta. When the water level is less than 300 cm, the census area still covers about 20% of the water bodies in the Inner Niger Delta. The low-lying position of the Debo-Walado-Korientze-lakes explains why most waterbirds become concentrated here; elsewhere in the Delta, only few, mostly stagnant, permanent lakes remain. Obviously, the monthly counts cannot be used to show seasonal variations in numbers. Thus, while the census area is not representative for the Delta, the counts reliably monitor annual fluctuations in waterbird numbers (as illustrated for Caspian Tern, Black-winged Stilt and Eurasian Marsh Harrier).

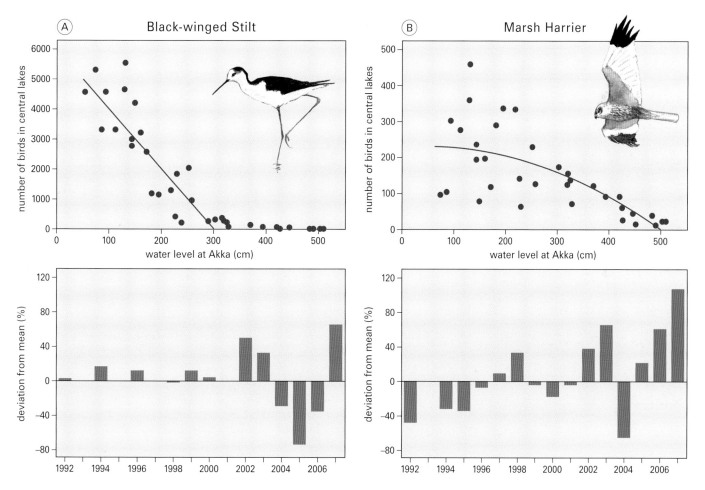

Fig. 55 Counts of Black-winged Stilt and Eurasian Marsh Harrier in the central lakes of the Inner Niger Delta (Fig. 53). Upper panels: number as a function of water level in October–February; lower panels: annual presence, calculated as average percentage deviation from the line shown in upper panel. Note: no count in 1993.

Inner Niger Delta

Year	1992	1994	1995	1996	1997	1998	1999	2000	2001	2002	2003	2004	2005
Akka, cm	86	75	239	113	303	55	44	97	21	181	196	230	150
African Darter		59	20	38	52	14	18	82	106	109	119	14	18
Long-tailed Cormorant		4460	13521	4349	9337	2562	6681	12185	5957	5616	1014	2829	509
Grey Heron		795	885	3475	52	4167	3998	471	5663	2574	3501	2246	1847
Purple Heron	848	850	1630	1523	766	795	2177	4171	1880	1913	1242	974	1425
Black Heron		80	90	252	20	68	390	82	320	236	130	11	0
Cattle Egret		3250	1600	1771	5054	1184	1054	351	625	339	99	624	978
Little Egret		3520	1178	2831	226	3001	5285	4002	10915	2178	2418	5723	5340
Intermediate Egret			309	35	450	510	98	1501	926	253	65	70	64
Great Egret		1850	2469	612	19	1809	1205	5534	2386	2131	1158	1958	840
African Spoonbill	0	238	11	0	86	1	430	19	186	0	140	1	3
Eurasian Spoonbill	0	38	0	0	0	0	7	4	0	0	51	38	75
Glossy Ibis	15375	4585	612	6125	131	6907	3149	1620	10651	6122	3554	252	8295
Sacred Ibis	133	129	6	71	0	150	188	25	518	20	86	44	54
White-faced Whistling Duck	340	1860	22	157	0	0	28	2	44	0	83	14	6643
Fulvous Whistling Duck	9500	5750	43	472	25	2	14	2	2	0	0	0	148
African Pygmy Goose	30	133	664	130	68	0	68	2	8	0	6	0	0
Knob-billed Duck	991	9124	27	129	0	50	13	12	458	902	0	2	178
Spur-winged Goose	82	105	2247	260	75	930	1951	1662	710	2001	208	685	110
Egyptian Goose	19	403	34	780	0	649	530	86	217	27	0	12	60
Northern Pintail	140000	150100	130	0	0	10	0	0	0	0	0	0	0
Garganey	95700	273000	4500	10000	10155	5316	49	144	3770	0	336	60	4525
Eurasian Marsh Harrier	116	196	151	215	175	302	231	195	193	275	336	64	222
Black Crowned Crane	50	3	0	3	0	0	32	4	4	0	0	1	0
Common Moorhen		17	0	0	124	0	41	9	35	59	0	0	186
Purple Swamphen		49	0	0	11	296	176	103	122	10	90	18	375
Black-winged Stilt	3291	5299	181	3306	300	4573	2758	371	2998	1078	1134	1775	3930
Egyptian Plover		68	100	107	63	57	59	71	139	95	71	57	48
Collared Pratincole	6270	8254	597	1136	2540	954	1417	1972	1515	1125	5257	538	20904
Spur-winged Lapwing		166	58	295	145	462	256	200	571	283	506	230	724
Common Ringed Plover	251	911	534	2037	20	6073	3664	4696	3070	1048	4136	2694	1595
Black-tailed Godwit	25136	23561	16973	16759	300	20126	26852	7074	22444	29195	20261	9136	37650
Eurasian Curlew	164	339	69	129	4	258	245	137	212	87	91	140	57
Little Stint	6900	13956	3165	6106	53	17666	11629	10962	10653	7472	6090	13032	4101
Curlew Sandpiper	426	714	530	2979	23	2770	2315	3754	1196	268	2475	1356	675
Wood Sandpiper	5	31	5	15	7	73	55	75	285	27	58	35	96
Ruff	90470	39050	17898	45958	1338	45075	26341	7671	47281	39646	26785	4262	31833
Spotted Redshank	1776	3108	1040	419	11	3318	4431	807	1011	1956	2266	4557	1153
Common Greenshank	297	421	413	685	34	892	1701	1811	1571	1364	403	2513	1416
Grey-headed Gull	170	242	181	40	88	60	12	2	19	0	1	0	0
Gull-billed Tern	465	1745	466	1564	100	1948	2364	2179	3344	1420	3086	1635	2080
Caspian Tern	2585	2519	1596	1583	311	2685	2783	3193	3334	2999	2469	2586	2781
Little Tern	325	160	172	136	84	133	140	346	283	86	51	31	179
Whiskered Tern		1810	1721	1703	2403	2453	3494	3227	3146	2059	2046	2217	6679
White-winged Tern		102	1410	1239	5874	3104	3241	2233	2902	1906	4009	2588	2212

2006	2007	Akka	year	b1	b2
130	132	R	R	b1	b2
20	67	-0.04	0	0.34	0.339
1369	1569	0.19	-0.68	-0.233	-0.257
1699	2202	-0.7	0.31	0.354	0.119
1213	946	-0.24	0.13	0.22	0.012
182	132	-0.56	-0.25	0.288	0.065
572	378	0.35	-0.68	-0.341	-0.375
4318	6176	-0.61	0.47	0.377	0.278
10	125	-0.3	-0.43	0.131	0.262
1369	843	-0.27	0.12	0.234	0.175
12	100	-0.39	0.17	0.047	0.051
61	55	0.12	0.62	0.672	0.681
3679	2403	-0.57	-0.05	0.034	-0.063
32	110	-0.67	0.12	0.116	0.118
0	14	0	-0.18	0.146	0.213
0	2	-0.27	-0.74	-0.468	-0.394
0	0	0.31	-0.79	-0.366	-0.438
0	0	-0.24	-0.55	-0.306	-0.356
120	0	0	-0.33	0.089	-0.099
17	0	-0.56	-0.38	-0.088	-0.172
0	0	-0.28	-0.73	-0.448	-0.351
0	3846	-0.25	-0.59	-0.435	-0.408
307	379	-0.19	0.35	0.513	0.393
0	0	-0.39	-0.51	-0.202	-0.243
210	9	0.16	0.28	0.373	0.404
147	9	-0.43	0.42	0.423	0.338
4343	5203	-0.61	0.22	0.208	0.119
44	51	-0.17	-0.48	0.056	-0.112
6315	947	-0.03	0	0.277	0.321
300	163	-0.34	0.43	0.528	0.394
1925	549	-0.44	0.32	0.355	0.221
26851	18252	-0.46	0.17	0.261	0.118
45	83	-0.74	-0.16	0.002	-0.067
9175	2021	-0.61	0.05	0.171	0.013
832	254	-0.39	0.05	0.178	0.013
31	59	-0.51	0.59	0.401	0.388
39098	55858	-0.56	0	0.076	-0.048
1193	3204	-0.24	0.23	0.333	0.166
1445	2899	-0.19	0.6	0.644	0.477
1	0	0	-0.9	-0.656	-0.651
3099	3510	-0.48	0.59	0.656	0.557
2165	2542	-0.67	0.27	0.369	0.198
157	92	-0.55	-0.41	0.119	0.022
5408	3104	-0.19	0.64	0.652	0.585
767	526	0.35	0.12	0.311	0.317

Table 6 Maximum number of birds counted in Lake Debo and Walado in December - March 1991-2007 (1-4 counts per season; blanks if not counted). The water level in Akka (2nd line) refers to the last count before 1 April, when usually the highest numbers were recorded. The last four columns give statistical details. R is the correlation between maximum number counted in a year and either water level in Akka or year (n=15). A multiple regression was performed on the 35 counts available for the period December-March to analyse the long-term change (number against year), taking into account the relationship of bird numbers with water level (3rd degree polynomial) and maximum flood level in that season (linear function); b1 is the standardised coefficient of number against year (thus without water level) and b2 the same coefficient but including the four variables describing the impact of water level. The significance level is indicated: P<0.05 red; P<0.01 green, P<0.001 blue.

Purple Heron in high *bourgou* vegetation. Niger, January 2008.

Inner Niger Delta

White-faced Whistling Duck, Garganey, Black-winged Stilt, Black-tailed Godwit and Ruff.

Caspian Terns arrive in the Inner Niger Delta in September-October, but in the central lakes large numbers are only present from January until their departure in March (Fig. 54A). The first Caspian Terns enter the central lake area when the water level drops below 300 cm (Fig. 54B). With a further decrease of water level, numbers increase. From the moment that the water level drops to 200 cm (it keeps reducing steadily), their numbers remain constant until departure. All Caspian Terns wintering in the Inner Niger Delta are by then concentrated in the central lakes, where they roost on the available levees and sandbanks and feed in nearby shallow water. The variation in numbers at a water level below 200 cm is a reliable reflection of total numbers present (and having survived so far) in the Inner Niger Delta in any one year (Fig. 54C). Until 2002, counts made at water levels below 200 cm show a gradual increase in numbers, followed by a decrease in later years (Fig. 54C); this will be explained in Chapter 31. In general, numbers are higher when the preceding months showed a high flood, indicating higher survival during high floods.

Black-winged Stilts show up in the central lakes at water levels below 250-300 cm. In contrast to Caspian Terns, numbers steadily increase with receding water level, to a peak level of 5500 birds (Fig. 55A). This is about half of the 12 000 birds maximally counted in the entire Inner Niger Delta from the air in 2006 and 2007 (Table 4). Departure from the Inner Niger Delta starts in early March. Stilts feed in shallow water and can be found in depressions and other temporary lakes during the receding flood until they dry up. The number of temporary lakes containing water gradually decreases between November and March, which explains why the numbers in the central lakes continue to increase during this period. For each cm of decrease in water level from 300 to 0 cm, Stilt numbers in the central lakes on average increase by 20 birds (Fig. 55A). To check for annual variations in numbers, we calculated the difference between counted and predicted numbers relative to water level during the

count. The wintering population was remarkably stable between 1992 and 2000, with above-average numbers in 2001-02 and 2007, and below-average numbers in 2004-06 (Fig. 55A). The 3258 Stilts counted in 1991 by Tinarelli (1998) at a water level of 122 cm represent an 8% above-average data point (not included in Fig. 55).

The story of water level-dependent numbers in the central lakes is – as expected – about the same for many other waterbirds and related species, including the **Eurasian Marsh Harrier**, a predator

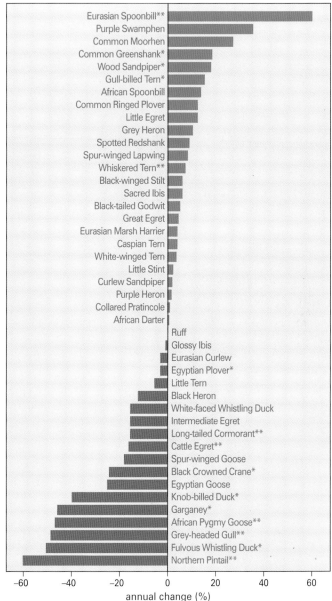

Fig. 56 The average change in waterbird numbers counted in Lakes Debo and Walado between 1992 and 2007 (data from Table 6). The significance of this function is given as: $P<0.05^*$, $P<0.01 = ^{**}$, $P<0.001=^{***}$.

Box 7

The significance of the lower floodplains for Glossy Ibis and *Corbicula*-eating waders

The water level on the floodplains of the Inner Niger Delta changes continuously between flooding and deflooding, an endless cycle of a dust-laden world transformed into wetland and gradually back into dry land. These extremes embrace intermediate stages that temporarily spell optimal foraging and resting conditions for water-associated birds. Waterbirds follow the retreating shoreline and consequently their distribution alters from day to day (Fig. 57). For some species, a change in distribution coincides with a change in diet. However, most bird species have a flexible, patch-related prey choice, induced by changes in distribution caused by the retreating water. For instance, the diet of Black-tailed Godwits and Ruff changes from rice grains, when available in inundated rice fields, to molluscs, when reduced flood level forces the birds into the lowest floodplains. Several mollusc species are taken, the most important one being a cockle-like sturdily-shelled bivalve 3-14 mm long, *Corbicula fluminalis*. Less important are *Caelatura aegyptica*, a larger, thin-shelled bivalve (<30 mm) and a small snail, *Cleopatra bulinoides*. These prey are taken not only by Black-tailed Godwits and Ruff, but also by Glossy Ibises and Black-winged Stilts. We determined prey selection and daily food intake in these birds and compared their combined predation pressure with the total food stock available at different water levels (van der Kamp et al. 2002b, van der Kamp & Zwarts unpubl.).

The birds swallow *Corbicula* intact and crack the hard shell in the gizzard. Ruff have a small gizzard and appear to select the smaller *Corbicula*, at least when they swallow the prey whole (Fig. 58), but many, like other waders, take the flesh from the gaping valves, an easy alternative since *Corbicula* die after emersion. Most Ruff eating dying bivalves are females, presumably because their gizzard is not strong enough to crack the shells. Black-tailed Godwits and Glossy Ibises ignore the small *Corbicula* taken by Ruff (Fig. 59).

Feeding starts early in the morning at 7:00 and continues until 18:00 (sun rise: 6:45; sunset 18:15). By counting the number of prey taken by individual birds, the daily consumption of the different species could be derived: 5600 *Corbicula* by an average Godwit and 9500 *Corbicula* by Glossy Ibis. Although *Corbicula* shells are crushed in the gizzard, the umbo (the part of the shell where the two valves are attached) remains intact. The width of the umbo and the length of the shell are closely related, showing a good exponential fit between size and flesh weight. Using this relationship, it was possible exactly to determine size selection by predators and to derive prey weights from the umbos found in droppings. By combining these data, individual daily food consumption could be estimated.

The numbers of Ruff and Black-tailed Godwit have been stable at 40 000 and 30 000 birds respectively, whereas Glossy Ibis crashed from 30 000 in the early 1980s to 3000 in 2007 (Chapters 21, 27 and 29). Consequently, the combined predation pressure decreased from 650 million *Corbicula* per day in the 1980s to 450 million *Corbicula* per day 25 years later; expressed in terms of dry weight, from 3.3 to 2.2 ton per day. Since *Corbicula* is the main diet for these bird species in the last month before their departure to the breeding quarters, their total annual predation

Fig. 57 How different bird species exploit the Inner Niger Delta relative to water height (From: van der Kamp et al. 2002b).

amounts to about 100 tons dry weight. How does this relate to the total amount on offer?

Corbicula live close to the surface and filter food from the overlying water. In the Inner Niger Delta, the species is found only on those floodplains that are covered by water for at least 6 months a year. Highest density is achieved on plains covered by water for 8-9 months. This clam species is a major prey for fish (*Tilapia*), and, during deflooding, for birds as well. When fully emerged, they die. The few individuals surviving the dry period produce the offspring for the next flood. Based on 1311 samples taken during three years (1999-2001), we estimated total food supply (dry flesh) on the floodplains around Lake Debo in January-March at 26 000 tonnes of *Corbicula*, 750 tonnes of *Caelatura* and 650 tonnes of *Cleopatra*. The birds consume 100 tonnes/year, i.e. less than 0.4% of the available flesh mass. This does not imply, however, that food is always exploitable, even when abundantly available. For example, Glossy Ibis feed in shallow water less than 20 cm deep, Black-tailed Godwit prefer a water depth of 15 cm and the still smaller Ruff forages in water less than 5 cm deep. Food therefore varies in accessibility during

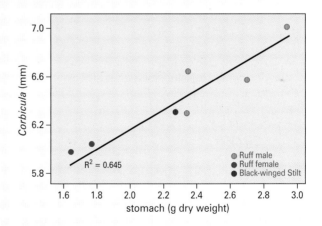

Fig. 58 Size of *Corbicula* found in six Ruff and one Black-winged Stilt as a function of the weight of their gizzard.

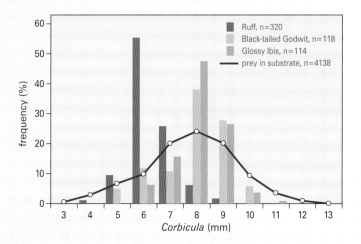
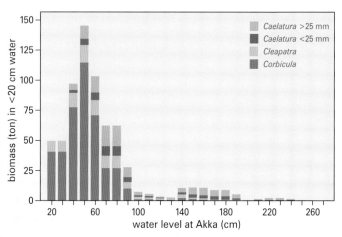

Fig.59 The size classes of *Corbicula* selected by Ruff, Black-tailed Godwit and Glossy Ibis compared to the frequency distribution of the bivalves present in the substrate.

Fig. 60 Biomass (tonnes dry flesh) of prey as a function of water level (gauge at Akka). *Caelatura* has been split into two size classes, since bivalves > 25 mm are too large to be eaten by Glossy Ibis.

higher water levels, or dies when the plains have dried out. To estimate daily available food supply, we need to know the food supply in the zone of shallow water < 20 cm. The digital elevation map (of which Fig. 46 gives a simplified version) shows that the shallow zone around Lake Debo measures 5 km² at a water level of 200-250 cm at Akka. This zone increases to 23 km² at a water level of 150 cm, but decreases substantially to 5 km² when the water level drops below 90 cm.

The average biomass of food available for mollusc-eating birds, such as that found in the shallow zone with <20 cm water, is determined by surface area and prey density. When the first floodplains become exposed during deflooding, the density of molluscs is still very low. Even at a flood level of 100 cm, prey density is not higher than 15 *Corbicula*/m², 9.5 *Cleopatra*/m² and 3.8 *Caelatura*/m², in total 0.4 g/m². With water level decreasing even further, density increases rapidly to reach its peak at a level of 50 cm: 2400 *Corbicula*/m², 281 *Cleopatra*/m² and 19 *Caelatura*/m², together 16.9 g/m². Despite the wide shallow zone at a water level of about 150 cm, available biomass is still very poor due to low prey density (Fig. 60). Food is most abundant when the dense *Corbicula* banks found at 40-60 cm come within reach of the birds. Given a daily consumption of about 3 tonnes a day, the combined daily predation pressure varies from 63% at a water level of 100-200 cm, to only 1.6% at 50 cm of water. However, water level decreases at about 3-5 cm a day. Thus, when the first Glossy Ibises start feeding in 20 cm-deep water, it will be another 4-7 days before Ruff are able to exploit the same site. When this is taken into account, predation pressure remains very low when the large *Corbicula* banks are exploited. However, with the water level down to 100 cm, the calculated predation pressure is higher than the estimated food supply. This suggests that either food supply is underestimated, or predation pressure overestimated. In any case, we are convinced that the birds deplete prey stocks entirely at this stage of the flood, because our sampling

Ruff, Black-tailed Godwit and Glossy Ibis feeding on *Corbicula* (centre left), *Caelatura* (centre middle). The faeces (right) were analysed to reconstruct prey selection of the three bivalve-predators.

Ruff and Black-tailed Godwits feeding on *Corbicula*.

Glossy Ibis feeding on *Corbicula*.

along transects in 20-60 cm deep water showed that prey were completely absent on the exposed floodplains, where a fortnight earlier high densities had remained. These data further suggest that mollusc-eating birds, after arrival on the shores of Lake Debo at a water level of 150-200 cm, face food shortages for at least 1-2 weeks. These observations raise several questions.

First, is the achieved food intake sufficient to fatten up, and if so at what rate? We estimated daily gross food intake of Black-tailed Godwit at 650 kJ; 20% of the energy intake is not assimilated, and net energy intake is therefore 520 kJ per day. From captive Godwits we know that, at constant body mass, the bird needs 360 kJ, so 160 kJ is available to build up body reserves necessary for migration, enabling them to in-

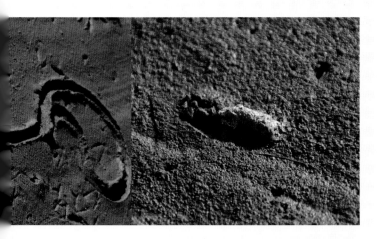

crease their body mass with an estimated 1.3% per day (A.-M. Blomert & L. Zwarts, unpublished). This rate of body increase is similar to that found in other waders on the Banc d'Arguin, living at the same latitude at the same time of the year (Zwarts *et al.* 1990). Bar-tailed Godwits flying non-stop from the Banc d'Arguin to the Wadden Sea have depleted their entire pre-migratory body reserve of 40% upon arrival in the Wadden Sea (Piersma & Jukema 1990). Assuming that Black-tailed Godwits in the Inner Delta also fly the 4000 km non-stop to the Dutch breeding areas

(Chapter 27), they need about a month to attain the necessary departure weight. Our estimated rate of fattening up refers to Godwits feeding on *Corbicula* banks in February 1992. The daily consumption was lower in the previous two months, when the birds consumed mainly *Caelatura*. Thus, it is not unreasonable to assume that, at a higher flood level, the rate of fattening up would be lower.

Secondly, is *Corbicula* essential for Glossy Ibis, Godwit and Ruff to fatten up in the central Inner Niger Delta? At the prevailing flood levels of the last 10-20 years, the birds exploit the *Corbicula* banks from February until their departure in March. How do they fatten up at much higher flood levels, such as those in the 1950s? This is a difficult question to answer, as we have no data on the distribution of *Corbicula* at different flood levels. It may be that at higher floods *Corbicula* also occurs higher up the floodplains and thus may remain the basic prey for mollusc-eating birds. If not, the birds have to fatten up on a different diet, *e.g.* rice grains, as they do in the coastal rice fields of Guinea-Bissau (Chapter 27). But how can they fatten up at much lower flood levels when *Corbicula* banks become available early in January, but hardly any molluscs are available in February-March? So far, we have no indication that the birds do advance their fattening up prior to departure from the Inner Niger Delta. Instead, extreme droughts create serious bottlenecks, as was shown in 1984/1985, the driest year of the century in the Sahel. Although female Ruff captured in late March were desperately looking for food, they lacked fat reserves – in fact, the body mass of these birds was even less than lean weight. A substantial cohort of females was unable to depart from the wintering grounds (males depart earlier than females) and must have perished that year (van der Kamp *et al.* 2002b; Chapter 27).

Thirdly, the last question boils down to a more general phenomenon, discussed in Chapter 43: is the timing of departure invariable or related to year-to-year variations in food supply which are determined by flood level or rainfall? The main conclusion from all the above observations is that, even in a large dynamic wetland such as the Inner Niger Delta, where birds seem to have so many opportunities to search for and exploit optimal feeding conditions, food supply still may be a limiting factor in some cases.

which, in the Inner Niger Delta, shows a sex-specific habitat choice, with females and juveniles preferring wetlands and males the drier surroundings (Chapter 25). Harrier numbers on average increase with declining water level (Fig. 55B). About 100 Marsh Harriers were present in the central lakes in the early 1990s and this gradually increased to 470 birds in 2007, with distinct year-to-year differences (Fig. 55B). The overall increase is consistent with population trends in Europe (Chapter 25).

Bird numbers in the central lakes may thus be used – after correction for water level – to determine long-term trends (Fig. 54C, lower panel of Fig. 55). However, there are several complications. First of all, the measured water level describes the flooding of the hydrosystem by the river, but the inundation of water bodies isolated from the river is a different matter. The higher the flood, the more depressions remain water-filled during deflooding (Fig. 38). Consequently, the number of birds arriving in the central lakes at a water level of 200 cm also depends on the peak level of the flood some months previously. Fewer alternative sites are available during low floods, and eventually a higher proportion of birds will be pushed into the central lakes. Examining this complication more closely, flood levels did indeed have some impact on Stilt numbers in the central lakes (a species that needs shorelines to feed), but no such effect was found in Caspian Terns. For the latter species, small isolated waterbodies presumably offer inadequate foraging opportunities.

Secondly, the counts in the central lakes produce many anomalies when compared with Delta-wide censuses. This is particularly evident in ducks and geese, which show erratic between-years fluctuations, partly governed by flood level. Ducks normally concentrate on the lakes along the NE and NW side of the Delta. Only when these lakes remain dry during droughts are the birds forced to move to the central lakes. Even so, their presence and numbers, and hence their chance of being detected and counted, largely depend on human disturbance, which is higher in the central lakes than elsewhere in the Inner Niger Delta. The impact of human disturbance on cormorants, herons and waders is supposedly smaller than on ducks and geese. In this regard, the significant declines of Cattle Egret and Long-tailed Cormorant in the central lakes stand out in comparison with their breeding numbers in the Inner Niger Delta, which are essentially stable. Does this imply that the central lakes, and their food supply, have changed for the worse for these species, as also suggested by the mass mortality of Long-tailed Cormorants in 2004-2005?

Even when considering the many pitfalls of counting birds in large and dynamic ecosystems, the central lakes show a clear dichotomy in trends between Palearctic and African species: many more Palearctic species are stable or increasing than African species, the latter often showing – steep – declines (Fig. 56). Whereas the seasonal presence of European species in the Inner Niger Delta is, for the most part, restricted to the northern winter, and consequently depends on the flood height in any one year, African species face local conditions year-round. African species may evade droughts by intra-African movements to regions less severely – or not at all – affected, but there is no escape from the human-induced habitat loss that is particularly serious in wetlands, and exacerbated by the steep human population growth (Fig. 28). A recent survey of trends in African waterbird species showed that 43% of all species for which estimates were available were in decline (Dodman 2007). The results from the central lakes are consistent with this overall picture.

Density counts

So far, counts have focussed on large, conspicuous birds, often concentrated in well-defined areas. Even considering the variety of possible biases, we feel confident that our attempts at counting fall within the realm of real figures. It is an entirely different story for secretive, small or widely dispersed bird species. A single look across the large floodplains west and east of the borders of the Diaka suffices to appreciate the magnitude of the problem. Vast expanses of seemingly identical grasslands stretch towards the horizon, only to turn into an amalgamation of habitats as soon as the area is penetrated on foot. The density of birds may seem insignificant at any one place. But the sheer vastness of the Inner Niger Delta accounts for stunning totals of many species, and this becomes obvious when clouds of Yellow Wagtails or Sand Martins start heading for roosts in the late afternoon. Yet these species may be encountered only occasionally in small numbers, or even not at all, during the day. The huge floodplains seem to spawn endless streams of birds around dusk. Any attempt at counting absolute numbers is futile.

Instead, we adopted an alternative census method, *i.e.* accurate bird counts in small plots of known size. As random sampling was impossible logistically, we opted for the next best solution: stratified sampling in the major habitats of the Inner Niger Delta. Habitat diversity within the Inner Niger Delta is low, permitting a geographically non-random set of plots to represent the entire Delta. Habitats were defined with the use of vegetation maps (based on satellite images) and information obtained from the digital flooding model. Large sections of the Inner Niger Delta could not be reached, or only peripherally so, but habitat types in those parts were adequately sampled elsewhere. Whenever possible, we chose >1 km long strip transects perpendicular to the river (Niger, Bani, Diaka). Where vegetation height, vegetation density and water depth prevented such selection, we devised intersecting transects in subplots. Plot lengths and widths were measured by a calibrated laser beam in adapted binoculars, or calculated from GPS-readings (checked by counting steps of known length in the field). This enabled an accurate calculation of plot surfaces. We recorded habitat type, vegetation height, vegetation density and water depth for each plot on pre-printed forms, and also date, local time, coordinates, census takers and census type.

Plot sampling was based on the assumption that all birds in the plots had to be recorded. To meet this criterion, several techniques were used, often accompanied by shouting, hand-clapping and throwing mud into the vegetation to assure that all birds were flushed and recorded. This was achieved by using a suite of techniques:
• Two or three people walked at separation distances of 20-50 m in parallel along the transect. Separation distances depended on vegetation density and height.
• One observer would criss-cross a plot while the other watched from the side and recorded the birds.
• After encircling a small plot, we would flush birds by means of noise and mud-throwing.
• Boating transects were necessary in water deeper than about

Table 7 Estimated number of birds in Lake Télé in March 2003 and March 2004, based on density counts in 92 plots averaging 4.5 ha in area.

Species	Count
Grey Heron	543
Purple Heron	1202
Squacco Heron	17646
Cattle Egret	3144
Great Egret	344
Intermediate Egret	3611
Glossy Ibis	291
Purple Swamphen	3843
Lesser Jacana	5765
African Jacana	4115
Black-winged Stilt	875
Greater Painted Snipe	545
Collared Pratincole	71
Little Stint	516
Little Ringed Plover	142
Wood Sandpiper	35166
Ruff	10060
Crested Lark	202
Yellow Wagtail	71280
Sedge Warbler	2490

Fig. 61 Locations of 1617 sites within the Inner Niger Delta where bird density counts in plots were carried out. The scene is based on a true colour composite (Landsat, 16 October 2001, water level in Akka 429 cm), showing an area of 183 x 342 km.

1.2 m, when we would apply distance sampling in bands of varying width (20->100 m) for the various bird species. Obviously, a Great Egret can be located further away than a Sedge Warbler.

While plot-sampling, a keen eye was kept on neighbouring fields because the effect of flushing birds was never confined to the plot being worked. In order to count birds reliably in the near-distance, we had to keep track of birds which we had flushed previously, and also to account for birds flushed prematurely from the next plots in sequence. The Inner Niger Delta constitutes a flat environment of wide vistas, simplifying the tracking of flushed birds. When a particular area had been disturbed by counting teams, they would traverse hundreds of metres in silence without counting to resume plot-sampling where all birds were still present and none had settled from disturbed plots. We may possibly have introduced an inadvertent bias when we found large stationary flocks ahead, such as resting herons or Collared Pratincoles. We normally steered clear of such groups, which may have led to under-sampling.

For each plot, we made a subjective assessment as to whether all birds present had been recorded or not. Density counts were designated incomplete when plots could not be entered, the water being more than waist-deep or the vegetation being impenetrable; numbers from these plots were not used in the calculation of bird

Table 8 Average density per km² of 46 species in the Inner Niger Delta within six habitat types. *Mimosa* = *bourgoutière* with *Mimosa pigra* bushes; *bourgou*= *bourgoutière*, rice-c = cultivated rice, rice-w = wild rice, bare/grass = unvegetated (including water) or grass-covered, stagnant = Lakes Télé and Horo. Reliability of the calculated average densities is low with a high standard error (SE >60% relative to mean; marked red) and high with a low SE (<30% relative to SE; marked green). Counts in plots on dry ground were omitted (bird density very low). Anthus sp. refers to Plain-backed or Grassland Pipit.

Habitat	Mimosa	Bourgou	rice-c	rice-w	bare/grass	stagnant
n of counts	15	790	137	67	93	86
Long-tailed Cormorant	0	85	0	1	1	34
Grey Heron	0	24	4	49	0	8
Black-headed Heron	0	0	30	0	0	0
Purple Heron	19	49	3	4	1	30
Squacco Heron	30	148	6	2	2	255
Cattle Egret	0	105	306	169	14	78
Little Egret	93	30	4	5	36	35
Intermediate Egret	0	14	1	0	2	73
Great Egret	0	10	0	0	1	12
Little Bittern	0	1	0	0	0	0
Great Bittern	0	5	0	0	0	0
Glossy Ibis	0	24	0	0	0	37
Common Moorhen	6	0	6	0	0	0
Lesser Moorhen	56	1	0	0	0	0
Purple Swamphen	0	11	0	0	0	58
Lesser Jacana	0	13	1	5	0	104
African Jacana	0	52	13	13	0	15
Black-winged Stilt	0	68	51	0	197	152
Great Snipe	20	1	5	0	10	0
Common Snipe	0	6	3	3	0	7
Greater Painted Snipe	0	3	15	0	0	1
Collared Pratincole	0	86	319	39	150	6
Spur-winged Lapwing	0	28	38	19	100	3
Kittlitz's Plover	0	55	17	0	117	0
Common Ringed Plover	0	36	1	0	315	1
Black-tailed Godwit	0	111	6	26	19	4
Little Stint	0	207	66	4	665	205
Curlew Sandpiper	0	7	0	0	385	5
Green Sandpiper	0	2	0	0	0	0
Wood Sandpiper	87	271	94	74	180	823
Spotted Redshank	0	1	6	0	6	0
Ruff	0	415	531	12	750	2215
Common Greenshank	0	4	6	0	15	0
Marsh Sandpiper	0	5	0	0	11	0
Crested Lark	94	97	138	45	35	6
Red-throated Pipit	0	1	0	0	0	0
Anthus sp.	0	2	2	3	0	0
Yellow Wagtail	649	661	241	186	294	1786
Bluethroat	245	16	14	2	0	0
Savi's Warbler	0	1	0	0	0	0
Sedge Warbler	760	260	0	4	4	3437
Aquatic Warbler	0	1	0	0	0	0
Subalpine Warbler	40	0	0	0	0	0
Chiffchaff	424	0	0	0	0	0
Prinia sp.	238	3	0	2	0	0
Zitting Cisticola	0	29	76	17	2	3
Total	2818	3012	2006	684	3327	9461

densities per habitat. All density counts were performed between 1 November and 15 March 2001/2002, 2002/2003 and 2003/2004. Altogether 1617 counts were made in plots averaging 2.99 ha in size, nearly all situated in the southern half of the Inner Niger Delta and around the central lakes (Fig. 61). Although plots were not randomly distributed across the Delta, sampling covered a stratified cross-section of major habitat types, water depths and vegetation types available in the Inner Niger Delta (van der Kamp *et al.* 2005a).

We also carried out density counts in two northern lakes, Télé (57 km²) and Horo (139 km²), in March 2003 and 2004. It was estimated that 56 km² of Lake Horo was covered by floating vegetation, comprising a large species locally known as *kouma* (*Polygonum senegalense*) and a small one known as *loubou* (*Ludwigia stolonifera*). Only Sedge Warblers were seen feeding on the emerging stems, reaching extremely high densities here (as already noted by Jarry *et al.* 1986): on average there were 103 birds per ha or an estimated total of 575 000 Sedge Warblers (95% confidence interval: 393 000 –756 000; n = 29) in this habitat alone.

During both visits the SW corner of Lake Télé was covered by floating vegetation of two grass species, locally known as *horia* and *garsa* (27.5 km²). Yellow Wagtails were the most common bird species (25.6/ha), but other species were abundant as well: Wood

Sandpiper (12.4/ha), Squacco Heron (6.4/ha), Lesser Jacana (2.1/ha), African Jacana (1.5/ha) and Purple Swamphen (1.4/ha). Sedge Warblers did not reach as high a density as in *Polygonum* vegetation (0.9/ha). Apart from the 51 counts in floating grass fields (water depth >20cm), we carried out another 26 counts along the water's edge where water depth was <20 cm, amounting to a strip of 25 km long and 20 m wide (covering 0.5 km²). Taking all data together, we arrived, via habitat-specific extrapolation, at an estimated total of 162 000 birds for Lake Télé (Table 7).

Bird numbers on the floodplains of the Inner Niger Delta were estimated in a similar way. Water depth and vegetation type were registered for each plot. Water depth is a major determinant of presence and abundance of many species, ranging from dry ground species such as Crested Lark to feeders on floating vegetation in deep water, such as Squacco Heron. However, most bird species reach their highest density in shallow water (Fig. 62). All density counts were classified according to five flood level bands (based on measurements at the gauge of Akka: 0-100 cm, 100-200 cm, ... 400-500 cm). For each flood level we used the digital flooding model described above to estimate the surface areas of dry ground, areas covered by up to 20 cm of water, 20-40 cm of water, and so on. Unfortunately, the flooding model did not allow a distinction between "moist ground" and "dry ground", ecologically a highly relevant difference. We tackled this problem by assuming that the width of the wet zone along the water's edge was 5 m, on average. The total surface area of moist ground can then be calculated, since the total length of shoreline can be derived from the digital flooding model. The higher the flood level, the longer the shoreline, but shore length increases even further during deflooding, reaching a maximum of 80 000 km due to the presence of isolated lakes and pools. The many temporary lakes at a water level of 200–300 cm at Akka explain why the relation between length of shoreline and water level at Akka is not straight but curved. The surface area of moist ground is calculated whilst accounting for this relationship.

Vegetation is another major factor that determines bird density. We distinguished 14 vegetation types, which in the final analysis were reduced to six major habitat types. *Bourgou*, *didere*, *poro* and water lily were collectively named '*bourgoutière*'. Areas with short vegetation or lacking it, *i.e.* open water (without *bourgoutière*, rice or water lily) and two types of grasslands were grouped. Thirdly, all habitat types in the stagnant Lakes Télé and Horo were put in one group. Most density counts were performed in *bourgoutière*, wild

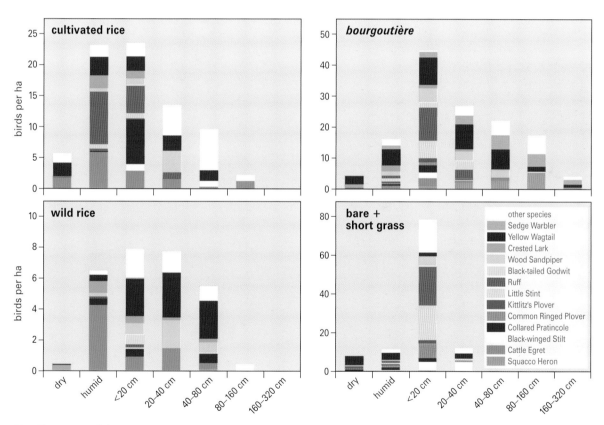

Fig. 62 Density of the 13 commonest bird species, and summed density of a further 49 species across the gradient from dry ground to deep water in four main habitats: cultivated rice (n=168 plots), *bourgoutière* (n=945), wild rice (n=160) and bare ground + short grass (n=95). All data were collected on the floodplains of the Inner Niger Delta (see Figs. 48 and 61). Note the different scales.

rice, cultivated rice and in areas with short or no vegetation. Bird density was remarkably low in wild rice, high in *bourgoutière* and exceptionally high in stagnant water (*i.e.* emerging *Polygonum* vegetation) (Table 8). Waders reached a high density in plots with short grass or no vegetation. A large part of this variation may be attributed to water level (Fig. 62).

In areas with no, or short, vegetation, birds reached a high density in shallow water, while they were absent from areas covered by deep water. In contrast, a variety of bird species fed in deeper water on the floating stems of wild rice and even more often on floating *bourgou* (explanation given in Box 8). Some species, like Yellow Wagtail, occurred in all habitat types, whilst others, like Sedge Warbler, were observed almost exclusively in *bourgoutière*. Species composition differed considerably between the four habitats, with more herons in rice fields and *bourgoutière* and mainly waders in shallow water lacking floating vegetation. Cultivated rice fields no longer covered by water, but not yet completely dried out, attracted many Ruff and other rice-eating birds.

Bird density is determined by more than water depth and vegetation. For instance, the presence of cows attracted Cattle Egrets and caused an increase in density from 0.9 birds per ha (cattle absent) to 4.8 birds per ha (cattle present). A similar difference was found in the density of Yellow Wagtails (2.9 vs. 12.1 birds/ha). Other variables may explain variations in bird density, but these were not quantified. Densities relative to water depth and vegetation type were used to extrapolate density counts into total numbers present on the floodplains. The vegetation map of Zwarts *et al.* (2005b) was used to determine the surface area of the main vegetation types. Since *Mimosa pigra* is mixed with wild rice or *bourgoutière*, it could not be distinguished as a separate habitat type.

The vegetation map refers to the lower inundation zone (<360 cm at Akka), an area of 5855 km², which can be divided into:
- 1040 km² of cultivated rice (17.8%),
- 1433 km² of wild rice and wild rice with water lily (24.5%),
- 1543 km² of *bourgoutière* (*bourgou*, *didere*, sometimes mixed with water lily) (26.4%),
- 1105 km² of bare ground when flooded, but changing into grassland after exposure (18.9%),
- 735 km² of water without vegetation at a water level of 86 cm (12.6%).

Combining the digital elevation model and the vegetation map allowed us to calculate the area per category of water depth separately for each vegetation type. This calculation was performed for each of five flood levels: <100, 100-200, 200-300, 300-400 and 400-500 cm. The outcome is an estimate of the total number of birds present in the lower 5855 km² of the floodplain of the Inner Niger Delta. Since we did not perform any density counts on the lakes, we assumed that bird density here was zero, which would undoubtedly be correct for most species. It is not true, however, for five tern species, Eurasian Marsh Harrier and some other bird species feeding above the lakes. This is not relevant, since these species were not included in density

Box 8

Herons on floating vegetation

Fish-eating herons capture prey by surprise from a standing position, by stalking or by chasing. For each species, tarsus length determines the water depth they can exploit. Because tarsus length differs between species, we would expect some habitat segregation between the six heron species inhabiting the Inner Niger Delta. Great Egrets should be able to forage in the widest depth range, from shallows up to 35 cm, Grey Herons up to 27 cm, Purple Herons up to 22 cm, Little Egrets up to 19 cm and Squacco Herons

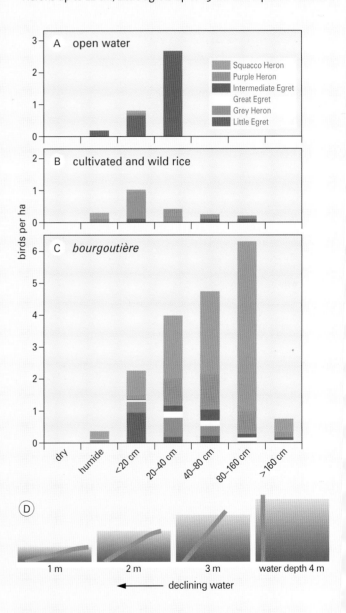

Fig. 63 Feeding density per ha of five heron and egret species as a function of water depth in (A) open water, (B) wild or cultivated rice and (C) *bourgoutière* (*bourgou* or *didéré*). All density counts were performed on the floodplains of the Inner Niger Delta, between 15 November and 1 March. Panel (D) visualises the changing position of *bourgou* stems with declining water level.

Purple Herons feeding in *bourgou* with about 1 m of water.

up to 11 cm. Density counts performed in open water (Fig. 63) show that herons feed in shallow water, usually less than 40 cm deep, but exploit deeper water if floating vegetation permits walking. The density is less on floating rice than on floating *bourgoutière* (Fig. 63). Great and Little Egrets do not, or rarely, stand on *bourgou* and rice stems.

Why is heron density higher on *bourgoutière* than on floating rice? The answer lies in the physical properties of these plant species. *Bourgou* grows in deep water and forms stems on average 4 m long. When the flood level declines, the thick *bourgou* stems gradually flex into a more horizontal plane. When the water recedes from 4 to 2 metres, the buoyancy of the loosely packed stems is apparently insufficient to support a heron. When the flood covers the *bourgou* fields with less than 1 metre of water, the stem mass is sufficiently dense and entangled to cope with the weight of a heron, thus offering the opportunity of exploiting this otherwise inaccessible zone (see inset). Floating rice is less attractive for herons than *bourgou*, since wild and cultivated rice grow in shallower water and form thinner stems which are only about 2.5 m long. The buoyancy of floating stems of rice is probably not great enough to support a heron. On top of that, floating rice often forms a dense mat that covers large stretches of water; locating swimming prey visually is therefore impossible. The open tangle of rather thick *bourgou* stems allows herons to locate fish more easily.

Unfortunately, we have no data on fish density beneath floating vegetation. Thus, we cannot exclude the possibility that floating rice harbours fewer fish than floating *bourgou*. We may assume that, as found in several studies (*e.g.* Wanink 1999, Wanink *et al.* 1999), fish density decreases in shallow water and only very small fish venture into the shallowest water. This may explain why heron density decreases when the water column drops below 40 cm. An exception is the Little Egret. In contrast to the other five heron species, Little Egrets usually feed communally to facilitate the capture of fish swimming in shoals.

If it is true that the buoyancy of loosely-packed floating stems explains why herons avoid *bourgou* in deep water, we would expect heron species to be zonally differentiated, depending on body mass rather than tarsus length. Our classification of birds feeding in various water depths is rather crude, but even so it is evident that Squacco Herons (less than 300 g) reach their maximum density in *bourgou* floating on a water column of 80-160 cm depth. Purple Herons (800-1000 g) are most common in *bourgou* with 40-80 cm of water, while Grey Herons (1400 g) are commonest in shallower water (20-40 cm). However, body mass is not the only determinant of where and when herons can feed. Compared with egrets and Grey Herons, both Squacco and Purple Herons adroitly walk across floating stems. In this regard, it is telling that the toe of a Squacco Heron (67 mm) is 18% longer than its tarsus and that of a Purple Heron 8% longer; in the four other species the toe is about 30% shorter than the tarsus (Cramp & Simmons 1977). Longer toes permit a better grip on buoyant vegetation and allow the bird's weight to be spread across a wider area and a greater number of stems.

By feeding on floating vegetation, herons may enlarge their potential foraging range considerably. Take, for example, the short-legged Squacco Heron. In the absence of floating vegetation, feeding is confined to shallow water less than about 10 cm deep. During high floods (>15.000 km^2 of the Inner Niger Delta flooded) the shallow zone with less than 10 cm of water covers about 1000 km^2. This surface area gradually decreases to 100 – 150 km^2 with floods as small as 3000 km^2. The surface area of the *bourgoutière* zone at water depths of 40-160 cm varies from 700 km^2 during a high flood to 300 km^2 during a low flood. By exploiting floating stems of *bourgou* or *didere*, Squacco Herons effectively double the size of their potential feeding area.

Table 9 Maximum number of waterbirds in November – February in the central lake area (460 km²; Fig. 53), and average numbers on the lower floodplains of the entire Inner Niger Delta (5855 km²; Fig. 39). All numbers refer to floodplains covered by water at a flood level of 350 cm. Numbers were calculated from density counts at four flood levels (0-100, 100-200, 200-300, 300-400 cm at Akka), then averaged for "all floodplains". The standard error SE shows the variation in numbers depending on flood level. For the central lakes the highest number in any of these four categories is given, usually coinciding with the lowest flood level.

Species	Central lakes	All floodplains	
	max	Mean	SE
Grey Heron	3195	23915	8131
Purple Heron	7179	50417	19273
Squacco Heron	41218	183390	56496
Cattle Egret	19118	314560	40276
Little Egret	7611	37389	10039
Intermediate Egret	2461	13009	4958
Great Egret	1611	8879	3088
Great Bittern	793	3852	1046
Purple Swamphen	1692	10844	4792
Lesser Jacana	2216	22310	11566
African Jacana	16320	112664	55496
Great Snipe	109	2061	589
Greater Painted Snipe	396	3667	2289
Spur-winged Lapwing	7222	72503	3736
Little Stint	46243	164576	66468
Green Sandpiper	243	558	122
Wood Sandpiper	38198	199448	52394
Ruff	77021	276834	70188
Marsh Sandpiper	891	2651	1138
Crested Lark	19593	193619	28879
Yellow Wagtail	100999	960785	53807
Bluethroat	2772	7030	2038
Sedge Warbler	41027	248879	77393

counts anyway, but were counted in absolute numbers.

Our estimate of bird numbers is confined to the floodplains below 360 cm. Maximal floodplain size varied between 10 000 and 20 000 km² annually. The number of birds, and of which species, that occurs on the higher floodplains depends on flood level. At a flood level below about 360 cm, all birds feeding on humid ground and in water are concentrated in the lower 5855 km² covered by our density plots. Cattle Egret, Kittlitz's Plover, Yellow Wagtail, Crested Lark and other bird species which also feed on dry ground (Fig. 62) may be found on the upper floodplains as well. If the water level

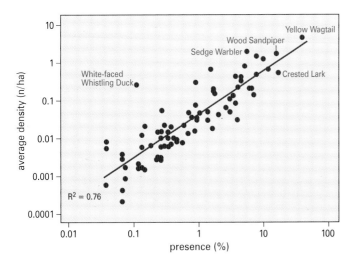

Fig. 64 Relationship between average density per ha and the proportion of sampled plots in which 78 bird species were present during 2999 density counts in the Inner Niger Delta and Senegal Delta.

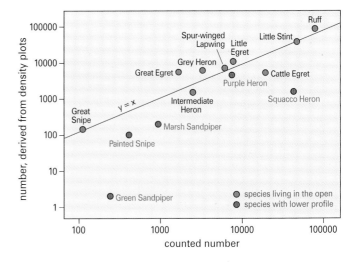

Fig. 65 Maximum number counted in the central lakes (Table 4) compared with the maximum number derived from density counts (Table 6). Species are indicated separately.

Hooklines are used to catch birds (three Collared Pratincoles). Low-flying birds are particularly vulnerable at night, when the touch of a single flight feather suffices to get caught by the razor-sharp hooks. Where hooklines are strategically positioned near water edges, many birds are captured during a single night and their distress calls can be heard over long distances.

exceeds 360 cm, part of the species community feeding in shallow water will move to the higher floodplains. As a consequence, our estimated totals for the lower floodplains (5855 km²) are too low for flood levels surpassing 400-500 cm. To demonstrate the effect of flood level on bird numbers, we estimated numbers for four flood levels: <100, 100-200, 200-300 and 300-400 cm. The standard errors given in Table 9 refer to these four estimates.

For density counts to be used in extrapolations, bird species should be present in a fairly large proportion of the plots (Fig. 64). The Yellow Wagtail was the commonest bird species in terms of presence (39.6%), as well as average density (4.7/ha). Crested Lark and Wood Sandpiper were both present in about 16% of the plots, but the average density of Crested Larks (0.5/ha) was lower than that for Wood Sandpipers (1.7/ha). Sedge Warblers had a higher density (2.0/ha) than Wood Sandpipers, but were present in only 5.5% of the plots. The variation in frequency of presence is partly habitat-related, some species being more selective than others. Also, species feeding in flocks may have an extremely patchy occurrence. The White-faced Whistling Duck, belonging to the 15 most common bird species, is a case in point: it was found in only 3 of the 2999 density plots in the Inner Niger Delta and similar habitats in the Senegal delta. Density counts are clearly unreliable for birds feeding in flocks. This limits our analyses (see below) to 22 out of the 78 registered species.

For several species, the estimated total derived from density plots may be compared with numbers obtained via counting methods mentioned previously. Our extrapolation of 960 000 Yellow Wagtails is hardly surprising, given counts of many hundreds of thousands at roosts. But how do we validate an estimate of 183 000 Squacco Herons, for instance? As it happens, the bird counts in the central lakes offer precisely that possibility. We have 62 reliable and complete counts of this well-defined region, offering an independent check of bird numbers in the same region as derived from density counts (Fig. 65). In general estimated totals corresponded quite well with counted numbers for species living in the open, but less so for species which were not so visible.

Via extrapolation, the density counts produced tentative totals for the entire Inner Niger Delta (Table 9), including standard errors. Thus, the estimate of 314 000 Cattle Egrets must be read as 274 000–355 000 (close to the total actually counted, *i.e.* 335 000 on the roosts). As expected, the SE is smaller in well-distributed common species than in scarcer, patchily distributed ones. To improve standard errors - in other words, to improve the reliability of the calculated totals - would have required an inordinately larger plot-sampling effort and a fully randomized set-up of plots. This was impossible in the available time and with the available manpower.

Direct counts showed that Little Stint and Ruff are the commonest wader species in the central lakes (Table 5), but plot sampling revealed that the less conspicuous Wood Sandpiper (150 000-250 000) is as common as the Little Stint (100 000-230 000). Density counts also show that the Ruff is the commonest wader on the floodplains of the Inner Niger Delta (200 000-350 000), a much higher figure than obtained from aerial surveys (100 000-200 000; Table 4). Ruff are easily missed during aerial counts; they are widely distributed across the floodplains, and inconspicuous in flight. On the other hand, our extrapolation for Ruff may be too high. Ruff do not simply select shallow water with humid zones to feed, but instead concentrate where small *Corbicula*-bivalves are common (Box 7). Since *Corbicula* is especially common on floodplains with some current, Ruff may reach higher densities on floodplains along the river than on the borders of isolated, temporary lakes. This is difficult to test with our data set, but if true, our estimate would be too high, since

Hippos near Dire, the only large wild mammal in the Inner Niger Delta having survived the competition with man for space (January 2009).

Most flooded forests with *Acacia kirkii* disappeared from the Inner Niger Delta in the 1970s and 1980s. The remaining forests are of prime importance for colonial breeding birds, notably Long-tailed Cormorant, Darter, herons and storks, African Spoonbills and Sacred Ibis. Yellow-billed Kites profit from such colonies (discarded prey, dead nestlings) and may reach high densities. Akkagoun, February 2005.

comparatively more plots were sampled along the rivers and near the central lakes. The same reasoning applies to fish-eating birds, although we have no indication that the density of fish-eating birds differs between floodplains connected to the river and those that are not. This all remains speculative. The totals in Table 6 are the best possible estimate.

Threats and protection

The Inner Niger Delta is a huge floodplain which attracts millions of waterbirds, not only from Europe but also, as will be shown in the following chapters, from Asia as far as eastern Siberia. From a hydrological point of view, the floodplain system may be considered relatively unspoiled. Nevertheless, the flooding of the Inner Niger Delta has been gradually reduced during the second half of the 20th century (Fig. 41), partly due to irrigation near the Markala dam and water withheld in the Selingue reservoir. A further decline in flooding is to be expected following the construction of the Talo dam in the Upper Bani, possible dams near Djenne, in the Upper Niger (Fomi) and downstream of the Inner Niger Delta (Taoussa), and an extension of the irrigation area of *Office du Niger*. Together these schemes seriously reduce flooding and harm the ecological significance of the floodplains. Flood level has a direct bearing on the survival of migratory birds spending the northern winter in the floodplains, being poor in years with low floods and vice versa. Purple Heron and Caspian Tern are points in case (see Chapters 16 and 31), but many more species are affected by flood level in the Inner Niger Delta. Further permanent reductions of the floodplains will lead to irreversible losses in populations of Palearctic and African bird species which depend on floodplains for part of their life cycle. The series of counts in 1992-2007 shows that the numbers of the migratory bird species (except Pintail) have so far remained stable, but that many African bird species are in decline (Table 4 & 6; Fig. 52 & 56).

Not only flooding is pivotal to the ecological significance of the Inner Niger Delta, but also human exploitation. Over the decades, the Inner Delta has been converted into a semi-natural habitat by the local inhabitants. For instance, the tenfold extension of the area of cultivated rice between 1920 and 2000 (Zwarts & Kone 2005b) caused a decline in the area of wild rice. Birds may have profited from this shift, however, especially waterbirds; judged by densities found in both habitats, cultivated rice is preferred over wild rice (Fig. 62). Rice farmers were also responsible for the removal of flood forests, heralding the loss of breeding sites for herons, egrets, ibises and storks. Colonial waterbirds are nowadays concentrated in the few remaining flood forests (Fig. 49), which makes them vulnerable to human predation and excludes their use of potential feeding areas too far from the colony. The decline in the surface area of flood forests, and of forests on levees and around the floodplains, must also have contributed to the decline of passerines concentrated here during the northern winter; see Box 3 and Chapter 37 (Common Redstart). The area is grazed by an increasing number of cows, sheep and goats, 6 million altogether in 2000 (Chapter 5). Grazing on this scale converts the floodplains into grassy pastures of a uniform low vegetation that is removed before seeding can commence; see Chapter 32 (Turtle Dove). The resulting lack of cover and food shortage are likely to have an impact on birds in the Inner Niger Delta, although this is not yet quantified.

Fishing in the Inner Niger Delta poses another problem, albeit multifaceted. Birds may have profited from overexploitation, which led to a reduction of average fish size during the past 30 years and hence to more fish becoming available in the size classes preferred by birds. However, many birds are accidentally killed by nets and hooklines. Hooklines are used to catch birds intentionally for food, as are disused fish nets on a grand scale. Most birds are captured, deliberately or otherwise, when water levels are low, *i.e.* when birds and fishermen concentrate in the last remaining water-filled vestiges (Kone *et al.* 2002). During a low flood this situation is reached by January, exposing the birds to heavy exploitation for another 2-3 months before return migration commences to their northern breeding grounds. This period of vulnerability is much shorter during normal floods. The threat is small, sometimes almost non-existent, in years of maximum floods, as happened before the 1970s, when the birds remained widely dispersed across the entire Inner Niger Delta until they departed for the breeding grounds. This must be, in combination with food availability, the overriding explanation for the link between annual mortality and flood level; for further explanation, see Chapters 23 (Garganey) and 29 (Ruff).

The relationship between annual bird mortality and flood level may have changed since the 1990s, especially in the light of the intensification of fishing activities. Bird exploitation is also likely to have increased. Where birds formerly were captured mainly for local consumption, nowadays trade is booming, facilitated by improved roads (expanding the area over which trapped fish and birds are transported, even as far as Bamako) and increased usage of ice to freeze and preserve food. Middlemen have professionalised the trade infrastructure. The first step in this process was already taken in the 1960s when inexpensive nylon nets enabled the large-scale capture of birds such as Garganey and Ruff.

It is still possible to counteract the anticipated decline of birds in the Inner Niger Delta. The management of irrigation schemes and water reservoirs in the Upper Niger could be adjusted to reduce ecological and socio-economical damage downstream. Much would be gained if vulnerable sites were protected effectively, as has been achieved in the Senegal Delta (Chapter 7). The Inner Niger Delta is not entirely without protection, though. The two largest flood forests, Akkagoun and Dentaka, are managed through local management committees, but the agreements between the surrounding villages are often violated by people from outside, who frequently visit the huge bird colonies to collect eggs or nestlings (Beintema *et al.* 2007). This kind of disturbance and exploitation has a large negative impact on colonial waterbirds (Box 12). For the same reason, it is likely that the imminent, and perhaps actual, extirpation of the Black Crowned Crane in the Inner Niger Delta could have been prevented if law-enforced protected zones had been designated for the species to breed safely.

Whereas local breeding birds can profit from the creation of safe havens, it is less likely that migratory bird species will profit likewise. Migrants are distributed across the entire Inner Niger Delta and their distribution changes continuously in relation to flood level. It is crucial that they have access to food without running the risk of being shot, captured or disturbed. The 1-2 months prior to the return to Eurasia are critical, when fattening-up coincides with declining food supplies in the wake of the receding flood. With climate change ahead (less rain expected in the longer run), the greatest threats to migrants are a further reduction of flood level due to events outside the Inner Niger Delta, and an increase of bird exploitation in the area itself.

Fishermen as well as waterbirds get concentrated during receding water in the last remaining water bodies; Lake Debo, February 1994.

Senegal Delta

For time immemorial, the Senegal River has been a lifeline for local communities, offering a rich source of fish, fertile soils for flood recession agriculture and grazing grounds for nomadic pastoralists. Early on in the European colonial history of the region, the French had shown particular interest in developing the river valley, but it was not until the 1960s that the necessary infrastructure for irrigation was put into place; from then on, cultivation of the delta was rapid. The dam construction programme in the 1980s was a pivotal change in the history of the delta, affecting livelihoods and lifestyles profoundly, in a way that generally was detrimental to nature. Large scale development of the delta went ahead regardless, placing intense pressure on traditional flood recession agriculture, grazing systems and local fisheries, with resultant socio-economic impacts on local communities. It is less than surprising that the extent of irrigation and artificial flooding achieved fell short of that projected. At the time of writing, the transition of the once-estuarine delta into an irrigated area for the cultivation of rice and other crops is in full swing. Unfortunately this process has had the serious concomitant complications characteristic of the development stages of large-scale irrigation schemes in the tropics: salinisation, unsanitary conditions for many local communities and the ubiquitous presence of invasive plant species that outcompete local flora. Nevertheless, the delta remains of socio-economic importance to the peoples of Senegal and Mauritania, its rice output comprising a substantial part of the region's annual rice production. The taming of the Senegal

River, and the transformation of its delta, resulted in the partial loss of a major stronghold for migratory and resident waterbirds in the Sahel; other biological assets were also affected adversely. In the early 1960s, French ornithologists recognised the sword of Damocles hanging over the delta and pleaded for the creation of reserves (de Naurois 1965a). Yet another call for mitigation measures was made prior to the dam construction schemes (GFCC 1980). Thanks to such efforts, and the actual designation of protected areas, the delta has retained a high biodiversity and accommodates important numbers of waterbirds.

Introduction

The Senegal River rises in the Fouta-Djallon mountains in Guinea and flows in a northwesterly direction through Mali where several tributaries join the main stream. In its central section, the valley floor is some 25-35 km across. About 150 km before entering the Atlantic Ocean, the Senegal crosses a flat alluvial plain where the river created a huge estuarine delta, roughly between Richard Toll and Saint-Louis. Unlike other Sahelian floodplains, the Senegal Delta has an open access to the ocean. In the past, salt water would penetrate the delta during the dry season, resulting in a variety of habitats and high biodiversity, but the construction of dikes and dams has undermined this dynamic system; at present, only the estuarine part around Saint Louis is still in contact with the Atlantic Ocean. The historical development of the Senegal River, including the lower delta, has been described at length (*e.g.* Bernard 1992, Crousse *et al.* 1991, Engelhard & Abdallah 1986). Each of these studies identified the construction of dams in the 1980s as *the* pivotal event in recent history. To understand the distribution and abundance of birds it is necessary to understand the major hydrological developments (Triplet & Yésou (2000) and Hamerlynck & Duvail (2003)). We provide a brief overview.

The delta before the construction of dikes and dams The Senegal discharges in a dynamic area where ocean currents, seasonal swells, and littoral drifts created sand barriers along the coast. The resulting strip of dunes was constantly being breached by the river, whose main embouchure has moved southward in recent geological history (Barusseau *et al.* 1995, 1998). The Chat Tboul lagoon in the north is an example of a former river mouth; other former connections to the ocean can be found further south. The river mouth was close to Saint-Louis early in the 19th century, but its position changed continuously. The shifting embouchure, varying channels and a shallow bar at the river mouth hampered navigation in the 19th century and the development of commerce upstream (Bernard 1992).

Behind the narrow ridge of dunes a vast natural estuarine floodplain developed over an area of about 3400 km². In the 19th century the French colonial administration developed plans for its cultivation, but significant hydrological engineering remained in abeyance until the mid-20th century. Highly dynamic conditions were characteristic of the natural estuarine floodplain – a freshwater flood occurred from August to October and sea water intruded as the flood receded, and so the local wetlands had long adapted to these fluctuations – within and between years – along with the variable salinity. The pristine floodplain was extremely rich in birds and fish (de Naurois 1965a, Voisin 1983). Ethnic groups exploited these natural

Rainfall in June-August rapidly transforms the dusty, brown Sahelian landscape into a brilliantly green scene. Almost as suddenly, this green blanket withers after the rains have passed, but not before birds have profited from the blooming and – later on - fructification of plants and trees.

resources but the lower delta remained only sparsely populated. Heavy flooding and the silty, often saline, soils prevented the development of flood recession agriculture, in contrast to the long-established recession agriculture employed by the peoples further inland along the river's mid-course.

The flood would begin rising in July, fresh water arriving in the delta from August onwards. The rising water would start to fill Lakes R'kiz (Mauritania) and Guiers (Senegal), both connected to the river by small streams and depressions. Both lakes, having a combined surface area of *c.* 285 km^2 (Bernard 1992), were important regulators of the floods. Downstream of Richard Toll, the flood entered the delta via a labyrinth of streams and channels bordered by levees, filling low-lying basins (*cuvettes*) in between (Fig. 66). During high floods fresh water covered the entire delta up to and including the Ndiaël basin in the south and reaching the Aftout el Saheli in the north. The salt water tongue in the Senegal moved upstream as soon the river flow decreased, mostly from November onwards, travelling as far as 200 km inland during the dry season (Fournier & Smith 1981).

Controlling the flood The first plans for water management aimed at improving the navigation of the river and developing cotton cultivation to support the French textile industry. In 1824 the first irrigation scheme was attempted near Richard Toll, but the hostile attitude of the local population, amongst other things, led to its abandonment.

The Manantali dam supports one of the largest hydro-power plants in the region, generating electricity for Mali, Mauritania and Senegal. Water retention to fill Manantali's reservoir significantly reduces the river discharge of the Senegal.

Fig. 66 The Lower Senegal Delta, showing Lake R'kiz in the north, Lake Guiers in the south and the main watercourses. The map indicates the maximum inundation at high floods in the past. From: 1:200 000 IGN-maps (based on aerial photographs of 1954 and ground surveys in 1957).

Then, in 1906, a small area near Richard Toll was irrigated to test the cultivation of cotton. Similar initiatives were undertaken along the river's mid-course. In July 1916, the Taoueye was dammed by a sluice whereby Lake Guiers lost its direct connection to the Senegal River. Although the lake might still be filled by the flood, the water could no longer escape at lower flood levels, thus preventing saltwater intrusion. An improvement of this sluice planned in the 1920s faced opposition of local inhabitants who asserted that the stagnant water in the lake had led to the invasion of '*roseaux*' (Bernard 1992), probably meaning stands of Reed *Phragmites australis* and Cattail *Typha australis*. [1] This early protest was a harbinger of one of the present major ecological bottlenecks in the delta. Lake Guiers became - and remains - the main potable water source for Dakar.

In 1937, *la Mission d'Aménagement du fleuve Sénégal* (MAS) was created, following several exploratory assessments by French engineers. It was estimated that 85 000 ha would be suitable for irrigation (Bernard 1992), but it was not until after the Second World War that the first implementation could be realised. Near Richard Toll, a rice growing complex of 6000 ha was developed; others were developed along the river's mid-course. After Senegal became independent in 1960, the OAD (*l'Organisation Autonome du Delta*) superseded the MAS, constructing an 84-km dike along the Senegalese side of the river, completing it in 1963-1965. Hamerlynck & Duvail (2003) considered that this embankment had led to higher floods on the Mauritanian side of the river. In 1965 the *Société Nationale d'Aménagement et d'Exploitation des terres du Delta du fleuve Sénégal* (SAED) was founded, its aim being to irrigate another 30 000 ha of rice cultivation within a decade, but by the mid-1970s only about a third of this target had been realised (Crousse *et al.* 1991). From then onwards, cultivation was undertaken in large separate perimeters of about 1000 ha or more, each with its own pumping station and canal network to distribute water and control drainage. Along the river's mid-course, much smaller perimeters (of about 20 ha) were created for management by the villagers themselves.

In 1972, the newly-created *Organisation pour la Mise en Valeur du fleuve Sénégal* (OMVS) developed plans to irrigate 375 000 ha,[2] of which 240 000 ha was to be realised in Senegal, 126 000 ha in Mauritania and 9000 ha in Mali (Crousse *et al.* 1991). To achieve this goal, two dams were planned to control the river discharge, one in the lower delta near Diama and another in the upper basin near Manantali (Fig. 14). The Diama dam was intended to prevent salt intrusion and to provide a reservoir for gravity-fed irrigation. A high water level in the reservoir would also facilitate navigation between Saint-Louis and Kayes in Mali and ensure the inflow of fresh water for Lake Guiers. The Diama dam became operational in July 1986. The Manantali dam, upstream in the Bafing tributary and completed in 1987, was designed to generate hydro-power and to supply the Diama reservoir with water during the dry season. However, it took another five years to finalise the Diama reservoir scheme; dikes had to be constructed on the Mauritanian side of the Senegal River and raised on the Senegalese side. During this time, further rice perimeters were developed on each side of the river (Fig. 67). In addition to these engineering works in the delta, Lake R'kiz was dammed in the 1980s to control the water inflow for agronomic purposes, and a small dam was built in 1986 at Foum-Gleïta in Mauritania in the

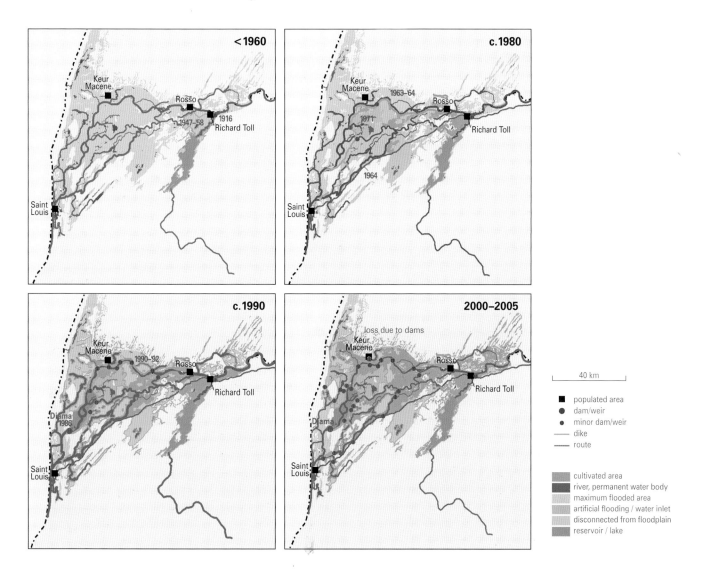

Fig. 67 The Senegal Delta showing dikes and dams and, approximately, years of completion. From: IGN-maps and sources mentioned in the text. The map also shows the area under cultivation. From: IGN-map (survey 1957), Fournier & Smith (1981) and Wulfraat (1993); data and maps provided by OMVS and SAED.

During a very high flood in October 2003, the dune ridge south of Saint-Louis was breached artificially to make a shortcut to the ocean and to avoid large-scale inundations of the city. The new outlet rapidly broadened from a width of about 50 m in October 2003 to about 1400 m in 2006.

Gorgol Noir valley, a small tributary of the Senegal River. Depending on rainfall, this small reservoir holds between 0.2 and 0.6 km³ of water to serve the irrigation of 1950 ha of rice paddies.

Flooding ceased to occur on a large part of the Mauritanian floodplain of the Senegal River after the dams and the Diama reservoir had been completed, leading to desertification and hyper-saline conditions in and around the remaining water bodies (Hamerlynck & Duvail 2003). In view of the expected negative environmental impact of the dams, GFCC (1980) recommended the creation of an artificial estuary in the delta; by 1994, embankment of the Bell and Diawling basins had restored the flood, albeit artificially, via water inflow from the Diama reservoir (Hamerlynck et al. 2002, Hamerlynck & Duvail 2003). The current hydrological management of the basins includes the simulation of a dry period and the inflow of salt water in the dry season at Bell. The rehabilitation measures were successful (Hamerlynck & Duvail 2003), not only restoring biodiversity (see last Section of this Chapter), but also allowing utilisation of natural resources by local communities.

In summary, the taming of the floods caused the Senegal Delta to lose much of its character as an estuarine floodplain, bringing far-reaching environmental, ecological and socio-economic consequences (e.g. Salem-Murdock et al. 1994, Verhoeff 1996, Bousso 1997, Adams 1999, Jobin 1999, DeGeorges & Reilly 2006). Furthermore, at the river mouth the dams affected the geomorphological processes, such as the Diama dam causing the river mouth to silt up (Barusseau et al. 1998), a circumstance that probably is a key factor in the recent (1999 and 2003) excessive flooding of the lower part of the delta, when large parts of Saint-Louis were inundated (Kane 2002). To restrict damage during the 2003 flood, an artificial breach was made in the dune ridge of Langue de Barbarie, forming a shortcut to the ocean about 5 km south of Saint-Louis and 20 km north of the natural embouchure. The new outlet rapidly broadened from about 50 m wide in October 2003 to about 1400 m in 2006. The sudden increase of marine influences changed the tidal amplitude, sedimentation and salinity in the delta downstream of Diama (Mietton et al. 2007); a large part of the lagoons south of Saint-Louis became exposed and dried out in December-February (P. Triplet in litt). The ecological consequences of this artificial shortcut on the environment of the Nthiallack and Gandiol basins (mangroves, fish)

Several stages of rice production in the Senegal Delta., from left to right (1) Seedbeds are used to nurse rice plants before planting the fresh shoots in paddies (July 2006); this tactic reduces the loss of seed to birds and rodents, (2) Rice paddies with full-grown rice (November 2004) and (3) just after harvest (December 2005); Palearctic waders such as Black-tailed Godwits and Ruff are most common in rice fields immediately after the harvest when some water is still present and spilled rice has not yet been depleted by the multitude of seed-eaters, (4) After thrashing, the rice is transported to the villages.

may well be profound, but have yet to be ascertained (Mietton *et al.* 2007, but see Triplet & Schricke 2008). Mangroves covered parts of the tidal zone and fiddler crabs *Uca tangerii* were abundant in December 2008 (pers.obs. E.Wymenga).

Cultivation of the delta   The left bank of the Senegal Delta was embanked in the early 1960s but it took about four decades to cultivate the delta (Fig. 67), starting with the left bank under direction of the SAED. On the right bank, in Mauritania, cultivation got under way after completion of the Diama reservoir in the early 1990s, although in the 1970s some pilot schemes had already been carried out by *Sociéte Nationale de Développement Rural* (SONADER). By 1993, the area under cultivation on the Mauritanian side was estimated at 13 600 ha, but the system degraded rapidly due to the drainage system malfunctioning and lack of maintenance (Peeters 2003).

From the start of the cultivation of the Senegal Delta, irrigated rice fields dominated, though large areas near Richard Toll were converted to sugar cane plantations in the early 1970s. Until the 1990s, the maximum area of rice on the left bank amounted to some 10 000 ha, but it covered *c.* 15 000 ha in 2005, with substantial annual fluctuations (Fig 68). This area comprises only a part of the potential rice paddies in the delta and the river valley – problems of salinisation and insufficient labour are proximate causes. For example, in the 1995-1996 season, the potential area in both these areas for rice cultivation covered 69 769 ha, whereas only 29 822 ha (42%) comprised irrigated crops (SAED 1997 *in* Peeters 2003). Developmental bottlenecks during the early cultivation of the delta were low productivity, salinisation, an inadequate drainage system and inefficient financing (Crousse *et al.* 1991, Engelhard & Abdallah 1986). Pivotal to cultivation success in the delta is regular flushing of the rice fields to counter the shallow and hypersaline groundwater that otherwise quickly leads to high soil salinity, which when it occurs means that rice fields have to be abandoned after a few years (Raes *et al.* 1995). In the most recent period (2000-2005), much of the delta in Mauritania was under cultivation, even though large parts have not been used for rice growing for a number of years. Abandoned areas are located mostly in the western part of the delta (Peeters 2003, see Fig. 76), where salinisation is more predominant than elsewhere. In the 2001-2005 period, the (once) cultivated area in the Senegal Delta covered *c.* 102 000 ha, of which approximately 35% was used annually to grow rice, an estimate based on satellite maps. This may however be substantially higher in some years, given the large annual fluctuations (Fig. 68) and shifts in the cultivated area.

In the 1990-2000 period, rice production in the Senegalese part of the delta and along the river's mid-course amounted to 14 000 tons annually (Data: SAED). The yield fluctuated between 4 and 5 tons/ha. On the left bank, about 5000 ha is used annually for off-season rice, which became possible since the completion of the Diama reservoir in the 1990s (Fig. 68). The large perimeters near Rosso and Richard Toll contain a greater crop diversity, including tomatoes, onions and other vegetables. East of Saint-Louis, several areas of horticulture are located on higher grounds. The sugar cane plantations near Richard Toll covered *c.* 11 000 ha in 2000-2005.

People   Before the advent of irrigation, large sections of the lower Senegal Delta were unusable for any kind of farming. Temporary

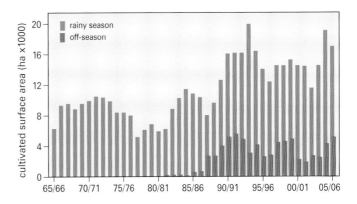

Fig. 68 Annual surface area (ha) of cultivated rice during the rainy and off- (dry) season in the Dagana Region in the Senegalese part of the Senegal Delta. Source: SAED.

Senegal Delta

Fish landing in Saint-Louis, where supply and demand meet on a daily basis.

residents may have dwelt on the higher vegetated dunes and the natural levees, particularly fishermen and pastoralists, but most of the estuarine floodplain was uninhabited. The dynamic flooding and salt intrusion further restrained the development of flood recession agricultures. However, the inundation of the alluvial plains along the river's mid-course valley (*waalo*) and the cultivation of the river banks (*falo*) supported a relatively dense population (Jobin 1999). The larger settlements in the Senegal Delta, such as Saint-Louis, Rosso, Richard Toll, remained small until exploitation of the delta could be sustained. Saint-Louis was the exception, playing a major role in its early days as capital of French West Africa (1895-1902) and of Senegal (1902-1958). Saint-Louis acted as the national administrative centre as well as attracting fishermen, pastoralists and traders. The poor navigability of the Senegal River and its dynamic embouchure probably prevented Saint-Louis' development as a national port giving access to the interior (Bernard 1992), a function fulfilled by Dakar since at least the late 19th century.

Following the embankments, new settlements were established in the delta near the larger rice paddies, such as Diadiem and Boundoum around the Djoudj. Villages like Ross-Béthio and Rosso developed into large rural centres in the 1980s. From 1960 to 1988, the annual population growth in the region was modest but did show a clear increase between 1976 and 1988 (>3.8%). About the same time, Richard Toll was booming (annual population growth 16%), linked to agronomical development and the busy sugar cane factories. Saint-Louis remains the main centre in the region; after slow development through the 19th and early 20th centuries, the community rapidly spread after 1950. In 2001, the council censuses of Saint-Louis, Richard Toll and Dagana counted 154 000, 70 500 and 25 142 inhabitants (www.gouv.sn/senegal/population_chiffres.html) respectively. Most of the delta (in 2008) currently is as an agricultural area experiencing a high level of human activity. Protected areas, in particular the National Parks, are quieter and form a refuge for wildlife.

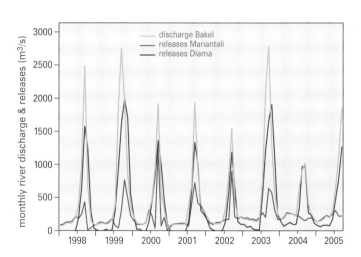

Fig. 69 Discharge at Bakel and water release from the Manantali and Diama reservoirs in 1998-2005. From: OMVS.

Flooding and river discharge

River discharge and water levels   The river discharge upstream at Bakel has been measured since 1950. In a rainy season in the Guinean mountains (from May to September), the river discharge starts to mount in July. In 48 of 56 years (1950-2005), the maximum discharge at Bakel occurred in September; in 6 dry or wet years it was in August and October, respectively. The flood takes about a month to reach the delta at Richard Toll. The river discharge is subject to large annual fluctuations; very low flows occurred during the droughts in the 1970s and 1980s (Chapter 3). The rainfall recovered in the 1990s, but the Manantali reservoir, in operation since 1987, affected the river discharge negatively. Water retention reduced the maximum monthly river discharge by an average of 11%, but it has reduced the discharge by 36% in dry years (Fig. 15). The current discharge of the Senegal is therefore determined by rainfall and by releases at Manantali. Releases by Manantali generate an artificial flood to enable

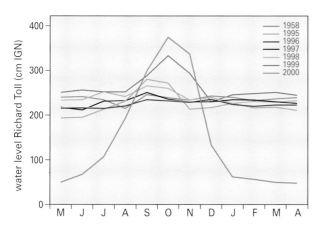

Fig. 70 Water level at Richard Toll in 1995-2000 relative to sea level (cm IGN), after the completion of the Diama reservoir. The water level in 1958 shows a relatively wet year before the construction of dikes and dams. From: OMVS.

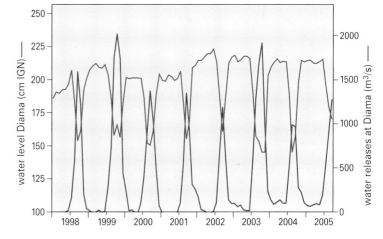

Fig. 71 Monthly releases and water levels upstream the Diama dam in 1998-2005. From: OMVS.

flood recession agriculture to be maintained along the river's midcourse. An increasing demand for hydro-electric power may prevent such flood management in the near future (Salem-Murdock et al. 1994, Duvail et al. 2002, DeGeorges & Reilly 2006), if the withholding of the flood peaks in 2004 and 2005 is a reliable predictor (Fig. 69).

Before the construction of dams, the flood level in the delta followed the pattern of the river's discharge, reaching a maximum in October, after which water levels dropped rapidly in the receding flood. Since the completion of the Diama reservoir, the water level has remained high and stable throughout the year (Fig. 70). The reservoir's water level is kept at about 200-250 cm (IGN), as measured at Richard Toll. The high water table is needed for gravity-induced irrigation and for double crop cultivation. Just upstream of Diama, the water level is maintained at c. 150 cm (IGN) during the flood and at 200 cm (IGN) during the rest of the year (Fig. 71). During the flood peak, the water level is lowered to release water and to avoid too much pressure on the dam.

Reconstructing the flood Unrestrained flooding of the low-lying basins in the delta could occur up to the mid-1960s, although by the early 1960s inundation of the Ndiaël basin had already been hampered by the Saint-Louis to Rosso road acting as a low dike (de Naurois 1965a). In the 1980s and 1990s the area subject to natural flooding rapidly declined (Fig. 67). Today (2008) only the Ntiallakh and Gandiol basins north and south of Saint-Louis function as estuaries in the former floodplain, although even there the river is strictly controlled. These basins cover c. 33 000 ha, some 9.7% of the original lower delta floodplain (3400 km²). [3]

Since its conversion into farmland, inundation of the Senegal Delta floodplain no longer is related to river discharge. Quantitative data on the flooded area before the building of the first embankments are scarce. According to Voisin (1983) and Triplet & Yésou (2000), the delta between Richard Toll and the ocean covered c. 320 000 ha, of which at most 190 000 ha were subject to flooding. Large floodplains (up to 676 500 ha) occur along the river's midcourse between Podor and Bakel (Diop et al. 1998). [4] Since flood extent is a function of river discharge (Chapter 3), flood extent in the pre-dam era had to be reconstructed using satellite images of the flooded area and maximum discharges (Fig. 72), but unfortunately, such imagery from the pre-dam era is scarce. We used 1984 as the year representative of a very low natural discharge. The maximum inundation zone was digitised from the 1: 200 000 IGN map, surveyed in 1957 (Fig. 66). With the exception of Lakes Guiers and R'kiz, and the coastal lagoon of Aftout es Saheli in the north, the delta between Richard Toll and the ocean comprises 290 000 ha of low-lying areas, of which 162 000 ha is on the left bank of the river (Senegal). This area was flooded only during exceptionally high floods, as in 1950. We obtained another usable image for 29 October 1963 (monochrome, stereo medium; http://www.edcsns17.cr.usgs.gov/EarthExplorer). Though the data set is very limited, the relation between discharge and flooding is strong (Fig. 72).

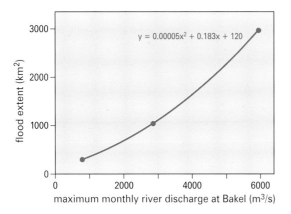

Fig. 72 Flood extent in the Senegal Delta plotted against river discharge at Bakel in a natural, uncontrolled situation. Using the equation in this figure, and the data on river discharge in Chapter 3, we reconstructed the annual flood extent of the Senegal Delta (Fig. 73). Sources: see text.

The loss of flooded area (Fig. 73, red bars) is the combined effect of dams (Fig. 15) and reclamation (Fig 67). The surface area that would have been flooded without dikes and dams was estimated proportionally to the extent of the river discharge. Similarly, we took increased flooding into account after the flood had been artificially restored in the Djoudj National Park (since 1971) and the Diawling National Park (Bell and Diawling basins) in Mauritania (since 1994) (Fig. 73, light blue bars). These protected areas today (2008) contribute significantly to the actual flooding of the Senegal Delta (Fig. 73).

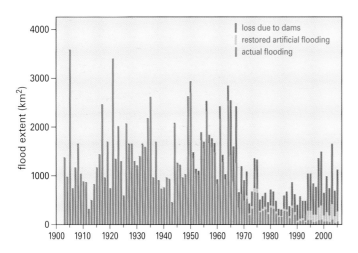

Fig. 73 Reconstructed flood extent in the Senegal Delta, based on the river discharge, the impact of dikes and dams and the partial restoration of the floodplain. The Diama reservoir and inundated parts within the embanked areas are not included in the part of the delta artificially flooded.

Our reconstruction as presented in Fig. 73 does not reflect reality fully. First, part of the river discharge is used for irrigation of rice paddies, which further reduces the flooded area in the remaining floodplain, although, the embankments may have resulted in additional flooding during the flood peak, as observed for some years after the Diama dam was built (Triplet & Yésou 2000). Secondly, in the restored parts of the floodplain (Djoudj NP, Diawling NP) the annual inflow of water may not be proportional to the maximum river discharge. Thirdly, during each flood, areas within the embankments are still inundated to some extent, partly the result of seepage and discharge of drainage water from the rice paddies. Lastly, some low-lying basins may act as catchment area for local rainfall.

To get an impression of the actual flood extent, we estimated the surface of inundated areas – including embanked areas – from satellite images that recorded the flood peak. For this purpose, we used 17 "quicklooks" (www.glovis.usgs.gov). Though these images are

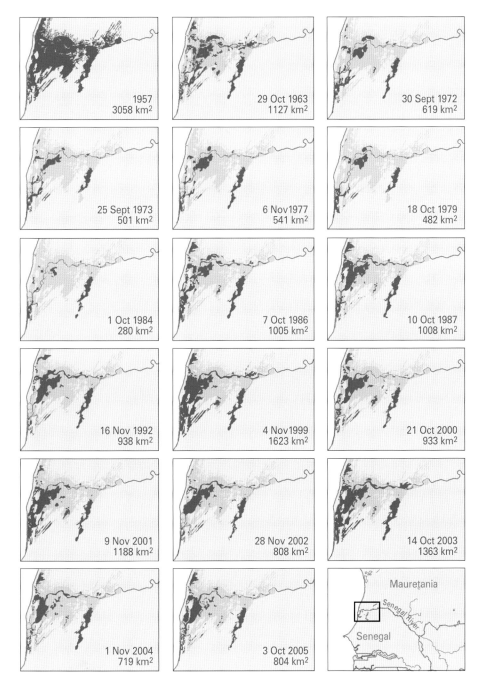

Fig 74 Waterbodies and inundated surface area in the Senegal Delta as derived from satellite images nicknamed "quicklooks" (Landsat 1-7. Source: www.glovis.usgs.gov). For each quicklook, date and surface area (dark blue) are given; the maximal flood extent in 1957 is shown as light blue background in all other panels. Irrigated rice paddies are not indicated.

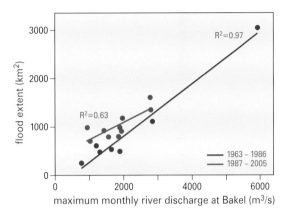

Fig. 75 Inundated (wet) surface area during the maximum flood extent in the Senegal Delta (excluding Lake Guiers, Aftout el Saheli and irrigated areas), plotted against river discharge for the years 1963-1986 and 1987-2005 (derived from satellite images, Fig. 74).

partly obscured by vegetation cover, they offer a fairly accurate impression of the annual variation in flooding and the extent of inundated areas and shallow water bodies in the delta (Fig. 74, which omits irrigated rice polders, shown in Fig. 76). The images clearly show the droughts in the 1970s (the 1972 and 1973 images) and the 1980s (the 1984 image). During the droughts of 1983 and 1984 almost the entire delta remained unflooded and water was scarce (Altenburg & van der Kamp 1985). The 1987, 1999 and 2003 images illustrate the flooding during wet years, when more water is also present within the embanked basins, possibly due to the combined effect of local rainfall and the higher inflow during the flood. The excessive flooding of the Mauritanian part of the delta in 1987 was caused partly by the impoundment of water near the Diama dam, which had become operational in 1986, when the dikes on the Mauritanian part of the reservoir were still incomplete. The images from 1999 onwards clearly show the effects of the 1994 rehabilitation of the Diawling and Bell basins in Mauritania. Note also that Trois Marigots – formerly an important delta wetland – has become progressively wetter since 1990 (Triplet & Yésou 2000). The close correlation between flood extent and river discharge in the past (Fig. 72) has remained partly intact despite the change in the hydrological regime (Fig. 75), but the hydrological engineering has seriously impacted plant communities in the wetlands (see next Section).

Wetlands

Originally, the alluvial plain of the Senegal Delta comprised a huge wetland zone interconnected with other wetlands such as Lake Guiers, Lake R'kiz and the Aftout es Saheli. De Naurois (1965a) described the area as a vast system of lakes, creeks, gullies and flooded

Estuarine habitats – mangroves, tidal mudflats and sandy plains – as found downstream of the Diama dam. The sandy plains in the northern part of the estuary only become inundated during high floods.

marshes in the grip of considerable water level fluctuations – within and between years – and a continuously shifting estuarine gradient from the coast inland, by which, in the dry period, the brackish zone moved from far inland (as far as Richard Toll) to close to the coast when the river flow was maximal. Much of the temporarily inundated marshes, in particular around the Grand Lac (in the current Djoudj NP), comprised flooded grasslands of *Sporobolus robustus*,

Vossia cuspidata, *Oryza longistaminata* and *Vetiveria nigritana*, typical of Sahelian floodplains. The flooded pastures, grazed in the dry season, extended southwards up to the Djeuss, Lampsar and Ndiaël basins. Coastal lagoons and marshes hugged the dunes. Flood forests of *Acacia nilotica* could be found along the river, from Richard Toll to Diama, but downstream of Diama were replaced by mangrove forests of *Avicennia* and *Rhizophora*.

In the early 1960s, the delta lost a significant amount of its wetlands. The Ndiaël – home in the 1950s to tens of thousands of staging waders such as Black-tailed Godwit and roosting ducks (Roux 1959b) – became partly dry in 1962 to 1963. This loss of wetland habitats followed the implementation of hydrological engineering schemes and subsequent cultivation of the delta (Fig. 67). Within a few decades, most flood forests and *Sporobolus*-grasslands had been converted into cultivated areas or dry, saline plains, which received water only during the rainy season. On the completion of the Diama dam, the northern part of the estuary changed into a desert-like environment. The flooded marshes in the central delta had received protection through the designation of the Djoudj as a National Park in 1971, but its flooding since then has been artificially so, with water from the Diama reservoir. Flood restoration in the Bell and Diawling basins in Mauritania led to the recovery of the *Sporobolus* habitat and the reappearance of floating *Echinochloa colona* meadows (Hamerlynck et al. 1999, Hamerlynck & Duvail 2003). The restoration of the estuarine gradient, by fresh water discharge from the Bell and Diawling basins into the Ntiallakh basin, initiated the regrowth of mangrove forests downstream. Both National Parks – Djoudj and Diawling – are typical, albeit managed artificially, of the original wetlands in the Senegal Delta (Box 9).

Wetland habitats The delta consists of a wide array of habitats, including an estuary, artificial floodplains, marshes, lakes, the Diama basin, embanked basins with irrigated agriculture, dried-out saline *sebkha*-like plains and patchy marshland in former tributaries and depressions (Fig. 76). The marshes are inundated temporarily or used as drainage basins for irrigation agriculture. Most wetland

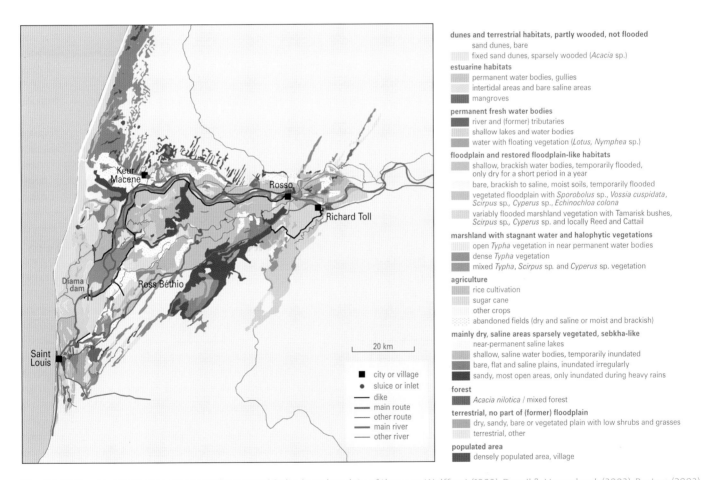

Fig. 76 Wetland habitats and land use in the Senegal Delta, based on data of the SAED, Wulffraat (1993), Duvail & Hamerlynck (2003), Peeters (2003), satellite images and field observations in 2003-2005 (E. Wymenga unpubl.). Habitats are distinguished on the basis of hydrological features and vegetation communities.

In the Djoudj NP a variety of marshland habitats can be found in the central part, frequently with Tamarisk bushes.

habitats in the delta are degraded in one way or another through insufficient flooding, invasive plants or human interference.

Estuary (35 000 ha) The estuary downstream of the Diama dam, in the Ntiallakh and Gandiol basins, encompasses the main river, tidal gullies, tidal mudflats, lagoons, vegetated salt marshes, mangroves and – in the northern part - some *Sporobolus* grasslands. Tamarisk *Tamarix senegalensis* bushes pepper the levees and the more elevated parts of the inundation zone in southern Gandiol. The tidal influence reaches as far as the Diama dam. The sandy plains to the north, just south of the Bell basin, are flooded only during high river floods (Fig. 74). Here, small mangrove forests fringe the main gullies. In the past, mangroves covered a much larger area, but these largely vanished in the 1980s and 1990s due to hydrological changes in the delta. The artificial flood releases from the Bell and Diawling basins revitalised the estuarine gradient in the Ntiallakh basin, allowing a partial recovery of mangroves (Z.E. Ould Sidaty, pers. comm.). The recent creation of a new river mouth south of Saint-Louis increased the influence of the ocean (Mietton *et al.* 2007), impacting on the estuarine environment. A comparison of counts in January of the five years before and the four years after the breach, revealed a significant decline in Black-winged Stilt, Little Ringed Plover, Kentish Plover, Redshank and Little Stint, and an increase in Knot, Sanderling and Ruff (Triplet & Schricke 2008).

Artificially flooded plains and marshes (34 000 ha) The Djoudj and Diawling National Park, situated across from each other on the left and right bank of the river, have been being flooded artificially since 1971 and 1994 respectively (Fig. 67), a process that imitates natural flooding. During the flood, the basins are fed from the Diama reservoir, mostly in July and August (Djoudj) and July to October (Diawling). The inlets are closed during the rest of the year, resulting in a gradual desiccation until in June the basins are almost completely dry. In the low-lying basins of the Djoudj, the extent of desiccation is usually much less, because the water level in the Diama basin tends to prevent their drainage (Diop & Triplet 2006).

The lakes and gullies lack vegetation in their deeper parts, but shallower sections are vegetated with floating and submergent plants like Common Hornwort *Ceratophyllum demersum*, Pondweed *Potamogeton* sp., Water Spinach *Ipomoea aquatica*, Water Lily *Nymphea lotus* and *N. maculata*. Floating assemblages, *Echinochloa stagnina* and *E. colona*, cover only small areas. Around the Grand Lac in the Djoudj and in the Diawling and Bell basins, extensive grassy floodplains – *Sporobolus* sp. and *Vossia cuspidata* – containing sedges *Cyperus* sp. and rushes *Scirpus*[5] sp. occur. *Sporobolus robustus* is often the dominant species, but *Scirpus littoralis*, *S. maritimus*, *Eleocharis mutata* and *Cyperus* sp. may abound whenever soil type, grazing, salinity or flood duration permit (Wulffraat 1993). In the Djoudj, marshland vegetation are usually dominated by *Sporobolus* and *Scirpus*, occasionally also by Cattail and Reed. Tamarisk and other shrubs are present on drier ground, accompanied on the higher levees by Toothbrush Tree *Salvadora persica*, whose berries form an

important food source for pre-migratory passerines (Stoate & Moreby 1995). While the landscape of the Diawling National Park has an open character, the Djoudj comprises a mosaic of open plains and denser vegetated marshes and levees.

Lake Guiers and the Diama reservoir Lake Guiers and the Diama reservoir have a more or less high and stable year-long water level (Fig. 71). Lake Guiers serves as a reservoir of potable water for Dakar and is continuously fed by fresh water from the river. The borders of the lake are vegetated with dense Cattail stands; the lake itself holds hardly any aquatic vegetation. An outburst of the invasive aquatic Kariba Weed *Salvinia molesta* in the 1990s was successfully suppressed through biological control (Pieterse *et al.* 2003). The Diama reservoir comprises the main river and adjacent riverbanks and after it was created, the shallow banks were rapidly colonised by Cattail, which at present forms dense stands over much of that area. Stands of Cattail outcompete other plant species, except for Reed and the floating Kariba Weed.

Wetlands in embanked basins Beyond the irrigated paddies (121 000 ha, 35% of the delta), the arid eastern and southern saline areas of the delta lost their wetland character after the embankment work. They can be inundated still after exceptionally heavy rains, but for only a short period; normally they remain dry (Fig. 74). The arid and saline plains containing *Salsola baryosma* and Glasswort *Salicornia* sp. constitute the outlying and more elevated parts of the original floodplain, whose open aspect is maintained by salinity and by grazing. Within the irrigated fields, a scattering of pools with Water Lily and Cattail provide (temperate) shelter to birds, amphibians, Monitor Lizard *Varanus niloticus* and mongoose. Wetlands within embanked areas are fed permanently or temporarily by fresh or brackish water: these include water inflows from the Diama reservoir, dammed streams, drainage basins and natural depressions. Most of the former tributaries and depressions used for irrigation or drainage have been invaded by Cattail. Before the flood was tamed in Trois Marigots in the southern delta, Cattail vegetation developed in the deepest parts with standing water, which is typical of permanent waterbodies in the delta; in some cases Cattail is mixed with Reed and *Scirpus* sp. (Peeters 2003). Patches of open water may be occupied by submergent or floating vegetation such as Bladderwort *Utricularia stellaris* and Water Lilies.

In temporarily inundated brackish basins, Cattail is less dominant. In the Ndiaël, a former core wetland in the delta in the 1950s, only its northern part holds water annually. The basin is fed by drainage water and serves as catchment basin for precipitation on higher ground. The brackish nature of the water is reflected in the mixed vegetation of rushes, sedges and Cattail. Many former depressions, inundated temporarily or not at all, are highly saline; as a typical example, the salty depressions east of Saint-Louis lack any vegetation. Bare and salty plains are found in parts of the Ndiaël, Trois Marigots and in the northwestern Mauritanian part of the delta.

Abandoned fields often are arid and saline, vegetated sparsely by *Salsola baryosma*; Tamarisks grow in the surrounding dikes. Low-lying parcels of land near marshes, if inundated regularly, may regain their wetland character. Abandoned plots becoming permanently inundated are overgrown rapidly by Cattail; only the presence

Flooding of the Diawling NP, through inlet of water from the Diama reservoir, was restored artificially in 1994. The subsequent recovery of floating meadows of *Echinochloa colona*, the return of *Sporobolus* on the floodplains and the shallow lakes that came into being created favourable conditions for waterbirds.

Salinisation and other constraints led to the abandonment of significant areas of cultivated rice fields. Unexploited low-lying rice paddies regain a wetland character, others turn bare and saline. December 2005.

of disused banks and dikes distinguishes them from Cattail marshes. The difficulties encountered in cultivating the delta resulted in the abandonment of many fields. From recent satellite images we estimate the surface area of abandoned fields at 35 000 ha (source: SAED, Peeters 2003), apart from many small abandoned parcels of land within large areas of cultivation.

Forests (225 ha) Forests have all but vanished from the delta. The flood forests, or *gonakeraies*, consisted predominantly of *Acacia nilotica* whose canopy was up to 20 m high. They were inundated for at least 15 days, and at most two months per year. Flood control, grazing and tree-felling successively depleted the forests, just as happened outside the delta. According to Camara (1995, *in* Peeters 2003), the surface area covered by forests in the region of Podor, just east of the delta, decreased by 74%; 64% between 1954 (9055 ha) and 1986 (3294 ha) and by another 10% until 1991 (2385 ha). In former times, the river's mid-course valley held extensive riverine woodlands of *Acacia nilotica*. Tappan *et al.* (2004) estimated that between 1965 and 1992 the surface area of these woodlands declined from 39 357 ha to 9070 ha (-77%). Significant forests nowadays (2008) are confined to protected areas (*forêts classées*); such forests are vital to species such as Turtle Doves for roosting and loafing (Chapter 32). In the delta, wooded cover is now extremely scarce and predominantly occurs along the main channels in the Djoudj National Park. In the Guembeul Nature Reserve, south of Saint-Louis, the original *A. tortilus raddiana* woodland is preserved as habitat for reintroduced game.

Invasive plants Human-induced changes to wetlands worldwide, direct or indirect, pave the way for non-indigenous plant species, and

The construction of the Diama dam resulted in a stabilisation of water levels and prevented salt from infiltrating the hinterland. These conditions provided a red carpet welcome to non-indigenous plant species. In particular, Cattail (left) and Kariba Weed (right) rapidly invaded open water, creating impenetrable floating mats up to one metre thick.

Extensive and dense Cattail vegetation in the Senegal Delta (Diama reservoir) in December 2004. Transect counts were performed along the fringes of Cattail stands, while mist-netting was carried out perpendicularly to the vegetation edges.

the Sahelian wetlands in general and the Senegal Delta in particular are no exception (Keddy 2002, Zedler & Kercher 2004). Water Hyacinth *Eichhornia crassipes* and Water Lettuce *Pistia stratiotes* are aggressive invaders that may rapidly blanket aquatic habitats. They form dense floating mats, outcompete indigenous plant species and negatively impact botanical and animal biodiversity (Zedler & Kercher 2004, Frieze et al. 2001). As long ago as the early 20th century, the people living near Lake Guiers protested against a sluice controlling the water inlet of the lake, as it had resulted in the invasion of Cattail (Bernard 1992). The Senegal Delta is a classic example of a disrupted wetland infested with invasive plants (Fall *et al.* 2004), a circumstance that has spawned much research (Boubouth *et al.* 1999, Fall *et al.* 2004, Hellsten *et al.* 2002, Pieterse *et al.* 2001, Peeters 2003, Kloff & Pieterse 2006).

The construction of the Diama dam stabilised the water table and prevented salt intrusion. This radical change in the ecosystem was immediately followed by explosive growth of Cattail in the reservoir's shallows. The dense, monotypic stands of Cattail were said to form 'an almost impenetrable wall between the river dikes and the open water' (Kloff & Pieterse 2006). The Diama reservoir was not the only source of infestation; most permanent freshwater habitats were invaded too.

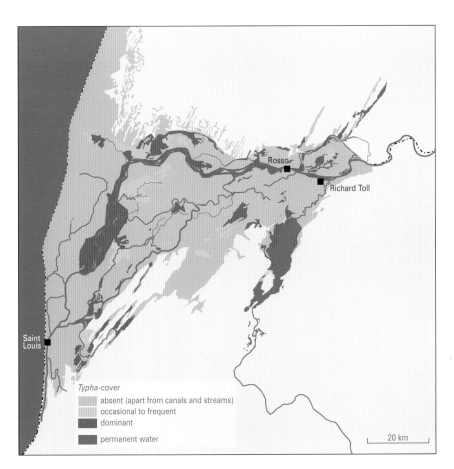

Fig. 77 Distribution and relative abundance of Cattail in the Senegal Delta, based on data of L. Manding (SAED, *in litt.*), OMVS, Peeters (2003), satellite images and field observations from 2003-2005 (E. Wymenga unpubl.). **NB** In areas where Cattail is indicated as absent, it is actually omnipresent in non-saline, permanent shallow waterways.

Senegal Delta

Dense Cattail clogged waterways and impeded access to fishing grounds. Moreover, pools in Cattail stands provided favourable habitat for snails and molluscs, which act as a host for parasites and waterborne diseases (Boubouth *et al.* 1999, Jobin 1999, Kloff & Pieterse 2006).

Of the other non-indigenous plants proliferating in the Sahelian floodplains, Water Lettuce, Kariba Weed and Water Hyacinth (Fall *et al.* 2004) probably present the greatest threats. Water Lettuce had become abundant after the mid-1990s, but Water Hyacinth had been much less so. Kariba Weed was accidentally introduced in the delta in 1999, but in under a year it had invaded all open freshwater habitats in the Diama reservoir, forming dense mats, expanding from Rosso to the Diama dam and clogging irrigation canals. By 2000, Kariba Weed had formed up to 1 m thick floating mats along Cattail stands near sluices and dams (Kloff & Pieterse 2006), but in June and July 2000, the army were called in to help remove it manually in the larger water bodies affected. This measure worked for only a few months (Pieterse *et al.* 2003, see also Triplet *et al.* 2000, Diop & Triplet 2000, Diouf *et al.* 2001). A subsequent campaign of biological control employing the neo-tropical weevil *Cyrtobagous salviniae*, which feeds exclusively on Kariba Weed, proved to be more successful (Pieterse *et al.* 2003). In 2005 and 2006, Kariba Weed seemed to be under control and, although it is still abundant in most freshwater habitats, but at much lower densities.

The domination of Cattail   At present (2008) Cattail rules the delta (Fig. 77), for it covers at high density an estimated 53 000 ha and is abundant in another 21 600 ha, a total area representing *c*. 50% of the delta wetlands. It is also dominant in the waterways of irrigated agriculture. Several measures have been tried to reduce Cattail in the delta; these involved manual or mechanical clearance, dredging of important waterways, mowing or burning and, locally, chemical control with glyphosate (Hellsten *et al.* 1999, Peeters 2003), but none offer a long-term solution. The GTZ investigated the possibility of carbonising Cattail, aiming at a profitable use for local communities (Henning 2002 in Elbersen 2006), but so far this study yielded no method for a wide-scale applicable use.

In the delta, the prolific growth of Cattail is rooted in its ecology. Cattail is a pioneer in shallow marshlands. It grows on waterlogged soils on peat, clay and sand, up to a water depth of 1.5 m. The seeds do not need oxygen to germinate. Its tolerance of anoxic conditions and its large ecological range (fresh to brackish, mesotrophic to hypertrophic) are the key to the rapid colonisation and often its dominance of a wide range of marshland habitats. In the once estuarine Senegal Delta, reduced salt intrusion, eutrophication (via fertilisation of irrigation cultures) and increased stagnancy of water levels upstream of Diama offered well-tailored conditions for Cattail and other invasive plants. The human-induced simplification of the intricate biological relationships in a variety of dynamically stable ecosystems provides ideal niches for invasive pioneers at the expense of a loss of biodiversity. The Senegal Delta is far from unique, for worldwide, cattail species have invaded such wetlands (Frieze *et al.* 2001, Keddy 2000).

The installation of protected areas, in particular the Djoudj National Park (1971) and the Diawling National Park (1991), safeguarded a significant part of the biological assets of the delta. The necessary management, surveillance and maintenance, being the core business of actual protection, requires a constant input of labour and finances.

Women from neighbouring villages collect *Sporobolus* in the Djoudj NP and Diawling NP to make their traditional mats and baskets, a well-known artisanal product from the region.

National Parks in the Senegal Delta: Djoudj and Diawling

In the early 1960s, de Naurois (1965a) anticipated the imminent danger that large-scale irrigation plans posed to natural habitats by proposing to safeguard core areas in the Senegal Delta. The Ndiaël wetland had already by then turned into a semi-arid landscape. A major step was the creation of the Djoudj National Park (PNOD, *Parc National des Oiseaux du Djoudj*) in 1971, as a national bird sanctuary of 13 000 ha, enlarged to 16 000 ha in 1976. The park became a stronghold of breeding Afrotropical waterbirds. Thanks to artificial flooding, the park also attracted huge numbers of migratory Palearctic ducks and waders. Over the years, local communities increasingly became involved in the management of the park (Beintema 1989, 1991), including the promotion of small-scale ecotourism. In this regard, the breeding site of Great White Pelicans was – and is – particularly eye-catching. Increasingly, however, artificial flood management is causing ecological problems. Discharge of water, to lower the water level in the park, is impossible because the artificial lake is always much higher than the central lakes of the park (Diop & Triplet 2006). This implies that near-dry conditions can not be simulated anymore, inadvertently creating the perfect habitat for invasive plants.

The establishment of the Diawling National Park (PND, *Parc National du Diawling*) in Mauritania provided another boost to the restoration of natural habitats in the Senegal Delta. It took more than ten years of debate before the park became reality (GFCC 1980, Hamerlynck 1997). PND was designated in 1991 and covered 16 000 ha, including the Bell and Diawling Basins. Embankment of the basins and water inflow from the Diama reservoir facilitated flood restoration, starting in 1994 (Hamerlynck & Duvail 2003). The success of this project is a reminder of the resilience of Sahelian ecosystems. Invasive plant species remained at bay thanks to flood dynamics (managed artificially) and the inflow of salt water in the Bell basin during the dry season, circumstances that resemble the natural situation. The restoration of the estuarine gradient in this part of the lower delta encouraged the recovery of mullet migration, leading to renewed fishing. Also, the *Sporobolus* plains in the Diawling have recovered, enabling local women to reintroduce their production of mats and other traditional products. Local communities were closely involved in the restoration process, which was crucial for gaining their support (Hamerlynck 1997, Hamerlynck & Duvail 2003). Because the success of the park attracts many people from elsewhere, the use of natural resources (fish, shrimps, grasslands and forests) needs to be carefully balanced and demands permanent attention. The same is true for grazing in the Djoudj NP (Touré *et al.* 2001) and hunting in the border zones (annually 5000-6000 Garganey, 500-1000 Whistling Ducks and a few hundred waders are shot; Peeters 2003). International investments in national conservation schemes often act as a magnet to people seeking opportunities associated with park-related funding, which jeopardises the effectiveness of conservation measures. A similar example in the Sahel relates to the reflooded Waza Logone (Scholte 2003; Chapter 9), and higher human population growth rates near protected areas have been evident across ecoregions, countries and continents (Wittemyer *et al.* 2008).

The two National Parks cover c. 10% of the total delta and hold c. 75% of the water (excluding the tidal part) in the dry period. The artificial flooding restored estuarine gradients, and regenerated shallow brackish lakes and floodplains with *Scirpus maritimus, S. littoralis, Sporobolus robustus* and *Echinochloa colona*. Most waterbirds in the delta are now concentrated in the National Parks. The Parks host a diverse fauna of insects, fishes, amphibians, reptiles (Nile Crocodile *Crocodylus niloticus*, Monitor Lizard *Varanus niloticus*, Python *Python sebae*), and both small and larger mammals (like Patas Monkey *Cercopithecus patas* and Warthog *Phacochoerus aethiopicus*). However, species such as Dorcas *Gazella dorcas* and Red-fronted Gazelle *G. rufifrons*, which may have been common in former days, remain on the verge of extinction (Peeters 2003).

Waterbirds

Birdlife in the Senegal Delta impressed naturalists in the past (Neumann 1917) and still attracts many birders. Half a century later, French naturalists started to quantify the ornithological importance of this part of the West African coast (de Naurois 1965b, Roux 1959a&b, Morel & Roux 1966a&b, Morel & Roux 1973).

Bird counts   In the early 1970s, coinciding with the start of the Great Drought, the *Centre de Recherches sur la Biogéographie des Oiseaux* (CRBPO) set up a bird monitoring scheme involving aerial counts in January, concentrating on wintering Palearctic ducks and geese in the Senegal Delta (Roux 1973a, 1973b, Roux & Jarry 1984). Aerial censuses in the 1974-1978 period were combined with a study on their population dynamics (Roux *et al.* 1976, 1977, 1978). From the 1980s onwards, starting with a Dutch Spoonbill expedition (Poorter *et al.* 1982), many surveys of migrant waders and passerines have been carried out in the Senegal Delta (*e.g.* Altenburg & van der Kamp 1986, Meininger 1989a, OAG Münster 1989b, Tréca 1990, Beecroft 1991). Since the 1989, the waterbird censuses have been pursued by the French organisations of *Office National de la Chasse et de la Faune Sauvage* (ONCFS) and *Oiseaux Migrateurs du Paléarctique Occidental* (OMPO) in collaboration with *Direction des Parcs Nationaux* (DPN, Senegal). Unlike the Inner Niger Delta in Mali, from which year-round data are available (Chapter 6), censuses in the Senegal Delta were confined mostly to midwinter (January) in collaboration with the International Waterbird Census (*e.g.*. Triplet & Yésou 1998, Schricke *et al.* 2001, Dodman & Diagana 2003 ; Box 10).

Information on Afrotropical (water) birds is limited, apart from studies on the exploitation of rice fields by ducks (Tréca 1999, Triplet *et al.* 1995). An impression of the ornithological diversity and importance of the Senegal Delta can be obtained from Morel & Morel (1990), the checklists of birds recorded in the Djoudj (Rodwell *et al.* 1996, Triplet *et al.* 2006) and reviews of (breeding) waterbirds in the delta (Roux *et al.* 1976, Triplet & Yésou 1998, Schricke *et al.* 2001). We focus on the impact of the transformation of the delta, from an estuarine floodplain into a cultivated area, on waterbirds.

Waterbirds breeding in colonies   The few delta-wide surveys of breeding waterbirds have been largely confined to colonies and large species; additional information relates mostly to anecdotal observations.

Distribution   The estuarine floodplain of the Senegal Delta is an attractive breeding ground for colonial waterbirds. Next to the floodplain proper, the adjacent coastal zone offers productive feeding grounds and safe breeding sites on islets or in the mangroves. De Naurois (1969) recorded 12 colonies of cormorants, herons, ibises and spoonbills in 1961-1965 (Fig 78). A large colony of Great White Pelicans was situated in the Aftout es Saheli, north of the delta, where De Naurois (1965b) also discovered a colony of Lesser Fla-

Box 10

Censusing waterbirds in the Senegal Delta requires a great effort

Patrick Triplet & Vincent Schricke (OMPO & ONCFS)

Of all the large floodplains in the Sahel, the Senegal Delta is probably the least difficult to survey. Aerial surveys are essential to cover the much larger Inner Niger Delta, or the Chad Basin, whereas it is possible to count the birds across the Senegal Delta simply by using cars and boats, although to do so demands much time and manpower. Since 1989, the midwinter count of the delta has been organised annually by *Direction des Parcs Nationaux* (DNP), *Office National de la Chasse et de la Faune Sauvage* (ONCFS) and *Oiseaux Migrateurs du Paléarctique Occidentale* (OMPO). Data are available from previous surveys carried out since the late 1950s, which makes the delta probably one of the best monitored sites in Africa.

In the Senegal Delta, counts have been carried out in the following main sub-sections:

- Djoudj National Park.
- Djeuss hunting zone.
- Langue de Barbarie National Park.
- Gueumbeul Special Reserve and the surrounding lagoons and ponds.
- Ndiael Reserve and the Trois Marigots.
- Lake Guiers (only one marsh, close to the lake, has been counted: N'Der).
- The Diawling National Park and its surrounding wetlands.

Given the size of the total area, counts are organised in a hierarchical way, starting with the Djoudj NP and Diawling NP. During our series of twenty January counts, the Djoudj NP was the major waterbird wintering site, mainly Anatidae during the day. However, the Diawling NP did become increasingly attractive to waterbirds, such as Great White Pelicans and Lesser Flamingos, which moved back and forth between them, the very reason why both parks are counted simultaneously on 15 January. All the other sites are counted during the following two days.

Apart from census errors (see Chapter 6: Inner Niger Delta), practical problems abound, either human-caused or posed by the environment. Logistic problems are manifold, ranging from car breakdowns, impassable tracks or high flood levels preventing teams from reaching important counting positions in the delta. Some former counting routes and observation points in the delta may have disappeared due to physical changes such as rice field development, prolific growth of Cattail and variations in the water regime. During the January count in particular, drifting sand is a major problem. Sometimes visibility is reduced so badly

that counting is impossible for a couple of days. Most frustrating is when a clear view of the area is restricted to just the first few hours of the day (between 07:30 and 09:30). Imagine having to count around 400 000 ducks in just two hours on the Grand Lac in the Djoudj NP! That is certainly no sinecure and would be absolutely impossible without the help of many volunteers.

To cover the Djoudj NP fully during a survey, we adopted a number of standard procedures, starting with meeting all the participants (about 50) at the Djoudj NP Biological Station. It takes seven teams to count birds in the Djoudj. Each team is composed of an experienced ornithologist and several (3-6) assistants. The team leader is responsible for the survey in his section, and must be acquainted with the local terrain and the expected distribution of birds. At present (2008), only a few people are sufficiently trained to count waterbirds on the big lakes (Grand Lac, Lac du Khar) in the central Park area, where most ducks roost during daylight. It takes 27-30 people to accomplish a complete count of the Djoudj NP; six or seven cars and three boats are also required.

The most complicated site to count is Lamantin Lake, where the team has to walk through water for three to four hours before arriving at the preferred observation point. Since the late 1990s, dense Cattail has thwarted any attempt to reach the lake or, more precisely, the last few remaining hectares of the lake proper. Another challenge, sometimes almost a nightmare, is the Grand Lac. The lake is very difficult to cover in the two hours after sunrise when visibility is adequate; the heat and the wind frustrate any attempt at counting later in the day, but it is even more difficult when there has been no wind at all the previous night. In such an event, and we were never able to figure out why, the ducks roost in remote and inaccessible ponds on their return. Duck numbers, in particular of Pintail and Garganey, are then inevitably underestimated, making it necessary to undertake repeated counts over several days to obtain a reliable estimate of numbers.

The water level varies constantly in the Trois Marigots, the greater part of which has become overgrown by Cattail and Reed, making counting birds more and more difficult. Nevertheless, annual monitoring is desirable because this unprotected site hosts a small group of African Pygmy Geese, a species considered to be a reliable indicator of wetland quality.

The lagoons near Saint-Louis attract large numbers of coastal waders, which can be counted from only two observation points: the first involves the dike separating the lagoons from Saint-Louis, and is a suitable counting site when the flood is sufficiently high to flood the local area; the second is near the small village of Leybarboye, where two people can count, but only during the mornings, because for most of the day the birds remain at considerable distance from the observers. Counting here is further complicated by the presence of mixed-species flocks.

During the 1989-2008 period of 20 years of systematic counting, more than one hundred people - rangers, students and ecoguides - were trained and became involved in the census program. However, turnover among observers was extremely high, more so than in the average wader population! Some people simply did not turn up, others had quit or had transferred to jobs in other protected areas, and some had died. As of 2008, the team leaders at the more difficult sites are still going strong, but new recruits are lacking and so there is no guarantee that monitoring the numbers of wintering and staging waterbirds, particularly in January when efforts are greatest, will continue over the next 20 years at the same levels of accuracy attained so far. We can but hope that young Senegalese ornithologists will take up the challenge and continue this unique nonstop series of waterbird counts in West Africa.

By joining forces waterbirds in the Senegal Delta are counted in midwinter (January) already two decades in a row. The annual census requires 30-40 participants, organised in seven teams each counting a subsector of the delta. This huge effort is possible through a co-operation between the DPN, ONFCS and OMPO.

Senegal Delta

Situated on an islet in Marigot du Djoudj, the breeding site of Great White Pelicans in the Djoudj National Park is subject to heavy erosion (Diop & Triplet 2006, I.S. Sylla pers. com.).

mingo. This location is the only, though irregular, breeding site of Lesser Flamingos in this part of Africa (Trolliet & Fouquet 2001). During the 1960s 17 colonially breeding bird species were recorded in the Senegal Delta, including Great White Pelican but excluding gulls and terns restricted to the coastal zone. Despite the embankment construction in the early 1960s, the delta was still largely natural (de Naurois 1965a). Most colonial waterbirds bred in trees, especially in Tamarisk and *Acacia nilotica* in the eastern and northern part of the delta, and in mangroves (*Avicennia* sp., *Rhizophora* sp.) in the lower delta.

A second survey in the early 1980s revealed large changes in numbers and distribution of colonies (Fig. 78). While De Naurois (1969) found no colonies in the central part of the delta, up to five colonies were present in the Djoudj in 1979-1980: one of Purple Herons breeding in a Reed marsh and the others of mixed species composition in Acacias. In 1975, the main mixed heronry was located in the Djoudj NP, totalling 8000 pairs (Roux 1974). The absence of colonies in this part of the delta in the 1960s was explained by the lack of favourable feeding conditions and the absence of trees for breeding, the latter being associated with heavy grazing (Voisin 1983). The establishment of the National Park in 1971 resulted in the removal of livestock, reflooding and protection against poaching. These measures turned the Djoudj NP into a core area for breeding waterbirds in the delta. While colonies elsewhere in the delta continued to disappear, the Djoudj hosted important breeding sites with at least 15 species of colonial waterbirds in 1985-1987 (Rodwell et al. 1996).

From 2000 to 2006, the breeding colonies in the delta were confined mostly to protected National Parks (Fig. 78), although in July 2006 indications of Cattle Egret breeding were obtained near Richard Toll (J. van der Kamp, unpubl.). The distribution of colonies overlaps with favourable feeding grounds (shallow marshes, estuary) and sites suitable for breeding (protected sites on islets, trees and bushes). In the Djoudj 2-4 mixed colonies of cormorants, egrets and herons are present, as is a colony of Great White Pelicans in the centre of the park. [6] The other breeding sites in the delta are in the Diawling NP, where, after the restoration of the flood in 1994, colonial waterbirds showed a remarkable upsurge in numbers (Hamerlynck & Duvail 2003, Diagana et al. 2006; Table 10). From 2004 to 2006, the Diawling NP harboured 4-5 mixed colonies, located in low Tamarisk bushes or mangroves. The colonies of Great Cormorant, African

Table 10 Estimates of the breeding population of colonial waterbirds in the Diawling National Park (Diawling, Ntiallakh, Tichlit basins) in 1993, 1999 (Hamerlynck & Duvail 2003) and in 2004 and 2005 (Direction Parc National du Diawling, Diagana et al. 2006). Key: + = present, - = no breeding recorded.

Year	1993	1999	2004-2005
African Darter	-	>25	1150
Long-tailed Cormorant	<50	>250	1300-1400
Great Cormorant	<30	>500	4500-5000
Great White Pelican	-	1400	
Pink-backed Pelican	-	5-10	
Black-crowned Night Heron	<5	>100	20-30
Squacco Heron	<50	>150	
Cattle Egret	<10	>250	50
Little Egret	<25	>200	100-320
Great Egret	-	>200	800
African Spoonbill	-	5	50-600
Sacred Ibis	-	+	50-70

The breeding site of Great White Pelicans was – and is – particularly eye-catching, and a magnet for tourists visiting the Djoudj National Park.

Darter and Great Egret are among the largest known in West Africa.

Population trends   The establishment of the Djoudj NP in 1971 was vital to the preservation of breeding waterbirds in the delta. The increase of colonial and ground-nesting waterbirds in both the Djoudj and Diawling can be attributed to: (1) Flood rehabilitation (new feeding habitats) in combination with a partial recovery of rainfall and flooding. (2) Return of trees and bushes after the area had been closed to livestock (new breeding habitats). (3) Prevention of poaching (safe haven).

The quantitative information on breeding colonies in the Senegal Delta for the 1960s and early 1980s must be considered as rough estimates (Table 11); data from the 1990s and subsequently are incomplete for the Djoudj. **African Darter** and **Long-tailed Cormorant**, widespread in the 1960s, faced a strong decline during the 1970s,

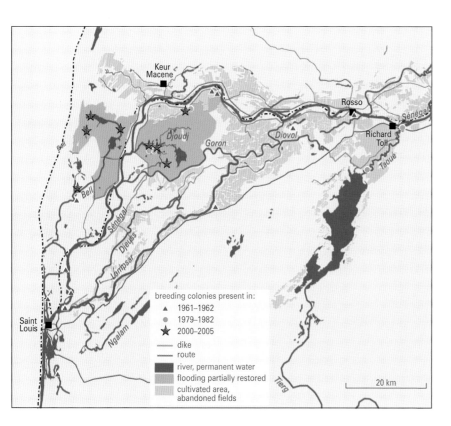

Fig. 78 Location of breeding colonies of pelicans, cormorants, herons, egrets, spoonbills and ibises in the Senegal Delta in the 1960s, 1980s and in the 2000-2005 period. Data from: de Naurois (1969), Voisin (1983), Diagana *et al.* (2006). The delineation of the areas subject to artificial flooding (green) largely overlaps with the National Parks of Diawling and Djoudj. [6]

Senegal Delta

African Darters, vulnerable breeding birds in the Djoudj NP and Diawling NP, depend on undisturbed breeding sites with fishing grounds nearby.

1980s and 1990s, related to drought and reclamation of the floodplain (Voisin 1983, Yésou & Triplet 2003; Table 11). From 1999 onwards, both species showed a spectacular recovery, stimulated by the flood restoration in the Diawling and probably also by a series of high floods. **Great Cormorant** was recorded as having only a few pairs in the 1960s, but has increased since then (Yésou & Triplet 2003). In 2004-2006 large colonies of this species were found in the Diawling NP. The breeding success of African Darters and cormorants in the Diawling was high (Diagana et al. 2006), but not high enough to explain this fast increase. This hints at a shift from other areas, such as the Djoudj, but Long-tailed and Great Cormorants may also have come from littoral breeding zones in the north; on the Banc d'Arguin for example, where the breeding numbers of Long-tailed Cormorant more than halved between 1997 and 2004 (Veen et al. 2006). [7]

In the 1960s, De Naurois (1969) recorded 11 species of herons and egrets, which were reconfirmed by Voisin (1983). Cattle Egret, Great Egret, Black-crowned Night Heron and Squacco Heron were by far the most numerous species in those days, ranging from sev-

Table 11 Numbers of colonial waterbirds in the Senegal Delta (Djoudj NP, Diawling NP and other sites combined) in the 1961-1965 period (de Naurois 1969), 1979-1981 (Voisin 1983 and 1987-1993 Rodwell et al. 1996,). Recent data from the Djoudj NP are lacking; for the Diawling NP see Table 10. The location of the colonies is given in Fig. 78.
Key: breeding pairs; + = a few to tens, ++ = hundreds, +++ = thousands.

Period	1961-1965	1979-1981	1987-1993
African Darter	400-500	> 60	++
Long-tailed Cormorant	>600 (>1500)[1]	>1200[2]	50 – 250
Great Cormorant	>10	?	660
Great White Pelican	1500-2000[3]	8500[4]	+++
Pink-backed Pelican			c. 10
Grey Heron	+	+	-
Purple Heron	50	+	+
Black-crowned Night Heron	>>500	>>50	>1375[5]
Squacco Heron	>>500	>>190	230[5]
Western Reef Egret black	>25	?	-
Black Heron	>25	+	+
Cattle Egret	>1500	>800	1960[5]
Little Egret	>50	+	>800[5]
Intermediate Egret	>100 (>300)[1]	>45	+
Great Egret	>250 (>480)[1]	>600	1450[5]
Green-backed Heron	+	+	+
Yellow-billed Stork	50?	>30	>>20
African Spoonbill	>100	>50	+
Sacred Ibis	>125	>26	+
Total estimate	**>> 6500**	**>> 12000**	**?**

Notes:
[1] Morel (1968) reports breeding colonies near Rosso in 1964 with 1500 breeding pairs of Long-tailed Cormorant, 480 Great Egret and 300 Intermediate Egret.
[2] Roux (1974) mentioned 3000 pairs for the Djoudj NP.
[3] Breeding in the Aftout es Saheli (de Naurois 1969b); in 1972 1500 pairs attempted to breed in the Djoudj NP and in 1976 2000 pairs bred successfully in the park (Roux 1974).
[4] Counted in 1982 in the Djoudj (Data from: BirdLife International 2008, no primary source mentioned).
[5] Based on figures mentioned by Kushlan & Hafner (2000) for November 1988.

In the Senegal Delta the value of shallow, brackish lakes, such as Grand Lac and Lac du Khar in the Djoudj NP and the Bell basin in the Diawling NP, depends largely on their protection of waterbirds from exploitation by man.

eral hundred to a few thousand breeding pairs. Apart from species that occurred in low numbers (insufficient information), other species seem to have increased by the end of the 1980s (figures from November 1988, Kushlan & Hafner 2000; Table 11). This increase may be attributed to protection measures. We lack delta-wide estimates for the 2000s, although some species resettled in the Diawling (Table 10). Most heron and egret species were still breeding in the Djoudj in the mid-2000s, except Black Heron and Western Reef Egret (Triplet et al. 2006); the status of Purple Heron is uncertain.

The **Black-crowned Night Heron** bred in considerable numbers in the Djoudj in 1988, as recorded by Rodwell et al. (1996) for 1985 and 1987 ('several hundred nests' for the heronries at Poste de Crocodile and Lac du Khar in the Djoudj NP). The population of **Squacco Heron** was never found to exceed some 500 pairs. **Little Egret** and **Great Egret** are increasing in the Senegal Delta, the latter especially in the Diawling. **Cattle Egret** bred with a few thousand pairs in the Djoudj by the end of the 1980s (Kushlan & Hafner 2000). Morel & Morel (1990) noted numerous colonies in the Senegal Delta, but many were destroyed or otherwise disappeared (see above).

Breeding numbers of storks, ibises and spoonbills are small, as they were in the 1960s. **Yellow-billed Stork** has been recorded breeding in the Djoudj in small numbers only, but in January of 2002 to 2004, between 100 and 450 non-breeding birds resided in the Djoudj (data from: P. Triplet and V. Schricke). **Sacred Ibis** numbers dipped in the 1980s and 1990s; the recovery in the 2000s can be attributed largely to the increase in the Diawling. The population increase of **African Spoonbill** in the Diawling zone must originate from colonies elsewhere in West Africa; numbers in the past rarely exceeded 50-100 breeding pairs. Although seen regularly in the past (Rodwell et al. 1996), and recorded breeding in December 1981 (Dupuy 1982), observations of **Marabou Stork** are lacking for the past decades (Triplet et al. 2006, P. Triplet in litt.); the species is mentioned incorrectly as a recent breeding bird by Groppali (2006).

The **Great White Pelican** holds a prominent position in the Senegal Delta. Along the West African coast it is found breeding only in Mauritania (up to 3800 pairs in 1997; Isenmann 2006) and Senegal (Veen et al. 2006). The birds breeding in the Djoudj probably originate from the Aftout es Saheli (de Naurois 1969). Irregular breeding in the Aftout still occurs (Peeters 2003, Diagana et al. 2006), as was the case in the Diawling in 1995 and 1999 (Hamerlynck & Duvail 2003). Great White Pelicans in the Djoudj settled in 1976 (I.S. Sylla pers. com.). The colony has been estimated at 8500 pairs in the 1980s (Table 11) and comprised in the order of 4000-5000 pairs from 2000 to 2008 (P. Triplet in litt., I. Ndiaye in litt.). Also, a small number of **Pink-backed Pelicans** uses the breeding site. After the breeding season, the total population of Great White Pelicans in the Senegal Delta during the January counts is on average 22 500 (1999-2004; Table 12). Pelicans use the Djoudj and Diawling as a single feeding unit, judging from the flights back and forth. Since the reflooding of the Diawling, the number of pelicans using these basins has increased significantly (Diawara & Diagana 2006), reflecting the apparent improvement of feeding conditions.

	1972	1973	1974	1975	1976	1979	1983	1984	1985
Great White Pelican	8500	6580	4997	18904	8000	5380	10	35	
Pinked-backed Pelican	270								
Glossy Ibis	1150	346	92	200		40	1090	145	
White-faced W. Duck	43000	11781	6800	54312	27000	13195	1970	2100	600
Fulvous Whistling Duck	400	754	100	6430	5000	2800	2410	100	400
African Pygmy Goose	4								
Knob-billed Duck	650	872	544	4580	1020	2050	2	30	62
Spur-winged Goose	1170	480	256	8514	4205	1378	1575	296	254
Egyptian Goose	250	1798	24	678	40	70	175	210	5
Northern Pintail	55000	114716	90900	217675	78500	26930	1110	700	3908
Garganey	135000	35000	86400	239977	120250	67570	9050	2800	17215
Northern Shoveler	2300	8000	2700	31840	7250	6040	16200	3800	2400
Black-crowned Crane	280	741		17			1	4	
Black-tailed Godwit	10500	23888	200	5600	7990		100	3000	
Ruff	500000	343960	20050	18450	5000	850	37000	500	80000

	1995	1996	1997	1998	1999	2000	2001	2002	2003
Great White Pelican	4536	10193	3349	10320	18251	20961	28368	18335	8888
Pinked-backed Pelican	87	23	161	127					
Glossy Ibis	42	430	305	427	397	512	177	146	467
White-faced W. Duck	21530	12160	32273	36890	25204	51090	83207	51147	120683
Fulvous W. Duck	2458	3208	3029	1782	3612	1682	16096	3259	1995
African Pygmy Goose				8	17		78	2	39
Knob-billed Duck	41	132	1894	1675	733	1381	1529	799	1023
Spur-winged Goose	218	1598	1792	1556	1130	1846	707	545	709
Egyptian Goose	270	183	126	782	779	498	256	433	563
Northern Pintail	87500	44993	53430	127608	107501	150127	133235	38266	32176
Garganey	19940	61758	153394	269144	232017	296191	150000	93424	73240
Northern Shoveler	10829	23202	18229	12821	23948	13241	18992	8862	16672
Black-crowned Crane	549	171	148	79	69	384	110	111	152
Black-tailed Godwit	2210	6944	2495	2473	4000	5000	3800	1200	1200
Ruff	124060	38150	135929	160063	88515	22910	30307	1158	794

The Senegal Delta and the adjacent coastal depression of Aftout es Saheli constitute the only West African wetlands where breeding of **Lesser Flamingos** has been proven. De Naurois (1965a) was the first to record this species as a breeding bird in this region; in July 1965 he located a colony with some 800 nests in the Aftout. Breeding attempts of Lesser Flamingos were apparently triggered when the depression held sufficient water, which in its turn depended on the discharge of the Senegal River. In years with a high flood (in the 1950s and 1960s), the species may have bred regularly (de Naurois 1965a). During the Great Drought in the 1970s and 1980s suitable breeding conditions failed to materialise as inundations were insufficient to flood the breeding grounds. However, in 1999, when the flood was high again, breeding was suspected in the Diawling (Hamerlynck & Messaoud 2000). In 2001, successful breeding was observed in the Chat Tboul (Peeters 2003). Some 3200 birds, including 350 immatures, were seen the next year. Breeding has been infrequent, long non-breeding intervals occurring between attempts. In most years, the staging population along the entire West African coast does not exceed 10 000-15 000 birds (Dodman & Diagana 2003). However, in 1990 exceptionally large aggregations were seen

1986	1987	1988	1989	1990	1991	1992	1993	1994
			2234		12400	9162	4820	14328
						8	175	69
			186	258	633	370	330	257
66410	43960	9015	22744	11960	4690	7064	10492	13649
30806	37760	942	1091	88	500	213	379	1165
398	1910	38	38	40	247	787	477	100
1370	945	2469	1059	1518	140	380	148	676
668	729		87	224	2662	365	388	602
247354	99132	39125	210000	44890	73801	72621	50000	77233
125550	183684	83417	85000	52011	77303	100990	118670	128631
8800	34236	437	15657	14185	19130	8740	11893	10024
			27	119	400	200	233	515
					332	5501	11110	4090
					178000	190000	192500	32072

2004	2005	2006	2007
39406		21925	22875
187			
113830			91609
2937			5388
314			
841			
715			
835			269
18778	68542	157456	177466
47314	188776	218182	41398
18560			
186		146	142
1000	2500	3191	5613
3457		7078	11423

Table 12 Midwinter (January) counts of 15 waterbird species in the Senegal Delta (excluding Aftout es Saheli and Lakes Guiers, R'kiz & Aleg and the river valley east of Podor). Data from: Roux 1973, 1974, Roux et al. 1976, Roux & Jarry 1984, Schricke et al. 1990, 1991, Girard et al. 1991, 1992, Trolliet et al. 1993, Triplet & Yésou 1994, Yésou et al. 1996, Triplet et al. 1995, 1997, Schricke et al. 1998, 1999, Dodman & Diagana 2003, Trolliet et al. 2007, 2008 and complemented by data from the AfWC-database of Wetlands International. The 2007 count is published in Bull. African Bird Club 14: 223-4.

Great White Pelicans fishing in groups, an efficient technique to chase schools of fish into shallow water where they can be simply scooped up.

The Eurasian Spoonbills breeding in The Netherlands, NW France and Spain form a discrete population, which moves up and down the Atlantic coast to and from wintering sites in the Banc d''Arguin (Mauritania) and in the Senegal Delta (Mauritania/Senegal). The shallow coastal waters provide an abundance of food throughout the year. First-winter birds remain in West Africa for the next three or four years until they are sexually mature. This population has little or no contact with the central and eastern populations. Among 48 000 records of Dutch-ringed Spoonbills, less than 1% were reported from the Mediterranean (mostly Tunisia) (Smart et al. 2007).

pressions in Mauritania (e.g. Mare de Mahmouda, R'kiz and Aleg) may hold large concentrations of waterbirds (notably ducks and Ruff) when conditions are favourable (van Wetten et al. 1990, Dodman & Diagana 2003). The abundance of waterbirds in the Senegal Delta is a reminder of the scarcity of suitable wetlands in this part of the Sahel. Waterbird counts in January (Table 12) do not necessarily reflect maximum wintering numbers. Palearctic migrants may have already started their migration (Hötker et al. 1990; Chapter 27, Black-tailed Godwit), and many Afrotropical waterbirds may have left the area in response to the gradual desiccation of flooded areas.

in the Djoudj (45 000 to 46 500 birds), while in June 2003 36 000 Lesser Flamingos were counted in the Diawling. Given the sporadic breeding and large fluctuation in numbers, an exchange between the populations in West Africa and those in East and southern Africa is likely (Trolliet & Fouquet 2001).

Apart from colonial breeders, a variety of other waterbirds breeds in the delta, including waders [8], ducks, rails and terns, as well as Arabian and White-bellied Bustard (Rodwell et al. 1996, Triplet et al. 2006). Another large ground-nesting bird, the **Black Crowned Crane**, used to be a widespread species, with an estimated thousand birds in the Senegal Delta. Morel & Morel (1990) mentioned a steep decline from the early 1980s onward, but by 1980 the numbers already had decreased to 200 over the entire delta (Poorter et al. 1982). Mid-winter counts in the past decade yielded on average 152 birds (January 1997-2007, Table 12). In the Djoudj NP the crane population is small, but the Mauritanian population has increased since the flood rehabilitation in the Diawling, from 2 breeding pairs 1993 to more than 30 in 1999 (Hamerlynck & Duvail 2003).

Wintering and migratory waterbirds   From July to April, the Senegal Delta is an important staging area for non-breeding Afrotropical and Palearctic waterbirds. This may come as a surprise considering the hydrological changes in past decades, and the ensuing loss of wetland habitats favoured by waterbirds. However, alternative wetlands in this part of the Sahel are almost non-existent, although large de-

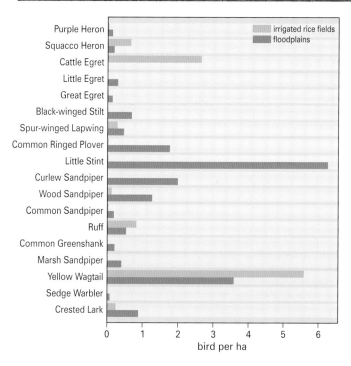

Fig. 79 Densities of selected waterbird species in rice fields (113 plots) and restored floodplains (36 plots in Djoudj and Diawling) in the Senegal Delta in February and December 2005.

Large concentrations of Palearctic ducks, in particular Garganey, Pintails and Shovelers, and Afrotropical ducks, mostly White-faced and Fulvous Whistling Ducks, use the Djoudj NP as a resting place during daytime. After sunset, the birds disperse to feeding sites in the wider surroundings, including rice fields.

Aerial counts, complemented by ground surveys from 1989 onwards (Box 10), concentrated on gregarious and large species (Table 12). The smaller, secretive or thinly distributed bird species were covered by 149 density counts in February and December 2005 (Fig. 79), a small sample compared with the Inner Niger Delta (Chapter 6, also for methods). Densities of waterbirds on restored floodplains were much higher than on rice fields, and diversity was greater. This outcome complies with the findings in the Inner Niger Delta (Wymenga *et al.* 2005, van der Kamp *et al.* 2005a). Most herons, egrets, waders and marshland passerines are abundant in natural habitats, and much scarcer in rice fields (as found elsewhere in West Africa, Chapter 11; Bos *et al.* 2006). Exceptions are Cattle Egret, Yellow Wagtail, Ruff and Squacco Heron.

The Senegal Delta is an important wintering site for **Black-crowned Night Herons**. High counts in January included 5300 birds in 1993 (Trolliet *et al.* 1993), 5208 in 1997 (Triplet *et al.* 1997) and 7610 in 2002 (P. Triplet & V. Schricke pers. comm.). The proportion of Palearctic birds is unknown, but may have increased in the 2000s given an increase in the European breeding population (Chapter 17). **Grey Heron** and **Purple Heron** are encountered mostly along watercourses and in shallow marshlands, averaging 0.25 Grey Herons and 0.13 Purple Herons per hectare (Fig 79). In December and January the delta offers 12 000-18 000 ha of suitable brackish marshes [9], indicating wintering populations of at least 3000-4500 and 1560-2300 birds respectively. These estimates are conservative, because both species also occur in vegetated water bodies and along watercourses throughout the delta. The same holds for **Little Egret**, estimated at >4000-6000 birds including local breeding birds.

The **Cattle Egret** is the typical heron of rice fields (Morel & Morel 1990). In 1985, Mullié *et al.* (1989) counted on average 3.3 birds/ha in a sample of 21 rice fields. In February and December 2005, we arrived at an average of 2.64 birds/ha (113 plots, SE = 0.70; n = 96, excluding seedbeds (only a few plots) and abandoned cultivated land (hardly any birds)). Given a surface area of >30 000 ha of rice fields annually under exploitation in the delta, the post-breeding population could comprise up to 55 000 birds.

Small numbers of **Glossy Ibis** (on average 309 birds annually in 1994-2004), **Black Stork** (35, 1990-1998) and **White Stork** (34, 1994-2004; sources in Table 12) inhabit the more or less natural habitats of the delta in midwinter, notably in depressions (Ndiaël), shallow

Patchy Cattail habitat, interspersed with temporary mudflats and pools with open water and water lilies, hosts an array of bird species (particularly waders and ducks), contrasting markedly with the scarcity of birds in homogeneous Cattail stands.

lakes and *Sporobolus* and *Scirpus* marshes. The status of the **Great Bittern** is difficult to assess, but it may be a common Palearctic winter visitor in wet grassy marshes around Grand Lac; Flade (2008) frequently flushed them from shallow ponds during fieldwork in January and February 2007, and estimated the wintering population tentatively at tens or even hundreds. This would make the Senegal Delta second to the Inner Niger Delta as a primary wintering site in West Africa.

The Palearctic fauna is well represented by ducks, especially **Pintail** and **Garganey**, and to a lesser extent **Shoveler**. In the 2000s, Pintail and Garganey numbered up to 200 000 and 150 000 birds respectively (Table 12), which means that the Senegal Delta harbours a substantial part of the West Palearctic breeding population (Delany & Scott 2006). Numbers were much smaller during the Great Drought (notably 1973, 1984, 1985), but both species recovered when rainfall improved again. Palearctic ducks roost in daytime in open, shallow and brackish waters in the Djoudj NP (Grand Lac, Lac du Khar) and, since the reflooding in 1994, also in the Bell and Diawling basins. About 90% of the Palearctic ducks once congregated in the Djoudj NP, but the restoration of the Diawling caused a redistribution in favour of the latter (Triplet & Yésou 2000). Also, the northern part of Ndiaël and Trois Marigots may function as roost and feeding site.

Afrotropical ducks are represented mainly by **White-faced Whistling Ducks** (Trolliet *et al.* 2003), showing an increase from tens of thousands to >100 000 over the years (Table 12). **Fulvous Whistling Duck** is less abundant, totals in January normally being of the order of 1500-3000 birds. Other Afrotropical ducks and geese occur in much smaller numbers (Table 12).

Along with ducks, waders are widely distributed and numerous across the Senegal Delta. Until the late 1990s, Ruff numbers during midwinter ranged up to 190 000 (on average 126 000 in the 1990s; Table 12), but since then have been considerably lower (11 000, on average, in the 2000s). Similarly, the **Black-tailed Godwit** has much declined from tens of thousands in the 1960s and 1970s to only 1000-5000 birds remaining in the 2000s (Chapter 27). Black-tailed Godwit and Ruffs were mostly counted when they commuted between the rice fields and the large wetlands (Djoudj NP, Ndiael) for drinking and resting. Apart from population changes (Chapter 27), the observed declines over the past decade (1997-2007) may be linked to changes in the way rice is harvested (P. Triplet, V. Schricke in litt). Triplet & Yésou (1998) reviewed the occurrence of waders in the Senegal Delta in mid-winter. Next to Ruff and Godwits other wader species with relatively high numbers are Black-winged Stilt (313-1137), Avocet (1342-4940), Kentish Plover (1624), Little Stint (5000-10 000), Curlew Sandpiper (2000-4000) and Spotted Redshank (1200). In the estuarine part of the delta (Gandiol and Ntiallakh basin, downstream Diama), various waders from marine coastal areas are present (*e.g.* Grey Plover, Bar-tailed Godwit, Red Knot and Whimbrel). As a result of the recent breach south of Saint-Louis, both wader numbers and species composition are changing (Triplet & Schricke 2008; see above).

In the open shallow floodplains, some wader species may reach high densities, like Little Stint (4.50 birds/ha), Curlew Sandpiper

(1.56 birds/ha) and Wood Sandpiper (1.21 birds/ha). Conservatively estimating the surface area of these habitats at 7500 – 11 000 ha [10], total numbers would then range between c. 10 000 (Curlew Sandpiper and Wood Sandpiper) and 50 000 birds (Little Stint), which is one or two orders of magnitude higher than assessed in the delta during mid-winter counts (Triplet & Schricke 1998, 2008).

The variegated ecosystem of brackish marshes, vegetated plains and rice fields attracts a plethora of wetland-related Palearctic birds other than ducks and waders, some omnipresent, others, such as **Little Crake** and **Baillon's Crake**, living a secret life in dense vegetation beyond the sight of birders. Apparently they are not uncommon in the delta marshland (Morel & Morel 1990, Triplet *et al.* 2006), but habitat choice and numbers were anybody's guess, until in 2007 the use of cage traps serendipitously uncovered the presence of Baillon's Crake in wet *Scirpus/Sporobolus* marshes (Flade 2008), presumably in large numbers. The intention had been to trap ground-dwelling passerines, but the five trapping sites (each with 15 traps) in *Scirpus/Sporobolus* yielded 1-4 Baillon's Crakes. If we assume a density of 1 bird/ha in *Scirpus* marshes, and that this habitat type occupies a surface area of 12 000-18 000 ha in the Senegal Delta, we can theoretically account for the larger part of the European population wintering here (Flade 2008). Little Crake, on the other hand, was captured in smaller numbers, but in contrast to Baillon's Crake, also inhabited Cattail vegetation as one of the few Palearctic species to occupy it.

Harriers are among the most widespread raptors in the Senegal Delta. In the winter of 1992/1993 Arroyo & King (1995) found two major roosting sites of **Eurasian Marsh Harriers** (each holding 300-350 birds, some 4 km apart), and 3-5 smaller roosts (<100 birds) within the same area. All roosts were located in dry rice fields, where dense rice plants were some 80 cm high. The wintering population must have exceeded 1000 birds, given the average density of >10.6 birds per km² for the area visited (94 km², including part of the Djoudj NP and surroundings). Only 15% of 131 aged birds were adult males. More recently, a census along the Senegal from Débi towards Rosso produced >130 birds leaving a single roost (14 December 2005), and several dozens hunting in other parts of the delta between 11 and 17 December 2005 (obs. I Ndiaye, D.Kuijper & E. Wymenga). A census in the rice field complex near Rosso on the same date yielded 2.6 birds per km². Arroyo & King (1993) found that Marsh Harriers spent about as much time over marshland as over rice fields when excluding roosts and pre-roost gatherings. Data on fluctuations in wintering numbers are lacking, but it can be assumed that the population increase in Europe would be reflected by a steady increase in the Senegal Delta (although not to the same extent, given the increasing numbers wintering in southern Europe; Chapter 25). Such a trend contrasts with that of **Montagu's Harrier**, whose numbers in the Senegal Delta may have dropped in past decades, with steep variations between years on top of that. Referring to the 1950s and early 1960s, Morel & Roux (1966a) considered the

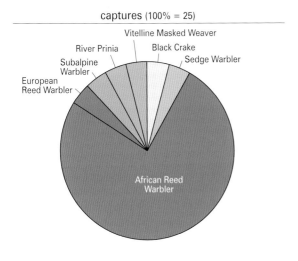

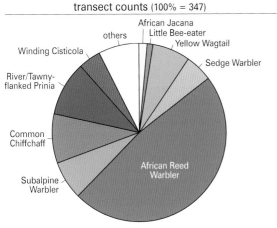

Fig. 80 Pie chart illustrating the bird species composition in Cattail vegetation, based on captures and transects. Birds associated with Tamarisks in the border zones are excluded. From: Bruinzeel *et al.* (2006).

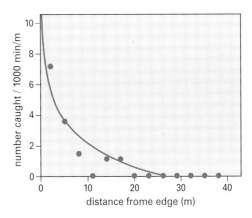

Fig. 81 African Reed Warblers were recorded within some 20 m of edges of Cattail stands; the vast uninterrupted Cattail beds lacked any warblers away from the edges. From: Bruinzeel *et al.* (2006).

Montagu's Harrier 'very common' in the Senegal Delta, with 'very large numbers' near Richard Toll. Actual estimates ranged from 200-250 in 1988/1989 (Cormier & Baillon 1991) to 150 in 1992/1993 (Arroyo & King 1995); a single roost east of Débi held 160 birds on 15 February 1992 (Rodwell et al. 1996). These numbers are rather small compared with single-roost counts of 1000 birds between M'Bour and Joal south of Dakar in 1998/1989 (Cormier & Baillon 1991), and of 1300 birds near Darou Khoudoss, some 80 km NE of Dakar, on 2 February 2008 (Chapter 26).

The vegetated floodplains and higher levees in the Djoudj NP and Diawling NP that have tall grasses and bushes (especially *Salvadora persica*) are important wintering sites for a range of Palearctic passerines (Morel & Roux 1966b, Beecroft 1991, Morel & Morel 1990, Flade 2008). The **Yellow Wagtail** is among the most abundant Palearctic passerines here, reaching densities in excess of 5 birds/ha. In the combined surface area of 70 000 ha of floodplains and abandoned rice fields, the population may exceed hundreds of thousand of birds, which is in accordance with records of single roosts of 100 000-250 000 birds in November 1991 and December 1992 (Rodwell et al. 1996), and the estimate of 50000-200000 in *Scirpus* marshes in January to February 2007 (Flade 2008). Similarly, **Sand Martin** roosts are known to number up to 2 million birds (Rodwell et al. 1996), and this order of magnitude was found to forage over the floodplains of Grand Lac in January to February 2007 (Flade 2008). The Senegal Delta also harbours significant numbers not only of **Bluethroat, Grasshopper Warbler, Savi's Warbler, Sedge Warbler**, and **European Reed Warbler**, based on density counts (Fig. 79) and trapping (Beecroft 1991, Flade 2008), but also of **Aquatic Warblers** (Box 11).

The vast Cattail marshes in the delta form an important habitat for insects, (small) fish and, potentially, for birds. In Europe and North America, where many natural marshes have been drained, Reed and Cattail marshes offer alternative habitat for a wide range of marshland fauna. Up till recently, little was known about avian biodiversity in West African Cattail vegetations. Dense stands of invasive plants are assumed to reduce both plant and animal diversity (Zedler & Kercher 2004, Frieze et al. 2001). For the Senegal Delta, Peeters (2003) mentions Cattail vegetation as feeding habitat for herons, rails and passerines, and as roosting site for Sand Martin and Yellow Wagtail.

An exploratory study in February 2005 in the Djoudj and Diama Basin showed that dense Cattail harbours few birds in comparison with other freshwater habitats in the Senegal Delta (Bruinzeel et al. 2006). For example, patchy Cattail habitat interspersed with temporary mudflats and pools with open water and water lilies, host an array of species (particularly waders and ducks) that are not present in dense and monospecific Cattail stands. Bird species composition in dense Cattail stands in the Senegal Delta is limited, as observed by mistnetting and transect counts (Fig. 80). Bird densities are highest along the edges (Fig. 81). In dense Cattail stands, African Reed Warblers were only caught within about 20 m of the edges; additional observations failed to reveal the presence of other bird species. In January to February 2007, the Aquatic Warbler Conservation Team sampled dense Cattail stands on a much larger scale (Flade 2008) and they also found low densities. However, their results indicated that Cattail vegetation may be of substantial importance for European Reed Warbler, Great Reed Warbler and Little Crake.

Impact of large-scale habitat changes   Many studies have tried to elucidate the impact on waterbirds of habitat changes in the delta (Lartiges & Triplet 1988, Tréca 1992, Triplet et al. 1995, Triplet & Yésou 2000). Since the 1960s, c. 90% of the original floodplain was embanked and cultivated, resulting in a major change to habitat composition and proportion (Fig. 82), but few were straightforward, however. For example: (1) Flood extent varies annually, low flood levels resulting in proportionately larger parts of the estuary and the natural floodplains remaining dry. (2) A substantial part of the cultivated area was abandoned and became either dry and saline or Cattail-dominated wetland. (3) After local rainfall or when drainage basins for irrigated fields appear, small wetlands may be formed that are independent of river discharge.

From the habitat changes depicted in Fig. 82, concomitant changes in bird numbers seem inevitable. Counts of staging waterbirds in the 1950s and 1960s (Roux 1959b, de Naurois 1965a, 1969, Morel & Roux 1966a&b, Roux & Morel 1966) indeed suggest that many waterbird species were (much) more numerous than in the 1990s and subsequently. However, the declining numbers of waterbirds in the intervening period cannot be attributed unequivocally to habitat changes. Many more factors have played a dominant role in the (wider) Senegal Delta, often operating in concert, and still do so:
1. The 1950s and 1960s were characterised by high floods of the Senegal River (Fig. 15), in itself a major determinant of waterbird numbers;

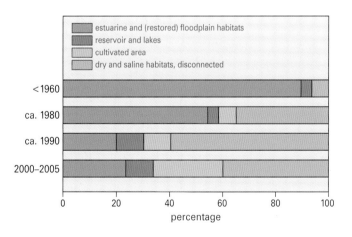

Fig. 82 Relative presence of four major habitat types in the Senegal Delta (3400 km²), based on Fig. 67. Note that the maximum potential flood extent is indicated of estuarine and floodplain habitats, but that the actual inundation depends on annual flood performance (Fig 73).

Box 11
Aquatic Warblers in the Djoudj and the Inner Niger Delta

In 2007, the Djoudj NP was found to be a (or the?) core wintering area for the Aquatic Warbler (Flade 2008). In the Djoudj, the species favoured ungrazed floodplain habitat dominated by *Scirpus* sp. and *Sporobolus* sp. The absence of livestock grazing is crucial for grasslands remaining suitable for wintering Aquatic Warblers. For example, the extensive floodplains of the Inner Niger Delta, as speculated by Schäffer *et al.* (2006) possibly to be of greater importance than the Djoudj as a wintering site for this species, are probably unsuitable in the main: during flood recession, livestock in their millions strip the available vegetation and turn the Inner Niger Delta into a dry 'wasteland'. Despite annual fieldwork throughout the Inner Niger Delta, especially in the surroundings of the central lakes, since 1991 (Fig. 48), we recorded Aquatic Warbler just once during density counts in 1191 plots, *i.e.* on 22 January 2005 near Akka (R.G. Bijlsma & W. van Manen, unpubl.). This site was 100% flooded (up to 20 cm deep) at that time, overgrown with didere *Vossia cuspidata* and poro *Aeschynomene nilotica* (height 10-25 cm, density 70%) and even then already being grazed by cows. This observation can be interpreted in several ways. If representative of *bourgoutière* (floating grass), it would make for an average density of 1 Aquatic Warbler/km² (Table 8, Chapter 6). *Bourgoutière* covers about 1540 km² in the Inner Niger Delta, which might theoretically translate into some 1500 Aquatic Warblers. However, as a foraging habitat *bourgoutière* is available for only a short period of time during flood recession, and Aquatic Warbler would have to be a habitat generalist to account for the successive removal of vegetation types by herbivores. However, the Senegal observations do not suggest that its habitat choice is catholic, but quite the contrary. It is therefore more likely that the

An Aquatic Warbler in one of its main breeding areas: the Sporava fen mire in Belarus.

record of the lone Aquatic Warbler in the Inner Niger Delta was but a chance observation, rather than indicating that we had found another important wintering site in the Sahel. Flade (2008) estimated the wintering population in the Senegal Delta at several thousand birds, which would comprise the major part of the breeding population in Poland, Ukraine and Belarus.

Extensive *Sporobolus* floodplains in the Djoudj NP, December 2006. This seemingly homogeneous vegetation turned out to be of vital importance for wintering Aquatic Warblers and Eurasian crakes.

African Spoonbills, not to be confused with balsaci Eurasian Spoonbills (which, by the way, are not Eurasian but African breeding birds confined to Mauritania), mingle in the Senegal Delta with Eurasian Spoonbills (in fact, NW and SW European breeding birds) in the Senegal Delta.

2. The Great Drought in the 1970s and 1980s caused steep population declines in many Sahelian-dependent (water)birds, independent of human-induced habitat changes in the wintering areas;
3. Breeding populations of several Palearctic migrants were much larger in the mid-20th century, as for Black-tailed Godwit (Chapter 28). However, the reverse may apply to certain species with much smaller breeding populations, thus accounting for changing numbers in Eurasian Spoonbill (Box 26), Eurasian Marsh Harrier (Chapter 25) and Osprey (Chapter 24);
4. Loss of one habitat type (such as natural floodplain and estuary) was countered partly by species switching to newly created habitats (*e.g.* rice cultivation; Tréca 1994 for Ruff);
5. Artificial reflooding partly nullified the negative impact of floodplain loss (see above), while restoration of wetlands in southern Mauritania boosted waterbird numbers (Table 10);
6. The designation of Natural Parks (Djoudj in 1971, Diawling in 1994) created safe breeding and roosting sites, removed livestock grazing and improved protection.

This amalgamation of factors may account for the Senegal Delta continuing to host a wide diversity of birds and in large numbers, despite the substantial human-induced changes it has suffered. For example, Triplet & Yésou (2000) failed to find obvious trends in the presence of ducks between 1972-1996, except for a decline in Knob-billed Duck and an increase in Shoveler. Overall, numbers of Palearctic and Afrotropical ducks showed large annual fluctuations, largely related to feeding conditions relative to rainfall and flooding (in 1972-1976; Roux *et al.* 1977) and to shifts to other Sahelian wetlands (Mauritanian lakes Mal, Aleg and R'kiz). Several duck species have profited from the cultivation of rice in the former floodplain, though mostly in poorly-maintained fields (Tréca 1975, 1977, 1981, 1988, 1992). The combined effect of the changes does not balance out for all species, however. Densities of Wood Sandpiper, for example, average 1.21 birds/ha in restored floodplain habitats but next to zero in rice fields (Fig. 79). If we take the surface areas from Fig. 82, and assume the recorded densities ($\pm$ SE) are representative (which, over the years, is probably not the case), Wood Sandpipers would have numbered 106 000 ($\pm$ 45 000) before effective flood control, but only 13 600 ($\pm$ 7550) from 2000 to 2005, a reduction of 88%. A similar calculation for the Purple Heron yields numbers of 11 400 ($\pm$ 7040) in the 1960s and 1740 ($\pm$ 1190) in the 2000-2005 period. Across the delta, densities are not uniform, being dependent on water depth and vegetation cover (Chapter 6). The high standard errors clearly show the need for more density counts (across more habitat types), a more accurate habitat map and detailed information on water levels to substantiate the calculations.

Conclusions

Estuarine floodplains like the Senegal Delta are scarce, in Africa and elsewhere in the world. In West Africa, the Senegal Delta is unique in its combination of riverine floodplains and estuarine habitats. The flat landscape, rich soils and presence of fresh water led to its conversion into irrigated farmland where natural flooding is kept at bay with dikes and dams. The associated degradation and loss of floodplains (-90% between 1950 and 2005) affected biodiversity, fish production and grazing grounds adversely. Reduced flood dynamics and the prevention of salt water intrusion created niches for invasive plant species that reduced the biodiversity even further. The Great Drought in the 1970s and 1980s eclipsed these large-scale changes. When rainfall increased again in the 1990s, a full, delta-wide recovery of bird populations was thwarted by the extent of cultivation and by embankments constructed in previous decades. However, the overall changes in waterbird numbers since the 1950s have been smaller than anticipated from the large-scale habitat changes since that time. The gradual improvement in numbers, starting in the 1990s after drought- and habitat-induced losses of the 1970s and 1980s, coincided with recoveries both of rainfall and flooding.

The Senegal Delta still is of paramount importance to waterbirds. The key to the rich avian biodiversity of the Senegal Delta is the presence of protected areas (*c.* 10% of the delta is National Park or Nature Reserve). These areas offer safe breeding, feeding and roosting sites for a multitude of waterbirds and other fauna. The protection of wetlands also prevents overexploitation by livestock grazing. Hunting is prohibited in protected areas, but duck hunting at dawn and dusk along the fringes of the National Parks partly circumvents protection and causes disturbance.

New threats to the Senegal Delta are in the making, and existing ones may become larger. Until 2005, the water regime in the Manantali reservoir had been fine-tuned in favour of flood recession agriculture in the river valley. The mounting demand for energy may result in increased water retention in the reservoir, which would limit the ability of the flood control system to provide ideal flood dynamics. A continued spread of Cattail, with its associated problems, is expected. However, the successful restoration of the Diawling shows the potential resilience of Sahelian floodplains, a characteristic that should serve well in future rehabilitation of the lost floodplains of the abandoned rice fields and of the former wetlands at Ndiaël and Trois Marigots.

Endnotes

1   According to Fall *et al.* (2004) the cattails in West African floodplains consist mainly of *Typha domingensis*, and to a lesser extent of the broader-leafed *Typha latifolia*. Others mention *T. australis* as the main species (Pieterse *et al.* 2003). *T. domingensis* and *T. australis* are similar-looking, small-leafed species occupying the same habitats. Durand & Levêque (1983) and Hepper (1968; Flora of West Tropical Africa) use *T. australis*, which is also followed in this book.

2   Crousse *et al.* (1991) mention a potential surface area for irrigation of 375 000 ha, while others use different figures (DeGeorges & Reilly 2006): 255 000 ha by GFCC (1980) and 355 000 ha by Bosshard (1999).

3   Note that this is 200 km^2 larger than the size mentioned by Voisin (1983) and Triplet & Yésou (2000); these differences are due to a varying surface area of dry areas being included.

4   The maximum flooded area was digitized using "quicklooks" of the high flood of 1999, date 26-09-1999 (source: http://glovis.usgs.gov/). The floodplains in the middle valley have a different character, because they have a shorter flood duration and have been cultivated for a long time.

5   In botanical literature, the genus *Scirpus* is named alternatively *Bolboschoenus*.

6   The mixed breeding colonies in the Djoudj in the 1980s and early 1990s were situated in Tamarisk bushes around the larger streams and lakes: Marigot du Djoudj, Grand Mirador (small cormorant sites), Canal de Crocodile and Lac du Khar (Rodwell *et al.* 1996). In recent years (2005-2008), the breeding colonies of herons, egrets and cormorants along Canal du Crocodile are still present, but breeding is not confirmed in the central part of the NP (pers. comm. I. Ndiaye). The breeding site of pelicans is on a small islet at the end of Marigot the Djoudj, a site subject to heavy erosion (Diop & Triplet 2006, pers. com. I.S. Sylla).

7   The breeding numbers of Long-tailed Cormorants on the Banc d'Arguin were: 2460 pairs in 1997, 2883 in 1998, 893 in 1999 and >622 in 2004 (Veen *et al.* 2006).

8   In 2004 the Black-winged Stilt was recorded breeding for the first time in the Mauritanian Senegal Delta (two pairs, pers. comm. Z. E. Ould Sidaty). In July 2006 it was recorded breeding on the Senegalese side of the river, and also in the Diawling. Two adults defended two fledglings feeding in a salty pool at the Djoudj's entrance (obs. I. Ndiaye, I. Diop, J. van der Kamp). Ten days later all birds had disappeared. In early August three nests with freshly laid eggs were found in the freshwater basins of the Diawling (obs. Z. E. Ould Sidaty, J. van der Kamp). Nesting here had apparently been triggered by the water inflow that started on July 1st.

9   The total surface area of floodplain habitats in the delta, largely comprising the restored habitats in the Djoudj NP and Diawling NP, covers 34 000 ha (based on Fig. 76). For calculating densities of waders and herons, dry areas and (deeper) open water (lakes in the Djoudj NP and large basins in the Diawling NP) have to be subtracted. The remaining surface area of vegetated floodplain habitats (brackish marshes) is 21 000 ha. If we add *open* and *mixed* cattail-vegetation (where we recorded numerous birds during our field work, for example in the Ndiaël in December 2005) the total area amounts to 24 000 ha. This area gradually dries out in the dry season to become unsuitable for waterbirds. We estimate that in December-January 50-75% of this area still holds enough water to be suitable as foraging area, some 12 000-18 000 ha. Density counts (Fig. 79) were performed in these habitats.

10 Waders forage in more or less open, shallow water bodies in the delta, while rising marshland habitats and cultivated fields are largely avoided (except for Ruff; Fig. 79). The surface area of these suitable habitats in the delta totals up to 15 000 ha, excluding estuarine habitats. In December-January 50-75% of this area (some 7500-11 000 ha) still functions as foraging area (damp conditions or shallow water), but these habitat rapidly dry out later in the season.

Hadejia-Nguru floodplains

Hadejia-Nguru is famous, not so much among ecologists or conservationists, but among socio-economists. It figured prominently in an extensive study on the hidden, economic values of wetlands. The costly irrigation schemes, made possible by construction of upstream dams, have been computed as economically just feasible: its net value amounted to 20-31 $/ha of irrigated land, or 0.03-0.04 $ per million litres of water. This calculation was inaccurate, however, because it disregarded the reduction of the river flow and the consequent loss of income for fishermen, farmers and other people living downstream. Barbier & Thompson (1998) concluded that "the increased irrigation benefits can only partially replace the lost agricultural, fishing and fuelwood values from reductions in floodplain downstream" and therefore "further expansion of large-scale irrigation should be avoided". However, the political reality is that "the focus on birds that constitutes one element of the HNWCP (Hadejia-Nguru Wetland Conservation Project) has proven an important stick to beat almost any project set up in the region. Typically, Government officials accuse externally funded programmes of being interested only in European birds and not in the welfare of resident populations. Such dogma has no empirical support from the published and grey literature emerging from these projects but as political rhetoric it plays extremely well to a Nigerian audience".[1]

Hydrology

The Hadejia-Nguru floodplains are situated in the Komadugu-Yobe Basin in NE Nigeria, along the fringes of the Hadejia and Jama'are rivers (Fig. 83). The rivers rise on the Jos plateau and the hills around Kano. In 1960-2000, the annual rainfall in the catchment area varied between 250 and 850 mm. The rivers are seasonal, with little flow in the dry season between October and May. Depending on rainfall, the floodwater level rises between 3 and 5 m from early July to mid September, followed by gradual decline from October to February. The maximal extent of flooding in September once was as much as 2500-3000 km² in the wet 1960s and early 1970s, but declined to 700-1000 km² thereafter, only 300 km² being flooded in the dry disaster year of 1984 (Fig. 84). The annual extent of the Hadejia-Nguru floodplains varies in synchrony with the flooding of the Inner Niger Delta. [2]

Upstream of the Hadejia-Nguru floodplains, there are 20 dams, of which in the Upper Hadejia River two that are large. The Tiga dam (1974) was constructed to provide water to the city of Kano and to source the Kano River Irrigation Project (670 km² planned, of which by 2000, 140 km² had been completed). The Challawa Gorge Dam (1992) was built to facilitate irrigation 200 km downstream in the Hadejia Valley (125 km² planned, of which 75 km² had been realised in 2000). Both reservoirs are large compared to the natural annual flow of the Hadejia River, even when the wet 1960s are taken as a reference point (2.7 km³ at Wudil): the storage capacity is 1.99 km³ for the Tiga (but for safety reasons reduced to 1.43 km³ in 1992) and 0.97 km³ for the Challawa. The Tiga dam reduces the peak flow of the Hadejia River by 30-40% (Goes 2002). As a consequence, the Hadejia-Nguru lost 200 to 500 km² of their floodplains (Hollis & Thompson 1993). The relative impact was highest in dry years. In 1984, for instance, the flooded area was only 300 km² when otherwise 600 km² would have been inundated had it not been for the Tiga dam. The water stored in both reservoirs is partly released in the dry period. Under natural conditions, only 2% of the annual runoff would have passed the town Hadejia between November and May, but this increased to 16% once the Tiga Dam was on line, and to 32% after the Challawa Dam had been completed (Goes 2002).

At present, the Hadejia-Nguru floodplains depend largely on the runoff of the near-natural Jama'are River, especially since the hydrological interventions in the Upper Hadejia. The proposed construction of the Kafin Zaki Dam, intended to facilitate a large irrigation scheme near Katagum, will have a considerable impact on flow of the Jama'are River.

The hydrology of the Hadejia-Nguru wetlands has changed markedly in the aftermath of dam building. Water management of the two newly created reservoirs reduced peak flood level, but increased water level in the dry period. Seasonal wetlands have become much smaller, to be replaced by permanent marshes densely vegetated by Cattail, which caused blockage of the Hadejia River in the floodplains, thus further enlarging the extent of stagnant

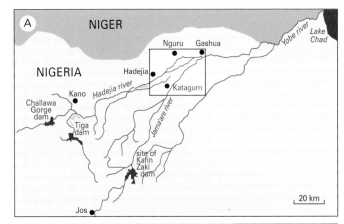

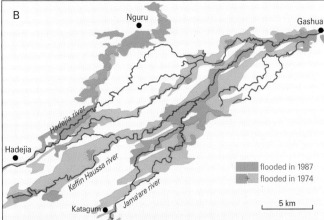

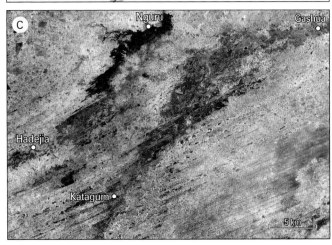

Fig. 83 The Rivers Hadejia, Jama'are and Yobe in NE Nigeria with the Challawa and Kafin Zaki dam is planned (map at top). The flood extent of the Hadejia-Nguru floodplains is shown on map, middle, for years with high (1974) and low river discharges (1987). From: Hollis *et al.* (1993). The Google Earth satellite composite shows the same area as the top map during peak flooding (24 September 2003). Note the parellel (NE-SW) ridges of fossil sand hills between which the rivers meander, and the dry floodplains along the Kafin Hausa, in the past normally water-covered even in relatively dry years.

water, increasing the flow in the direction of the city of Nguru and reinforcing the reduction in flooding in the centre of the Hadejia-Nguru wetlands.

The annual flow of the Yobe River at Gashua, just downstream of the floodplains, represented 24% of the combined inflow of the Hadejia and Nguru rivers into the marshes (Hollis & Thompson 1993).³ Since the early 1990s, the amount of water entering the floodplain through the Hadejia River alone has been insufficient to add to the outflow, which is dependent entirely on the inflow of the Jama'are River (Goes 2002). Without regular flooding, the ground water table is 5-10 m lower, affecting the whole region of the Hadejia-Nguru complex (Hollis et al. 1993). The two dams lower the ground water table in this region by a further metre, another adverse impact of the Tiga and Challawa dams on economic and ecological values downstream (Acharya & Barbier 2000, Barbier 2003).

People and land use

The Hadejia-Nguru floodplains and their immediate surroundings (5100 km²) are densely populated: in 1997, 1.22 million people, which represents a total population density for the wetlands of 239 people/km², and for the rural population, 168 people/km² (Thompson & Polet 2000). By established practices, local people reshaped the landscape, turning the wetlands into a semi-natural habitat. In the 1980s, half the area was cultivated, the upland planted with millet and sorghum (34.8%), the inundated zone with rice (13.8%), and irrigated areas with wheat and vegetables (2.1%). The area of irrigated cropland increased after the dissemination by the State of Kano to local farmers of 70 000 portable irrigation pumps (Hollis & Thompson 1993; Lemly et al. 2000). Rangeland grazed by

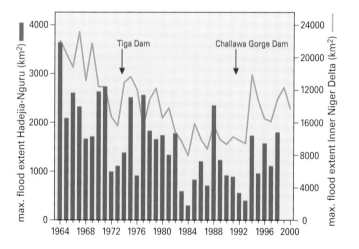

Fig. 84 The maximum flood extent of the Hadejia-Nguru floodplains and the Inner Niger Delta (Fig. 41). From: Barbier & Thompson (1998) and Goes (2002).

cattle, sheep and goats covered 21.8%. Even in the late 1980s, 27.5% of the land comprised more or less natural habitats, i.e. permanent, open water (2.4%) and wooded savanna (25.1%). Red acacia *Acacia seyal* and buckthorn *Ziziphus sp.* are tolerant of some flooding, and hence occur on the higher floodplains; other tree species, such as Doum Palm *Hyphaene thebaica*, Baobab *Andansonia digitata* and Tamarind *Tamarindus indica*, grew in the highlands. Trees in more accessible parts of the wetlands have been removed relentlessly, chiefly as fuel wood for the surrounding cities. In the 1980s after the drought that had killed many trees, people began to fell also in the Baturiya Reserve (340 km²) along the Kafin Hausa River (Adams 1993, Thomas 1996). There are still forest remnants in the central and least accessible parts of the inundated areas.

As in other Sahelian wetlands, fishermen use gill nets, clap nets, seine nets, baskets, fish traps and hook lines. Fish species composition and biomass are dominated by catfish *Clarias*, cichlids *Tilapia*, silversides *Alestes* and Nile Perch *Lates niloticus* (Neiland et al. 2000). Captured fish are eaten fresh locally (constituting only 1% of the fish market (Thomas et al. 1993)), or stored for trade by drying (59%), smoking (34%) or frying (6%). Captures are minimal in July and August at 5 tonnes per day, but increase to 33 tonnes in December-February when the fish concentrate in the last remaining waters. The annual catch in 1989/1990 was estimated at some 6300 tonnes (Thomas et al. 1993). By the late 1980s fishermen were complaining of declining fish stocks (Thomas et al. 1993), most likely due to reduced flooding (Fig. 84). Despite the reduction in flood extent, the numbers of fishermen have increased. As in the Inner Niger Delta, higher fishing intensity produced a reduction of the fish size and a decrease in fish variety.

Birds

Elgood et al. (1966) mentioned the presence of large numbers of Palearctic waterbirds in the Hadejia-Nguru wetlands, but gave no figures. Ash (1990) counted several thousand Squacco Heron at

Tens of thousands of White-faced Whistling Ducks and Garganey may be present in the Hadejia-Nguru wetlands, but only if the flood extent is large. Whistling Ducks, *nomen est omen*, are highly vocal, and their sibilant 3-note whistles lend a special flavour to African wetlands, as does the honking of geese in NW European wetlands. Whistling Ducks and Garganey are particularly vocal during nocturnal flights to and from feeding areas, an impressive auditory avalanche of rushing wings, whistles and strident croakings.

Matara Nuku in November 1987. In winter, Sand Martin is common in flocks of up to 2000 birds (Elgood *et al.* 1994). Since January 1988, systematic midwinter bird counts were performed for eleven years (Table 13). The low numbers from the first few counts were due in part to the counters' lack of knowledge of the area. Common species include Garganey, Northern Pintail, White-faced Whistling Duck, Ruff and Black-tailed Godwit (Table 13). The steep decrease of Ferruginous Duck is remarkable: 2100, 2120, 1830 and 2000 were present in January of 1969-1972 respectively (AfWC database; Wetlands International), there were 1600 in 1988, but only few birds remained in the late 1990s (Table 13).

The number of wintering waterbirds appears to be correlated with the variation in flood extent, as concluded by Polet (2000). Seasonal variations in numbers, if any during the winter period, are not quantified, because censuses are restricted to January. It is therefore not clear whether low numbers during a dry January

Table 13 Counts of waterbirds in the Hadejia-Nguru floodplains in January. From: Polet (2000) for geese and ducks, and AfWC-database of Wetlands International for Glossy Ibis and waders.

	1988	1989	1990	1991	1992	1993	1994	1995	1996	1997	1998
Extent of flood (km2)	700	?	?	860	893	545	387	1728	967	1567	1107
Knob-billed Duck	225	2784	1763	678	1330	948	1163	2196	1287	1601	563
Spur-winged Goose	143	1682	1039	2177	1448	2265	1983	7332	1479	2022	1386
Egyptian Goose	0	3	0	0	0	0	1	0	0	0	175
African Pygmy Goose	11	0	17	35	16	0	3	31	6	27	8
Fulvous Whistling Duck	274	1234	1248	1160	4080	2565	197	3551	965	9510	237
White-faced Whistling Duck	1608	3279	6329	6498	5019	10888	12112	47879	28430	58613	30053
Northern Shoveler	700	324	79	175	5	0	210	5	82	84	57
Northern Pintail	13937	1745	20618	17323	20820	5950	25083	13520	12905	12565	34866
Garganey	9235	8538	22458	11812	12395	36319	13271	98689	74570	145462	147563
Ferruginous Duck	1594	15	280	150	8	0	102	17	0	2	0
Glossy Ibis	186	275	359	?	?	425	640	446	553	2565	567
Black-tailed Godwit	1609	324	530	?	?	150	845	615	2577	6474	357
Ruff	8529	27562	8676	?	?	3976	11405	70940	69317	59925	47618

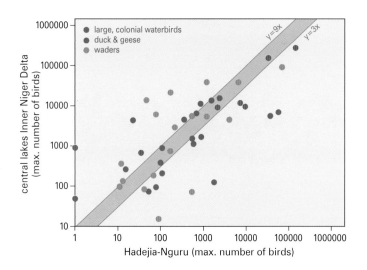

suggest an exodus of birds before then (to areas unknown, but nearby Lake Chad might be a likely candidate; Coulthard 1991), or smaller numbers of birds being present in the first place.

When we compare the figures of the Hadejia-Nguru wetlands with those from the central lakes in the Inner Niger Delta, and correcting for surface area, the Hadejia-Nguru numbers lie in the same order of magnitude (Table 14, Fig. 85)[4], which is remarkable, since the Hadejia-Nguru wetlands are eight times more densely populated than the Inner Niger Delta and are subject accordingly to

Fig. 85 Maximum numbers counted in the Hadejia-Nguru and the central lakes of the Inner Niger Delta; data from Table 14. Note the use of logarithmic scales. Grey field indicates expected relationship: 3 to 9x more birds in the Inner Niger Delta as in the Hadejia-Nguru.

Table 14 Maximal number of birds counted during 11 mid-winter counts (Table 13) and 3 counts performed in July (1995-1997) in the Hadejia-Nguru wetlands (HNW) compared with the maximum counted in the central lakes of the Inner Niger Delta (IND). From: AfWC-database of Wetlands International and Table 5.

Species	HNW	IND
Long-tailed Cormorant	1511	13521
Great White Pelican	23	4300
Grey Heron	678	6218
Black-headed Heron	79	94
Purple Heron	355	4606
Squacco Heron	870	1667
Black Heron	102	390
Cattle Egret	37832	5571
Little Egret	845	11100
Intermediate Egret	549	1503
Great Egret	518	5534
Yellow-billed Stork	110	210
White Stork	1774	130
African Spoonbill	110	900
Eurasian Spoonbill	52	75
Glossy Ibis	2447	15375
Sacred Ibis	572	1160
White-faced Whistling Duck	58613	6643
Fulvous Whistling Duck	9510	9500
African Pygmy-Goose	35	664
Knob-billed Duck	2196	9124
Spur-winged Goose	7332	11481
Egyptian Goose	1	900
Northern Pintail	34866	150000
Garganey	147563	273000
Osprey	12	5

Species	HNW	IND
Eurasian Marsh Harrier	99	458
Black Crowned Crane	1	50
Black-winged Stilt	1210	5536
Pied Avocet	42	86
Great Snipe	13	140
Greater Painted Snipe	11	102
Collared Pratincole	170	20904
Spur-winged Lapwing	551	5829
Black-headed Lapwing	92	16
Kittlitz's Plover	48	13887
Common Ringed Plover	80	6073
Black-tailed Godwit	6473	37654
Eurasian Curlew	12	372
Little Stint	1231	38362
Wood Sandpiper	168	755
Common Sandpiper	533	74
Ruff	70845	90470
Spotted Redshank	4065	4557
Common Greenshank	214	3031
Marsh Sandpiper	70	185
Grey-headed Gull	269	242
Gull-billed Tern	174	3896
Caspian Tern	8	3545
Little Tern	430	346
Whiskered Tern	988	6679
White-winged Tern	134	5874

Yellow-billed Storks are distributed across sub-Saharan Africa, but have disappeared from several West African wetlands. The species is confined to shallow water where they feed mainly on fish.

more intense human exploitation. Such exploitation of waterbirds in the Hadejia-Nguru explains a decline of the larger species (storks, pelicans and cranes; (Akinsola *et al.* 2000). Smaller species, such as Ruff, are captured with hook lines and snares (Ezealor & Giles 1997). The Black Crowned Crane, the national bird of Nigeria and once common in the north, "formerly ... in the valleys of great rivers ... with flocks up to 200" according to Elgood *et al.* 1994, has now disappeared from the Hadejia-Nguru – 14 counts produced but a single observation.

A count in 1991 found 7500 remaining breeding pairs of larger bird species (5575 Cattle Egret, 661 Long-tailed Cormorant, 493 Intermediate Egret, 291 Squacco Heron, 332 Little Egret and 137 Great Egret; Garba Boyi & Polet 1996).

Conclusion

Since 1964, the extent of flooding of the Hadejia-Nguru wetlands has varied between 300 and 3600 km^2, mainly due to variable rainfall. Dams and irrigation schemes upstream reduced the flooded area by 300-500 km^2, but at the same time covered the lower floodplains with permanent water. The wetlands are now heavily exploited by local people, who have converted the floodplains into a semi-natural, mostly treeless, habitat. Nevertheless the floodplains still harbour large numbers of waterbirds, at least in wet years.

Endnotes

1. R. Blench: http://www.old.uni-bayreuth.de/afrikanistik/mega-tchad-alt/Bulletin/bulletin2003/programmes/wetland.html
2. The flood extent of the Hadejia-Nguru is 0.112x the flood extent of the Inner Niger Delta (R^2=0.458). Without dams, the correlation between the flood extents improves (R^2=0.522) and the ratio increases: 0.128x.
3. Hollis & Thompson (1993) derived a water balance for the Hadejia-Nguru in the period 1964-1987. The average annual inflow was 4.39 km^3, the rain input 0.57 km^3, the evaporation 2.83 km^3, the ground water recharge 1.46 km^3 and the outflow 1.06 km^3. The groundwater storage decreased in this period from 10 to 5 km^3, because the annual water output (5.35 km^3) was larger than the input (4.96 km^3) during these 23 years.
4. The dam-affected flood extent of the Inner Niger Delta and the Hadejia-Nguru differ by a factor of 8.9 (see endnote 2 above), but since 30-100% of the bird numbers in the Inner Niger Delta concentrate in the central lakes during receding water, we expect the bird numbers in the central lakes of the Inner Niger Delta to be about 3 to 9 times as high as in the Hadejia-Nguru. On average, the numbers in the Inner Niger Delta are 4.93 times as high when all species are combined. For waders the average is 6.76, for ducks and geese 5.55 and for herons and other large colonial waterbirds 3.85 (Fig. 85). In other words, the Hadejia-Nguru wetlands harbour relatively many waders (*i.e.* Ruff), duck (mainly Garganey and White-faced Whistling Duck), and relatively few colonial waterbirds (except for plentiful Cattle Egret).

Lake Chad Basin

With contributions of Paul Scholte

A shrinking African lake, which still covered 26 000 km² in the 1960s (almost the size of Belgium), and now threatens to disappear in the near future, appeals to the imagination. The international attention mounts whenever the level of Lake Chad dips, but no such outcry is heard when the water level rises again in subsequent years. Consequently, the general public is convinced that Lake Chad is shrinking in size – in fact dying – and for one obvious reason: climate change. The fact is that Lake Chad has always shown large variations in size and the clear decrease in the 1970s and 1980s was partly due to irrigation along the rivers responsible for filling it. For the local people living along the border of Lake Chad the seasonal and annual variation in water level is part of life, but this elusive phenomenon appears to be difficult to accept by national and international decision makers. Lake Chad must be tamed, with predictable variations in water level, enabling large irrigation schemes along the borders of the lake. They should know better. Long-existing irrigation projects in Nigeria and Chad were destroyed by high water in the wet 1950s, while large irrigation schemes built in the wet 1960s could not be used in the many dry years since then. The ambitious South Chad Irrigation Project (with a planned 670 km² of land irrigated) used a supply canal originally extending 24 km into the lake, but during the dry mid-1980s the water's edge was up to 70 km away from the inlet. Local farmers and fishermen followed suit in the 1980s, as they, and their ancestors, had done during previous droughts. But times are changing.

Lake Chad and its surroundings have been divided over four states in the 20th century. After Nigerian fishermen founded a village in the dry years on the other side of the border (in Cameroon), military skirmishes were the result and, finally, they had to withdraw after a judgement from the International Court of Justice in The Hague. Little has been learnt: there is even a grotesque plan to change Lake Chad into a lake with a more or less constant water level by diverting the Ubangi River from the Central African Republic to Lake Chad.

Introduction

Lake Chad Basin in North-Central Africa can be considered as a cuvette between the Niger and Nile valleys (Fig. 20), on the bottom of which lies Lake Chad. About 7000 years ago, when the Sahara was still grassland (the 'Mega-Chad' period), its shoreline was several hundred km further inland (Leblanc *et al.* 2006, Kröpelin *et al.* 2008). For thousands of years the basin has been predominantly desert, transitioning into Sahelian grasslands and wooded savanna in the south. These arid lands sharply contrast to the basin's oases of life: the wetlands. In addition to Lake Chad, Lake Fitri and Lake Iro also contain water throughout the year. The basin holds extensive floodplains: the Hadeija-Nguru floodplains in northern Nigeria (300 km west of Lake Chad; see chapter 8), the extensive floodplains along the Chari and Logone Rivers in Cameroon, Chad and the Central African Republic, and the borderlines of the three lakes cited above. These floodplains generally dry up in December – January, offering the perfect wintering site for migratory birds. In addition, there are many small and isolated ephemeral wetlands formed by rainfall and local run-off, especially in the transition zone of Sahel and Sudan savanna. For a few weeks or months, especially after the rainy season from August till October, they contain water and vegetation, allowing both livestock and wildlife good foraging and water supply when the other wetlands are inundated. The various wetlands complement each other in the timing of flooding and drying up, the key to survival of the basin's inhabitants, including birds.

Lake Chad

Hydrology Lake Chad is a closed lake, with a single surface outlet under exceptionally wet circumstances into the northeast, the lowest part of the basin, the old 'Mega Chad'. By the time of the first rains in June, the water level in Lake Chad starts to rise 5-6 cm per month, owing to variation in local rainfall between 9.4 and 56.5 cm per year (Vuillaume 1981) and an increasing inflow of the Chari and Logone Rivers (Fig. 86). In a dry year, with low riverine input, the water level in the lake already reaches its peak in November, but, if the inflow is large, the water continues to rise until January or even February. When, after the rainy season, the river discharges largely cease, the level of the lake drops gradually by up to 6-7 cm per month in April-May. The water level varies seasonally by around 90 cm, in some years by even more than 150 cm.

In a wet year the annual inflow exceeds the annual evaporation,

The Logone floodplains are covered by some 50 cm of water during flooding. After flood recession, grazing turns the green marshlands into a scorched, almost bare steppe within a couple of weeks. It takes another flood before the cycle is to be repeated.

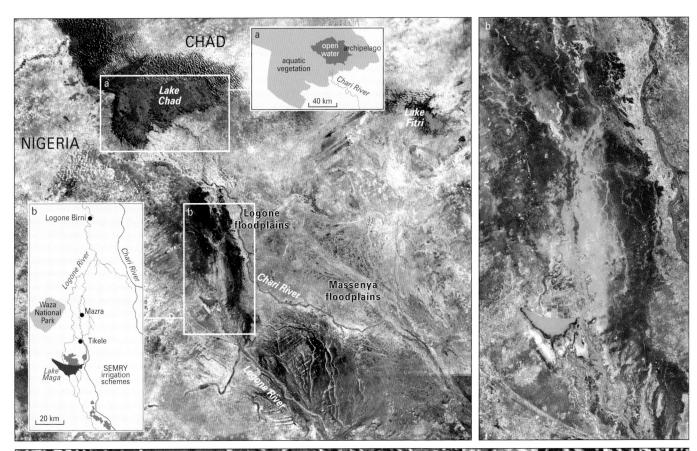

allowing the lake to expand in size. In contrast, during a drought period, evaporation surpasses the inflow, causing a gradual decline of the water level. The water level in Lake Chad, measured since 1908 in Bol (Fig. 87), was high in the relatively wet 1950s, but dropped five metres between 1962 and 1987, only to increase slightly again in the 1990s.

Before 1973, the maximum surface area of Lake Chad varied annually between 15 000 and 25 000 km² but has dropped below 10 000 km² since then. Each year the lake was about 2000 km² smaller in May than in the few preceding months, but this amounted to about 3500 km² since 1973. Thus, a decline of the total surface area of Lake Chad went along with a substantial increase of the seasonal floodplains along the borders of the Lake (an explanation for which is given below). The changes are even larger than these figures suggest. At high water levels Lake Chad is without islands, but thousands of sand dunes emerge along the eastern side of the lake when the water level drops, forming a large archipelago (Fig. 86), and also named as such.

Lake Chad measures 24 000 km² when the water level at Bol reaches 478 cm, but it covers only 9000 km² at 50 cm. Using these correlates, the measurements in Bol could be used to estimate the lake size at various water levels were it not for one complication. The northern and southern half of the lake get separated when the water level drops below 100 cm, as was the case in 1973. Also, in the following years, the floods did not pass the "Great Barrier" between Baga Sola and Baga Kawa (Fig. 88), and the northern basin dried out. [1] At the same time, the average seasonal fluctuation in water level increased from 90 cm to 150 cm in the southern basin (Fig. 87A), expanding the surface area of the seasonal floodplains from 2000 km² to a maximum of 7000 km² (Fig. 89). The minimum lake size between 1977 and 2006 varied between 6000 and 7000 km², and the maximum lake size between 7000 and 15 000 km² (Fig. 87B). [2] During the relatively high floods of 1994 and 1999, the water level again surpassed the barrier between the northern and

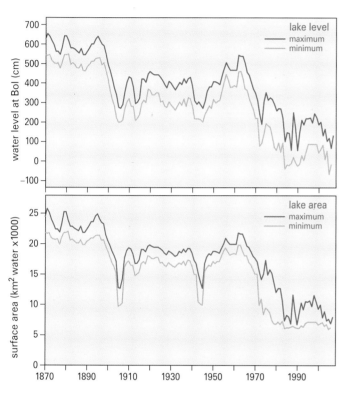

Fig. 87 (A) The annual variation in the highest and lowest water level of Lake Chad between 1870 and 2006. Since the level of Lake Chad and the flow of the Nile river are correlated, the flow data of the Nile (available since 1870) were used to estimate the water level in Lake Chad before 1908 when the actual measurements started at the hydrological station of Bol; the data for 1980-1983 and 1987 were also estimated. The hydrological data from Olivry et al. (1996) (1870-1994) and Coe & Birkett (2004) (1992-2000) have been combined with satellite radar altimetry of Lake Chad since 1992 (http://www.pecad.fas.usda.gov/cropexplorer/global_reservoir; Crétaux & Birkett 2006). (B) The surface area of Lake Chad, as derived from functions describing the relationship between lake level and lake size (Vuillaume (1981).

Fig. 86 (previous page) Lake Chad, Lake Fitri and the Logone floodplains. The Logone floodplains and the southern part of Lake Chad are shown enlarged (see inset maps); Google Earth satellite composites 27 December 2004.

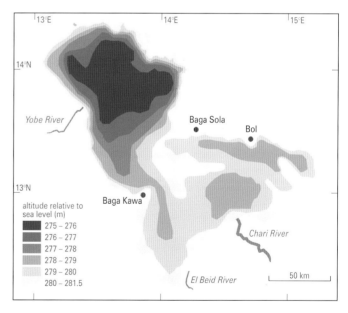

Fig. 88 The altitudes of Lake Chad relative to mean sea level; 0 cm at the gauge of Bol corresponds with 277.87 m relative to sea level. Note the relative high area south of Baga Sola, which functions as a barrier between the southern and the northern basin when the level of the lake is low. From: Carmouze & Lemoalle (1983).

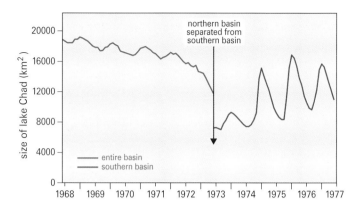

Fig. 89 Monthly variation in the size of Lake Chad between May 1968 and May 1977. In May 1973 the water level dropped to such an extent that the northern and southern basin became separated. Within two years the northern basin dried out (water surface not shown), but the seasonal variation in water level in the southern basin became more punctuated. From: Olivry et al. (1996).

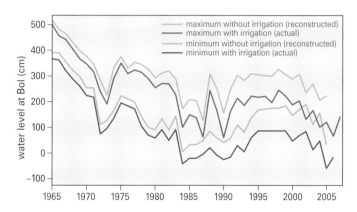

When the northern basin gradually dried up after 1973, one of the first plants to colonise the bare ground was Ambatch *Aeschynomene elaphroxylon*, a soft-stemmed scrub with cork-like wood. Local fishermen in other floodplains (such as these children in the Inner Niger Delta) collect *Aeschynomene nilotica* stems to be used as floats for their nets.

Fig. 90 Annual maximum and minimum water level in Lake Chad, as observed, compared with simulated lake levels without irrigation along the Chari and Logone Rivers. The impact of irrigation on the river flow is shown in Fig. 21.

southern basin, but not by enough to fill the northern basin (Lemoalle 2005).

Impact of irrigation The decline of the water level since 1962 can only partly be attributed to the reduced rainfall in the catchment areas of the Chari and the Logone Rivers. Irrigation schemes have been taking water from the rivers since the early 1960s and the amount has increased since 1992 (Fig. 21). The impact of this reduced inflow has been modelled, based on monthly measurements of river inflow, local rainfall, evaporation and seepage (Vuillaume 1981; summarised in Olivry et al. 1996)[3].

We used the close relationship between lake level and inflow to quantify the impact on lake size of irrigation along the Chari and Logone Rivers (Fig. 90). Irrigation had only a small effect on Lake Chad before 1985, but this has increased since then until 1992. This is partly due to the increased water intake (see Fig. 21), but also to evaporation exceeding the reduced river inflow. Irrigation schemes have lowered the water level by about one metre (Fig. 90). Vuillaume (1981) used his hydrological model to simulate a reduced river flow caused by irrigation between 1954 and 1977, when the amount of irrigation was still rather minor. Nevertheless, he concluded that in 1977 the lake level would have been 37 cm higher were it not for the irrigation. Coe & Foley (2001), simulating the level of Lake Chad from 1954-95, found that about half of the observed shrinkage of Lake Chad since 1975 was due to human water use. Without irrigation, Lake Chad would have measured 12 000 to 16 000 km^2 in the 1990s and 2000s, instead of the observed 7000-11 000 km^2 (Fig. 87).

Vegetation Until 1973, Lake Chad was a large shallow lake with up to 20 000 km^2 of open water. About 2400 km^2, or 12% of the lake surface, was covered by aquatic vegetation before that year (Iltis & Lemoalle 1983). Where the Chari River flows into the lake, extensive aquatic meadows of *Vossia cuspidata* (about 200 km^2) occurred, a plant species locally known in West Africa as *didere* and confined to water columns of one to three metres deep. In more shallow waters Papyrus *Cyperus papyrus* and Reed *Phragmites australis* were found, dominating the eastern border of the lake, abounding in the region of the Great Barrier and forming floating islands. As salinity in Lake Chad is low (1‰) but increases further north, Papyrus is replaced by Cattail *Typha* sp. in the northern part of the lake.

The northern basin gradually dried up after 1973. Open water turned into dryland, with temporary pools during the rainy season. One of the first plants to colonise the bare ground was Ambatch *Aeschynomene elaphroxylon,* a soft-stemmed scrub with cork-like wood . The Mesquite *Prosopis juliflora*, an invasive tree species from Central America, has increased rapidly. During the 1970s and 1980s it was widely planted in the struggle against desertification. It fixes mobile dunes and colonises dry, sandy grounds, but the species has started to invade more fertile areas. A forest of 10 ha planted in 1977, just north of Lake Chad, has spread across the northern basin and was already covering 3000 km^2 by the early 2000s (Geesing et al. 2004).

Since 1973 the southern basin has changed into a relatively small area of open water (c. 1500 km^2), fringed with large seasonal floodplains (3500-4000 km^2). Dense Ambatch vegetation developed on the drier sediments, while the increased seasonal fluctuation in water level from 50 cm to 150 cm in the floodplains favoured *didere* at the expense of Reed and Cattail. Similarly, massive floating mats of Water Lettuce *Pistia stratiotes* and Water Lily *Nympheaea lotus* came into existence.

People The human population in the Chad Basin has been estimated at about 10 million in 1999, of which most live in the southwest (Fig. 35B). The population density around the Lake itself varies from less than 5 inhabitants per km^2 in the north to 25-50 inhabitants per km^2 in the south.

There are at least 20 000 professional fishermen in the lake basin; fishing is a part-time activity for at least another 300 000 people. Before 1960, fishermen used traditional fishing techniques which could only be applied in rivers, creeks and shallow waters. On the lake proper hardly any fishing occurred, but that changed after 1963 with the introduction of nylon nets (Carmouze et al. 1983). The annual fish capture, estimated at 65 000 tonnes in 1969, increased to 227 000 tonnes in 1974 (Durand 1983). Taking into account the concomitant decline in lake size from 19 000 to 9300 km^2, the increase in yield was even larger, from 34 to 245 kg/ha. The high captures in the mid-1970s were due to the gradual desiccation of the northern basin, which enabled the fishermen to capture entrapped fish in the remaining waters. Durand (1983) concluded that by 1977, with the exception of the temporary peak in the mid-1970s, the yield of fish in Lake Chad stabilized at about 100-120 kg/ha. Consequently, the size of the lake must have repercussions for the total capture of fish. The data given by Neiland et al. (2004) for 1978-2001 do indeed show that the annual total capture varied between 101 000 tonnes in 1978 and 22 000 tonnes in 1982, in concordance with the size of the lake. The yield between 1978 and 2001 averaged 86 kg/ha, i.e. somewhat smaller than predicted by Durand (1983).

Farmers in the Sahel cultivate their land during the short rainy season, but in floodplains and along the borders of Lake Chad, they also use exposed moist soils to grow sorghum and, to a lesser degree, millet, and also maize and beans in the south. Since time immemorial, farmers have constructed temporary, small dams to regulate the water level in the valleys between the islands. Satellite images show that by the early 2000s large-scale 'polders' had been made with concrete dams.

Groups of nomadic pastoralists from a variety of ethnic backgrounds (Arabs, Kredda, Borno) follow the receding lake and, during the dry season, let hundreds of thousands of their cattle and sheep graze on the vegetation of the exposed lake bed. A rapidly declining number of the endemic Kouri cattle, descendants of *Bos taurus* (the common humpless cattle) with huge 'floating' horns, are permanently present in the flooded pastures. No livestock counts have been conducted over the entire area, but we may assume that

African Elephants still occur in the floodplains of the Waza-Logone, where *Acacia seyal* flood forests are important food sources (as are *Piliostigma reticulatum*, *Combretum* sp. and *Balanites aegyptiaca*). The number of trees killed by them doubled in the 1990s and early 2000s, but the main effect of Elephants in Waza National Park is on *Acacia* dynamics and structure (Tchamba 2008).

The Waza-Logone National Park attracts more tourists than any other National Park in Central Africa. In the dry season, wildlife, birds and tourists concentrate near remaining pools.

livestock numbers have increased since the last Great Drought, not only because of the overall trend in the Sahel (Fig. 33), but also because new grazing grounds came into existence in the expanding seasonal floodplains.

Other wetlands

Lake Fitri is often referred to as a 'mini Lake Chad' (Fig. 86). Depending on the inflow of the Batha River, its maximum size, generally reached in November, varies between 420 and 1300 km² (Keith & Plowes 1997); it was completely dry in 1973 and 1984. The Batha flows for only 3-4 months of the year, with a seasonal variation in water level of about 150 cm (http://www.pecad.fas.usda.gov/cropexplorer/global_reservoir; Hughes & Hughes 1992). The flooded fringes are covered by *Echinochloa stagnina* (known in West and Central Africa as *bourgou*), *didere* and *Ambatch*. The open water is partly covered by floating vegetations of mainly Water Lily and Pondweed *Potamogeton* sp. The lake is intensively used by the local people. The annual catch of fish has been estimated at some 3000 tonnes (Hughes & Hughes 1992). Farmers cultivate their land when waters recede and the floodplains are grazed by cattle belonging to nomads.

Little is known about the floodplains along the Chari, estimated by Lemoalle (2005) to reach 7000 km² for the area around Massenya and even 50 000 km² for the Salamat, the main tributary of the Chari. The Salamat River flows across the Zakouma National Park (3000 km²), one of the last strongholds for wildlife in West-Central Africa (Chapter 5), and Lake Iro, a little-known lake, slightly smaller than Lake Fitri.

In contrast, there are many studies on the Yaérés, the floodplain areas south of Lake Chad, covering some 8000 km² (Fig. 86). Most studies have focussed on the southwestern part, the Logone floodplains in North Cameroon, and especially on the impact of the Maga dam (Fig. 86). The Maga dam, built in 1979, created a reservoir, Lake Maga (with a volume varying between 0.28 and 0.68 km³ and surface area between 100 and 300 km²), allowing gravity irrigation, and thus reducing costs of rice cultivation. Since embankments were built along the Logone River to prevent flooding, the area downstream from the irrigated rice fields has no longer been immersed and the ground water table has gone down by 3-10 m (Ngatcha et al. 2005). This is also visible on satellite images (Fig. 86): the area between the rice fields and Tikele are yellow instead of green. The effect of the Maga dam was notable as far as the Waza National Park (1700 km²), 40-50 km downstream, where the perennial grasses on the former floodplains (such as wild rice) were replaced by less productive annuals (Scholte 2007) and where the population of Kob antelopes dropped from 20 000 to 2000 (Scholte et al. 2007).

The negative impact of the Maga dam on the lower floodplains was particularly serious because the inflowing water that could not be stored in the reservoir was drained directly into the Logone River rather than into the floodplains. The surrounding desiccated area could be reflooded, without doing much damage to the rice culture, by making an opening in the dike downstream of the rice complex. This occurred in 1994, after many studies and consultation rounds, allowing a flow of 20 m³/s. A second opening was made in 1997 (7-10 m³/s). The reflooded area averaged 200 km² (Loth 2004). That a relatively small release could have such a large effect hinges on the fact that the Logone plains are submerged by only 50-100 cm. The floodplain rehabilitation was, however, not the success it initially seemed to be. First, the Maga dam was constructed in 1979, a relatively dry year, while the opening in the dike took place in 1994, a wet year; the large changes in

The drop in antelope numbers in northern Cameroon, from 20 000 in the 1960s to 2000 in the 1980s (slightly increasing to 5000 in the mid-1990s), is better explained by antelope-livestock contacts (provoking the transmission of diseases such as rinderpest during droughts) than by poaching, which has remained rather constant. The increasing livestock numbers pose a serious threat to antelope populations (Scholte et al. 2007a).

flooding were therefore partly due to variations in rainfall. Moreover, the reflooded areas attracted so many people from elsewhere that the restoration of the protected area was jeopardized. For instance, the number of cattle tripled between 1993 and 1999, severely limiting the availability of foraging for the Kob and other wild herbivores, which did not therefore increase in numbers (Scholte 2003).

Birds

In the 1970s Lake Chad changed from a large waterbody locally fringed with marshland into scrub- and woodland in the northern section and a relatively small lake surrounded by extensive seasonal floodplains in the southern part. The repercussions for birds must have been considerable.

When the lake was still large, waterbird counts by Vielliard (1972) showed that Northern Shoveler was the commonest Palearctic duck species (8000 in 1969 and 11 000 in 1971). Thousands of Garganey and Pintails, together with African Pygmy Geese, were concentrated on the flooded grasslands, where also several 10 000s of White-faced Whistling Duck were present. These counts must have underestimated real numbers. Aerial counts of large waterbirds in Lake Chad

The Waza-Logone floodplains during flood recession. This stage of the flood cycle offers excellent feeding opportunities for birds and other wildlife, but often for a short time span only, especially during droughts. January 1999.

The number of Black Crowned Cranes in the Lake Chad Basin was estimated at 5000 birds in the early 1990s (Scholte 1996), compared with less than 50 in the Inner Niger Delta, 150-380 in the Senegal Delta and 0 in the Hadejia-Nguru (Chapter 6-8). Human activities, via poaching and disturbance of breeding grounds, are the main cause of decline throughout West Africa.

Based on fieldwork in 1979, Jarry et al. (1986) characterised the avifauna of Lake Chad in five habitat types. African passerines and possibly rails abounded in dense vegetation of Papyrus, Cattail and Reed, but Palearctic bird species (except for a single Savi's Warbler or Bluethroat) were absent. Few birds used the farmland along the border of the lake or in the archipelago. The open water of the Lake was only visited by some Grey-headed Gulls, pelicans and terns. Most bird species were present, and in large numbers, in the submerged meadows of *didere* and Water Lily and on emerging mud and sand flats (Dejoux 1983). This suggests that wintering conditions for most Eurasian bird species may have improved after Lake Chad changed in 1973 from a large stagnant lake with predominantly open water into a seasonal wetland. In contrast, wintering conditions for most Eurasian birds in the 'complementary' floodplains elsewhere in the Lake Chad Basin (Logone, Lake Fitri) have most likely deteriorated.

Vielliard (1972) had already noted that the distribution of waterbirds varied seasonally, with a shift from the Logone floodplains to

in January and December 1999 and in December 2003 revealed 95 000 (on a single day), 388 000 (over 6 days) and 537 000 waterbirds (over 2 days) respectively. The available counts are difficult to compare for methodological reasons, but it is remarkable that only a few hundred Shovelers were counted in 1999-2001, compared with 200 000 Garganey and 38 000 Pintails (data from the AfWC database; Wetlands International). When compared to Vielliard's survey in 1969-1971, the large increase of Garganey and Pintails must have been due to the expansion of floodplain size since then, as this is their preferred habitat.

Another survey method was employed by Van Wetten & Spierenburg (1998), who counted 9046 waterbirds (49 species) in three plots in floating meadows along the southern fringe of Lake Chad in January 1993. Their plots totalled 17 km² with an average density of 5.3 bird per ha. Part of the plots must already have been exposed, as evidenced by the presence of relatively large numbers of Cattle Egrets, Spur-winged Lapwings and Collared Pratincoles, all species which avoid areas covered by water. This is possibly also the explanation for the relatively low density of waterbirds. In the Inner Niger Delta, where waterbird density (excluding passerines) in floodplains amounts to 10-16 birds per ha, density drops to 1-3 birds per ha when the floodplains have dried up (Table 7; Fig. 62). Taking counting differences into consideration, i.e. large plots in Lake Chad (where small and secretive species may be partly overlooked) and small plots in the Inner Niger Delta (where counts are more precise but flock-feeding birds may be underestimated), the species composition of Palearctic species on the floodplains in Chad and Mali match rather well, with many Glossy Ibis, Black-winged Stilt, Black-tailed Godwit, Ruff and Little Stint in both areas.

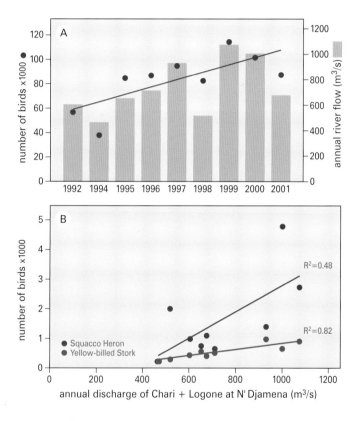

Fig. 91 (A) Number of waterbirds in January-February counted during ground surveys in the Logone floodplains (Cameroon; western bank) compared to the annual discharge at N'Djamena (Logone and Chari Rivers combined) in the preceding year. The correlation is high: R = +0.84. (B) Relationship between counted numbers of Squacco Heron and Yellow-billed Stork and the river flow. From: Scholte (1996); AfWC database of Wetlands International.

Collared Pratincoles winter in large numbers near floodplains in the Sahel. One typically encounters large groups keeping a low-profile on dry ground (only noticed when disturbed), or insect-hawking groups at high altitudes. The latter are often so high in the air that they resemble mosquito swarms, and it is the contact calls that make the observer aware of the hundreds of birds.

Lake Chad during receding water, between November and March. Similarly, Van Wetten & Spierenburg (1998), responsible for the first systematic waterbird count of the Logone floodplains in January 1993, were aware of the impact of receding water on birds using the floodplains. They recorded, for instance, 12 000 Black-tailed Godwits on a single roost in November, where some months later no more than 371 birds were counted in the entire region.

Aerial counts of Lake Chad and the Logone floodplains confirm that waterbirds leave the Logone floodplains in dry years, and move to Lake Chad. The counts in 1984, 1986 and 1987 coincided with dry years (annual discharge of the Chari 484-545 m^3/s), and those in 2000 and 2004 with wet years (discharge 998 and 891 m^3/s). In the three dry years 312 000 – 553 000 Garganey were counted in Lake Chad and 200 – 6800 on the Logone floodplains. In contrast, in the two wet years, Lake Chad held 400 000 Garganey, and the Logone 200 000. The story is similar for Pintail (<1% on the Logone in the dry years and 41-47% in both wet years) and other waterbirds.

An overview of waterbird counts in the Cameroon part of the Logone floodplains in January-February between 1992 and 2000 indicate an increase in the numbers of Afrotropical waterbirds, but strongly fluctuating numbers of Palearctic waterbirds without any clear trend (Scholte 2006). The increase of Afrotropical birds may be partly attributed to the floodplain rehabilitation near the Maga dam, resulting in 200 km^2 of newly flooded area and another 600 km^2 where flood levels have been raised (Scholte 2006). This impact is, however, difficult to separate from the concomitant increase in rainfall and river discharge. Birds are always present in larger numbers at a higher river flow, as shown for two species in Fig 91. There are two exceptions: Collared Pratincole and White Stork. These species prefer dry floodplains and avoid the flooded parts. The twice higher number of waterbirds in wet years can be explained by the generally better feeding conditions and reduced mortality. For the local breeding birds, reproduction is probably better. Another pivotal factor is the distribution of birds: many stay in the Logone floodplains during high floods but disperse to Lake Chad when the areas dry up.

Two ground surveys of waterbirds in the Logone floodplains, including the Chadian part on the eastern bank of the Logone River (Fig. 86), carried out in February 2000 and 2001, clearly showed that large numbers may concentrate here (Table 15). A comparison with the Inner Niger Delta (Table 5) reveals that numbers in the Inner Niger Delta are three times higher than in the Logone valley. This was not unexpected, given the three- to fourfold difference in size. Some species, predominantly resident Afrotropical species, are remarkably numerous in comparison with the Inner Niger Delta and the Senegal Delta:

- 1400-2200 Black Crowned Cranes were counted in the Logone floodplains. Nowhere further west in the Sahel can such large numbers of Cranes be seen. Further east in the Lake Chad Basin, however, similar numbers are found (Lake Fitri and Zakouma National Park; Scholte 1996). This also applies to the Sudd floodplain (Chapter 10).

Table 15 Two ground surveys of waterbirds in the Logone floodplains (east and west bank combined) and six bird counts of Lake Fitri from the air. From: Ganzevles & Bredenbeek 2005; Beeren et al. 2001 (Logone), AfWC database; Wetlands International (Fitri).

Site	Logone; ground		Lake Fitri; aerial count					
Date	Feb 00	Feb 01	Feb 84	Feb 86	Feb 87	Feb 99	Dec 99	Dec 03
Long-tailed Cormorant	6304	2898				2	3013	13
Great White Pelican	105	151		21	202	450	80	33
Pinked-backed Pelican	24	201	9	4	203		63	
Grey Heron	790	1096	17	50	86	60	1085	184
Black-headed Heron	3956	12323			1		15	141
Purple Heron	161	125	4	2	19	11	507	21
Black-crowned Night Heron	926	1896						
Squacco Heron	7338	23211	85		400	3060	3546	373
Black Heron	7264	858			20	105		35
Cattle Egret	20895	24004		500		1130	9416	3378
Little Egret	8643	2920						203
Intermediate Egret	140	470						
Great Egret	808	689				150	3626	1542
Yellow-billed Stork	1077	1055			145	10	30	42
African Openbill Stork	1466	1705					100	17
White Stork	535	4242				1615	2331	61
Woolly-necked Stork	489	21						
Saddle-billed Stork	30	8	5		2			
Marabou Stork	1784	1860	17			20		6
Glossy Ibis	1683	1382	21	93	243	113	4154	628
Hadada Ibis	174	62						
Sacred Ibis	3126	3616	5	1056	148	138	1796	2270
African Spoonbill	43	304		1	500	900	696	1072
White-faced Whistling Duck	41942	6275	16000	4400	24800	10740	95238	43945
Fulvous Whistling Duck	132	2	6600	350	11000	340	5469	1140

- Very large numbers of Black-headed Heron, Black Heron, Sacred Ibis and White-faced Whistling Duck were observed, and also large numbers of African Openbill and Marabou Stork.
- Squacco Herons are difficult to count in dense vegetation. Density counts in the Inner Niger Delta suggest that numbers actually present may be 25 times higher than observed during a regular waterbird count (Table 5 and 9). However, counted numbers in the Logone may well almost approach actual numbers, as the floodplains are mainly dry in January-February, with the birds concentrating in the remaining small depressions, creeks and pools.
- Large numbers of Collared Pratincoles (up to 11 353, Table 15) were counted in the Logone floodplains, highlighting the importance of dry floodplains in January-February.

- Many species, mainly Palearctic migrants, seem to be relatively rare in the Logone floodplains during January-February: large numbers of Purple Herons were expected but the Logone surveys revealed only 100-150 birds, as opposed to 4600 in the Inner Delta. However, transects in the Logone floodplain conducted earlier in the year (October – November) in 1994, suggest the presence of larger numbers ('a thousand').
- Not many Garganey and Pintails were recorded, but we know from aerial counts that one million ducks can be concentrated on day-time roosts at Lake Chad (Chapter 22 and 23).
- Apart from Ruff, Black-winged Stilt and Collared Pratincole, wader numbers were generally low, as in the Hadejia-Nguru (Chapter 8). This contrasts with the Inner Niger Delta, where the wetter conditions attract larger numbers of waders (Chapter 6).

Site	Logone; ground		Lake Fitri; aerial count					
Date	Feb 00	Feb 01	Feb 84	Feb 86	Feb 87	Feb 99	Dec 99	Dec 03
African Pygmy Goose	175	114	2		28	7		6
Knob-billed Duck	5939	1899	3350	4300	5200	2970	8295	2044
Spur-winged Goose	2407	1461	1450	116	1400	2195	1849	1295
Egyptian Goose	0	39	5	101	200	13	172	380
Northern Shoveler	452	1777	80	35	300	100	658	3165
Northern Pintail	6213	271	32000	56000	35300	3380	36865	19760
Garganey	3314	23869	17000	83000	21600	4040	97332	17100
Ferruginous Duck	0	0	40	300	500	1	3800	8450
Eurasian Marsh Harrier	652	934		9	19	22	21	
African Jacana	2560	3876						
Black Crowned Crane	1478	2313	300	20	51	134	441	164
Black-winged Stilt	3489	9147	11	76	48	48	1471	2770
Egyptian Plover	378	246						
Collared Pratincole	3693	11535						
Spur-winged Lapwing	2444	2753				35	557	45
Black-tailed Godwit	952	2318	4		200	220	950	2990
Little Stint	1353	2212						
Wood Sandpiper	1212	2926						
Ruff	72544	146343	550	500	1200	1000	10503	4195
Spotted Redshank	155	882						
Common Greenshank	649	332						
Grey-headed Gull	906	56						
Gull-billed Tern	131	6						
Whiskered Tern	956	1204						
White-winged Tern	1391	88						

- Few terns were observed at any time of the year, possibly because the height of the flooding rarely exceeds 1 m of water (Scholte 2007).

Apart from waterbirds, the Logone floodplain and its surroundings are also noted for their high numbers of Palearctic raptors, especially Eurasian Marsh and Montagu's Harriers (Thiollay 1978). Palearctic raptors, with the exception of Common Kestrel, have remained relatively stable in numbers since the 1970s, compared to a dramatic decline of resident vultures and the larger eagles (Thiollay 2006a).

The six counts of Lake Fitri (Fig. 86) were aerial, so smaller and less conspicuous bird species have been overlooked or underestimated. Lake Fitri harbours many waterbirds, especially Afrotropical ducks and storks. Up to 8450 Ferruginous Duck were observed in the last two counts. This species was not recorded during aerial surveys of Lake Chad, but Gustafsson et al. (2003) saw 500 and 1000 Ferruginous Ducks on 19 and 21 October 2000 along the Nigerian coast of Lake Chad.

Mullié et al. (1999) and Brouwer & Mullié (2001) showed the importance of the estimated 818 relatively small (average 54 ha) isolated wetlands in Niger, including the western part of Lake Chad Basin. Most only held water for a few months after the rainy season, enough to support possibly up to one million Palearctic and Afrotropical waterbirds. A comparison with numbers on the permanent Niger river (west of the Basin) showed clear complements: Palearctic ducks tended to disappear from the permanent Niger River after a good rainy season, apparently preferring to stay at the isolated wetlands where there was a concomitant increase in numbers (Mullié

& Brouwer, pers.comm.). It is expected that the permanent water of Lake Chad and the main rivers play a similar role.

Gustafsson *et al.* (2003) compared their observations from the Nigerian part of Lake Chad in 1997-2000 with those published by Ash *et al.* (1967), Dowsett (1968, 1969) and Elgood *et al.* (1994) up to 30 years previously. They failed to observe Skimmers and Ospreys, and recorded fewer Yellow Wagtails and Sand Martins and far fewer Quails. It is unlikely that their observations, referring to the western coast of Lake Chad, can be extrapolated to the rest of the lake. Ospreys, for example, are still widely present along the southern edge of Lake Chad. On the other hand, Skimmers used to be common along the Logone in the late 1950s, but have not been observed in the relatively well-studied Waza-Logone since 1980 (Scholte *et al.* 1999).

The waterbird counts clearly show that Lake Chad Basin still holds important waterbird concentrations. The high numbers of large, Afrotropical waterbirds, such as Marabou Stork, Sacred Ibis and White-faced Whistling Duck, are striking in comparison with their near-absence in the Inner Niger Delta and the Hadejia-Nguru wetlands (Scholte 2006) . The number of Black Crowned Crane in the Lake Chad Basin was estimated at 5000 birds (Scholte 1996), compared with less than 50 in the Inner Niger Delta, 150-380 in the Senegal Delta and 0 in the Hadejia-Nguru (Chapter 6-8). Equally, vultures are more common than elsewhere in the western Sahel (Scholte 1998). A possible explanation would be the smaller human population density, but that is hardly likely, given the proximity of large cities such as Kousseri (and N'djamena). In and around the Logone floodplains some 800 000 people are living (Mouafo *et al.* 2002), a population density higher than in the Inner Niger Delta. Most likely, local differences in human predation rates, based on varying cultural habits, may play a role. Large-scale commercial exploitation of waterbirds, as is the case in the Inner Niger Delta (Kone 2006), is not known for the Chad Basin (although for a number of years, Black Crowned Cranes have been captured for export as pets). Also, large colonial waterbirds are not safe in Logone (see Box 12), although their predicament seems to be better than in the western Sahel.

Numbers presented in this chapter are generally based on counts from the mid-dry season and do not reflect the dynamics of birds moving between various wetlands in the Lake Chad Basin. The fluctuating Lake Chad reaches its maximum water level late in the dry season when all surrounding areas, including the floodplains, are drying up. Neither do the dry season counts show the importance of ephemeral wetlands which contain water earlier in the dry season. We speculate on these interactions as follows (Scholte *et al.* 2004). During the rainy season from June to September, hundreds of small and mostly ephemeral wetlands, dispersed over the Lake Chad Basin, harbour large numbers of Afro-tropical ducks and waders, many breeding at this time of the year. A few weeks later, after the termination of the rainy season, many of these birds move into the floodplains along the Chari, Logone and Komadougou Yobe riv-

Box 12

The significance of effective bird protection

Counts annually conducted in January showed that the Black-headed Heron became ten times more numerous in the Logone floodplains between 1992 and 2001. Also the breeding colony in the village of Andirni went up from 742 nests in 1993 to 2479 nests in 2003. Elsewhere in West Africa, colonies of Black-headed Herons never exceed 150 pairs. How can the exceptionally large colony in Andirni be explained? The background story is simple. The herons breed in the home village of National Park guides, whose presence creates a safe haven, amidst a generally dangerous environment with high human predation. Even in Waza National Park, with a moderately effective management (see chapter 5), breeding colonies of Marabou Stork, Yellow-billed Stork and Pink-backed Pelican are regularly destroyed by people taking (large) young from the nests.

This situation is reminiscent of the persecution of herons in Europe during the first half of the 20th century, as can be illustrated with two

telling examples. In 1958 more than half of the total French population of Grey Herons bred in a single, well-protected colony with 1300 nests. With decreasing persecution, this large colony split up into several new ones in the surrounding area. And in The Netherlands in 1926, the Grey Heron population (6500 pairs) was distributed over 130 heronries (settled in very high trees only), compared with 10 000 pairs over 500 colonies in the year 2000. Average heronry size in The Netherlands declined from 50 to 20 pairs, whilst large colonies of 500-1000 pairs have not been found since the 1950s. In this respect, the sheer size of the two breeding colonies of large waterbirds in the Inner Niger Delta (Chapter 6) reveals more about the protection status and available breeding sites, than about the ecological conditions for these waterbirds.

These observations may be generalised to all bird species in Africa, especially those species which are worthwhile for local people to collect. Persecution explains much of the ongoing decline of storks, herons, darters, cranes, ducks and geese, but also of large raptors (Thiollay 2006a, 2006b). The bad news is that it will be difficult to convince local people to stop harvesting birds. The good news is shown by the heronry in Andirni: high bird numbers can be achieved again with effective community-based protection. From: Scholte (2006).

White-faced Whistling Ducks are very common in Lake Chad and surrounding wetlands.

ers, which dry up in the early- and mid-dry season (December-February). Migratory Eurasian birds, such as White Stork, egrets, ducks and waders, join the Afrotropical birds. Later in the dry season, birds probably disperse to the Lake Chad area, which by then has reached its maximum size, before returning to Eurasia. Afrotropical waterbirds will rest there till the onset of the next rains. Species such as Garganey and Black-winged Stilt seem to prefer ephemeral wetlands, whereas White-faced Whistling Ducks have a preference for the larger lakes, including Lake Chad and riverine habitats.

Conclusion

Lake Chad and its associated wetlands have witnessed enormous changes in size over the millennia, changes that continue to the present day. Before 1973, Lake Chad used to encompass an area of 20 000 km^2, of which 12% was covered with a dense vegetation of mainly Reed, Papyrus and Cattail. After 1973 lake size declined to 7000-12 000 km^2; half of this decline may be attributed to reduced rainfall, the other half to irrigation along the Chari and Logone Rivers. The northern basin fell dry in 1973 and turned into a woodland with temporary pools. The seasonal variation in water level in the southern basin increased from 50 cm to 150 cm, enlarging the surface area of seasonal floodplains from 200 to 5000 km^2. Since the bird density is generally low on open water and in the dense swamp vegetation, but high on seasonal floodplains, we assume that after 1973 Lake Chad became more attractive for many waterbird species. Counts from the air do indeed show that huge numbers of ducks and waders are present on Lake Chad. Waterbird counts reveal that Lake Fitri and the Logone floodplains also contain high numbers of waterbirds, though probably fewer than in the past when these wetlands were larger.

Endnotes

1. The annual discharge of the Yobe river entering into the NW corner of Lake Chad amounted to 0.5 km^3 in 1957-77 (Vuillaume 1981), equivalent to 1% of the entire river flow into Lake Chad. This is much too low to inundate the northern basin of Lake Chad, when it is separated from the southern basin at a low lake level.

2. Lake size (y, km^2) is a function of water level in Bol (x, in m): $y = 62.99x^3 - 418.4x^2 + 3658.x + 8065$. If the northern half of Lake Chad becomes isolated from the southern basin, another function has to be used to predict the surface area of the southern basin: $y = 205.0x^3 - 65.55x^2 + 935.5x + 6479$. The functions have been derived from graphs presented by Vuillaume (1981).

3. The hydrological variables show a good correlation. For instance, maximum water level can be predicted precisely when we know the inflow and the maximum water level in the preceding year: take 63% of the previous maximum (in cm) and add 15% of the inflow (m^3/s) to get the predicted maximum level in the next year. In the same way the minimum water level can be derived from the inflow and the minimum water level in the preceding year. In our analysis, however, we used curvilinear functions, to get a better prediction. The maximum water level in Bol (cm; usually in November-January) is a function of the annual inflow of the Chari and Logone Rivers (m^3/s; mainly dependent on the flow in July-September) and the maximum lake level during the preceding year (max$_{-1}$, cm): $0.01238 max_{-1} + 0.000983 max_{-1}^2 + 0.3063 flow - 0.000078 flow^2 - 20.391$ ($R^2=0.95$, N=72; P<0.001; 1934-2005). The fit is even better for the minimum lake level: $0.5117 min_{-1} + 0.000543 min_{-1}^2 - 0.1015 flow + 0.000013 flow^2 - 54.679$ ($R^2=0.97$, N=72; P<0.001; 1934-2005).

The Sudd

The River Nile drains a huge area (3 million km^2, or about 10% of Africa), but its annual flow (88 km^3/year) is that of a much smaller river, such as the Rhine. The Nile is a real "desert river" and so its function as a lifeline is even more pronounced than for "dry-land rivers" such as the Senegal or Niger in the western Sahel. The Nile's flow is relatively small for several reasons: its water suffers a high evaporation rate in the southern half of its basin, no tributaries add water to it further north as it traverses the easternmost Sahara and its natural flow is reduced greatly by dams and large-scale irrigation. However, the Nile is more than just a desert river. In particular, the White Nile causes the flooding of large areas in southern Sudan. These floodplains, known as the Sudd, belong to the world's largest wetlands, which are of prime importance to birds and wildlife. Why, then, is southern Sudan not as famous as other African wonders of the natural world, such as the Serengeti. In the first instance, probably because it is a harsh, unhealthy environment, where the first rains in April or May turn the dry black clay bottom into "the most slippery kind of mud" (Howell *et al.* 1988) for seven to eight months, transforming the area in the words of Cobb (1981) and Nikolaus (1989) into "underwater, or, hardly better, a sea of black mud", where "fieldwork is hampered by ... disease and a variety of unpleasant biting insects". Travellers in earlier times experienced the Sudd as an insurmountable barrier, where water creates an ever-changing network of channels which, if navigable at all, are often blocked by floating vegetation. Its general inaccessibility is the main reason why little is known about its bird life, but, in addition, since the 1950s, southern Sudan has been plagued by periods of civil war and famine (1956-1972 and 1984-2005). As these human tragedies tend to have their own specific and usually opposite impacts on wildlife, there is a strong need to shed light on the Sudd's present state and the bird numbers involved.

The Nile Basin

The Nile has two principal branches: the Blue Nile draining the western Ethiopian highlands and the White Nile, whose sources are on the upland plateau of East Africa around Lake Victoria. The two tributaries merge near Khartoum to form the Nile proper (Fig. 92). Rainfall in the mountainous upper reaches of the drainage basins of the White and Blue Nile averages 1000–1200 mm per year. The remaining 80% of the Nile basin is semi-arid (the northern half of Sudan) and arid (Egypt). On average, 86% of the water of the Nile descends from the Ethiopian highlands, the remaining 14% from the equatorial lakes. Rainfall in the Ethiopian highlands is confined to a single season (July-September), causing the Nile to flood in late summer. The runoff from the equatorial lakes shows little seasonal variation, because the large volume of water stored in the lakes evens out the effects of variations in regional inflow and rainfall. As a consequence downstream, the White Nile is the more important source of water supply during the low-flow stage from November to June – Fig. 93 shows the monthly flow of the White Nile at Mongalla, where the Sudd marshes begin. Only 300 km further downstream, the outflow has decreased by almost 50% of the inflow rate, due to evaporation and transpiration from the marsh vegetation (see below).

During the 20th century, the average annual flow of the Nile at Aswan amounted to 88 km³, but varied between 52 and 121 km³. Most of this variation was due to the variable discharge of the Blue Nile, which was on average 50 km³, but varied between 21 and 80 km³. The lowest river flows were registered in 1913 (20.6 km³) and 1984 (29.9 km³) when rainfall in the highlands was exceedingly poor: respectively 20.0% and 17.7% below the long-term average for the 20th century (Conway 2005, Conway *et al.* 2005). In the Sahel, 1913 and 1984 were extremely dry years as well (Fig. 9). In contrast, the 1950s and 1960s were relatively wet in Ethiopia, *i.e.* similar again to the western Sahel. In fact, the Sahel rain index (Fig. 9) and rainfall in the Ethiopian highlands are correlated (R=+0.51; calculated for 1955-2003 only, because the number of gauges in Ethiopia before 1955 was limited).

The flow of the White Nile depends on the water level in Lake Victoria, and hence on rainfall in the preceding years (Sutcliffe & Parks 1999). Between 1900 and 1960 the water level in Lake Victoria varied by about half a metre, but due to the extremely high rainfall in 1961, an El Niño year (Birkett *et al.* 1999), the water level in 1961 rose 107 cm and another 45 and 52 cm in 1962 and 1963, reaching a level 2 m above those of the 1920s and 1940s (Fig. 94). The storage volume of Lake Victoria increased in these three years by 151 km³. The outflow in 1961 (20.6 km³) was hardly higher than in 1960 (20.3 km³), but in the following three years the outflow steadily increased (38.7,

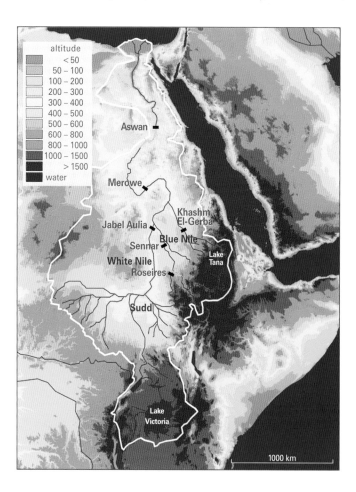

Fig. 92 The Nile Basin, showing tributaries and six major dams and irrigation schemes; the Merowe dam is still under construction.

44.8, 50.5 km³ in 1962-64), only to decrease slowly again from 1965 onwards (Sutcliffe & Parks 1999). After 1961, the rainfall continued to decline and finally dropped below the long-term average in the 1980s (Conway 2005). As a consequence, the level of Lake Victoria gradually decreased by nearly 150 cm between 1965 and 1995 (Fig. 94). 1997 was the first good rainy season for many years, but this has only slowed the steady decline of the lake level. High rainfall, such as in 1961, must be rare, since a historical study yielded only one other year, 1878, with a similar hydrological impact in the last century and a half (Sutcliffe & Parks 1999).

The annual outflow from Lake Victoria (Sutcliffe & Parks 1999) is determined by its water level (Fig. 94). [1] Fig. 94 therefore also illustrates the variation in the flow of the White Nile. The flow was maximal in 1964 with 51.1 km³ and at its lowest level in 1922 (11.8 km³).

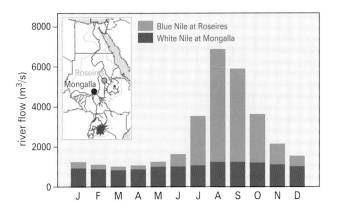

Fig. 93 Average monthly flow of the Blue Nile at Roseires, and of the White Nile at Mongalla, calculated for the same period (1912-1982). From: GRDC.

The consequences for the Sudd marshes are discussed below.

The flow of the Nile is affected by several dams and irrigation schemes. Dams created deep lakes, whose total surface is 1000 km², while most of the water from the storage reservoirs is tapped for irrigation. Three dams were constructed before 1933 in Egypt to allow irrigation. The original Aswan dam, the largest, had a lake whose area was 45 km² and volume 5 km³, but in the 1960s the Aswan High Dam was built 6.5 km upstream, creating a huge reservoir, Lake Nasser, 163 km³ in size, a volume roughly equivalent to twice the total average annual flow of the Nile. As Lake Nasser is situated in the hot desert, its evaporation is high, at 7 mm per day, which amounts to an annual water loss of 10 km³. The Aswan High Dam was intended to generate hydro-electricity on a large scale, but the control of water flow not only heralded the end of annual flooding of the lower Nile valley, which experienced diminishing replenishment of fertile soils because much less material was in suspension, but also permitted the extension of irrigation schemes along the river. In 1997, Egypt started to pump 5 km³ from Lake Nasser annually to the Toshka depression 100-250 km west of Aswan, to create "a second Nile valley" in the middle of the desert.

The Hamdab Dam near Merowe, Sudan, is also intended to produce electricity. It will be another large reservoir, some 850 km upstream of Aswan. Its lake will be 350-800 km² in area and will contain up to 12.5 km³ of water. Further upstream, four other dams allow Sudan to irrigate substantial areas: the Khashm El-Gerba (1964) in the Atbara River, the Jabal Awlia (1937) in the White Nile 50 km southwest of Khartoum, the Sennar (1926) in the Blue Nile 300 km south of Khartoum and the Roseires (1950) in the Blue Nile 500 km southeast of Khartoum. The FAO estimated the surface of the actual irrigated

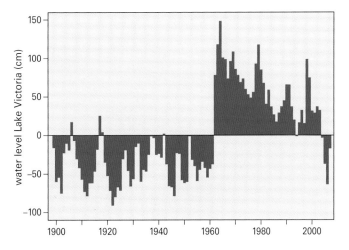

Fig. 94 Fluctuation of water level in Lake Victoria, given as cm deviation from the 20th century average. From: Conway (2005), and for recent data: Topex/Poseidon/Jason satellite altimetric datasets (given on the 'Crop Explorer' site of the USDA Foreign Agricultural Service).

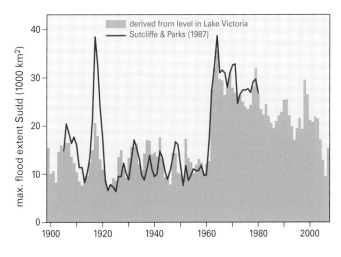

Fig. 95 The maximally flooded area of the Sudd, according to the hydrological model of Sutcliffe & Parks (1987) calculated for the period 1912-1980. Also shown is the flooded surface (1894-2007) as derived from the relationship between the flooded area of the Sudd and the water level of lake Victoria.

land in Sudan in 1995 at 19 000 km² and the total water withdrawal at 17.8 km³ (http://www.fao.org/docrep/W4356E/w4356e0s.htm).

In the 1930s a proposal had already been made to create a channel for the White Nile which would bypass the Sudd marshes in order to expand the water flow of the Nile in the deserts of Egypt and northern Sudan. In 1974, these two countries agreed to construct the Jonglei Canal along the Sudd. Although work started in 1980, it halted at the outbreak of the civil war in 1983, at which time 260 of the 360 km had been completed, but since then no attempt has been made to finish the project.

The Sudd

Hydrology The Sudd is a riverine floodplain, but due to the small seasonal variation of inflow (Fig. 93), the within-year flood level varies usually only by about 50 cm (Suttcliffe & Parks 1987). The between-year annual variation in flooding is much larger. When in the mid-1960s the Sudd inflow trebled compared to that of preceding decades, the water level on the local gauge at Shambe, in the centre of the Sudd, rose 3 m.

Sutcliffe & Parks (1987) developed a water balance model to es-

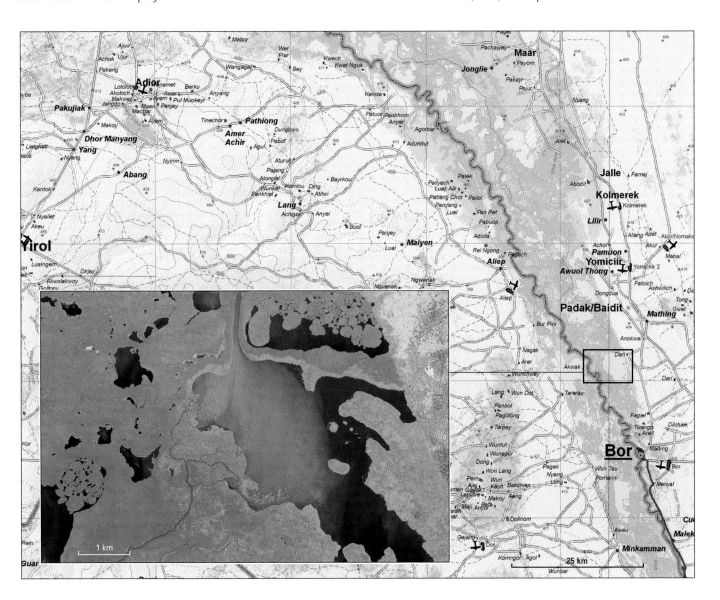

Fig. 96 Topographical map of the southern third of the Sudd. Inset photo shows satellite image (10 November 2002) made available by Google Earth. The circular forms on the image are floating swamp islands. The map (scale 1:500 000) was released in October 2005 by the Geoprocessing Unit at the Centre for Development and Environment, University of Bern (www.cde.unibe.ch).

Patrick Denny (left), one of the few botanists who have studied the vegetation in the Sudd, among Papyrus *Cyperus papyrus* up to 4-5 m high (Lake Naivasha, March 2001). Papyrus is a herbaceous perennial native to Africa, tolerant of temperatures of 20°-30°C and a pH of 6.0-8.5. In deep water it forms floating, tangled masses of vegetation known as *sudd*. Papyrus attracts few Palearctic migrants, but is home to a suite of African habitat specialists which need undisturbed swamps (Owino & Ryan 2006). Papyrus also offers an important protection from eutrophication and supports fish larvae recruitment (Kiwango & Wolanski 2008). Papyrus swamps are hazardous to penetrate on foot, especially when standing in deep water, and are also inhabited by crocodiles.

timate the seasonal and annual variation in the flooded surface in the Sudd. Their estimate is primarily based on the monthly data of the flow into the Sudd at Mongalla, the outflow at Malakal (minus the flow of the Sorbat River), local rainfall and evaporation. They conclude that between 1900 and 1960 the flooded surface varied mostly between about 7000 km² in May and 12 000 km² in November. In this period 60% of the Sudd comprised permanent and 40% seasonal swamp. The maximally flooded area increased from less than 10 000 km² in 1960 to nearly 39 000 km² in 1964 (Fig. 95). Unfortunately, the hydrological model cannot be used to predict the size of the Sudd since 1983 because the required hydrological data are lacking. The predicted surface of the Sudd depends mainly on the measured inflow at Mogalla [2], but since this represents 76% of the outflow of Lake Victoria, the size of the Sudd floodplains may

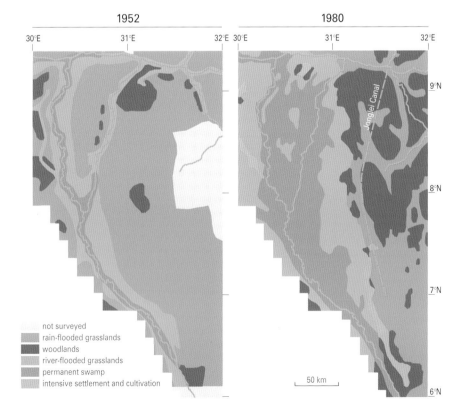

Fig. 97 (left) Vegetation of the Sudd in 1952 (low flood) and in 1980 (high flood). From: Howell *et al.* (1988).

Fig. 98 (right) The distribution of settlements (huts), cattle, Tiang (an antelope better known as Topi) and Shoebill in the Sudd and its environs during wet and dry seasons, based on aerial surveys in 1979-1981. From: Howell *et al.* (1988).

also be derived from the water level in Lake Victoria (Fig. 94). The size of the flooded area in the Sudd has gradually decreased from 39 000 km^2 in 1964 to 10 000-15 000 km^2 in 2005-2007, thus returning to levels normal for the 1890-1960 period (Fig. 95).[3]

Vegetation Like the Inner Niger Delta and, before its river was dammed, the Senegal Delta, the Sudd is a floodplain, but with one essential difference: a much larger part of its area is permanently flooded. In this respect the Sudd resembles other African floodplains such as the Okavango Delta (Sutcliffe & Parks 1989), the Kafue plains (Howell et al. 1989) and the post-dam Senegal Delta (Chapter 7). Denny (1984) described the permanent swamp vegetation zones of the Sudd, and in 1983 a vegetation map of the Sudd and its environs (69 000 km^2 in total) was produced by the 'Range Ecology Survey (RES)' and the 'Swamp Ecology Survey (SES)', based on field work from the 1979-1983 period. All RES and SES data are summarised by Howell et al. (1988). The following description is based on these sources.

The rivers and lakes within the Sudd (1500 km^2) are partly covered by Water Hyacinth *Eichhornia crassipes*, a free-floating plant

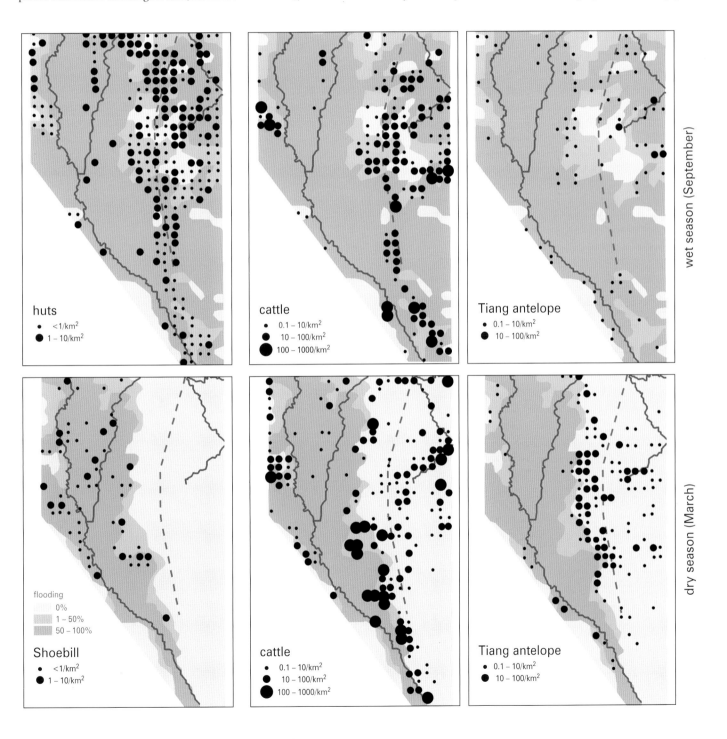

species which appeared in the Sudd in 1957 and by the late 1970s had covered 112 km^2 (Freidel 1979). It has largely replaced the Water Lettuce *Pistia stratiotes*. The deep open water zone is surrounded by other plant species. *Bourgou (Echinochloa stagnina)*, a key species of the Inner Niger Delta growing in sites that flood to 3-5 m depth (Fig. 43), is restricted in the Sudd to deep, often seasonal, pools in the grasslands. In contrast, shallow-flood species such as *Vossia cuspidata* (known as *didere* in West Africa and hippo grass in East Africa) are common in the Sudd, but restricted to the areas with a fluctuating water depth, which occur mostly in the south, covering some 250 km^2. Papyrus *Cyperus papyrus*, growing in up to 2 m depth of water, covers a larger area, 3900 km^2, but is absent from areas whose water levels vary. The same is true for Cattail, which grows in shallower water, being found mostly in the central and northern part of the Sudd, covering 13 600 km^2. These giant swamp herbs form dense vegetation and bulky clumps, which not only prevent human access but also block any view from the river.

The permanent swamp vegetation is surrounded by seasonal grasslands. Howell *et al*. (1988) distinguished two grassland types being seasonally flooded, one dominated by Wild Rice *Oryza longistaminata* (13 100 km^2), flooded 5-9 months a year and the other by Antelope Grass *Echinochloa pyramidalis* (3100 km^2), flooded 4 months a year or less. These seasonally river-flooded grasslands are in turn surrounded by three types of seasonally rain-flooded grasslands (20 000 km^2), dominated by Cockspur grass *Echinochloa haploclada*, Whorled Dropseed (or Rats Tail Grass) *Sporobolus pyramidalis* and Jaragua (or Thatching Grass) *Hyparrhenia rufa*, which are all widespread in Africa and elsewhere in the tropics. Although 90% of grasslands are burnt yearly in the dry season, woodlands remain on the higher floodplains and in the surrounding land; Red Acacia *Acacia seyal* (5400 km^2) is more flood-tolerant than the Desert Date *Balanites aegyptiaca* (5300 km^2).

The above vegetation types within the Sudd floodplain have been calculated as comprising 16 900 km^2 of permanent and 16 200 km^2 of seasonal swamp, but these figures refer to the late 1980s when the floodplains were very large (Fig. 95). The swamp composition was different prior to 1960, when the Sudd was much smaller (Fig. 97). The high floods of the early 1960s saw the Red Acacia and Desert Date succumb at sites that previously had been flooded, at maximum, for only some months a year. Not only did the maximum flood level rise suddenly in the early 1960s, but the minimum flood level also increased. Consequently, much of the seasonally river-flooded grasslands became permanent swamp. The higher flood level allowed Cattail to become more widely distributed than any other plant species, from 6700 km^2 in 1952 to 19 200 km^2 in 1980. Seasonally river-flooded grassland increased at the expense of seasonally rain-flooded grassland.

People, livestock, fish and wildlife   The Sudd and its environs held about 300 000 to 350 000 people between 1950 and 1980 (appendix 6 *in* Howell *et al*. 1988). The largest city in the area is Malakal, which grew from 12 000 inhabitants in 1954 to 50 000 in 1981, but in this period, most people remained in the countryside. The high floods of the 1960s were disastrous for local people, especially because they coincided with civil war. "Flooding rendered large areas uninhabitable and war made habitable areas unsafe" (Howell *et al*. 1988: Chapter 11).

Aerial counts between 1979 and 1981 showed that herders held nearly half a million cattle and 100 000 sheep and goats, but seasonal influxes from the highlands increased the totals to 800 000 cattle and 180 000 sheep and goats in the dry season. The herders move seasonally with their cattle, but do not enter the permanent swamps (Fig. 98).

Farmers grow mainly sorghum, but also maize, cowpeas, groundnuts, sesame, pumpkins, okra and tobacco. The cultivated land, about one ha per family, lies surrounding the settlements, because the crops must be continuously protected from birds, mainly seed-eating weavers, but also doves. The distribution of the settlements is shown in Fig. 98; obviously there are only a few settlements within the actual floodplains, a fact which is confirmed by high-resolution satellite images (Google Earth).

Herders and farmers cannot benefit from the flooded area, but fishermen, in their need to live close to their fishing grounds, usually camp on the river's high levees. Fishing is a rather recent development in the Sudd – before the nets became available in the early 1950s, open water was rarely fished (Howell *et al*. 1988: Chapter 14). From that time, fishermen as far away as West Africa began to settle in the area. They sun-dried and salted the fish to store and sell. Local people did learn to fish, not at first as full-time fishermen, but when the high floods of the 1960s turned their traditional grazing lands into swamp, many farmers and herders had no choice but to start fishing. However, fewer than 8000 canoes were counted during the three aerial surveys in the early 1980s. Howell *et al*. (1988) estimate the annual catch in the Sudd at about 27 000 tonnes, although that might be doubled if the Sudd experiences the level of fishing pressure affecting other African floodplains. [4]

Much of the Sudd, especially the permanent swamp, is rarely visited by local people, which is probably why the area still harbours remarkably large numbers of herbivores. Aerial counts during 1979-1981 revealed the following maximum numbers: 460 Bushbuck, 10 200 African Buffalo, 3900 Elephant, 66 000 Mongalla Gazelle, 6000 Giraffe, 3500 Hippo, 11 600 White-eared Kob, 2500 Uganda Kob, 32 000 Nile Lechwe, 6000 Oribi, 33 400 Reedbuck, 4100 Roan Antelope, 1100 Sitatunga, 360 000 Tiang (or Topi), 8900 Waterbuck and 3900 Zebra (Mefit-Babtie Srl 1983). Most of these species are seasonal migrants, entering the flooded areas during receding water and returning to high ground at the start of the rainy season; as an example Fig. 98 shows the distribution of the Tiang during the wet and the dry season. According to Howell *et al*. (1988) most species are confined more to the swamps than they once were, due to human population pressure. The importance of inaccessible swamps as safe havens has certainly increased since then (see Box 4).

African Openbill Storks are food specialists, eating snails and freshwater mussels in aquatic habitats. Very large numbers have been counted in the Sudd. When opportunities are favourable, as during desiccation of rivers and lakes, large flocks may arrive out of the blue, deplete local food stocks and disappear equally suddenly. This picture epitomises this very moment for the Boteti River near Maun, Botswana, after the water level had dropped to wading depth in January 1994. Up to 172 Open-billed Storks (here part of the fishing group on 15 January 1994) depleted the local stock of bivalves in less than two weeks, whereas up to 770 Great White Pelicans did the same with fish.

Table 16 Three aerial counts of large bird species in the Sudd within the area shown in Fig. 97 and 98 in 1979-1981. From: Howell et al. 1988. Status given is derived from Nikolaus (1989): U = uncommon, FC= fairly common, LC = locally common, C = common, VC = very common.

Month	mid-Sept	Nov-Dec	late March	
Season	wet	early dry	late dry	Status
Ostrich	1486	4961	6240	U
Long-tailed Cormorant	232	8883	6006	C
Great White Pelican	0	0	5643	R
Pink-backed Pelican	3649	6110	11187	C
Grey Heron	984	0	0	FC
Black-headed Heron	1652	1460	1716	C
Purple Heron	2587	2091	5049	LC
Goliath Heron	0	3819	3234	C
Squacco Heron	3845	9402	18414	C
Cattle Egret	172 359	65 253	86 724	VC
Great Egret	75 806	9530	19 074	FC
Yellow-billed Stork	0	3775	11 154	
African Openbill	13 469	288 536	344 487	VC
White Stork	0	16500	0	U
Abdim's Stork	0	0	858	
Woolly-necked Stork	1350	2475	1485	R
Saddlebill Stork	3640	4017	4158	FC
Marabou Stork	196	359 719	194007	C
Shoebill	6407	5143	4938	
Glossy Ibis	787	1695 240	8778	U
Hadada Ibis	697	429	231	C
Sacred Ibis	16 201	4419	17 688	VC
White-faced Whistling Duck	7150	0	51 810	C
Fulvous Whistling Duck	0	0	8775	FC
Knob-billed Duck	394	9611	9075	LC
Spur-winged Goose	1153	88 220	150 216	FC
Black Crowned Crane	36 823	22 715	14 685	C
Arabian Bustard	299	945	728	FC
Black-bellied Bustard	944	396	297	FC

Birds Howell et al. (1988) describe the seasonal grasslands between November and April as "an extremely favourable habitat for birds, particularly aquatic ones ... as well as being essential to some million intra-African migrants. Ripe grass seed, crustacea, molluscs, aquatic insects and enormous numbers of fish, all contribute to the diversity and high productivity". Some bird species were counted in the 1979-1981 aerial surveys; e.g the Shoebill (Fig. 98) a species confined to permanent swamps.

Fishpool & Evans (2001) give a table with the maximum bird numbers censused during the three aerial surveys in 1979-1981, but the ornithological handbooks of Brown et al. (1982) and Del Hoyo et al. (1992) make no reference to these data. These counts may been overlooked, but possibly they have been ignored, because the numbers of some species seem unrealistically high (Table 16). To interpret the numbers, knowledge of the count methodology is essential. As described in detail by Howell et al. (1988: appendix 5) the area was divided into a 10 x 10 km grid. From the aircraft, two observers counted all animals, canoes and settlements, within each grid square in narrow strips 145 m wide on either side of the aircraft, effectively an aerial line transect method. Large herds of animals were photographed in order to check the field estimates. This method of aerial line transects, some 700 km long, covered 3% of the area of the Sudd. The total estimates were extrapolations for the entire 69 000 km².

Large herbivore counts were similarly extrapolated. Howell et al. (1988) give the standard errors for these estimates. For instance, the standard error for the Tiang (average 360 000) was 95 000, and for other counts of herbivore species, it was approximately 30 to 50% relative to the mean. Unfortunately, the standard errors are given only for large herbivores, not for bird species, which makes it difficult to evaluate the bird counts.

The single-most important site for Shoebills in Africa is the Sudd, where several thousands are said to occur. Nowhere else are numbers even remotely that large. It is an impressive, man-sized bird associated with swampy grasslands bordering papyrus. Bangweulu swamps, Zambia, March 2008.

We can assume that the counts were more accurate for large, conspicuous species (pelicans and storks) and less so for those that are difficult to detect (Glossy Ibis); for the same reason, the numbers of skulking species, such as Squacco and Purple Heron, must have been largely underestimated. The numbers of common but small species as Ruff, Black-tailed Godwit and Garganey were not counted at all. We can also assume that extrapolation was more accurate for widely distributed species (e.g. Sacred Ibis, Black Crowned Crane) and less so for species feeding in flocks in only a few locations, such as Glossy Ibis (of which only five large concentrations were ascertained in December; Mefit-Babtie Srl 1983). Similarly scattered are those species which spend the day in large roost assemblages (ducks).

A comparison of the three counts reveals that comparable numbers were counted for some resident, and especially for solitary, bird species (e.g. 5000 – 6000 Shoebill, 2000 Woolly-necked Stork, 4000 Saddlebill Stork, 1500 African Fish Eagle). For other species the three counts gave totals varying during the season: Ostrich, Openbill and Marabou entered the area during deflooding, as did White Stork, although the latter had left the area by late March. In contrast, the largest number of Black Crowned Crane was counted in September, but decreased as the water receded.

The recorded maximum number of 37 000 Black Crowned Cranes and certainly of 1.7 million Glossy Ibis are extremely high. Even if by the merest chance, all Glossy Ibis in the area were concentrated in the 3% covered by the strip counts, at least 50 000 Glossy Ibis would still have been present, 1.5 times more than has ever been counted in the Inner Niger Delta (Chapter 6). Nikolaus (1989), who did fieldwork in S Sudan between 1976 and 1980, mentioned the species as "rather uncommon winter visitor in small numbers to large swamps". For other species, the status assessed by Nikolaus agrees with the counted numbers; Table 16, last column.

Although the Shoebill is another species of which incredible numbers were counted, about the same numbers were tallied during the three counts (4900-6400), and so the figures seem reliable. According to Brown et al. (1982) the African population "should not exceed 1500 birds", but Del Hoyo et al. (1992) arrived at a higher estimate of 11 000 birds, of which 6000 were in the Sudd. The latter figure is based on Nikolaus (1987) who mentioned this figure with-

Dense Cattail and Papyrus vegetations provide the ideal habitat for rails and some specialist passerines, but are largely avoided by most waterbirds, unless patches of open, shallow water within these dense marshes are available. A Goliath Heron disturbed along the edge of a papyrus swamp in Lake Navasha, Kenya (upper right), and a Long-tailed Cormorant roosting in papyrus in the Okavango Delta, Botswana (upper left). Foraging Grey Herons, Great Egrets, African and Eurasian Spoonbills, Glossy and Sacred Ibis, Black-winged Stilts and Black-tailed Godwits in Typha marshes in the Senegal Delta.

The Sudd

Large birds are particularly prone to human predation, chemical pollutants and habitat degradation. For example the Ostrich, once widespread in West Africa, is now extinct west of Chad. Along with vultures, eagles and bustards, Secretary Birds (above) have also markedly declined in the northern Sahel and in the savanna woodlands of West Africa, by at least 50-60% between 1968-73 and the early 2000s (Thiollay 2006a-c). The decline may have been smaller in the eastern Sahel, where human pressure has not yet reached the West African levels, although actual trend information is lacking. Goliath Herons (below) are still quite common in eastern Africa; here seen foraging in water overgrown with Water Lily bordering papyrus in Lake Naivasha.

out further reference, but it could be that he took it from the aerial survey results. The Shoebill is restricted to permanent swamps in the Sudd (Fig. 98) and similar habitat elsewhere in Africa (del Hoyo *et al.* 1992).

Del Hoyo *et al.* (1989) considered the African Openbill as probably the commonest stork in Africa, the largest known colony (5000 pairs) being in Tanzania; the aerial surveys of the Sudd suggested about 300 000 birds in the dry season. Nikolaus (1987) also agreed that this species is very common in southern Sudan.

The aerial survey suggests the presence of 15 000 – 36 000 Black Crowned Cranes in the Sudd, which numbers have never been matched elsewhere in Africa. The total African population (western and eastern subspecies combined) was estimated at some 80 000 – 100 000 birds in the 1980s, of which half were in Sudan, but since then the population declined steeply, although current data for the Sudan are lacking (Beilfuss *et al.* 2007).

It is obvious that the Sudd remains an extremely important bird area. Most remarkable is the presence of large numbers of pelicans, egrets, Goliath Heron, Black Crowned Crane and seven stork species. It seems that these birds still can find remote places in the impenetrable swamp vegetation to breed, unlike in the floodplains of West Africa, where due to the present lack of safe areas, the same species are rare or absent (*e.g.* Chapter 6). It should be noted, however, that the Sudd counts were performed around 1980, when the flood extent was at its maximum (Fig. 95), so numbers were most likely lower before 1965 and also in the 1990s.

Most of the species in Table 16 are residents and so we can only speculate about the significance of the Sudd for migrant duck, waders and other Eurasian wetland species. Large numbers could be expected in the Sudd, given its size and the presence of a large variety of wetland habitats. On the other hand, much of the Sudd is covered by Cattail, which in Senegal is a habitat without waterbirds, except for moorhen, rails and a few warbler species (Box 11) and by Papyrus, which from our own observations in Kenya, Botswana and Zambia, is just as sparse in bird life. Although these extensive monotonous stretches of vegetations are poor bird habitats, many birds are attracted to the edges. Satellite images clearly show that there are many pools, lagoons and creeks relieving the monotony of Cattail and Papyrus (Fig. 96, although open water has gradually decreased in extent between 1973 and 2002, due to the lower flood level (Petersen *et al.* 2007a).

We do not really know which species could be expected to winter in the Sudd and we can only make guesses about long-term trends. One reasonable guess is that the long-term variation in the flooded surface (Fig. 95) must have had repercussions on the population trends in bird species which concentrate in the Sudd. If so, the trends of the populations wintering in the Sudd differed from those elsewhere in the Sahel between 1960 and 2000, because the high floods in the Sudd in the 1970s and 1980s coincided with a dry epoch in the Sahel.

Conclusion

The Sudd is about twice as large as the Inner Niger Delta, but the Sudd holds only a third of the human numbers present in the Delta. The presence of many large herbivores and huge numbers of resident large bird species is an obvious sign of the much lower human pressure on the Sudd habitats. The few available data from 1980 suggest the presence of large numbers of species such as Goliath Heron, seven stork species and Black Crowned Crane. However, observations from other African floodplains suggest that bird densities in the permanent swamps are probably low. Nevertheless, we predict the presence of many migrant waterbirds, due to the sheer size of the Sudd and the presence of large seasonal floodplains. The Sudd was large in the 1970s and 1980s (and probably attracted more birds), unlike the rest of the Sahel, which was dry in that period.

Endnotes

1 The annual outflow from Lake Victoria at Jinja is a linear function of the level of Lake Victoria: km^3 = 0.164 level + 26.67 (R^2=0.979; calculated over the period 1956-1978; level defined as cm deviation from 20th century average).

2 The maximally flooded surface of the Sudd (km^2), such as predicted by Sutcliffe & Parks (1987) is highly correlated with the annual flow (km^3; data from the GDRC database) of the White Nile at Mongalla: flooded area =708 flow -5791 (R^2=0.949; calculated over the period 1912-1980). A similar regression with the level of Lake Victoria (cm; data from Fig. 95) gives: 103 cm +17520 (R^2=0.479; calculated over the period 1912-1980). The low correlation is mainly due to three deviating years, 1917-1919, when the flooding derived from Lake Victoria was 46% lower than that derived from the water balance study.

3 Sutcliffe & Parks (1987) base their hydrological model on a monthly evaporation rate varying between 150 mm in June and 200 mm in December-January, in total 2150 mm per year. According to Mohamed *et al.* (2004) this should be 1355 mm, since the assumption of Sutcliffe & Parks that the evaporation for open water and swamp would be equal was incorrect (but see Sutcliffe 2005). As a consequence, the flooded area would still be larger than indicated by Sutcliffe & Parks. The error cannot be large, however, because the size of flooded areas, according to the hydrological model of Sutcliffe & Parks, fitted well with the surfaces derived from air photography in 1930/1931, satellite imagery of February 1973, and a vegetation map indicating permanent and seasonal swamp, combining field work, aerial photographs and satellite images (Sutcliffe & Parks 1987).

4 Using the data of Laë & Levêque (1999), Zwarts & Diallo (2005) calculated the relationship between the annual fish catch (ton) and the maximum size of the flooded area (x, km^2): ton/year = 52.986 $x^{0.6462}$; R^2=.615). Using this relationship, the predicted annual catch in the Sudd could have been 20 000 tonnes before 1960 and up to 50 000 tonnes in the late 1960s and the 1970s.

Rice fields

Irrigation along Sahel rivers is often detrimental to natural floodplains. The large irrigation scheme of *Office du Niger* in Mali annually takes so much water from the Niger River that the flood extent of the Inner Niger Delta, just downstream, is reduced by some 300 km^2. Floodplains in the Logone valley and the Senegal Delta have been embanked and converted into irrigated land. Conversely, irrigated areas and reservoirs might be considered artificial wetlands, offering some compensation for the loss of floodplains. All large-scale irrigation schemes in western Africa were constructed in the second half of the 20th century, but small-scale irrigation has existed in Africa since well before then. The farmers who grow rice in areas reclaimed from the mangrove belt between The Gambia and Sierra Leone store the rainwater in embanked areas, enabling them to wash away the salts which enter from the estuary, and to prevent salt from rising by capillary action from deeper soil layers. They cannot drain their diked fields too much, however, if they are to prevent the soils from turning acid due to the presence of dissolved iron and aluminium. Acidified soils are among the most hostile environments known, making farming impossible. Thus, to grow rice successfully, farmers have to control salt and fresh water levels, as well as soil acidity and soil salinity. However, if all individual farmers were to attempt to create the ideal growing circumstances on their own parcels of land, they would often undo the work of their neighbours. Farmers solve the problems by mutual agreement and that makes the water management system of the Jola in the Casamance and the Balanta in Guinea-Bissau even more impressive. When the first Portuguese reached West Africa in the 15th century, they witnessed people growing rice in polders. However, African rice *Oryza glaberrima* was already domesticated from wild rice *Oryza barthii* some 2500 years ago, by people living in what is today the Inner Niger Delta. The annual production of rice fields on the floodplains and in the coastal zone amounts to 1-2 tonnes per ha, compared with 4-6 tonnes in the large-scale irrigated areas. While the costs of traditional rice growing are next to nothing (apart from manpower), the costs of the modern system of irrigation, although partly hidden, are high. Farmers in irrigation schemes pay a rent, but this is not sufficient to cover the costs of maintenance, let alone the construction of the irrigation schemes. The production of large-scale irrigated rice is thus heavily subsidised. Nevertheless, rice imported from Asia is cheaper than locally grown rice. The Sahel countries safeguard their own farmers, by charging tax on imported rice (being 32.5% in Mali in the early 2000s). The general policy is to achieve food-security, in dry years as well, and to keep the country self-supporting. This also explains why the governments are eager to implement still more and ever larger irrigation schemes, certainly as long as external donors will pay for these large prestige projects.

Rice in floodplains

The farmers on the floodplain use a rice variety that is well-adapted to grow along with the flood during raising water, but rice farming is not easy in floodplain areas. Ideally the seed should have germinated before the flood arrives. This means that the farmers have to sow the rice grains before the first rain, in the hope that the rain will come before the flood. Rice plants are able to grow 3-4 cm a day and thus keep up with the water level during the flood, increasing in size by several cm a day. The stems may be as long as 5 m, but usually they are less than 2 m long. After a flooding period of about three months, the rice can be harvested during receding water. Much can go awry in such a system:
- If there is no rain before the flood covers the floodplains, the seed does not germinate before the area is covered by water.
- If there has been sufficient rain for rice to sprout, the rice still needs water to grow. That is why the flood must arrive not later than a fortnight after the last rains.
- If timing and amount of rain have been good, but the flood low, rice plants will grow, but the yield will be poor due to the short growing season. A minimal flooding of three months is required.
- If there has been enough rain, but the flood is higher than expected, the production will be smaller. The optimal water depth is about 2 metres.
- If the growing of rice has been successful, the ripening grain must be protected later on against seed-eating birds.

A production of 65 000 tonnes of rice is necessary to feed all 800 000 people in the Inner Niger Delta, but this level was not reached in 4 of the 16 years between 1987 and 2002, when the annual rice production varied between 40 000 tonnes in dry years and about 150 000 tonnes in wet years (Zwarts & Kone 2005b). To enhance rice production, dikes and sluices have been constructed to stem the receding water, if necessary. Dams and sluices were useless, however, because the floods were not as high as anticipated in most

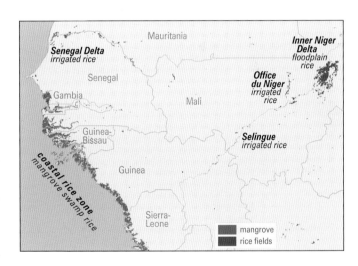

Fig. 99 Bird density in rice fields was measured in five areas: the coastal rice fields between Gambia and Guinea, the rice fields on the floodplains of the Inner Niger Delta and in the irrigated rice fields in the Senegal Delta, near Lake Selingue and in the area of *Office du Niger*.

Rice fields 171

In 2002, the flood in the Inner Niger Delta was sufficiently high to produce an above-average rice harvest in January 2003. Much work in rice fields is manual, involving large numbers of people working the fields during planting, crop protection in the ripening stage, and harvesting.

years, and, consequently, the embanked areas (680 km² managed by *Opération Riz Mopti* and *Opération Riz Segou*) were usually only partly flooded between 1970 and 2005. The total surface area cultivated for rice in the Inner Niger Delta has increased from 160 km² in 1920 to 3400 km² around 1990, of which 140 km² is actively irrigated (Zwarts & Kone 2005b).

Coastal rice

Rice polders are located in the tidal marshes and were reclaimed from mangrove habitat in the zone from The Gambia to Sierra Leone (Fig. 99). The total surface area of the mangroves may be estimated at some 8800 km², of which 700 are in The Gambia, 1400 in Senegal, 2500 in Guinea-Bissau, 2600 in Guinea and 1600 in Sierra Leone (Bos *et al.* 2006). Rice is grown between August and January, but natural

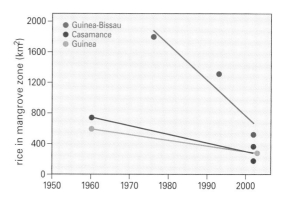

Fig. 100 Extent of rice fields in lowlands and mangrove swamps in the Casamance (southern Senegal), Guinea-Bissau and Guinea between 1960 and 2003. From: Bos *et al.* (2006); van der Kamp *et al.* (2008). The decline was probably smaller than indicated (see text).

Coastal rice fields hugging the tidal and mangrove zone, here shown for the rainy season with wet rice fields between wooded highlands and a tidal inlet in the Casamance (September 2007; left), and for the dry season with desiccated rice fields after the harvest, with mangroves in the background fringing a side branch of the Rio Mansoa in Guinea-Bissau; notice the small tidal flights along the inlet in the centre (March 2005). Both sites clearly show the high diversity and small scale of ecotopes, resulting in a high diversity of birdlife and good wintering grounds for Palearctic migratory birds.

succession results in a dense vegetation of grasses and rushes in the rainy season. Long vegetation is cut and deposited on small dams surrounding the parcels. A part of the sod is used to improve the dams, and the remainder turned on the spot. The ditches are deepened and the mud is laid down on the ridges. As a result the ground is completely bare before the rice is planted. Subsequently ditches may need to be deepened and ridges (re)constructed.

Rice is planted at intervals of 10-13 cm, usually on 35-40 cm wide ridges with 3-4 plants next to each other. The small ditches in between are about 35-45 cm wide. Rice fields are encircled by low dams of some 30-50 cm wide. (These dams were particularly helpful when performing density counts, just as the local people used them to cross fields without trampling the rice.) The estimated rice density lies in the range of 20-40 plants/m². Planting one ha takes about 200 hours, but this represents a fraction of the total amount of work to obtain a harvest of 1-2 ton/ha (500-700 hours per year per ha; de Jonge et al. 1978). Obviously, rice farmers work hard for their living.

Bos et al. (2006) arrived at a figure of 1128 km² of rice cultivation in coastal West Africa, of which 88 were in The Gambia, 183 in southern Senegal, 530 in Guinea-Bissau, 287 in Guinea and 40 in Sierra-Leone. This estimate was based on remote sensing combined with field work and referred to active cultivations. Earlier estimates gave the total rice-growing area, including fields in fallow. Van der Kamp et al. (2008) estimated that about half of the total rice extent in the Casamance in 2007 was actually in cultivation. During our own field work in the 1980s we did not attempt to quantify how much of the land laid fallow or was abandoned, but it must have been less than a quarter. The decline in surface area of rice cultivations since the 1960s may therefore be smaller than suggested in Fig. 100. Nevertheless, the rice extent in the 1980s was probably twice as large as in the early 2000s. Since traditional subsistence rice-growing systems are time-consuming, a reduction in rice farmers inevitably causes a reduction in the rice-growing area. As elsewhere in Africa, people are leaving the countryside in favour of cities. Locally, rural depopulation is the rule, e.g. in the northern half of the Inner Niger Delta (Zwarts & Kone 2005a). This must also have been the case in many coastal rice regions, due to political instability since the early 1980s.

Abandoned rice fields in the tidal zone may transform back into mangrove or change into bare sand- and mudflats or flats covered with low saline vegetation of rushes and grasses. Abandoned rice fields in the rain-fed zone above the high-tide line may convert into shallow backwaters covered with Water Lilies.

Irrigated rice

Rice needs a lot of water. Rice growing in the Sahel is therefore something of an anomaly. The climate of the Sahel is obviously more suited to grow drought-resistant and heat-tolerant millet and sorghum, as has been the case for centuries. Nevertheless, rice has become the staple food of people in the Sahel (Chapter 5), made possible by import and irrigation schemes along the Rivers Senegal,

Prior to planting, women clear the vegetation from the fields with a grub hook that looks like a slender pickaxe (*ebaraye* in Diola); men use a large spade (*kadiandou* in Diola). This system of rice growing is delicately attuned to the restrictions set by local conditions and manual labour.

Irrigation along Sahel rivers is not only detrimental to natural floodplains, but also to natural forests. The expansion of rice fields in the zone of *Office du Niger* in Mali led to widespread clearance of local forests; June 2005. Small-scale clearances are often carried out manually, which may seem less detrimental at first sight, but, in the long run, results in the same outcome: a man-made landscape with few natural forests.

Niger, Nguru and Logone (Chapters 6-9). The *Office du Niger* in Mali, the largest irrigation scheme in West Africa (740 km²), annually produces 333 000 tonnes of rice (state of affairs in 2001; Wymenga *et al.* 2005), being 40% of the national rice production in Mali. It is the rice granary of Mali with a more or less secured production, independent of rainfall and river flow, where 250 000 people can make a living (Bonneval *et al.* 2002). The down side of the coin is the extraction of water by *Office du Niger* from the Niger, before entering the Inner Niger Delta, limiting – among other things – profitable rice cultivation on the floodplains. Independent of the river discharge, *Office du Niger* takes 2.5 km³ per year, being equivalent to 6% of the annual flow in a wet year (1995) but as much as 16% in a dry year (1990). Consequently, due to upstream irrigation, the risk of crop failure in the Inner Niger Delta has increased, especially in dry years.

The cultivated rice zone of *Office du Niger* has expanded, on average, by 2.3% annually between 1983 and 2001. Water intake has remained stable over the same period but the yield has trebled to 6 tonnes per ha. Water use has declined from a staggering 45 000 l per kg rice in the mid-1980s to less than 10 000 l/kg in 2000. The development plan launched by *Office du Niger* in 1998 envisages a further expansion of at least 140-230 km², but possibly 300-400 km², by 2020 (Keita *et al.* 2002). Given hydrological constraints, this would only be feasible if the water use per ha can be reduced by another 20%.

Irrigation schemes are normally characterised by rectangular fields and straight canals, diverting water into smaller canals, and finally into irrigation ditches among rice parcels. However, the *Office du Niger* uses former river branches, called *fala*, to direct water to the various irrigation schemes. The *falas* and adjoining forelands have been turned into permanent wetlands, mainly covered by a Cattail vegetation of varying density. Cattail also abounds along canals and ditches, while the shallow water is covered by Water Lily *Nymphea* sp., Water Fern *Azolla* sp. and Knotweed *Polygonum senegalense*, and by recent invaders like Water Hyacinth *Eichhornia crassipes* (since

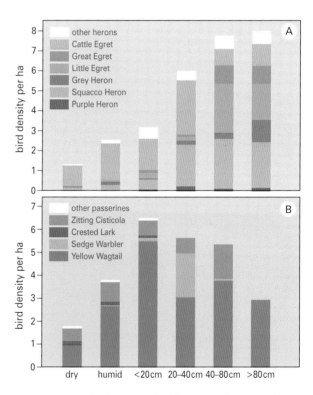

Fig. 101 Average bird density of (A) herons and egrets and (B) passerines as a function of water depth in rice fields. Data collected in November-February 2002-2006, combined for floodplains, irrigation schemes and coastal rice fields in Mali, Senegal, Guinea-Bissau and Guinea.

Rice farmers use seedbeds to germinate rice seed. Although it takes more time, it prevents seed predation (*e.g.* by Black-tailed Godwits). The fresh shoots are transplanted to rice fields where they are planted on bare ridges in the rainy season. Black-tailed Godwits prefer to feed on these freshly tilled paddies and cause some damage by trampling the tender rice shoots.

the early 1990s) and Kariba Weed *Salvinia molesta* (since the late 1990s). The overall impression of this irrigation scheme is therefore more 'natural' than of many other schemes.

Birds in rice fields

The bird density in cultivated rice fields has been measured in floodplains, in irrigation schemes and in the mangrove and tidal zone (Fig. 99). These data have been used to estimate the number of waterbirds present in:
- rice fields cultivated in floodplain areas (3300 km² in the Inner Niger Delta), based on 169 density counts (Chapter 6);
- irrigated rice fields, of which there are 900 km² in Mali (*Office du Niger*, Inner Niger Delta, Selingue) and 350 km² along the Senegal River), based on 933 density counts in *Office du Niger* (Wymenga *et al*. 2005), 432 density counts in Selingue (van der Kamp *et al*. 2005b) and 113 density counts in the Senegal Delta (Chapter 7);
- coastal rice fields (1130 km²), based on 1308 density counts in The Gambia, 80 in southern Senegal, 1108 in Guinea-Bissau and 558 in Guinea (Bos *et al*. 2006).

Altogether, 4701 density counts are available for November-February, with another 317 counts in July (irrigated rice fields only).

The field work was performed in 2002-2006. The plots measured 0.72 ha, on average (see Chapter 6). For each plot, vegetation type was noted as well as vegetation height and density, and water level. The analyses showed that, apart from vegetation type, water level was the most important variable explaining the variation in bird density (Fig. 101). For example, herons and Little Egrets prefer a water depth of at least 20 cm, but Cattle Egrets avoid rice fields with more than 40 cm of water (Fig. 101A). Sedge Warblers were found in rice fields with shallow water, which is also the preferred habitat of Yellow Wagtails, but the latter can also be observed in dry fields and in rice standing in deep water (Fig.101B). Waders typically reached their highest density (4-6 birds/ha) in shallow water.

Only 6% of rice field plots in the floodplains were dry in November, but this gradually increased to 35% in January and 62% in February. The trend is the same in the coastal zone, although less marked: from 6% in November, to 16% in December, 32% in January and 38% in February. No such trend is visible in irrigated fields: between November and February 50% of the plots were dry (but only 24% in July). Deep water is rare in rice fields: less than 1% of the plots had a water depth of more than 80 cm, and only 3% of 40-80 cm. In the analyses, we lumped the 1387 counts of dry rice fields and 3321 counts of humid and wet rice fields.

The density of waterbirds averaged 11 birds/ha in wet coastal

Table 17 Average bird density per ha in three types of rice fields, either dry or wet (= humid or covered by water), between November and February 2002-2006 (see also Fig. 101). Average values with a high standard error (SE, > 60% relative to mean) are marked red and values with a low SE (<30% relative to SE) are marked green; the category between has no colour. Number of counts given in the bottom line.

Birds per ha	Floodplain Wet	Floodplain Dry	Irrigation Wet	Irrigation Dry	Coastal Wet	Coastal Dry
Grey Heron	0.037	0	0.015	0	0.158	0.042
Black-headed Heron	0.302	0	0.003	0	0.061	0.030
Purple Heron	0.031	0	0.001	0.002	0.052	0.012
Squacco Heron	0.060	0	0.293	0.006	0.700	0.212
Cattle Egret	3.060	1.801	2.613	0.927	1.421	1.047
Little Egret	0.040	0	0.130	0	0.630	0.080
Intermediate Egret	0.006	0	0.258	0	0.022	0.006
Great Egret	0	0	0.022	0	0.189	0.055
Green-backed Heron	0	0	0	0	0.228	0.017
Hamerkop	0	0	0.062	0	0.093	0.030
African Jacana	0.131	0	0.167	0	0.181	0
Black-winged Stilt	0.513	0	0.277	0	0.186	0
Spur-winged Lapwing	0.384	0.248	1.652	0.466	0.742	0.272
African Wattled Lapwing	0.012	0	0.589	0.083	0.378	0.429
Little Stint	0.658	0	0.147	0.002	0.010	0
Green Sandpiper	0	0	0.015	0	0.070	0
Wood Sandpiper	0.937	0	3.668	0.001	0.618	0.006
Common Greenshank	0.061	0	0.061	0	0.300	0.015
Ruff	5.311	0	0.182	0.149	0.226	0
Spotted Redshank	0.060	0	0	0	0.095	0
Common Sandpiper	0	0	0.023	0.001	0.371	0.034
Crested Lark	1.379	0.104	0.261	0.429	0.030	0.016
Yellow Wagtail	2.409	2.194	8.784	1.522	2.767	0.411
Sedge Warbler	0	0	0	0	0.298	0
Zitting Cisticola	0.759	1.273	0.138	0.218	0.933	0.764
Other species	3.493	0.049	0.414	0.033	0.436	0.157
Total	19.641	5.670	19.781	3.852	11.235	3.638
N	112	31	598	560	2332	718

Table 18 Average bird density per ha rice field in Sahelian irrigation schemes in November-February (gradual desiccation of rice fields) and in July (at the end of the dry season). A selection is made of species present in July. Average values with a high standard error (SE, > 60% relative to mean) are marked red, values with a low SE (<30% relative to SE) green; a high SE indicates poor reliability of the figure). Number of counts given in the bottom line.

Species	Nov-Feb	July
Purple Heron	0.002	0.005
Great Egret	0.012	0.125
Cattle Egret	1.852	0.189
Squacco Heron	0.154	0.480
Green-backed Heron	0.000	0.190
African Jacana	0.086	2.155
Spur-winged Lapwing	1.077	0.742
African Wattled Lapwing	0.344	0.051
Total	3.691	3.977
N	1160	318

rice fields, but only half as much in dry rice fields (Table 17). Such densities are in the same order of magnitude as in floodplains and irrigated rice fields. The species composition was about the same in the various types of rice field, but densities varied relative to presence/absence of water and type of rice field. The most ubiquitous species across all types of rice field was the Yellow Wagtail. In coastal and irrigated rice fields, Yellow Wagtails were common in wet habitats, but scarce when rice fields had dried out. On the floodplains, however, Yellow Wagtails abounded in dry rice fields as well. Here, the flies accompanying grazing cattle offer plenty of food for Yellow Wagtails (Chapter 6).

When rice fields successively dry out in the course of the northern winter, waterbirds may concentrate in the remaining wet areas (unless they depart). Indeed, densities of Cattle Egrets and Squacco Herons, for instance, doubled between November and February. On the other hand, most Wood Sandpipers left the rice fields in the course of the winter months. The density of Yellow Wagtails, however, remained similar in wet rice fields between November and February.

The concentration of waterbirds in the few remaining waterlogged areas can also be illustrated by comparing irrigated rice fields in November-February with those in July, at the end of the dry season (Table 18). The density of African Jacanas, for instance, was 25 times higher in July than in winter. Higher densities in July were also evident in other species, even in Purple Herons and Squacco Herons, whose numbers are boosted by an influx of Palearctic birds in November-February. Neither Cattle Egrets nor lapwings showed higher densities in July, presumably because these species are not dependent on water to feed. For most waterbirds though, irrigated rice fields are among the very few wetlands available in the Sahel at the end of the dry season.

Compared to floodplains and irrigated rice fields, coastal rice fields showed relatively low bird densities in combination with a

Farmers, or rather their children, spend much time in crop protection, notably when harvest time is approaching and the ripening rice grain constitutes an attractive food source for seed-eating birds, such as Village Weavers (and many other species). Shouting, catapults, slings and other contraptions... the resourcefulness of rice farmers in protecting their crop almost defies description.

high bird diversity (Table 17). Coastal rice fields are interspersed with fallow land, and adjoined by mangroves, vegetated upper tidal flats and ponds with aquatic vegetation. This mix of habitats attracts many more species than those typical of rice fields proper (such as Sedge Warblers). Moreover, rice fields are used as high tide roosts by birds feeding on tidal flats during low tide; conversely, mangroves function as breeding and roosting sites for birds feeding in the rice fields. The whole is more than the sum of the parts.

The densities given in Table 17 may be used to estimate the numbers of birds present in the three types of rice fields. The irrigated rice fields in the Inner Niger Delta measure 140 km^2, those on the floodplains 2000 km^2. Given these multipliers, we estimate that the total number of Yellow Wagtails inhabiting these rice fields varies between 460 000 (when dry) and 600 000 birds (when wet). The difference is much larger for the other species (Table 17). Given the gradual decline in the surface area of wet rice fields from 94% in November to 36% in February, a decline of bird numbers present on rice fields is to be expected. Later in the winter season most waterbirds get concentrated in *bourgou* fields, which remain covered by water for much longer (Chapter 6). The numbers calculated to be present in the irrigated rice fields of *Office du Niger* (550 km^2 + 130 km^2 in the immediate surroundings) and in the coastal rice fields (1120 km^2) amount to 0.73 and 1.16 million respectively (Table 19). The figures should be read with caution, as standard errors are quite large. Our estimate of 76 000 Black-tailed Godwits, for example, must actually be read as 33 000 - 119 000. The reason for this wide range is that Black-tailed Godwits feed in flocks. The species was encountered in only 20 plots, meaning that 99% of the counts were devoid of Godwits. Estimates based on density counts are much better for species which feed scattered across the area. The error of estimate is, for instance, only 8% in Yellow Wagtails.

Rice fields a substitute for natural wetlands in the Sahel?

The high bird numbers in rice fields (that is to say: if wet) may indeed suggest an ecological compensation for the loss of floodplains. However, such a conclusion should be interpreted with caution. As discussed by Wymenga *et al.* (2005), rare and endangered species are completely missing among the birds feeding in cultivated rice fields (see also Table 19). Floodplain loss affects a much wider range of bird species. In the Inner Niger Delta, for example, the construction of the Selingue dam and the irrigation scheme of *Office*

Box 13
Rice fields as a substitute for natural wetlands: a European perspective

The value of rice fields as a substitute for natural wetlands in southern Europe has been studied by Fasola et al. (1996), Fasola et al. (1996) and Sànchez-Guzman et al. (2007). They conclude that rice fields are an important feeding area for herons, duck, waders, gulls and terns. In several lowlands and deltas, rice fields constitute the only feeding habitat left for waterbirds, since 80-90% of the original Mediterranean wetlands have been lost. There are still many natural wetlands in the Camargue, beside extensive rice fields. This offered the opportunity to compare habitat choice of waterbirds if both wetland types are available (Tourenq et al. 2001). Natural marshes attracted many more birds of a greater variety than rice fields. Where rice had been cultivated for longer time periods, even fewer birds were present due to the negative impact of soil management and pesticides (Tourenq et al. 2003). On the other hand, as natural marshes are mostly flooded in the winter months, flooded rice fields may offer alternative feeding grounds during the breeding season and during migration perods, especially in dry years. Rice fields in southern Europe may also be kept flooded during the winter months to facilitate shooting ducks and other waterbirds which feed on spilt rice on harvested fields. Black-tailed Godwits, returning from Africa in January, use the rice fields in Portugal and Spain for some weeks as a refuelling station. During this short stay, their diet consists entirely of spilt rice (Chapter 27).

du Niger resulted in the loss of 12% of the wintering waterbirds (but calculated to increase to 44% if the proposed Fomi dam comes into existence). The newly created irrigation zones compensated for a tiny fraction of these losses (Wymenga et al. 2005: 217), i.e. slightly more than 12% (but declining to a mere 4% with the Fomi dam).

The extent of West African mangroves has not declined much in recent decades, except in southern Senegal during the drought in the 1980s (Bos et al. 2006), but the cultivated rice area in the coastal zone seems to be in decline, although trends differ throughout the

Coastal rice fields attract a variety of large waterbirds, such as locally breeding Sacred Ibis and African Spoonbills (above). Without exception heronries are in the highest trees, relatively safe from human predation. The African Spoonbills nesting on Ilha de Papagaios, Bissau, also breed on huge trees (centre); to take this picture, one must have a good head for heights and no fear of Green Mambas or other venomous snakes which are particularly abundant in heronries. The high trees in Bissau, Ziguinchor and other West African towns close to coastal rice fields also harbour large colonies, for example, of African Darters and Great Egrets (shown on picture below; Ziguinchor, September 2007).

Table 19 Estimated total number of waterbirds present in the rice fields of *Office du Niger* (and other irrigated rice schemes in the immediate surroundings; 680 km²; Wymenga *et al.* 2005) and in coastal rice fields (1120 km²; Bos *et al.* 2006) in November-February. + = species present but estimate not reliable (standard error >60% relative to the mean).

Species	Office du Niger total	SE	Coastal rice Total	SE
Long-tailed Cormorant			2927	876
Grey Heron	702	347	14605	5326
Black-headed Heron			6003	1709
Purple Heron			4839	1191
Great Egret			17632	4453
Intermediate Egret	5806	3214	2033	611
Little Egret	2019	951	56961	13435
Cattle Egret	84295	19897	149257	21729
Squacco Heron	4903	1706	65547	9640
Green-backed Heron			20012	11831
Hamerkop	+		8731	1835
Sacred Ibis			445	257
White-faced Whistling Duck			9775	5277
African Jacana			15490	5478
Lesser Jacana			506	329
Black-winged Stilt	12189	4621	15913	3329
Collared Pratincole	7163	4565	1527	736
Painted Snipe			1404	665
Spur-winged Lapwing	87336	21612	70684	10190
African Wattled Lapwing			43684	8608
White-headed Lapwing			3746	2545
Common Ringed Plover			4270	1316
Little Ringed Plover			1632	1067
Spotted Redshank			8140	3800
Marsh Sandpiper			1336	859
Common Greenshank	+		26109	4864
Green Sandpiper	+		5951	1838
Wood Sandpiper	82082	18312	53105	6987
Common Sandpiper			32676	3340
Common Snipe	2058	1078	21369	3207
Little Stint	+		855	468
Ruff	+		19382	7727
Black-tailed Godwit			76008	43249
Pied Kingfisher			7961	1814
Yellow Wagtail	398082	88986	247708	19868
Red-throated Pipit			844	588
Plain-backed Pipit			11684	2122
Yellow-throated Longclaw			5552	2017
Sedge Warbler			25531	13034
Savi´s Warbler			901	487
Bluethroat			931	506
Crested Lark	27155	8747	3000	1032
Zitting Cisticola	9933	3014	100047	9873
Prinia sp.	3222	1736	230	145
Total	732019		1168382	
Number of counts	716		3051	

region (Fig. 100). The bird numbers estimated for the coastal rice fields (Table 19) are based upon a surface area of 1120 km², *i.e.* probably half of the area available in the early 1980s. Large losses would have dire consequences for migratory birds wintering in sub-Saharan coastal rice fields, such as West European Black-tailed Godwits (Chapter 27). African bird species may suffer as well, such as, for example, Black Crowned Cranes, which are still present in relatively large numbers (Beilfuss *et al.* 2007).

Conclusion

Rice fields, especially when wet, constitute an important habitat for wintering Palearctic birds. The species composition is not very different in rice cultivated on floodplains, in irrigation schemes or in coastal rice fields in the mangrove and tidal zone. Coastal rice fields harbour a larger variety of species than the other types of rice fields, and this is associated with the proximity of tidal flats and mangroves. At the end of the dry season, in July, irrigated rice fields offer water in an otherwise largely desiccated Sahel. Large numbers of waterbirds are then concentrated in this man-made habitat. Most bird species in irrigated rice fields belong to the (very) common species. Conversion of natural floodplains into irrigated rice fields results in declining bird numbers, as the creation of new habitat (rice fields) only partly compensates for losses incurred in floodplains.

Wetlands in the Sahel: a summary

The wetlands in the Sahel have much in common, but they also differ in several important respects (Chapters 6-11). The differences mainly depend on flood dynamics and whether or not natural resources are exploited by local people. The timing and extent of flooding have been substantially altered by the creation of reservoirs in rivers responsible for the flooding of the Senegal Delta, Hadejia-Nguru, Logone and Inner Niger Delta (Fig. 102). Flooding and human exploitation have an impact on vegetation and fauna, including birds.

Floodplain size The seven wetlands described in the previous chapters nicely illustrate the various fortunes of flooding, vegetation and bird life in the past decades (Fig. 103). A comparison of maximal flood extent (Fig. 103A) in the 1960s and 2000s shows that the Sudd became the largest wetland in Africa, after some years with an extremely high discharge of the White Nile in the mid-1960s (when flood extent more than tripled in just four years). It took over 30 years before the Sudd was back again at its pre-1960s size of 10 000 km^2. Lake Chad, on the other hand, measured about 25 000 km^2 before 1973, but has been about twice as small since then. The flood extents of the Inner Niger Delta, Hadejia-Nguru, Logone and Fitri showed large annual fluctuations in response to variations in rainfall and river flow, but – with the exception of Lake Fitri – were also permanently reduced in size due to the construction of hydrological infrastructures upstream. The annual flood extent in the Senegal Delta varied even more than in the other floodplains: between 500 and 3500 km^2 before the mid-1960s (still a natural river system) and between 250 and 500 km^2 since 1990 (dikes and dams have gradually reduced floodplain size). Without the artificial flooding of the National Parks Djoudj and Diawling, flood extent would have been reduced by another 200 km^2. All together, the Sahelian floodplains and lakes lost more than half of their surface area between the 1960s and 2000s, mainly due to reduced rainfall, but also exacerbated by the construction of dams and dikes. Floodplain loss in the Waza-Logone, resulting from the implementation of an irrigation scheme, was partly rectified by reopening the dike, which enabled reflooding of some 200 km^2.

Water column The depth of maximum flooding is highest in the Inner Niger Delta, varying from 4-6 m. The Hadejia-Nguru held up to 3.5 m of water before upstream reservoirs started to reduce floodplain size. The taming of the Senegal River substantially reduced the annual fluctuation of water level in the Senegal Delta, from 3.5 m to 0.5 m. In contrast, the seasonal variation in water level trebled in the southern basin of Lake Chad, but the northern basin fell dry. The water level varied less in Fitri (up to 150 cm), Logone (100 cm) and the Sudd (50 cm). (Fig. 103B).

Seasonality of flooding The rainy period for all wetlands falls between June and October, but the timing of flooding and deflooding differs (Fig. 103C). The rivers feeding the Hadejia-Nguru system are short with a steep gradient, and peak flooding is therefore already reached by September. The Senegal Delta also has a short hinterland. Discharge of the Niger River normally reaches its maximum in November, but peak flow is one month earlier with a low discharge, and two months later with a high inflow. Moreover, the Inner Niger Delta is so large that when the water in the southern part has already receded, the northern delta still has to be flooded.

Floodplain size, permanent and seasonal The surface area of the immersed floodplains is six to ten times as large as the surface area permanently covered by water (Fig. 103D), except for Lake Chad and the Sudd, where the difference is much smaller. In the Sudd, more than half of the wetlands are permanently flooded, independent of flood extent. In Lake Chad, the surface area of the seasonally flooded fringe of the lake increased from 2000 to 5000 km^2 after 1973, due to a drop of the lake's level. Dikes and dams in the Senegal River changed a large part of the Senegal Delta from an estuarine floodplain into a lake and, beyond the dikes, into drylands or irrigated areas. The presence of permanent lakes in the Inner Niger Delta depends on flood level: with low floods the lakes in the northern half of the area are not reached and dry out, but the lakes are refilled during high floods.

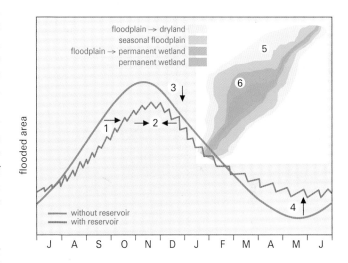

Fig. 102 Upstream reservoirs have an impact on floodplains in several ways. When the reservoir is filled at the start of the rainy season, it retards flooding (1) and lowers the peak level (3). This also reduces the flood extent (5). Some months later, when the reservoir is emptied, the low water level is raised (4), causing an extension of the permanent marshes (6). These man-induced changes turn the higher floodplains into drylands (5) and the lower floodplains into permanent marshlands (6); the remaining seasonal floodplains contain less water for a shorter period (2). The irregular water releases from the reservoir cause erratic short-term fluctuations in water level, unlike the normally gradual daily increase and decrease of water level associated with unhampered flooding and deflooding.

Vegetation The vegetation in Sahelian wetlands primarily depends on water depth and seasonal fluctuations therein. Another factor is human exploitation, notably grazing. Reed, Cattail and/or Papyrus do not grow in seasonal floodplains, as they are unable to survive the dry period when floodplains turn into semi-arid environments. As soon as floodplains are permanently covered by water, Cattail starts to colonise the shallow waters. This happened on a large scale in the Senegal Delta and locally in the Hadejia-Nguru. The reverse trend could be observed in Lake Chad, where the dense vegetation of Cattail, Reed and Papyrus disappeared when fluctuations in water level increased. These helophytes were replaced by grasses, adapted to extreme flood dynamics and characteristic of Sahelian floodplains (see below). Floating water plants do not dominate the vegetation in the floodplains, apart from Water Lily. Permanent lakes may be covered by Water Cabbage and - since the 1960s - by Water Hyacinth, or by emergent water plants such as Pond Weed. The Senegal Delta is a classic example of how the loss of flood dynamics profoundly changed the ecology of the floodplain, offering Cattail and invasive plants like Kariba Weed a red-carpet welcome.

Seasonal floodplains are mainly covered by grass species that can cope with variable inundation depths and flood durations (Fig. 103E &F). *Bourgou* grows where water depth reaches at least several metres in areas which are flooded for more than half of the time. The species is common in the Inner Niger Delta, and was found locally in the Hadejia-Nguru before dams reduced the flooding. *Didere* or Hyppo grass is widespread in African floodplains, where

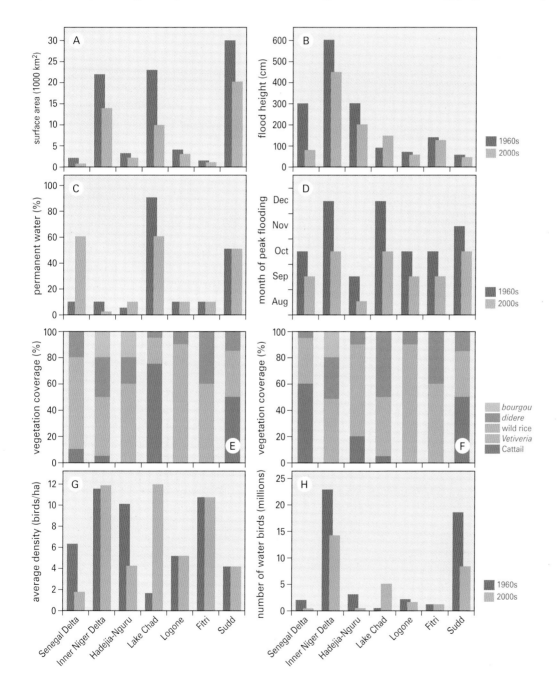

Fig. 103 Hydrological and ecological characteristics of seven Sahelian wetlands, ranked from west to east, in the 1960s and 2000s. (A) Maximum annual surface area of the floodplains and lakes; (B) Percentage of wetland permanently flooded during a year; (C) Height of the flood relative to the average lowest water level; (D) Month during which the flood reaches its peak level; (E) Coverage of major vegetation types in the 1960s; (F) Coverage of major vegetation types in the 2000s; (G) Overall average bird density in floodplains (given an average density of 20 birds/ha in bourgou and didere, 5 birds/ha in rice, 2 birds/ha in Vetiver Grass and 0 birds/ha in Cattail, Papyrus and open water); (H) Numbers of birds present, obtained by multiplying bird density and surface area of the floodplains.

The Black Crowned Crane is a majestic bird, sporting a regal perch in its territory, from which in early morning the rich unison call – in fact a carefully orchestrated duet performed by the pair – is uttered for all to hear and to rejoice in. The demise of this species in much of West Africa is strongly felt when visiting areas where the species still abounds, such as Guinea-Bissau and the Casamance: the loud bugling here is almost otherworldly. Casamance, September 2007.

water depth varies between one and three metres. However, in floodplains with less water (*e.g.* Logone), the species is rare and replaced by wild rice (in about 1 m of water); in even shallower zones, other grasses, such as Rats Tail Grass *Sporobolus pyramidalis* and *S. robustus*, Antelope Grass and Black Vetiver Grass take over. In part of the Logone floodplain the perennial Vetiver Grass was replaced by less productive annual grass species after the Maga dam reduced the already shallow flooding to zero.

Bird densities Highest bird densities in the Inner Niger Delta were found in *bourgou* and *didere* vegetations with about 30 birds/ha, with lower numbers in wild rice (7 birds/ha) and in the Vetiver zone (2 birds/ha; the latter estimate is based on few counts). The few counts in dense Cattail fields in the Senegal Delta indicate that bird density here is close to zero. Extrapolating these densities to aquatic vegetations in other floodplains, separately for the 1960s and 2000s, a decline in overall density of birds in the Senegal Delta and Hadejia-Nguru is apparent, associated with a concomitant loss in flood dynamics, resulting in an increase of Cattail and a decline of *didere* and *bourgou* (Fig. 103G). Between the 1960s and 2000s, we estimate that the number of wetland-associated birds in the seven Sahelian wetlands has declined by some 40%, given the known habitat changes in the interim period and average bird densities per habitat type (Fig. 103H). This decline is mainly caused by a reduction in floodplain size (Inner Niger and Senegal Deltas, Sudd) and vegetation dynamics (more Cattail in the Senegal Delta).

These estimations can only be, at best, indicative. First, we do not know whether bird densities in the zones with *bourgou* and wild rice differ between Sahelian wetlands. Secondly, are bird densities independent of the surface areas of the various habitats? This question is particularly relevant in the face of reductions in habitat availability in the wake of the construction of upstream irrigation schemes or storage reservoirs. For instance, the Fomi Dam, planned in the Upper Niger River, will reduce the flood extent of the Inner Niger Delta by 10%, but *bourgou* (associated with deep water), with its high bird numbers, will face a loss of 60% of its area. Can *bourgou* sustain even higher densities, off-setting the loss of habitat, or are densities already at a maximum and will the anticipated loss of *bourgou* fields result in a decline of *bourgou*-associated birds? We believe the latter assumption the most likely, as shown in the following chapters. Population trends of several Palearctic bird species are correlated with flood extent, strongly indicating that numbers of some bird species are regulated on the wintering grounds. Habitat loss may, however, be compensated by the creation of new habitats. For example, the irrigation system by *Office du Niger* upstream of the Inner Niger Delta reduced the flood extent of the floodplain, but also created 700 km^2 of habitat for waterbirds. The same holds for the Senegal Delta where 200 km^2 of floodplains were turned into rice fields. The average bird density in

Box 14
Kafue plains and Okavango Delta

The seven Sahel wetlands described in this chapter can be ranked according to flood dynamics, ranging from relatively small (the Sudd) to large (Inner Niger Delta). Other African wetlands are equally different in dynamics.

The Okavango Delta in Botswana is, just like the Sahel wetlands, situated in a semi-arid zone with a high evaporation rate; the flooding of the system also depends on river inflow. The rainy season is more prolonged than in the Sahel and covers another part of the year (November-April). The outflow from the marsh amounts to only 5% relative to the water inflow, very small compared with the large Sahelian wetlands, where about half of the water is 'lost' to evaporation and seepage. Sutcliffe & Parks (1989) used a water balance model to calculate the monthly extent of the floodplains in the Okavango. The water level in the Okavango varies less than 100 cm and the annual flood extent ranges between 6000 and 8000 km². Hence, the major part of the Okavango is a permanent swamp. In the context of wetlands in the Sahel, it mostly resembles the Sudd, being also covered for a large part by Papyrus and Cattail.

The Kafue Flats cover some 6500 km² of floodplains and wetlands in Zambia. The gradient of the Kafue River is less than 5cm/km, and it takes two months before the water has passed the full 250 km of the Kafue Flats. A dam in the Kafue Gorge, downstream of the wetlands, and another dam 270 km upstream, profoundly changed the hydrology of the system. The peak flow was reduced from 700 to 450 m³/s, while the flow in the dry season increased from 50 to 200 m³/s. Both peak and trough of water level were mitigated (Fig. 102). *Bourgou*, wild rice and water lily in the seasonal floodplains were replaced by permanent marsh vegetation of Reed and Cattail; *Mimosa pigra* and Ambatch colonised huge areas (Mumba & Thompson 2005).

The Okavango Delta is an alluvial fan with a very low elevation gradient, resulting in low flooding dynamics. Some 30% of the area is permanent swamp, characterised by extensive stands of *Cyperus papyrus* and *Miscanthus junceus*. This habitat is poor in birds in comparison with Reed swamps. Three boating transects of 5 km each on 13 January 1994, when this picture was taken, revealed 292 individuals of 36 waterbird species in Reedbeds *Phragmites communis*, 208 individuals of 15 species in Okavango Water Fig *Ficus verruculosa* mixed with *Phragmites*, and only 28 individuals of 8 species in Papyrus with water ferns (R.G. Bijlsma, unpubl.).

rice fields amounts to 11 birds/ha, showing the potential partly to off-set losses incurred in floodplains.

The reasoning so far hinges on the influence of human-induced changes in flooding and natural vegetations on bird populations. However, natural floodplains are also heavily exploited by local people. Farmers in the Inner Niger Delta removed wild rice and flooded forest to cultivate rice. Many forests surrounding the floodplain were cut in the dry 1970s and 1980s. Grazing intensity by the millions of cattle, sheep and goats has increased tremendously in the 20th century. Fish exploitation has also intensified, as indicated by the plethora of fish traps, nets, baskets and hook lines; large fish have become scarce, and average fish size is still declining.

Impact on bird populations Many Palearctic and Afrotropical species depend on Sahelian floodplains for their survival. The marked changes in the wetlands of the Sahel in the 20th century, whether man- or climate-induced, have had a big impact on a variety of African and Palearctic bird species. Waders using the Senegal Delta showed a marked reduction in numbers after the seasonal floodplains were reduced from 3000 to 1000 km² and the coverage by Cattail increased from 0 to over 50%. On the other hand, neither ducks nor fish-eating birds showed a decline in numbers, possibly because ducks started to exploit the irrigated rice fields. Also, the artificial flooding of the Djoudj and – especially – the Diawling created estuarine habitat that attracted ducks and fish-eating birds.

Black-winged Stilts, with White-winged Terns, egrets and a Reef Heron, forage in shallow waters, sometimes forming feeding associations when temporary food bonanzas are available.

In this particular instance, the wider Senegal Delta maintained its waterbird population at the level from before the immense reduction of the natural floodplains. When the maximum numbers from Table 20 are converted into densities, using the surface areas shown in Fig. 103A, the Senegal Delta has an extremely high bird density compared to other Sahelian wetlands.

The impact of the changes mentioned above differs per species, but is, generally speaking, larger on local breeding birds than on winter visitors. The former loose feeding grounds and are exploited when breeding unless protection is enforced. The only Sahelian wetlands with strictly protected areas are the Senegal Delta (Djoudj National Park, Diawling National Park and other protected zones), and, to a lesser degree, the Logone floodplains (Waza National Park) and Lake Fitri. The Djoudj harbours the only remaining large Great White Pelican colony in the western Sahel (apart from the Banc d'Arguin). Large bird species in general offer a profitable and easy source of protein for local people, especially when breeding in colonies. This effect is clear-cut in densely-populated wetlands (Inner Niger Delta, Hadejia-Nguru), but is presumably less devastating in areas with lower human pressure (Sudd). The latter is exemplified by the ground-nesting Black Crowned Crane, which is still common in the Sudd (as are wild herbivores), Lake Fitri and the Logone but (virtually) lost from other Sahelian wetlands. The species may tolerate human pressure when refugia for breeding are available, such as estuarine habitats - including mangroves - bordering coastal rice fields in Senegal and Guinea-Bissau.

Resilience of Sahelian wetlands Taming the floods changed the ecology of Sahelian wetlands, often with negative effects on bird life. However, Sahelian floodplains have always been dynamic systems with large interannual variations in flooding, resulting in plant and bird communities adapted to marked changes in local conditions. This is borne out by the remarkable recovery of colonial breeding birds and Black Crowned Cranes following floodplain rehabilitation in the Diawling National Park, and after protection measures were taken in the Senegal Delta. Ecological restoration can be successful, provided that protection from exploitation (collecting young, capturing and shooting birds) is guaranteed. Also, protection measures and restoration can be nullified or endangered when new colonizers – with different cultural backgrounds and often not rooted in local customs – are opportunistically attracted to newly created farmland and grazing grounds

Site	Maximum number								Population X 100				
	IND	SD	HNW	Logone	Fitri	Sudd	ON	coast	W Africa	Africa	Europe	W Asia	Total
count type	mixed	mixed	ground	ground	aerial	aerial	dens	dens					
Great White Pelican	4300	39406	23	151	450	5643			600	2300			2900
Pink-backed Pelican	72	270		201	203	11187				500-1000			1000
Long-tailed Cormorant	50875	2837	1511	6304	3013	8883		2927	>1000	>1000			2000
Grey Heron	23915		678	1096	1085	984	702	14605		10000	4900		14900
Black-headed Heron	94		79	12323	141	1716		6003	3000				3000
Purple Heron	50417		355	161	507	5049		4839		880	2300	250	3430
Black-cr. Night Heron	12000	5300		1896		36823				1200	790	1000	2990
Squacco Heron	183390		870	23211	3546	18414	4903	65547		4500	640	1000	6140
Green-backed Heron	200					20012				10000			10000
Black Heron	390			7264	105					1000			1000
Cattle Egret	335377		37832	24004	9416	172359	84295	149257		14100	2800	1000	17900
Little Egret	37389		845	8643	203		2019	56961	1000	5800	1300	580	8680
Intermediate Egret	13009		110	470			5806	2033	1000				1000
Great Egret	8879		518	808	3626	9530		17632		3000	470	1000	4470
Yellow-billed Stork	210		110	1077	145	11154				890			890
African Openbill Stork	2			1705	100	344487				4000			4000
Woolly-necked Stork	22			489		2475				1000			1000
Saddle-billed Stork	1			30	5	4158				250			250
Marabou Stork	102			1860	20	359719				3500			3500
Shoebill						6407				65			65
Sacred Ibis	1160	293	572	3616	2270	17688		445		3300			3300
Glossy Ibis	25800	1150	2447	1683	4154	1695240				15000	570	1000	16570
Eurasian Spoonbill	75	6396	52							65	230		295
African Spoonbill	900	322	110	304	1072					1000			1000
White-faced Wh. Duck	73647	120683	58613	41942	95238	51810		9775	6500	16500			23000
Fulvous Whistling Duck	72700	37760	9510	132	11000	8775			750	3400			4150
Knob-billed Duck	19818	4580	2196	5939	8295	9611			650	3900			4550
Spur-winged Goose	18519	8514	7332	2407	2195	150216			750	4000			4750
Egyptian Goose	1829	2662	1	39	380				75	3600			3675
Northern Pintail	384685	247354	34866	6213	56000							15000	15000
Garganey	899916	296191	147563	23869	97332							25000	25000
Northern Shoveler	13907	34236		1777	3165							4900	4900
Ferruginous Duck	15066	230		0	8450							1500	1500
Black Crowned Crane	50	711		2313	441	36823			150	570			720
Purple Swamphen	10844								250	1100	250		1600
Black-winged Stilt	5536	1137	1210	9147	2770		12189	15913		1000	1270		2270
Pied Avocet	86	4940	42							1200	1200	250	2650
Egyptian Plover	753			378					350	700			1050
Collared Pratincole	20904		170	11535			7163	1527		1000	430	1000	2430
Spur-winged Lapwing	72503		551	2753	557		87336	70684		4000	1000		5000
Common Ringed Plover	6073		80					4270			730	10000	10730
Kittlitz's Plover	13887		48						350	3000			3350
White-fronted Plover	791								130	800			930
Black-tailed Godwit	55060	23888	6473	2318	2990			76008			3000	1000	4000
Spotted Redshank	4557	1200	4065	882				8140			900	1000	1900
Common Greenshank	3031		214					26109			2300	10000	12300
Marsh Sandpiper	2651		70	649				1336			270	750	1020
Little Stint	164576	50000	1231	2212				855			2000	10000	12000
Wood Sandpiper	199448	10000	168	2926			82082	53105			10500	20000	30500
Ruff	276834	500000	70845	146343	10503			19382				12500	12500
Gull-billed Tern	3986		174	131						50	130	630	810
Caspian Tern	3545		8						530	555	225		1310
Whiskered Tern	6679		988	1204							1280		1280

Box 15
Waterbird numbers in large sub-Saharan wetlands

% of African-Eurasian population							
IND	SD	HNW	Logone	Fitri	Sudd	ON	coast
1	14	0	0	0	2		
0	0		0	0	11		
25	1	1	3	2	4		1
2	0	0	0	0	0	0	1
0	0	0	4	0	1		2
15	0	0	0	0	1		1
4	2	0	1		12		
30	0	0	4	1	3	1	11
0					2		
0			7	0			
19	0	2	1	1	10	5	8
4	0	0	1	0		0	7
13	0		0			6	2
2	0	0	0	1	2		4
0	0		1	0	13		
0	0		0	0	86		
0			0		2		
0			0	0	17		
0			1	0	103		
0					99		
0	0	0	1	1	5		0
2	0	0	0	0	102		
0	22	0					
1	0	0	0	1			
3	5	3	2	4	2		0
18	9	2	0	3	2		
4	1	0	1	2	2		
4	2	2	1	0	32		
0	1	0	0	0			
25	16	2	0	4			
36	12	6	1	4			
3	7		0	1			
10	0		0	6			
0	1		3	1	51		
7							
0	0	0	1	0		1	1
0	2	0					
1			0				
9		0	5			3	1
15		0	1	0		17	14
1		0					0
4		0					
1							
9	6	2	1	1			19
2	1	2	0				4
0	0		0				2
3	0		1				1
14	1	0	0				0
7	0	0	0			3	2
22	40	6	12	1			2
5	0	0					
3	0						
5		1	1				

The significance of Sahelian wetlands, including the rice fields of *Office du Niger* in Mali and the coastal rice fields from Gambia up to and including Guinea, shows in the peak numbers counted, when compared to population sizes in W Africa, sub-Sahara Africa (including W Africa), Europe (including the Mediterranean and hence N Africa) and W Asia (Table 20).

The table is only indicative. Estimates of population sizes are often based on incomplete data, certainly so for the species breeding in Africa. Maximum numbers for the wetlands are also variously obtained. Density counts and extrapolations were used for coastal rice fields, aerial counts for Lake Fitri, density counts and aerial surveys for the Inner Niger Delta; for the latter area ground surveys and roost counts are also available. The numbers for the Sudd are based on just three systematic density counts from a plane. Moreover, maximal numbers of the Logone are based on not more than two complete ground surveys, as opposed to dozens for the Inner Niger Delta. We believe, however, that the table gives a fair impression of the significance of the Sahel wetlands for large waterbirds. Especially noteworthy are the following points:

The maximum numbers for the Sudd are (nearly) equal to, or even surpass, the total population size for African Openbill Stork, Saddle-billed Stork, Marabou Stork and Glossy Ibis, indicating that either the world population is underestimated or the estimate for the Sudd is too high, or both. Delany & Scott (2006) assumed that the 1.5 million Glossy Ibises breed locally, but that is unlikely, since this number was counted in midwinter, while hardly any birds were present in September and March. Obviously, the Sudd must be an important area for waterbirds, but actual data are of utmost importance. The present estimate of the world population of Black Crowned Cranes (72 000), for instance, still hinges upon the 37 000 birds from one of the three aerial counts of the Sudd in 1979-81.

The Inner Niger Delta harbours a quarter or more of the world population of Long-tailed Cormorants, Squacco Herons, Northern Pintails, Garganeys and Ruffs, and 10% or more of seven other waterbird species.

The counts in the much smaller Hadejia-Nguru, Fitri and Logone suggest that, of some ten bird species, 1% of the African-Eurasian population may be present, or even more. The three areas combined, together with Lake Chad (not given in the table), are as important, or even more so, than the Inner Niger Delta.

The Senegal Delta is about ten times smaller than the Inner Niger Delta, but Northern Pintails and Garganeys (and Ruffs in the past) were present in numbers equal to those found in the Inner Niger Delta. It is an important wintering site for Eurasian Spoonbills and contains the largest known breeding site of Great White Pelicans.

Irrigated rice areas attract many waterbirds, as shown for the area of *Office du Niger* in Mali. Even more important are the coastal rice fields behind the mangroves between Gambia and Sierra Leone. See especially Black-tailed Godwits, Spur-winged Plovers and Squacco Herons.

Table 20 Maximum number counted in Inner Niger Delta (IND; table 4, 5 & 9), Senegal Delta (SD; Table 12), Hadejia-Nguru (HN; Table 14), Waza-Logone (WL; Table 15), Fitri (Table 15), Sudd (Table 16), the irrigated rice fields of *Office du Niger* (ON; Table 19)) and the coastal rice fields (coast; Table 19), compared with population sizes for Africa, Europe and West Asia (Delany & Scott 2006). The last columns give the maximum numbers as percentages of the total population size for Africa, Europe and West Asia combined.

The Sahel as wintering area for Eurasian bird species

The relatively narrow latitudinal strip of land that forms the southern edge of the broad Sahara was by no means terra incognita before it featured widely in the media in the early 1970s. Ernst Hartert was only 24 when he took part in the famous Flegel expedition in 1885-86 to Nigeria, then still largely uncharted faunistically and botanically (Hartert 1886). He collected extensively in northern Nigeria, despite often ailing from fever or losing his gunpowder in river accidents. Subsequently, many more naturalists explored the region, and by 1947 Guichard (1947) had produced a wide-ranging overview of the birds of the Inner Niger Delta. However, once standardised monitoring of breeding bird populations in Europe had been established, the results enabled the changes in Sahelian bird numbers to be detected, raising the concomitant questions of 'how and why?'

Glue (1970), Berthold (1973, 1974) and Winstanley et al. (1974) were among the first to suggest a relationship between the variations in breeding numbers of Whitethroat in Europe and the circumstances in the Sahel in the preceding winter. They reported sudden declines in German and British populations of several trans-Saharan migrants from 1969 onwards, i.e. at the early onset of what later was known as the Great Drought, a prolonged period of little rainfall (1972-1993), a circumstance that had not occurred for over 40 years. Among the bird species most afflicted were Common Sand Martin, Common Redstart, Sedge Warbler and Common Whitethroat. Other studies went on to cement this provisional conclusion into a solid Palearctic-African framework, in which Sahelian hydrology acts as underlying trigger of population changes. For example, Cowley's 1979 analysis of Sand Martin vicissitudes in Nottinghamshire (Great Britain) showed a clear relationship with Sahel rainfall, as did those of Szép (1993, 1995a, 1995b) for Sand Martins breeding along the River Tisza in Hungary. A more detailed analysis of the Nottinghamshire data (Cowley & Siriwardena 2005) revealed that survival rates of this particular population were also driven by rainfall in the breeding area in the previous summer, whereas rainfall during the breeding season impacted productivity. For an entirely different species, the Purple Heron, den Held (1981) and Cavé (1983) had found that numbers in several Dutch colonies fluctuated in synchrony with the combined discharge of the rivers Niger and Senegal. A plethora of studies followed, increasing in scope, number of species, geographical range and use of various hydrological and climatic features. Bruderer & Hirschi (1984), for instance, took the number of ringed birds in Switzerland as a measure of population fluctuations and compared these trends with Sahelian rainfall, river discharge of the Senegal and water level in Lake Chad. Svensson (1985) compared Finnish trends for 13 bird species with rainfall in the Sahel, an exercise extended by Marchant (1992) and Baillie & Peach (1992) to 15 bird species breeding in Great Britain, Sweden, Denmark, Finland and Czechoslovakia. These studies and many others will be discussed in the following chapters.

The Palearctic hordes between the Sahara and the rain forest

Playing with numbers and making lists have long been favourite pastimes of birdwatchers and scientists. Figuring out to what extent Africa is swamped by which migratory Palearctic bird species is no exception. Moreau (1972) tallied 187 Eurasian species migrating to Africa (K.D. Smith *in* Moreau added another 73 species, mostly marine, coastal and vagrant species). A reappraisal by Walther & Rahbek (2002) resulted in a list specifying 351 Palearctic species, of which 240 cross the Sahara.[1] Of the 240 sub-Saharan migrants, 40 are restricted to marine habitats (seabirds and coastal waders) and some of the remaining 200 landbirds must be considered as vagrants. The Sahel and Sudan zone receive a fair number of all the Palearctic species that winter in Africa (Fig. 104). From west to east, the proportion of Palearctic species in West African avifaunas ranges from 29% in Senegambia (Morel & Roux 1973, Barlow *et al.* 1997), to 21% in Liberia (Gatter 1997b), 18% in Ghana (Grimes 1987), 18% in Togo (Cheke & Walsh 1996) and 19% in Nigeria (Elgood *et al.* 1994). These unadorned statistics attest to the rich Palearctic element in avifaunal West Africa.

The numbers of Palearctic, and in particular Eurasian, species in sub-Saharan Africa decrease from north (open landscapes) to south (wetter and more densely-vegetated areas), a cline largely related to rainfall and habitat (Moreau 1972, Alerstam 1990, Newton 1996a). The European birdwatcher visiting an African savanna or floodplain during the northern winter can relate more easily to their avifauna than to that of the rainforest, where the small numbers of Palearctic species are swamped by African species and in any case are difficult to see in dense canopy. This general picture is typical for West and Central Africa, but does not apply in East Africa, where two rainy seasons, topography and a much more complex amalgam of habitats attract Palearctic migrants further south and in larger numbers (Fig. 104).

For our purposes, we reduced the number of Palearctic bird species to those whose lives are – in one way or another – closely intertwined with conditions in the vegetation zones of western and central Sahel and of Sudan (Fig. 105)[2]. Our final list of 84 species covers a range of habitats between 10° and 20°N, from semi-desert, savanna and scrubland, to floodplains, and slightly further south almost to 15°N to woody savanna. Twenty four bird species are found mainly in semi-desert and savanna, 17 in scrubland and woody savanna. In addition to these 40 dry-land species, wetlands hold another 35 species, of which 12 occur mainly in floodplains. Eight species could not be restricted to any of these categories, such as omnipresent generalists like Yellow and White Wagtail or aerial feeders like Pallid Swift and Barn Swallow. Not all species depend upon the Sahel to the same degree. For some, a substantial part of the population winters north of the Sahara, whereas others winter largely south of the Sudan vegetation zone or strictly along the coast. Only 27 species can be considered as dependent completely on the Sahel, *i.e.* they cannot avoid Sahelian conditions, which, when adverse, act metaphorically as a bottleneck narrowing the lifeline of the species during the northern winter. Palearctic migrants showing long-term and consistent declines often are from this dependent category, and we discuss this subject at length in Chapter 44. To provide some perspective, we selected (Chapter 15-41) an array of species to identify events pivotal to numerical fluctuations, such as migration strategies, distribution and habitat selection in Africa and changing conditions on the breeding grounds.

Three general questions need to be addressed before we explore the fortunes of different bird species wintering in or near the Sahel:

- Do birds (individuals or species) from western and eastern Europe mingle in the Sahel, and if so, to what degree? This question is pertinent not only when survival conditions differ between eastern and western sections of the Sahel, but also when escaping adverse conditions is easier along the eastern latitudinal axis, where topography and habitat allow access to the intertropical rain belt across the equator, than along the western axis, where southern Africa can be accessed only by crossing the rainforest barrier.
- What are the conditions in the Sahel that are of vital importance to survival?
- How do we estimate annual winter mortality in Africa, and is it possible to derive winter mortality in the Sahel from monitoring schemes in Eurasia?

Sub-Saharan distribution and migratory pathways

The distribution of Palearctic birds across the Sahel is not simply a matter of occupation of certain habitats, resource partitioning or throwing a dice. The chances of a British Sedge Warbler turning up in the Chad Basin are pretty slim, just as a record of a Ukrainian bird in the Senegal Delta would seem strange. Even when Palearctic birds use broad-front migration, as most do, rather than concentrating at short sea crossings, western European birds tend to occupy the western and central Sahel, and eastern European the central and eastern parts. With few exceptions, winter distributions in the Sahel reflect those in the breeding quarters. This of course is the general picture, mainly as it emerged from ringing recoveries (Fig. 106), but

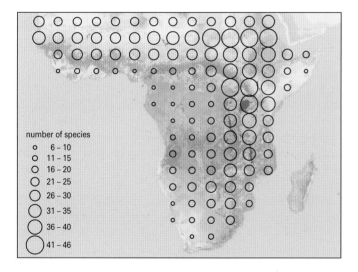

Fig. 104 Numbers of Eurasian insectivorous species bird species wintering in sub-Saharan Africa; land-cover taken from Fig. 32. From: Alerstam (1996) who based his Fig. 70 on the distribution maps of Moreau (1972).

Box 16
Playing with the numbers

Numbers, or as Moreau was taught in primary school, 'figures' are irresistible (Moreau 1972: 45). Present-day scientific texts revolve round 'figures'. Some numbers are factual, others are not or even imaginary, or perhaps not entirely so. Estimating the number of Palearctic migrants heading for Africa is a daunting enterprise, one tackled by Moreau using humour and juicy prose. In his own words:

"Let us now toss a few figures into the air. I use this lighthearted expression deliberately, because the last thing I would wish is to be taken too seriously in this connection and for the figures I shall put forward to be taken as "hard" and quoted as such. Whatever I may say in deprecation, experience teaches me, I am afraid, that they will be. Nevertheless, the temptation to numerical speculation is irresistible and one can, I think, be reasonably sure that at least the order of numbers involved can be ascertained."

His prognosis, that his estimate of 5000 million Palearctic migrants in Africa would be taken literally, actually was reasonably accurate, not missing the mark by much. What is all the more remarkable is that his calculation was based on numbers that were wrong. His starting point, the detailed study by Einari Merikallio (1958) on distribution and numbers of Finnish birds, was a good enough choice, even though it was the only one available at the time. Finnish ornithologists had been censusing breeding birds since the early 20th century, long before the rest of Europe, and using (semi-)standardised mapping techniques in homogeneous plots or transects stratified according to habitat (Palmgren 1930, Koskimies & Väisänen 1991). Merikallio's study provided country-wide estimates for all Finnish breeding birds, calculated mostly from quantitative habitat-stratified transects walked in 1941-1956 across Finland. Moreau used this singular study to apply its methodology to estimate the number of Palearctic migrants breeding per km^2 in Finland, then to extrapolate the results across the Western and Central Palearctic. To this end, he summed all estimates of Finnish bird species wintering in Africa, assumed Finland to comprise some 250 000 km^2 of land, and subsequently arrived at an average density of 100 pairs of migrant species to each square kilometre of land surface. Expressed in bird numbers, he multiplied the 100 pairs by four (pair plus on average two young per pair), then decided that 400 birds/km^2 was too high and so 200 birds/km^2 would be a more reasonable figure. The basis of this downward estimate was the Willow Warbler, for which the total of 5.7 million estimated pairs in Finland was felt to be over-weighted in comparison to the figures for the rest of Europe. Using 200 birds/km^2 as the multiplication factor, he arrived at 5000 million Palearctic-African migrants from the entire West and Central Palearctic (*i.e.* 25 million km^2 of which 0.8 million km^2 lie in N Africa, 10.4 million in Europe and some 14 million in Asia west of 90° E).

Curlews are scarce in West African floodplains. The long bills of these two birds clearly hint at their Asian origin. Curlews breed in Eurasia between 5°W and 120°E. Within this vast expanse, bill length increases gradually from west to east, from 115 to 140 mm in males and from 135 to 180 mm in females. Average bill length of Curlews captured in West Africa is 137 mm and 176 mm, in males and females respectively (Zwarts 1988, Wymenga et al. 1990).

For several reasons, the above calculations were flawed. For a start, Finland comprises 305 470 km² of land. Secondly, summing all Palearctic-African migrants in Merikallio's study produces an average density of 44 pairs/km² (=176 birds including young, or rather 88 birds when following Moreau's principle of halving the number), resulting in a corrected total number of 27 million birds (305 500 × 44 × 1/2 × 4), rather than his estimate of 50 million (250 000 × 100 × 1/2 × 4). Now, is this really important enough to create doubt? That depends. On a Palearctic scale, Moreau's guess may not have been wide off the mark by much, even when we take into account the vast differences in bird assemblages within Eurasia. Remember that Moreau did say, "We can, I think, be pretty sure that the order of number is correct", and these figures are of the same order. Number estimates should not be taken as gospel. It pays sometimes to recalculate earlier estimates and it shows that humour in science is a blessing and that more recent estimates like 3500-4500 million Palearctic migrants (B. Bruderer in Newton 2008) and 70 000-75 000 million African birds (Brown et al. 1982) should be viewed in a just as relaxed manner as Moreau did for his own estimates. Incidentally, the present Finnish population of Palearctic-African migrants is estimated at 21.1 million pairs, i.e. 42.7% of the total breeding population (Koskimies 2005), not the 13.5 million assumed by Moreau. The difference is based largely on improved census methods.

even the classic textbook example of a migratory divide, the White Stork (Schüz 1971), turned out to have a more versatile migratory strategy than previously thought: east and west do meet (Berthold et al. 2001a, 2002, 2004, Brouwer et al. 2003), sufficiently so that the eastern flyway may be used during outward migration, and the western flyway for the return migration (admittedly illustrated so far only by a single bird, perhaps not representative of White Storks in general; Chernetsov et al. 2005).

To investigate overlap across the entire width of Africa from Atlantic Ocean to Red Sea, 6089 recoveries of Eurasian birds in Africa between 37°N in the north (covering most of Africa north of the Sahara) and 4°N in the south (up to the rainforest) were divided into seven longitudinal categories corresponding to the east-west gradient of breeding in Eurasia (Fig. 106). Only 142 recoveries originated from breeding areas east of 30°E, and these were grouped in the category of 25-30°E. To have an indication of the degree of overlap in wintering range for the seven European longitudinal classes, we calculated the average longitude of recovery in Africa and its associated standard deviation. This exercise showed that the east-west distribution in sub-Saharan Africa could be largely attributed to a similarly zoned arrangement of the breeding areas of Eurasia, but not entirely. For example, Baltic and Ukrainian Caspian Terns overlap fully in their winter distribution, both concentrating in the Inner Niger Delta and along the West-African coast (Fig. 107). The bias of the amount and location of ringing effort within Eurasia is also likely to corrupt comparisons in species whose breeding grounds extend east of 25°E. The low explained variance in Black-tailed Godwits, for instance, is an artefact: almost all recoveries came from The Netherlands; few godwits have been ringed in eastern Europe, and the comparison is therefore flawed. Similarly, reporting rates are not evenly distributed across Africa, being generally much better in West Africa than in the central or eastern Sahel. Higher population density, prosperity (wider use of guns), bird research and absence of war are prerequisites for recoveries to be reported.

The degree of overlap in the Sahelian east-west gradient appears to be indifferent to correlates such as body size, non-passerine versus passerine, or waterbird versus landbird (Fig. 107). The evidence points at separate wintering grounds for western and eastern European birds, but with a varying degree of overlap in the central Sahel. The only species in Fig. 107 squeezing its Africa-directed flight into narrow corridors on either end of the Mediterranean, the White Stork, shows a very small degree of overlap. This is consistent with earlier ideas based on ring recoveries (Schüz 1971), and is not really contradicted by the recent findings of satellite-tracked birds (Berthold et al. 2004) which show merely that eastern and western wintering populations may meet in Chad, northern Nigeria and northern Cameroon. Incidentally, precisely the same pincer-like movement across the Sahel has now been found for Black Storks, which follow the same routes as White Storks (Bobek et al. 2006). For waterbirds that depend on just four major wetlands across the Sahel, overlap is also likely to be small. The unique flooding regime in the Sudd, water level in the Lake Chad Basin and flooding in both floodplains of the western Sahel are asynchronous and have very different year-to-year impacts on bird species concentrated in these types of wetlands.

Our analysis is restricted to recoveries between 4°N and 37°N (Fig. 107), but were the southern limit to be shifted south, many more recoveries from eastern Europe would have become available to us from Kenya through to South Africa (see for instance recoveries of birds ringed and recovered in southern and eastern Africa: Dowsett 1980, Dowsett et al. 1988, Dowsett & Leonard 1999, Pearson et al. 1988, Underhill et al. 1999). This latter cohort originates overwhelmingly from eastern Europe, Finland and Russia, supplemented by a small number of species from Western Europe that follow a southeasterly course.

Annual variation in wintering conditions

Understanding annual variations in living conditions of birds spending the northern winter in the Sahel requires quantitative information on habitat quality and food supply for the entire region. What, for instance, is the temporal and spatial variation in biomass and exploitability of fish, fruits and insects? Such data are not available, except for small areas and short time-intervals. In some cases we can circumvent the lack of direct measurements by using sub-

Fig 105 The geographical distribution of 84 bird species along a north-south gradient, between Sahara in the north and tropical rainforest in the south. Colours indicate main habitat choice (based on Borrow & Demey 2004). Also indicated the degree to which these populations spend the non-breeding season north of the Sahara ('N'), south of the equator ('S') and along the coast ('C'). Dependence on the Sahel indicated with different fonts: full (bold), moderate (italic), little (grey). The last column gives for 27 species the Chapter number.

stitutes for food availability. Locusts and grasshoppers are an important food resource for large bird species and even passerines (Chapter 14). Using data on locust outbreaks, Dallinga & Schoenmakers (1989) were able to partly explain population fluctuations of the White Stork. Another example is the quantity of fish captured by fishermen in the Inner Niger Delta, which has been registered since 1966. These data, with certain restrictions, provide a baseline for the annual variation in fish biomass in an area where a considerable proportion of the fish-eating birds in the Sahel is concentrated.

Variations in fish captures in the Inner Niger Delta are closely related to the surface of the flooded area (Chapter 6), whereas scale, timing and location of locust breeding and migration may be predicted from rainfall patterns and annual changes in the green vegetation. Put simply, rainfall, vegetation and flooding are vehicles that enable us to make educated guesses of variations in overall food supply. For a fish-eating species like the Caspian Tern, of which the Baltic population to a large extent winters in the Inner Niger Delta, this approach translates into a close fit between survival and flood level. For species such as many warblers that are widely distributed across the Sahel except in floodplains, rainfall

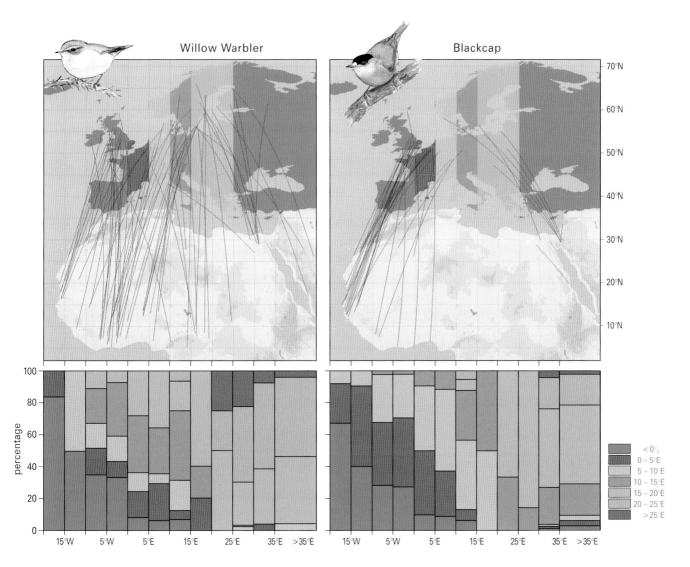

Fig. 106 Longitudinal split of European breeding origins as reflected in ringing recoveries in Africa north of 4° N, here illustrated for Willow Warbler (n=362) and Blackcap (n=1919). The graphs are based on these large numbers but, in order not to clutter the figures, no lines are shown on the map for the many African recoveries north of 30°N. More than 80% of the Willow Warblers and nearly 70% of the Blackcaps recovered along the Atlantic coast in Mauritania and Senegal (15-20°W) are from Europe west of Greenwich, but birds recovered in Libya, Egypt, Chad and Sudan (>20°E) are mostly from (north)eastern Europe (>15°E).

should be the better indicator of wintering conditions in the Sahel, or more accurately, not rainfall *per se* but its derivate, green vegetation (expressed as NDVI; Chapter 4), which relates to these species' food supply in the form of insects and fruit. Since NDVI data are available only from 1981 onwards, to link Sahel conditions with breeding bird trends that started earlier than 1981, we have been compelled to use rainfall statistics.

Vegetation clearly depends on rainfall in the preceding months, but also on the amount of rainfall in the preceding year(s). Consequently, green vegetation cover shows smaller fluctuations than rainfall does (Fig. 26), not to mention a distinct delay in greening

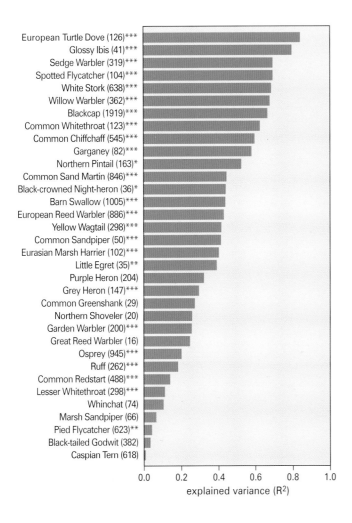

Fig. 107 The west-east distribution in Africa (4-35°N) for birds ringed in seven longitudinal zones in Europe (see map in Fig. 106) varies between complete overlap (Caspian Tern) and hardly any overlap (Turtle Dove). Shown as the explained variance in one-way analyses of variance, based on recoveries and recaptures (number between brackets). Levels of significance are * = P<0.05, ** = P<0.01 and *** = P<0.001; not significant = P>.05.

relative to the actual timing of rainfall. We have described a similar time-lag for flooding (Chapter 3), which in the Inner Niger Delta depends on river flow (Fig. 40), which in turn depends on rainfall in the Upper Niger during the preceding months and the preceding years (Figs. 18 & 19). This interrelated series of events explains why the hydrological drought always lags behind the meteorological drought (Fig. 108).

Over the past century, rainfall and flooding in the Sahel alternated between extended periods of drought and abundance (Fig. 108), as had been the case since time immemorial. Ample evidence exists to show that such periods are instrumental in shaping bird populations via differential mortality. To that extent, the long-term decline in Sahelian rainfall, and flooding of the Inner Niger Delta, Senegal Delta and Hadejia-Nguru floodplains since the 1970s, offer a 'natural experiment' in population dynamics. Unlike for previous droughts, the present wealth of ornithological data enables an educated assessment to be made of the impact on birds of large-scale events, whether or not human-induced.

The difference in survival rates between birds depending on floodplains and those using the drier parts of the Sahel may not be large, since flooding of the Inner Niger Delta and rainfall in the same year are closely correlated (Fig. 109). However, floods nowadays are irreversibly smaller than in the past, due to irrigation and the presence of dams. Birds depending on the Inner Niger Delta have lost 25% of their habitat (Chapter 6). The loss of floodplains due to irrigation and damming is proportionately even larger for the Hadejia-Nguru wetlands (Chapter 8) and the Waza-Logone floodplains (Chapter 9). Furthermore, the floodplains in the Senegal Delta have been embanked and turned largely into irrigated rice fields (Chapter 7). The present river flow of the Senegal no longer has an impact on flooding in this region (Fig. 109). However, even before the Senegal River was tamed, the correlation between flooding and rainfall anomaly, such as calculated for the entire western Sahel, was already much poorer than for the Inner Niger Delta.[3]

To all appearances, the Sahel may seem a desiccated world (100-500 mm rainfall per annum), which indeed it is during the northern winter after the effects of rainfall (July-September) gradually have worn off. During the transition from green to yellow, the few large wetlands within the realms of the Sahel are of vital importance to half of the Palearctic bird species residing in the Sahel (Fig. 105). Some species entirely depend upon these 'wet havens', others not at all. The distribution pattern of ring recoveries south of the Sahara provides a rough indication of habitat use of wintering Palearctic birds, and – at the species level – of the degree of dependency on wetlands (Fig. 110). Taking two extremes, incidentally from the same family, it appears that Sand Martin is a typical floodplain species, whereas Barn Swallow favours neither floodplains nor rice fields. Ruff tends to concentrate in floodplains, but is sparsely distributed in the coastal or quasi-coastal rice fields of Senegal, Guinea-Bissau and Guinea-Conakry (substantiated by counts, Chapter 11), where Black-tailed Godwit mostly appears. Not surprisingly, ducks and herons are typical floodplain species, whereas warblers are distributed mostly across the entire Sahel zone.

All Sahelian wetlands are hotspots for ringing recoveries, with a glaring exception: the Sudd, a blank in the Euring database. The lack of recoveries here, despite the huge numbers of waterbirds present (Chapter 10), is testimony to the fact that reporting rates must be smaller than elsewhere in the Sahel, even taking into account the very much smaller proportions of Russian and Asian birds that carry a ring. The same considerations apply to the Lake Chad Basin, where recoveries are few notwithstanding overwhelming numbers of wintering Palearctic birds. Despite the manifold pitfalls of using ring recoveries at face value, the results regarding temporal and spatial distributions do not digress too far from impressions gained and counts carried out in the field.

Winter mortality

Intuitively, it makes sense that mortality of birds wintering in the Sahel depends on rainfall and flood size. Anyone having first-hand experience of the Great Drought of 1985 will retain the retinal image of dying cattle and can recall the taste of dust. The 27 female Ruff that were captured on 23 March 1985 by Albert Beintema and Buba Fofana were emaciated rather than carrying pre-migratory fat, and their chances of returning to the breeding quarters were nil. Mortality must have been high. Despite the obvious link between mortality and Sahelian conditions, we are aware of only one study where this was shown directly: in years of low floods in the Inner Niger Delta, more waterbirds appeared to be captured by local people (Kone *et al.* 2002, 2005; Chapter 23).

Winter mortality in the Sahel may also be estimated indirectly from an analysis of ring recoveries. Many European ringing schemes have added their recoveries to the Euring database, but this scheme lacks information on numbers ringed per annum, a prerequisite for calculating mortality. Some ringing schemes have digitalised this kind of information (for example, Denmark), but otherwise it is nec-

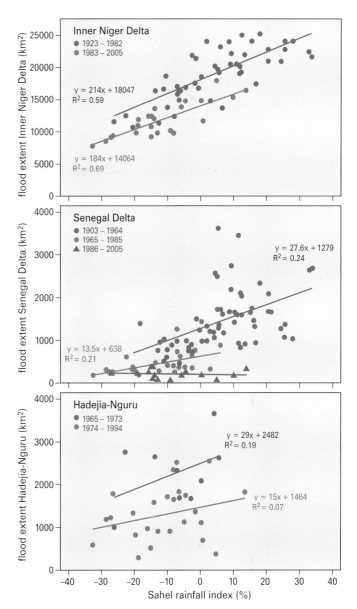

Fig. 109 Relationship between flood extent of the Inner Niger Delta, the lower Senegal Delta and the Hadejia-Nguru floodplains and annual rainfall in the Sahel, expressed as % departure from the average calculated over the 20th century. Subdivision into temporal categories well illustrates the impact of newly-constructed dams and irrigation schemes on flood extent.

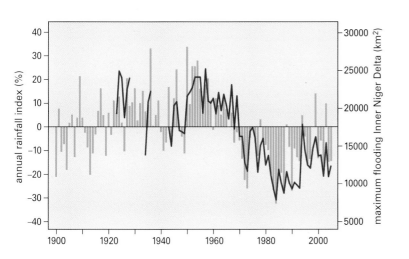

Fig. 108 Annual rainfall in the Sahel (1900-2005), expressed as % departure from the average calculated over the 20th century (columns, left-hand axis; data from Fig. 9) and the flooding of the Inner Niger Delta (line; right-hand axis; data from Fig. 40).

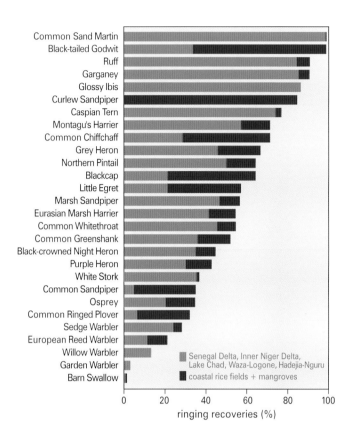

essary to consult the annual reports of the various ringing schemes to extract the number and age of birds ringed per annum. For three species (Black-tailed Godwits from The Netherlands (43 000 birds ringed), Glossy Ibis from the Ukraine (5000) and Caspian Terns from the Baltic (65 000), the annual number of ringed birds was known, allowing a recovery analysis, with Sahel rainfall and flood size of the Inner Niger Delta as covariates.

A simple alternative method of investigating whether mortality in the Sahel is drought-driven, is to compare the number of recoveries from the Sahel as a fraction of the annual total of recoveries from Europe and Africa. However, using recoveries as a measure of mortality is fraught with pitfalls. A major problem is the possible variation in reporting rate over time. For example, it has been suggested that since the 1960s reporting rates of German-ringed passerines from Italy have declined (Bezzel 1995, Schlenker 1995). Furthermore, changes in legislation and public awareness of conservation issues have possibly resulted in reduced shooting and catching (Tucker *et al.* 1990, McCulloch *et al.* 1992, Barbosa 2001), although as mentioned earlier, declining willingness to report shot birds (especially for species protected by law) to the ringing stations cannot be ruled out as an underlying factor in this trend. It is to be expected that the reporting rate will drop to zero in war-stricken regions and countries that face increasing poverty and anarchy, which tragedies have since the 1980s occurred in several countries in Africa, though not really in countries of the western Sahel. We believe, or rather hope, that the Sahel countries have had a more or less stable reporting rate due to the input of birders and researchers from abroad, who often question the local people for killed or found birds with rings.

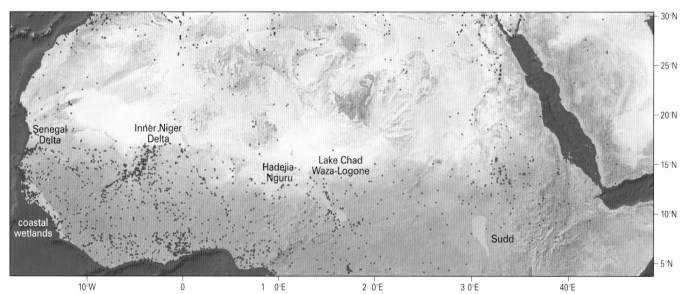

Fig. 110 Distribution of ringed birds found dead, captured or shot in Africa between 4° and 30° N. The six large wetlands are indicated. The graph shows the proportion of birds captured, shot or found dead in floodplains or in coastal rice fields (light blue); if neither, then in totally different habitats. Only species with more than 10 recoveries selected. From: EURING.

Another fundamental problem is the question whether recoveries may be used as a proxy of winter mortality in the Sahel. The Euring database contains 3375 recoveries from the Sahel proper, of which 2475 have a known cause of death: 51.2% shot, 19.6% captured and 16.3% accidentally killed (mainly fish-eating birds such as Osprey, Caspian Tern and Purple Heron). Fifty to ninety percent of the larger bird species are reported as being shot. More than half of the smaller birds had been captured (Fig. 111). Only a small fraction of the reported birds has not been killed by people. During droughts we expect mass mortality due to starvation, especially amongst the smaller species, but very few of these will be found, making them grossly underrepresented in the recoveries. Annual variations in recoveries in the Sahel therefore tell us more about the risk of being shot or captured than of starvation.

Winter mortality in Africa can also be derived from data collected in the European breeding areas. If sufficient breeding birds are captured and recaptured, the recapture rate translates into an inverse measure of mortality. This method was applied, for instance, in Sedge Warblers (Peach *et al.* 1991, Cowley & Siriwardena 2005) and Common Sand Martins (Szép 1995a & b). The annual winter mortality rate of the White Stork (Kanyamibwa *et al.* 1990, 1993; Barbraud *et al.* 1999; Schaub *et al.* 2005) could even be based on annual re-sightings of individually marked individuals.

When this kind of information is not available, the relative change in population size between two years may be cross-compared to one of the Sahel variables during the intervening winter. A direct comparison of population size in the breeding season with circumstances in the Sahel in the preceding winter is a less useful way of esti-

Over the past decades, large numbers of Osprey chicks have been ringed in Europe. In Finland alone, in the period 1913-2005, voluntary bird banders ringed 38 900 Ospreys, currently 1200-1400 annually (Saurola 2006). Almost 50% of the Finnish population nowadays breed on artificial nests erected by the bird banding groups in order – among other reasons – to facilitate (relatively) easy and safe access to nests in a species that is known for its tendency to use dead trees ('widow-makers') for nesting. Not surprisingly, this huge ringing effort has resulted in a plethora of recoveries, not only in Europe but also in Africa, where the species may even be specifically targeted because of its size. In some regions, such as the floodplains of the Inner Niger Delta, few Ospreys survive a prolonged stay because of human predation.

Shooting birds, especially small ones, is too expensive for most people in Africa. Lead shot costs 0.46 € (Mali and Senegal in 2007), which is the equivalent of the price for a bird weighing 0.5 kg on the market. Consequently, a bewildering variety of cheap catching devices have been developed, using discarded fish nets and tubes, hook lines, snares, traps, catapults and slings. This clay tablet with nylon snares is used in the Inner Niger Delta to capture small birds. Heavier tablets with larger snares are employed to catch herons, jacanas, Great Bitterns and Purple Swamphens in the dense marshland. The large number of wings found in fishing villages testifies to the efficiency and profitability of these catching methods. In fact, man's superior hunting skills are much better exemplified by small children using their catapults and snares than by the puffed-up shooting fraternity in the Mediterranean.

During the northern winter, the West African avifauna has a rich Palearctic element: between 18% (Ghana and Togo) and 29% (Senegambia) of the species are of Palearctic provenance. Among the passerines, many accumulate in the Sahel and Sudan zones of West Africa, where they arrive after the rains and remain throughout the dry season under deteriorating conditions before embarking on the return trip to the breeding quarters. Some species, such as Red-backed Shrikes heading for the Kalahari, use the Sahel as a stopover only, and continue their flight to southern Africa (usually via eastern Africa). Yellow Wagtails occur throughout Africa except in rainforests, with some geographical separation between subspecies (*flava* shown here). Whinchats, on the other hand, largely remain in the Sahel and Sudan zones. Northern Wheatears are typically confined to the Sahel, where they are very common. Wheatears from West Greenland and Canada, belonging to the 30-gram subspecies *leucorhoa*, cross the Atlantic Ocean in a single 4000-km flight from their breeding grounds to West Africa (Thorup *et al.* 2006).

mating winter mortality, because recovery from excessive mortality is spread across at least two and often more years due to limits of reproductive capacity and opportunity. Therefore, relative changes in the breeding population comprise the preferred indicator of winter mortality, but this approach suffers two disadvantages; because mortality here is described as the relative change between two suc-

cessive years, the derived estimate is based on *two* estimates, which increases the error of estimate,[4] and secondly, bird species differ in age of first breeding and in lifespan. A species like the Common Sand Martin breeds at one year of age, and most individuals have a lifespan of under 2 years. In this species, winter mortality in the Sahel can be derived directly from the annual change in the population. In contrast, the long-lived Caspian Tern starts breeding only when it is four years old. Changes in the breeding population are then not only dependent on winter mortality in the preceding year but just as much on winter mortality during earlier non-breeding years.

Based on the expected differences in impact of flooding, rainfall and vegetation on bird species, we expect winter mortality to vary accordingly. Floodplain-dependent migrants should thrive when floods are high, but survival of migrants inhabiting the drier parts of the Sahel, should be governed more by vegetation cover and rainfall. However, there is more to add to this simple prediction. Birds

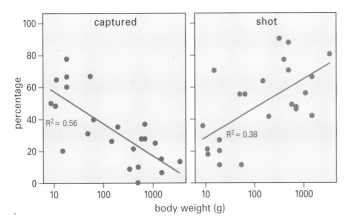

Fig. 111 Proportion of 23 bird species reported as shot or captured (with >10 birds for which cause of death is known) recovered in Africa between 4° and 25° N as a function of body weight. From: EURING.

Palearctic birds dominate in the West African floodplains during the northern winter. They are much more common than the African species occurring there: Black Herons, Kittlitz's Plovers, Spur-winged Lapwing, African Fish Eagles, African Wattled Plovers, Sacred Ibis, Egyptian Plovers and African Jacanas.

from Western, Central and Eastern Europe follow parallel migration routes to their African wintering grounds, and thus mostly, because there is some overlap, end up in separate parts of the Sahel. Because fluctuations in annual rainfall are similar in western and eastern Sahel (Fig. 7), there is no reason to expect a systematic difference in survival rates for bird populations breeding separately in western and eastern Europe, with the exception of species dependent on floodplains. The lower Senegal in the west has been losing its natural floodplains to reclamation since the early 1960s (Chapter 7), whereas the much larger Inner Niger Delta in Mali so far has lost a relatively small part of its floodplains to upstream dams and irrigation (Fig. 109). We therefore expect that Western European species concentrated in the Senegal floodplain will show a stronger decrease during low floods, or a less complete recovery during high floods, in comparison with their congeners from Central and Eastern Europe wintering in the Inner Niger Delta, Chad Basin or Sudd. For all species wintering in the Sahel, whether depending on floodplains or not, we have checked breeding trends for independence of flooding (lower Senegal and Inner Niger Delta), rainfall and the Normalized Difference Vegetation Index (NDVI) (western Sahel). For instance, birds concentrated in Sahelian woodland or using fallow farmland may be impacted by the long-term effects of deforestation or intensification of farming, and not just by rainfall patterns and green vegetation cover. In Chapters 15-41, we take into account factors on the breeding grounds that may overrule those in the winter quarters.

Small variations in rainfall make a big difference in the Sahel. This is especially evident in wetlands where depressions may remain dry and dusty or be filled with rain water. Temporary wetlands are green and abound in insects and waterbirds. Whether sites are green or dusty depends on rainfall and since this varies from year to year, there is also a large annual variation in the wintering conditions for migratory bird species. In contrast, deforestation and overgrazing generate a gradual change in the Sahelian landscape irrespective of rainfall, causing a long-term and lasting impact on birds (Senegal Delta, December 2008).

Endnotes

1 We added two species to the list of Walther & Rahbek (www.macroecology.ku.dk/africamigrants): Red Knot and Curlew Sandpiper.

2 We arrived at 84 bird species for which the Sahel-Sudan zone is an important wintering area. From the list of Eurasian migrants to Africa, we skipped the vagrants, but also some common species, such as Eurasian Teal, Common Kestrel, Common Snipe and White Wagtail, because only a minor fraction of their populations crosses the Sahara. Furthermore, we skipped species wintering at sea, such as Black Tern, and coastal waders, such as Sanderling, Grey Plover and Whimbrel, but also other waders, such as Common Ringed Plover and Curlew Sandpiper, for although relatively large numbers of these may be recorded in Sahelian floodplains (see Chapter 6 and 7), they pale into insignificance when compared with the large numbers present in the intertidal zone of the Banc d'Arguin (Altenburg et al. 1982, Zwarts et al. 1998) or Guinea-Bissau (Zwarts 1988). After removing these and other species from consideration, 133 species remain, of which a significant proportion of the populations winters south of the Sahara. This total includes 30 species spending the non-breeding season mostly in tropical or southern Africa, occurring only in the Sahel or Sudan zones during spring or autumn migration: European Honey Buzzard, Lesser Spotted Eagle, Steppe Eagle, Red-footed Falcon, Eleanora's Falcon, Eurasian Hobby, Corn Crake, Black-winged Pratincole, Common Cuckoo, Common Swift, Alpine Swift, European Bee-eater, European Roller, Common House Martin, Thrush Nightingale, Eurasian River Warbler, Marsh Warbler, Olive-tree Warbler, Icterine Warbler, Barred Warbler, Garden Warbler, Blackcap, Wood Warbler, Willow Warbler, Pied Flycatcher, Collared Flycatcher, Spotted Flycatcher, Eurasian Golden Oriole, Red-backed Shrike and Lesser Grey Shrike. We also removed Little Ringed Plover, Temminck's Stint, Green and Common Sandpiper, since the numbers wintering in the Sahel are low compared to the total in Africa. When we further confine our list to species occurring in the western and central Sahel and Sudan zone, we lose 11 more species: Great White Pelican, Steppe Buzzard (*vulpinus*), Levant Sparrowhawk, Common and Demoiselle Crane, Pied and Cyprus Wheatear, Rüppell's Warbler, Ménétries's Warbler, Masked Shrike and Cretzschmar's Bunting. Of Common Cranes migrating through western Europe, small numbers cross the Mediterranean and none the Sahara; in eastern Africa, though, large numbers migrate in a broad front across Egypt to winter in southern Sudan (Goodman & Meininger 1989). The majority of the Great White Pelican breeding in eastern Europe migrates to eastern Africa, where the Sudd is an important wintering area. The Great White Pelicans in the western Sahel originate from breeding colonies in the Senegal Delta and on the Banc d'Arguin. Finally, we skipped Spotted, Little and Baillon's Crake, since hardly anything is known about their winter distribution, although it is likely that significant fractions of these populations are concentrated in the Sahelian marshes.

3 The low correlation between the Sahel rainfall index and the flood extent of the then untamed Senegal River may be attributed to: (a) deviation between the Sahel rainfall index and the rainfall in the catchment area of the Senegal River; both series are correlated, but the fit is far from perfect ($R^2=0.59$), (b) the river flow, and thus the flood extent, depend not only on the rainfall in the catchment area but also on the rainfall in the preceding year or even two years. This long-term impact is relatively large in the Senegal River (Fig. 15; Table 1).

4 Suppose that in two subsequent years, the actual population is 100 and 80, but that we also know that the error of estimate is 10. In this example, the population may be underestimated in the first year at 90 (100-10) and overestimated in the second year also to 90 (80+10), by which the observed annual change is nil. It may, however, also be other way round: overestimated at 110 in the first year and underestimated at 70 in the second year, thus a decrease with 40. Consequently, although the average error of estimate of the population size is only 10, the error made in the derived year-to-year change in the population is automatically much larger.

Birds, locusts and grasshoppers

Wim C. Mullié

It is still pitch dark when we step into the dugout pirogue which will take us to the roost, discovered twelve months previously by Philippe Pilard of LPO's *Mission Rapaces*. After the water crossing, we still have several kilometres of walking through the mud ahead of us. Tracks of Spotted Hyenas are everywhere. We take up a strategic position and wait for first light. Then the spectacle begins... Thousands of Lesser Kestrels ascend from the roost and disappear in a northeasterly direction. Even bigger clouds of Swallow-tailed Kites rise and resume position in the top of the Baobabs in which they have roosted. At sunrise some 70 000 raptors have already passed overhead, an unforgettable sight. But another surprise is waiting. Under the roosting trees we find a carpet of pellets, literally untold millions of them. When we give these a closer look we see that the predominant prey items are grasshoppers. Two days later we have located the falcons' main hunting grounds. Despite the seemingly unfavourable conditions of the dry season, we are surprised to see massive numbers of grasshoppers... and their predators: not only thousands of Lesser Kestrels, Montagu's Harriers, Northern Wheatears and Abyssinian Rollers, but also White Storks and Marabou Storks. Flocks of up to 500 Cattle Egrets move rapidly through the dry vegetation, chasing grasshoppers in front of them. Hundreds of Lesser Kestrels hover above the moving flocks and grab the orthopterans flying away: a profitable feeding association indeed. Still later we locate the biggest roost of Montagu's Harriers ever found: over 1300 individuals. It does not take us more than 15 minutes to collect 500 pellets, also exclusively consisting of grasshoppers. All these birds feed on sedentary grasshopper species which spend the dry season as non-reproducing adults, and not on migrating locusts with their erratic occurrence, as European ornithologists have stated for decades. It's time for a reappraisal.

Introduction

Predation of locusts and grasshoppers by vertebrates, and birds in particular, has attracted attention since time immemorial. Nevo (1996) cites numerous historic sources which report predators of locusts and grasshoppers in the eastern Mediterranean basin, such as Eusebius of Caesarea (*c.* 260–340 AD, Bishop of Caesarea) who stated that the Egyptians honour the ibis for its destruction of snakes, locusts and caterpillars. Pliny (23-79 AD) explains that predators arrived in response to prayers offered to Jupiter by the people in the region of Mt. Cadmus (Jebel el Akra in today's Turkey), when locusts were attacking their fields. According to his information, migratory birds destroyed the invading locusts (Nevo 1996).

In the early 20th century the Committee of Control of the South African Central Locust Bureau considered bird predation as an important control mechanism of Red Locusts *Nomadacris septemfasciata* and Brown Locusts *Locustana pardalina*. In their annual reports special emphasis was given to this phenomenon (Lounsbury 1909). The word "Locust Bird" or "Sprinkhaanvoël" in Afrikaans (Van Ee 1995) was commonly used for predators that were considered important in destroying locusts, such as White Storks and Black-winged Pratincoles (Lounsbury 1909) and Wattled Starlings (Meinertzhagen 1959). Elsewhere in Africa (Moreau & Sclater 1938, Hudleston 1958, Schuz 1955) and India (Husain & Bhalla 1931, Singh & Dhamdhere 1986), birds were also seen as important natural enemies of locusts and grasshoppers. In the second half of the 19th century Indian Mynahs *Acridotheres tristis*, acridivores from the Indian subcontinent, were even introduced to Madagascar to control Red Locusts, which, however, they failed to do (Franc 2007). Rosy Starlings have been mentioned as the major predator of the locusts which used to occur in the steppes of the Cis-Caucasus, *i.e.* Migratory Locusts *Locusta migratoria* (Belik and Mihalevich 1994) and probably Moroccan Locusts *Dociostaurus maroccanus*. Outbreaks were particularly heavy in the 1920s, reputedly due to the vast stretches of newly abandoned fields during and after the Civil War (Znamensky 1926). Wild Boars *Sus scrofa* living in the Kizlyar Steppe and the "Caspian reeds" fed entirely on locusts during locust invasions; up to 1.5 kg of locusts have been found in a single stomach (Heptner *et al.* 1989). After the frequency of locust upsurges slumped Rosy Starling numbers dropped sharply, only to increase again during outbreaks of Italian Locusts *Calliptamus italicus* (Belik and Mihalevich 1994).

The widespread use of chemical pesticides for locust and grasshopper control since the Second World War removed the focus on "economic ornithology" (Kirk *et al.* 1996). National crop protection services, locust control centres and international organizations lost interest in natural factors contributing to population regulation of

Desert Locust hopper bands in the early morning in the Tamesna, Niger, October 2004. The yellow individuals are 5th instars, the pink individuals are immatures.

Swarm of Migratory Locusts Locusta migratoria capito on the Horombe plateau in Madagascar, May 1999, accompanied by Black Kites and an occasional Sooty Falcon.

Swarm of Desert Locusts, between Rosso and Tiguent, Mauritania, August 2004.

Acrididae (also called acridids), the family of locusts and grasshoppers to which most economically damaging species belong. More important, local knowledge on the role of these factors was lost as well. Chemical locust control in Africa has been held responsible for avian population declines in Africa and Europe (African Swallow-tailed Kites: Thiollay 2006a,b,c, White Storks: Dallinga & Schoenmakers 1984, 1989). However, since the use of organochlorine compounds, such as dieldrin, in locust control was abandoned in favour of acutely toxic but non-persistent organophosphorous insecticides, impacts on avian populations are considered to be less likely, though not impossible.[1]

Most published accounts on predation of acridids are of birds present during major outbreaks of locusts. This makes for good reading. The sight of a locust swarm (adults) or of hopper bands (nymphal stages) is certainly impressive, as witnessed by Meinertzhagen (1959): "The hoppers were at Middelburg (Cape) and so numerous that trains were held up owing to squashed bodies greasing the lines so that engine wheels would not bite the rails...". It is therefore not surprising that several authors concluded that the efficacy of (avian) predators in dealing with swarms during upsurges or plagues was rather limited (Dean 1964, Stower & Greathead 1969). The relationships between birds and grasshoppers (acridids not showing phase changes from solitary to gregarious and vice versa) have been studied much less in Africa.

Grasshoppers, in particular the species surviving the dry season as non-reproducing (diapausing) adults, contrary to those producing diapausing eggs, are a much more reliable food source for (mi-

Fig. 112 Seasonal occurrence of Desert Locusts in Africa from 1980 to 2008 per latitude between 10° and 20°N. Based on data from *Desert Locust Bulletin*, FAO.

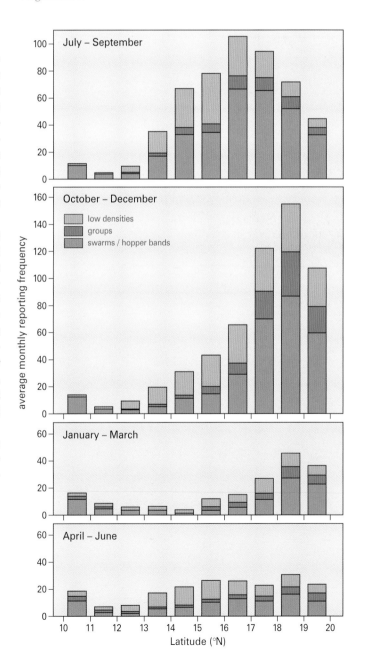

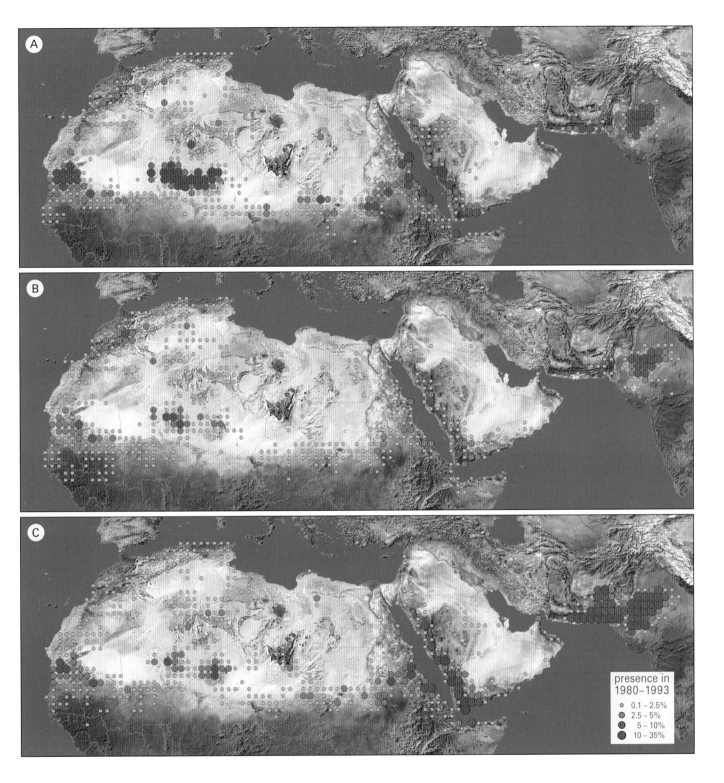

Fig. 113 Occurrence of Desert Locusts in Africa and SW Asia per square degree. (A) Swarms and hopper bands, (B) Groups, (C) Low densities and solitary locusts. Based on data from *Desert Locust Bulletin*, FAO.

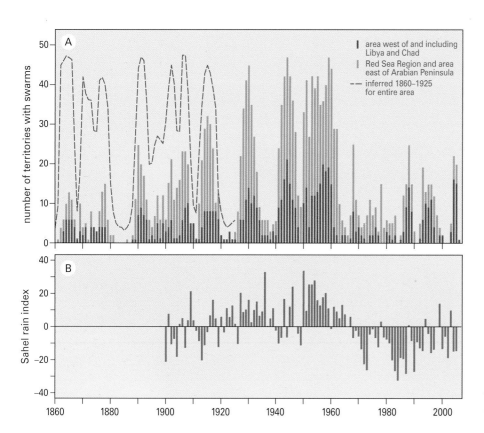

Fig. 114 (A) The upper graph shows territories reporting Desert Locust swarms from 1860-2006 (After Waloff 1976; modified from Magor et al. (2007). Data courtesy FAO, Joyce Magor in litt.). (B) Sahel rain index (Fig. 9).

In January 2007, Philippe Pilard of the *Ligue pour la Protection des Oiseaux* discovered a mixed roost of Lesser Kestrels (28 600) and African Swallow-tailed Kites (16 000) in Senegal. The exquisite grace of a Swallow-tailed Kite defies description, watching thousands entering and leaving this roost must have been mind-boggling. This delicate insectivorous bird is still largely an enigma. Very few accounts of its life history, all of them anecdotal and none recent, have been published; the latest relates to its behaviour in response to high grasshopper densities in Niger (Mullié *et al.* 1992). The pellets underneath the roosting trees in Senegal exclusively contained remains of grasshoppers. (March 2008, Senegal).

gratory) birds wintering in the Sahel than migratory locusts. However, it was not until recently that this concept found its advocates (Mullié *et al.* 1995, Jensen *et al.* 2006). The importance of *Ornithacris cavroisi* as a major prey species during the dry season in the Sahel has meanwhile been described for Swallow-tailed Kites (Mullié *et al.* 1992, Pilard 2007), White Storks (Mullié *et al.* 1995), Fox Kestrels (Brouwer & Mullié 2000), Abdim's Storks (Falk *et al.* 2006), Montagu's Harriers (Chapter 26) and Lesser Kestrels (Pilard *et al.* 2008).

This chapter will give a short introduction to the ecology of locusts and grasshoppers in the Sahel and the northern Sudan zone, and describes the role of birds as their predators. The possible relationship between recent land cover changes in Senegal, resulting in the creation of favourable habitats for acridids and the presence of important populations of acridivorous birds in this area, will be discussed. Alternative management options for locusts and grasshoppers, utilising combined natural and biological control mechanisms, are also put forward.

Locusts

As stated above, locusts relevant for the scope of this book belong to the orthopteran family Acrididae. The information in this paragraph is mostly based on Steedman (1990) and Symmons & Cressman (2001). Locusts are usually large insects which have the capacity to change habits, behaviour and physiology when they occur in large numbers in dense concentrations. From being solitary, occurring in low densities, they become gregarious. This shift is accompanied by changes in behaviour, colour and reproduction. Winged adults and immatures form swarms and migrate over large distances. The wingless larval stages, called nymphs or hoppers, form dense bands. A hopper band is a cohesive mass of insects which persists as long as meteorological conditions and predator pressure permit; the group moves in a single direction. Grasshoppers, unlike locusts, do not generally show this behaviour. Several locust species occur in the Sahel, such as Desert Locusts *Schistocerca gregaria*, African Migratory Locusts *Locusta migratoria migratorioides* and Tree Locusts *Anacridium melanorhodon*. Desert Locusts usually occur in arid and semi-arid areas, whereas African Migratory Locusts in the Sahel are usually found under more humid conditions, such as those in the Lake Chad Basin and in the Inner Niger Delta (Mestre 1988).

Desert Locusts have clear seasonal patterns in their distribution across the Sahel (Fig. 112). The lowest frequencies are recorded from January till June, whereas the highest frequencies occur from July till December. This is caused by the circular migration pattern of Desert Locust swarms which originate from the gregarisation areas in Mauritania, Algeria, Mali, Niger, Chad and Sudan. Gregarisation is the process by which initially solitary locusts develop gregarious behaviour, when weather conditions are simultaneously favourable over large areas (FAO 1980). Desert Locusts develop into gregarious populations which invade the Sahel during the rainy season. When conditions continue to be favourable, they move into agricultural lands and may inflict serious damage. During the last upsurge in West and NW Africa from 2003-2005, some 13 million ha were treated with chemical insecticides to reduce or prevent damage to crops and pastures (Brader *et al*. 2006).

During recessions (periods without widespread or heavy infestations by swarms), or remissions (periods of deep recession marked by the complete absence of gregarious populations (FAO 1980), Desert Locusts occur in very low densities in their favourite arid habitats (Fig. 113). Their densities dwindle so much (<25 individuals per hectare), and their behaviour and colouration become so cryptic, that vertebrate predators are no longer a serious

Lesser Kestrel with grasshopper prey, Khelcom, Central Senegal, March 2008.

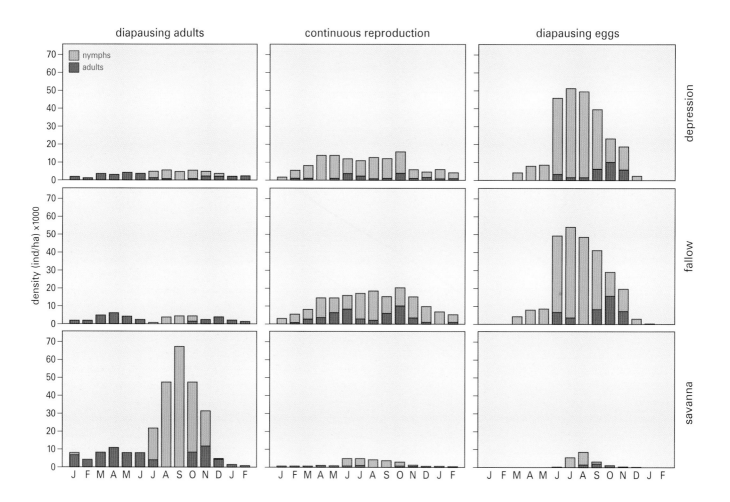

Fig. 115 Annual cycle of the grasshopper community at Saria, Burkina Faso, in three distinct habitat types: sub-humid depression, fallow land and semi-arid savanna and three different life-history traits: diapausing adults, continuous reproduction and diapausing eggs. Based on Lecoq (1978) and M. Lecoq (*in litt.*).

Acorypha clara (left) *Ornithacris cavroisi* (right). *O. cavroisi* is the preferred prey of many bird species during the dry season.

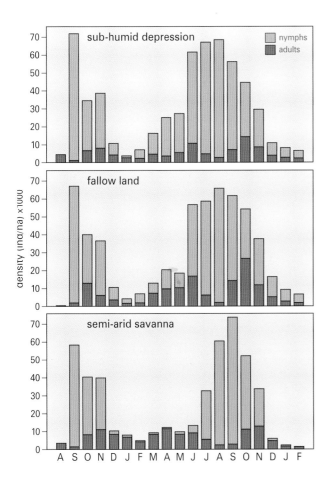

Fig. 116 Total monthly densities of grasshoppers for combined life history traits at Saria, Burkina Faso, in three distinct habitat types: sub-humid depression, fallow land and semi-arid savanna. Based on Lecoq (1978) and M. Lecoq (*in litt.*).

threat to their survival. It has been postulated that high levels of predation by birds will result in strong selective pressures for the evolution of crypsis and other anti-predator defences in acridids, such as micro-habitat selection, which facilitate crypsis or active escape (Schulz 1981).

During upsurges, resulting from successful breeding over a number of generations by an initially small population (Symmons & Cressman 2001), Desert Locusts start migrating at the end of the rainy season. In the Western Sahel they migrate to the Maghreb. In the Red Sea Region migration heads for the Arabian Peninsula, Ethiopia, Eritrea, but sometimes as far as India and Pakistan or Kenya and Tanzania, where they continue to reproduce. In the Western Region they return to the Sahel in late June, where they peak at latitudes between 17°N and 20°N from October till December. No clear pattern has been found in the frequency of outbreaks of Desert Locusts, for which data have been recorded since 1860. In one or more Desert Locust regions, upsurges start with successive seasonal rains, which are more widespread, more frequent, heavier and longer lasting than normal (van Huis *et al.* 2007). In the last two decades, upsurges in the Western Region were recorded in 1992-1994 (FAO Desert Locust Bulletin) and in 2003-05 (Brader *et al.* 2006). A major plague was recorded in 1986-1988 (US Congress OTA 1990). During a plague there are widespread and heavy locust infestations, the majority of which are bands or swarms (Symmons & Cressman 2001).

Starting in the 1960s, the global wind systems reversed to a system that prevailed before 1890 and which was associated with a more restricted north-south movement of the Intertropical Convergence Zone (ITCZ; Magor *et al.* (2007). As a result, the geographical distribution and duration of plagues of Desert Locust nymphs after 1965 are much reduced compared with the pre-1965 period (Magor *et al.* 2007; Fig. 114).

Densities that have been recorded during upsurges vary from 20 to 150 million individuals/km² for adults. Maximum densities found for 1st larval stages, or instars, are 30 000/m² and for 5th instars 1000/m². However, average densities over larger surface areas are much

Woodchat Shrike with grasshopper prey, Diouroup, Senegal, February 2008. Laniidae and Coraciidae are specialized acridivores with over 85% of all species within the genus in Africa feeding on acridids.

lower, probably 50-100/m² for late instar bands (Symmons & Cressman 2001). These latter densities are in the same range as grasshopper nymph densities that can be encountered under favourable conditions in the Sahel.

Grasshoppers

Most Sahelian grasshopper species also belong to the Acrididae, although several other families occur, most notably the Pyrgomorphidae. Senegalese Grasshoppers *Oedaleus senegalensis* are known to show incipient gregarious behaviour and to form hopper bands. What has been described in the acridological literature as "swarms" of Senegalese Grasshoppers must have been clusters of high numbers of individuals during important outbreaks (Maiga *et al.* 2008). In general, grasshoppers do not display coordinated gregarious behaviour like locusts, nor do they undergo massive diurnal movements (Maiga *et al.* 2008). The Senegalese Grasshopper is a species with three generations per year and extensive north-south migrations, triggered by rainfall.

Grasshoppers have life history traits that allow them to cope with the harsh Sahelian environment. Species can have continuous re-

Fig. 117 Minimum surface area infested with grasshoppers, predominantly Senegalese Grasshoppers, in Senegal, Mali, Niger and Sudan between 1989 and 2007. Based on data provided by Idrissa Maiga (Agrhymet Regional Centre, Niamey, *in litt.*) and National Crop Protection Directorates. No data were available for Sudan before 1999, and data for Mali are incomplete. Surface areas treated with pesticides against grasshoppers were used as a proxy for surface areas infested in case of missing data.

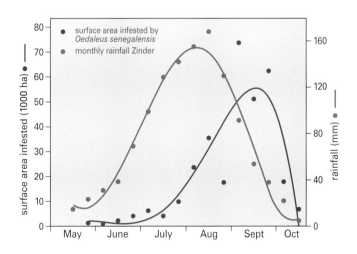

Fig. 118 Average monthly area infested by Senegalese Grasshoppers in Niger in relation to average monthly rainfall. Data courtesy Idrissa Maiga, Agrhymet Regional Centre, Niamey, *in litt.*

Lanner Falcon hunting on Desert Locusts chased by grazing camels in dry *Schouwia thebaica* vegetation, Aghéliough, Northern Niger, November 2005. Right composite photograph: Lanner in flight dismembering a captured Desert Locust.

production (CR), lay diapausing eggs (DE) at the end of the rainy season, or survive the dry season as diapausing adults (DA). Furthermore, the species with continuous reproduction can have two to four generations per year, DE-species one to three and DA-species one or two. A study of the annual cycle of the grasshopper community at Saria (Burkina Faso: 12°16' N 02°09' W) by Michel Lecoq (CIRAD) between 1975 and 1977 showed that, after maturation, most grasshopper species which survived the dry season as immature adults (DA) laid eggs after the first significant rains (*c.* 20 mm), and then died. DA species preferred savannas for reproduction, whereas the other species inhabited depressions or fallow land (Fig. 115). At the same time, eggs laid at the end of the previous rainy season hatched; from June till September the population was dominated by nymphs. Towards the end of the rainy season the survivors had developed into adults. Their numbers peaked between September and November when the major part of the population consisted of species which will lay eggs and die. Only populations with continuous reproduction and those which will survive as non-reproducing individuals will do so until the next rainy season.

The timing of egg laying or hatching and migration in the Sahelian zone depends on the position of the ITCZ: the farther north, the shorter the rainy season because the ITCZ arrives later and returns

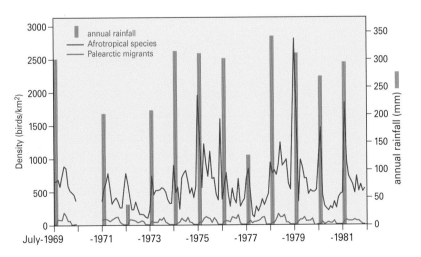

Fig. 119 Monthly bird densities of Afrotropical species and Palearctic migrants at Fété Olé, Northern Senegal, between 1969 and 1982. Data courtesy G. and M.-Y. Morel (*in litt.*).

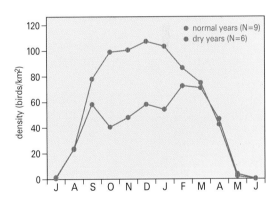

Fig. 120 Average monthly densities of Palearctic migrants during dry and normal years at Fété Olé, Northern Senegal, from 1960-1982. Based on Morel & Morel (1978) and G.J. and M.-Y. Morel (*in litt.*).

Table 21 Number and percentage of bird species per family recorded as taking locusts or grasshoppers as food in Africa. Bird families have been grouped according to the proportion of species being recorded as acridivorous within a family.

Afric : Number of species in each family known from Africa (following The Birds of Africa, Vol. I-VII)

Acrid : Number of bird species within each family known to be acridivorous, at least for part of its diet

Pal : Number of bird species within **Acrid** presumed to be of Palearctic origin

% : **Acrid** as a percentage of **Afric**

Rel Imp : Average number of acridid species recorded as food of bird species within each family

Text in orange: Bird families of which individual species have at least 1.5 different acridid species in their diet, on average.

Non-Passeriformes					
Family	Afric	Acrid	Pal	%	Rel_Imp
75.0-100 %					
CORACIIDAE	8	7	1	87.5	2.00
BURHINIDAE	4	3	1	75.0	1.00
50.0 - 74.9 %					
NUMIDIDAE	6	4	0	66.7	1.75
OTIDIDAE	21	13	0	66.7	1.31
MEROPIDAE	19	12	0	63.2	1.50
CICONIIDAE	8	5	2	62.5	2.60
BUCEROTIDAE	24	13	0	54.2	1.31
CAPRIMULGIDAE	27	14	1	51.9	1.20
FALCONIDAE	24	12	5	50.0	2.17
25.0 - 49.9 %					
GRUIDAE	7	3	0	42.9	1.31
CHARADRIIDAE	34	14	7	41.2	1.14
PHOENICULIDAE	8	3	0	37.5	1.33
ALCEDINIDAE	22	8	0	36.4	1.88
THRESKIORNITHIDAE	11	4	1	36.4	1.50
TROGONIDAE	3	1	0	33.3	1.00
TURNICIDAE	3	1	0	33.3	1.00
GLAREOLIDAE	13	4	2	30.8	2.75
STRIGIDAE	46	14	1	30.4	1.14
ARDEIDAE	29	8	3	27.6	1.38
LARIDAE	26	7	4	26.9	1.29
CUCULIDAE	38	10	0	26.3	1.50
PHASIANIDAE	46	12	1	26.1	1.08
TYTONIDAE	4	1	0	25.0	2.00
0.1 - 24.9 %					
ACCIPITRIDAE	90	20	9	22.2	2.30
CAPITONIDAE	41	9	0	22.0	1.22
RALLIDAE	37	4	1	10.8	1.25
SCOLOPACIDAE	55	4	4	7.3	1.00
APODIDAE	29	2	1	6.9	1.00
COLUMBIDAE	46	2	0	4.3	1.00
ANATIDAE	68	2	1	2.9	1.00
Families with <3 species					
STRUTHIONIDAE	1	1	0	100.0	1.00
SAGITTARIIDAE	1	1	0	100.0	2.00
ROSTRATULIDAE	1	1	0	100.0	1.00
RECURVIROSTRIDAE	2	2	2	100.0	1.00
UPUPIDAE	1	1	1	100.0	2.00

earlier than at more southerly latitudes (Fig. 5). Species with a wide distribution may reproduce continuously in the Guinea savanna, whereas they may have either diapausing eggs or adults in the Sahel and in the northern Sudan savanna (as in Burkina Faso; Fig.115) because they have to cope with a long dry season. Large numbers of grasshoppers with continuous reproduction or diapausing adults will move southward with the retreating ITCZ and the lowest densities are usually encountered in January and February. As we have seen in the previous paragraph, Desert Locusts are also largely absent from the Sahel and Sudan zone during this period and the food supply for acridivorous birds may become critical.

Thus, when Palearctic bird migrants arrive south of the Sahara, they may sometimes find over 300 000 grasshoppers per ha. Those which stay in the Sahel zone will be confronted with rapidly declining acridid numbers, with the lowest level in January and February

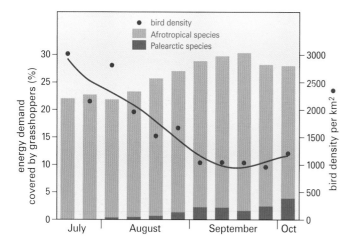

Fig. 121 Diet of bird species between July and October 1989 in the Western Ferlo, Northern Senegal, based on gizzard contents (Mullié, unpubl.).

Passeriformes Family	Afric	Acrid	Pal	%	Rel_Imp
75.0-100 %					
LANIIDAE	18	16	5	88.9	1.56
PRIONOPIDAE	8	6	0	75.0	1.00
50.0 - 74.9 %					
DICRURIDAE	8	4	0	50.0	1.25
25.0 - 49.9 %					
MOTACILLIDAE	41	19	4	46.3	1.21
MALACONOTIDAE	46	21	0	45.7	1.19
CAMPEPHAGIDAE	13	5	0	38.5	1.00
ALAUDIDAE	74	28	1	37.8	1.46
CORVIDAE	19	7	0	36.8	2.43
TURDIDAE	134	49	14	36.6	1.18
CISTICOLIDAE	94	34	0	36.2	1.03
STURNIDAE	52	17	1	32.7	1.59
MONARCHIDAE	19	6	0	31.6	1.00
EMBERIZIDAE	17	5	0	29.4	1.00
PLOCEIDAE	112	30	0	28.8	1.10
PASSERIDAE	24	6	1	25.0	1.50
0.1 - 24.9 %					
PLATYSTEIRIDAE	29	7	0	24.1	1.00
MUSCICAPIDAE	38	9	1	23.7	1.00
TIMALIIDAE	40	8	0	20.0	1.13
PYCNONOTIDAE	63	12	0	19.0	1.17
SYLVIIDAE	150	16	7	10.7	1.06
FRINGILIDAE	54	3	0	5.6	1.00
HIRUNDINIDAE	38	2	1	5.3	1.50
ESTRILDIDAE	78	1	0	1.3	1.00
NECTARINIIDAE	87	1	0	1.1	1.00
Families with <3 species					
PICATHARTIDAE	2	1	0	50.0	1.00
BOMYCILLIDAE	1	1	0	100.0	1.00

Cattle Egret with *Acorypha clara*, Khelcom, Central Senegal, March 2008.

(Fig. 115). This is perhaps somewhat surprising, as we might expect the lowest numbers to occur at the end of the dry season, in May. Most likely, this is the result of resource-driven movements rather than natality or mortality. Because the ITCZ moves from south to north and back, these migrations are usually in either direction, depending on the position of the ITCZ. Some species show more distinct movements than others. Senegalese Grasshoppers have already been mentioned for the rainy season. During the dry season, and depending on latitude, *Ornithacris cavroisi* and *Harpezocatantops stylifer* have been shown to make extensive movements, but a species like *Acorypha clara* shows less pronounced movements (based on Lecoq 1978). Between January and April, the grasshopper community of savannas at Saria in Burkina Faso consisted of 38.2 ± 7.5% *O. cavroisi*, the most important species in terms of biomass at this locality (Fig. 116). Also, at more northerly latitudes *O. cavroisi* reaches high densities in favourable habitat during the dry season. *Ornithacris* has a preference for evergreen broad-leaved shrubs and trees, such as Combretaceae. We hypothesise that this feeding behaviour makes the species less vulnerable to food shortages during dry years. It also explains why many bird species favour *O. cavroisi*: it is a secure and predictable food source.

As with locusts, infestations of grasshoppers vary from year to year (Fig. 117). In Niger the difference between low and high infestation rates is as much as a factor of 118. Apparent upsurges in the Sahelian countries appear unrelated, probably the result of inconsistencies in scouting, reporting and archiving activities, as a result of insecurity in the field, lack of fuel or decentralisation. For grasshopper species surviving as immature adults, such information is not available, but it can be safely assumed that such large inter-annual fluctuations are the rule (*e.g.* Lecoq 1978). The extent of the infestations shows that annual chemical grasshopper control may have a larger overall environmental impact than the control of locusts during upsurges.

In the years before 1965, when Desert Locust outbreaks were very

Montagu's and Marsh Harriers, Lesser Kestrels and vultures (mostly White-backed) were observed drinking at this waterhole, Khelcom, Central Senegal, March 2008. Drinking is frequently observed in the Sahel, but rarely so in temperate regions.

frequent and received most attention, grasshoppers were considered to be a local problem only (M Lecoq, pers. comm.). It was not until the rainy season of 1974, when widespread crop damage occurred two years after the major drought of 1972/1973, that grasshoppers started to be seen as a regional pest to crop production in the Sahel (Launois 1978). The most important species, both as a pest to agriculture and as a prey for birds, is the Senegalese Grasshopper. Between 1974 and 1989 outbreaks of Senegalese Grasshoppers, and of several other species showing the same ecological requirements and life history traits, in the central Sahel, occurred in 1974, 1975, 1977, 1978, 1980, 1985, 1988 and 1989 (Maiga et al. in press). Populations of Senegalese Grasshoppers in Niger peak on average one month after the peak of the annual rain (data courtesy Idrissa Maiga, Fig. 118). [2]

Bird-acridid relationships

Bird densities and distribution   We are well informed about the seasonal occurrence of birds in Sahelian savannas, thanks to a unique series of transect counts in northern Senegal by Gérard J. and Marie-Yvonne Morel (Morel 1968; Keur Mor Ibra, 16°20'N, 15°25'W), Morel & Morel (1973, 1978, 1980, 1983; Fété Olé, 16°13'N, 15°05'W, Fig. 119) and G.J. Morel (in litt. 1990 and 2008). Their data were collected in two periods, 1960-1962 and 1969-1982, covering nine years with average rainfall and six dry years. When these studies were conducted, tree density in their plots was considerably higher than nowadays (e.g. tree density on top of dunes in Fété Olé 296/ha in 1976 (Poupon 1980), 147/ha in 1995 (Vincke 1995) and 109/ha in 1999 (Danfa et al. 2000)). Grasshopper populations, in particular those associated with grassland habitats and degraded wooded savanna, were probably less important. Gillon & Gillon (1974) measured above-ground arthropod biomass in Fété Olé in 1971 (rainfall 202 mm; dry year) and found grasshopper biomass in September 1971 to be 2.66 g dry weight/100 m^2. Based on data presented in Balança & De Visscher (1990) it can be calculated that average grasshopper biomass in September 1989 near Keur Mor Ibra (rainfall 184-314 mm; average 235.5 mm) was c. 38 g dry weight/100 m^2, or at least 14 times as high as in Fété Olé. [3] This shows that, apart from changes in the presence of grasshopper populations due to alterations in habitat (see below), there are also important natural fluctuations in prey availability, mainly as a function of annual rainfall (hatchability of eggs, extent of migration) and vegetation development (survival of nymphs and adults).

Palearctic migrants show distinct reactions to drought (Fig. 120). Upon arrival in July and August, bird densities are similar between dry and normal years, but by September a difference already becomes visible. From October till February densities of Palearctic migrants are significantly reduced by a factor of two in dry years. From March till May densities in dry and normal years again do not differ,

sources conducted for this chapter revealed that 537 bird species of 61 families are currently known to prey upon locusts and grasshoppers in Africa (Table 21).

Several bird families in Africa, such as the Coraciidae and Laniidae, appear to be specialised acridivores. A wide range of acridid species is recorded as food for Ciconiidae, Glareolidae and Corvidae. Individual bird species which can be associated with a diverse acridid diet are *e.g.* Black Kites and Pied Crows, but also White Storks and Abdim's Storks, Lesser Kestrels and Montagu's Harriers. Based on extensive fieldwork in Niger, Petersen *et al.* (2008) and Falk *et al.* (2006) found that pre-migratory movements of Abdim's Storks are in synchrony with the seasonal movements of Senegalese Grasshoppers.

From the available data, Elliott (1962) concluded that predation during real infestations, in which swarms of over 5 billion insects frequently occur, may be between 0.25 and 6% of the acridid biomass. Based on the data presented by Smith and Popov (1953), Greathead (1966) calculated that about 875 000 locusts out of a swarm of "some tens of millions" were taken daily by avian predators. This would represent a *daily* predation rate of about 4%. The authors considered this to be exceptional. Stower & Greathead (1969) calculated that – over a 23-day period – bird predation amounted to about 4% *in total* of a population of 1.2 million locusts (0.17% daily), a low predation rate. Cheke *et al.* (2006) studied the impact of predation on an isolated Desert Locust population by Common Kestrels and Lanner Falcons during a field trial with the entomopathogen *Metarhizium anisopliae* var. *acridum* in North Niger. They found that the predation rate increased from 0.05%/day to 3.8%/day in 36 days, when the initial population of Desert Locusts had decreased from about one million individuals to 13 000 individuals, due to the combined impact of *Metarhizium* and predation by falcons. The cumulative impact was calculated to be at most 26% of the locust population (Cheke *et al.* 2006). Interestingly, falcons continued to prey on locusts, even when recorded locust densities had dropped from an initial 1700 to 20 per ha at the end of the study. However, it is known that locust densities calculated from transect counts may be 25-50% too low (Magor *et al.* 2007).

Desert Locust hopper bands may face much higher losses from bird predation. Ashall and Ellis (1962), studying Desert Locust populations in Eritrea, estimated that one population of about 15.2 million hoppers was reduced to 5.2 million individuals in just 14 days, of which birds alone took 8 million individuals. This corresponds to 52% of the initial hopper population. In Mauritania and Niger, Wilps (1997) found that birds, particularly Cream-coloured Coursers, Golden Sparrows and Desert Sparrows, took 97.5-99.5% of Desert Locust hopper bands (2nd-4th instar) of 130 000 – 500 000 individuals in 4-11 days, and 95% of a hopper band of 1.1 million individuals. Culmsee (2002) followed a hopper band of 20 000 individuals in Mauritania and found a reduction to less than 5000 in ten days due to bird predation, a daily predation rate of 7.5%. High predation rates on Desert Locust nymphs, particularly in field trials

reflecting the northward migration of birds which have wintered further south, where conditions were more favourable.[4]

The food of vertebrate predators in N Senegal was studied in 1989 by means of gizzard contents of birds, mammals and reptiles (Fig. 121), and their energy demands were calculated (Mullié, unpubl.). Grasshoppers were found to be a major part of the diet for a range of species. Interestingly, nightjar species qualified as acridivores, which has also been found in Niger (Brouwer & Mullié 1992). The proportion of the total avian energy demand satisfied by grasshoppers increased from 22 to 30%, starting mid-August. This coincided with an increase in grasshopper numbers on the study plots, in particular an influx of Senegalese Grasshoppers (from an average of 28 700 ind./ha on 5 September to 54 000 ind./ha on 16 September) on their southward migration, following the descent of the ITCZ (Balança & De Visscher 1990). At the end of September, grasshopper numbers decreased rapidly as the herbaceous vegetation started to dry out (Niassy 1990). The energy demand of Palearctic migrants satisfied by grasshoppers was at maximum 4% of the energy demand satisfied by grasshoppers for the entire bird community. Highest densities of Palaearctic migrants at this latitude were not reached until December (Fig. 120), *i.e.* several months after the grasshopper peak.

Birds as predators of locusts and grasshoppers Elliott (1962) mentioned more than one hundred bird species of 34 families as locust predators in Eastern Africa. Husain & Bhalla (1931) did the same for the Indian subcontinent and came up with 35 species, while more than 200 bird species were found to prey on grasshoppers in the USA (Metcalf 1980). A survey of published accounts and other relevant with the entomopathogen *Metarhizium anisopliae* var. *acridum*, have also been found by Kooyman & Godonou (1997) and Kooyman *et al.* (2005, 2007). Hopper bands in Algeria and Mauritania, containing some tens to hundreds of thousands of individuals, were generally eliminated by birds in less than a week. In the latter trial Golden Sparrows were among the prime natural enemies of 4th instars of Desert Locusts (data W.C. Mullié & C. Kooyman). Golden Sparrows are considered as pests on ripening rice and millet in the Sahel, and, as such, are killed in their millions with organophosphate pesticides. The same controversy holds for Village Weavers, generally considered an agricultural pest, but, at the same time, one of the prime consumers of Senegalese Grasshoppers in Niger (Axelsen *et al.* in press). Baddeley (1940) in "*The rugged flanks of Caucasus*" recounts that Rosy Starlings were called *Birds of Mahomet* or *Holy Birds* because of their locust-devouring qualities, but *Devil Birds* when they devastated fruits in orchards and vineyards two months later.

Kooyman (in Bashir *et al.* 2007) calculated that 1500 Yellow Wagtails foraging on second instar Desert Locust nymphs consumed 225 000 hoppers per day at Aitarba, 260 km south of Port Sudan, Sudan, along the Red Sea coast. Assuming an underestimation of 40%,[5] the birds would have taken about 3% daily of the estimated 12 million hoppers present. The total toll of hoppers taken by birds between hatching and fledging was estimated by Kooyman (in Bashir *et al.* 2007) in the order of 30%, which, based on the above calculations, might be higher as well.

Recent work in Niger has shown that Cattle Egrets reduced local grasshopper densities (initially 2.5-5 ind./m^2) by 55-85% (Petersen *et al.* in Kooyman 2006). Similar work in northern Nigeria found that Cattle Egrets removed 45% of grasshoppers at a density of 1.33 ind./m^2 (Amatobi *et al.* 1987). In areas visited by Abdim's Storks, reductions were even more pronounced: from 9.9 to 1.2 ind./m^2 at high densities and from 3.6 to 0.8 ind./m^2 at low densities (Petersen *et al.* 2008a). Although reductions are partially caused by the insects fleeing from foraging flocks of birds, it is obvious that birds deplete grasshopper densities.

It should be noted that the importance of predation cannot be assessed from predation rates alone in periods when food is in short supply for the grasshoppers (Belovsky *et al.* 1990). Under those conditions intra- and interspecific competition may occur and preda-

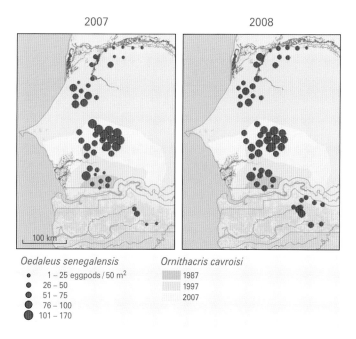

Fig. 122 Range extension between 1987 and 2007 of *O. cavroisi* into the Eastern part of the "peanut basin" in Senegal (after Alioune Beye, DPV, in Mullié (2007)) and the Senegalese Grasshopper egg pod densities in Senegal in 2007 and 2008 (data courtesy Kemo Badji, DPV).

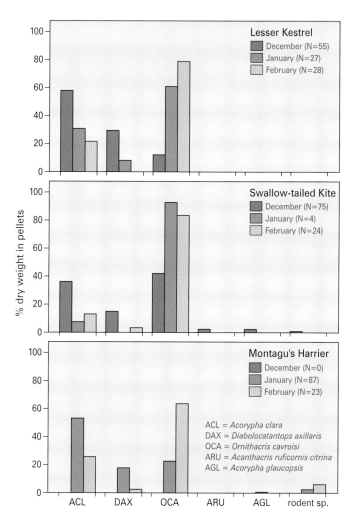

Fig. 123 Diet of Lesser Kestrels, Swallow-tailed Kites and Montagu's Harriers in Senegal, based on analysis of regurgitated pellets collected from night roosts. Data LPO (France), SWGK (The Netherlands) and Mullié (unpubl.).

Black Kites and Pied Crows feeding on grasshoppers chased away by a bush fire, Khelcom, Central Senegal, November 2006.

tion may be compensatory, *i.e.* the grasshoppers killed by predators would have died anyway by competition or lack of food. A similar situation occurs at the end of the rainy season, when the grasshoppers which have laid diapausing eggs will die anyway. Bird predation on these individuals will not have any influence on their population levels.

Montagu's Harriers can serve as an example for birds which increase in density when grasshopper or locust densities are high (numerical response), and increase their intake rate of acridids at high locust densities (functional response). During the Desert Locust outbreak of 1988 more than 1000 Montagu's Harriers were reported from a locust-infested area between Mbour and Joal in Senegal (Cormier & Baillon 1991). A year later, when locusts were absent, only up to 120 individuals were present in the same area (Baillon & Cormier 1993). This area is still being frequented by Montagu's Harriers, and in 2006/2007 their number was estimated to be at least 150 individuals (Claudia Puerckhauer pers. comm.). Food remains in regurgitated pellets collected in central Senegal contained up to 95% acridids; alternative prey, such as small songbirds, mammals and lizards, was widely available but rarely taken. The harriers showed a preference for large grasshopper species among the dozen or so of species available (W.C. Mullié own obs.). This is notably different from their food choice during the breeding season in western Europe, where prey choice is diverse and vertebrates, voles in particular, are important food sources (Arroyo 1997, Underhill-Day 1993, Salamolard *et al.* 2000, Koks *et al.* 2007). However, it is similar to the food taken in southern Spain (Corbacho *et al.* cited in Salamolard *et al.* 2000) and in India (Ganesh & Kanniah 2000).

Montagu's Harriers in their wintering quarters show foraging bouts during mid-morning hours - when temperatures rise and grasshoppers become active - and at the end of the afternoon, just before moving to their night roosts. During the hottest hours of the day most harriers either loaf in the shade of low shrubs or spiral up in thermals. They are often seen at water holes (Cormier & Baillon 1991, W.C. Mullié own obs), a behaviour probably associated with the need for water under high temperatures, as it is not recorded in their European breeding range (Ben Koks pers. comm.). High grasshopper capture rates have been observed in N Senegal (1 prey/4.5 min; Arroyo & King 1995). At this rate of food intake, daily energy requirements could be met within three (*Ornithacris*) or four hours (*Schistocerca*) of active foraging per day. Just before their autumn migration, Abdim's Storks in eastern Niger satisfied 95% of their daily energy needs (excluding pre-migratory fattening) in just one hour of active foraging on grasshoppers during a year with high densities. When grasshopper densities were 60-65% lower than in the year with high densities, 70-75% of their energy demands were met in two hours of active foraging (Petersen *et al.* 2008a). Although, theoretically, energy needs may be met in a short period of time, the digestive bottleneck (Kenward and Sibley 1977) would prevent a single foraging bout, as the digestive tract can only hold a limited number of prey items and the birds need time to digest their prey.

Experimental studies Several experimental exclusion studies showed the potential of avian predation in regulating grasshopper populations. In a two-year study in grasslands in N Dakota, USA, Fowler *et al.* (1991) were unable to detect differences between plots with birds and plots where birds were excluded at low (*c.* 1 ind./m^2) densities during a dry year; however, grasshopper densities were 33% lower in a year of normal rainfall, when densities averaged *c.* 3 ind./m^2. In another experiment Joern (1986) found a reduction of 27.4% of grasshoppers in plots in Nebraska, USA, subjected to 40 days of

Box 17
The limits of growth?

Land degradation is a continuous, and, in some cases, an irreversible process if not halted in time. It seems to have created - at least temporarily - favourite habitats for the development and increase of grasshopper populations in the latest extension of the "peanut basin" in central Senegal, where 55 400 ha out of 73 000 ha agropastoral reserve and protected forest were cleared between 1991 and 2004 for groundnut production (Schoormaker Freudenberger 1991, Mullié & Gueye in press). Not surprisingly, this has become an important region for wintering Lesser Kestrels and Montagu's Harriers, perhaps even for Northern Wheatears. However, it is to be expected that, in the long run, conditions will become less favourable; major soil erosion from wind and water will occur when the current land use continues unchanged. Soils in this area are among the most vulnerable in Senegal for Aeolian erosion (CSE 2003). A reduction in fallow periods, a trend apparent everywhere in the Sahel, will have a negative impact on biodiversity. The process bears resemblance to that of eutrophication. A slight increase in nutrient levels increases total biodiversity and biomass, but when a certain threshold is reached, species richness starts to decrease and the system collapses in favour of a few generalist species. Land cover of the older parts of the "peanut basin", which have been cultivated for a much longer period, may serve as an example of what might be expected. It is tempting to think of the succession in the grasshopper-predator complex as a function of the age of the area brought under cultivation. Unfortunately, our knowledge is still rudimentary. The question of how close we are to the limits of growth is, however, pertinent.

bird predation, or 0.7% per day, whereas Joern (1992) reported that, over a 3-year period, the average bird-induced reduction was 25%. Under dry conditions no statistically significant difference between plots with and without birds could be detected, due to movement of grasshoppers from the plots.

In a 4-year exclusion field experiment, birds limited grasshopper numbers in ungrazed semi-arid grassland in SE Arizona, USA (Bock et al. 1992). By the final year of the study, adult grasshopper density was >2.2 times higher, and mean nymph density >3.0 times higher, on plots from which birds were excluded. Despite their impact, Bock et al. concluded that birds still failed to qualify as keystone predators, at least in the short term.

The first results from a simulation study of the role of birds and egg-pod predators on Senegalese Grasshoppers (Axelsen et al. 2008) showed that the combined natural predation caused reductions of between 60 and 75%, of which some 25-30% were by birds (Jørgen Axelsen, pers. comm.). These reductions are of the same magnitude as those found in the above-mentioned experimental studies. The model included various taxa, such as plants (natural Sahelian Gramineae, Millet), grasshoppers (Acrididae, i.e. Oedaleus senegalensis), insect predators (Bombyliids, Tenebrionidae) and vertebrate predators (24 species of birds). The parameters of the model were constructed using field data collected in Senegal and Niger in 2003 and 2004.

Land cover change in Senegal: has grasshopper availability increased?
The main agricultural belt of Senegal, also called the "*bassin arachidier*" (peanut basin), suffers from pressure on its natural resources by the increasing human population. Land cover changes in Senegal between 1988 and 1999 were assessed using satellite imagery (CSE 2003). One of the study areas, covering 515 673 ha of the eastern part of the Department of Kaffrine within the latest agricultural extension area of the peanut basin (Tappan et al. 2004), showed that cropland increased from 138 396 ha to 215 725 ha (from 26.8 to 41.8% of the total land cover); denudated soils increased from 62 to 2038 ha and steppe from 25 to 39 671 ha. On the other hand, surface areas of savanna decreased from 370 267 ha to 256 954 ha; only 37 659 ha, or 10.2 % of the savanna in 1988, remained without change.

In line with land cover changes, the distribution pattern of grasshoppers changed simultaneously. Some species, such as *Ornithacris cavroisi*, have gradually increased and invaded areas where the species formerly was uncommon, particularly in a NE direction (Mullié 2007; Fig. 122). The same holds for other species with similar life-history traits, such as *Acorypha clara* and *Acanthacris ruficornis citrina*. The clearing of dense stands of woody vegetation in favour of fallow land or low bushes (mainly Combretaceae and Acaciaceae with an occasional Baobab tree), has created favourable habitat for grasshopper species with a preference for low bushes on grassland or for more xeric conditions of bare soils. The transformation of formerly wooded savanna into steppe, interspersed with cropland (groundnuts, millet, sorgho and maize) also appears to have created favourable habitat for grasshoppers surviving the dry season as diapausing eggs, such as *Oedaleus senegalensis*, *Kraussella amabile* and *Cataloipus cymbiferus*. The combined average densities of both

Feeding associations of Cattle Egrets and Lesser Kestrels at Khelcom, Central Senegal, February 2008, are a common feature during the season when feeding is profitable at medium to high grasshopper densities.

Birds found dead after Desert Locust control operations in Niger and Mauritania. From left to right Abyssinian Roller, White-throated Bee-eater, Golden Sparrow and Fulvous Babbler.

adults and nymphs from August till October now often approaches 40 ind./m^2, and occasionally up to 120 ind/m^2 (W.C. Mullié and Y. Gueye, unpubl.), which is higher than the economic threshold for treatments, as applied by the Senegalese Crop Protection Directorate (DPV) (Mullié 2007). Recent counts of egg pods of Senegalese Grasshoppers show that this is more than a hypothesis. Egg pod densities found in the field in this area during the dry seasons of 2007 and 2008 were among the highest found in Senegal (Fig. 122).

Increases in grasshopper densities have also been reported elsewhere, in association with meteorological/ecological or anthropogenic disturbances, such as drought (Launois 1978), ungulate grazing (Merton 1960, Bock *et al*. 1992, Moroni 2000), bush fires (Thiollay 1971) or land cover changes, especially reduction of tree cover and transition towards mosaic habitats (Merton 1960, Idrissa Maiga *in litt*. 2008).

Of particular importance is the study of Franc (2007), who was able to show a causal relationship between recent deforestation in NW Madagascar and the development of a new gregarisation and breeding area for Red Locusts *Nomadacris septemfasciata*. Red locusts used corridors, newly created by deforestation, to exploit biotopes which were favourable for egg laying. Franc (2007) also showed that, under the prevailing climatic conditions, succession of the newly created habitats might eventually make these areas less attractive for breeding. Drastic conservation measures in Thailand, where a similar relationship between acridid development and deforestation had been found, showed that this ended outbreaks which had occurred over a 25-year period (Roffey 1979).

There are indications that vertebrate predators, birds in particular, have increased in numbers. We hypothesise that this involves a response to increased prey availability. Our surveys in central Senegal since late 2006 showed that the newly created steppes and grasslands with up to 15%, and occasionally 40%, of low bushes, dominated by *Guiera senegalensis* and *Combretum glutinosum*, attracted very high numbers of acridivores, including Lesser Kestrels (25 000-30 000 on a single roost, but >>10 000 elsewhere), Swallow-tailed Kites (35 000-40 000) and Montagu's Harriers (7000-10 000 +) [6] dispersed over a feeding area of *c*. 3500 km^2. In an area of approximately 550 km^2 Abyssinian Rollers and Northern Wheatears each reached maximum densities of *c*. 100 ind./km^2, while Cattle Egrets (4000-5000), White Storks (3500), Woolly-necked and Marabou Storks (tens) were common. Densities of Northern Wheatears are in the same range as reported from northern Nigeria (Jones *et al*. 1996), whereas numbers of foraging Montagu's Harriers (>5000 birds; confirmed during roost counts) are among the highest reported in the litterature. Buffalo Weavers and White-throated Bee-eaters in September-October, White Storks with Black Kites and Lesser Kestrels in November and Cattle Egrets and Lesser Kestrels from January till March often formed feeding associations, sometimes following cattle herds (W.C. Mullié, own obs.). Feeding associations in response to high densities of acridids are well known from the literature (*e.g.* Smith & Popov 1953, Schüz 1955, Meinertzhagen 1959, Triplet & Yésou 1995).

Analyses of food remains in regurgitated pellets collected on roosts from November 2007 until April 2008 showed that Lesser Kestrels, Montagu's Harriers and Swallow-tailed Kites predominantly consumed grasshoppers (Fig. 123). In the course of the dry season, the diet of all three raptor species shifted from the smaller *Acorypha clara* and *Diabolocatantops axillaris* to the bulkier *Ornith-*

acris cavroisi. We have insufficient data on grasshopper density and distribution to explain these changes. Experimental studies indicate that birds specifically select the larger species of grasshoppers, or, within a species, the larger sex (females), in line with foraging decisions related to energy maximisation (Kaspari & Joern 1993). Birds did indeed preferentially kill large-bodied grasshoppers (Belovski & Slade 1993), although males were found to be more vulnerable than females because of their more exposed behaviour (Belovski *et al.* 1990).

Integrating bird predation in an alternative approach to locust and grasshopper control Looking at Desert Locust invasions since the 1960s in relation to the period before, and taking into account the informed opinion of renowned acridologists (Michel Lecoq, CIRAD, and Joyce Magor, FAO, *in litt.*), we can summarise the situation as follows. The introduction of plague prevention measures in the 1960s, in particular by using the organochlorine insecticide dieldrin in so-called barrier treatments, proved to be effective against the build-up of gregarious populations. Dieldrin was also involved in population declines of various bird species in Europe and North America, and was suspected to have contributed to population declines of Palearctic and Afrotropical species (wintering) in Africa. During the 1980s, dieldrin was replaced by organophosphates and other compounds, but these products were rather short-lived and therefore costly. Full-cover repeated sprayings were needed to achieve the same result. With the introduction of fipronil a rather persistent product became available for barrier treatments, but it soon became clear that fipronil had some major environmental drawbacks (Danfa *et al.* 2000, Mullié *et al.* 2003, Peveling *et al.* 2003). At present, it is no longer recommended for full-cover treatments of Desert Locusts (PRG 2004). Upsurges since 1965 have been attacked by massive use of mainly organophosphate insecticides and it has been claimed that this has ended upsurges and prevented further plague development.

However, distribution and abundance of Desert Locusts in the last 40 years can also be explained by changes in the global wind circulation and subsequent changes in the extent of the movement of the ITCZ, which, in turn, has caused the geographical extent and duration of Desert Locust invasions to decrease sharply, without affecting the upsurge frequency as such (Fig. 114). While some experts claim that the massive use of pesticides ended the 1986-1988 and 2003-2005 plagues, others state that unusual and extreme weather conditions were responsible for their termination. With current knowledge it is not possible to separate the different factors involved reliably, but the data provided by Magor *et al.* (2007) indicate that climate and weather most likely play a bigger role than hitherto considered.

For birds the consequences are unequivocal. Migrating locusts, such as Desert Locusts, have never been a reliable food source. Birds will attack swarms and hopper bands, but most of the time Desert Locusts have a limited distribution and do not occur in the Sahel when Palearctic migrants are present. Furthermore, after 1965, reproduction in outbreak areas is geographically much reduced when compared to the period before (Magor *et al.* 2007). Therefore encounter probabilities have decreased, reducing the importance of locusts as a food source for birds. Similarly, opportunities of feeding on sprayed locusts have become less frequent.

In the same time frame, grasshopper populations developed into a pest to subsistence farming in the Sahel. Before 1974 grasshoppers were considered a local problem. This changed after the massive outbreak in 1974 and, ever since, grasshoppers have been indicated as a major factor in crop production in the Sahel. There is evidence, outlined in this chapter, that land cover changes have contributed to the creation of favourable conditions for the development of grasshopper populations, and for changes in species composition, at

During flood recession, the Inner Niger Delta is swamped with an estimated half-million Cattle Egrets, feasting on the plethora of grasshoppers and locusts. When the floodplain is inundated, the birds concentrate in wild rice fields, where their diet also consists of acridids. November 2008.

places where previously they were not considered as a pest. Due to a rapidly increasing human population, savanna is cleared and new crop lands are developed. Grasshopper species, in particular the species surviving the dry season as diapausing adults, are present throughout the year and form a reliable food source for birds.

Grasshoppers, like locusts, are treated with chemical insecticides and their annual occurrence over vast areas (Fig. 117) makes grasshopper control potentially more environmentally damaging than locust control. A recent review of the entomopathogen *Metarhizium anisopliae* var. *acridum* (Green Muscle®) has shown that it may be a viable alternative to broad-spectrum chemical insecticides, at least for controlling a range of grasshopper species and nymphs of Desert Locusts in a preventive control strategy (van der Valk 2007). This strategy should be actively promoted to reduce the negative impact on birds of conventional control methods. Although *Metarhizium* has been shown to be effective in the field against adult Desert Locusts (Cheke *et al.* 2006), the relatively slow speed of action, in particular when the daily ambient temperature range is unfavourable for pathogen development, will prevent it from being used against swarms and chemical control methods will continue to be used in the foreseeable future. However, as preventive control hinders swarms from building up, the need for later chemical control will be reduced.

Based on available information, Mullié (2007b) found that synergy of bird predation and biocontrol of locust and grasshoppers is the rule. This results from a few fundamental differences between chemical and biological control mechanisms. Chemical control with broad-spectrum insecticides kills a large number of insects, target and non-targets alike, within a short period of time, usually from a few hours to less than one day. This will deprive insectivorous birds almost instantly of their food source and, as a result, massive numbers of birds will leave the area. Mullié and Keith (1993) found a reduction of around 50% in bird numbers the day following treatment. Dead insects are not usually an attractive prey and they simply decay. Biopesticides do not kill instantly, but rather weaken the targets until they become sluggish and are easy to capture. Furthermore, biopesticides are selective and do not kill predators and parasitoids. Birds quickly learn that easy prey is available, and, rather than leaving the plots, they are attracted to them. These findings are important to understand insect pathogen-predator relationships in order to develop integrated pest management strategies that take into account numerical responses of avian predators and other natural enemies (*e.g.* Thomas 1999).

Despite cases mentioned in the text where avian acridivorous species also cause damage to agriculture in the Sahel, field records and experimental studies have provided overwhelming evidence of birds suppressing acridid populations at low and medium, and sometimes at high population levels. As advocated by Kirk *et al.* (1996), this "ecological service" should be valued. It is therefore questioned if chemical treatment of larval stages of locusts and grasshoppers *under high predation pressure* always makes sense, in particular at low densities, or of hopper bands up to several million individuals.

Acknowledgements

Wim Mullié is particularly grateful to the "Fondation Agir pour l'Education et la Santé" (FAES, Senegal, President Mrs Viviane Wade; First Lady of Senegal) for having funded the writing of Chapter 14. He dedicates this chapter to Mrs Wade. He kindly acknowledges the following persons for having contributed unpublished data or other information to, or for having commented upon, earlier versions of Chapter 14. They are, in alphabetical order: Drs Mohamed Abdellahi Ould Babah (CNLA, Mauritania), Kemo Badji (DPV, Senegal, Project PRELISS), Amadou Bocar Bal (Agrhymet Regional Centre, Niger), Sébastien Couasnet (Fondation AES, Senegal), Harry Bottenberg (USAID, Ghana), Moussa (Baba) Coulibaly (DPV, Niger), Yves Dakouo (OPV, Mali), Meissa Diagne (DPV, Senegal), Fakaba Diakité (UNLCP, Mali), Ousseynou Diop (DPV, Senegal), Helena Eriksson (SLU, Sweden), Ali Gado (OPV, Mali), Hans Hut (Foto Hut, The Netherlands), Rabie Khalil (Locust Control Center, Sudan), Gerrit van de Klashorst (The Netherlands; ex-FAO, Dakar), Ben Koks (SWGK, The Netherlands), Moussa Konaté (DPV, Senegal), Christiaan Kooyman (Fondation AES, Senegal, and ex-IITA, Benin), Michel Lecoq (CIRAD, France), Joyce Magor (FAO, Italy), Idrissa Maiga (Agrhymet Regional Centre, Niger), Gérard J. and Marie-Yvonne Morel (France; ex-ORSTOM, Senegal), Bo Svenning Petersen (Orbicon, Denmark), Philippe Pilard (LPO France), Claudia Puerckhauer (Bayerisches Artenhilfsprogramm Wiesenweihe, Germany), Toumani Sidibé (UNLCP, Mali), Leen Smits (SWGK, The Netherlands), Amadou Demba Sy (CNLA, Mauritania), Harold van der Valk (The Netherlands), Larry Vaughan (Virginia Tech, USA).

Endnotes

1 An avian risk assessment carried out for earlier Desert Locust and grasshopper control campaigns in Senegal showed significant risk for avian mortality in 55% of the areas treated during the 1986-1989 Desert Locust campaign and in 50% of those treated during the 1993-1995 campaign, both due to the use of fenitrothion (Mullié & Mineau 2004).On the other hand, a similar assessment for the 2003-2005 Desert Locust campaign showed that 35-41% of the areas treated was associated with bird mortality (data W.C. Mullié).

2 It is interesting that the mathematical models of Holt & Colvin (1997) and Colvin & Holt (1996) of the interaction between the migration of Senegalese Grasshoppers, their predators and a seasonal habitat indicate that severe and prolonged drought may induce outbreaks due to a reduction in natural predator activity. However, our regression analyses of Senegalese Grasshopper outbreak years did not produce significant results when compared with either rainfall index for the Sahel or national annual precipitation figures. This is not entirely surprising since Senegalese Grasshoppers are a migratory species, whose population dynamics partially depend on ecological conditions found outside the areas where the first generation hatches. Therefore such an analysis ought rather to be performed on a regional than on a national scale, taking into account the movements of the species. Nevertheless, recent progress in modelling Senegalese Grasshopper populations with rainfall-models, and taking into account multiple generations, migrations and natural control mechanisms (Fisker et al. 2007, Axelsen et al. in prep., Jørgen Axelsen pers. comm. 2008), may eventually lead to more reliable predictions.

3 A study at Khelcom, central Senegal, in October 2008 showed that maximum grasshopper biomass was 600-1200 kg wet weight /ha, which is equivalent to 180-360 kg dry weight/ha. The average biomass was 45-105 kg dry weight/ha (data W.C. Mullié and Y. Gueye unpubl.). Total rainfall at Khelcom was 931.5 mm in 2008, which was exceptionally high (M. Sy, Direction de la Météorologie Nationale, Poste Khelcom, pers. comm.).

4 Recent studies that have estimated bird densities in N Senegal did so mainly in the rainy season (July-October: Keith & Mullié 1990, Mullié & Keith 1991, 1993; July-August: Petersen et al. 2007b). Densities generally fall in the range that Morel & Morel (1973, 1978, 1980, 1983) found in Fété Olé, but they were much higher in July and August 1989 due to large numbers of Golden Sparrows on transects (Keith & Mullié 1990). When densities were compared along two precipitation gradients in Niger (between 600 and 200 mm isohyets) and Senegal (between 900 and 300 mm isohyets), from south to north, densities were surprisingly similar and ranged from 475 – 762 (average 663) ind./km^2 in Senegal and from 477 – 862 (average 639) ind./km^2 in Niger (Petersen et al. 2007b). In a study in N Nigeria (c. 13°N) in December 1992 – January 1993 Jones et al. (1996) found densities ranging from 1630-7890 birds/km^2 of which 360-450 per km^2 were Palearctic migrants. These densities are much higher than those usually found in the northern Sahel, at latitudes of about 16°N (Fig. 119 & 120). In central Senegal, Mullié & Gueye (unpubl.) found densities of 247-4308 (average 1728) birds/km^2 between December 2008 and February 2009, of which 43-412 (average 106) birds/km^2 were Palearctic migrants. These densities are in the intermediate range between N Senegal and N Nigeria.

5 The calculations of Kooyman were based on an intake of 9 g wet weight per day per bird, which is conservative as their corrected field metabolic rate (following Nagy 2005) is 100 kJ/day, corresponding to a food intake rate of 4.5 g dry weight (15 g wet weight) of grasshoppers per day. Yellow Wagtails have minimal fat reserves during wintering (Lundwall & Persson 2006), therefore no correction for pre-migratory fattening has been made. Based on these assumptions we estimate that the real consumption could have been 375 000 hoppers per day.

6 Regularly repeated density counts (N=12) on 19 km of transects at Khelcom, central Senegal, from September 2008 till February 2009, showed that combined densities of Montagu's and Marsh Harriers in a 550 km^2 study area gradually increased from 3.9 ind./km^2 in September and 8.3 ind./km^2 in October to 14.2 ind./km^2 in January and 15.8 ind./km^2 in February. Simultaneous counts at dusk on three night roosts in the same area in January 2009 showed that at least 5000 Montagu's Harriers were present (Wim Mullié kindly acknowledges the help of Ben Koks, Leen Smits, Jean-Luc Bourrioux, Thierry Printemps and Benoît van Hecke with counting). Marsh Harriers constituted 10-15% of the numbers observed in September, but this was reduced to 2-5% by the end of November and to 1-2% in February. The majority of the Marsh Harriers only used Khelcom as a staging area during their southerly migration in September and October.

Grey Heron
Ardea cinerea

Hunched beside a frozen watercourse, its bill sealed with lumps of ice, and waiting for death to come, this is the typical picture that comes to mind during severe winters in much of Europe, when resident Grey Herons are decimated. Not all Grey Herons share this fate, however, because the species is partly migratory, and those which escape the frosts by moving south trade starvation for the dangers of migration. Long-distance migrants, in particular, encounter telephone wires, power cables and a myriad of shooters, but also have to endure the stresses of crossing vast expanses of water or sand. The Grey Herons which reach the sub-Saharan floodplains reap the reward of temporary food bonanzas so enormous that foraging takes only a small part of the day. The typical sighting of Grey Herons in West Africa is that of dozing flocks standing on dry land, away from the nearest water. Danger here comes from man, either from guns or from the scatter of thousands of hooklines. It may pay to escape harsh winter conditions in Europe, but when mild winters predominate it pays even more to remain in the breeding quarters. Climate change may increase the resident proportion of Grey Heron populations in their western breeding range.

Breeding range

The Grey Heron occurs throughout Europe, and is particularly abundant in floodplains and lake districts in marine and continental temperate climate zones between 45° and 60° N. The European population in the 1990s was estimated at 210 000-290 000 pairs, mostly in the United Kingdom, The Netherlands, Germany, France, Ukraine and Russia (BirdLife International 2004a). Many populations are resident, dispersive or partly migratory.

Migration

On average, northern- and easternmost breeding birds tend to migrate over longer distances than the more southerly and westerly breeders (recoveries December-January, >75 km): 1141 km for Kaliningrad, 917 km for Sweden and 841 km for Denmark, as opposed to 454 km for The Netherlands and 413 km for France (Rydzewski 1956).

The standard direction taken by long-distance migrants from western Europe is southwest in autumn, except for those from Norway, which fly west to Great Britain (Rydzewski 1956). The standard direction of the birds from eastern Europe is more variable. Consequently, there is some segregation of eastern and western populations in sub-Saharan Africa (Fig. 124). Considerable overlap is found in the Inner Niger Delta, where birds from westernmost (France, Denmark, The Netherlands) and easternmost origins (Ukraine) mingle. Some birds from Central Europe head for East Africa (Fig. 124), but low reporting rates in East Africa may obscure the frequency of this phenomenon. For Egypt, Mullié *et al.* (1989) and Meininger *et al.* (1994) came up with eight recoveries of Grey Herons ringed in the former USSR between 1948 and 1977, mostly originating from colonies along the Volga and the eastern Caspian Sea. This complies with data in Skokova (1978), showing that Grey Herons from the Volga, Crimea, eastern Black Sea and Caspian Sea head for the Nile Delta and from there may penetrate southwards, following the Rift Valley up to Lake Victoria.

A typical view of Grey Herons in West African wetlands: loafing birds along the water line (or even on dry land, far away from the nearest water), here seen together with roosting Caspian Terns and feeding Ruff. Grey Herons in Sahelian floodplains are rarely seen feeding, an indication of high food abundance (disregarding the possibility of nocturnal feeding). Inner Niger Delta, February 2007.

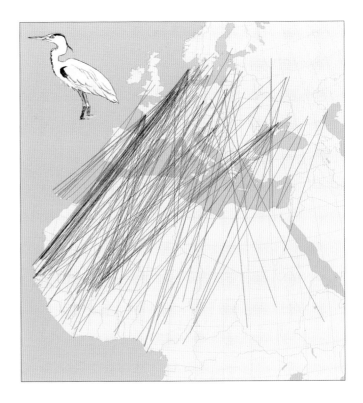

Fig. 124 European ringing locations of 145 Grey Herons recovered in Africa between 4° and 30°N and from Egypt. From: EURING, and Mullié *et al.* (1989) and Meininger *et al.* (1994) for Egypt.

Grey Heron *Ardea cinerea*

Distribution in Africa

The overwhelming majority of European birds never even attempt to reach Africa during migration, but nevertheless large numbers have been counted in the Inner Niger Delta (about 24 000 birds; Chapter 6) and the nearby rice fields of Office du Niger (550 birds; Wymenga *et al.* 2005), in coastal rice fields between Senegal and Guinea (an estimated 14 000; Bos *et al.* 2006; Chapter 11), and in the Waza Logone in northern Cameroon (300-700 counted in 1992-2000; Scholte 2006). For the entire Chad Basin, the estimate stands at 5000 birds in 1993 (van Wetten & Spierenburg 1998). This makes for some 45 000 Grey Herons in these areas alone, not to mention their more scattered presence in smaller wetlands and coastal lagoons throughout West Africa. The total European population in the 1990s amounted to 600 000 – 900 000 birds (see above), indicating that perhaps about 10% of all Grey Herons may winter in West Africa. The age-distribution of 116 Grey Herons recovered in sub-Saharan Africa indicates that long-distance flight is not exclusively a juvenile prerogative. Birds of 1, 2, 3, 4 and 5-17 years old were represented by 34%, 26%, 11%, 10% and 19% of the total number of African recoveries. Grey Herons defer returning to their breeding colonies by between one and four years (Fernández-Cruz & Campos 1993), and the high proportion of these age categories in African wintering quarters seems to be consistent with delayed breeding and avoidance of competition with more experienced older birds. Similarly, 31 out of 36 Grey Herons recovered freshly dead in sub-Saharan Africa between April and August were 1-4 years old (EURING). Whether summering of younger age classes in Africa improves survival is questionable, however. Grey Herons are large birds, dependent on wetlands; this makes them an easy, profitable prey for local people.

Two Grey Herons feeding together with a Great Egret and Reef Herons in a ditch in a coastal rice field. Guinea Bissau, October 1989.

Population change

Wintering conditions in Africa Grey Herons in the Sahel are concentrated in the few floodplains, where the annual number of recoveries is clearly related to flood level. When all years are combined and categorised into dry (<12 000 km² flood extent), intermediate (12 000-16 000 km²) and wet years (>16 000 km²), the number of recoveries in dry years is four times larger than in wet years, 4.7/year versus 1.1/year. However, these figures may be biased, because the total number of ringed birds, and thus the number of recoveries, has varied considerably over the decades – the EURING files contain data concerning 20 329 dead Grey Herons, over the 1947-2004 period

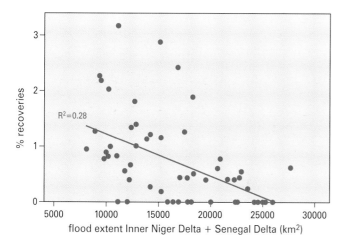

Fig. 125 Annual number of Grey Heron recoveries from Africa (4°-20°N) as a percentage of the total number recovered across Europe and Africa (dead birds only) between 1946 and 2004 (n= 59 years, P<0.001). From: EURING.

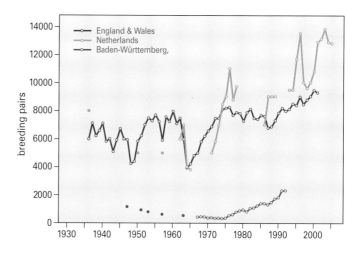

Fig. 126 Fluctuations in the number of breeding pairs of the Grey Heron in England & Wales (Marchant *et al.* 2004), The Netherlands (Bijlsma *et al.* 2001, SOVON) and Baden-Württemberg, Germany (Kilian *et al.* 1993).

Grey Heron taking a sun-bath, amidst feeding Black-winged Stilts. Inner Niger Delta, January 1994.

Grey Herons wintering in Europe run the risk of dying from starvation during cold spells. Throughout the 20th century, each decade experienced between 1 and 5 cold, very cold or severe winters (Hellmann index >100, The Netherlands), a sufficiently high frequency to keep the migratory tendency in the Grey Heron population alive. However, from 1998 onwards, up to and including 2008, all winters, except for two normal ones, were mild, very mild or extremely mild. This may eventually result in a declining proportion of Grey Herons wintering in Africa. Netherlands, December 2004.

varying between 64 and 644 recorded annually. From the latitudinal zone between 4° and 20ºN, covering Africa south of the Sahara, a total of 126 recoveries is available (including 10 birds of unknown ages).

The 126 West African recoveries represent 0.62% of the total number of recoveries, well below the estimated proportion of Grey Herons wintering in West Africa (10%). There are two possible explanations. First, only 1049 Grey Herons ringed in eastern Europe were recovered, *i.e.* only 5.19% of the European total. Nevertheless, 46 of 126 African recoveries came from eastern European countries. Conversely, the remaining 19 280 western European recoveries yielded only 80 Grey Herons from Africa, *i.e.* 0.41%. Taking into account population sizes for eastern Europe and the rest of Europe (BirdLife International 2004a), multiplied by 3 (pair plus one young), this would produce some 16 200 trans-Saharan migrants from eastern Europe (91%) and 1600 from western Europe (9%). In reality, numbers are about five times higher than that (see above), indicating that reporting rates in Africa are much lower than in Europe.

Notwithstanding the lower reporting rate of Grey Herons in Africa, year-to-year variations in the number of recoveries from West Africa may be used to estimate annual variations in winter mortality. Winter mortality is indeed significantly higher in dry years (Fig. 125), even when taking into account that a wide variation is to be expected from the small number of recoveries available in separate years (11 at most).

Breeding conditions European Grey Heron populations have shown a fivefold increase since the 1970s, then levelled-off in the 1990s (Hagemeijer & Blair 1997, BirdLife International 2004a). The impact of harsh winter conditions is highest in more or less resident populations (as in Britain; Marchant *et al.* 1990), but less so in Central European populations (Marion 1980, Bauer & Berthold 1996).

During the first six decades of the 20th century, breeding numbers in Europe remained well below threshold capacity, mainly because of relentless persecution exacerbated by contamination of the food chain in the 1950s and 1960s. Reproduction and survival improved substantially following legal protection (Marion 1980, Kilian *et al.* 1993, but see Campos *et al.* 2001 for Iberia) and banning of persistent pesticides from widespread use in agriculture (Marchant *et al.* 2004). Present-day fluctuations are relatively small and mediated by events, such as severe winters in Europe, and density-dependent variables (Lande *et al.* 2002). The Sahel impact may be larger in the eastern European population, from which relatively more birds spend the winter in the Sahel, but unfortunately long-term series data are lacking.

Conclusion

Annually, large numbers of juvenile, immature and adult Grey Herons reach sub-Saharan Africa from their European breeding quarters. High numbers concentrate in floodplains, notably the Inner Niger Delta and the Senegal Delta. The overwhelming majority of Grey Herons wintering in sub-Saharan Africa is likely to originate from eastern Europe, where population size is greater and migratory tendency stronger. Annual winter mortality is low in wet Sahel years, but high in dry years. On the population level, however, any effect of poor flooding in West Africa is overridden by factors present in Europe (persecution, pesticides, severe winters), because only 10% or less of Grey Herons winter in West Africa.

Purple Heron
Ardea purpurea

Purple Herons reach high densities in West African floodplains, where well-scattered individuals stand on the floating vegetation (locally known as *bourgou*) along edges and in small open spaces. A casual observer may not see more than a handful of Purple Herons at any one time. Scanning *bourgou* fields for slender necks or pointed bills protruding from this buoyant dense vegetation may be more rewarding. When foraging Eurasian Marsh Harriers weave their meandering flights across *bourgou* fields, the real number of Purple Herons becomes especially apparent, as they take flight to evade the raptors. One particular floodplain, the Inner Niger Delta in Mali, holds tens of thousands of Purple Herons during the non-breeding season. This Sahelian wetland functions as the single most important wintering ground in West Africa (see below), and Purple Herons are therefore highly susceptible to changes in this floodplain, whether natural or human-induced.

Breeding range

In the western Palearctic, the Purple Heron breeding distribution is disjunct and largely confined to freshwater marshes and reedbelt-fringed lakes south of 53°N, but it is rather more continuous from central Europe eastwards, from the Danube towards south Ukraine and south Russia (del Hoyo et al. 1992, Bankovics 1997). The European population is estimated at 29 000-42 000 pairs, of which some 20 000 breed in Russia and Ukraine (BirdLife International 2004a). The entire population is migratory and winters mainly in western tropical Africa and East Africa from Sudan southwards (see below). The populations breeding in India and SE Asia and in East Africa are well separated from the western Palearctic population (McClure 1974).

Breeding in West Africa is restricted to very small numbers in Senegal (80 nests in 1974, Djoudj; Dupuy 1975) and Mali (Lac Aougoundou; Lamarche 1980). Breeding here is erratic, though, as exemplified by the Inner Niger Delta, where nesting was not recorded for many years until, in 1994, a high flood (the first high one since 1972) triggered 2-10 pairs into breeding (Chapter 6); in the following years breeding ceased again. The number of 1475 in Niger, in Kushlan & Hancock (2005) and based on Brouwer & Mullié (2001), refers to dry season counts of local and migratory non-breeding birds, not to pairs.

Migration

The distribution pattern of birds ringed in The Netherlands was significantly different from that of birds ringed elsewhere in Europe (Fig. 127). Birds recorded in Senegal, The Gambia and Mauritania originated exclusively from The Netherlands and France. On the other hand, Mali is a melting pot of all European populations of Purple Herons, but as shown by Voisin (1996), Purple Herons from eastern Europe have not yet been recorded from Senegambia. We consider this to indicate on average a slightly more easterly distribution of eastern European Purple Herons within West Africa. Many Purple Herons from Russia move through Greece and Turkey, through Egypt and Eritrea, to winter in eastern Africa, particularly from Sudan southwards (Kushlan & Hancock 2005). However, Russian birds have been recovered as far west as Nigeria (Elgood et al. 1994) and Benin (Cheke & Walsh 1996), and it is likely that a substantial portion of Russian birds winter in West Africa.

Distribution in Africa

At least 91 Purple Herons ringed in Europe have been shot, captured or found dead in West Africa, south from 17°N to the coastal regions at 4°N; the majority of these records have been extracted from the EURING database (Fig. 127). The recoveries are distributed across the Sahelian, Sudano-Sahelian and Sudanian vegetation zones (Chapter 4), i.e. covering grassland, cropland, savanna, woody savanna, forest, and inland and coastal wetlands. 22 birds were recovered from a single area, the Inner Niger Delta in Mali, and 11 from the coastal wetlands (including wet rice areas) from South Senegal through Sierra Leone. Only 4 recoveries came from another floodplain area, the Senegal Delta. The recovery distribution pattern shows that, within the Sahelian zone, most Purple Herons were found in the floodplains, the remainder being more or less evenly distributed across other wet habitats in the Sudanian vegetation zone.

The fact that most recoveries came from south of 10°N might lead to the idea that Purple Herons mostly winter in the Sudan zone, rather than in the Sahel. We believe this to be an artefact of distribution and density of the human population (and hence shooting): 45% of the sub-Saharan recoveries shown on the map refer to birds being shot. This proportion differs, however, per country, being particularly high in Liberia (86%, n=7) and Sierra Leone (75%, n=8). In other countries along the Guinean Gulf, 50% of the reported birds were shot (n=18). In the sparsely populated Sahelian countries of Mauritania, Senegal, Mali, Niger and Burkina Faso, 6 birds were shot, compared with 13 reported as being found dead (including trapped by accident). Hence, we may conclude that Purple Herons

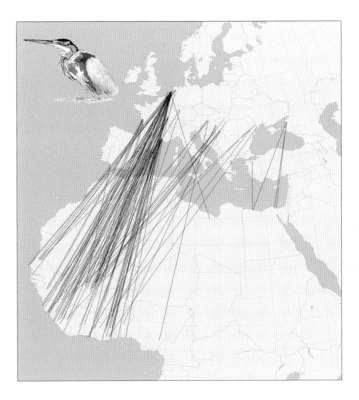

Fig. 127 European origins of 102 Purple Herons recovered in Africa between 4° and 30°N and from Egypt (EURING), 23 from Voisin (1996), 1 from Grimes (1987)) and 4 from Mullié et al. (1989).

Purple Heron *Ardea purpurea*

Box 18

Purple Herons on the move in Africa?

Jan van der Winden, Martin J.M. Poot & Peter van Horssen

So far, ring recoveries have revealed only limited information about migration routes and the behaviour of Purple Herons in Africa. Although the correlation between flood size in Sahelian floodplains and Purple Heron survival is clear, the actual movements of Purple Herons within African wetlands are still an enigma. A major question is whether Purple Herons are Sahel-bound or range more widely across West Africa necessary. To be more precise: do Purple Herons disperse further south when the Sahel is stricken by drought? The rather high recovery rates of Dutch birds in the Sudan and Guinea zones of West Africa (Fig. 127) suggest either regular wintering south of the Sahel or southward dispersal within West Africa during the northern winter. The range covered by wintering Dutch herons is at least 80 times the size of their breeding area. Within this vast area floodplains clearly stand out as regular wintering sites, but the limited ringing effort in The Netherlands hampers the identification of other important wintering areas.

In the summer of 2007, five adult and three juvenile Purple Herons were equipped with satellite transmitters. We partly hoped to remedy our ignorance of what the birds were doing in the wintering quarters, and also to collect information on habitat use and daily rhythms whilst feeding, roosting and migrating. We used two transmitter types: GPS transmitters providing accurate and regular data on locations (for studies on habitat use), and a PTT transmitter to study intra-African movements. All birds started migration after breeding, but two juveniles first dispersed into northern France before starting migration a month later. Losses during migration were high: two adults died in sand storms in the Sahara and a juvenile was shot in Morocco. Another juvenile was disoriented and died on the Atlantic Ocean not far from Brazil. Although unfortunate, these events emphasised some of the many dangers that Purple Herons encounter during migration.

The migration routes revealed by tracking are used in conjunction with the information based on ring recoveries. All birds moved southwest across the Iberian Peninsula, mostly passing the Pyrenees along the west side, continuing through Morocco or Algeria into the Sahara. Most birds crossed the Mediterranean Sea at the southern point of the Iberian Peninsula, where the sea passage is shortest. Four juvenile Purple Herons from the Camargue, equipped with satellite transmitters in 2004, partly used the same route during their southbound migration (Jourdain et al. 2008).

Two of our adults and one juvenile reached the African wintering areas in the Senegal or River Niger basins. Both adults headed further south after one or more short stop-overs in the Sahel, finally reaching Ivory Coast and Guinea respectively. The one in Ivory Coast wintered along a small river valley and nearby artificial lake. Unfortunately its transmitter failed in November 2007. The other adult navigated directly into Guinea and wintered in a mangrove-fringed rice field complex close to the coast. The juvenile bird started wintering in the wetlands along the Niger in Mali and continued southwards to Ivory Coast in January-February. This shift coincided with the seasonal decline in floodplain size of the Inner Niger Delta.

Too few birds were tracked to draw firm conclusions. The wintering range of Dutch Purple Herons may indeed be huge, as assumed from ring recoveries. And perhaps the juvenile Purple Heron "Lena", which reached tropical Africa in late winter, does support the hypothesis of intra-African movements in dry periods.

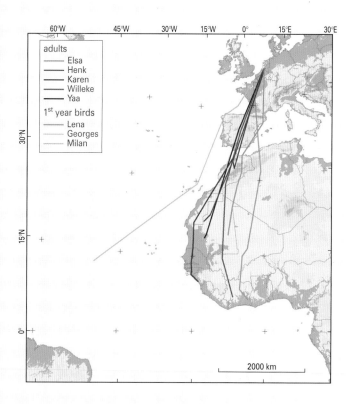

Fig. 128 The migration routes of eight Purple Herons equipped with satellite transmitters.

are shot more often in the Sudan zone than in the Sahel, and therefore run a higher chance of being reported in the Sudan zone. The relative scarcity of Purple Herons in the Sudan zone is corroborated by the description of being "not uncommon" in Liberia (estimated at 1000-1500 wintering birds; Gatter 1997b), Ghana (Grimes 1987), Togo (Cheke & Walsh 1996) and Nigeria (Elgood *et al.* 1994). In comparison, we estimate the wintering population in the Inner Niger Delta (Mali) at 30 000-70 000 birds (Chapter 6). This is by far the most important wintering area of Purple Herons in West Africa.

Population change

Wintering conditions On average, 2 birds were recovered per year, from the 4°-20°N latitudinal zone, the variation being 0 to 10 annually. To correct for ringing density, we calculated the % African recoveries relative to the annual total. There are 860 recoveries in the EURING database, of which in total 8% are from Africa. The annual percentage in years with a large flood extent is always below 10%, but it can rise to 20-80% in years when the flooded areas are small (Fig. 129). This is an obvious indication that the winter mortality depends on the flood extent.

Den Held (1981), Voisin (1996) and Gatter (1997b) suggested that Purple Herons wintering in West Africa moved further south in dry years. This is not supported by our analysis of the ringing data. Compared with earlier analyses, we have a larger data set, covering a longer time series, at our disposal. Limiting the analysis to the period before 1980, we also find, as did Den Held (1981), more recoveries in the Sudan vegetation zone in relatively dry years. However, this can be ascribed to the 7 Purple Herons shot in Liberia in 1973, an extremely dry year indeed. We cannot exclude completely the

Purple Herons in the Inner Niger Delta feed on floating meadows of *bourgou* grasses. With an average body mass of about 1 kg, the dense *bourgou* mats can easily hold the weight of a Purple Heron, especially because the long toes evenly distribute the weight across the thick stems floating on the water.

suggestion that Purple Herons may temporarily shift from drought-stricken regions to less affected regions further south; such a strategy is specifically mentioned for Liberia, where J. v.d. Linden (in Gatter 1997b) recorded greater numbers of Purple Herons in the very dry winters of 1983/84 and 1984/85 than in the winters preceding and following. Whether or not Purple Herons shift their wintering grounds to the south in dry winters, and to what extent, remains unsolved. Neither ring recoveries nor direct observations as yet have sufficed to answer this question (but see box 18).

Since the pioneering study of Den Held (1981) on the impact of drought in the wintering area on breeding Purple Heron numbers in The Netherlands, further studies essentially came to the same conclusion (Cavé 1983, Fasola *et al.* 2000, Barbraud & Hafner 2001). We have extended earlier analyses with longer time series and with trends in breeding numbers for as many countries/regions as available (Fig 130).

Population trends in European countries are rather well synchronised.[1] This suggests that population fluctuations are determined only partly by changes in local circumstances (see below) and that there is a common denominator that has an overriding impact on overall population level, i.e. wintering conditions in the floodplains of West Africa. The reduction in the extent of the floodplain of the Inner Niger Delta in the early 1970s, the recovery in subsequent years and renewed poor flooding from 1983 onwards are clearly reflected in the numerical vicissitudes of the Purple Heron. However, the figures also show that the number of Purple Herons did not decline any further during a decade of consistently low flood levels from the mid-1980s to 1994.

Purple Herons in West Africa are not entirely concentrated in the floodplain areas (see also ring recoveries above). Hundreds, if not

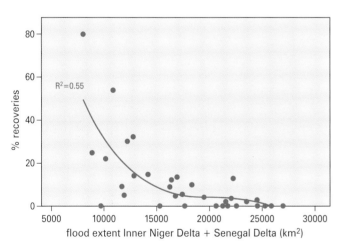

Fig. 129 Annual number of recoveries from Africa (4°-20°N) as % of the total number in the same year (dead birds only) between 1953 and 1987 (n= 35 years, P<0.001). From: EURING.

thousands, winter along river margins, shores of freshwater lakes, reservoirs and lagoons (Grimes 1987, Elgood et al. 1994, Cheke & Walsh 1996, Gatter 1997b). These habitats are much more prone to annual variations in rainfall than floodplains – and hence vary in suitability for Purple Herons in winter. A statistical analysis [2] clearly suggests that for Purple Herons wintering in West Africa, the floodplains are far more important than temporary rain-fed lakes and ephemeral wetlands; how important is explained below.

The close relationship between population size of the western European Purple Heron and the flood extent must imply that more birds die during reductions in the extent of floodplains, as was indeed found in the analysis of the recoveries (Fig. 129). The change in breeding numbers between two consecutive years may be used as another proxy for mortality. The population does indeed increase during high floods, but population changes are also related to the size of the population: the population tends to decrease when at a high level, and to increase if the breeding population is small.[3] If the annual estimates of the population are less than accurate, sampling errors will then automatically generate such a negative correlation. Although such an error cannot be ruled out entirely, the Dutch – and increasingly, other European - population estimates are actually very precise. Hence we conclude that the tendency of a low population to increase and a high population to decrease should be interpreted as a density-dependent mechanism. Simultaneously, the population is known to increase during high floods in the West African floodplains (see above). Annual mortality, such as deduced from the population

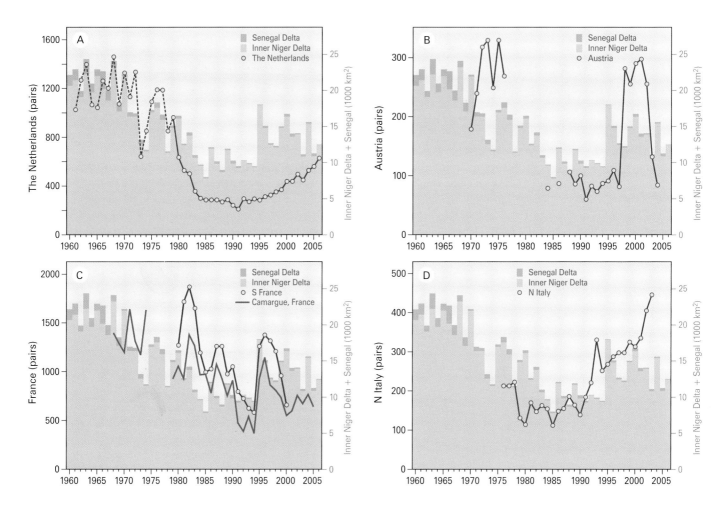

Fig. 130 Population trends of the Purple Heron in parts of four European countries related to the flood extent of the Inner Niger Delta and the Senegal Delta in the preceding year. From: den Held 1981, van der Kooij 1982-2007 (The Netherlands), Grüll & Ranner 1998 for 1970s; Dvorak pers.comm., since 1984 (Austria, Neusiedlersee; numbers in 1984-1997 were under-recorded), Barbraud & Hafner 2001, Kayser et al. 2003, tourduvalat.org (France), and Fasola in Kushlan & Hafner 2000 and in BirdLife International 2004b (NW Italy). All Dutch colonies have been counted since 1980, but only two major colonies had been counted (Nieuwkoop and Maarssen) in the 1961-1979 period. Because these two colonies accounted for 41% of the Dutch population in later years, we have used this ratio to reconstruct the overall Dutch population in 1961-1979 on the basis of Nieuwkoop and Maarssen.

change, is density-dependent, but clearly is also working in concert with the surface area flooded in African floodplains.

Breeding conditions The all-important interaction between flooding of wetlands in West Africa and breeding numbers in Europe does not exclude the possibility that local factors in or near the breeding colonies can also have an impact on numbers. Many examples of such locally induced variations in colony size are available, of which loss of breeding habitat to reclamation, industrial development, tourism and urbanisation are the most detrimental because of their permanent nature (Kushlan & Hancock 2005). Short-term declines have been reported following excessive use of pesticides (Spain, Ebro Delta, 1960s; Prosper & Hafner 1996), manipulation of water levels and commercial reed exploitation (France, Camargue; Deerenberg & Hafner 1999), and sedimentation of reedbeds (central France; Broyer et al. 1998).

Many other factors influence the quality of breeding and feeding grounds (review in Kushlan & Hafner 2000). In The Netherlands, for example, Purple Herons breed mainly in reed marshes in the low-lying peat district, where they forage in ditches in the neighbouring grasslands (rarely in shallow marshland). The loss of natural water level dynamics (high in winter, low in summer) and the industrialisation of farmland have resulted in negative changes in nesting and foraging habitats. Purple Herons have adapted by habitat-switching and by changes in diet and nest site choice (van der Kooij 1991a, 1995a, 1997a, van der Winden & van Horssen 2001, van der Winden et al. 2002).

Local improvements in such conditions may lead to larger than average increases, as happened in the 1990s in the Zouweboezem colony, where, due to specific conservation management, the population boomed from 5 pairs in 1990 to 174 pairs in 2006 (30% of the Dutch population); in contrast, the entire Dutch population – including the Zouwe – only doubled during this time interval (van der Winden et al. 2002, van der Kooij in series) (Fig. 131). The steep growth of the Zouwe colony cannot entirely be explained by local reproduction rates, because birds from other Dutch colonies must have moved to the Zouwe (van der Winden et al. 2002). The main question is: what would the size of the Dutch population be without the creation of the Zouwe? The simplest answer would be that the birds had the opportunity of colonising a new habitat, which, if created proportionately in other suitable areas, would allow the Dutch population to grow by 30%. However, if the population size of the Purple Heron is determined primarily by conditions in the wintering area, the creation of a new breeding site leads only to a redistribution of the birds, not to an enhanced population. We now know that, although the population is largely regulated by the wintering conditions, improvements to breeding areas have an influence and so the truth is most likely to be somewhere between the two sets of circumstances.

Conclusion

Trends in European populations of Purple Herons are largely determined by climatic conditions in the main winter quarters, i.e. the floodplains of Sahelian West Africa. High floods correlate with high numbers, low floods with declines. This is particularly evident during long-lasting droughts, as in the 1970s and 1980s. In most European countries, the devastating effects of the droughts in the 1970s and 1980s were only partly ameliorated in the 1990s. This major trend is superimposed on smaller variations in breeding numbers caused by local changes in conditions in the breeding areas.

Endnotes

1 The trends in the three longest series are well correlated: Austria vs. The Netherlands: R=+0.62 (n=26 years), and France vs. The Netherlands: R = +0.63 (n=34).
2 When the size of the Dutch population is plotted against rainfall in the Sahel, the correlation is, although significant (R^2 = 0.285), much less evident than for flooding (R^2=0.688). When rainfall and flooding are entered in a regression equation, flood appears to be the dominant factor; the added contribution of rainfall is small (R^2=0.042) and weakly significant (P=0.017).
3 A multiple regression analysis shows that the population change in the Dutch population is significantly related (P< 0.001) to population size in the preceding year (ß=+0.820) and to flood extent (ß=-0.817); R^2=0.604.

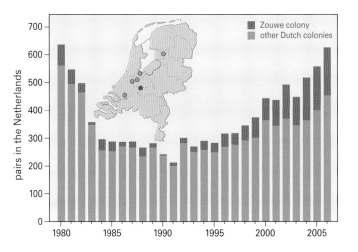

Fig. 131 Population trend of Dutch Purple Herons, with the Zouwe colony shown separately. From: van der Kooij 1982-2007, SOVON.

Black crowned Night Heron
Nycticorax nycticorax

Silence reigns in the Inner Niger Delta, accentuated only by the steady rhythm of wood against wood from passing pirogues – a greeting, then silence again. Solitary egrets and herons wade along the water courses; some gather where fishermen have built barrages to trap fish when water levels fall. Rarely do we encounter a Night Heron in daylight, and when we do, it is likely to be a juvenile, hooked by one of the many thousands of hooklines that litter the water's edge – even the slightest touch suffices to become hooked, as we know from painful experience. By day, the Night Heron is enigmatic, but its mystery begins to resolve around dusk, when, from many directions, long silent lines of diurnal herons and cormorants start homing in to roosts, phalanxes of birds slowly undulating low in the skyscape and disappearing against the darkening horizon. Soon after, the silence of the delta is fractured by the eerie croaking of large flocks of Night Herons leaving their day-time roosts. There is so little light that this mass movement would be easily missed if the birds did not call. Thousands of Night Herons spread out through the delta at night, wading along the same watercourses that egrets and diurnal herons have just vacated. Most of the birds exploiting the abundant food resources in this watery landscape originate from Europe, their stay in the delta but a sojourn.

Breeding range

In Europe, the Black-crowned Night Heron is distributed across low altitude wetlands, occurring rather patchily in small colonies along the northern edge of its range and most commonly in large colonies in the lower catchments of rivers like the Rhône, Po, Danube, Dnestr, Volga, Terek and Kura. The European population is estimated at 50 000-75 000 pairs (Marion *et al.* 2000), Italy holding particularly high numbers (fluctuating between 14 000 and 20 000 pairs in 1985-1990, and between 12 000 and 14 000 in 1995-2002; Brichetti & Fracasso 2003). There are also significant populations in Russia (10 000-15 000 pairs), Romania (decreased 47% between 1974-1975 and 1987-89 to 3100 pairs in 1986), France (4176 pairs in 1994) and Ukraine (3645-5000 pairs in 1986). Despite strong fluctuations, populations were considered more or less stable in the early 2000s (Kushlan & Hancock 2005).

Breeding by Black-crowned Night Herons in West Africa is uncommon. For the Inner Niger Delta, Skinner *et al.* (1987a) mention up to 10 breeding pairs, nesting between August and March. In the Dentaka Forest, they may breed semi-annually in very small numbers, as evident from observations of downy-headed juvenile birds in February and March. In February-March 1995, 100-300 pairs of Night Herons were recorded in Dentaka Forest; in March 1995 several dozen recently fledged juveniles proved that a few pairs had been successful (van der Kamp *et al.* 2002c). Since then, breeding numbers have reverted to 1-10 pairs annually in the Inner Niger Delta (Chapter 6). Elsewhere in sub-Saharan Africa, breeding colonies usually are small (no more than a few hundred) and widely scattered, as, for example, in Nigeria (three colonies, up to 200 pairs; Elgood *et al.* 1994), Togo (not uncommon resident from Lomé north to Dapaon; Cheke & Walsh 1996), Ghana (*c.* 40 pairs at Weija, not annually; Grimes 1987), and Liberia (small numbers resident at Harper power plant; Gatter 1997b). The species is commoner in the Senegal Delta, where in the Djoudj National Park up to 1375 nests were counted in November 1988 (Kushlan & Hafner 2000); in 1985 and 1987 several hundred nests with eggs were recorded for only Poste de Crocodile and Lac de Khar (Rodwell *et al.* 1996).

Migration

The distribution of ringing recoveries in West Africa shows a concentration in the Inner Niger Delta, a metropolis for Night Herons from France to Ukraine. The present number of recoveries is insufficient to detect any segregation of European breeding populations within West Africa. Central European birds, from Hungary to the Balkans, show a scatter of recoveries in directions between west and southeast (Schmidt 1978, Nankinov 1978). Eastern birds on average winter further east in Africa than those from southern Europe (Sapetin 1978b). Some ringed Night Herons from colonies along the northern and eastern Black Sea, and on the Caspian Sea, were recovered in Africa, from Lake Chad through Congo-Brazzaville to the Nile in Sudan (Fig. 132). Birds ringed in Azerbaijan were recovered during migration in Italy, Lebanon and Iraq, which indicates that this population also exhibits broad front migration to wintering areas in Central West Africa and East Africa (Sapetin 1978b, Patrikeev 2004).

Of the larger floodplains in the Sahel, only the Inner Niger Delta stands out with 7 recoveries; the Senegal Delta and the Waza Logone in northern Cameroon each are represented with a single recovery. The recoveries in Algeria, Libya and Egypt indicate that Night Herons cross the Sahara on a broad front; the birds along the Nile in Egypt and Sudan may have been heading to wintering grounds in East Africa (Fig. 132).

The cause of death of 18 birds found in Africa is known: 9 shot, 6 trapped (4 of which were by hookline in Mali) and 3 from natural causes (EURING). Night Herons are particularly vulnerable to human predation on their wintering grounds, not only because roosting sites are few and traditional and attract large numbers of birds, but also because their foraging strategy of wading along shallow water edges makes them vulnerable to hooklines, particularly if they are baited.

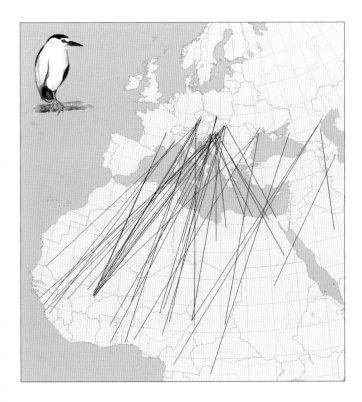

Fig. 132 European ringing locations of 46 Night Herons recovered in Africa between 4°N and 30°N (and including Egypt). From: EURING (31), Wetlands International (Mali, 5), Mullié *et al.* 1989 (Egypt, 6), Sapetin 1978b (Black Sea and Caspian Sea, 4), J. Brouwer pers. comm. (Niger, 1).

Black-crowned Night Herons in southern Europe prey upon the non-indigenous Red Swamp Crawfish *Procambarus clarkii*. This prey species thrives in disturbed habitats, including rice fields. Its present abundance in southern Europe has boosted breeding and wintering in a wide range of large fish-eating birds, including Great Bitterns (relative abundance of crayfish explained 56% of the inter-annual differences in Bittern density in southern France; Poulin *et al.* 2007), herons and White Storks (Chapter 20). Albufera, Spain, May 2008.

Distribution in Africa

Less than 1% of the European population of Black-crowned Night Heron winters in southern Europe. The main wintering areas lie south of the Sahara, mostly in the floodplains of the Sahel and in West-African wetlands. The wintering population in West Africa is estimated at 70 000-100 000 birds, with high concentrations in the Senegal Delta (10 000, year not specified, Kushlan & Hancock 2005; 5200-7600 in 1993-2002, Chapter 7), the Inner Niger Delta (11 000-12 000 in the early 2000s and 21 000 in 2008, based on counts at roosts; Chapter 6), and Waza-Logone in northern Cameroon (up to 4779 in 2000; Scholte 2006). A high proportion of juveniles may remain there for up to several years (Kushlan & Hancock 2005), which emphasises the importance of Sahelian wetlands for this species even further.

Population change

Wintering conditions Linking ring recoveries to three categories of flooding of the Inner Niger River shows that Night Herons are more likely to be recovered in dry years, and less so in wet years: 0.36 recovery/year when more than 16 000 km^2 was flooded, 0.91 when less than 12 000 km^2 was flooded and 0.81 for interim flood extents. This trend, albeit based on small numbers, applies to both the Sudan vegetation zone and the Sahel. We should find that flooding extent relates to mortality rates, as indicated by the levels of breeding populations in Europe, or, more precisely, we would expect a delayed effect, because Night Herons enter the breeding population only when 2-3 years old (Kushlan & Hafner 2000). This hypothesis is testable by data from southern France, for which breeding population the longest time series is available: the breeding population reflected closely the annual variation in water level in the floodplains of West Africa (Fig. 133A).[1] Because wintering Night Herons are concentrated in the floodplains, it was presumed that flood extent would be much more important than rainfall, but the direct impact of the preceding flood on the breeding population half a year later came as a surprise. Apparently, during dry winters adult mortality is sufficiently large to have an immediate effect on breeding numbers. Over the 34 years of monitoring in the Camargue, the population has shown clear rises and falls, but not clear increasing or decreasing trends (Fig. 133A).[2]

Monitoring programmes should cover at least several decades in order to depict trend directions reliably, so that local short-term changes in numbers can be assessed in context. They should also take into account that data from a single site can be misleading: for example, data about Night Herons in colonies along the Dnestr in Ukraine (Fig. 133B). Ukrainian birds appeared to show the same pattern as French birds, with dips in the mid 1970s and 1980s and higher numbers in between, but the trends differed. However, in the late 1980s and early 1990s, when local breeding conditions had improved after the negative effects of dam-building had been mitigated (Schogolev 1996b), the local population recovered. As had already been found by Fasola et al. (1996), the size of the Ukrainian population and the Sahelian rainfall index are significantly correlated (see also Fig. 133B), with indications of a delayed effect of 1-2 years.[3] This contrasts with the immediate response of the Camargue population.

Other time series, such as those from the Albufera Natural Park, Spain (39°20'N, 00°20'W, 1982-93; Prosper & Hafner 1996) and NW Italy (1976-97; Fasola in Kushlan & Hafner 2000), show consistently low population levels in the 1980s, coinciding with drought conditions in West Africa. The colonies in the Ukrainian Danube Delta also showed a distinct dip in 1985 (very dry conditions in the Sahel), a recovery in 1986-88 and lower numbers throughout the 1990s, with an upsurge in 2002 (Platteeuw et al. 2004). In the adjacent Romanian Danube Delta, breeding numbers were low in the 1980s and started to increase in the 1990s. Taken altogether, the Danube Delta held 2140 pairs in 2001 and 2964 pairs in 2002 (Platteeuw et al. 2004). According to Kushlan & Hafner, the Danube Delta population

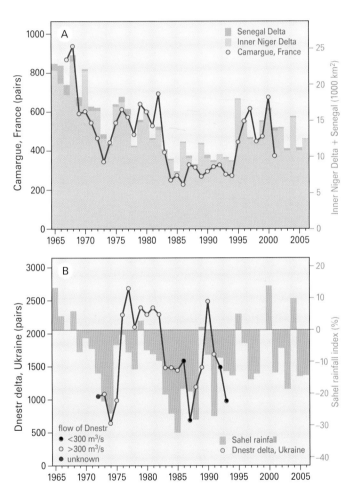

Fig. 133 (A) Night Heron breeding population changes in the Camargue, France, related to the flood extent of the Inner Niger Delta and Senegal Delta in the preceding year. From: Kayser et al. 2003 and www.tourduvalat.org. (B) Night Heron breeding population changes in the Dnestr delta, Ukraine related to the Sahel rain index in the preceding year. From: Schogolev 1996b.

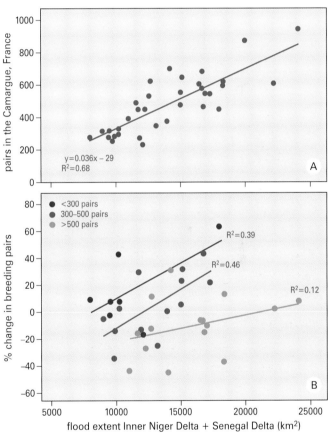

Fig. 134 Size of the breeding population in the Camargue in 1967-2000. (A) number of pairs. (B) % change in the population as a function of flood extent of the Inner Niger Delta and Senegal Delta in the preceding year. Same data as Fig. 133A, subdivided for three population levels.

Thousands of Night Herons roost by day in a Tamarisk forest in the Djoudj National Park, Senegal Delta, January 2008. Undisturbed roosts are as important as undisturbed breeding sites.

numbered 3100 pairs in 1986 and declined by 47% between 1974-75 and 1987-89, coinciding with droughts in the Sahel.

The relationship between size of breeding populations and the extent of flooding in their African wintering areas (Fig.134A) implies that annual winter mortality increases when the floodplains of the Inner Niger Delta and the Senegal Delta are relatively small (see also Fig. 133). The numerical change in a population in successive years, used as a rough measure of mortality, revealed no relationship with the size of West African floodplains. However, further analysis showed that population size does play an intervening role. Numbers in the Camargue tend to remain stable if the population is large (>500 pairs); under such conditions flood extent seems to have little or no impact on population size. In contrast, when the population is small (<300 pairs), numerical increases are significantly larger if floodplain size had been large in the preceding winter (Fig. 134B). [4] The same conclusion may be drawn from Fig. 135, showing the annual population change, but now as a function of the population size.

Figs. 133 - 135 clearly demonstrate that the extent of flooding of West African floodplains largely determines the size of the population in the following breeding season. The breeding population size in the Camargue tends to level off and remain stable at a level of 600 pairs at a large flood extent, indicating density-dependent effects in

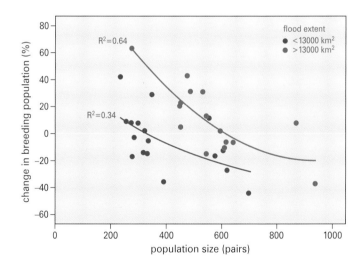

Fig. 135 % change in the Camargue population as a function of population size in the preceding year, subdivided for years during which the Sahelian flood extent was above or below 13 000 km² in the preceding year. The data from Fig. 134B have been re-plotted in this graph to show the density dependency and the Sahel effect in another way.

the wintering areas (provided the Camargue represents an unbiased sample of Night Heron populations).

Breeding conditions As in the Camargue, trends elsewhere in France depict a marked low in 1974 (Voisin 1994), variously followed by further declines or, locally, steep increases [5] Strong fluctuations, but generally with a declining trend, are also known for Spain (Fernández-Cruz et al. 1992), Croatia (D. Munteanu in Kushlan & Hafner 2000) and the Ukrainian Danube Delta (Platteeuw et al. 2004). Many of these fluctuations are caused by local variations in hydrological conditions, especially in southern Europe, where the suitability of sites for breeding and foraging depend upon very dry or wet seasons. Similarly, as demonstrated for the Dnestr colonies (see above), the construction of dams may permanently destroy prime breeding sites and negatively affect major food resources, unless remedied by a regime that permits natural flooding (Schogolev 1996b). On the other hand, dams sometimes create new feeding and breeding opportunities for Night Heron (e.g. Leibl 2001).

Conclusion

Despite the scarcity of long-term census data from European breeding quarters and of ringing recoveries from West Africa, the information available suggests that: (1) The Black-crowned Night Heron winters mostly in the floodplains of the Sahel; (2) The extent of Sahelian floodplains, determined by rainfall, during the northern winter largely determines the size of the population in the following breeding season; (3) The impact of local breeding conditions on population size is small in comparison with that of protracted periods of droughts or wet conditions on the wintering grounds; the latter have a significant bearing on the extent of feeding areas, and hence survival.

Endnotes

1. The scatter around the regression line shown in Fig.134A cannot be attributed to flood extent in West African floodplains in the two or three years preceding the breeding season, nor to rainfall in the Sahel. Although these variables are significantly related to breeding population size (for instance, rain: $R^2=0.41$, $P< 0.001$), none adds significantly to the variance explained by flood level in the preceding year.
2. A multiple regression analysis shows that the Camargue population depends entirely on flood extent in West Africa ($R^2=0.68$), and not on year (enhancing R^2 with 0.1). Thus, the small scatter around the regression line shown in Fig.134A cannot be attributed to long-term changes.
3. Sahel rainfall in the preceding year has a significant impact on the size of the Ukrainian population ($R^2=0.28$). The explained variance increased to $R^2=0.54$ if the rainfall in the two years before the preceding winter is added to the analysis. Surprisingly, the Ukrainian numbers are only weakly related to flooding of the Inner Niger Delta ($R^2=0.11$).
4. A multiple regression analysis shows that the percentage population change of the Night Heron (NH%) is significantly correlated ($P< 0.0001$) with population size (P), and with the flood extent of the Inner Niger Delta (IND) ($R^2=0.52$): NH% = -6.40 - 0.131P + 0.005IND.
5. There were declines (Les Dombes: 400 pairs in 1968, 215 in 1974, 81 in 1986, 0 in 1989), increases (Loire, Allier and Cher: 256 in 1974, 502 in 1978, 680 in 1982, 450 in 1985; Aquitaine: >80 in 1974, 99 in 1980, 313 in 1989; Vendée: 15 in 1974, 51 in 1989), and steep increases (Durance and Rhône Valleys: 0 in 1974, 25 in 1981, 248-253 in 1989; and especially the Midi-Pyrénées in central southern France: 275 in 1974, 1000 in 1981, 1665 in 1989, about 2000 in 1991).

Squacco Heron
Ardeola ralloides

Imagine the Darb el Arba'in Desert, the hyper-arid core of the eastern Sahara (<5 mm annual rainfall), as far as the eye can see, a blanket of sand, occasionally interspersed by small sandstone outcrops (Goodman & Haynes 1992). Vast numbers of birds cross this inhospitable area twice yearly, mostly unseen even by the local Lanner Falcons, whose favoured perches are the detritus of war, left behind from military operations by the British Army and the Sudan Defence Force in 1941, when lorry convoys were resupplying the recently captured oasis of Kufra in Libya. This 20 to 30 km-wide swathe, starting at Wadi Halfa in Sudan, traverses the border between Egypt and Sudan and remains littered with abandoned convoy markers, lorries and other vehicles, camp sites, fuel tins and drums. The Lanner begins nesting when few bird migrants are passing through (early February) and then feeds almost exclusively on locusts. Later, with young to feed, it can prey on exhausted or resting migrant birds, mostly Quails and Turtle Doves but also Squacco Herons.

Squacco Herons are often difficult to detect when feeding in dense marshland, unless standing in full view on top of floating water plants such as Water Lily and Water Cabbage. Senegal Delta, December 2005.

Breeding range

This rarest of European herons is thinly distributed in southern Europe, where some increase has taken place since the late 1990s (Kushlan & Hancock 2005). Higher numbers are present in eastern Europe, but large colonies in the major breeding areas are even further east in Turkey, Iran and Transcaucasia. In Azerbaijan, for instance, the population in the late 1980s and early 1990s was estimated at 15 000-18 000 pairs, but as many as 84 500 pairs were counted in 1964 in the Kizil Agach Reserve alone (Patrikeev 2004). The European population has been estimated at 14 300-26 800 pairs (Kushlan & Hancock 2005), mostly nesting in Russia (35%), Turkey (32%), Bulgaria (11%) and the Danube Delta in Romania (10%).

The species breeds in many West African freshwater marshes, from Mauritania, Senegambia, Sierra Leone, Mali, Ghana and Nigeria to Chad. Small numbers breed in the Senegal Delta (Chapter 7). Breeding numbers are apparently stable in the Inner Niger Delta in Mali, where Skinner *et al*. (1987a) reported 550-650 pairs in 1985; 500 pairs were found in 1999-2001, mostly in the Dentaka forest but also, in November 2000, at Akkagoun (90 nests) and in the degraded forest of Pora (some dozens of nests) (van der Kamp *et al*. 2002c, Chapter 6). In comparison with its early 20th century breeding status, the species' present distribution in West Africa is wider and its abundance greater (Kushlan & Hancock 2005).

Migration

Small numbers winter in the Mediterranean region, but most Squacco Herons are fully migratory and cross the Sahara to winter in the sub-Saharan band, from southern Mauritania, Senegambia and Guinea-Bissau eastwards. The EURING files contain only five recoveries of birds ringed in Europe: two birds from the former Yugoslavia, ringed in 1912 and recovered in Nigeria in 1914; one from France, ringed in 1968 and recovered in Sierra Leone in 1969; another ringed in 1969 and recovered in Guinea in 1970; the final one from Greece ringed in 1992 and recovered in Ghana in 1993. Rodwell *et al*. (1996) mention the observation of a wing-tagged bird from the Camargue, France, in January 1990 in the Djoudj, Senegal. For Nigeria, Elgood *et al*. (1994) mention a third Squacco Heron from the former Yugoslavia, as well as a bird from Bulgaria. Migration supposedly occurs in a broad front across the Mediterranean Sea and Sahara, as deduced from the extent of visible prey remains of Lanner Falcons in the Egyptian desert.

Nondescript on the ground, but quite conspicuous when taking flight, Squacco Heron counts in marshlands and floodplains are likely to be biased, depending on the counting method.

Distribution in Africa

Squacco Herons are abundant in the floodplains of West Africa. Ash (1990) mentioned the presence of several thousand in just one location within the Hadejia-Nguru floodplains (Matara Uku in November 1985. Although their cryptic coloration may conceal them well in suitable habitat, they are prominent when flushed. Our plot sampling in the Inner Niger Delta permitted an estimate of some 41 000 Squacco Herons in the central lakes Debo and Walado in the early 2000s and 200 000 in midwinter for the entire Inner Niger Delta, including Lake Télé (Table 6 and 8).

Apart from visiting floodplains, Squacco Herons are generalists, occurring wherever fresh or brackish water is found and sufficient cover is available. In most West African countries, it is considered a common or locally common Palearctic migrant from October to March, found at lakes, water reservoirs, sewage farms, rice fields, ponds, coastal wetlands and lagoon edges (Grimes 1987, Elgood *et al.* 1994, Cheke & Walsh 1996). In Liberia, wintering numbers are estimated at 1000 for the interior and a few hundreds along the coast (Gatter 1997b).

The importance of rice fields for wintering Squacco Herons was highlighted during stratified plot sampling in Senegal, The Gambia, Guinea-Bissau and Guinea-Conakry in the winters of 2004 and 2005 (Bos *et al.* 2006; Chapter 11). The density in rice fields extrapolated into some 71 000 Squacco Herons (95% confidence interval 50 000 – 92 000) for the 120 000 ha of this type of habitat in the above countries. Using the same type of counts in Mali in 2002-2004, van der Kamp *et al.* (2005b) arrived at 200 Squacco Herons in the irrigated rice fields near Lake Selingue, and Wymenga *et al.* (2005) at 4000

in the irrigated rice scheme of *Office du Niger* during the winter (see also Chapter 11). The density plots revealed that in June-July both areas harboured four to five times more Squacco Herons than in December-February, i.e. 900 in Selingue and 21 000 in *Office du Niger* respectively. Many birds in the latter area were in nuptial plumage (J. van der Kamp, pers. obs.), with some remaining to breed in the former river branches (*Falas*) and *Acacia* habitats (Wymenga *et al.* 2005). Most Sahelian wetlands dry out in the rainless period of November to June, which is probably why so many Squacco Herons are concentrated in the wet rice fields in June-July.

When combined, the winter counts in the floodplains and rice fields in West Africa produced a total of some 350 000 Squacco Herons in the early 2000s. This number exceeds any estimate for Eu-

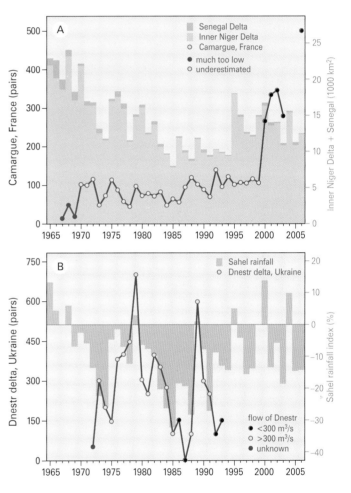

Fig. 136 (A) Breeding population of Squacco Herons in the Camargue, France related to the flood extent of the Inner Niger and Senegal Deltas in the preceding year. Coverage of French colonies improved after 2000 and was inadequate in 1967-69. From: Kayser *et al.* (2003), www.tourduvalat.org. (B) Breeding population in the Ukrainian Dnestr delta related to the Sahel rainfall index. From: Schogolev (1996b).

Squacco Heron along the shallow water edge with Cattail and an inevitable flipflop (omnipresent in most, even remote, African wetlands). Senegal Delta, December 2005.

rope and Africa, indicating not only serious underestimations of breeding populations (the species is indeed difficult to census on breeding grounds; Hafner *et al*. 2001), but also a substantial influx of birds coming from Asian breeding areas.

Population change

Winter conditions Monitoring of nesting pairs in the Camargue shows fluctuating numbers in the 1970s and 1980s, followed by an increase to >100 pairs since the early 1990s and even to 500 pairs in 2006 (Fig. 136A). Census methodology changed twice (Hafner *et al*. 2001), and counts before 2000 are now considered as underestimates (www.tourduvalat.org). In the 1970-98 period, the population increased annually at 1.4% on average. The year-to-year population change varied, however, from -47% to +50%. This variation weakly correlated with the extent of the West African floodplains. [1]

The Ukraine population in the Dnestr delta showed steep fluctuations, partly in conjunction with local river discharges and the construction of a hydroelectric dam 700 km upstream in 1983; the negative impact of the dam became less in 1988 when natural floods were allowed again to inundate the delta (Schogolev 1996b; see black dots in Fig. 136B). The Dnestr trend follows, with some delay, the flooding of the Inner Delta, but it correlates better with the Sahel rainfall index. [2]

In addition to the two series discussed so far, there are several country-wide and regional censuses, with larger sample sizes, from Italy (Brichetti & Fracasso 2003), Spain (Prosper & Hafner 1996) and Azerbaijan (Patrikeev 2004), collected intermittently over longer time intervals. These data, although possibly based on underestimations, show that hydrological conditions in West Africa are of major importance in shaping Squacco Heron populations. In particular, severe droughts are reflected in marked population declines on the breeding grounds. Between 1940 and 1960, the Western Palearctic population increased by some 30% (Kushlan & Hafner 2000). In 1970-90, the period characterised by severe droughts in West Africa, two-thirds of the European breeding populations decreased by some 20%. This decline was particularly prominent in eastern Europe and Russia. In Greece, for example, the national total amounted to 1400 pairs in 1970, declining to 1100 pairs in 1970-84 and a mere 201-377 pairs in 1985-86 (Crivelli *et al. in* Kushlan & Hafner 2000). Similarly, at Kopacki Rit, a breeding site in Croatia, numbers declined from 478 pairs in 1954 to 190 pairs in 1970 and <50 in the mid-1980s (Mikuska *in* Kushlan & Hafner 2000). An even steeper

Some 350 000 Squacco Herons spend the northern winter in West Africa, of which 200 000 are in the Inner Niger Delta; wherever shallow (vegetated) water is available, the species is omnipresent. January 2003, Diaka River, Inner Niger Delta.

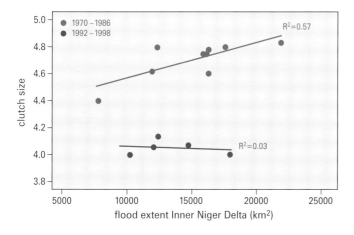

Fig. 137 Mean clutch size of Squacco Herons breeding in the Camargue as a function of the flood extent of the Inner Niger Delta in winters preceding the breeding season, for the episodes 1970-1986 and 1992-1998. Number of nests varied between 15 and 57 (on average 29). Data from: Hafner et al. (2001).

decline was recorded in the Volga Delta, where 7000 pairs bred in 1970, reducing to a maximum of only 300 pairs in the 1990s (Kushlan & Hancock 2005). However, since the late 1990s, southern European populations have again showed an increasing trend. Numbers in Italy increased from 270 pairs in 1981 to c. 400 in 1985-86 and 550-650 in 1995-2002 (Brichetti & Fracasso 2003), and in Spain from 204 pairs in 1986 to 822 in 1990 (Díaz et al. 1996).

Breeding conditions In the Camargue population, which shows a long-term increase, the increases and reductions in breeding numbers correlated with spring rainfall in southern France (Hafner et al. 2001). Despite numerical increase, clutch size and brood size decreased substantially somewhere between 1986 and 1992, and remained low in the 1990s. This decline occurred in parallel with a steep increase in the number of tree-nesting Cattle Egrets, whose habitat and food choice overlap to a high degree with those of Squacco Herons. Although this correlation has not been proved to be causal, further testing is required (Hafner et al. 2001). It is also possible that declining clutch size is associated with the increased rice production in the Camargue. The timing of this change in land use parallels the increase in Cattle Egrets, commonly known to feed in rice fields. Squacco Herons feed less frequently in rice fields, and may have been impacted negatively by the loss of their preferred feeding grounds and by contaminants associated with rice production, which affect prey populations outside this habitat. Neither of these hypotheses is as yet substantiated (Hafner *et al.* 2001).

We offer a third possible explanation: at low flood levels in Africa, birds which arrive in poor condition in their breeding area may produce smaller clutches (Fig. 137). The correlation between flood extent of the Inner Niger Delta and clutch size in the following breeding season is highly significant, but only for the period before 1987. Clutch size has been much smaller in more recent years, and the correlation with flooding disappeared, which suggests that clutch size depends on more than a single factor.

Except in France, rice fields are used by Squacco Herons in proportion to their availability in southern Europe. Food intake rates of adults foraging in rice fields were higher in NW and NE Italy than in the Ebro Delta (Spain), Axios Delta (Greece) and Rhône Delta (France) (Fasola *et al.* 1996). As dry-ground cultivation of rice is becoming increasingly popular in Italy and Spain, the availability of foraging areas for Squacco Herons is likely to decline, with negative consequences at the population level (Fasola *et al.* 1996).

Densities of Squacco Herons can be high during the early stages of flood recession, when the surface area of shallow water is still small; this particular stretch of floating didere vegetation (about 20x30 m) holds 17 birds (November 2008, Inner Niger Delta).

Conclusion

Density counts suggest the presence of at least 350 000 Squacco Herons in West Africa. Because this figure is many times larger than the European and West African populations taken together, the majority of these birds must originate from Asian breeding grounds, even if we assume that the European population has been largely underestimated. Long-term monitoring of local populations in southern France and Ukraine provided evidence for the impact of flooding or rainfall in West Africa on population fluctuations. Clutch size in southern France was correlated with flood level in West Africa throughout the mid-1980s, but not so subsequently, when clutch size declined for reasons unknown. Studied on a larger scale, but using incomplete time series, the overall vicissitudes of European populations showed a steep decline in the 1970s and 1980s, followed by an increase since then. These fluctuations are in close synchrony with flood levels in West Africa, where in particular the Inner Niger Delta is of crucial importance. Conditions on the breeding grounds may have substantial impact on numbers locally, as evidenced in the colonies of the Dnestr, Ukraine, where the construction of a dam led to severe drought conditions and steep declines.

Endnotes

1 The annual population change was weakly correlated with flood level (R^2=0.18, P=0.02), with decreasing numbers during low floods and increases during high floods. However, the birds are difficult to count and so possibly the expected relationship is masked partly by variable underestimations of the population.

2 When population size is regressed against Sahelian rainfall, the explained variance increased from R^2=0.34 to R^2=0.47 if years of low summer flow rates in the Dnestr River (<300 m^3/s) are omitted.

Little Egret
Egretta garzetta

Communal fishing is not just a human prerogative. In the floodplains of West Africa, at least seven bird species do so routinely, although partly by exploiting different niches. Long-tailed Cormorants and Great White Pelicans are particularly adept in chasing fish in waist-deep water. Cormorants sometimes form flocks of hundreds of birds. The frenzy almost immediately attracts the attention of Whiskered and White-winged Terns which adroitly flutter above the mass of cormorants, awaiting the chance of stealing a share. When the water level drops even more, and only shallow water remains, the paradise materialises for Little Egrets, Black Herons, Greenshanks and Spotted Redshanks. Little Egrets are normally confined to water edges or shallow water with submergent vegetation (where fish have fewer chances to escape from solitary egrets), but the drying floodplain offers them excellent opportunities to hunt communally on freshly-spawned fish in water less than 10 cm deep. Furthermore, the West African floodplains are almost devoid of kleptoparasitic gulls, which, elsewhere, would be the first to harass such communally hunting egrets. The inevitable handfuls of marsh terns which turn up take only the smallest fish. Usually, the modest uproar from the fishing birds is quickly over – either the fish have dispersed to safer waters or the shoal has been depleted. The Little Egrets begin to forage on their own again, but keep an eye on their neighbours, just in case.

Breeding range

The Little Egret has a wide but discontinuous distribution in southern Europe (south of 48°N), from Iberia eastwards well into Ukraine and southern Russia. Catchments of large rivers are favoured, the highest density occurring in large areas of rice fields in northern Italy (Hagemeijer & Blair 1997). Recently, the species has been spreading northwards along the Atlantic coast of France (>8000 wintering in 2000/01; Voisin et al. 2005) towards Belgium (33 pairs in Zwin in 2006, >200 birds in June-August in Flanders; Natuur.oriolus 73: 66), The Netherlands (94 pairs in 2005; van Dijk et al. 2007), Ireland (12 pairs in 1997, 32 in 1999, 55 in 2001 and 112+ pairs in 2003; www.birdwatchireland.ie) and southern England (>1650 individuals present in September 1999, Musgrove 2002; 354-357 pairs in 2004, British Birds 100: 337-338).

The European population was estimated at 68 000-94 000 pairs in 1990-2000 (of which 15 000-16 000 were in Italy and 10 000-20 000 in Spain; BirdLife International 2004a), but now is certainly in excess of this estimate, given the increases in Spain, France, Italy and NW Europe (Brichetti & Fracasso 2003, Kayser et al. 2003, Voisin et al. 2005).

In North Africa, Little Egrets breed at scattered locations along the northern borders of Morocco (200-750 pairs in the 1980s, Thévenot et al. 2003), Algeria (>140 pairs, Isenmann & Moali 2000) and Tunisia (hundreds of pairs, Isenmann et al. 2005). Breeding numbers in Senegal and Mauritania exceed 1000 pairs, with several thousand birds (equating to an unknown number of pairs) breeding in the Saloum (Kushlan & Hancock 2005), 250 pairs in Diawling National Park in 2005 (Chapter 7) and an unknown number in the Djoudj (Chapter 7). Breeding numbers in the Inner Niger Delta of Mali in the period from 1985 to 2005 varied between a low of 500-1000 pairs in 1994-96 and a high of 1500 pairs in 2005 (Chapter 6). Much smaller numbers breed irregularly in Ghana (Grimes 1987) and Nigeria (17 nests in February at Malamfatori; Elgood et al. 1994).

Migration

The species is resident, dispersive and migratory, with substantial numbers crossing the Mediterranean to winter in Africa. Taking into account the range expansion, increasing numbers and the milder winters (on average) in western Europe in the 1990s and 2000s, proportionally larger numbers of Little Egrets remained in Europe during winter (Hafner et al. 1994, Bartolome et al. 1996, Voisin et al. 2005).

Ring recoveries south of the Mediterranean are too few to delineate clear wintering grounds in West Africa (Fig. 138). Nevertheless, recoveries in the Senegal Delta, coastal rice fields, intertidal zones and mangroves between Senegal and Guinea, Inner Niger Delta and Mauritania coincide with regions and habitats where field observations have shown the presence of large numbers of Little Egrets (Chapters 6 & 11).

Seventeen birds ringed in the Camargue in southern France fanned out across West Africa, from the Canary Islands to southern Chad, but mainly in Senegal (7 birds) and Mali (4). These birds may have partly followed the southern Mediterranean coastline westwards through Spain to end up tracking along the Atlantic coast further south. Others apparently went the other way, via Italy, from where the crossing was made to Tunisia (Voisin 1985). These birds either stayed along the southern borders of the Mediterranean (from Tunisia eastwards to Libya and Egypt) or continued on a southbound course straight through the Sahara towards the Inner Niger Delta (Voisin 1985). A trans-Saharan passage of Little Egrets is substantiated by numerous observations and captures in autumn (oases in Algeria, such as Beni-Abbès and El Golea, as well as near Adrar in the inhospitable Grand Erg in early November 1970; Voisin 1985) and in spring (6 on 31 March 1971 near Amenas and 20 at the salt lakes of Ouargla on 2-3 April 1971; Haas 1974).

It may be that Little Egrets breeding in the Balkans also leave Europe via Italy. The bird ringed in Serbia on 16 June 1912 and shot in Nigeria in January 1914 was one of the earliest African recoveries of a bird ringed in Europe. As chance would have it, a Squacco Heron from the same Serbian heronry ringed a day later, 17 June, had been

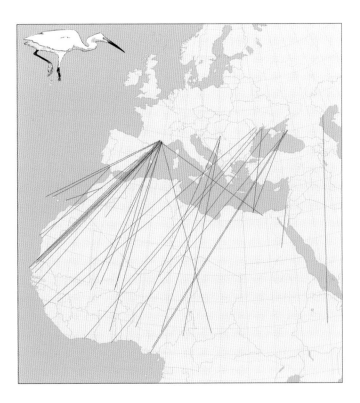

Fig. 138 European ringing locations of 37 Little Egrets recovered between 4° and 30°N and from Egypt. 26 from EURING and 11 from Grimes (1987), Nankinov & Kistchinskii (1978), Sapetin (1978a), Mullié et al. (1989) and Meininger et al. (1994).

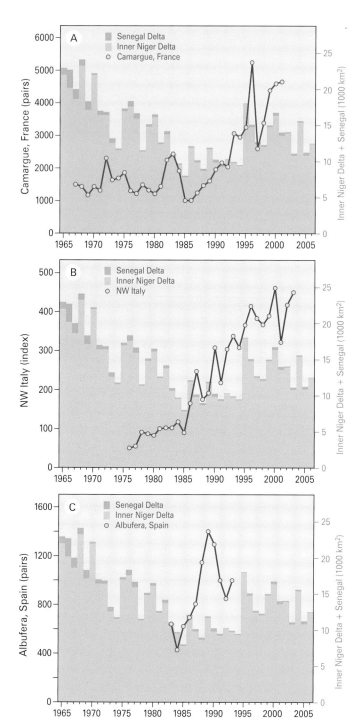

Fig. 139 Population trends of the Little Egret in (parts of) three European countries compared with the flood extent of the Inner Niger Delta and the Senegal Delta in the preceding winter. From: Kayser et al. 2003, www.tourduvalat.org (France), Prosper & Hafner 1996 (Spain), Fasola in Kushlan & Hafner 2000 and in BirdLife International 2004b (Italy).

shot at the same Nigerian site on 12 November 1912! Other Little Egrets from the Balkans have been recovered from Sierra Leone and Liberia, indicating a SW flight direction (Nankinov & Kistchinski 1978).

Birds from the colonies along the Dnestr and Volga on the northern Black Sea and Sea of Azov follow directions between S and SW and have been recovered along the Niger in Mali, elsewhere in Niger and in Guinea, Ghana and Nigeria (Sapetin 1978a; Fig. 138). Little Egrets breeding east of the Ukraine, for example those in Azerbaijan and along the western Caspian Sea, seem to follow a rather more southerly course, with recoveries coming from Pekhlevi in Iran and the marshes near Basra in Iraq (Patrikeev 2004, Sapetin 1978a), Central Saudi Arabia (Sapetin 1978a) and Ethiopia (Fig. 138). These recoveries suggest parallel flight paths for the various Eurasian populations, except those from France, which disperse more widely across West Africa and the southern Mediterranean.

Distribution in Africa

Although censuses of non-breeding egrets often do not distinguish between Little Egrets and white morphs of Western Reef Herons *Egretta gularis,* less than 1% of the nominate Western Reef Herons occurring in Senegal, The Gambia and Mauritania are white, thus making this problem hypothetical in the West African context (Dubois & Yésou 1995).

In the Sahel region, Little Egrets are confined to wetlands, rivers and irrigated farmland. Censuses in January 1993 in the floodplains of the Waza Logone and along the southern fringe of Lake Chad were used to calculate a total number of 10 700 for the entire Lake Chad (van Wetten & Spierenburg 1998). For the Inner Niger Delta in Mali, we used density counts from the early 2000s to arrive at a total of 37 000 birds during the northern winter, which we presume is a gross underestimate (Chapter 6). The overall wintering population in the Inner Niger Delta has increased during this period (Table 6; Fig. 56). A total of 1650 birds has been estimated for the irrigated rice

Little Egrets in the Inner Niger Delta regularly feed collectively in association with other fish-eating species, usually other egrets, Long-tailed Cormorants, Black Herons, White-winged and Whiskered Terns, Spotted Redshanks and Greenshanks. Little Egrets can also be found profiting from easy fish access near nets and fish cages.

fields of *Office du Niger* in Central Mali (Chapter 11). Little Egrets are ubiquitous in this region, occurring wherever shallow water is available, and although they are never in large concentrations, the total number for Mali must surely be higher than the sum of the above regional totals.

The wintering population in the Senegal Delta is tentatively estimated at 6000 birds (Chapter 7). Density counts in the coastal rice fields of Senegal, The Gambia, Guinea-Bissau and Guinea arrive at about 8200 Little Egrets (Bos *et al.* 2006; Chapter 11). On the Banc d'Arguin in Mauritania 2700 Little Egrets were counted in January-

Little Egret *Egretta garzetta*

February 1980 (Altenburg *et al.* 1982), 3400 in January-February 1997 (Zwarts *et al.* 1998) and 3912 in January 2000 (Hagemeijer *et al.* 2004). Between 3800 and 10 150 (mean 7166) local and migratory Little Egrets have been estimated to occur in the wetlands of Niger in January-March; this estimate is based on censuses in 1994-97 (Brouwer & Mullié 2001). In nearby Burkina Faso, the species is well represented in small numbers during the northern winter (Weesie 1996), but is much scarcer during the rainy period (near Ouagadougou, rare in July-August and absent in September; Thonnerieux *et al.* 1988).

For West African countries south of the Sahal, an estimate is available only for Liberia: 3000-4000 in midwinter, mostly along lakes and in rice fields in the interior (Gatter 1997b). In Ghana it is a "not uncommon resident" at lagoons, saltpans and reservoirs along the coast and near irrigation dams (Grimes 1987). Essentially the same status has been described for Togo ("common resident and abundant after influx of Palearctic migrants in August"; Cheke & Walsh 1996) and for Nigeria ("common in all types of wetlands"; Elgood *et al.* 1994). The influx of Palearctic migrants starts in late July in the north, and becomes noticeable in the south by October. Most Palearctic visitors have left by early April.

Taking all information together, and considering the lack of quantitative data from many West African countries, a minimum of 90 000 Little Egrets may well be present in midwinter, some 80% of which are in the Sahel. The local population may not exceed 6850 pairs, amounting to some 20 000 birds (assuming on average one young per pair). This indicates an influx from the Palearctic of about 70 000 birds, or 18-35% of the European population. Since the 1990s, the proportion of Little Egrets remaining in Europe to winter has increased (*cf.* Thibault *et al.* 1997 for the Camargue).

Population change

Wintering conditions The population of the Camargue has been monitored since 1968. Up to the early 1990s, the overall trend was stable with marked fluctuations. After 1991, numbers trebled (Fig. 139A), a trend also noted elsewhere in southern Europe (Kuslan & Hancock 2005) and in NW Europe (see above). The increase in some Italian colonies, for example, also started in the early 1990s and resulted in a doubling of numbers in less than a decade (Fig. 139B). The Spanish series is too short to show any relationship (Fig. 139C). The increase of the studied populations in Italy and France was highly significant. Independent of this overall trend, flooding of the Inner Niger Delta had a significant impact on fluctuations in France, but was non-significant in NW Italy. The impact of the floodplain extent in West Africa on population size in southern Europe concurs with the importance of the Inner Niger Delta as a wintering site for Little Egrets in West Africa, and with our calculation that at least 18-35% of the European population winters there (see above).

Breeding conditions Following the severe winter of 1984/85 in Europe, the breeding population of Little Egrets in the Camargue almost halved – Hafner *et al.* (1994) concluded that this particular winter had caused the decline. By coincidence, however, that same winter was characterised by severe drought in the Sahel, and it is more likely that European and African winter conditions worked in concert to produce such a steep decline, which was followed by a constant rate of recovery. In this population, the number of nests correlated strongly with the reproductive output of the previous year (but not of earlier years), where output had been affected by local rainfall during the early breeding season (Hafner *et al.* 1994). Wet years were productive years. The number of locally wintering Little Egrets was considered a good predictor of the size of the breeding population the following year. Few Camargue birds were thought to migrate over long distances (Hafner *et al.* 1994), as found in Spain (Bartolome *et al.* 1996), but this may be a rather recent development consistent with climate change. Residency and short-distance dispersal are at present supported by findings from the 1990s in western France, where severe winter weather did not result in southerly movements, but instead led to dispersal into the surrounding areas, in search of alternative feeding sites (Voisin *et al.* 2005). Since the early 1990s, the steadily rising temperatures and milder climate in much of western Europe have produced a radical shift in the distribution and temporal abundance of Little Egret. Wintering in Europe, even as far north as Britain and The Netherlands, is now commonplace (Musgrove 2002, Voisin *et al.* 2005). This strategy implies that severe winter weather is likely to cause high mortality, as indeed occurred in the winters of 1984/85 and 1996/97 in western France (Voisin *et al.* 2005), and in Britain in 1996/97 (Musgrove 2002). The effect of such crashes is short-lived, provided that milder winters predominate (Hafner *et al.* 1994). A study of colour-ringed Little Egrets in the Camargue in 1987-95, which period lacked severe winters, did indeed show a constant annual survival rate in adults, but marked variations in yearling survival, independent of severity of winters or trophic conditions at fledging (Hafner *et al.* 1998).

Hydrological conditions in colonies and nearby feeding grounds are important determinants of reproductive output. Drought conditions, either through lack of rainfall (Hafner *et al.* 1994) or via human-induced change (as along the Dnestr in Ukraine, where damming upstream prevented the Dnestr Delta from being flooded in spring; Schogolev 1996b) may cause food shortages and stress, resulting in low productivity (as also found in New South Wales in Australia; Maddock & Baxter 1991).

Finally, habitat destruction, hunting and over-exploitation are still threatening the very existence of heronries, including those of Little Egret, when laws are poorly enforced (Albania, see Vangeluwe *et al.* 1996).

Conclusion

Sahel droughts have an immediate negative impact on populations of the Little Egret in southern Europe; when followed by wet winters in West Africa populations recover from such declines. The population increase, possibly due to the milder European climate in recent years, has resulted in a concomitant, fast expansion of the numbers wintering in Europe. Particularly severe winters then cause higher mortality than normally would have occurred. Local factors may explain short-lived variations in abundance and reproduction.

White Stork
Ciconia ciconia

White Storks are conspicuous birds, weighing 3.5 kg and with a wingspan of 2 m. Each year they fly up to 23 000 km between Europe and Africa. To avoid extended sea crossings, most White Storks tackle the Mediterranean Sea by crossing the Strait of Gibraltar or by diverting through Turkey and the Middle East, where half a million White Storks (and one million raptors) squeeze themselves into the bottleneck of Israel (Leshem & Yom-Tov 1996). This is an impressive sight, not only for birdwatchers but for anybody witnessing flocks of storks gaining height in thermals and gliding off in their preferred direction.

In the words of David Lack (1966), a White Stork is "the most cherished of European birds. It breeds on houses and towers, also on platforms specially put up for it, and throughout its present breeding range in northern Europe it is protected not only by law but by universal sentiment". Yet this could not stop the disastrous decline of breeding numbers in NW Europe during the 20th century. The species became extinct in Belgium (1895), Switzerland (1949), Sweden (1954), Italy (1960), The Netherlands (1991) and Denmark (1998). Considerable effort has been put into reintroduction efforts. Birds bred in captivity were released as young birds to breed in the wild. The first 'breeding stations' were established in Switzerland in 1948, leading to the first free-ranging breeding pair in 1960. Breeding stations were also implemented in Belgium (1957), Alsace, France (1962), Baden-Württemberg, Germany (1968) and The Netherlands (1970). The reintroduction programmes were successful, but also elicited heated debates as to whether this was the best way to safeguard endangered species. One hostile argument was: why invest in reintroduction programmes when birds suffer high winter mortality in Africa? Nowadays, we know that conservation measures should ideally be applied wherever a bird spends significant time in its life-cycle, but biased towards where it suffers high mortality rates. Although the majority of White Storks still migrate annually to Africa, for other reasons increasing, albeit still small, numbers, no longer cross the Sahara – they winter in the Mediterranean Basin or have become resident there. Nevertheless, the future of the species continues to hinge on what happens thousands of km away from the nest platform.

Breeding range [1]

The breeding range of White Storks is shaped rather like a large triangle, encompassing much of Europe west of the line from St. Petersburg to Moscow and the Crimea, but including NW Africa, much of the Middle East and Turkey. Since at least the mid-19th century, range expansions have been noticed in the species' eastern and northeastern distribution (Schulz 1998), but without the reintroduction programmes, the species would have disappeared from NW Europe in the late 20th century. In 2000, the estimated European population was 200 000 pairs, strongholds being in Portugal and Spain (22 000 pairs), Poland (45 000), Ukraine (30 000) and the Baltic states (25 000) (Bird Life International 2004a).

Migration

Since 1901, more than 300 000 White Storks have been ringed, resulting in a plethora of recoveries that by the 1950s had already been used to discover the general migration pattern (review in Schüz 1971), the migratory divide between eastern and western migration routes ('Zugscheide', Schüz 1953) and the distribution in Africa during the northern winter (reviews in Schüz 1971, Bairlein 1981). Satellite tracking studies since 1991 have produced information to fine-tune these findings (e.g. Berthold et al. 2001a, 2004).

The migratory divide between western and eastern populations runs from The Netherlands to Switzerland (Fig. 140A). The 120 000

Without so much as a wingbeat, this sequence of climbing and gliding is repeated over and over again, enabling the birds to cover hundreds of km every day in an energy-efficient manner. A wave of gliding White Storks, 2043 to be exact, heading North above Kasanka National Park in Zambia, late March 2007 (bottom) and Black and White Storks gaining height in Israel (top).

western birds fly via Spain and Morocco to the western Sahel, where they mingle with White Storks from NW Africa. The eastern population flies via Turkey and the Middle East to eastern and southern Africa (550 000 birds). Another 2000 birds cross the Mediterranean between Tunisia and Italy.

Birds equipped with satellite transmitters confirmed the general migration pattern as known from ring recoveries, but – not surprisingly – also showed wide individual variations. One such example relates to a Russian bird which crossed the Mediterranean from France to Tunisia where it spent its first winter and second summer. In its second winter, the bird moved to an area near Lake Chad, then switched to Spain in its third summer and winter, moved on to Poland in its fourth summer, to return to Lake Chad in its fourth winter (Chernetsov *et al.* 2005). This bird successively used the Central, Western and Eastern Mediterranean flyway, thus exhibiting a stunning flexibility in migration behaviour.

With the average daily migration distance being 200-300 km, the 4000 km between NW Europe and the Sahel may take 2-3 weeks, and the 11 000 km between Europe and southern Africa about two months, but these periods become more prolonged when adverse weather calls for stopovers. The southbound migration lasts longer than the return flight, since the birds routinely use more stopovers for longer intervals during the northern autumn. The eastern population, for example, uses the eastern Sahel as a stopover site for 4-6 weeks before embarking on the trip to southern Africa.

White Storks leave the breeding area between early August (southern range) and late August (northern range). Western European birds arrive in the western Sahel mostly before December and leave in February. Arrival on the breeding sites steadily progresses from south to north, *i.e.* in February in the Maghreb, in February/March in Iberia and in the first half of April in NW Europe. The birds following the eastern flyway spend less than three months on their wintering grounds, but altogether stay more than six months in Africa. On average, passage in Israel peaks around 21 August and 28 March. Satellite tracking studies show that the duration of stopovers in the east-

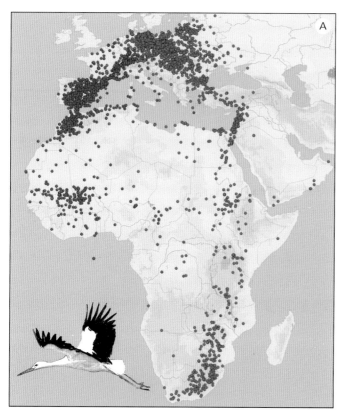

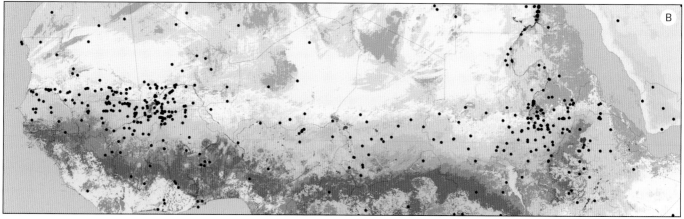

Fig. 140 (A) Recoveries of dead White Storks, shown separately for birds ringed in western Europe (n=2757; Belgium, France, Switzerland, Portugal, Spain) and in eastern Europe (n=3592; former East Germany, Hungary, and countries further east). The birds ringed in Denmark, Sweden, The Netherlands and West Germany are not shown; they may use either flyway (n=4620). (B) Recoveries of dead White Storks in the Sahara, Sahel, and the Sudan and Guinea zones. The birds avoid forest (green) and scrubland (brown) and are mostly found in grassland (yellow) and cropland (pink); for the full legend for land cover designation, see Fig 32. From: EURING.

Being riders of thermals, White Storks migrate only during daylight hours. Stopovers during adverse weather conditions can ground them in any place, be it dry fishponds in Israel (below) or desert in Sharm-el-Sheikh, Egypt (22 February 2002).

ern Sahel during autumn migration varies from year to year.

Increasingly, White Storks are wintering north of the Sahara, including in Europe (Gordo & Sanz 2008, Archaux *et al.* 2008). Before the 1950s, neither southern Europe nor northern Africa had produced any ring recoveries in December and January. After 1970, however, midwinter recoveries in Morocco and Algeria became as common as during the rest of the year (16% of the total, n=503). Similarly, the proportion of White Storks recovered in Portugal and Spain during December and January increased to 6.3% (n=1374) after 1970, and to 5.4% in France, Switzerland, Belgium and The Netherlands (n=1064) and 2.2% in Germany (n=2120). Monitoring in Spain, Portugal and France indicates that increasingly larger numbers remain to winter: up to 3000 White Storks were counted in the early 1990s in southern Spain and 1100 in southern Portugal and more than 1000 in France in 2004, of which 50% were near refuse tips (Merle & Chapalain 2005). Total numbers are surely higher. For example, Tortosa *et al.* (2002) recorded 31% (15 of 49) of their marked population in southern Spain during the winter months; if this is applicable to the entire Iberian peninsula, totals in winter must be high indeed. White Storks in southern Iberia also make extensive use of refuse tips (Blanco 1996, Tortosa 2002) and irrigated fields in which to feed; a complete survey of such sites and habitats has not

yet been attempted. Irrigated fields have become particularly attractive because the non-indigenous Red Swamp Crayfish *Procambarus clarkii* has proliferated to the extent that it has become a major food source for White Storks. For example, the ground beneath stork

White Stork *Ciconia ciconia*   255

roosts in eucalyptus near Montijo, Portugal, was littered with a red carpet of pellets (25 December 1991 – A.-M. Blomert & R.G. Bijlsma, unpubl.); this testifies to the overriding importance of that crustacean to local White Storks.

Like their western congeners, birds of the eastern population have shown a tendency to winter further north than usual. There were no records of wintering birds in Israel until the late 1950s, but about 3000 did so in the early 1990s. However, by the late 1990s only a small percentage at most of the eastern population was wintering in SE Europe, Turkey and the Middle East, a much smaller fraction than the 10% or more of the western population that winters in NW Africa and SE Europe.

Distribution in West Africa

Excluding an estimated 20 000 birds wintering north of the Sahara, but including the 40 000 breeding birds from the Maghreb population, 140 000 White Storks spend the winter in the western and central Sahel. The Inner Niger Delta is reputedly one of the species' main wintering areas and certainly there have been many ringing recoveries from Central Mali (Fig. 140B), but our ground surveys of the central Inner Niger Delta never arrived at numbers higher than 208; aerial counts of the entire floodplain revealed a maximum of only 3657 birds. Duhart & Descamps (1963) mentioned flocks of 10 to 100 on the deflooded short-grass areas, concentrating where locusts were abundant. Since White Storks seemed to avoid the flooded areas proper, the birds may have been feeding in the area beyond our surveys' coverage, a circumstance that probably applies also to other Sahel floodplains (data: African Water Bird Census database of Wetlands International). The Hadejia-Nguru floodplains usually harbour a mere 10-200 White Storks (an outlier population of 1747 birds was present in January 1997). Higher numbers were observed in the Lake Chad Basin: there were 4349 in December 1999 (of which 2331 were around Lake Fitri) and 6090 in January 1999 (of which 2745 were in Waza and 1615 around Lake Fitri). The overview of Mullié et al. (1995) indicates the presence of thousands of White Storks in wetlands in Chad, Niger and Cameroon before 1970, but only a few hundred appeared in the 1980s and 1990s. Decreases were also noticed for the Senegal Delta (from over 4000 in the late 1950s; Morel & Roux 1966) and Nigeria (Fry 1982).

Standard waterbird count methodology by itself is inadequate to assess wintering numbers of White Storks in Africa, particularly because the birds feed on dry land possessing short vegetation, a widespread habitat. The species does use wetlands, however, to rest and to drink, and to allow thermoregulation during the hottest part of the day. The many permanent or semi-permanent wetlands scattered across the Sahel may therefore play a key role in the life cycle during the northern winter (Goriup & Schulz 1991, Mullié et al. 1995, Brouwer et al. 2003).

The distribution of ringing recoveries (Fig. 140B) suggests that

White Storks in Africa roam around in flocks, and single ones seem a bit outplaced, especially when surrounded by Black Crowned Cranes and White-faced Whistling Ducks. (North Cameroon, 1992).

White Storks are less common in the central part of the Sahel, but field data (Mullié et al. 1995, Brouwer et al. 2003) showed otherwise. This was confirmed by Berthold et al. (2001a), who compared the staging areas of 26 tagged birds (11 in Chad and 15 in Sudan) with the distribution of ringing recoveries (4 in Chad and 155 in Sudan). The lack of ringing recoveries from the central Sahel may therefore be attributed to the low reporting rate of birds shot or captured in Chad and Niger. The same holds for the Sudd; this important staging area (with 16 500 White Storks; Table 16) produced only three recoveries.

Berthold et al. (2002) tracked 120 individual White Storks. The birds did not show site fidelity to stopovers during migration or to wintering sites. One bird, followed for nine years, sometimes migrated as far as Tanzania, but in other years strayed as far south as Cape Province in South Africa (Berthold et al. 2004). This pattern is corroborated by South African counts of wintering White Storks, the numbers varying between very few and 200 000. Only small numbers reach South Africa in years when their staple food, the armyworm *Spodoptera exempta*, a caterpillar eating grass and cereals, abounds in East Africa, an example being when in 1987 100 000 White Storks (and 40 000 Abdim´s Storks), were counted in northern Tanzania (Schulz 1998).

In eastern Europe, White Stork nests lining streets, or on roof tops, are nothing out of the ordinary; this street held ten occupied nests in a row. (Poland, June 2008).

The preference for dry lands in Africa is associated with the distribution and abundance of locusts, grasshoppers and other large insects. Migratory locusts may be an important prey in years when they are abundant, but grasshopper numbers tend to be more regular, forming a more reliable food resource, especially during the rainy season (Chapter 14), which ends just as the birds arrive in the Sahel (from September onwards). The western population must find alternative food locally during the dry months, but in eastern Africa, there is no significant geographical barrier to prevent birds moving to alternative habitats available in southern Africa in years when there is a poor supply of food in the Sahel.

Africa is not exactly the land of milk and honey as far as White Storks are concerned. Food supply is highly erratic, both temporally and spatially. White Storks are expedient in their search for feeding areas – for example, they follow rain fronts, stay close to livestock and wild herds and will profit from insects flushed by bush fires – but these foraging strategies may not always meet daily energy requirements. The fact that wintering White Storks have large fat reserves suggests that they are prepared for prolonged fasting periods. Berthold et al. (2001b) and Michard-Picamelot et al. (2002) found, as did Hall et al. (1987) and Michard et al. (1997) before them, clear seasonal changes in body mass.[2] Birds held in captivity in Europe, and free-ranging birds in Europe and Israel, weighed about 3.5 kg in June-July but 4.5 kg in December-January; in the latter case, the mass included up to one kg of fat. Berthold et al. (2001b) and Michard-Picamelot et al.(2002) assume that body mass is endogenously regulated. When subjected to the photoperiod prevailing in midwinter in African wintering quarters, captive birds in Europe showed the same body mass as free-living birds. Data on body mass of White Storks in the Sahel are not available, but Berthold et al. (2001b) have noted that wintering White Storks in South Africa and Tanzania were indeed very fat. This is confirmed by the observations of L. Hartmann (in Schüz 1937), who noticed that the region near Iringa in Tanzania (7°48'S, 84°50'N) had a 'sprinkling' of White Storks in the northern winter of 1937, and that . . . "the local people constantly brought speared Storks, also young ones which were dead beat from the flight; the latter sometimes recovered, sometimes died. The local people remarked: very much to eat, very good, and a lot of fat. They seemed to consider the Storks as manna from heaven."

To build up body reserves, White Storks must increase their daily food intake and start doing so upon arrival in Africa in October, at the end of the rainy season, when large insects are still relatively abundant. Interestingly, Holger Schulz, who studied White Storks during this period in Sudan, was surprised to see that "...even when locusts were abundant, much more time and energy had to be expended to catch them than is required to obtain food in Europe. In Sudan, the Storks had to work continuously for most of the day to obtain their food requirements whereas in West Germany Storks foraged for periods lasting only 20-120 minutes, with long breaks for resting and preening" (Goriup & Schulz 1991). Most likely, the observed birds were not having any difficulty meeting their daily energy requirements but were in the process of enhancing their daily food intake to build up fat reserves. We assume that White Storks, after arriving in the Sahel, increase their body reserves to have a safeguard against possible adverse conditions later in the year. The wintering months may be critical for White Storks, but the preceding fattening-up period is equally crucial for survival. The Sahel functions as an essential refuelling station for White Storks, not just for the western population that remains there throughout the northern winter, but also for the birds which go on to spend the northern winter in eastern and southern Africa.

Population trends

Wintering conditions   The European White Stork population increased from 136 000 pairs in 1984 to 166 000 pairs in 1994/1995 (+22%) and to 230 000 pairs in 2004/2005 (+39%) (Schulz 1999; K-M. Thompson in http://bergenhusen.nabu.de). The overall increase was particularly spectacular in the western population: +75% in the first period (from 16 000 to 28 000 pairs) and +89% in the second 10-year period (from 28 000 to 53 000 pairs). A large part of the western population breeds in Spain, where the population boomed between

1984 (6750 pairs) and 2004 (33 200 pairs). The small, peripheral populations in NW Europe recovered very fast from near-extinction in the 1980s, partly due to immigration (e.g. Barbraud et al. 1999); by the late 1990s numbers were higher than had ever been counted in the 20th century. The increase in the eastern population was still sizeable; from 120 000 pairs in 1984 to 138 000 pairs in 1994-1995 (+15%) and 177 000 pairs in 2004-2005 (+28%).

The ups and downs in the western population might be attributed to rainfall in the Sahel but, as shown in Fig. 141, the relationship is far from clear-cut (Fig. 142A). To some degree, outbreaks of Desert Locusts coincided with a population increase of White Storks in Alsace (Fig. 142B); this has also been shown by Dallinga & Schoenmakers (1984, 1989) for other populations.

The increase of White Storks in Baden-Württemberg and The

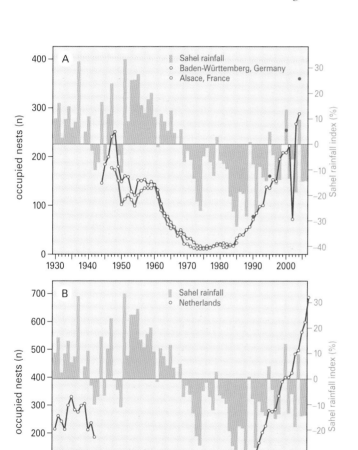

Fig. 141 Population trends of White Storks in (parts of) three European countries (lines, left axis) compared with Sahel rainfall (bars). From: Bairlein 1991, NABU (Baden-Württemberg), Bairlein 1991, LPO (Alsace, France), van der Have & Jonkers 1996, SOVON (The Netherlands).

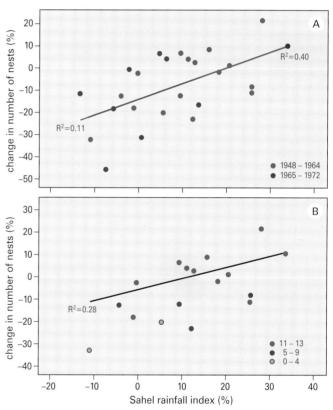

Fig. 142 Annual change in the Alsace population of White Storks as a function of rainfall in the Sahel in the preceding year, (A) for two periods, (B) for three density levels of Desert Locusts (i.e. the number of Sahelian countries with locust plagues, 1948-1964), with positive changes when locusts occurred widely in the Sahel. Alsace data from Fig.141, locust data from Fig. 114.

These White Storks resting in plain desert seemed to be exhausted; when approached they took wing for only some 100 m before landing again (Amenas, Algeria, 1 April 1971; left). White Storks captured alive are 'stored' by breaking wing bones; for the time being they also function as playmates (Ghardaia, Algeria, 31 March 1966; centre). Two mummified and sandblown White Storks, probably 1-3 months dead but still holding some moisture and living dermestid larvae. (Tanezrouft, 17 April 1973). From: Wilfried Haas (pers. comm.).

Netherlands since the 1980s seems unrelated to variations in Sahel rainfall. Initially, annual population growth was fast (+22% in Baden-Württemberg, +25% in The Netherlands) and gradually slackened to +7% in Baden-Württemberg and +9% in The Netherlands, probably due to saturation effects on the breeding grounds. This fast recovery coincided with extremely dry years in the Sahel, contrary to expectations. The reverse, *i.e.* wet Sahel years having a positive effect on trends, was not found either (after grouping time series in 10-year blocks).

The lack of correlation between population trends and Sahel

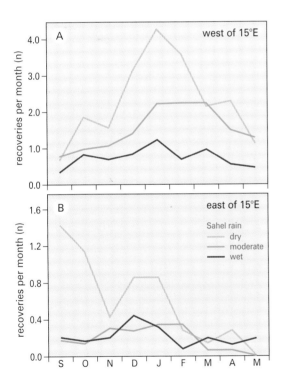

Fig. 143 Monthly distribution of recoveries of White Storks between 4°N and 20°N 1953-2005, split between the west (Lake Chad and further west) and the east. The averages are given separately for the 7 driest years (rain index <-20%), for 26 years with moderate rainfall and for the 20 wettest years (index >0%). Data from: EURING.

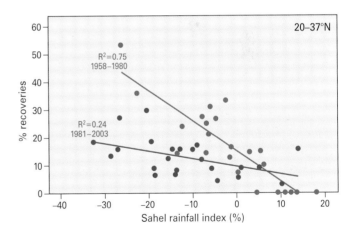

Fig. 144 Annual number of recoveries from the western Sahara (20°-37°N, west of 15°E) in January-May, as % of the total number of recoveries in Europe and Africa between July and June, plotted against Sahel rainfall. The data refer to dead birds ringed as nestlings in France, Portugal and Spain (n=1924). From: EURING.

White Stork *Ciconia ciconia*

rainfall does not necessarily imply that wintering conditions are irrelevant to survival. Indeed, White Storks are more often recovered from the sub-Saharan belt between 4°N and 20°N during dry years than during wet years (Fig. 143). This finding holds for western and eastern populations, although the seasonal pattern – notably in dry years – differs. In the western Sahel, most Storks are recovered between December and February, but in the eastern Sahel many recoveries date from September and October. The eastern Sahel functions partly as a stopover site for Storks heading for wintering areas further south, unlike the western Sahel where Storks stay throughout the northern winter (Fig. 140). In the eastern Sahel, a higher number of recoveries during the northern midwinter is typical for dry years (Fig. 143), presumably because mortality among birds which did not continue their migration to southern Africa is higher than usual. Eastern birds wintering in the Sahel suffer as much from droughts as do western birds. Apart from mortality through starvation, droughts have an emaciating effect on storks, making them even more vulnerable to human predation. Using such simple techniques as spears, slings, stones, boomerangs, snares, quarterstaves or bare hands, thousands of storks are bagged annually across the Sahel by the local people, either to diversify their diet (Sudan; Schulz 1988) or to obtain at least some protein-rich food (Niger and Nigeria; Giraudoux & Schüz 1978, Akinsola et al. 2000). Shooting storks for sport is also a widespread phenomenon in the Sahel, especially by rich Arabs (G. Nikolaus in Schulz 1988) and by Europeans on safari (Giraudoux & Schüz 1978). The available information is too erratic to permit any correlation between human-inflicted mortality and rainfall in the Sahel.

Compared to other species, storks (and broad-winged users of thermals in general) fall victim to powerlines relatively frequently (Bevanger 1998). Of the 5624 EURING recoveries with known cause of death, a staggering 46.5% of the White Storks were electrocuted and 9.9% died from a collision with wires.

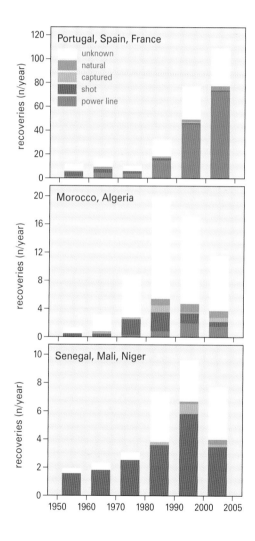

Fig. 145 Average annual number of recoveries of dead White Storks during six 10-year periods (the last of which is 2000 up to and including 2005). Note that different scales are used in the three panels. From: EURING.

The annual variation in the number of spring recoveries from the western Sahara is correlated with rainfall in the Sahel in the preceding months (Fig. 144); this effect was less clear-cut after 1980. [3] The cause of death is unknown for most recoveries from northern Africa, but we assume most birds were shot. The apparent higher mortality in NW Africa after dry Sahel winters may be the outcome of poor condition, caused by food shortage upon departure from the Sahel. Unfortunately, data on body condition of migrating Storks are lacking. However, White Storks are known to arrive later on the breeding grounds after a dry Sahel year (Dallinga & Schoenmakers 1989), implying a slower migration pace, more stopovers of longer duration, a later departure from the wintering area (although recoveries from Morocco do not show a shift in the average passage date in relation to rainfall in the Sahel), or combinations thereof.

The number of birds shot in the Maghreb shows a decline since the 1980s, which is all the more remarkable since the total (ringed) population has increased steadily during the same period of time (Fig. 145). The question remains in what degree this is related to an actual decline in shooting or in the willingness to report a shot bird carrying a ring (or stating a politically correct cause of death). In the 1950s, none of the recoveries from Morocco and Algeria was reported as a "natural death", but 12% of the recoveries in the 2000s were labelled as such. A similar shift in the reported causes of death has been evident in Portugal, Spain and France, where, in the second half of the 20th century, the proportion of shot birds among the recoveries declined from 44% to 2%, and 'natural deaths' increased from 1.8% to 10.5%.

Schaub & Pradel (2004) estimated that annually 25% of the juveniles and 7% of the adults of the Swiss White Stork population died after crashing into powerlines. Garrido & Fernández-Cruz (2002) arrived at equally stunning numbers for Central Spain in 2000: an average rate of collision of 3.9 birds per km powerline, on top of electrocutions at a rate of 0.39 birds per pylon, mostly outside the breeding season and near landfill sites. Of the annual mortality rates of 50-60% in juveniles and c. 25% in adults (Fig. 146B), the collisions represent respectively about half and under one third of the totals.

The production of electricity in western Europe has tripled between 1975 and 2006, an increase of 3.5% per annum. In the same region, the total length of powerlines has increased in the same proportion, for example, from 176 000 km to 217 000 km in the 2000-2007 period. [4] It is no wonder that powerlines have become such a major cause of White Stork fatalities in such a short time, despite efforts to reduce mortality by lining them with spheres.

Annual population changes and the number of recoveries may be used as proxies of mortality, but direct measurements of annual mortality, based on sightings of marked birds, are also available for birds breeding in Alsace (1957-1995; Kanyamibwa et al. 1990), three German sites (1957-1969; Kanyamibwa et al. 1993), SW France (1986-1995; Barbraud et al. 1999), Poland and eastern Germany (1983-2002; Schaub et al. 2005) and The Netherlands (Doligez et al. 2004) The studies showed that the annual mortality is dependent on Sahel rainfall; for the Swiss Storks the Normalized Difference Vegetation Index (NDVI), a measure of the greenness of the African vegetation, had been used instead of rainfall (Schaub et al. 2005). The incorporation of NDVI data from eastern and southern Africa explained little of the temporal variation in mortality, presumably because White Storks wintering in those regions are nomadic and not nearly so dependent on limited and declining food stocks as in the Sahel proper (Schaub et al. 2005). Our analysis of Schaub's data, using rainfall in the Sahel, eastern Africa and southern Africa instead of NDVI, arrived at the same conclusions: annual survival is lower when rainfall in the Sahel is poor (shown for adult birds in Fig. 146A) and the impact of rainfall in eastern and southern Africa on survival is insignificant. [5] The latter finding is deceptive, for a careful analysis of Schüz (1937) and Lange & Schüz (1938), based rather on month-by-month on-the-spot data of rainfall and Stork vicissitudes as supplied by a multitude of correspondents, showed that Storks in southern and eastern Africa – despite their nomadic behaviour – can still run into trouble. Almost 70 years later, Sæther et al. (2006) confirmed these overlooked studies.

The decline in survival rate between the 1950s and the 1980s be-

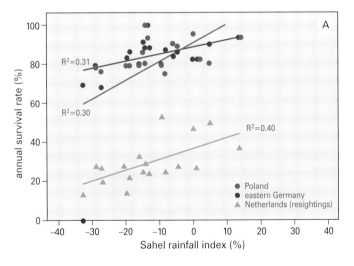
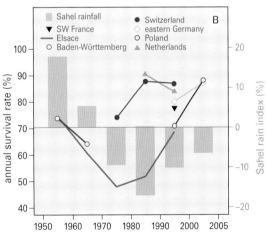

Fig. 146 (A) Annual survival of adult White Storks (>1 year old, ring recoveries) from Poland and E Germany between 1985 and 2002 and from The Netherlands between 1983 and 2000 (resightings, mostly at the nest), as a function of Sahel rainfall in the preceding year. (B) Average annual survival in 10-year periods in seven European regions compared to the average Sahel rainfall during the same periods. Survival data, based on ring recoveries, from: Schaub et al. 2005 (eastern Germany and Poland), Bairlein 1991 (Baden-Württemberg, Alsace), Kanyamibwa et al. 1990 (Alsace), Barbraud et al. 1999 (SW France), Schaub et al. 2004 (Switzerland), Doligez et al. 2004 (Netherlands).

gan long before the droughts in the Sahel in the 1970s and the 1980s. The recovery in survival rate after the 1980s is also much stronger than could be expected from the slight recovery of rainfall in the Sahel (Fig. 146B). Obviously, the regulation of White Stork populations depends on the interplay of a suite of factors (Fig. 148).

Carry-over effects The higher mortality of White Storks in NW Africa during spring migration (Fig. 143) in years when the preceding northern winter in the Sahel had been very dry, is a good example of a carry-over effect: harsh conditions in the wintering area influence the performance of the surviving birds during the subsequent migratory period. In fact, carry-over effects were known, and had been described by Ernst Schüz for White Storks, long before this term was coined in the scientific literature. In 1937, for example, the arrival of White Storks on European breeding sites was delayed by 3-4 weeks, many nests remained unoccupied, and the number of chicks per nest was well below average (Schüz 1937). According to Schüz's informants, the weather in eastern and southern Africa had been extreme, either much wetter than normal (February) or much drier (January), resulting in high mortality and very late homebound mass migration (if at all, as hundreds of birds stayed behind).

Many long time series of the number of breeding and non-breeding pairs, and their reproductive success, are available. The data collected by, for example, A. Schierer for Alsace (1947-1984), by R. Tantzen for Oldenburg in Niedersachsen (1928-1963) and by the doyen of White Stork research, E. Schüz for Central Europe, were the bulwark of analyses by Lack (1966), Bairlein (1991) and others. For a number of European populations, Dallinga & Schoenmakers (1984, 1989) matched *Störungsjahren* (disastrous or upset years) and several breeding parameters with rainfall and food supply in Africa. They found that White Storks in Alsace arrived a week later than normal after a dry Sahel winter and, on average, were delayed another five days in years without locust plagues in the Sahel. In years of high locust abundance in the Sahel, White Storks in Oldenburg arrived at the nest sites early in April, but in years when locust numbers were poor, they returned later in April or in May, and sometimes as late as June. Dallinga and Schoenmakers' analysis almost precisely matched the findings of Schüz (1937), half a century earlier.

The analysis of long-term White Stork data has been much facilitated by having free access to annual breeding parameters from German populations, as provided by the Naturschutzbund Deutschland (NABU, http://bergenhusen.nabu.de). We used these data to check whether the annual variation in the proportion of unsuccessful pairs was synchronised across the various regions, which was the case for the *Länder* of Schleswig-Holstein, Niedersachsen, Sachsen, Sachsen-Anhalt and Mecklenburg-Vorpommern, but not for Baden-Württemberg. [6] Except for the birds of Baden-Württemberg which migrate to the western Sahel, all the others are part of the population that reaches eastern and southern Africa by the eastern route. This suggests that the synchronisation dichotomy cannot be explained by invoking a common denominator for an Africa-induced failure to breed; in the eastern population, such failure may not operate in synchrony with the western population. For instance, 1997 was a normal year for the western population. Eastern birds not only were delayed by more than a month, but half the usual number of pairs refrained from breeding. This delay was caused by extremely cold weather in Turkey, which also compelled birds to back-track temporarily (Berthold *et al.* 2002).

A regression of the percentage of unsuccessful breeding pairs against rainfall in the Sahel, eastern and southeastern Africa produced remarkably consistent trends for all German federal states: fewer Stork pairs commenced breeding after a dry year in the Sahel, but no such relationship was found with annual rainfall patterns in eastern and southeastern Africa. [7] This outcome held for the eastern population (here exemplified by Schleswig-Holstein; Fig. 147) and the western population (Baden-Württemberg). The proportion of non-breeding pairs in Schleswig-Holstein, or elsewhere in Germany, did not systematically increase or decrease in the course of the 20th century.

The number of fledglings per pair was not related to rainfall in the preceding winter in eastern and SE Africa, and, in most time series, the impact of Sahel rainfall was also non-significant. The variation in reproductive output of successful pairs was highly synchronised for the various German sub-populations, although least so for Baden-Württemberg. [8]

That Sahelian rainfall impacts on reproduction, as evidenced in arrival dates, laying dates and the respective proportion of non-breeding pairs in both eastern and western populations, is beyond doubt, but what mechanisms explain it? During dry years, White Storks wintering in the western Sahel may have failed to store sufficient body reserves to cross the desert successfully (Fig. 146), or to arrive in timely fashion on the breeding grounds. This reasoning is easy to understand as applying to the western population, which after all remains in the Sahel throughout the northern winter, but not for the eastern population, for which the Sahel is only a stopover in September, October and November *en route* to the eastern and southern African winter quarters. [9] The carry-over effects of rainfall in eastern and southern Africa on reproduction are not very pronounced, probably because the birds can switch easily between different zones to benefit from local conditions. During spring migration, they cross the Sahel zone relatively quickly, implying that any Sahel impact on the eastern population occurs in autumn. More detailed field work in Africa is clearly needed (as described for Montagu's Harriers in Chapter 26).

Breeding conditions Apart from coping with the burden of wintering conditions, White Storks must also face local conditions on the breeding grounds. The NABU data show that a higher proportion of pairs starts breeding when May and June are warm and dry, which conditions are also conducive to a higher number of young to be raised per pair. The impact of local rainfall and temperature on reproductive parameters remains significant, after correcting for other variables (shown in Fig. 148) operating simultaneously. More im-

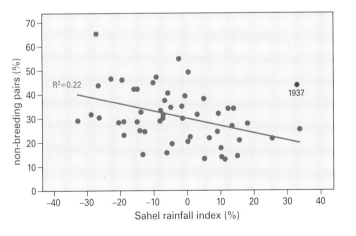

Fig. 147 Fewer White Stork pairs in Schleswig-Holstein (1930-2005) breed when rainfall in the Sahel in the preceding year has been poor. From: http://bergenhusen.nabu.de.

The 1937-outlier (high rainfall, yet high proportion of non-breeding) has been elucidated by Schüz (1937) and Lange et al. (1938). Rainfall in East and southern Africa was then either well below or well above average; local droughts and downpours interfered with foraging and migration. Moreover, locusts were recorded as scarce or lacking over much of the eastern and southern African range. First signs of storks in trouble were noted in February and March, when large numbers in poor condition were recorded from South Africa to Kenya. Hundreds, perhaps thousands, were captured locally (Limpopo region, normally few birds here), mainly because the birds were extremely tame (in hindsight: probably emaciated). The number of recoveries of ringed birds in 1937 was much higher than usual (mostly juveniles). Large numbers remained in Africa south of the Equator through May and June. Apparently, such a suite of exceptional conditions does not occur very often.

White Stork *Ciconia ciconia*

portant, and independent of other variables, the number of fledglings per successful pair has declined since the 1930s, not only in Schleswig-Holstein but also in Niedersachsen, Baden-Württemberg, The Netherlands (Dallinga & Schoenmakers 1984) and NE France (Bairlein 1991), but not in eastern Germany.[10] Such a difference is not unexpected, since the quality of the breeding habitat in NW Europe has deteriorated due to agricultural intensification, unlike in eastern Europe where traditional grasslands, farmlands and flooded river valleys offered prime Stork feeding habitat, containing a large variety of prey species, each prey category remaining sufficiently abundant for at least part of the breeding season to guarantee a good food supply from arrival to departure (For a pair with young, food demand during this 30-day period is estimated at 179.4 kg; Kosicki *et al.* 2006). In Poland, invertebrates were a major source of food, alternative diet comprising earthworms, fish, amphibians and small mammals as available (Kosicki *et al.* 2006). White Storks studied in the Obra valley produced more fledglings in areas possessing high densities of Common Voles *Microtus arvalis* (Tryjanowski & Kuzniak 2002); a guaranteed vole diet provides a high food intake rate (Böhning-Gaese 1992). Intensification of farmland, as observed throughout western Europe, has led to impoverished and prey-depleted stork habitats. For example, Dutch White Storks studied in the 1980s were found to depend largely on earthworms (van der Have & Jonkers 1996), normally a marginal prey due to their low gross energy intake rate (on average 461 kJ/h, compared to 1431 kJ/h if voles are consumed; Böhning-Gaese 1992).

A recent example of the effect of deteriorating habitat quality is provided by the land changes in western Poland since the mid-1990s, such as the increase of arable monocultures at the cost of grassland, through mechanisation and intensification, which re-

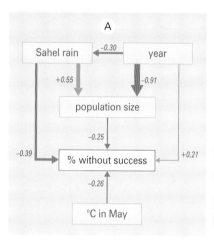
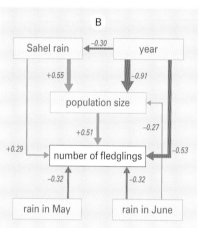

Fig. 148 Population dynamics of White Storks in Schleswig-Holstein, northern Germany, showing the correlations between relevant variables and: (A) the proportion of unsuccessful pairs; (B) the number of fledglings per successful pair. Arrows show assumed causality, red and green indicating negative and positive correlations, respectively. The degree of correlation is indicated by the boldness of the arrow. Correlations are calculated over 53 observation years between 1930 and 2005, and given in italics if non-significant (P>0.05); lines are omitted if R<0.20. From: http://bergenhusen.nabu.de. Explanatory notes: Many pairs were unsuccessful when the Sahel had been dry during the previous winter, and when the weather in May had been cold. The proportion of unsuccessful pairs increased over the observation period, but this non-significant effect disappeared when the Sahel becoming simultaneously drier is taken into account. Fewer pairs were unsuccessful when the population size was larger, which is precisely what one would not expect, given the competition between pairs for the best breeding sites. The correlation is non-significant, however, and the partial correlation becomes very small (R=-.06) after correction for the impact of Sahel rainfall (which is positively correlated with population size).

The number of fledglings per successful breeding pair has declined during the observation period. Independent of this decline, rain in May and June has a negative impact. Two correlations are counterintuitive: the larger the population or the drier the Sahel, the more young are raised. This reverse density-dependent effect disappears, however, when we control for year as an intervening variable (partial correlation becomes: R=+0.13), and the same holds for Sahel rain (partial correlation: R=+0.02).

Conclusion: The % of unsuccessful pairs is reduced after high rainfall in the Sahel in the preceding year and when the breeding area has warm weather in May. The number of fledglings per successful breeding pair increases with dry weather during the breeding period. Independent of weather, the reproductive output (chicks per successful pair) has declined significantly between 1930 and 2005. Other impacts are lacking and, if present, assumed not to be causal.

sulted in the loss of 16-24% of White Stork breeding sites in Silesia and Lubuskie; in other parts of Poland, so far having escaped similar agricultural intensification, White Stork populations have remained stable or have increased (Kosicki *et al.* 2006). The decline in the number of fledglings per pair and increased mortality have steepened the decline of the Stork population in NW Europe up to the 1980s at least. Considering that the impoverishment of farmland has since then continued unabated, or even accelerated, the fast recovery of local populations since the 1980s comes as a big surprise. Several factors may have contributed to this recovery. First, reintroduction programmes boosted numbers and provided food in a food-depleted environment. Secondly, annual survival rate has improved and nowadays compensates for a decline in chick production (Schaub *et al.* 2004). Thirdly, dispersal from core breeding areas enabled rapid colonisation of areas up to 200-300 km away (Barbraud *et al.* 1999, Skov 1999, Bijlsma *et al.* 2001). Finally, an increasing proportion of White Storks have ceased to migrate at all, becoming residents (like the Swiss birds, although these birds did not have a higher reproductive output than migrants; Massemin-Challet *et al.* 2006), or have learned to interrupt their migratory flight in the Mediterranean Basin where they profit – at least for the moment (this may change with stricter EU regulations) – from waste disposal on landfill sites. All these changes, partly human-induced and partly behavioural, have led to the counter-intuitive result of European populations booming in intensive farmland, otherwise considered an ecological wasteland.

Conclusion

The White Storks breeding in SW Europe and in a small part of NW Europe, comprise what is known as the western population, which winters in the western Sahel. Storks breeding east of the migratory divide winter mostly in eastern and southern Africa; for these birds, the eastern Sahel functions largely as a refuelling station. Both populations suffer high winter mortality in years of poor rainfall in the Sahel. Survival rates have improved since the 1980s, mostly because fewer birds have been shot during migration. However, mortality from collisions with powerlines is high and has steadily increased as a consequence of the steeply rising demand for energy in Europe. After dry Sahel years, many more White Storks from the western population die whilst crossing the Sahara than do so after wet years; dry Sahel years also delay arrival dates on the breeding grounds, when fewer pairs will commence egg-laying. This carry-over effect is present in both the eastern and western populations. In NW Europe, the number of fledglings per successful pair has steadily declined since at least the mid-20th century, mainly due to habitat deterioration. Several reasons explain the recovery of this peripheral population: reintroduction programmes, an improved adult survival rate and a relatively low level of immigration from core populations in SW and E Europe.

Endnotes

1. If no reference is specified, the statement in question is based on the review of Schulz (1998).

2. White Storks do not deposit fat before migration, nor enlarge their flight muscle (Berthold *et al.* 2001b, Michard-Picamelot *et al.* 2002). Soaring is an energetically-efficient way of travelling, and the Storks' low wing-loading and high glide ratio permit large distances to be covered without flapping when strong thermals are available. Wing-loading (bodymass/wing area) is 6.3 kg per m^2 in White Storks (Shamoun-Baranes *et al.* 2003) and the glide ratio (the equivalent of the horizontal gliding distance that the bird attains for every 1 m loss of altitude) 15:1 (Alerstam 1990).

3. The correlation between the Sahel rainfall and the relative number of recoveries from the western Sahara (Fig. 144) was highly significant before 1980 (P<0.001) and weakly significant after 1980 (P< 0.02).

4. Data were extracted from the Statistical Yearbooks of the "Union for the Co-ordination of Transmission of Electricity"; see www.ucte.org.

5. The annual survival rate of adult birds (Polish and eastern German birds combined) was significantly correlated with the Sahel rainfall index (Fig 9; R=+0.491, P=.002), but not correlated with rainfall in SE Africa (Fig. 9; R=-0.06; P=0.760) or in eastern Africa (Hulme *et al.* 2001; R=0.228; P=-0.226). A multiple regression analysis showed that the Sahel rainfall remained significant and that the contribution of the other variables was next to nothing.

6. The % of unsuccessful pairs in the different federal states was highly correlated. For instance, the correlation of Mecklenburg-Vorpommern with Niedersachsen was R=+0.91 (n=23), Sachsen R=+0.64 (n=14), Sachsen-Anhalt R=+0.85 (n=15) and Schleswig-Holstein R=+0.80 (n=23). The correlation of Baden-Württemberg with these *Länder* was non-significantly negative (*e.g.* R=-0.19 with Niedersachsen (n=13) and R=-0.36 with Schleswig-Holstein). The average correlation between the series (except Baden-Württemberg) was R=0.45.

7. The % of unsuccessful pairs in the different federal states was correlated with Sahel rainfall in the preceding year: Baden-Württemberg (R=-0.43, P=0.03, n=25), Bayern (R=-0.12, P=0.56, n=26), Mecklenburg-Vorpommern (R=-0.423, P=0.05, n=23), Niedersachsen (R=-0.40, P=0.02, n=34), Sachsen (R=-0.11, P=0.71, n=14), Sachsen-Anhalt (R=-0.63, P=0.01, n=15), Schleswig-Holstein (R=-0.62, P=0.003, n=53).

8. The annual number of fledglings per successful pair in the different federal states was highly correlated. For instance, the correlation of Mecklenburg-Vorpommern with Niedersachsen was R=+0.58, Sachsen R=+0.68, Sachsen-Anhalt R=+0.53 and Schleswig-Holstein R=+0.27. The correlation of Baden-Württenberg with these *Länder* was non-significantly positive (*e.g.* R=+0.28 with Niedersachsen and R=+0.27 with Schleswig-Holstein. The average correlation between the series (except Baden-Württemberg) was R=+0.45; n given in note [v].

9. The Sudd may be the only eastern Sahelian wintering site, with up to 16 500 birds in midwinter; Table 16. The importance of the Sudd for White Storks and other birds badly needs a reappraisal, because recent information is completely lacking.

10. The annual number of fledglings per successful pair in the different federal states declined in Baden-Württemberg (R=-0.61, P=0.0005, n=24), Niedersachsen (R=-0.35, P=0.04, n=34) and Schleswig-Holstein (R=-0.53, P<0.0001, n=53), but the trends were non-significantly negative in Bayern (R=-0.22, P=0.30, n=24) and non-significantly positive in Mecklenburg-Vorpommern (R=+0.31, P=0.13, n=34), Sachsen (R=+0.11, P=0.71, n=14) and Sachsen-Anhalt (R=+0.35, P=0.15, n=19).

Glossy Ibis
Plegadis falcinellus

Flat, wet and silent – the overall impression one gains of the Inner Niger Delta. Even the omnipresent fishermen, herdsmen and villagers, over a million strong, are lost as if in space, their conversations fading away in the light breeze. The silence contrasts strongly with the din that all too often exists in many other large wetlands, particularly in the Mediterranean region, where the shooters claim glory in filling the air with blasts; the banned lead shot is still being deposited in the wet landscape where it will find its way to the top of the food chain. The crack of a shotgun in the Inner Niger Delta seems especially deafening, because it happens so rarely. Upon investigation it is quite likely that a local villager with an old-fashioned single-barrel gun had stalked a flock of Glossy Ibis. Glossies usually forage in rather tight flocks, and tend to concentrate in the central lakes as water tables fall in the delta. Weighing 500-700 g, one bird makes for profitable hunting, and especially so if more than one is hit with one shot. Is this perhaps why so many Glossy Ibises are recovered from the Inner Niger Delta, despite the rarity of (illegal) shooting? Of nine ringed birds with known cause of death in the Inner Niger Delta, six were reported as shot, a telling figure.

Breeding range

The breeding distribution of the Glossy Ibis in Eurasia is discontinuous and mostly restricted to Romania, Turkey, Russia and Ukraine. These countries held >95% of all 21 000-23 000 European pairs in 1978-1998 (Heath *et al.* 2000), moderately declining to 16 000-22 000 pairs in 2000 (BirdLife International 2004a). Much smaller numbers, usually tens or less, have been recorded in other eastern and southern European countries. Glossy Ibis disappeared as a breeding bird from southern Spain in the 1950s, but reappeared in the 1990s and, henceforth, markedly increased to at least 1100 pairs in 2004 (García-Novo & Marín 2006). In Africa, breeding is mostly confined to the eastern part of the continent. In West Africa, intermittent breeding in small numbers is known for the Inner Niger Delta (Mali), with 150 pairs in 1994-1995 (Chapter 6).

Migration

The recoveries, mainly derived from birds ringed as chicks in the colonies in the Dnestr Delta along the western Black Sea (Schogolev 1996a), show a sparse scattering around the Mediterranean Sea, and a concentration in Sahelian floodplains (Fig. 149). The great majority of birds apparently winter in the Inner Niger Delta, some 4600 km from the colonies in the Ukraine.

The recoveries in Europe and northern Africa refer mostly to migrants. Adults from the Dnestr colonies usually depart in July-August, heading to wintering areas in West Africa via the eastern Mediterranean. During their first three months of life, juveniles first fly 200-250 km southwest to mingle with juveniles from the Danube Delta. Some recoveries indicate that these birds continue in part flying west, to cross the central Mediterranean via Italy, enter Africa at Tunisia and Algeria, and then head southwards across the Sahara, eventually reaching the same West African wintering quarters used by the adults (Fig. 149; Schogolev 1996a). However, more intensive ringing is needed to confirm this strategy, and to evaluate its relative significance compared with a route across the eastern Mediterranean. Observations in the 1970s-1990s in Tunisia (Isenmann *et al.* 2005), Algeria (Isenmann & Moali 2000) and Morocco (Thévenot *et al.* 2003) indicate that: (1) hundreds pass during autumn migration; (2) numbers observed during autumn migration are generally higher than during spring migration; (3) numbers increased since the late 1980s, at least in Morocco, where it has become an occasional resident since 1994 (Thévenot *et al.* 2003). The latter development is probably linked to the marked increase in Spain, as shown by the observation of six different birds colour-ringed in Spain (C. Bowden *in* Thévenot *et al.* 2003).

Interestingly, some older data indicate that Glossy Ibis routinely wintered in NE Africa and the Middle East. In the 1970s and 1980s, up to 500 wintered in Israel before the local breeding population had increased from the 50-100 pairs of the 1980s to >300 pairs in

Seen on 24 April 2004 on Lesbos, Greece, together with some Black-winged Stilts, these Glossy Ibis may have spent the northern winter in one of the Sahelian floodplains. During the 20th century the population in Greece was subject to a decline, as elsewhere in SE Europe and West Asia but unlike that in Spain.

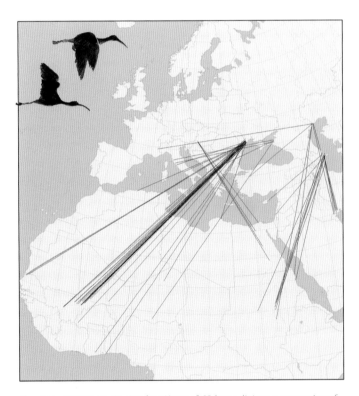

Fig. 149 European ringing locations of 62 long-distance recoveries of Glossy Ibis (17 were extracted from the EURING database; 25 from Schogolev (1996a), 17 from Sapetin (1978c) and a total of 3 others from Mullié *et al.* (1989) (Egypt), Thonnerieux (1988) (Niger) and Wetlands International (Mali).

Glossy Ibis follow the retreating shoreline during the deflooding of the Inner Niger Delta. They feed on small bivalves which they swallow whole; the hard shell is cracked in the gizzard (see Box 7).

1992 (Shirihai 1996). On the other hand, regular wintering in large numbers could not be confirmed for Egypt (Goodman & Meininger 1989), where up to 6 birds have been recorded in December and January. The large nesting population of Azerbaijan (tens of thousands of pairs in the 1950s and 1960s) was said to winter in NE Africa (Vinogradov & Tcherniavskaya *cited in* Patrikeev 2004). However, the scattering of recoveries from birds ringed in the large colonies of the Kizil Agach Reserve (Azerbaijan), *i.e.* from the Lenkoran Lowland (Azerbaijan), Astrakhan Region and Dagestan (Russia), Iran (3 shot), southern Iraq and Israel to Sudan, points at an eastern – rather than NE - African wintering range (Patrikeev 2004). In Sudan a strong autumn passage has been noted along and east of the Nile, but most pass to winter further south (Nikolaus 1987). In Saudi Arabia, the Glossy Ibis is a passage migrant in autumn and spring, but 300-400 birds have been recorded wintering in the southwest of the country (Rahmani & Shobrak 1992); these birds may originate from colonies near the Caspian Sea. Similarly, birds from the Volga Delta (northern Caspian Sea) have been reported in eastern Africa (mainly Sudd) and in India (Sapetin 1978c); the latter involved two birds ringed in 1931 and 1941, and recovered 8 months later on either side of northern India (McClure 1974).

Distribution in Africa

The large floodplains of West Africa attract Glossy Ibis like magnets. The majority of European birds concentrates in the Inner Niger Delta, which held up to 35 000 in 1981. Two aerial counts of Lake Chad and surroundings (AfWC database) recorded 23 000 Glossy Ibis in December 1999 and 4300 in December 2003; the latter survey also revealed 630 birds in Lake Fitri. In 1988-1998, the wintering population in the Hadejia-Nguru wetlands varied between 200 and 2600 birds (Table 13). The 23 mid-winter counts of the Senegal Delta suggest a decline: highest numbers were counted in 1972 (1150) and 1983 (1090), but in 1984-2003 usually just 100-500 (Table 12), but occasionally larger numbers were recorded, such as *c*. 1000 in the Djoudj in 1991-1992 (Rodwell *et al.* 1996) and 2600 at Lac de Guiers in February 1984 (Sauvage & Rodwell 1998).

The importance of floodplains is corroborated by ring recoveries (Fig. 149). Away from the floodplains, small parties of Glossy Ibis can be found wherever wetlands provide shallow water bordering muddy shores where they can forage, but even so still mostly in the Sahelian zone. Near Garoua in northern Cameroun, for example, flocks of 10-20 birds appear after the first rains in late May-June, and 50 or more visit the area during falling water levels in September-October (Girard & Thal 1996). In Niger, at the mouth of the Mékrou and along the Niger River, small flocks of several dozens are encountered commonly in December-February, but there were up to 160 on 4 January 1983 and 280 on a roost near Saga in March-April 1983. Numbers in Niger show steep rises and falls; highest numbers were recorded during droughts, as in 1983 (Giraudoux *et al.* 1988), likely due to concentration. In Burkina Faso, Oursi is the most important site, normally holding 100-250 wintering birds, but there was an outlier of 691 birds in 1998 (AfWC). In Central Chad, the Glossy Ibis is a regular visitor in the wet season in the Quadi Rime – Quadi Achim Reserve, where the first birds arrive after mid-August and numbers peak at 300-400 between 13 and 30 September, as they concentrate on the larger stretches of water (Newby 1979). Relatively large numbers were reported in the midwinter counts of the lakes in southern Mauritania: in Lac Mal usually fewer than 100 in 1975-2001 (but 460 in 2000) and in Lac Aleg 100-600 (but 1040 in 2001 and 2380 in 2000). Counts of Lake Mahmouda show that large numbers may be concentrated here: 2600 in 1990 and 4800 in 1991 (AfWC). The Sudd might be, or might have been, an important wintering area (Chapter 10).

In the Sudan vegetation zone and along the Atlantic Coast, num-

years of life, the summer population would amount to 30-40% of the winter population. Numbers registered during counts in June 1999-2006 in the Central Lakes, where Glossy Ibis congregate during summer, were on average 28% of the numbers counted in the preceding January-February period (van der Kamp *et al.* 2002a, unpubl.). This proportion is lowest during droughts when birds are forced to leave the area after the desiccation of feeding grounds (see Box 7 for a description of feeding circumstances). If we remove the three driest years from the dataset, numbers during summer account for 40% of wintering numbers.

bers are generally much smaller, rarely exceeding several dozens at a time, as in southern Nigeria (Elgood *et al.* 1994), Togo (rare; Cheke & Walsh 1996), Ghana (mostly 1-4 in wetlands and saltpans; Grimes 1987), Ivory Coast (16 near Dabou, February; Thiollay 1978a) and Liberia (rare; Gatter 1997).

The combined data suggest that around 2000, the wintering population in West Africa may be estimated at 30 000 to 40 000 birds. There is no indication that the decline in the Inner Delta since the early 1980s is due to a shift to wetlands elsewhere in West Africa and so we must assume that the entire wintering population has decreased.

A significant part of the wintering population in the Inner Niger Delta remains there during the summer. The two recoveries from June and July refer to birds of one and two years of age. If it were the case that all birds remained in the wintering areas for the first two

Population change

Declining populations Glossy Ibis wintering in the Inner Niger Delta have shown a steady decline of about 90% in less than 30 years (Fig. 150). Counting logistics enabled full coverage of the southern and central Inner Niger Delta to be achieved, but not of the northern part; the overall trend, though, is reliable.

Within Europe, only a single monitoring scheme covered the Glossy Ibis continuously for >20 years, *i.e.* that in the Ukrainian Dnestr Delta, in the northwestern Black Sea (Fig. 151). From ring recoveries, we know that birds breeding here form a separate sub-population with those of the Danube Delta, Dnjepr Delta, Kuban Delta, Crimea and Tiligul Liman (Schogolev 1996b). Fluctuations in breeding numbers in the Dnestr Delta seem to be determined partly by local variations in water level (Fig. 151). Human-induced fluctuations notwithstanding, a long-term decline was apparent between 1972 and 1993. [1]

Other sites with important colonies, such as Kis-Balaton in Hungary, Evros Delta in Greece, various sites in Bulgaria, Danube Delta in Romania and those in Azerbaijan, have been counted intermittently during the past century. Even considering the species' low site fidelity, and hence the poor reliability of single-site - and single-year censuses (Bauer & Glutz von Blotzheim 1966), the overall impression is a decline throughout much of the 20th century. The number of pairs at Kis-Balaton increased from 50 in 1912 to 1000 in 1922, 1923 and 1926, then declined to 2-3 in 1953 and 0 in 1954 (Warga *in* Bauer & Glutz von Blotzheim 1966). In the late 1980s, up to 20 pairs started again to breed in dispersed settlements elsewhere in the country (Heath *et al.* 2000). The Greek colonies amounted to 1100-1500 pairs in 1971-1973. A steep decline was noted in the early 1970s, when numbers in the Evros Delta dropped from 1000-1200 in 1971 to 400-500 in 1973, but this has continued, to 50-71 pairs in 1984-86, breeding ceasing entirely in 1995 (Handrinos & Akriotis 1997). The colonies in the Romanian Danube Delta were estimated at 1200 pairs in 1976-1977, then increased to 2000 in 1995, 2055 in 2001 and 3340 in 2002 (Platteeuw *et al.* 2004). For Bulgaria, the number of breeding pairs ranged between 100 and 700 in the 1970s and 1980s; since then, the species has been in steady decline, with only 52-57 pairs left in 2006 (Shurulinkov *et al.* 2007).

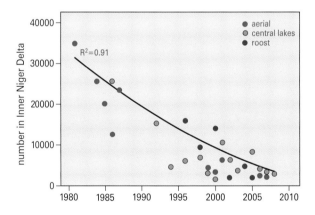

Fig. 150 Number of Glossy Ibis in the Inner Niger Delta, based on three different counting methods, in January or February. Data from Table 4 (aerial counts), Table 6 (ground surveys of Central Lakes) and van der Kamp *et al.* 2002 & 2006 (roost counts). Since each method may underestimate the numbers present, the trend line is based on the highest annual counts.

Glossy Ibis *Plegadis falcinellus* 269

In Azerbaijan, a marked decline was recorded in the 20th century (Patrikeev 2004). At Lake Aggel in the Kura-Aras lowlands, numbers dropped from >10 000 pairs in the mid-1960s to 8300 in the late 1960s and 6000-8000 pairs in 1988-1990. In the southeastern lowlands, Lake Mahmudchala was abandoned in the 1940s, then recolonised in the late 1980s (11 000 pairs); some decline was evident thereafter (5500-6000 in 1990). In the Kizil Agach Reserve perhaps as many as 50 000 pairs bred in the mid-1950s, then numbers declined steeply during the 1960s to reach 450 pairs in 1972. Since then, the number of pairs somewhat recovered to 1500 in 1973, 2500 in 1975, 2000 in 1976 and 900-3000 in 1982 (Patrikeev 2004).

The colonisation of southern Spain, and the subsequent increase, in the 1990s stands in stark contrast to the declining populations in eastern Europe.

Wintering conditions   The work of Schogolev (1996a), who ringed 5000 nestling Glossy Ibis between 1972 and 1982 in the Dnestr colonies (mostly in 1977), permitted a comparison between actual and expected number of recoveries per year, based on the age distribution of 45 recoveries. The oldest Glossy Ibis in this series reached the respective ages of 15, 17 and 22 years, but most birds shot or found in the Sahel did not exceed 4 years of age. Of Glossy Ibis still alive on 1 September in their first year of life, 75% would be still alive after 1 year, 64% after 2, 55% after 3 and 49% after 4. Average adult mortality amounts to 10% per annum, assuming a constant survival rate for adults. We used this information to estimate the expected number of recoveries per annum south of the Sahara, and compared these figures with the observed number of recoveries (Fig. 152). Deviations between expected and observed annual number of recoveries were correlated with flood level in the Inner Niger Delta. Of 31 recoveries, 40% came from the eight driest years (flooding <10,000 km^2), much more than the expected 22%. In eight years with intermediate flooding levels (10 000-12 000 km^2), expected and observed recoveries were similar at 30%, whereas expected numbers were smaller than observed in wet years (>12 000 km^2; inset in Fig. 152). Irrespective of this clear impact of droughts on numbers, the wintering population has declined steadily, i.e. also in wet years (Fig. 150). This decline might be related to increased human pressure (hunting) in an environment where human living conditions are harsh and deteriorating (Chapter 6): the 121 and 165 Glossy Ibises registered for sale on local bird markets in the Inner Niger Delta in 1999 and 2000 respectively (Kone et al. 2002) underestimate the number of birds shot annually.

Breeding conditions   The Glossy Ibis is a difficult species to study on the breeding grounds, due to poor site fidelity. Nevertheless, the overall decline throughout the 20th century seems realistic. The recent increase in Spain is something of an enigma. Large-scale drainage of wetlands in Greece, as elsewhere, may have played an important role in the decline. The spread and intensification of agriculture and fisheries in the lower reaches of deltas deprived birds of major feeding and breeding sites (Tucker & Heath 1994, Schogolev 1996b, Platteeuw et al. 2004). In the Dnestr Delta, Ukraine, the building of a hydroelectric dam 700 km upstream in 1983 was particularly eventful; it stabilised flood levels, caused drought during the breeding season and affected major food resources until semi-natural flooding was restored in 1988 (Schogolev 1996b).

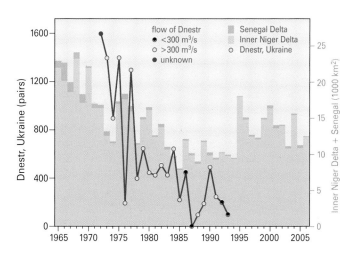

Fig. 151 Breeding population of the Glossy Ibis in the Dnestr Delta, Ukraine (left-hand axis) compared with the flood extent of the Inner Niger and Senegal Deltas in the preceding winter (blue bars, right-hand axis). From: Schogolev (1996b). Years with low discharges of the Dnestr River (<300 m3/s) are indicated with black symbols.

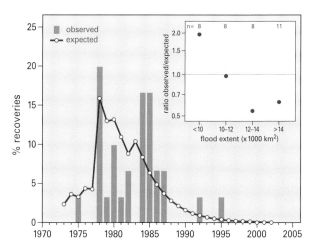

Fig. 152 The % recoveries of Glossy Ibis per year (n=31), as recorded in sub-Saharan Africa, compared with the number expected from the average survival rate as calculated for the same cohort. The inset shows – on a log scale - the ratio of observed and expected number of recoveries for years with different flood extents of the Inner Niger Delta; n = number of years.

Flying Glossy Ibis are a wonder to behold. The alternation of flapping and gliding flight at different levels creates a wave which ripples like a fluid continuum across the flock (January 2008, Niger).

Conclusion

Glossy Ibis showed a long-term decline in the 20th century, mostly caused by loss of breeding habitat through drainage. Most European breeding birds winter in the floodplains of the Sahel, where Inner Niger Delta and Lake Chad are of particular importance. Annual survival is closely linked with the extent of West African floodplains, *i.e.* lowest in dry years and highest in wet years. Short-term population fluctuations are governed largely by flood conditions in West African floodplains. The present wintering population in the Inner Niger Delta has declined by some 90% since the early 1980s. This decline is not compensated for by increases elsewhere in West African wetlands.

Endnotes

1. The decline over the years was linear ($R^2=0.59$, $P<0.001$). Breeding numbers also correlated positively with the flood extent of the Inner Niger Delta ($R^2=0.33$), but in a multiple regression analysis, flood extent does not add significantly to the variance already explained by year. The flow rate of the Dnestr appeared not to be related to the breeding numbers ($R^2=0.03$).

Northern Pintail
Anas acuta

Buss (1946) started his paper in *Auk* with a rueful sigh, "throughout the war sudden and mysterious radar signals rushed combat men to battle stations, sent fighter planes on 'goose' chases, prompted lookouts to report unidentified aeroplanes diving into the sea, gave rise to several E-boat scares, started at least one invasion alarm, and tested the vocabulary of many skippers." Apparently, during the Second World War, most Allied radar operators, usually sailors and airmen, neither knew of a British report from 1941, stating that "birds can reflect radio waves with sufficient strength to be detected by radar sets", nor of any guide that interpreted such responses (Lack & Varley 1945). Buss (1946) not only knew about radar responses from flying birds, but also saw the possibility of using radar to measure the flight speed of birds. He mentioned, for instance, a flock of 50 Northern Pintails flying over the sea at a steady speed of 29 knots. Now, half a century after the papers of Buss and Lack & Varley, anyone on-line can see the daily progress being made by individual Pintails flying between Alaska and California (www.werc.usgs.gov/pinsat). Since the 1990s, it has been common practice to radio-tag hundreds of Pintails to study nocturnal habitat choice, daily flight distance and even mortality rate (*e.g.*Cox & Afton 1996, Fleskes *et al.* 2002, 2005). Satellite telemetry has also made clear that the average migration ground speed, some 77 km/h (Casazza *et al.* 2005), is slightly higher than the 63 km/h as measured by Buss using more primitive technology. Who could have foreseen these research results? Certainly not Tom Lebret, who during the Second World War, whilst in hiding from the German occupiers, rowed night after night across inundated fields in the northern Netherlands to elucidate daily flight distances covered by wintering ducks (Lebret 1959).

Pintails prefer to feed in shallow water where they filter their food (*e.g.* larvae of midges, plant seeds) from the top layer. Sahelian floodplains offer particularly favourable foraging conditions, except during droughts.

Breeding range

The Northern Pintail is, together with Mallard and Teal, one of the most widely distributed ducks, breeding in a broad holarctic belt across Eurasia and America. In winter, the species shows a similarly wide distribution, albeit at a much lower altitude. The American population has declined from 5 to 7 million before the 1980s to fewer than 2 million birds since about 1985, since when the numbers have been more or less stable (Miller & Duncan 1999, Miller *et al.* 2001). The Eurasian population has been estimated at 2 million birds, of which 0.5-1 million birds are in European Russia, much smaller numbers being in the rest of Europe (Berndt & Kauppinnen 1997, Delany & Scott 2006).

Migration

Northern Pintails may fly 10 000 km or more between breeding areas and the sites where they spend the northern winter. Birds ringed while moulting in the marshes of the Volga Delta were recovered from a huge swathe stretching from southern India through the Arctic Circle, and from the Atlantic coast in Africa to the coast of the Pacific Ocean in eastern Siberia. Ringing and satellite telemetry show that birds wintering in Asia may be found in Alaska in summer, and birds that winter in North America may breed in Asia (Henny 1973, Miller *et al.* 2005, Dobrynina & Kharitonov 2006). Of 7555 recoveries in the Euring database of Pintails ringed in Eurasia, 90 came from Africa, *i.e.* 72 from north and 18 from south of the Sahara (Fig. 153A). The Euring database does not include the many Pintails ringed and recovered in Asia (but see Dobrynina & Kharitonov 2006).

The 18 recoveries south of the Sahara, with another 3 recoveries from birds ringed in Mali and Senegal (2 of which were shot in Eurasia, Fig. 153A), represent only 1% of all 2719 winter recoveries. Relatively few Pintails winter in Europe (60 000 according to Delany & Scott 2006), and the comparatively large number of recoveries of birds shot in France (n=531) and elsewhere in NW Europe (n=1252) bears testimony to the much bigger chance of being shot and reported in Europe than in Africa. Birds from the African wintering area pass largely through SE Europe in March-April, but post-breeding migration heads for West Europe, and from there to West Africa (Fransson & Pettersson 2001, Wernham *et al.* 2002). This migratory strategy seems to be corroborated by the pattern of recoveries, with relatively many recoveries in NW Europe in July-October (n=949) and low numbers in spring (March-April, n=149); for SE Europe it is the other way around (119 in spring and 46 in autumn). The many recoveries from Russia in April (n=530) and May (n=531) indicate that upon arrival on the breeding grounds Pintails are not yet free from shooters.

Pintails ringed in the Volga Delta (mainly July-August, 40% of all recoveries; Fig. 153B) and NW Europe (August-February but mainly September-October, 47% ditto; Fig.153C) show a clear dichotomy in recovery patterns. NW European birds ending up in Africa were found in Morocco (11), Algeria (6), Tunisia (2), Senegal (5), Ghana (1) *i.e.* exclusively in the western half of the continent. Birds ringed in the Volga Delta, on the other hand, ranged from Egypt (29), Tunisia

Fig. 153 (A) Eurasian origin of 90 Pintails recovered in Africa and the African origin of 2 Pintails recovered in Eurasia, shown with lines. (B) Distribution of recoveries of birds ringed in NW Europe (n=3563) and (C) Volga Delta (n=3026). From: EURING.

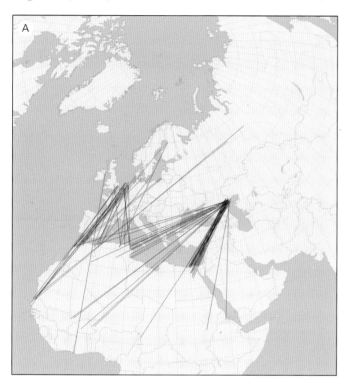

(11), Algeria (9) through Morocco (1), apparently spanning all of northern Africa, although dominant in the eastern half. This is also borne out by the recoveries south of the Sahara, *i.e.* Ethiopia (1), Sudan (1), Chad (1), Niger (2), Burkina Faso (1), Mali (2). Birds from the Senegal Delta migrate through NW Europe, but to some extent also via SE Europe to southern Russia; the latter route predominant for birds wintering in the central and eastern Sahel (Fig. 153B). The recoveries in Russia and Siberia also show that birds migrating via NW Europe generally breed to the north of those migrating via the Volga Delta. These EURING data seem to contradict the idea of flux, both on the wintering grounds (Pintails exchanging between various wet-

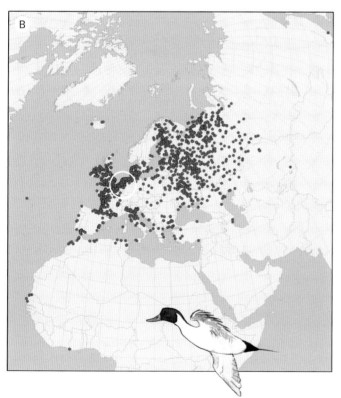

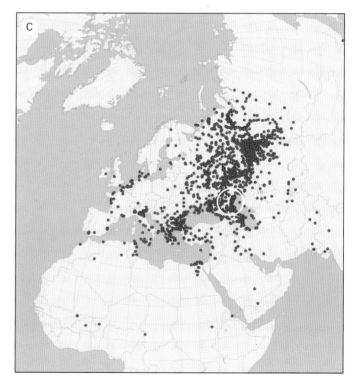

Pools in Sahelian floodplains and rice fields are typically covered with Water Lilies. They flower from September to January. From December onwards the seeds are an important food source for dabbling ducks, such as Garganey and Pintails. Stomach analyses show that – at least during November-December – Pintails in the Senegal Delta feed mainly on Water Lily seeds, which comprised 89% of the stomach contents of 34 birds, complemented with seeds of sedges (Tréca 1993).

lands in the Sahel, depending on flood size and precipitation; see Winter distribution) and on the breeding grounds (Henny 1973, Miller et al. 2005, Dobrynina & Kharitonov 2006).

Distribution in Africa

The Pintails wintering in West Africa are concentrated in only a few large wetlands (Fig. 154). They feed at night and are concentrated at roosts during daytime. Telemetry in North America showed that Pintails fly routinely between 8.7 and 24.4 km from roost to feeding areas (Cox & Afton 1996), and even up to 43 km (Fleskes et al. 2005). Such flight distances indicate that distant feeding sites can be covered from a single roost. In the Senegal Delta, for instance, the Grand Lac in the Djoudj National Park offers Pintails such a safe daytime haven, from where birds depart 10-20 minutes after sunset to feeding sites in southern Mauritania (region of Keur-Massène, some 10-12 km away), in the Diar wetlands of the Djoudj (12 km) or SW to floodplains beyond the boundaries of the Park, i.e. Djeuss and Lampsar (20 km) and Trois Marigots (30 km) (Roux et al. 1978, Triplet et al. 1995).

In the much larger Inner Niger Delta, Pintails use permanent and temporary lakes in and around the floodplain during daytime as roost (Lamarche 1980), from where flooded rice fields and other shallow waters can be reached by night. Similar variable flight strategies probably apply for the birds wintering in and near Lake Chad, the Hadeija-Nguru wetlands and Lake Fitri (Vieillard 1972).

In the major Sahelian wetlands, between-year variation in numbers is huge (Fig. 155). For instance, numbers in the Senegal Delta between 1972 and 2007 varied between 1000 and 247 000, and in the Inner Niger Delta between 11 000-385 000. When counts are tallied for both areas, numbers counted in the same year vary less, although still considerable, i.e. between 90 000 and 410 000. The number of Pintails in Lake Chad seems to have decreased since the 1980s, but the number of counts here is small.

Taking all large West African wetlands in the Sahel together (Fig. 155), the total wintering population has decreased from, on average, 600 000 in the early 1980s to 400 000 a quarter of a century later. The large between-year variation may partly be due to counting errors, but the dip in the drought-stricken mid-1980s (430 000 in 1984 and 480 000 in 1986 in the four largest wetlands combined) is likely to be real. Whether Pintails recovered from this drought-related dip as quickly as the count of 880 000 ducks in the three largest wetlands in 1987 suggests, is open to debate (see Chapter 23).

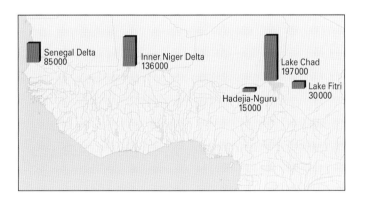

Fig. 154 Average number of Pintails (1972-2007) present in five wetlands during mid-winter aerial counts. From: data shown in Fig. 155.

Population change

Wintering conditions   The Finnish population (20 000-30 000 pairs) is considered to be in decline (Koskimies 2005), but was at a low ebb during the dry Sahel years and started to increase when flood levels improved (Fig. 156) Unfortunately, this series does not include the driest Sahel years, *i.e.* 1984 and 1985. We may assume that in 1986 the Finnish population already had recovered from the preceding Sahelian drought, which may have, according to the January counts in Senegal Delta and Inner Niger Delta, wiped out two-thirds of the Eurasian Pintail population (Fig. 155): 288 000 in 1983, 147 000 in 1984, 109 000 in 1985, 325 000 in 1986 and 311 000 in 1987. Pintails in the Inner Niger Delta are particularly vulnerable to capture when concentrating in the last remaining floodplains during deflooding. In dry years, higher numbers are captured by the local people because the birds have to concentrate in fewer sites to feed. The decline of the population in 1985 may be explained by human predation (Tréca 1989) but, just as in Garganey, also by high mortality due to starvation (see Chapter 23 for a discussion).

Breeding conditions   As in Eurasia, the Pintail is a favoured quarry species in North America, and has consequently been studied in great detail. Population fluctuations are thought to be dependent mainly on conditions in the northern breeding areas (*e.g.* Henny 1973, Miller & Duncan 1999, Miller *et al.* 2001). According to these studies, the large year-to-year variation in the wintering population may be attributed to rainfall patterns in the northern prairie, since

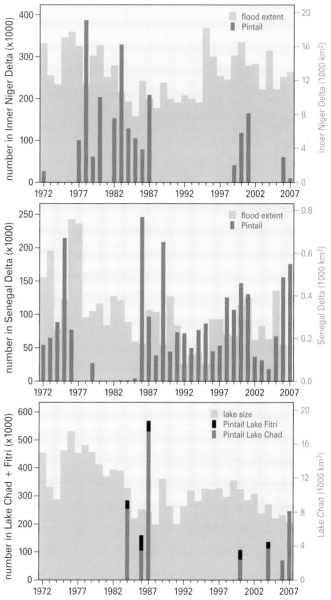

Fig. 155 Mid-winter numbers of Pintails in Senegal Delta, Inner Niger Delta, Lake Chad and Lake Fitri. Data from: AfWC database Wetlands International, Trolliet et al. 2007 (for details see Chapters 6-9). The blue background graph shows maximum flood extent of the Inner Niger Delta or Senegal Delta and maximum size of Lake Chad (taken from Chapter 6, 7 and 9).

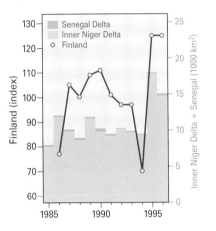 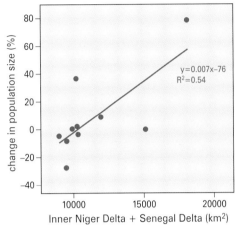 

Fig. 156 Population trend of Pintails breeding in Finland in relation to the flood extent of Inner Niger Delta and Senegal Delta (left). The population change (%) between two years is plotted against the total flood extent (right). Data from: Väisänen et al. (2005).

Pintails produce fewer young in dry years. When the prairie is dry, birds move on to areas further north, but this strategy does not result in an improvement of breeding success. The wet-dry cycle on the prairies cannot explain the long-term decline, for which the conversion of native prairie into cropland is thought to be responsible.

It is not likely that Eurasian Pintails have suffered a similar loss of breeding habitat as those in North America, although drainage and river regulation have had, at least locally, a large impact (e.g. Stanevicius 1999). Whether a wet-dry cycle on the Russian breeding grounds affected the annual variation in the wintering population is as yet unknown. Also, we can only speculate about the impact of shooting on the population. Shooting is unlikely to have a direct impact on the population level if birds are killed which would have died for other reasons before the next breeding season. However, many Pintails are also killed during spring migration and during arrival on the breeding areas, i.e. impacting the size of the breeding population.

Conclusion

The Eurasian population of the Pintail wintering in the western Sahel declined between 1980 and 2006 from 600 000 to 400 000 birds. It is unclear whether this decline is Sahel-driven. The population dipped in the dry Sahel years in the mid-1980s, recovered somewhat in the years to follow, and then plunged into a long-term decline. This trend concurs with that found in North American breeding grounds, where human-caused habitat change is thought to be responsible for the decline.

Garganey
Anas querquedula

Hundreds of thousands of Garganey breeding in Siberia spend the winter in the Sahelian floodplains, but to get there, they have to fly 10 000 km, or more. Why fly all the way to the Senegal Delta, the Inner Niger Delta and Lake Chad when seemingly equally attractive wetlands abound in southern and Central Asia (although large numbers do winter in S and SE Asia)? In Africa Garganey are associated with floodplains and lakes in the Sahel, where they concentrate in the larger wetlands, resting on disturbance-free open water during daylight and heading for nocturnal feeding areas 10-15 km away once the short dusk has arrived (Roux *et al.* 1978, Triplet *et al.* 1995). Wave upon wave of croaking Garganey speeding low over pitch-black desert provide an unforgettable experience in an otherwise silent world, where one might expect only the occasional yelp of a jackal. Dry years in their winter quarters land the birds in dire straits, because the scarcity of feeding sites on floodplains acts as a refuelling bottleneck for the long return trip to Siberia – the competiton for limited resources is intense. Normally during deflooding, Garganey concentrate in the steadily reducing number of shallow lakes and pools, where they gorge on water lily seeds, which makes catching the birds quite simple: in the evening, put up some old fishing nets where the shallows have retained water lilies; in the morning return to remove the captured birds. Bozo fishermen from the Djenné area in the southern Inner Niger

Delta have acted this way at least since the 19th century (Tréca 1989). The photos on page 286-7 tell the story in a nutshell. Bozo birdcatchers were engaged by Guy Jarry, Francis Roux (*CRPBO*)[1] and Bouba Fofana (*Eaux & Forêts*, Bamako) from January to March in 1977 to 1979 to catch and ring large numbers of Garganey, one of which was recovered in eastern Siberia, nearly 12 000 km from its ringing station. In February 2007, Nicolas Gaidet (*CIRAD*)[2] and his team again employed Bozo birdcatchers to catch Garganey, this time to fit them with satellite transmitters. Usually, local birdcatchers slit the throats of captured birds, the daily catch being sold to the local fish dealer, who transports the birds with iced fish to the market. During pre-migratory fattening, Garganey increase their body mass by some 40% during February, which process is parallelled by their price on the market in Mopti, rising from the equivalent of 0.62€ in January to 1.10€ around 1 March in the early 2000s. Such a high fat load may enable the birds to fly many thousands of kilometres without refuelling but it also makes them a highly desirable prize to the people of Mali.

Breeding range

The Garganey breeds in a broad belt across Eurasia from the Atlantic Ocean in the west to the Pacific Ocean in the east, roughly between 40° and 65°N. Within this huge area of some 10 000 × 2500 km, it reaches its highest density between 50° and 55°N and between 15° and 105°E, an area of 6000 × 550 km comprising eastern Germany, Poland, Lithuania, Belarus, Ukraine, S Russia, N Kazakhstan and N Mongolia (Farago & Zomerdijk 1997, Fokin *et al.* 2000).

Around 1990, the breeding population in European Russia was estimated at 570 000 – 960 000 pairs, with another 79 000-92 000 in the rest of Europe (30 000 in Belarus, 28 000 in Ukraine; Farago & Zomerdijk 1997). For the 1990s, Fokin *et al.* (2000) arrived at about half a million pairs in the former USSR, of which 100 000 are in the Asian part and 400 000 in the European part.

Migration

The EURING database holds 2347 recoveries of Garganey ringed in Eurasia, of which 68 come from Africa. A further 25 birds have been recovered of those ringed in West Africa, including 13 from Eurasia (Fig. 157A). The latter records have proven particularly informative. Four birds, all adult males, ringed between 30 January and 3 March 1978 in the Inner Niger Delta, were shot between one and nine years later in their presumed breeding areas (in May or June). One male was recovered in Belarus (at 30°E, 5367 km from the Inner Niger Delta), but the other three came from eastern Siberia (at 83°, 101° and 127°E, 8830 km, 10 144 km and 11 846 km respectively from their breeding areas), suggesting (as substantiated below) that birds wintering in Mali originate from eastern breeding areas.

Most recoveries referred to birds shot during migration (see Fig. 158). Recoveries in May-June are considered to relate to birds in their breeding quarters, and this coincides roughly with the regions where breeding density is recorded as high (Fokin *et al.* 2000). Recoveries during homeward migration (March-April) came mainly from southern and eastern Europe, contrasting with those during autumn migration (July-October) that cluster in western as well as eastern Europe (few from southern Europe). For instance, 69% of Italian recoveries are from March and only 10% from August-September (100%=486). In The Netherlands, however, none came from March and 79% from August-September (100%=163). This difference may be biased by the timing of open hunting seasons and by differences in hunting pressure, but counts and observations of migrating birds also suggest that, in March, birds from African wintering areas migrate via southeastern Europe (Vogrin 1999), whilst post-breeding movements pass on a large scale through western Eu-

One of the unforgettable impressions of West African floodplains is the Garganey, present in their hundreds of thousands. Hard to find except by airplane, and even then not easy to count, but unexpectedly materialising on the horizon like a cloud of mosquitos after some unknown disturbance.

Garganey *Anas querquedula*

rope towards West Africa (Impekoven 1964, Farago & Zomerdijk 1997).

The birds wintering in East Africa migrate through Egypt (Urban 1993). Observations along the northern Sinai coast revealed substantial passage in August and September (1970s and 1980s), for example >221 000 between 16 August and 24 September 1981 (Goodman & Meininger 1989). The few recoveries here indicate an origin from Russia or further east, the birds having been ringed just west of the Volga Delta (Mullié et al. 1989).

The majority of recovered Garganey had been ringed in early autumn in the Volga Delta (an important moulting centre; Fokin et al. 2000) and in NW Europe. Of the NW European birds, 34 had been recovered in Africa, 18 from Senegal, 2 from Mauritania and one from Morocco, representing 62% from westernmost Africa; the remaining 13 birds came from Mali (8), Algeria (4) and Burkina Faso (1) (Fig. 157B). Very few Volga birds were recovered in Africa, and none from the westernmost part: Mali (2), Nigeria (1), Chad (1) and Tanzania (1) (Fig. 157C). Two more Volga-ringed birds were recovered in Egypt, but lack details on the recovery site and date (Mullié et al. 1989). The wintering areas in Africa are largely separated into eastern and western populations, with some overlap in Mali. Whether Garganey show winter site fidelity is not known; most recoveries refer to birds being shot, which allows little chance of unravelling intricate life-histories. However, one of five Garganey ringed in Senegal in 1974 was captured in Mali ten years later.

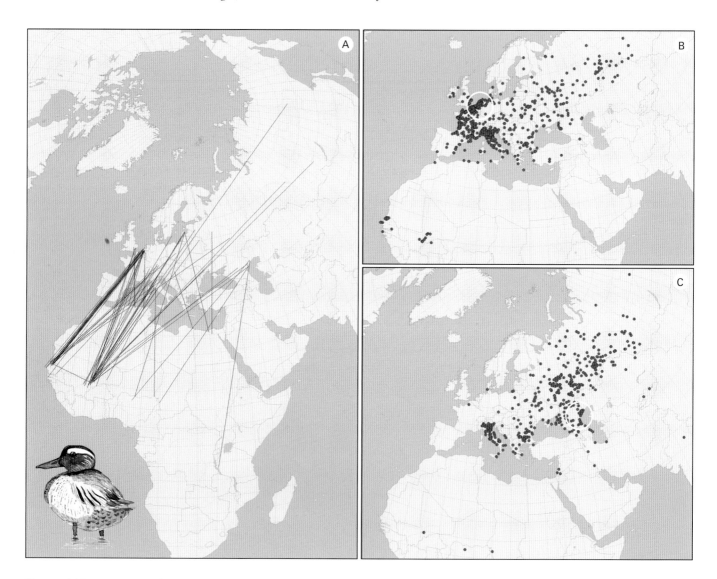

Fig. 157 (A) Eurasian origin of 68 Garganey recovered in Africa and African origin of 13 Garganey recovered in Eurasia, shown by lines, (B) recoveries of birds ringed in NW Europe (n=1101), (C) recoveries of birds ringed in the Volga Delta and surroundings (n=500). From: EURING.

Many birds breeding in Siberia moult in the Volga Delta. Those from Europe moult either in the Volga Delta or, more often, in NW Europe (Fig. 157C), although moult centres in Europe nowadays are presumably dispersed and small. In the past, however, large concentrations of drakes in NW Europe were known to occur in the former Zuiderzee, a shallow inlet (5000 km^2) of the North Sea in The Netherlands, that had attracted drakes from late May onwards, increasing to "very large numbers" after mid-June (ten Kate 1936). After the enclosure and embankment of the Zuiderzee in 1932, Garganey numbers dwindled in the remaining border lakes to some 10 000 in 1957-1962, 2500 in 1963-1969 and <200 in the 1970s and 1980s (Gerritsen & Lok 1986). Equally small numbers (a few hundreds, mid-May to July) are known to moult in Schleswig-Holstein in northern Germany (Kuschert & Ziesemer 1991) and in Niedersachsen in NW Germany, where flocks of up to 200 moulting birds have been noted in June-July (Seitz 1985). A patchy distribution of small numbers of moulting drakes per site is presumably typical throughout NW Europe. Recoveries of Garganey in May-June, supposedly corresponding with breeding sites, averaged a longitude of 60°E for birds which had been ringed whilst moulting in the Volga Delta (n=119), but for birds ringed in autumn in NW Europe, an average longitude of 38°E (n=47) was found, or 1500 km to the west.

The recovery of Volga birds in China (1), Kyrgyzstan (2), India (1), Iran (3) and the Middle East (2), indicates that the Volga Delta is a melting pot of birds wintering in Africa and Asia. Hundreds of Garganey ringed in India were recovered from Russia (not included in Fig. 157A), mostly from W Siberia between 60° and 90°E (McClure 1974, Dobrynina & Kharitonov 2006). This range overlaps with the breeding grounds of birds wintering in Africa, and raises the question whether exchange between Asian and African wintering grounds occurs.

Winter distribution

Garganey rarely spend the winter north of the Sahara, be it in Spain (Díaz et al. 1996), France (Yeatman-Berthelot 1994, Guillemain et al. 2004), Italy (Brichetti & Fracasso 2003), North Africa (Thévenot et al. 2003, Isenmann & Moali 2000, Isenmann et al. 2005) or Egypt (Goodman & Meininger 1989). Instead, the range of wintering sites covers the floodplains and lakes across the Sahel from Senegal through to Sudan, and in East Africa extends south into Kenya (where the wintering population was estimated at 20 000 in the early 1980s; B.S. Meadows in Lewis & Pomeroy 1989) and Uganda ("27 November, many thousands on Lake Wamala, and still 'good numbers' on 27 April"; G. Archer in Bannerman 1958). The latter observation indicates that East African wintering sites are used by birds from the species' easternmost breeding distribution (Central Asia, Primorie-Yakutia), where spring arrival dates average between 10 and 20 May (Fokin et al. 2000).

Urban (1993) arrives at an estimated 95 000-181 000 Garganey wintering in East Africa, mainly in Sudan (70 000-120 000), but many more are found in West Africa (Fig. 159). The Djoudj National Park offers a quiet daytime roost in the Senegal Delta. In the Inner Niger Delta, Garganey have to be more flexible and change their daytime roosting sites frequently in response to the activities of fishermen and to floodplain deflooding. The species' predictable use of nocturnal feeding areas is a habit on which birdcatchers capitalise in deciding when and where to put up their nets at various stages of receding flood levels. Based on 185 birds collected between 1973 and 1977, Garganey food choice in the Senegal Delta varies seasonally in relation to availability (Tréca 1981a, 1993). Grass *Echinochloa colona* seeds are preferred from October onwards, and upon depletion, are supplemented by Water Lily seeds in January, seeds of sedges in December and February) and wild and cultivated rice in March. In the Inner Niger Delta, Garganey feed mainly in Water Lily fields at lower flood levels; the patchy occurrence of such fields makes the species particularly vulnerable to human predation.

Our present knowledge of the distribution and numbers of Garganey in West Africa is based largely on the activities of, first, Francis Roux and Guy Jarry (Roux 1973, Roux & Jarry 1984) and, later, of several other French ornithologists (see Chapters 6, 7 and 9), who in the month of January for the past 35 years have been criss-crossing the extensive floodpains in small, low-flying aircrafts (Fig. 159). On average, a majority of the Garganey in West Africa (43%) was located in the Lake Chad area, which included the Waza-Logone floodplains, and about 37% was found in the Inner Niger Delta. Fewer birds used the Senegal Delta and relatively small numbers resided in the Hadejia-Nguru wetlands and Lake Fitri. Although numbers varied considerably between years, on average a tallied total of one million Garganey occupied the Sahelian wetlands, which result aligns rather well with the numbers from the only year in which all five wetlands were surveyed simultaneously – in January 2000 – when 1.3 million Garganey were counted (Fig. 160).

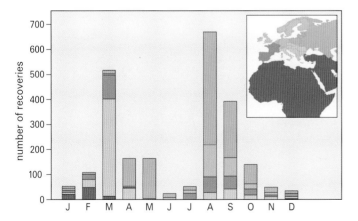

Fig. 158 Total number of recoveries per month for Africa (including Near and Middle East) and SW, SE, N and E Europe. Same data as Fig. 157.

Garganey in the Sahel feed at night and roost by day. Only quiet, often isolated and temporary lakes without fishermen or hunters will serve as roosting sites. (Lake Debo, Inner Niger Delta, January 1994).

Variations in numbers may indicate exchanges of wintering birds between large African wetlands. Counts suggest that this may be the case between Lake Fitri and the nearby Lake Chad.[3] Roux & Jarry (1984) further suggested that in 1984 Lake Chad may have experienced an influx of Garganey from the Senegal Delta and the Inner Niger Delta, where very low numbers had remained to winter. That major lateral movements occurred between these wetlands is also suggested by Monval & Pirot (1989), but they may possibly even range further east into the Nile basin (Urban 1993). Although the strategy of switching between various floodplains makes sense, because the sizes of floodplains and lakes vary asynchronously (see Fig. 160), we found no real evidence proving that such large-scale movements do indeed occur.[4]

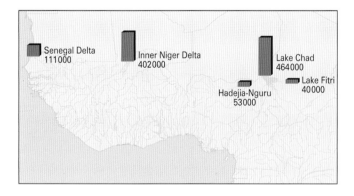

Fig. 159 Average number of Garganey present in five major wetlands during mid-winter aerial counts in 1972-2007. From: data shown in Fig. 160.

Box 19
Bird counts from the air

Garganey numbers counted in the Sahelian floodplains show wide variations between areas and years (Fig. 160). These differences are partly due to counting errors. Dervieux *et al.* (1980) concluded that flocks of water birds counted in the Camargue from a plane were underestimated by 20%, on average, but it is questionable whether this may be generalised to aerial counts of the Sahelian floodplains. Rappoldt *et al.* (1985) found that the counting error, independent of group size, was 17% for birds in flight and 37% for birds gathered in roosts. Garganey have the 'unfortunate' habit of congregating during daylight in a few large daytime assemblages; counting errors over many small flocks usually cancel each other out, but any counting error in one vast flock will have far-reaching consequences for the census overall. For instance, in January 2007 the Garganey present on Lake Chad were nearly all concentrated in a single flock covering a 5 km-diameter area holding around 600 000 birds (Trolliet *et al.* 2007). The error of estimate for Lake Chad, and in that year for West Africa as a whole, largely depended on this one count's error estimate. Such a vast flock is difficult to miss, but smaller Garganey flocks are not, especially in extremely dry and wet years when birds have a more scattered and variable distribution. Several, mainly early, counts in the Inner Niger Delta (1972, 1974, 1982) are known to be incomplete (Roux & Jarry 1984).

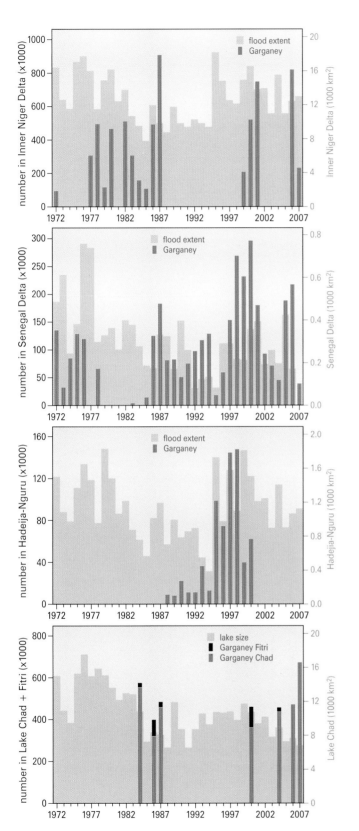

The counts from Lake Chad and Lake Fitri are too infrequent to be able to detect trends. Numbers in the Inner Niger Delta, varying between 100 000 and 800 000, seem more or less stable in the long run. In the Senegal Delta, numbers showed wide variations between 40 000 (dipping even lower during droughts) and 300 000, again without a long-term increase or decrease. Numbers in the Hadejia-Nguru wetlands are related to the size of this floodplain, with a gradual increase between 1988 and 1998, from 9000 to 148 000 Garganey, and lower numbers (about 50 000) in 1999 and 2000 (see also Chapter 8).

Combining simultaneous counts for the Senegal Delta, the Inner Niger Delta and Lake Chad, we arrive at the following figures (millions): 0.71 in 1984, 0.93 in 1986, 1.54 in 1987, 1.17 in 2000, 1.5 in 2006 and 0.94 in 2007.

Population change

Declining populations Very few long-term trends are available for Garganey, which is a rather secretive species in its breeding quarters. On the other hand, past and recent estimates of numbers allow a qualitative trend assessment for most Eurasian countries, as compiled by BirdLife International (2004a, and previous editions), Fokin et al. (2000) and many regional and national avifaunas and atlases (see for example Bauer & Berthold 1996 for a Central European review). These sources indicate that Garganey have declined, often markedly so, since at least the 1970s (examples in Fig. 161). The Eurasian population in the 1990s may be estimated at about 1.5 million birds. During the 1980s the population in the former USSR was estimated to have been about four times larger than in the 1990s, but Russian hunting statistics suggest that the decline must have already begun before the 1980s (Fokin et al. 2000), a conclusion consistent with trends from Belarus (significant decrease, despite fluctuations; Nikiforov 2003) and Lithuania (declining since the 1960s; Stanevicius 1999).

The wintering population may be estimated at 1.6-2.2 million birds, with about 1-1.5 million in West Africa, 0.1-0.2 milion in East Africa and about 0.5 million in South Asia (Brown et al. 1982, Lewis & Pomeroy 1989, Urban 1993, Delany & Scott 2006). The steep decline on the European breeding grounds (Fig. 161) is not reflected by similar vicissitudes in West Africa, suggesting not only that steep declines in well-studied small populations in western Europe do not reflect the overall Eurasian trend, but also that the assessment by Fokin et al. (2000) of "a general decrease in the second half of the

Fig. 160 Midwinter numbers of Garganey in the Inner Niger Delta, Senegal Delta, Hadejia-Nguru, Lake Chad and Lake Fitri. The sources of the separate counts are given in Chapters 6-9. The background blue graph gives the maximal flood extent of the three floodplain areas and of Lake Chad (taken from Chapters 6-9). The size of the Hadejia-Nguru floodplains was known for recent years and estimated for others (see Chapter 8).

Box 20
Pre-migratory fattening of Garganey in the Inner Niger Delta

In February-March 2007, Nicolas Gaidet of CIRAD and the women in Mopti market allowed us to weigh Garganey that had been captured the night before. We recorded weights of 206 females and 221 drakes (see graph). In mid-February, drake median body mass was 375 g, which gradually increased to 485 g by 15 March, an increase of 29%. Females showed a similar trend from 335 to 410 g, an increase of 22%. When the mid-March weights are compared with mid-winter weights (see below), the overall increases are even greater.

The birds captured in mid-March were from the last remaining birds, since the majority had already left the Inner Delta in the preceding fortnight. On 14 March, over 60% of the birds weighed more than 480 g. We assume that birds having attained this mass are ready to leave. Other birds that had been found having reached this peak weight earlier in the season were presumed to be birds that would accordingly start migration at an earlier date. On average, drake body mass increased from 340 g in midwinter to 480 g at departure, an increase of 40%. For females, we calculated the body mass increase from about 320 g in midwinter to 440 g at departure. (For further discussion of the pitfalls in calculating the rate of fattening, see Zwarts et al. 1990.)

Estuarine waders on the Banc d'Arguin (Mauritania) also increase their body weight by 40% (Zwarts et al. 1990), which enables them to fly to NW Europe non-stop, a distance of 4300 km (Piersma & Jukema 1990). Garganey do the same. Of the 400 Garganey captured in Senegal by Jarry & Larigauderie between 2 and 17 March 1970, 7 were shot in northern Italy that same month, one only 5 days after it had been ringed (Morel & Roux 1973).

Under normal Sahel conditions, Garganey departing from the Senegal Delta or Inner Niger Delta have stored sufficient body reserves to be able to fly directly to their breeding grounds in western and central Europe. At a continuous flight speed of 50 km/h (Bruderer & Boldt 2001), they cover some 1200 km per day. Assuming there are no head or tail winds, they take 3.5 days to cover this distance. The average date of arrival in the Camargue in southern France for 11 different years between 1954 and 1966 was 1 March (M. Guillemain unpubl.), indicating that the birds had left the Sahelian wintering grounds around 25 February. However, the weights of Garganey departing from Mali are insufficient to fly non-stop to the Siberian

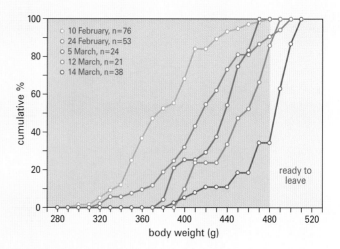

Fig. 162 Increase of body mass in Garganey drakes in the Inner Niger Delta between 11 February and 15 March 2007. Masses are given as a cumulative frequency distribution for 5 days within this period. Data collected by M. Diallo, J. van der Kamp & L. Zwarts.

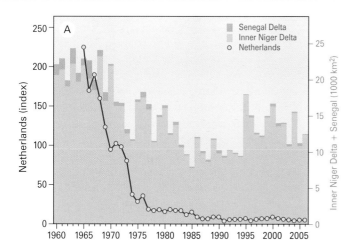
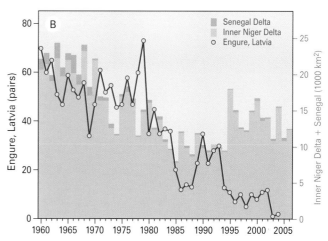

Fig. 161 Breeding population of the Garganey in The Netherlands (left-hand graph) and in Lake Engure (Latvia) (right-hand graph) in relation to the flood extent of Inner Niger Delta and Senegal Delta. Data from: Bijlsma et al. 1993, van Dijk et al. 2008 (The Netherlands), Viksne et al. 2005 (Latvia).

breeding areas. These eastern-breeding birds need refuelling stations in European Russia or in the Black Sea region.

The arrival dates at the Camargue offer the opportunity to investigate whether the birds adjust their timing of pre-migratory fattening in accordance with dry or wet years. Information from the wintering grounds suggests that Garganey in dry years have serious problems in gaining weight, or even in obtaining sufficient food at all (Tréca 1993). We should therefore expect a slower rate of fattening, or no fattening at all, in dry years. It is also possible, however, that birds leave earlier in a dry year, forced away by dwindling feeding areas and declining food stocks. The data-set from the Camargue (Guillemain et al. 2004; pers. comm.) suggested a retarded arrival in a dry Sahel year, but the relationship was far from significant. For the time being, we have to conclude that the time of departure is not related to the flooding.

Given the median departure weights we calculated in Mali, we have no reason to expect that Garganey will need to fatten up again in northern Africa, or even in southern Europe. From Garganey captured in March in the Camargue, Guillemain et al. (2004) estimated that their body masses (drake weight averaged 345 g) would allow the birds to fly 1800-2000 km beyond the Camargue before reaching starvation body mass levels. The body mass of birds captured thus in the Camargue did not show any significant weight gain trend with time, i.e. as days passed since the first migrants arrived; average body mass gain was only 0.27 g/day for 316 males and 0.68 g/day for 56 females (Guillemain et al. 2004). Assuming this result applies to all Garganey stopping over in the Camargue, it raises the question why the birds interrupt their migration in the Camargue at all? For Shovelers we believe we know the answer: the birds arrive unpaired from the African wintering areas and leave as pairs (Zwarts, unpubl. data from Camargue, February 1967). It is plausible that the Camargue also functions as a 'matrimonial market' for Garganey in early March, because their courtship activity is just as intense as that of Shovelers in February (see also Amat 2006, but contra J.-Y Pirot in Guillemain et al. 2004).

Garganey drakes captured in the Volga Delta in October, just prior to outward migration, weighed 400-600 g (Cramp & Simmons 1977, Fokin et al. 2000), about 100 g more than their departure weights before the return migration from Mali. We therefore would expect that Garganey leaving the Volga Delta undertake correspondingly longer nonstop flights, and indeed, they fly nonstop 5000 km to reach Lake Chad or 6000 km to the Inner Niger Delta.

20th century" in the former USSR needs substantiation. Whatever the Eurasian findings in the breeding quarters, numbers on the Sahelian wintering grounds since the 1970s do show distinct fluctuations, but these lack indications of long-term decline or increase (Fig. 160). Several of the early aerial winter counts have been incomplete (Box 19). Statements on long-term trends in wintering numbers are therefore tentative.

The Garganey is an esteemed game bird in Europe. The annual hunting bags have been estimated by Schricke (2001) at 37 500 – 62 500 in the European Community, while Rutschke (1989) arrives at more than 500 000 in the whole of Europe, particularly in Russia, Ukraine and Poland (see also Fokin et al. 2000). In contrast to the Pintail, most Garganey are shot in autumn and not in spring or in the breeding season (Fig. 158). Compared to these numbers, the totals killed annually by the people in Mali during drought years (see below) appear to be less significant.

Wintering conditions Irrespective of long-term population changes, droughts in the Sahel have a direct impact on population size. For example, in the record dry year of 1984, less than 5% of the normal wintering population of Garganey remained in the Senegal Delta. Only after flooding had improved during a succession of years since the mid-1990s, have the number of Garganey increased substantially here (Fig. 160; Trolliet et al. 2003). Drought-triggered declines may be due to dispersal to (neighbouring) areas less severely hit by drought (the available evidence is scarce, see above), but increased mortality from starvation and hunting may also play a more or less decisive role; for these two scenarios, there is some supporting evidence.

Although many Garganey have been ringed, there are only a few recoveries from sub-Saharan Africa (Fig. 157) and in 40% of the years since 1946 there have been no recoveries from Africa. To correct for the variation in the total number of recoveries, the Sahel numbers have been expressed as a percentage of the total (Fig. 163). Despite the small sample size, the trend is clear: there are more recoveries from dry years.

Deteriorating conditions during severe droughts, notably advanced desiccation of pools, lakes and shallow waters, reduce the availability of feeding sites and hamper fructification and production of seeds, the Garganey's main food. The species is then forced to widen its food choice (including less profitable foods, such as Water Lily tubers), to expand foraging activities into the daylight hours and to concentrate at fewer and fewer sites. These are the second best options, as recorded in the Senegal Delta for birds which showed average body masses of 345 g (n=3) and 359 g (n=11) in January and March respectively in 1973, a very dry year. Such low body masses compare unfavourably with averages of 419 g (n=19) and 452 g (n=3) in the same respective months of 1974 to 1978, which were years of normal flood levels (Tréca 1981a). Without pre-migratory fattening, Garganey would be unable to undertake long-distance flights to the breeding quarters (see also Box 20). A generally poor condition also reduces wariness in a species and therefore increases predation risk; in the dry season of 1972/1973 Tréca (1981a) was able to approach openly to within two metres of foraging Garganey.

In years with normal or greater flooding, Garganey wait till darkness before embarking on flights from their roosts to the foraging sites, from which they return in early morning darkness.[5] In the dry season of 1972/1973, food shortages must have forced birds into extending their feeding activity period well into daylight, even as early as 15:00-16:00 (Tréca 1989). Daytime feeding, and hence increased predation risk, has also been recorded in the Senegal Delta in 1985

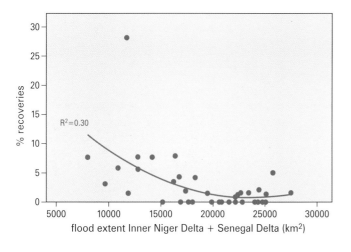

Fig. 163 During years with poor flooding, a higher proportion of Garganey was recovered from West African floodplains (4°-20°N) than in years with good flooding, suggesting better survival in years with high floods. The recovery rate is expressed as the percentage of recoveries in West Africa based on the total number recovered from Europe and Africa (dead birds only) in 36 years with >10 recoveries (1950-1985) ($p<0.01$). From: EURING.

By using old fishing nets mounted on poles, birdcatchers in the Inner Niger Delta can take more than 1000 Garganey per night. The birds are stored in ice and transported to Mopti market to be sold. The picture on the market was retouched since the lady asked to be unrecognisable.

Garganey *Anas querquedula*

(Tréca 1989) and in the Inner Niger Delta in dry years (January-February 1992 and 1994, until 16:00-17:00; J. van der Kamp & L. Zwarts unpubl.). Starvation during dry years must have been widespread, because declines of up to 50% were recorded on the breeding grounds after such droughts in the Sahel (Fig. 161).

The impact of hunting on numbers is less equivocal. Not only in Europe is the Garganey an esteemed quarry species (by shooting), but such is the case in parts of Africa too (by netting and shooting). The annual catch by netting in the Inner Niger Delta varied between 1800 and 27 000 birds (Kone *et al.* 1999, 2002, 2005, Beintema *et al.* 2002, Kone 2006). This tenfold variation can be explained by differences in flood level, with many birds being captured in dry years (1999 and 2005) and few birds in wet years (2000 and 2004). To capture Garganey in nets, the water level on the floodplain must be less than 200-250 cm on the gauge of Akka. Furthermore, after late February, birds begin departing for Eurasia and catches decline steeply. The window of capture is therefore restricted to the last fortnight of February in a wet year (2004), but covers three months in a dry year (2005). To complicate matters further, fewer birds are captured during full moon periods compared with the intervening dark nights.

The number of captured birds is based on counts in four *cercles* (Djenné, Mopti, Tenenkou, Youvarou) covering the southern half of the Inner Niger Delta (Fig. 164). Four interviewers visited hunters, dealers and markets as often as possible during the three relevant winter months, but the numbers they registered certainly underestimate the actual numbers captured in this huge area. To correct for missing days for each *cercle*, we substituted the average daily capture in that week (Fig. 165). The number of captured birds in a dry

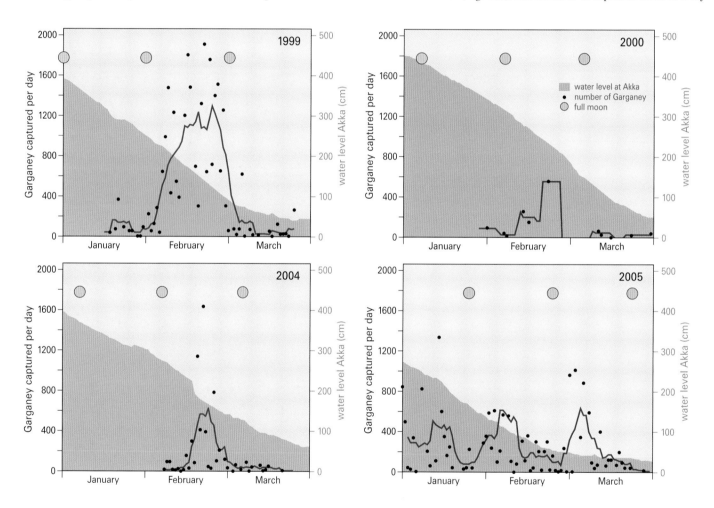

Fig. 164 Number of Garganey captured per day in the southern Inner Niger Delta in four different years during the period 1 January to 1 April. Direct comparisons between years are limited to the part-periods for which data are available. The line shows the 7-day running mean. The decreasing flood level is shaded in blue (values at the right-hand axis), and the period of full moon is indicated. If the water level is less than 200 cm at Akka and the night is dark, many more birds are captured than under a full moon; this circumstance applies until the birds depart in late February and early March. From: Kone *et al.* (2002, 2005), Kone (2006).

year (such as 2005) was six times higher than the number netted in a wet year (such as 2000).

The regression equation in Fig. 165 can be used to predict how many birds may have been captured annually since 1956, assuming that the water level in mid-February at Akka is a reliable determinant of the total catch. If correct, the numbers of Garganey captured in the southern Inner Delta would have varied between 0 in wet years and 60 000-70 000 in dry years (Fig. 166). These numbers are likely to be higher for two reasons: (1) The extrapolation is based on registered and marketed numbers, ignoring local consumption in the Inner Niger Delta; (2) The estimate is based on data collected in the southern Inner Niger Delta, ignoring the smaller numbers captured in the northern half of the delta.

Although it is perfectly possible to capture 1000 Garganey in hundred 12 metre-nets in a single dark night, this would be worthwhile only if the sale of the birds could be guaranteed and also that they could be transported quickly or stored in ice. Nowadays, fish traders can cope with such catches, but back in the 1980s the bird trade had not achieved anything like that level of organisation, which dates from about 2000. Prior to 1999, however, there were other reasons for high capture rates. First, in the dry 1980s, fish and rice production were very low, which forced local people to find alternative food supplies to avoid starvation, and, secondly, Seydou Bouaré (pers. comm.), who witnessed the capture and trade of Garganey in these dry years, estimated that annually possibly even more than the expected 70 000 (Fig. 166) were captured.

Presumably, Garganey were rarely captured in the 1950s and 1960s when flood levels were high (Fig. 166). Although the 1940s were relatively dry, catches of Garganey were likely to have been small because cheap lightweight nylon fishing nets became available in West Africa only from the 1960s onwards.

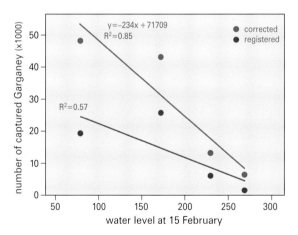

Fig. 165 Registered and extrapolated numbers of Garganey captured per year in the southern Inner Niger Delta in 1999, 2000, 2004 and 2005 as a function of water level on 15 February; see also Fig. 164 for registered numbers. The extrapolation corrects for missing days.

Bernard Tréca was another eyewitness of the Garganey captures in Mali in January and February of 1985, after the driest year of the century. He noticed that Garganey had switched from traditional feeding areas in the southern Inner Niger Delta to irrigated rice fields near Markala and Dioro, just SW of the delta. He estimated that in this area alone (128 km^2), 50 bird catchers captured 11 000-11 500 Garganey (and 3500-4000 Pintail) (Tréca 1989). Trapping in the delta itself must have continued meanwhile, as 200-300 standing bird nets were seen from the air during the mid-winter bird count, indicating that not all Garganey had left the floodplains.

The estimated 60 000-70 000 Garganey killed annually in the Inner Niger Delta between 1983 and 1994 corresponds to a third of the average number counted in those years in mid-January. Within the Sahel, massive trapping of Garganey seems to be restricted to the Inner Niger Delta. Netting is not employed in the Senegal Delta and not at a large scale around Lake Chad. The western wintering population of Garganey (Senegal+Mali) would therefore suffer an annual loss of 0-15% via netting, depending on hydrological circumstances in the Sahel.

The EURING recoveries also show that Garganey in the Sahel are hunted. According to Schricke (2001) more than 10 000 are annually killed in the Senegal Delta. For Niger, Abdou Malam (2004) arrives at an annual total of 1000-8000 Garganey being shot in 29 wetlands, given the presence of 12-80 hunters in four different winters and assuming that each hunter bagged 80 Garganey per winter. This represents a small proportion of the local winter population. For certain wetlands in Burkina Faso (Sâ-Sourou, Pama-Sud, Pama-Nord, Konkombouri, Oudalan-Fleuve, Béli, Nazinga), it is estimated that 300 Garganey (range 30–446) were shot annually in the winters of 2000/2001-2003/2004 (Kone 2006). These data do not suggest that shooting is an important mortality factor in the Sahel, but the impact may be larger than just removing birds from the population, because it causes substantial disturbance in the few wetlands suitable for foraging or resting Garganey. Although trapping in the In-

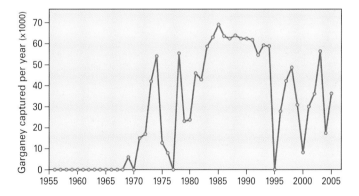

Fig. 166 Annual number of Garganey captured in the Inner Niger Delta, as derived from the regression equation of corrected numbers in Fig. 165.

ner Niger Delta occurs on a grander scale than hunting elsewhere, the associated human disturbance is smaller.

Two questions remain. The first addresses the netting of Garganeys in the Inner Niger Delta: will catching effort in the future increase – at similar flood levels – when methods of preservation and transportation further improve? Note that when fish captures per fishermen continue declining (Chapter 6), local people are likely to look for alternative sources of food and income. On 15 February 2007, we may have seen a glimpse of the future. That day we saw the arrival of 3500 Garganey at Mopti market. Even taking into account the prevailing ideal circumstances for catching Garganey (low water level and new moon on 17 February), this day's total was nearly twice the highest daily catch reported between 1999 and 2005 (Fig. 164). Continued monitoring of Garganey trapping would be necessary to track long-term changes in catch totals.

The second question refers to the combined impact of netting and starvation on winter mortality. As the highest capture totals were achieved in dry years, when conditions are poor anyway and mortality from possible starvation is high, the question is: if netting birds does increase the annual death toll, then to what degree? Our estimate is that in dry years about 30% of the local wintering population is harvested, but we have no clue as to how many birds in these circumstances do not return to their breeding area because, either they are unable to fatten up, or they die from starvation. It is likely that catches remove in part birds which otherwise would have died from starvation, but this fraction possibly is low because captures peak in the few weeks prior to departure. Our overall conclusion is that the capture process removes birds which have survived the dry

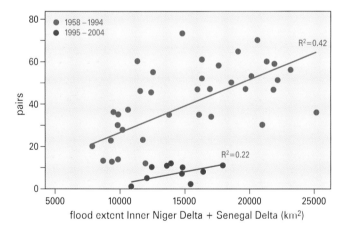

Fig. 167 Breeding population of Garganey in Lake Engure, Latvia (Viksne et al. 2005) as a function of the surface of the inundation areas in the Inner Niger Delta and the Senegal Delta, split into periods before and after 1994 (the year that predator control in Lake Engure was gradually banned and vegetation succession changed the favoured breeding habitat of ducks).

winter and are preparing for the flight back to the breeding quarters.

Admittedly, the recovery in the late 1980s from drought-induced declines showed that populations readily bounce back to previous levels when wintering conditions improve, but given the above discussion, will other factors intervene after amelioration following droughts in the future? On the other hand, the long-term decline on the breeding grounds, especially because it is concurrent with improved flood levels and rainfall in the Sahel since the 1990s, is a cause for concern.

Breeding conditions   Throughout the 20th century, most floodplains and rivers all over Europe have been regulated, taming seasonal floodplains and destroying the Garganey's preferred breeding grounds. Industrialisation of farmland (*e.g.* drainage) has led to further losses of breeding habitat. Nowhere is this better illustrated than in the low-lying Netherlands, where land improvement has caused a decline of >95% since the 1960s (Bijlsma *et al.* 2001). Such schemes were implemented throughout western Europe (Bauer & Berthold 1996), but also in eastern Europe where the impact on the species' world population is far greater (Fokin *et al.* 2000, Dobrynina & Kharitonov 2006).

Other local factors may include predator control (based on circumstantial evidence), and successional changes in the vegetation (meadows overgrown with reeds and shrubs), as mentioned for Lake Engure in Latvia (Viksne *et al.* 2005). Here, control of Eurasian Marsh Harriers, Ravens and American Minks *Mustela vison* was intensive in the 1975-1993 period, but gradually ceased from 1994 onwards. Garganey numbers declined in the 1980s and after 1994, and have remained at a low level since (to 2005). Interestingly, fluctuations in numbers before 1994 corresponded well with flood level in the Inner Niger and Senegal Deltas, but much less well thereafter (Fig. 167).

Proper management of nature reserves can reverse negative trends in the breeding quarters, as illustrated by the nature reserve of Vejlerne in NW Jutland, Denmark (60 km^2). Rising the water level on meadows in spring and summer, on top of controlled grazing and mowing, resulted in a marked increase from an all-time low of 4 breeding pairs/drakes in 1985 to 82-184 in 1998-2003 (Kjeldsen 2008). Other birds profited as well, including Black-tailed Godwit and Ruff (see also Chapter 44, Box 29).

Conclusion

Garganey in western Europe have been in steep decline for decades, and increasingly also in eastern Europe and Russia. A major cause of decline is breeding habitat loss, and this is greater in western than in eastern Europe. The western and central Sahel harbours 1.0-1.5 million Garganey. Surprisingly, wintering numbers in the Sahel seem to be relatively stable in the long run, contrasting with the species' fortunes on the breeding grounds. Steep between-year declines in breeding populations were associated with droughts in the Sahel, when Garganey congregated in diminishing floodplains and large lakes. During droughts, about 30% of the numbers wintering are trapped in the Inner Niger Delta, the likely impact being a mortality factor additional to starvation losses.

Endnotes

1   CRBPO: *Centre de Recherches par le Baguage des Populations d'Oiseaux.*
2   CIRAD: *Centre International de Recherche Agronomique et de Développement.*
3   Lake Fitri and the nearby Lake Chad have been counted simultaneously five times (Fig. 160). Numbers in Lake Fitri, varying between 4000 and 97 000 birds, were negatively related to numbers in Lake Chad (R=- 0.84), possibly due to exchanges between both areas, which together harboured about 500 000 Garganey during the five winters between 1984 and 2006 (Fig. 160).
4   Although the counted numbers in the Senegal Delta and Inner Niger Delta (which are positively correlated: R=+0.54, N=12) may suggest a negative relation to the Lake Chad numbers (N=7 and R=-0.57, and N=6, R=-0.45, respectively); neither correlation is significant.
5   In the Inner Niger Delta on 22/23 January 2005, for example, tens of thousands of Garganey passed overhead near Akka between 18:25-18:28 and 5:44-5:50 (civil twilight 18:59 and 5:58 respectively), when darkness was so deep that only birds immediately overhead could be discerned (R.G. Bijlsma unpubl.).

Osprey
Pandion haliaetus

When Kees Goudswaard got a job as fishery biologist in Ghana in 1991, he had to make many long boat trips across Lake Volta. To amuse himself, he started to count and plot all Ospreys residing along the shoreline. In this way, he discovered that for four winters in a row, the lake held 420 birds. Because individual Ospreys could be recognised, he realised that the birds were strictly resident in winter. The Ospreys of Lake Volta appeared to feed on Moustache Catfish *Hemisynodontis membranacea*. Normally catfish are bottom feeders, but like several other catfish species of the *Mochokiella* and *Synodontis* genera, it also has the habit of feeding inverted on insects and plankton at or near the water's surface, thus making life easy for local Ospreys. The evolutionary consequence is that these catfish species have reverse countershading, which matches the background when they are upside down (Chapman *et al.* 1994). The apparent benefit of concealment from predators from above or below is diminished, however, by their need at the air-water interface to employ aquatic surface respiration, which requires constant swimming movements to maintain position, an activity that cannot fail to attract the attention of Ospreys on their observation posts, drowned trees. Lake Volta is the world's largest man-made lake; its surface area is 8500 km^2 and its volume 148 km^3. In West Africa, there are many reservoirs and their economic and ecological impact is tremendous, though that of the latter is seldom positive, particularly when all downstream effects are taken into account. However, Ospreys certainly have benefited from the artificial creation of lakes and reservoirs, where drowned trees projecting above the surface provide perches in the middle of their feeding areas. But once the last tree has toppled or has been removed for fuel by the local populace, the species will have to work harder for its meals.

Ospreys and African Fish Eagles spend a large part of the day sitting on perches. In the words of Leslie Brown (1970), writing of African Fish Eagles: "... they are generally very idle birds that have no difficulty in catching their own food each day. Their idleness has suited my own mood, and fitted well with days spent reclining on cushions in a gently rocking boat, making notes at intervals."

Breeding range

The Osprey has a worldwide distribution across all continents except South America and the polar regions, provided that water teeming with fish is available. In the 19th and early 20th century, human persecution reduced both Osprey numbers and distribution in much of Europe, a circumstance aggravated by the organochlorine contamination of the food chain by agricultural practice in the 1950s-1970s period (Poole 1989). After persecution ceased (in the breeding areas) or abated (during migration; Saurola 1995) and the application of persistent pesticides on farmland was banned, Ospreys quickly recovered and recolonised lost breeding areas throughout Europe from the 1970s onwards. The European breeding population in 2000 was estimated at 7600-11 000 pairs (BirdLife International 2004a), of which 3400-4100 were in Sweden, 1150-1330 in Finland and 2000-4000 in Russia.

Migration

Satellite-transmitters have been used extensively for tracking Ospreys (Hake *et al.* 2001, Kjellén *et al.* 2001, Dennis 2002, Saurola 2002, Triay 2002, Thorup *et al.* 2003, Alerstam *et al.* 2006, Strandberg & Alerstam 2007). In some cases movements could be monitored closely, highlighting a bird's individual decisions on an almost hourly basis. The peregrinations of several Ospreys from Scotland and Finland have been published on the Internet, showing in detail how individuals navigate the landscape and their selection of stopovers, roost or hunting perches and how long they stay: www.roydennis.org, www.fmnh.helsinki.fi.

The broad front migration sweeps across Europe, the Mediterranean and northern Africa in directions varying from SW to SE. Individual flight paths are interrupted by a succession of stopovers along the way (Hake *et al.* 2001, Kjellén *et al.* 2001, Schmidt & Roepke 2001). The duration of stopovers may extend up to 35 days in autumn, but is reduced to 1-5 days in spring when birds are in a hurry to arrive in timely fashion in the breeding quarters. The 5800-7500 km flight between Sweden and West Africa takes on average 39 days in autumn (of which 25 whilst travelling 254 km/day) and 26 days in spring (22 days of travel, 294 km/day); long distances covered during a 24-hour period frequently require nocturnal flight, especially in spring. The annual flight paths of satellite-tracked birds followed during repeated journeys were often 120-405 km apart, the maximum east-west separation being 1400 km. This does not suggest route fidelity or the use of familiar landmarks in orientation, for these track separation distances deviated from earlier tracks by distances that far exceed the normal range of vision from normal migration altitude (Alerstam *et al.* 2006).

European Ospreys winter in sub-Saharan Africa, except for small numbers in the Mediterranean Basin. Most birds end up in West Africa, but those from Finland (Saurola 1994, 2007), Sweden E of 23°E (Fransson & Pettersson 2001) and Russia (Österlöf 1977) also winter in East and southern Africa (Fig. 168), where numbers are small compared with West Africa. Ospreys from various parts of Europe show extensive overlap in their wintering grounds, but an east-west cline is nevertheless prevalent. Ospreys from Finland on average

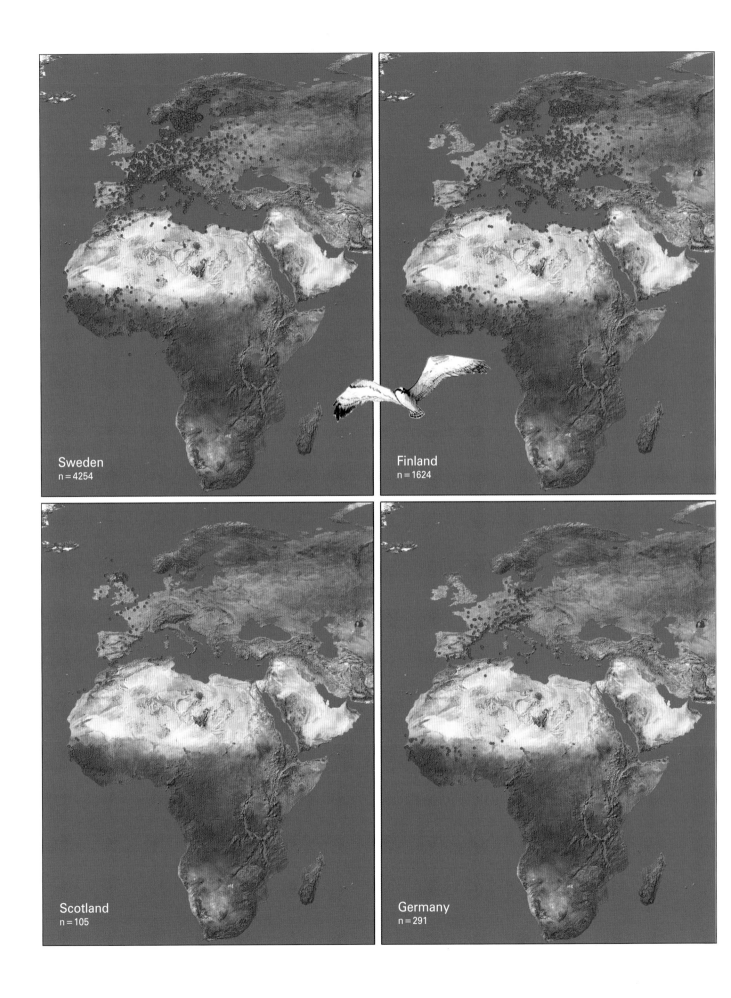

winter more to the east (1°57'E, n=531; equivalent to Nigeria) than those from Sweden (4°48'W, n=318) and Germany (8°43'W, n=65). Scottish birds are found in the westernmost parts of West Africa (14°17'W, n=36), mostly close to the Atlantic coast between Senegal and Ghana.

Distribution in West Africa

Counts [1] All major and many minor water bodies in West Africa, including salt and brackish water in the coastal zone, are occupied by Ospreys during the northern-hemisphere winter. Distribution and density in the wintering areas vary according to geographical location and habitat, ranging from low densities of birds scattered along exposed beaches to loose flocks in estuaries; in mangroves, the spacing is even. Ospreys in Africa are noted for their presence along salt and brackish waters, in stark contrast to the Eurasian breeding quarters where few are known to forage in saltwater even when available (Leopold *et al.* 2003, Marquiss *et al.* 2007). This difference is more likely caused by variations in water turbidity (being much higher in western European coastal areas), than by variations in surface temperature (as surmised by Marquiss *et al.* 2007).

Inland freshwater lakes have a sprinkling of Ospreys, albeit at lower densities than along the coast. Plotting wintering numbers against lake size reveals an average density of one Osprey per 12 km^2, irrespective of lake size (Fig. 169). [2] Using the regression line given in Fig. 169 to predict the numbers present in the 192 remaining reservoirs in West Africa, the combined surface area of 17 500 km^2 may theoretically hold 1127 Ospreys. [3]

Densities along rivers are usually low (Table 22), but an exceptionally high number of 153 Ospreys was recorded in March 1976 on the Niger River in Mali, between Macina and Mopti (175 km), but at the same time the next year a similar count produced just 13 birds (Lamarche 1980). In March, Ospreys should be on the return migration to the breeding grounds, but deviating from direct migratory tracks to forage at scarce water sources is commonplace (Strandberg & Alerstam 2007). This may explain temporary concentrations, such as Lamarche (1980) recorded along the Niger.

In comparison to numbers of Ospreys in coastal areas and at freshwater reservoirs, the species is relatively rare in floodplains, despite their plethora of fish when flooded. For example, in the 2000s, very small numbers were found in the largest Sahelian floodplain, *i.e.* the Inner Niger Delta; aerial counts showed a decline from 18, 23 and 14 respectively, in 1972, 1974 and 1978, to a mere 8, 1 and 6 in 1999-2001; furthermore, the annual ground surveys in the central lakes between 1992 and 2008 did not find more than 5 Ospreys. The much smaller Hadejia-Nguru floodplains harboured 7, 12 and 8 Os-

Fig. 168 Recoveries of Ospreys ringed as nestling in Sweden, Finland, Scotland and eastern Germany. From: EURING.

Jumping fish in shallow water (Bijagos Archipelago, Guinea-Bissau) at the approach of large predatory fish. Many coastal areas along the West African coast have crystal-clear water with high fish densities, raising the question as to why Ospreys do not remain in West Africa to breed.

preys respectively, in 1994, 1995 and 1998. Osprey numbers in the Senegal Delta, unlike in the Inner Niger Delta, have been highest in recent years, *i.e.* 21, 15, 6, 9, 25 and 14 annually in the 1999-2004 period. The construction of the Diama dam led to the conversion of upstream riverine floodplains into a lake, presumably improving conditions for Ospreys. Aerial surveys of Lake Chad and the nearby floodplains produced very few Ospreys (AfWC database). Consequently, the total of Ospreys in floodplains and Lake Chad Basin is, tentatively, below one hundred, a trifle compared with the numbers in lakes, rivers and coastal regions.

Ospreys wintering in Africa favour coastal habitats including estuaries and mangroves in preference to (and in declining order of usage) freshwater lakes, rivers and floodplains. Several explanations for habitat-related differences in density may apply, none of which is mutually exclusive: food supply, water turbidity, availability of perches and human disturbance. Bonga *Ethmalosa fimbriata*, being the main catch of fishermen using beach seines from Mauritania to Cameroon (Jallow 1994), was the major prey for Ospreys in Mauritania (K. Goudswaard, pers. comm.). Coastal Ospreys in Senegal took a variety of fish species, but at any one location clear preferences were recorded, involving both Flathead and White Mullet *Mugil cephalus* and *M. curema*, Sicklefin Mullet *Liza falcipinnis* and Bonga (Prévost 1982).

Coastal Ospreys spend very little time hunting. To meet their daily energy requirements, they need to spend only 30 minutes in capturing 1-3 fish whose total weight is 300-350 g; the energetics involved in hunting and flying amounted to 134 kJ/day (Appendix 3 *in* Poole 1989). Hunting success depends on fish type, tidal cycle, time of day, wind force, turbidity, social interactions and age (Simmons 1986, Poole 1989, Strandberg *et al.* 2006). Despite the relatively low hunting success of coastal and semi-coastal Ospreys in Africa

(Prévost in Poole 1989, Simmons 1986) – only 30-32% of attempts succeed – the species needs to spend only an insignificant 3-5% of daylight hours on hunting, which demonstrates clearly that it largely leads a life of loafing in coastal wintering quarters, as indeed is borne out by satellite-tracked birds (Triay 2002, Alerstam et al. 2006, www.roydennis.org, www.fmnh.helsinki.fi).

Fish diversity and productivity in freshwater lakes and rivers differ substantially according to connectivity (both are higher when exchange is possible; Adite & Winemiller 1997), whether rivers are natural or dammed (dams usually impact diversity negatively), the extent of shallow water in the littoral zone and the presence of drowned trees. The last-named is particularly beneficial for periphyton production, a major food source for invertebrates, and hence fish (www.dams.org). Lake Volta is fringed with dead trees, being exploited for fuel and building material when water levels are low, and thus are disappearing rapidly. Ospreys at Lake Volta prey on Moustache Catfish (P.C. Goudswaard unpubl.), found only in the River

Table 22 Densities of wintering Ospreys in salt, brackish and freshwater habitats in Africa, based on transects (km) or plot counts (km^2). From: Altenburg et al. 1992 (Banc d'Arguin, Mauritania), Prévost 1982 (Senegambia), Altenburg & van der Kamp 1991 (Guinea), wiwo 2005 (Sierra Leone), Schepers & Marteijn 1993 (Gabon), Gatter 1997 (Liberia), Lamarche 1980, J. van der Kamp, R.G. Bijlsma unpubl., AfWC database (Mali), Ambagis et al. 2003 (Niger), AfWC database (Cameroon).

Country	Habitat/river	Year	Month	Transect/plot	No.	Density
Salt and brackish water						
Mauritania	coastal	1980	Jan-Feb	265 km	50-100	0.19-0.38/km
Senegambia	estuary	1979-80	winter	26 km	70	2.69/km
Senegambia	mangrove	1979-80	winter	2650 km	179	0.1-0.3/km^2
Senegambia	coastal	1979-80	winter	457 km	221	0.42/km
Guinea	mangrove	1990	Jan	294 km	26	0.09/km
Sierra Leone	estuary	2005	Jan-Feb	337 km	30	0.09/km^2
Gabon	coastal	1992	Jan-Mar	90 km	28	3.21/km
Freshwater (rivers)						
Liberia	Cavalle, Dube, Cestos	?	winter	380 km	14	0.04/km
Mali	Niger (Mopti-Macina)	1976	Mar	175 km	153	0.87/km
Mali	Niger (Mopti-Macina)	1977	Mar	175 km	13	0.07/km
Mali	Niger (Tamani-Mopti)	1998	Nov	360 km	6	0.02/km
Mali	Niger (Bamako-Mopti)	1999	Jan	400 km	5	0.01/km
Mali	Niger (Mopti-Debo)	2004	Jan	150 km	1	0.01/km
Mali	Diaka	2005	Jan	18 km	0	0.00/km
Mali	Niger (Gourma-Bourem)	1978	Jan	180 km	11	0.06/km
Niger	Koro Gungu-Boumba	1995-99	winter	75 km	0-2	0.00-0.03/km
Niger	Mékrou	1995-99	winter	15 km	0-2	0.00-0.13/km
Cameroon	Logone (Pouss-Tikelé)	2001	Jan	23 km	3	0.13/km
Total (excl. Mali 1976)				**1418 km**	**49**	**0.035/km**

Fishermen using beach seines in Senegal to catch bonga and mullet, both important prey for Ospreys. Under clear water conditions, these 20-45 cm long (>300 g) and fat-rich species are normally well dispersed in deeper water during the day, but may form shoals over sand and mud bottoms in coastal regions and lagoons at the edge of the rising tide (Lowe-McConnell 1987, Abou-Seedo et al. 1990). Ospreys catch fast-moving mullet at the surface by flying close to the water and plunge-diving, feet-forward at shallow angles, but they locate sardines *Sardinella* spp. and flying fish *Cheilopogon* spp. offshore when circling at heights of 100-300 m above the water, sometimes several kilometres out to sea (Prévost in Poole 1989).

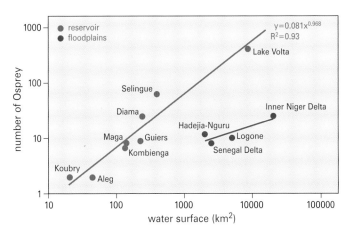

Fig. 169 Osprey numbers in reservoirs (Lake Volta in Ghana (1991-1995), Maga in Cameroon (1993), Guiers in Senegal (2003), Aleg in Mauritania (1997), Selingue in Mali (2003, 2004), Koubry (1988 & 1997) and Kombienga (2003) in Burkina Faso) and floodplains in West Africa (Inner Niger Delta (1978), Senegal Delta (1973), Hadejia-Nguru (1997) in January-February. Note the use of a double log-scale. From: P.C. Goudswaard (Volta; unpubl.), van der Kamp et al. 2005b (Selingue), AfWC database (other).

Volta and its four tributaries. In 1995-96, fishermen on Lake Volta exploited catfishes mostly under two years old, strongly indicative of over-fishing, given its normal longevity of about 5 years (Ofori-Danson et al. 2001). Fish populations in dammed lakes are usually depauperated, but some fish species have become more numerous, notably plankton feeders (www.dams.org). The waters of reservoirs constitute an important new habitat for Ospreys in Africa, but data are lacking on the negative impacts of dams on Ospreys using river systems (where Osprey densities are normally low, possibly because of high turbidity and high human-induced mortality).

The low densities of Ospreys in floodplains rather represent an enigma, given the huge level of fish production. The Inner Niger Delta is a case in point: in the 2000s fewer than 10 Ospreys occupied a floodplain measuring 400 by 80 km. Evidently, fish abundance *per se* is not the only factor determining Osprey densities. Floodplains differ from other water bodies in that the presence of water is seasonally restricted, perches are scarce, the mean size of fish is declining due to intensive human predation (average fish size as captured by fishermen declined to 10-20 cm in the Inner Niger Delta, compared with >30 cm in Lake Selingue; van der Kamp 2005b), the water is highly turbid and human population pressure has increased. Indeed, we have evidence that floodplains are risky habitats to Ospreys, acting as a population sink (see below).

Osprey *Pandion haliaetus*

Photographs play an important role, telling stories which are hard to convey in words alone. Many photographers, like Hans Hut, here in action in Niger, gave freely of their work.

Tens of thousands of nestling Ospreys have been ringed in Europe, notably in Fennoscandia. This ringed bird was observed in the Djoudj, Senegal Delta, in January 2007; the colour-ring enabled its identification as a bird from eastern Germany.

When Lake Selingue came into existence in 1992, 1.2 million trees drowned, of which many were still present in 2007, offering an abundance of perches for Ospreys.

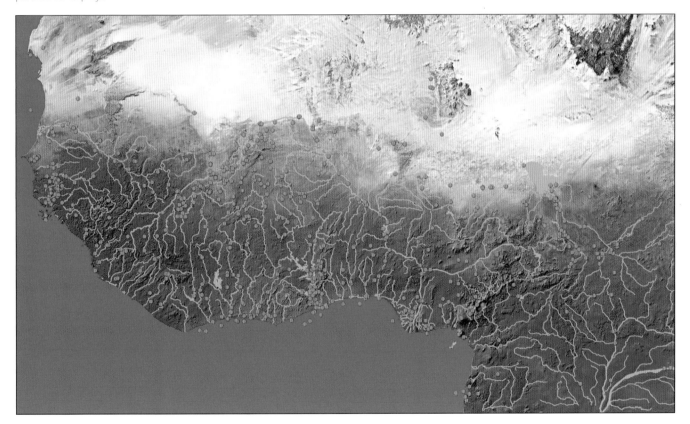

Fig. 170 Recoveries of Ospreys in West Africa between 4° and 20°N (n=1086). From: EURING.

Numbers involved In the 1990s, the Osprey's European population encompassed 7600-11 000 pairs, or some 22 800-33 000 birds (≡ pairs × 3). West Africa is the major wintering area for these birds, as reflected by the numbers observed relative to other parts of Africa (and by the distribution of recoveries; Fig. 170). Counts and extrapolations are available for a number of sites across West Africa. Highest numbers are found along the coast. On the Banc d'Arguin, Mauritania, numbers increased slightly from 92 in 1980 to 72 in 1997, 121 in 2000 and 123 in 2001 (Hagemeijer et al. 2004). On the Langue de Barbarie (N Senegal) Ospreys numbered between 12 and 43 in 2000-2004. Still more Ospreys are present in the Sine Saloum Delta, Senegal: 415 in 1997, 227 in 1998, 488 in 2000 and 326 in 2003. In the winter of 1977/78, Prévost (1982) estimated Osprey numbers here at 145, with loose flocks of up to 34 birds in January. For S Senegal, Altenburg & van der Kamp (1986) arrived at 'several hundreds' in late 1983, and similar numbers were extrapolated for Guinea-Bissau and Guinea (275-300 in mangroves) by Altenburg & van Spanje (1989) and Altenburg & van der Kamp (1991). In winter 2005, Van der Winden et al. (2007) counted 53 Osprey along the coast in Sierra Leone, and Gatter (1988) mentions 30-70 birds along the coast with another 300 along rivers in the interior of Liberia. The AfWC database holds dozens of other sites along the coast of the Gulf of Guinea with 1-10 Ospreys each. These counts together produce a minimum of 2500-4000 Ospreys along the Atlantic coast of Africa between Mauritania and Nigeria.

Another 1100 Ospreys are estimated to occupy inland reservoirs (see above). The numerical importance of rivers as wintering sites for Ospreys may amount to 1000 birds, considering the total length of rivers in West Africa [4] and an average density of 0.035 Osprey per km.

The total minimum estimate of 6000 Ospreys for West Africa is much too low since it represent only about 20% of the total European population. Most likely, we have underestimated the numbers along the coast and the West African rivers.

Recoveries By 2006, a total of 1086 recoveries from Africa were available, 90% of which came from between 4° and 30°N (20-50 recoveries annually in 1977-2005). Within that zone, most recoveries came from rivers (45.3%), coastal sites (26.6%), floodplains (10.9% in Inner Niger Delta, 2.8% in Senegal Delta, 0.7% in Hadejia-Nguru wetlands) and lakes (e.g. 6.8% in Lake Volta, 5.1% in Lake Chad). The distribution of recoveries across habitats is at variance with the actual distribution as observed from counts. For example, many more Ospreys in West Africa winter along the coast than are indicated from recoveries (about 75% versus 27%). Furthermore, dozens winter in the Inner Niger Delta but hundreds do so at Lake Volta; with so few wintering in the Inner Niger Delta, the large number of recoveries is surprising. Finally, recoveries from along rivers indicate a high presence, despite the thin distribution there (Table 22, Fig. 169).

For the Inner Niger Delta, where wintering numbers have declined since the 1970s, the annual number of recoveries increased to 3.7 in 1998-2006, compared with an average of 2.7 in 1977-1997. This increase reflects an improved reporting rate since 1998, associated with activities of Wetlands International in the region. A variable reporting rate is also the most parsimonious explanation for the discrepancies between distribution and recoveries as described above.

Summering in West Africa Almost 10% (89 out of 949) of the African recoveries in the EURING database refer to Ospreys reported in June-August, including three birds fledged that same year. The age-distribution of the 86 remaining birds was 52 first-year (60%), 13 second-year (15%), 9 third-year and 12 more than 3 years old, thus not only many first-year birds remain in Africa during summer (Österlöf

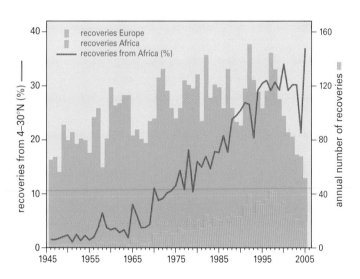

Fig. 172 The proportion of Ospreys recovered in Africa between 4° and 30°N (line & left-hand axis) relative to the total number of recoveries in the same year in Africa and Europe (histogram & right-hand axis). From: EURING.

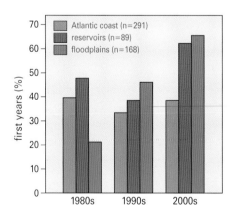

Fig. 171 Percentage of first-year Ospreys in recoveries from Sahelian floodplains, reservoirs and the coastal zone. Same data as Fig. 170.

Ospreys winter along the Casamance (left) and the Gambia (right), but they are only common in the estuaries (September 2008).

1977), but also some adult birds. The distribution of the birds in summer did not differ from the distribution found in winter, the highest number of recoveries being in Mali (22%), Senegal (16%) and Ivory Coast (11%).

Mortality in Sahelian wetlands Floodplains in the Sahel are of minor importance as wintering areas for Ospreys. At present, about as many Ospreys are recorded alive annually in the Inner Niger Delta and along the River Niger as are reported dead annually to the European ringing stations. Given the ratio of ringed to recovered being 10:1 (possibly even lower in Africa), the equal number of observations and recoveries suggests that this floodplain is a death trap for large numbers of Ospreys. The relatively high number of recoveries from Lake Chad, where very few other bird species have been recovered, hints at similar population sinks in other Sahelian wetlands. Of the 490 Ospreys recovered in the Sahel, 49% were in their first year of life. In reservoirs and along the coast, the proportion of juveniles among recoveries was respectively 43% and 38%. This proportion has changed considerably since the 1980s (Fig. 171), when only 21% of the birds recovered in the Sahel were juveniles (when winter-

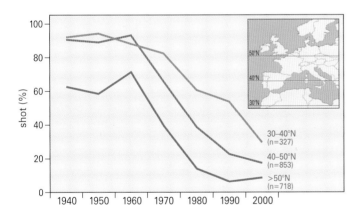

Fig. 173 Percentage of Ospreys shot in three latitudinal zones, relative to the total number of recoveries with known cause of death. From: EURING.

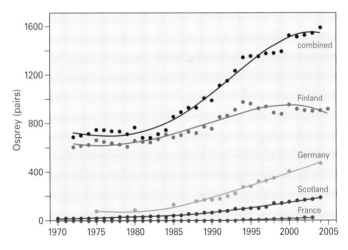

Fig. 174 Long-term population trends of Osprey in several European countries. The trend lines shown are second or third degree polynomials, explaining 94-99% of the variance in the different trends. From: Dennis & McPhie (2003) and annual reports in British Birds for Scotland; Saurola (2007) for Finland; Wahl & Barbraud (2005) for France; Schmidt & Roepke (2001) and Mebs & Schmidt (2006) for Germany.

Osprey *Pandion haliaetus*

Some rivers in Guinea-Bissau, like the Rio Geba (left), are fringed by extensive rice fields. Others, like the Rio Grande de Buba, the Rio Corubal and the Rio Cumbiiã (from left to right), have a tree-lined shore over large sections. Flow rate and turbidity differ between rivers and river sections, and within seasons. The detectability of fish varies accordingly; Ospreys can be found wherever local circumstances are favourable for catching fish (September 2008).

ing numbers in the Inner Niger Delta were larger), increasing to 65% in recent years. In the coastal areas the fraction of juveniles has remained stable at 38%. Do these data indicate that the importance of the Inner Niger Delta as a wintering habitat has further been marginalised? Possibly so, because mankind's intense exploitation of fish stocks has led to older and larger fish disappearing and to a concomitant decrease in fish size (Chapter 6). Hence the exploitation by Osprey of the present fish supply in the Inner Niger Delta might be less profitable than in the past. Moreover, Ospreys are remarkably faithful to wintering sites (Prévost *in* Poole 1989, Alerstam *et al.* 2006), and the high fraction of juveniles reported for the Sahel may then also suggest high mortality, as indeed borne out by the ratio of observed to recovered being about 1:1. Ospreys wintering in the Inner Niger Delta have a fair chance of taking fish from a hook line, just as an Osprey patrolling the Niger River runs a high risk of being shot. Excluding 228 birds whose cause of death is unknown, 39% of the Ospreys recovered between 4° and 30°N were shot, 20% trapped and 36% killed accidentally due to the hook lines. These proportions have not changed over time. Regional differences are large, however: of coastal birds 46% were shot, against 20% in Sahelian wetlands. For birds trapped, intentionally or by accident, it was the other way around: along the coast 39%, in the Inner Delta 63% and in the Senegal Delta and at Lake Chad even higher at 76%.

Most Ospreys winter along the Atlantic coast, and Sahel effects should therefore be limited, if any. Annual changes in breeding numbers did not, as predicted, differ between wet and dry Sahel years. The increasing proportion of recoveries from Africa (4°-30°N) since the 1980s (Fig. 172) does not mirror weather conditions in this region. Rather, it reflects reduced (reporting of) shooting in Europe, a tendency already pointed out by Pertti Saurola (1995) and also visible in the declining number of rings reported (Fig. 173).

Population trends

Throughout the Palearctic, Ospreys have increased in number since the 1970s or 1980s. In Sweden, numbers doubled between 1971 and the mid-1990s, from 2000 to 3400-4100 pairs, then remained stable (Svensson *et al.* 1999). Similarly, numbers in Finland were considered stable in the 1970s, increased at about 3% annually in 1982-1994 to 1200 pairs (numbers in Fig.174 refer to pairs checked annually) and again stabilised in later years (Saurola 2007). Other populations had a slower start, and have not yet reached the levelling-off stage. In Germany, for example, the number of pairs more than doubled between 1988 and 1998 (from 147 to 346 pairs; Schmidt & Roepke 2001, Fig. 174), whereas the Scottish population took a long time to gather momentum (Fig. 174). Recent colonisations, like the one in central France, may still have a long way to go (Wahl & Barbraud 2005).

The European population has doubled since the early 1970s. The increase in Finland was accompanied by improved reproduction (1.37 large chick/occupied territory in the 1970s, 1.65 in 1996-2005; Saurola 2007) and by reduced persecution in Central and South Europe since the 1970s (Saurola 1995). Conversely, reproduction in Germany did not improve in the 1988-1998 period and the increase here is attributed to reduced mortality during migration and the prolific supply of artificial nests (Schmidt & Roepke 2001).

Conclusion

It has been calculated that in the late 20th century, at a minimum 4100-6100 Ospreys wintered in West Africa, which figure accounts for about 20% of the European population (including Russia). This must be a large underestimation given the preponderance of ringing recoveries in West Africa, where about 55% reside along the coast, 20% near reservoirs, 20% along rivers and the rest in Sahelian floodplains. Survival of Ospreys in the Sahel is low, due to shooting and trapping, but because only a small fraction winters here, the impact on the population level is negligible. The increase in Osprey numbers since the 1970s is triggered by protection in Europe and the banning of persistent pesticides from the food chain, which factors enabled the species to re-colonise lost breeding areas, and to increase densities in core habitats. Improved reproduction and reduced human-induced mortality were instrumental in doubling the European population between 1970 and 2000.

Endnotes

1. Unless stated otherwise, all data are taken from the AfWC database (Wetlands International) and refer to counts done in January.
2. This outcome may come as a surprise. Because Ospreys are bound to the lake shore, one might expect that the fraction of the water surface being too far from the coast to be exploited will increase with lake size, thus implying that the overall density will be less in larger lakes. If lakes were circles or squares, one would expect the number of Ospreys (being bound to the shore) to be a function of the square root of lake surface. None of the reservoirs included in Fig. 169 are circles or squares, but are long and narrow former river valleys. Lake Volta, for instance, is 520 km long but only 6-12 km wide and contains a scattering of islands. Its total shoreline extends to 4800 km, thus providing ample opportunities for Ospreys to fish everywhere on the lake.
3. A full list of African dams can be found at www.fao.org/nr/water/aquastat/damsafrica/african_dams060908.xls. There are 192 reservoirs in West Africa; for 70 lakes, surface area (S) as well as volume (V) is given. Using the relationship $S = 1.199V^{0.839}$ ($R^2=0.914$), we estimated the surface of 123 small reservoirs of unknown surface from their volumes (adding Foum Gleita in southern Mauritania, and correcting for errors, such as the volume of Lake Shiroro into 2.5 km^3). Adding the measured and estimated surfaces we arrive at a total water surface area of 17 700 km^2.
4. The Niger is the longest river in West Africa, at 4200 km. The Bani (1100 km) and the Benue (1400 km) are its main tributaries. Other large rivers are: Senegal (1800 km), Black Volta (1350), White Volta (1150 km), Hadejia + Jama'a re + Komodougou (1100 km), Gambia (1100 km), Logone (1100 km) and Chari (1000 km). There are between the Gambia and The Niger Delta some 30 rivers each about 300 km long. Taking into account some 6000 km for other rivers and all tributaries, we arrive at a round figure of 30 000 km. Given an overall density of 1 Osprey per 28.6 km, some 1000 would be present along the West African rivers. The few available counts (Table 22) suggest a decline, between the 1970s and the 1990s and 2000s, in density of the Ospreys present along the Sahelian rivers.

Eurasian Marsh Harrier
Circus aeruginosus

For the present generations of birders and ornithologists, it is hard to imagine what Europe looked like to previous generations, even as recently as the 1950s. Many wetlands, still in prime condition, had yet to be consumed by relentless demands for farmland or urban sprawl. It may therefore come as a surprise that the extensive reedbeds of these wetlands harboured very few Eurasian Marsh Harriers, doubtless largely due to the persecution associated with "rational" game preservation, as proposed by Henning Weis (1923) with reference to Denmark, but aptly applicable to much of Europe. Continuous human persecution gave the Marsh Harrier a reputation as a secretive breeding bird. However, legislation to protect the species during the 20th century did not end covert killing. In The Netherlands, for example, gamekeepers shot and poisoned at least 400 Marsh Harriers in the Noordoostpolder in 1951, where numbers had boomed following its reclamation from the IJsselmeer in 1942 (Bijlsma 1993). Wilful extermination was exacerbated by large-scale cultivation of wetlands, reducing the available nesting habitat, and by the widespread application of persistent pesticides on farmland in the 1950s and 1960s. Marsh Harrier numbers remained at a low ebb in Europe until recently. Our generation is the first in over 100 years to see and hear sky-dancing Marsh Harriers, wherever suitable habitat is available. This higher abundance is likely to be reflected by higher densities in wintering areas, including the Sahel.

Breeding range

The lowlands of Europe, from the polders below sea level in The Netherlands, northern and western France, the North German Plain and Poland eastwards into the Baltic States, Belarus and much of Russia, are bespeckled with swamps, marshes and lakes, often fringed with dense reedbeds and other herbaceous vegetation. This lowland belt held about 80% of the European population of Eurasian Marsh Harrier in 1990-2000 (93 000-140 000 pairs). A country like Ukraine, with its large deltas, is equally favoured as breeding ground (13 800-23 600 pairs). North and south of this belt, densities decrease, except in southern Sweden, which has a sizeable population (1400-1500 pairs) (BirdLife International 2004a). Breeding in Africa is restricted to Morocco (widespread and common; Thévenot et al. 2003), Algeria (several dozen pairs; Isenmann & Moali 2000) and Tunisia (50-70 pairs; Isenmann et al. 2005), which all hold resident populations.

Migration

The migratory tendency of Marsh Harriers decreases from north to south, and from east to west across Europe. Breeding birds from Fennoscandia and eastern Europe abandon their breeding haunts completely in early autumn, mostly to winter in sub-Saharan Africa. Wintering is still rare in Germany, but in the United Kingdom (Oliver 2005) and The Netherlands (Zijlstra 1987, Clarke et al. 1993) this tendency is quite pronounced. The large population in western France is considered resident (Bavoux et al. 1992) and the same may apply to the small populations in Spain, Portugal (except birds breeding in the north; Rosa et al. 2001) and Italy. The frequency of wintering in western Europe seems to be on the increase in recent years (Oliver 2005), particularly of juveniles (probably mostly females; Clarke et al. 1993). Long-term censuses in The Netherlands show local and temporal variations in wintering numbers, respectively peaking in the early 20th century (Zuiderzee; ten Kate 1936), 1977-1982 (Flevoland; Zijlstra 1987), northern Delta (early 1960s, late 1980s and early 1990s; Ouweneel (2008), and southern Delta (increasing since 1990s; Castelijns & Castelijns 2008).

Harriers, including Marsh Harriers, cross the Mediterranean Sea in a broad front, as implied by the small numbers seen at migratory bottlenecks on either side of the Mediterranean (Bijlsma 1987). Direct observations in Italy (Agostini et al. 2003, and references therein), radar studies on the southern coast of Spain (Meyer et al. 2003) and satellite tracking (Strandberg et al. 2008a) have indeed shown that Marsh Harriers readily crosses the sea, even during twilight and at night, preferably with following winds and under stable weather conditions. With a predominantly southwesterly course from breeding grounds towards sub-Saharan Africa (as indicated by ringing recoveries; Fig. 175), some segregation between eastern and western populations should be detectable in the wintering quarters, which is precisely what the evidence of 95 recoveries south of the Sahara tells us. Western European birds winter mostly west of 5°W and easterly breeding birds mostly east of 0°E; intermediate breeding populations in winter are distributed mainly between these longitudes, although overlapping (e.g. in the Inner Niger Delta) considerably with the western wintering population. For example, of 35 Marsh Harriers ringed in eastern Germany, 9 were recovered from Mali and only 3 from Senegal, but of 19 Marsh Harriers ringed in The Netherlands, the reverse was the case: 3 from Mali as opposed to 9 from Senegal. The fact that these populations breed only some 400-500 km apart, yet are largely separated in their African wintering grounds, indicates that European breeding birds make segmented use of sub-Saharan wetlands.

Spring departure from the West African wintering grounds, as shown by satellite tracked Swedish birds, occurred between 11 February and 17 April for adults (Strandberg et al.2008a). This study showed that trans-Saharan migrants crossed the Mediterranean Sea between 18 March and 29 April. The average speed of spring migration was 219 km in males and 233 km in females, only marginally faster than during autumn migration (204 and 221 km respectively). Travel distances of juveniles in autumn were much smaller than of adults, i.e. on average 146 km (as was their total distance travelled,

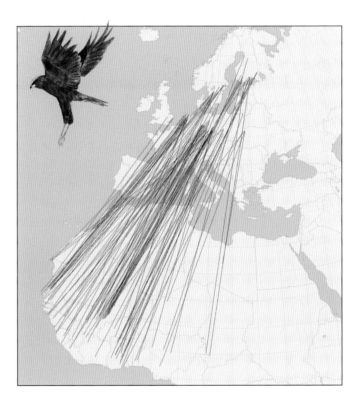

Fig. 175 European ringing locations of 102 Marsh Harriers recovered in West Africa between 4° and 20°N. From: EURING database, except one recent recovery from Mali (Wetlands International).

Marsh Harriers heading for their nocturnal roost in the Inner Niger Delta, 18 February 2007, 18.14 local time. The typical sight here, from one hour before sunset till dark, is of harriers purposely flying in straight lines from various directions towards the roost, very different from the more wanderig low foraging flights. Upon reaching the roost, the birds descend and settle out of view in the tall vegetation. Every now and then, all harriers present – as if on command – take wing again and start milling around. This caroussel-behaviour is not repeated in early morning, when the harriers take flight before sunrise and head straight for their foraging grounds.

viz. 1757 km in juveniles and 5093 km in adults). Many juveniles do not migrate all the way to sub-Saharan wetlands, but remain in Europe.

Distribution in Africa

Of the 95 recoveries south of the Sahara, many were recorded from large floodplains (21 Inner Niger Delta, 9 Senegal Delta), the extensive coastal wetlands between Senegal and Sierra Leone (notably Sine Saloum in Senegal, The Gambia) and from Keta Lagoon, Volta River and Oti Basin in Ghana (the latter extending into Togo; Cheke & Walsh 1996). At least 45 of the recoveries (47%) could be traced to large wetlands; the remaining 50 birds were distributed across the Sahel zone (n=20) and the Sudan zone (n=30), where the species uses a large variety of habitats, mostly associated with water (reservoirs, lakes, streams, rice fields) but also, for example, montane forests where above-canopy hunting has been recorded (in Liberia; Gatter 1997), and *Pennisetum*-covered gaps in high forest where Barn Swallow roosts are targeted (SE Nigeria; Bijlsma & van den Brink 2005).

The large floodplains hold high numbers in the northern winter. In 1992/1993, the wintering population in the Senegal Delta (94 km² including the Djoudj) was estimated at >1000 individuals (Arroyo & King 1995), and midwinter counts revealed 99 birds in the Hadejia-Nguru floodplains (Chapter 8), 22 birds in Lake Fitri and 237 birds in the Logone floodplains (Chapter 9). The much larger Inner Niger Delta may hold perhaps 1000-2000 Marsh Harriers, of which 470 in the central lakes (in 2007; Chapter 6). Systematic observations of hunting harriers in the Inner Niger Delta in January-February 2004 and 2005 showed that 77% of 387 Marsh Harriers hunted lake sides, *bourgou* fields and river banks (R.G. Bijlsma & W. van Manen, unpubl.). Hunting flights over dry ground were least evident in females and juveniles (respectively 14% of 35 birds and 12% of 214 birds), and more so in males (26% of 126 birds). Regardless of sex-specific differences in foraging habitats, the overriding preference for wetlands in Marsh Harriers stands out in comparison with Montagu's Harriers and Pallid Harriers, of which respectively 63% (n=72) and 71% (n=14) were recorded hunting over dry savanna-like habitats and dry rice fields (see Fig. 61 for the distribution of census plots throughout the Inner Niger Delta relative to wetlands). A similar harrier-specific habitat choice has been described for the floodplains of northern Cameroun between Léré and N'Djaména in the early 1970s, where Marsh Harriers occupied the wettest habitats and Pallid Harriers the driest (Thiollay 1978b).

In Sahelian wetlands, wintering Marsh Harriers roost communally in dry rice fields (Djoudj, Senegal; Arroyo & King 1995, Rodwell *et al.* 1996), wild rice and *didere* (Inner Niger Delta, Mali; R.G. Bijlsma & W. van Manen unpubl.) and dry grassland (Casamance, Senegal; 2 December 2005; R.G. Bijlsma unpubl.). Roost size varies seasonally and locally, from several tens up to 390 (Arroyo & King 1995, Rodwell *et al.* 1996, Strandberg & Olofsson 2007). Large roosts of Marsh Harriers are often single-species, apart from the occasional Montagu's or Pallid Harriers. Away from wetlands, Marsh Harriers lead a more solitary life, and often roost singly wherever the opportunity arises; near Afi Mountain in SE Nigeria, for example, a solitary

Marsh Harriers wintering in West Africa are largely confined to marshlands. Males (facing page) on average inhabit drier habitats than females and juveniles (this page). In the drier parts of the Sahel, where Montagu's and Pallid Harriers greatly outnumber Marsh Harriers, pools are frequently visited for drinking (Senegal, February 2007). Male Marsh Harriers, being slightly more agile than females, hunt for weavers, small mammals and grasshoppers in drylands and along edges of marshland (Inner Niger Delta, January 2005), whereas females and juveniles take bigger birds, fish, carrion and offal (here in *bourgou* in the Inner Niger Delta, January 2005; far right).

male roosted in elephant grass in forest gaps (Bijlsma 2001).

Adult Marsh Harriers tracked by satellite telemetry showed high site fidelity to wintering and post-migratory stopover sites, both within and between seasons. Post-migratory movements may be associated with local variations in food abundance, *i.e.* food shortage leading to trips of up to 811 km (when returning to the initial staging area). In fact, site fidelity of six adults tracked in 2004-2007 was even stronger for wintering sites than for breeding sites (Strandberg *et al.* 2008a).

Population change

Wintering conditions   Of birds recovered south of the Sahara with known cause of mortality (n=48), the majority were killed by shooting (21) and trapping (17) in relatively few areas. If the flood extent of the Inner Niger Delta was below 12 000 km², recoveries averaged four per year, only one annually in years with a flood extent above 16 000 km², and two in years of intermediate floods. To correct for the variation in the number of ringed birds, the sub-Saharan numbers of recoveries (0-8 per year) are expressed as a percentage relative to the total number of recoveries in the same year (varying between 8 and 84). As shown in Fig. 176, the Sahel mortality rate is higher in dry years. It is likely that the chance of getting killed is higher when birds are concentrated in the few remaining wet parts of the Sahel under dry conditions, rather than when they are distributed widely across the entire region when conditions are wet.

An upward trend in Marsh Harrier population size has been recorded in all European countries. Sooner or later during a recovery phase of a population, numbers will level off and will undergo greater levels of fluctuation than experienced in previous decades, albeit the mean level is often somewhat below the initial apex (Fig. 177). This pattern has become apparent in Sweden (since the early 1990s; Lindström & Svensson 2005), Denmark (after 1995; Fig. 177), The Netherlands (after 1995; Fig. 177), Germany (since the late 1980s; Kostrzewa & Speer 2001, Mammen & Stubbe 2006), France (since the late 1980s; Thiollay & Bretagnolle 2005), Spain (since the 1990s; Díaz *et al.* 1996) and Italy (since the late 1990s; Brichetti & Fracasso 2003). Conversely, the Marsh Harrier in the United Kingdom is still on the increase (Fig. 177).

After having recovered from pesticide-related crashes, fluctua-

Male Marsh Harrier crossing the Sahara, presumably an early migrant heading for the breeding quarters (Mauritania, February 2006).

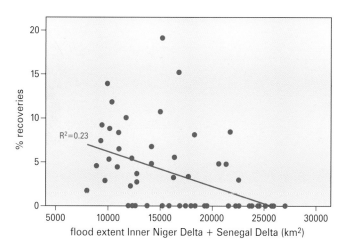

Fig. 176 Annual number of recoveries from Africa (0°-20°N) as % of the total number in the same year (dead birds only) between 1946 and 2004 (n= 58 year, P<0.001), plotted against flood extent. From: EURING.

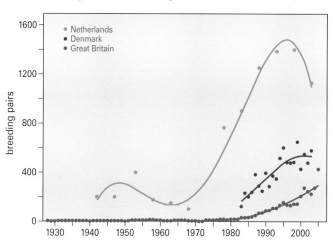

Fig. 177 Long-term population trends in NW Europe. From: Grell 1998 and Heldbjerg 2005 (Denmark), Underhill-Day 1984, 1998; Rare breeding birds in the UK, *in* British Birds in series (Great Britain), Bijlsma 1993, 2006b (The Netherlands).

Box 21
Strife and temporary respites

Up to the 1950s, the population level of Marsh Harriers in Europe was well below the carrying capacity of all countries with breeding birds. Loss of habitat and the species' classification as 'vermin', in the name of game preservation, were to blame. The struggling populations were hit even harder in the 1950s and 1960s when the massive use of organochlorine and mercury compounds in farming resulted in contamination of the food chain, causing reduced reproductive success and increased mortality (Odsjö & Sondell 1977). The already dwindling populations declined to precariously small numbers all over Europe. For decades, any impact of wintering conditions in Africa on breeding populations must have been swamped by human-induced declines on the breeding grounds.

Perhaps surprisingly, short-term upsurges were also noted during this long episode of struggle, particularly in The Netherlands, which, in 1900, was a country of only 32 000 km^2 in area, about half of which was below sea level. The demand for land was great during the 20th century boom, when the human population trebled to reach an average density of 466 inhabitants/km^2 of land, culminating in the enclosure of the Zuiderzee in 1932. In subsequent years, successive parts of this newly-created lake, named the IJsselmeer, were reclaimed, i.e. the Noordoostpolder in 1942 (480 km^2), Oostelijk Flevoland in 1957 (540 km^2) and Zuidelijk Flevoland in 1968 (430 km^2). After each area was reclaimed, it took up to 10 years before the virgin soil of the new polder was fully converted into farmland with infrastructure and towns. These years were typified by a luxurious growth of perennials, especially reed Phragmites australis that at times was sown by aircraft and helicopter to suppress the colonisation of weeds. The sudden availability of hundreds of square kilometres of suitable breeding habitat, in combination with booming Common Vole Microtus arvalis and Harvest Mouse Micromys minutus populations, attracted Marsh Harriers, which soon bred in numbers unprecedented in The Netherlands and the rest of Europe. In the Noordoostpolder, for example, some 1200 were present in August 1948 (Limosa 23: 306, 1950). Such high totals gave the local gamekeeper fraternity the excuse to kill at least 400 harriers in 1951. After the reclaimed land had become cultivated, the numbers of Marsh Harrier dropped virtually to nil in Noordoostpolder and Oostelijk Flevoland. The creation of a large nature reserve in Zuidelijk Flevoland, the Oostvaardersplassen, prevented a total demise, although numbers dropped considerably from >300 in 1977-1980 to 35-48 in the early 2000s (Fig. 178), mostly through loss of breeding and foraging habitats. It is of note that the steep upsurges in numbers during each reclamation period could not be accounted for by local reproduction; many recruits were probably immigrants, especially during the early years, perhaps even drawing breeding birds from nearby sites until some 5-7 years had elapsed (Dijkstra et al. 1995). In Zuidelijk Flevoland, net emigration came into effect from the late 1970s and early 1980s and surplus birds started to (re)colonise other breeding haunts elsewhere in The Netherlands (and possibly Great Britain; see Fig. 177). This premise is supported by regional trends in The Netherlands, where numbers failed to increase substantially until the early 1980s. The fact that earlier reclamations, like those in Noordoostpolder and Oostelijk Flevoland, did not result in similar long-term effects can be ascribed to culling and, particularly, to the negative impact of seed dressings and organochlorines on mortality and breeding success.

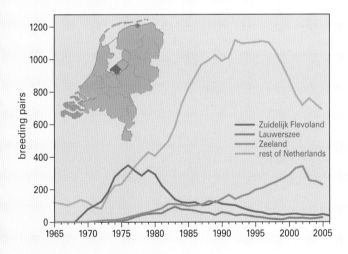

Fig. 178 Long-term population trends of the Marsh Harrier in several regions in The Netherlands compared with the trend in the rest of the country. Zuidelijk Flevoland and Lauwersmeer were reclaimed in 1968 and 1969 respectively, and subsequently cultivated. The estuarine habitats in Zeeland changed into freshwater marshes from 1969 onwards, due to damming. From: Bijlsma et al. 2001 (plus additions).

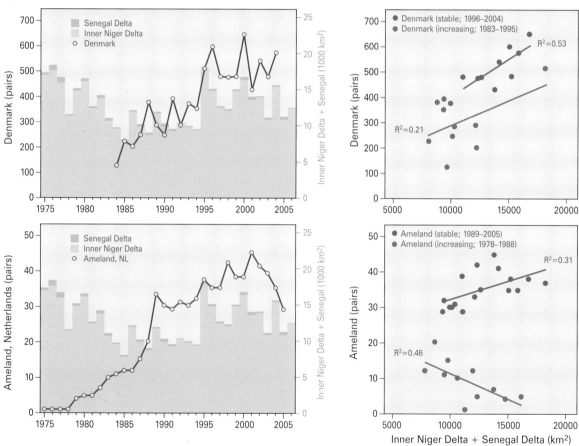

Fig. 179 Long-term population trends of the Marsh Harrier in Denmark and on the island of Ameland in the northern Netherlands related to Sahelian flood extent (left-hand panels); notice the better fit between population size and flood extent after numbers reached saturation levels on the breeding grounds (right-hand panels). From: Krol & de Jong *in* Bijlsma 2006 (Ameland) and Grell 1998 & Heldbjerg 2005 (Denmark).

A single Eurasian Marsh Harrier causes 33 000 Black-tailed Godwits to take flight, just after they had arrived on the rice fields near Samora, Portugal, where they feed on spilt rice during daytime. The birds leave the roost at twilight, and enter the feeding grounds when temperatures are still low (here -5°C). During the early morning, they are more jittery than later on, when feeding becomes more frenzied and passing harriers only startle the nearest godwits into flight (January 2008).

tions in populations have more or less reflected the extent of flooding in West African floodplains, as demonstrated for Denmark and Ameland (The Netherlands) (Fig. 179). This is consistent with the decline in ring recoveries during high floods in Sahelian wetlands (Fig. 176), indicating poorer survival during droughts. Sahelian conditions are particularly crucial for NW and N European populations which are fully migratory, except for several hundreds wintering in Belgium, The Netherlands and southern England (Zijlstra 1987, Bijlsma et al. 2001, Oliver 2005). Populations in W France and in S Europe are partly or largely sedentary (see above) and any impact of sub-Saharan conditions is likely to be small. As wintering in Europe and Africa is sex- and age-skewed, with females and juveniles on average wintering further to the north and in wetter habitats (Clarke et al. 1993, Bijlsma et al. 2001, Panuccio et al. 2005), droughts in Sahelian wetlands may impact sexes and ages differentially, given also the strong fidelity to wintering sites in West Africa (Strandberg et al. 2008).

Breeding conditions   Over time, man suppressed breeding numbers to levels well below carrying capacity (Bijleveld 1974). This changed in the 1980s. Legislation, law enforcement and successive bans of persistent organochlorines and seed dressings in Europe were instrumental in the species' rapid recovery from persecution and environmental contamination (Newton 1979).

The increase varied from a factor of 2.4 (Germany) to 14 (The Netherlands) between the minimum around 1970 and the peak and subsequent levelling-off in 1995-2000 (examples in Fig. 177 and 179). The populations of Sweden, Finland, Denmark (Jørgensen 1989), Czech Republic, Germany (Kostrzewa & Speer 2001), The Netherlands (Bijlsma et al. 2001), Flanders and France (Thiollay & Bretagnolle 2005) on average increased by a factor of 4.1 between the late 1960s and the late 1990s. The Marsh Harriers in the United Kingdom have fared even better: only 1-3 breeding females in 1969-72, but 429 by 2005 (Underhill-Day 1998, complemented with data from *Rare breeding birds in the UK* as published in series in *British Birds*).

Conclusion

Due to relentless persecution in the first half of the 20th century, and aggravated by widespread use of organochlorines and mercury on farmland in the 1950s and 1960s, European populations of Marsh Harriers approached extinction, reaching such a low population level that any effect of conditions in the wintering quarters was overridden by conditions in the breeding quarters. Sudden upsurges in The Netherlands in the 1950s and 1960s were short-lived, as temporarily favourable conditions followed reclamation work. The possible impact of floodplain size in the Sahel became apparent only after European populations had recovered from human-induced declines in the breeding areas, but we presume this impact to be greater on the fully migratory northern populations (of which adults show high wintering site fidelity in West Africa) than on those, largely sedentary, in southern Europe.

Montagu's Harrier
Circus pygargus
Christiane Trierweiler & Ben J. Koks

To talk of Montagu's Harriers, or Monties as they are known by harrier aficionados, is to conjure up farmland, protection and volunteers; or, in the case of The Netherlands, waving cereals, flowering oilseed rape and butterfly-covered lucerne. 'Empty' quarters which are not empty at all, but the remote corners of a small country otherwise packed with human beings. The protection of nests demands unfailing dedication on the part of both farmers and volunteers. Hundreds of people have helped us over the years and are immortalised in our hall of fame, or in the names of those Monties which carry satellite transmitters (www.grauwekiekendief.nl). A special place in our hearts is reserved for Rudi Drent. After his retirement as an animal ecologist at the Centre for Ecological and Evolutionary Studies (Groningen University), he continued to supervise our Monty research. Rudi was the ideal sparring partner, keeping sight of the larger picture amidst a multitude of details, pointing out promising avenues for further research and honing the data to perfection. His death on 9 September 2008 was a shock to all of us.

Breeding areas

Montagu's Harriers have a wide but patchy distribution across Europe, extending eastwards to the Caspian lowlands, Kazakhstan and the upper Yenisey at 93°E. The latest estimate of the European population, including Russia, stands at 35 000-65 000 pairs, with stable or slightly increasing numbers in the 1990s (BirdLife International 2004a, but see below).

Wintering areas

On the Indian subcontinent, notably NW India, large roosts have been found, containing several thousand individuals (Clarke 1996a, 1998). These birds are most likely to come from the eastern breeding grounds. The latter population may also, at least partly, winter in eastern and southern Africa, given the vast migration of Montagu's Harriers observed in Georgia during early autumn 2008 (B. Verhelst unpubl.), and the average migration directions attributed to birds recovered along the Dnepr, Don and east of the Volga (Mihelsons & Haraszthy 1985). If this interpretation is correct, the Sahel from Chad/Nigeria eastwards and eastern and southeastern Africa are a melting pot of Montagu's Harriers breeding from Fennoscandia (10°E) eastwards to at least 55°E. Three birds ringed in spring and autumn at Dzjambul in southern Kazakhstan (45°50'N, 71°25'E) were recovered in western Asia at distances of 674, 1209 and 1594 km to the NE-NNE (Mihelsons & Haraszthy 1985); these birds may have wintered in India, probably circumventing the impressive Tien Shan mountain range on the way.

In Kenya, the species is considered to be a fairly common visitor to the grasslands and other open habitats in the highlands and in the southeast (Lewis & Pomeroy 1989). Many birds venture further south into the semi-deserts, grasslands, savannas, pans and fallow lands of Tanzania and the eastern half of southern Africa, albeit in ever lower numbers the farther south they fly (Brown et al. 1982, Harrison et al. 1997). However, the birds from Europe are confined overwhelmingly to the narrow band of the Sahel in W Africa.

Migration strategies

By 2008, after almost a century of bird ringing, the EURING files contained 46 African recoveries of Montagu's Harriers ringed in Eu-

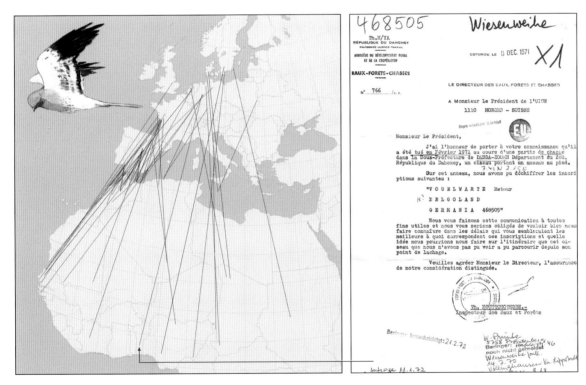

Fig. 180 European ringing locations of Montagu's Harriers recovered in Africa (n=47) and Malta (n=2), including 15 south of the Sahara. From: EURING and García & Arroyo (1998). The Montagu's Harrier ringed as a juvenile near Lippstadt, Germany by W. Prünte on 4 July 1970, and shot in the Dassa-Zoumé district in Benin in February 1971 (indicated by the arrow), is one of several reported well south of the wintering grounds in the western Sahel. The letter reporting the bird is a fine example of the many obstacles which have to be assailed in order to get the report onto the ringing centre's desk; no wonder that many rings from birds taken in Africa are never reported. From: Vogelwarte Helgoland, Wilhemshaven.

To capture adult Montagu's Harriers, we use a stuffed Northern Goshawk and a wide-meshed mistnet positioned close to the nest. This female (Karen) was captured near Ballum in Denmark in July 2008, and turned out to have been ringed during migration as a first calender-year in the Czech Republic in 2000. Interestingly, two of our satellite-tracked females travelled from western Europe via the Czech Republic during their outward migration (Fig. 181).

rope (Fig. 180). The picture emerging from these recoveries is one of migratory connectivity: birds from The Netherlands, Britain, France and Spain mostly ended up in de western part of the Sahel, birds from northern Germany and Sweden converged in the central Sahel. Notable exceptions occur, however, showing that birds from western Europe may find their way to the central Sahel (Fig. 180).

From the information derived from satellite telemetry, radar studies and visual observations in the Central Mediterranean, our perception of the migration of Montagu's Harriers has changed substantially from that provided by the ring recovery analysis by García & Arroyo (1998).

Autumn   Pre-migratory movements, following brood failure or shortly after the chicks have fledged, may occupy up to 73 days be-

Our Monty activities encompassed Europe from The Netherlands in the west to Belarus in the east. Raptorphiles, like Dmitri Vintchevski (holding bird leaflet and translating information from the farmer's couple in western Belarus), have been very helpful in directing us to habitats of Montagu's Harriers in their countries.

Ringing nestling Montagu's Harriers in western Europe is usually a social event, involving the farmer and his family (here proudly holding 'their' chicks), local birdwatchers and coordinators (represented by Lars Maltha Rasmussen of the Dansk Ornitologisk Forening) and 'Monties' (represented by Ben Koks and Aletta Buiskool from The Netherlands). This nest on arable land near Skærbæk in Denmark held five chicks, an above-average number. (July 2008).

This second calendar-year male Montagu's Harrier, photographed near Khelcom, Senegal, on 3 February 2008 (left), and again on the same spot on 28 January 2009 (right), was ringed and wing-tagged as a nestling near Deux-Sèvres, in western France, in 2007. The bird was one of the 1524 nestlings tagged in France in 2007, as part of a large European effort to improve our knowledge of dispersal and migration (www.busards.com).

fore true migration commences (own data, Limiñana et al. 2008). During pre-migratory movements, unless blocked by large stretches of open water, Spanish birds dispersed on random headings, short-hopping between sites that abound in food. They may also frequent conspecifics' breeding haunts up to 500-1100 km away (Trierweiler et al. 2007a, Limiñana et al. 2008). The latter behaviour may function as prospecting for next year's choice of breeding site, it may be due to wind drift or it may simply involve a brief return to the natal site.

In general, autumn migration follows broad parallel bands in an arc between SW and S. Despite some overlap, western birds winter on average in western Africa and eastern birds do so further east (Fig. 181; Limiñana et al. 2007). The previous idea that in order to cross the Mediterranean Sea, Montagu's Harriers converge at suitable short sea crossings, such as the Strait of Gibraltar, the numerous central Mediterranean islands, or the Near East (García & Arroyo 1998), is not substantiated by the course of satellite-tracked birds, nor is it apparent from radar studies in southern Spain (Meyer et al. 2003) and visual observations in the central Mediterranean (Panuccio et al. 2005). Radar observations from the southern coast of Spain, 25 km east of Malaga in autumn 1996, showed that 74% of the Montagu's Harriers continued their southbound flight without hesitation upon reaching the coastline. The Mediterranean Sea here is only 150 km wide, which in autumn would take about 4 hours to cross at an average ground speed of 11.6 m per second, or 42 km/h (Meyer et al. 2003). The extremely low wing loading of 2.05 (body mass in kg/wing area in m^2 = 0.300/0.1463), the lowest recorded among the 36 raptor species so far measured (Bruderer & Boldt 2001), and the high aspect ratio of the slender wings allow Montagu's Harriers to alternate between soaring, gliding and flapping-and-gliding flight and to migrate in less than favourable thermal and wind conditions (Spaar & Bruderer 1997). That Montagu's Harriers are able to make long sea crossings is proved from our satellite-tracked birds (see Fig 181, birds crossing the Gulf of

Catching Montagu's Harriers is but one activity, releasing them safely after taking measurements and fitting them with ring(s), tags or transmitter is the next step in a cascade of events related to the life of an individual bird, as illustrated by this colour-ringed adult male released by Assia Kraan, a volunteer from The Netherlands. Tagging and colour-ringing are used in the hope of increasing the number of sightings, something much more desirable than simply hoping for a recovery (which is usually synonymous with a dead bird). This beauty of a male was ringed near Cuxhaven, Germany (July 2008).

Biscay and the eastern Mediterranean near Greece and Crete).

During autumn migration, the average daily distance covered by our satellite-tracked birds was 153 km, a very similar distance to those of other long-distance raptor migrants.[1] One of our males made a nonstop flight from The Netherlands to northern Spain in 2006, showing that nocturnal flights are possible. Montagu's Harri-

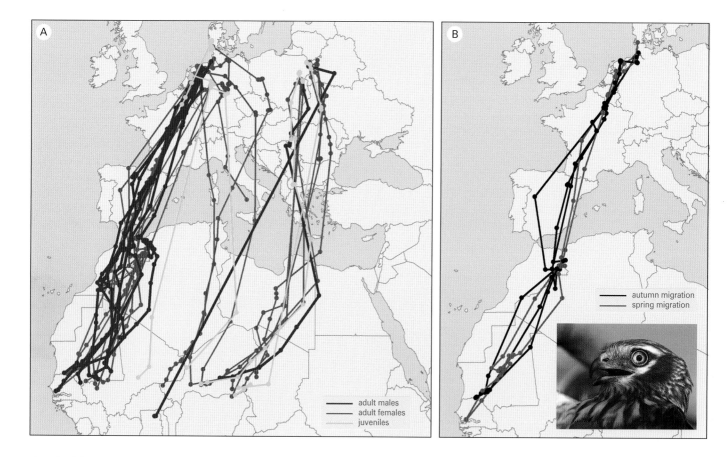

Fig. 181 (A) Representation of 34 flight paths of 25 satellite-tracked Montagu's Harriers during autumn migration in 2005-2008, showing birds from The Netherlands (10), Germany (5), Denmark (2), Poland (4) and Belarus (5); from Dutch Montagu's Harrier Foundation, Vogelwarte Helgoland, University of Groningen and Deutsche Bundesstiftung Umwelt. (B) Successive migratory routes in autumn 2006-2008 and spring 2007-2008, based on satellite telemetry of Merel, a female Montagu's Harrier named after 11-year old Merel Schothorst, our youngest Monty helper in The Netherlands. Lines represent the shortest distances between successive recorded positions (small dots) and not necessarily the actual route (Trierweiler *et al.* 2007a, 2008).

It is 5 May 2009. Female Cathryn has returned to her breeding grounds near Nieuw Scheemda in the northern Netherlands. She is some bird. First captured in 2004, when her nest failed during a spell of inclement weather, she was outfitted with a transmitter in 2006 when at least in her fifth calendar year. Since then, we have been able to closely track her whereabouts. In 2006 and 2008, it took her only 20 respectively 25 days to reach the wintering grounds, about twice as fast as the average Montagu's Harrier. In 2006-2008, she started her return migration like clockwork, *i.e.* around 19 March. Her site fidelity is also outstanding, both in Senegal and The Netherlands. Her consecutive breeding sites since 2006 were 0 km, 2.3 km and 3.5 km apart. In Senegal, she frequents the vicinity of M'bour, the mega-roost of Montagu's Harriers, which was located by tracking down the positions obtained via Cathryn's satellite transmitter. The year 2009 will prove to be a challenge: she picked a winter wheat field as her breeding site, again close to Nieuw Scheemda. However, Common Vole numbers are in a trough, and her breeding success will depend on the bird-hunting skills of her partner.

ers presumably speed up when crossing the Sahara; an adult female tracked in 2005 averaged 623 km in a day when crossing the Sahara (Trierweiler *et al.* 2007a). The satellite-tracked Spanish Montagu's Harriers apparently followed a different strategy, as their daily speed dropped considerably after the first stage of desert crossing in Morocco (more than 450 km in a day). These birds took up to two weeks to complete the remaining 1000-1500 km to the wintering grounds, with average daily distances ranging between 93 and 219 km; they did not fly at night (Liminaña *et al.* 2007). This strategy would imply feeding along the way, but how this could be accomplished in the Sahara is difficult to envisage.

Spring So far, most satellite-tracks have shown clearly that eastern European birds winter to the east of western European birds, with some overlap in Mali and Niger, and little movement along the length of the Sahel during the northern winter, with the possible

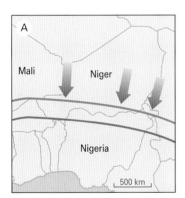 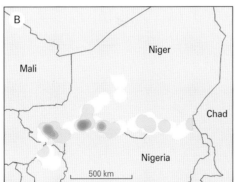 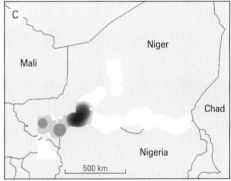

Fig. 182 (A) Schematic representation of winter movements of Montagu's Harriers in the central Sahel. Arrows show the incoming routes of Montagu's Harriers in autumn (based on satellite telemetry). The blue line represents the latitude with the highest concentrations during the early dry season (October-December), and the red line the southward shift in conjunction with the advancing dry season (mostly in January-March), based on data from satellite telemetry and field observations. Adapted from: Trierweiler *et al.* 2008. (B) Relative abundance of grasshoppers in prey transects and (C) of Montagu's Harriers on road transects in southern Niger and northern Benin in January-February 2007, represented as kernel densities; high colour tones reflect high densities (Trierweiler *et al.* 2007b).

A line transect in the magnificent landscape near Hombori, Mali, produced five Montagu's Harriers in second calendar-year plumage, 1 Pallid Harrier and some 100 Fox Kestrels in January 2008. This combination of avian predators is a clear indication that grasshoppers abound in the area.

exception of birds in the central Sahel (see below). Once settled in the Sahel, many birds remain there or reasonably close nearby (a

The satellite transmitter of Montagu's Harrier 'Franz', tagged in The Netherlands, produced such detailed coordinates that this roost near Mopti, Mali, could be located in January 2008; a careful survey yielded some 70 pellets.

few hundred km) for the rest of the winter. Eurasian Marsh Harriers were found to perform a small clockwise circular loop movement in West Africa, notably so between the latitudes of 20°N and 35°N, with the loop narrowing going northwards until it intercepted the autumn route in northern Spain (Klaassen et al. 2008b). In West Africa, the widest separation between autumn and spring routes was at 20°N, amounting to slightly more than 400 km. None of the Swedish Marsh Harriers showed any inclination to make the spring crossing of the Mediterranean in the central section (Cap Bon-Sicily), which accords with the spring migration of Eurasian Marsh Harriers which have wintered in West Africa following a more westerly flyway than in autumn (Klaassen et al. 2008b). So far, our NW-European Montagu's Harriers also refrained from using the central Mediterranean corridor on their return migration (www.grauwekiekendief.nl). Instead they retraced their outward route, except in West Africa, where they drifted slightly to the east or west of the autumn track. In spring 2008, eastern European birds, having wintered in the central Sahel, performed a clockwise loop which took them to the central Mediterranean on return migration. This loop was unrelated to locust movements; the Montagu's Harriers had been more or less sedentary in their wintering quarters where they fed mostly on resident grasshoppers (migratory locusts are prominent in the diet only during

Melanistic male (left, March 2008) and a bird with a Spanish ring (right, January 2009) near temporary pool at Khelcom, Senegal. Among the 1000 or so Montagu's Harriers on the roost near Dara (Khelcom), 30 melanistic birds were present.

outbreaks; see below). Furthermore, any significant movement associated with grasshopper supply occured along a north-south axis (as displayed by satellite-tracked Montagu's Harriers; Trierweiler *et al.* 2008), following the retreating Intertropical Convergence Zone (ITCZ). Whether a clockwise loop migration is consistent across years has yet to be proven. The suggestion by Klaassen *et al.* (2008b) that the clockwise loop migration of Eurasian Marsh Harriers in West Africa is influenced by the wind patterns in spring (predominantly from eastern sectors, and stronger in spring than in autumn; see also Chapter 42), may also hold for Montagu's Harriers (but see tracks of female 'Merel' at Fig. 181. In the central Mediterranean, visible migration of Montagu's Harriers in spring is more conspicuous than in autumn (unlike at Gibraltar, where migration intensity in both periods is about the same; Tables 3 & 4 in Finlayson 1992), but counts here are very small in comparison to the total volume of Montagu's Harrier migration (Panuccio *et al.* 2005). If this seasonal pattern indeed exists in the central Mediterranean, it must then involve Montagu's Harriers from northern and – especially – eastern Europe, which enter Africa via the eastern and central Mediterranean and return via a clockwise loop in spring. Indeed, five birds ringed at Cap Bon, Tunisia, in spring, were recovered in Hungary, Bulgaria (2), Ukraine and the Voronezh region (SW Russia), having taken directions between NNE and ENE (recovered in the same year, or up to six years later; Mihelsons & Haraszthy 1985).

Fidelity to wintering sites

During the northern winter, many European Montagu's Harriers, presumably the majority, are confined to the Sahel and northern Sudan zone (Fig. 181), but until recently next to nothing was known about their temporal distribution and movements within the wintering quarters. The predominance of insects in Montagu's Harrier diets would suggest a dynamic distribution related not only to outbreaks of locusts and grasshoppers but also to seasonality and movements of local grasshopper species triggered by the cyclic position of the ITCZ (see Chapter 2). In other words, birds may be forced to exploit different sections of the Sahel and northern Sudan zone each year and to perform seasonal movements within years (as suggested by Thiollay 1978c).

Between-year movements   Three tagged birds, which were followed for three consecutive seasons, returned to the same spots in Mali and Senegal (see Fig. 181 for an adult female). This sample is, of course, too small to conclude that fidelity to the wintering site is the norm for this species in Africa. On the population level, the high migratory connectivity demonstrated by ring recoveries and satellite telemetry of birds from western Europe strongly indicates the fidelity of Montagu's Harriers to restricted parts of the Sahelian and adjoining Sudan zones (García & Arroyo 1998, our data). A similar conclusion was derived from the ringing data of Eurasian Marsh Harriers (Chapter 25), a species for which satellite telemetry hinted at a stronger site fidelity to stopover and wintering areas than to breeding grounds (Strandberg *et al.* 2008a).

Within-year movements   The first results from our tagged birds indicate that upon arrival at the wintering quarters the birds remain for several weeks or months within a few km of the places where they first settle. A gradual southward shift begins as the dry season progresses, over some 200-250 km, a distance that bears testimony

Montagu's Harriers may form pre-roost gatherings on the ground, from which the birds depart collectively when the time has come to approach the actual roost site. (Fatick, Senegal, February 2008).

to several factors: the narrowness of the Sahel belt, the increasingly desiccating conditions in the Sahel that start in September and end with the first rains in May, and the change in abundance of birds and grasshoppers in conjunction with the southward shifting 'green belt' (Fig. 182A, see also Chapter 14, Jones 1999). This process would explain the Montagu's Harriers' movements into the adjacent Sudan vegetation zone.

Habitat use [2]

Montagu's Harriers in Africa are essentially birds of dry ground that is sparsely covered by trees. During road transect surveys in West Africa from 1967 to 1973, Thiollay (1977) found clear latitudinal gradients for the presence and density of Montagu's Harriers. Examples from his data are: absence of the species in the forested regions around 6°N in the Ivory Coast; near-absence in the well-wooded Guinean zone (0.02 birds/100 km); low densities in the wooded savanna of the Sudan zone (between 9.30°N and 14.30°N: 0.72-1.17 birds/100 km); the highest densities in the Sahel (which included the inundation zones of the Niger and Senegal: 0.79-3.11 birds/100 km; dry savanna is preferred over wet floodplains – see Chapter 25 for niche differentiation between harriers in the Inner Niger Delta); and, lastly, densities rapidly declined to zero in the northern Sahel close to the Sahara (near 20°N).

Within the Sahel, densities showed large variations, supposedly in relation to prey abundance (Orthoptera) and habitat. Our road counts in Niger in 2006-2007 revealed that large stretches of land were devoid of Montagu's Harriers; line transects often failed to come up with any grasshoppers there. Montagu's Harriers were most commonly encountered where grasshoppers were abundant (Fig. 182c).

In Niger, Mali and Senegal, Montagu's Harriers avoided severely degraded habitats (few remaining trees or shrubs and overgrazed), and regions with high tree cover, but favoured slightly degraded shrubland and cropland. The latter habitats, often having retained elements of more natural habitats, had the highest bird and grasshopper densities, a finding in concurrence with bird densities in northern Nigeria (Hulme 2007). In Niger, tiger bush is an important natural habitat for Montagu's Harriers, but such a patterned vegetation community, with alternating bands of trees and shrubs separated by bare ground or low herb cover (resembling the stripes on a tiger), is rapidly giving way to cultivation. The widespread replace-

Fig. 183 Cumulative number of Montagu's Harriers entering the roost near Darou Khoudouss, Senegal, on 3 February 2008 until 19:14 (local time); sunset by 18:32, civil twilight by 18:56, nautical twilight by 19:23.

ment of natural habitats by a mixture of degraded natural habitats (fewer trees, more scrubland, and low herb layer) and cultivated land (with some shrubs and trees) may have favoured Montagu's Harriers. However, such a modified landscape is not a steady state. Ongoing cultivation and human population pressure result in severe and irreversible degradation. Tree and shrub loss and impoverished bird and insect life are in evidence over much of the Sahel nowadays, a circumstance that has negatively influenced almost the entire raptor community in Mali, Niger and Burkina Faso, with Montagu's Harriers particularly affected (Thiollay 2006a, 2006c).

Roosts as treasure-troves [3, 4]

Daytime roosts   During the hottest part of the day, many Montagu's Harriers retreat into the shade of trees or shrubs, either singly or in small flocks. One such roost we accidentally encountered in January 2007 near Birni N'Konni in Niger. Around 14:00, we spotted twelve Montagu's Harriers, each in a different shrub without undergrowth, on the slope leading to a plateau. Underneath some shrubs, we found pellets and faeces, but in such small numbers that we suspected the site was used only during daylight. In February 2008, another typical daytime roost was found near Kaolack in Senegal, close to a salt lake. The day had been very hot, and we were desperately in search of some shade. Our bias towards shade-bringing bushes revealed an adult female Montagu's Harrier standing on bare ground in a small patch of shade beneath a shrub. This site produced some 20 prey remains of the grasshopper *Ornithacris cavroisi* and also some freshly-plucked Yellow Wagtail feathers. Because the only raptor species we observed here were Montagu's Harriers, we feel it safe to assume that the prey remains had indeed been left by this raptor.

Night roosts   The conventional way to find roosts is to watch for Montagu's Harriers flying in steady, straight lines before sunrise or just after sunset (Fig. 183). Usually, more than one bird will display this 'atypical' flight behaviour, not at all representative of that of foraging birds. The latter adopt a more roaming flight mode, meandering in low quartering searches across the terrain, switching direction and stalling frequently. Roost flights head clearly to or from a particular spot. However, to find the roosts one needs a number of well-spaced observers, preferably in contact with each other (walkie-talkie), to plot the general directions of the roost flight. Roosts can then be located either by triangulation of the extrapolated flight paths or by following the birds. This approach to finding roosts is most likely to be successful when roosts are large (the more harriers observed, the greater the chances of encountering roost flights).

By tracking our transmitter-carrying Montagu's Harriers, using the most up-to-date positions received each day, we were able to locate roosts in a completely different way that often pointed us towards areas which we would not otherwise have visited, because the logistic problems would have seemed too daunting. Furthermore, from the telemetry we detected that some birds, like an adult Polish female residing east of Niamey, Niger, daily switched between roost sites, some of which we managed to reach; at one such site we found the remains of a freshly-eaten grasshopper *O. cavroisi*. A tagged juvenile female, from The Netherlands, led us to find, near Niamey, a 700 ha plateau containing a roost of 2 or 3 Montagu's Harriers. A particularly spectacular discovery near Mopti, Mali, was a roost containing some 30 Montagu's and 5 Eurasian Marsh Harriers in a well-wooded agricultural area dominated by tall grasses, not exactly the type of habitat where we would have expected to find Montagu's Harriers. At some telemetry-indicated locations, we found pellets before we had actually seen a single Harrier, but on 24 January 2008,

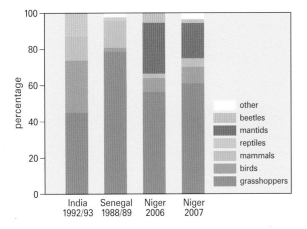

Fig. 184   Frequency distribution of prey categories in pellets of Montagu's Harriers in Gujarat, NW India (n=134 pellets; Clarke 1993), in Senegal (n=113; Cormier & Baillon 1991), and in Niger (n=41 in 2006, n=28 in 2007; Koks et al. 2006, Trierweiler et al. 2007b).

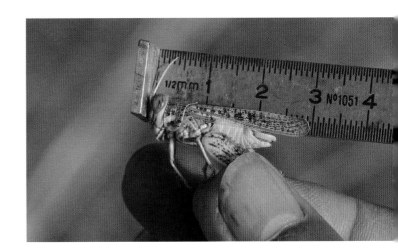

The relatively small grasshopper *Acorypha clara* was the main prey of Montagu's Harriers, Cattle Egrets, Lesser Kestrels, Wheatears and Woodchat Shrikes in central Senegal in February 2008.

The presence of predators, like Golden Jackals *Canis aureus*, may be one of the reasons why harriers prefer to roost in tall grasses (as found for Eurasian Marsh Harriers in India; Verma & Prakash 2007), or – when resting or drinking during daytime like this male Montagu's Harrier – prefer an unobstructed view. (Khelcom, Senegal, March 2008).

between 18:50 and 19:40, we recorded Montagu's Harriers heading inwards from every direction. Around 19:30, 'Franz', a seventh calendar-year male, came into view, the satellite transmitter clearly visible on his back. To watch this Dutch bird, whose nest in a field of lucerne we had successfully protected in the preceding summer, in a Malinese setting, made us feel we had received an accolade for our fieldwork. As a bonus, at this site we were able to collect 70 pellets, whose contents of small mammals and reptiles, passerine eggs and a few small grasshoppers were not that different from the average pellets in an average year in NW-Europe.

Satellite-generated data revealed the positions of three roosts in Senegal; two contained 'only' 100-200 birds, but the third was enormous in comparison. The existence of this mega-roost was first detected by Wim Mullié in the evening of 25 November 2006. Although its exact location remained hidden, 500-1000 birds were observed near Darou Khoudoss, just before sunset, flying to a roost. The surroundings of this site were revisited in the evening of 2 February 2008, when a pre-roost gathering of 90 birds on bare farmland was noted. Although the light was failing quickly, the actual roost was conservatively estimated as holding about 1000 birds. Between 17:40 and 19:10 on the next day we counted more than 1000 birds (Fig. 183). At 19:14 it was almost dark, but, as luck would have it, all the birds began to mill around again just before darkness prevented any further observations; at this time we noted 300 more birds on the side of a hill not yet surveyed. Our totals came to 1300 Montagu's Harriers and a few Marsh Harriers; perhaps 1500 birds might have been present. We could not be any more precise, because birds were still entering the roost moments before complete darkness descended and we had to leave the area for security reasons.

Until recently, in Africa, the largest roosts mentioned have not usually exceeded 70-160 birds for Senegal (Arroyo & King 1995, Rodwell *et al*. 1996) and well over 200 birds for Kenya (predominantly Montagu's Harriers; Meinertzhagen 1956). An apparent exception was the count on 8 February 1989 recorded for a roost in the area of M'Bour and Joal (delta of Sine Saloum, Senegal) of 800-1000 Montagu's Harriers; this large number was associated with an outbreak of Desert Locusts *Schistocerca gregaria* (Cormier & Baillon 1991). In February 2008, Montagu's Harriers on the Darou Khoudouss roost profited from an abundant supply of the medium-sized *Acorypha clara* (highest densities of 3-5 individuals/m^2); pellets contained this grasshopper species almost exclusively; in March, the larger *O. cravoisi* became the predominant item in the diet of Montagu's Harriers (Fig. 123; Chapter 14).

Since at least the mid-1980s in the Indian subcontinent, even larger roosts of Montagu's Harriers have been recorded by Clarke (1996), who found up to 2000 birds in Bhavnagar District in NW India. This roost was estimated as holding some 3000 harriers on 6 December 1997; 15-25% comprised Pallid Harriers and a few Eurasian Marsh Harriers, but the great majority were Montagu's Harriers (Clarke et al. 1998).

The role of resident grasshoppers

Both the general literature and the few real data collected in the field suggest that the locusts Locusta migratoria and Schistocerca gregaria are of crucial importance as food for acridivorous birds, including Montagu's Harriers (Brown 1970, Thiollay 1978c, Cormier & Baillon 1991, see also Chapter 14). These locust species are certainly abundant in some years at some sites, but the available evidence clearly shows that resident grasshopper species are of far greater importance to Montagu's Harriers, especially because their cumulative availability, usually in high numbers, is more stable within seasons and across years, thus representing a reliable food source for acridivorous species. The infrequent outbreaks of migratory locusts provide a stark contrast in annual availability, for, characteristically, locusts remain in very low numbers or are absent during the extensive periods of recession or remission. Furthermore, the frequency of such outbreaks has been much reduced since 1965 (Fig. 114), often occurring outside the window of presence of Palearctic migrants. Indeed the highest frequency of S. gregaria in the Sahel occurs from July until December; Chapter 14).

In Niger, both in 2006 and 2007, the most frequently consumed prey were Orthoptera, mostly comprising the resident grasshopper species O. cavroisi, but, surprisingly, mantids did form a large share of prey numbers (Fig. 184). Birds and mammals were insignificant in terms of numbers, but, individually weighing much more than a single grasshopper, were obviously more important in terms of biomass. A high frequency of O. cavroisi in Montagu's Harrier pellets was also recorded in 2008 (Niger, Senegal; analyses not yet fully completed), attesting to the importance of medium-sized (3-7 cm) and large (>7 cm) resident grasshopper species for Montagu's Harriers (and other acridivorous bird species, such as White Storks; Brouwer et al. 2003).

Montagu's Harriers wintering in the central and western Sahel have proved to be quite versatile in their choice of prey. In February 2008, for example, we found large differences in prey choice between regions, presumably reflecting local variations in food supply. Pellets collected in Niger, Mali and Senegal contained small insects (termites, beetles), small, medium-sized and large Orthoptera, rodents, passerines, eggs and reptiles (see also Fig. 123).

Our data suggest an interpretation that is a far cry from the notion of a diet predominated by locusts. Just as in NW India (Fig. 184), the Montagu's Harriers in the Sahel make do with whatever is available.[5] On the breeding grounds in Europe, Montagu's Harriers also forage on an assortment of prey species, predominantly passerines in Britain (Clarke 2002), voles and passerines in The Netherlands (Koks et al. 2007) and France (Millon et al. 2002) and birds and insects in Spain (Sánchez-Zapata & Calvo 1998). However, between-year differences are substantial, in Europe as well as in Africa. The predominance of locusts in the diet of Montagu's Harriers in Senegal in 1988/1989 perhaps reflected a dietary exception rather than the rule; even then, despite the abundance of locusts in the outbreak that year, rodents remained an important food source (Fig. 184). Clearly, the story of the importance of locusts to acridivorous bird species in Africa needs revising to account for the dietary prominence of resident grasshopper species (see also Chapter 14, for an elaboration on this issue) and alternative prey.

Population change

Wintering conditions   That Montagu's Harriers wintering in the Sahel have been in decline is shown by a comparison of road counts in Mali, Burkina Faso and Niger in 1969-1973 and 2003-2004 (down by 74% for unprotected areas and National Parks combined; Thiollay 2006a). The same effect had earlier been suggested for East Africa: "I have little doubt that some disaster has stricken the population that used to come to East Africa", wrote Leslie Brown (1970). Some support for that view for southern Africa may also be gleaned from observations in the Transvaal's Nyl River floodplain, where Tarboton & Allan (1984) recorded eight birds in 1959-1970, but none in 1975-1981. The information collated by Clarke (1996b) seems to suggest that – at least in East Africa – numbers since then may have recovered to some extent. However, according to Simon Thomsett (in litt.): "Brown's large roosts are no more." In southern Africa, the species is now relatively rare throughout the region except in Botswana, where it is fairly common in the north (Harrison et al. 1997).

How numerous Montagu's Harriers may once have been in East Africa, is apparent from the casual observations of Meinertzhagen (1956) in Kenya.[6] On 17 January 1956, he recorded 17 harriers (predominantly Montagu's) during a motor journey of 200 miles between Isiolo and Marsabit (5.3 birds/100 km), and another 11 individuals in February on another 140-mile motor journey in the Rift Valley (4.9/ birds/100 km). In the Sahel, such densities have rarely been encountered, and, if so, only in the very best habitats and many decades ago (Thiollay 1977). Our own road counts, for example, in southern Niger in January and February 2006 and 2007, yielded provisional densities of 0.43 (4172 km) and 0.52 birds/100 km (4950 km) respectively. The road counts by Thiollay (2006a) in the western Sahel in 2003-2004 produced between 0.7 and 0.9 Montagu's Harriers per 100 km.

The trends of European populations wintering in the western Sahel fluctuate independently of rainfall in the Sahel (Fig. 185), the variable affecting green vegetation and therefore the food supply

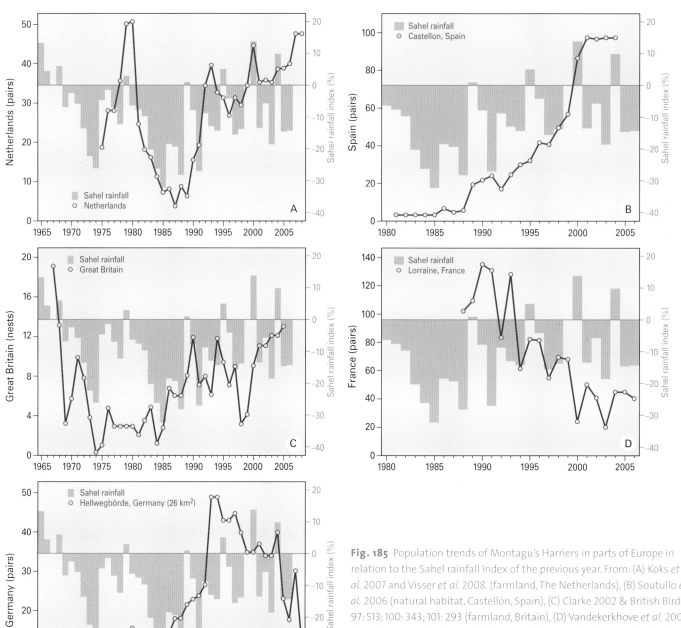

Fig. 185 Population trends of Montagu's Harriers in parts of Europe in relation to the Sahel rainfall index of the previous year. From: (A) Koks et al. 2007 and Visser et al. 2008. (farmland, The Netherlands), (B) Soutullo et al. 2006 (natural habitat, Castellón, Spain), (C) Clarke 2002 & British Birds 97: 513; 100: 343; 101: 293 (farmland, Britain), (D) Vandekerkhove et al. 2007 (farmland, Lorraine, France), (E) Hölker 2002 and Illner in litt. (farmland, Hellwegbörde, Germany). The graphs show a wide variety of trends, driven largely by local conditions on the breeding grounds. All studies refer to regions where nests are being protected when necessary.

of Montagu's Harriers. The trends of the Dutch Montagu's Harriers suggest an effect of rainfall in the Sahel on numbers, but this is an artefact of conditions on the breeding grounds (see below). Also, we were unable to replicate the significant positive correlation between the number of Montagu's Harrier nests found in Britain and the West African rainfall anomaly (Clarke 2002), using a longer time series and the relative change in breeding numbers from one year to the next. Despite large-scale habitat degradation in the Sahel (Thiollay 2006a, 2007) and elsewhere in Africa (Fishpool & Evans 2001), we have little evidence that the ups and downs in Europe are, as yet, triggered by such events. This may change with ongoing habitat loss.

Breeding conditions Montagu's Harriers are among the best-studied raptor species in Europe. Many of these studies started only in the

A male Montagu's Harrier quartering the windswept farmland of Groningen in the northern Netherlands, with the village of Noordbroek in the background (5 May 2009). In past years, our radio-tagged birds have shown that hunting is far from random, the birds showing a preference for set-aside and recently mowed fields where voles are more abundant and visible than in the surrounding farmland.

1970s or even later (examples in Fig. 186). Statements based on short-term trends (such as an increase in 1970-1990; BirdLife International 2004a, see Box 27 in Chapter 44) can be quite misleading when not viewed against the backdrop of historical data. In The Netherlands, for example, the population was estimated at 500-1000 pairs in the first half of the 20th century, but this had declined to a handful of pairs by the late 1980s (Bijlsma 1993); the subsequent increase to more than 40 pairs in the 2000s is a reminder that this recovery – heartening in itself – still represents only a small fraction of a once large population. The all-time dip in the Dutch population accidentally coincided with the Great Drought in the Sahel (1980s). A closer look at the data, however, reveals that the trends in The Netherlands are driven by local conditions. For example, the embankment of Zuidelijk Flevoland in 1968 created good breeding habitat and high food abundance, resulting in peak numbers around 1980, but the cultivation of that area in subsequent years caused loss of habitat, with a consequent steep decline of Montagu's Harriers to near-extinction in the early 1990s. However, a sudden upsurge in numbers came after fallow land had been introduced as a measure to counter overproduction in agriculture (part of the European Common Agricultural Policy; Pain & Pienkowski 1997). The ensuing increase in The Netherlands was actively assisted by nest protection and agri-environmental schemes in farmland (Koks & Visser 2002, Trierweiler *et al.* 2008).

In many regions in Europe, natural breeding habitats have disappeared, forcing Montagu's Harriers into farmland where, because of their late laying date, the onset of the cereal and lucerne harvests threaten nest survival (Corbacho *et al.* 1999). Nest protection is imperative to prevent large-scale nest failure, particularly so in western Europe, where the timing of the cereal harvest has much advanced, *e.g.* by about two weeks in Lorraine, France, between 1988 and 2006 (Vandekerkhove *et al.* 2007) and by a month in The Netherlands between 1968 and 2008 (Bijlsma 2006c, unpubl.). Without nest protection in farmland, reproductive output cannot sustain existing population levels in farmland populations (Koks & Visser 2002, Millon *et al.* 2002, Vandekerkhove *et al.* 2007). On average, 60% of the farmland nests would be destroyed in the absence of nest protection; typical estimates range from 41% to 98% in 14 regions in France, Portugal and Spain (Arroyo *et al.* 2002, Millon *et al.* 2002).

The human effort involved in nest protection throughout Europe is enormous. In France, for example, 40-50 groups in 60 districts have been active in nest protection each year; the combined effort safeguarded 11 000 nests from destruction, enabling 22 000 nestlings to fledge in 1976-2001 (Pacteau 2003). This massive involvement is estimated to cover between 7.5% and 17% of the French Montagu's Harrier population annually, but is apparently insufficient to stop the negative population trend in much of France (Pacteau 2003, Thiollay & Bretagnolle 2004).

Even in the tiny Dutch farmland population (16-48 pairs in 1990-

2008), whose nests are protected each year when necessary, nest protection and habitat improvement only just sustain a more or less stable population (Koks & Visser 2002). Colour-ringing has shown some exchange between breeding clusters in The Netherlands and northern and eastern Germany (Visser et al. 2008). Pre-migratory dispersal may be one of the mechanisms by which potential breeding habitats over a wide vicinity are being explored and tested (Limiñana et al. 2007, Trierweiler et al. 2007a). Food abundance, in particular of Common Voles *Microtus arvalis* (in Britain: Field Vole *M. agrestis*) and small passerines, may then serve as the trigger for settlement (Salamolard et al. 2000, Arroyo et al. 2007, Koks et al. 2007). The scarcity of formerly common prey species in Europe's industrialised farmland strains the already precarious status of European Montagu's Harriers. Nest protection should therefore be merged with improvement of harrier habitat in farmland (Millon *et al.* 2002, Koks *et al.* 2007), in addition to the preservation of natural breeding habitats, where reproductive output is – at least in Spain - better than in farmland (Limiñana *et al.* 2006).

Conclusion

The increase of Montagu's Harriers recorded in Europe between 1970 and 1990 represents only a small recovery from losses incurred earlier in the 20th century. Destruction of natural breeding habitats, and a subsequent shift to breeding in farmland, has had an overriding influence on the fortunes of this species in the 20th century. Harvesting often prevents pairs breeding in cropland from successfully raising chicks, unless nests are being protected. Modern farming has also had a devastating impact on prey abundance (notably voles and passerines). Without habitat improvement, nest protection in farmland cannot reverse negative trends.

Montagu's Harriers winter mostly in the Sahel, where the east-west distribution largely mirrors the longitudinal distribution of the breeding grounds. During that period they may be forced southwards by the desiccating conditions of the Sahel into the southern Sahel or northern Sudan zone. Their main food comprises many resident grasshopper species (with different phenologies and occurring in local outbreaks), which remain a reliable food source throughout the dry season, a diet complemented by passerines, mantids and small mammals; outbreaks of migratory locusts are but a rare dry-season opportunistic food source. Large harrier assemblages may occur wherever small mammals, resident grasshoppers or locusts abound. The present habitat degradation in the Sahel is likely to increase in extent, but for the time being may favour resident grasshopper species and hence Montagu's Harriers. The impact of Sahel rainfall and habitat degradation on Montagu's Harriers is overriden by the much greater land changes which have occurred in Europe in the 20th century.

Endnotes

1 Many medium-sized and large raptor species have now been tracked by satellite, providing information on migration speeds relative to sex, age and season (see table). On average, birds attain higher speeds during the return migration (but see Swainson's Hawk below), adults move faster than juveniles and immatures, and speeds over deserts are higher than over more hospitable land (European Honey Buzzard, Short-toed Eagle, Eurasian Marsh Harrier). Body weight is expressed in grammes (averaged for male and female; Dunning 1993). Distance relates to the one-way distance between breeding and wintering grounds in km (based on tracked birds), origin to the breeding site. Daily migration distances (in km) are given for the entire southbound or northbound migration period (including stopovers), with number of tracked birds in brackets. Notice that Wahlberg's Eagle is the only species which was tagged in the southern hemisphere, and for which the outward migration is to the north.

2 Inspired by the work of Jean-Marc Thiollay in West Africa, we adopted the method of road transects to detect spatial and temporal variations in abundance of Montagu's Harriers. Travelling at a maximum of 60 km/h, we counted all raptors systematically along roads, dirt roads and tracks while estimating the distance of each raptor from the road (Trierweiler et al. 2007b). Whenever a Montagu's Harrier was seen, we stopped to check whether there was more than one bird, but if so, they were not included in the road count itself. Since starting in SW Niger in the northern winter of 2005/2006, we have covered more than 15 000 km of road transects in Niger, Mali and Senegal. We also used the road counts to record habitat types (every 5 km) and habitat degradation (scored thus: no, little or much, the categories related to pre-determined values for the extent of undergrowth, erosion and tree-felling). In addition, we collected data on prey densities by walking line transects at least 30 m long, recording birds (species and numbers, within 20 m on either side), active burrows of small mammals (rodents, burrows <3 and >3 cm diameter, within 1.5 m on either side), reptiles (within 1.5 m on either side) and grasshoppers (lengths <3 cm, 3-7 cm or >7 cm, within 1.5 m on either side). Samples of grasshoppers were collected on the spot, to be identified at a later date and to be used as references. We used these data to calculate relative prey densities. Between 2006 and 2008, we collected data from more than 1100 prey transects (Trierweiler et al. 2007b, unpubl.).

3 'Roost' is used here in the sense of any place where one or more individuals spend time resting or loafing, at night or during daylight.

4 Finding roosts, and by doing so, finding pellets, is essential when investigating the diet of Montagu's Harriers in the wintering areas. Large roosts (hundreds of birds) are easier to find than small roosts (a few birds, sometimes only one). However, investigating only the large roosts will bias the results, because large roosts indicate areas of high food abundance, where the diet will usually be less diverse than in areas of lower food abundance. In the latter areas, Montagu's Harriers normally have a more diverse diet. The small body of literature on the food of wintering Montagu's Harriers is strongly biased towards large roosts in areas subject to outbreaks of *Schistocerca gregaria* (Cormier & Baillon 1991, Arroyo & King 1995); this bias is often exacerbated by small sample sizes and short sample periods. A second bias in dietary studies relates to the techniques used to study diets: pellet analysis, collection of prey remains, visual observations or

Species	Body weight	Distance	Origin	South	North	Source
Osprey	1486	6393	Sweden	162 (12)	244 (8)	Alerstam et al. 2006
Osprey	1486	5260	Scotland	168 (7)	236 (3)	Dennis 2008
Osprey	1486	4958	USA	241 (52)	-	Martell et al. 2001
European Honey Buzzard	758	6709	Sweden	148 (8)	-	Hake et al. 2003
Egyptian Vulture	2120	4160	France, Bulgaria	194 (3)	-	Meyburg et al. 2004a
Short-toed Eagle	1703	4365	France	234 (1)	-	Meyburg et al. 1996, 1998
Montagu's Harrier	316	5000	Europe*	153 (16)	-	C. Trierweiler et al. in prep.
Eurasian Marsh Harrier	628	4243	Sweden	127 (23)	161 (13)	Strandberg et al. 2008a
Broad-winged Hawk	455	6998	North America	69 (3)	105 (1)	Haines et al. 2003
Swainson's Hawk	989	12728	North America	188 (27)	150 (19)	Fuller et al. 1998
Lesser Spotted Eagle	1370	8725	Central Europe	164 (5)	177 (3)	Meyburg et al. 1995a, 2001, 2004b
Wahlberg's Eagle	640	3520	Namibia	214 (1)	185 (1)	Meyburg et al. 1995b
Hobby	240	9635	Sweden	151 (4)	-	Strandberg et al. 2008b
Eleonora's Falcon	390	8600	Italy, Sardinia	134 (1)	293 (1)	Gschweng et al. 2008
Peregrine Falcon	780	3841	Russia, Kola	190 (2)	-	Ganusevich et al. 2004
Peregrine Falcon	780	8436	North America	172 (22)	198 (7)	Fuller et al. 1998

* Consisting of birds breeding in Germany and The Netherlands (12), Poland (2) and Belarus (2).

video recordings at the nest. Each method in itself under- or over-represents certain prey categories (Schipper 1973, Simmons et al. 1991, Underhill-Day 1993, Sánchez-Zapata & Calvo 1998, Redpath et al. 2001, Koks et al. 2007), and so a combination of methods is usually considered to provide the smallest bias. Our Sahelian dietary studies are largely based on pellets.

5 An eye-witness account of Chris Magin may serve to illustrate that Montagu's Harriers in Africa also capitalise opportunistically on sudden outbursts of food. Returning to Addis Ababa from Lalibella, Ethiopia, in late January/early February 2008, he took the lowland route from Dese, descending into the Afar plains. "As we headed south to Awash NP we passed through Yangudi Rasa NP, which straddles the main road. The grassy plains were extremely well vegetated (i.e. the rains a few months previously must have been extremely good) and the grass was absolutely swarming with a super-abundance of small rodents. Every step you took seemed to cause one to scuttle for safety. I cannot say with certainty what species they were. The skies were alive with raptors, almost entirely harriers. I was so impressed with the numbers that at one point I attempted a rudimentary census, scanning a 90 degree arc very slowly with my 10x50 Zeiss Jenoptem binoculars and counting all harriers seen up to the horizon. I counted 125 in this quarter horizon, so estimated that there were around 500 present within the limits of my visibility. I could see quite far – although it was the middle of the day, this was the cool season, so relatively little heat haze – so probably I could pick out harriers up to 3-4 km away. As there were no locusts or large grasshoppers present and most of the harriers were quartering low to the ground, I assumed that they were congregating to feed on the 'exploded' rodent population. The harriers would have been between the hamlet of Gewane and the junction with the Dese road. I also assume that they would have been within the Yangudi NP limits, if only because the amount of good grazing would have attracted hordes of nomads and their livestock herds if it had been outside the NP boundaries. The harriers I saw must have been Pallid and Montagu's Harriers, which are common in the area."

6 The reputation of Richard Meinertzhagen, ornithologist extraordinaire in his time, as being a reliable source of information has crumbled in recent years (review in Garfield 2007). Painstaking research by Alan Knox, Robert Prŷs-Jones, Pamela Rasmussen and Nigel Collar has shown that "much that he left us cannot be taken at face value" (Knox 1993). His note on harriers in Kenya, however, has a ring of credibility, since it was published within months of the actual observation. This is also evident from the remark of Simon Thomsett (in litt.), born and raised in Kenya and intimately familiar with the raptors there: "I believe Meinertzhagen in this case." Garfield (2007) documents that most of the numerous fabrications occurred long after the event, but concedes that, "All the same, his bird writing, even if sometimes factually wrong, usually tends to be much more dependable and certainly more plausible – therefore less irritating – than his military or political memoirs."

Black-tailed Godwit
Limosa limosa

Under a cloudy sky scattered groups of farmers are tilling the land. A heavy shower has just blown past. Suddenly, the air is filled with a familiar sound, announcing a seasonal marker: Black-tailed Godwits are plummeting down in their characteristically dizzy way to alight nearby. After an absence of five months during which they bred in northwest Europe, they are back in their main non-breeding haunts along the West African coast. Here, in the rice fields of the Casamance, farmers cast resentful looks at returning godwits. They are considered a pest locally, because the birds may inflict damage to the recently-sown crop. At the other end of the Black-tailed Godwit flyway in The Netherlands, farmers welcome them as harbingers of spring. Wet grasslands below sea level became an attractive substitute for the rapidly dwindling natural habitats in flooded marshes, peat bogs and heaths, especially since artificial fertilizers have improved the food supply for Black-tailed Godwits. However, this ecological heaven has become a reproductive hell; in the late 20th century intensive farming practice, which has long replaced more traditional ways of farming, has altered their breeding habitat, making the seemingly attractive meadows a trap in spring – they still induce nesting and egg laying, but fledging success has steadily reduced because the date of first mowing has advanced by at least three weeks. Consistently poor breeding success and other problems associated with modern farming are the lot facing modern-day Black-tailed Godwits on their nesting territory.

Breeding range

The breeding areas of Black-tailed Godwits range from western Europe to eastern Siberia, where the species occupies temperate and boreal lowlands, roughly between 45° and 60° N (Cramp & Simmons 1983). The European population consists of two subspecies: the nominate *L. l. limosa,* which breeds almost entirely on the continent, and *L. l. islandica*, which has geographically separated populations on Iceland, (very) small numbers in Scotland (Shetland, Orkney) and in NW Norway (Thorup 2006). The population of Icelandic godwits is thriving and combines a steady increase in numbers with an expansion of the breeding range on Iceland (Gunnarsson *et al.* 2005). The population growth here is attributed to climate change from which the species benefits, mostly in its breeding range (Gunnarsson *in prep.*). In winter, Icelandic godwits are confined to coastal habitats in Ireland, Great Britain, The Netherlands, France and the Iberian Peninsula, while continental godwits winter mostly in sub-Saharan Africa (Fig 186; Beintema & Drost 1986, Gill *et al.* 2002). Black-tailed Godwits from E Siberia (east of the Yenisey River) are distinguished as a subspecies, *Limosa l. melanuroides*, and winter in coastal areas in India, SE Asia and Australia (Cramp & Simmons 1983, del Hoyo *et al.* 1997).

Continental godwits breed across much of lowland Europe, but their stronghold is The Netherlands (*c.* 45 000 -50 000 pairs in the early 2000s; SOVON), smaller numbers being found in northern Germany (*c.* 6000 pairs; Hötker *et al.* 2007). These godwits breed in open, moist and wet grasslands used for dairy farming. In Central and Eastern Europe, godwits predominantly inhabit semi-natural habitats such as wet pastures and floodplains. Significant numbers are found in Belarus (8500 pairs), Ukraine (7500-14 000 pairs) and western European Russia (15 000-32 000 pairs; Birdlife International 2004a, Thorup 2006). P. Tomkovich (*in* Thorup 2006) reports another population from Kazakhstan, where up to 5500 moulting birds have been recorded in the Tengiz-Korgalzhyn region (Wassink & Oreel 2007).

Distribution in Africa

Before Haverschmidt (1963) wrote his essay on Black-tailed Godwits, and before pioneering French ornithologists published their observations from West African wetlands (Roux 1959a & b, Morel & Roux 1966a & b), West European godwits were thought to winter in NW Africa in the Mediterranean basin (Witherby *et al.* 1940, Bannerman 1960). They did (and still do), but the overwhelming majority was soon found to stay in sub-Saharan Africa, where open, fresh to marine wetlands are favoured, and large flocks can be found in shallow waters, on emerging floodplains and in rice fields. Black-tailed Godwits are known to feed largely on rice grains (Tréca 1975, 1977, 1984), but they also take seeds of *Echinochloa* sp. and *Cyperus* sp. (Guichard 1947), and ostracods, small macrofauna and insects (Tréca 1984, Altenburg & van der Kamp 1985, van der Kamp *et al.* 2006). In the Inner Niger Delta pre-migratory weight gain depends mainly on the consumption of small bivalves (*Corbicula*, see Box 7).

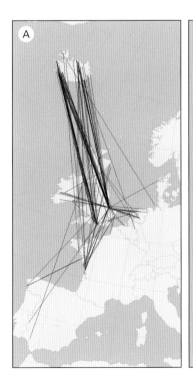
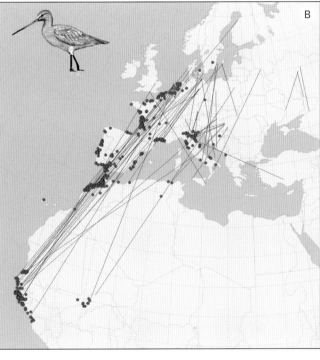

Fig. 186 Recoveries and ringing locations of: (A) 358 Black-tailed Godwits *L.l. islandica* (ringed mostly along the British coast in winter)[1], (B) Godwits *L.l. limosa* (mainly ringed as chicks in The Netherlands). In order to avoid cluttering the figure, lines for the 1582 Dutch birds are omitted; instead, dots indicate the recovery sites. The 224 recoveries of birds ringed outside The Netherlands are shown by lines. Data from: EURING, complemented by the Dutch Centre for Avian Migration & Demography for recent Dutch recoveries up to 2006.

The most important wintering areas are the rice fields in the coastal zone of southern Senegal (Casamance) and Guinea-Bissau, and in the large Sahelian floodplains of the Senegal Delta, the Inner Niger Delta and the Chad Basin. All EURING recoveries from sub-Saharan Africa, save 14, are from the West African coastal zone (Figs. 186, 187).

Black-tailed Godwits breeding in eastern Europe are thought to winter in the eastern part of the Sahel and East Africa (Glutz von Blotzheim et al. 1977, Beintema & Drost 1986). Accurate data on wintering numbers in central and eastern Africa are scarce. Moreau (1972) reports 'enormous numbers in October-December' in the Chad Basin. Mid-winter counts of godwits on the Logone floodplain varied between 50 and 2700, numbers depending on the area covered (OAG Münster 1991b, van Wetten & Spierenburg 1998, Dijkstra et al. 2002, Ganzevles & Bredenbeek 2005, Scholte 2006). Complete aerial surveys of the Chad Basin, including the Logone floodplain and Lake Fitri, revealed 13 976 birds in 1984, 30 365 in 1986 and 8 411 in 1987 (Roux & Jarry 1984; AfWC database). An aerial census in January 2007 by Trolliet et al. (2007) yielded 36 528 birds and an overall estimate of 40 000 birds (B. Trolliet in litt), the highest total so far. These recent numbers do not suggest a decline over time. As far as is known, large numbers do not winter in Sudan, where godwits are 'an uncommon visitor in small numbers; recorded on wet grassland' in the Sudd (Nikolaus 1989) and 'common on the northern rivers' (Cave & Macdonald 1955).

Other concentrations of Black-tailed Godwits in the central Sahel are known from the Hadejia-Nguru floodplains (6000 birds in 1997, see Chapter 8). In Niger, Nigeria and Burkina Faso the species is encountered generally in small groups of 10-100 birds in wetlands (Giraudoux et al. 1988, Thonnerieux et al. 1988), although the database of the African Waterbird Census of Wetlands International does hold some remarkable records such as 1075 birds at Mare d'Oursi (Burkina) on 30 March 2004 and 10 500 on 22 January 1984 in the

riverbed of the Niger near Niamey (Niger). The latter observation is of special interest because it indicates that during droughts birds may switch to alternative feeding areas not normally visited (also found in e.g. Garganey (Chapter 23) and Ruff (Chapter 29).

In the western Sahel, godwits concentrate in the floodplains of the Inner Niger Delta and the Senegal Delta. The maximum wintering population in the Inner Niger Delta fluctuated around 40 000 birds (Fig. 188) and, as in the Chad Basin, did not show any decline between 1971 and 2007. This finding contrasts with the crash in numbers in the Senegal Delta and in the coastal rice fields. The January census in the Senegal Delta did not exceed 5000 birds in most years since 1991. In the 1970s and 1980s such low numbers were recorded only under conditions of drought (Fig. 189A). Numbers had dwindled even further to 1000-3000 birds in 2001-2006 (Fig.

During the deflooding of the Inner Niger Delta, suitable feeding sites become ever more restricted in area, until most Black-tailed Godwits are eventually concentrated in the central lakes (here Lake Debo, 10 March 2005). It is a race against the clock, as the local food supply, bivalves buried in mud below shallow water, can be harvested for only a few weeks before the water has retreated completely (or the bivalve stock is depleted; see Box 7).

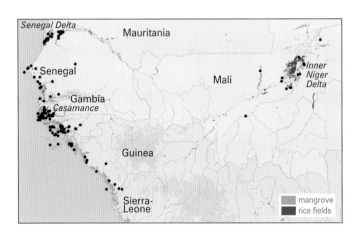

Fig. 187 Recoveries of Black-tailed Godwits in West Africa. For recoveries along the entire flyway see Fig. 186.

The plumage of Black-tailed Godwits in their wintering quarters is a composite of various tones of grey. This dull appearance, although beautiful in itself, is refracted the very moment that the bird takes flight and reveals its black-and-white tail to the world. The grey winter plumage is prevalent among godwits throughout the winter, as shown for the Casamance (September 2007, facing page) and the Inner Niger Delta (early February 1994, centre). Interestingly, godwits in Portugal start moulting into their red summer plumage as early as mid-January (1993, right); these birds will arrive on the breeding grounds in Western Europe in February. The timing of moult and return migration of the birds wintering in the Inner Niger Delta is much later, another indication that these birds belong to the eastern European population.

189A; Kuijper *et al.* 2006). These figures are a far cry from the tens of thousands recorded before 1973 (Morel 1973). Estimates were made of 15 000-20 000 birds in 1973-1979 (Tréca (1984) and *c.* 10 000 in December 1980 (Poorter *et al.* 1982). The largest concentration ever recorded for the Senegal Delta dates back to October 1958, when Roux (1959b) reported 'several hundred thousands'. This unusually high number was present for 2-3 weeks and may have involved birds profiting from the exceptional flood in that year during their southward migration (Roux 1959b, Tréca 1984).

The data above clearly demonstrate a progessive decline in the number of wintering Black-tailed Godwits in the latter part of the 20th century (Fig. 189A), even when disregarding the 1958-outlier (Trolliet & Triplet 1995). This trend is corroborated by a similar drop in ring recoveries from the Senegal Delta (Fig. 189B). The decrease started well before the construction of large dams in the Senegal River Basin in 1983 and 1987 (Tréca 1984, 1992) and must have been triggered at least partly by land reclamation in the early 1960s and exacerbated by recurrent droughts in the 1970s and 1980s; both events reduced the surface area of the inundation zone (Chapter 7). With the exception of the 11 000 birds counted in January 1993 (Trolliet & Triplet 1995), numbers remained low throughout the 1980s and 1990s. The continued decline in the 1990s and 2000s is not related to flooding in the delta (1991-2005: $R^2 = 0.13$) and therefore most likely reflects the decline of the western European breeding population. Possibly there was also a shift towards more southerly wintering areas in the Casamance and Guinea-Bissau, instigated by habitat loss in the Senegal Delta; we have no data to substantiate such a shift.

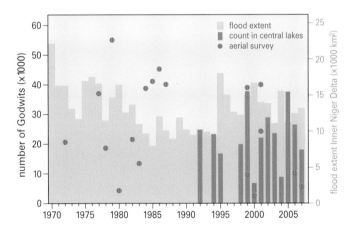

Fig. 188 Number of Black-tailed Godwits in the Inner Niger Delta counted from the air (entire delta) or from the ground (central lakes only). During high floods many birds are missed during ground surveys in the central lakes (2000, 2004), but counts from the air may also be incomplete. Hence we conclude that 35 000 - 45 000 godwits spend the winter in the Inner Niger Delta. Further explanation is given in Chapter 6.

The importance of Guinea-Bissau for Black-tailed Godwits was first highlighted by Poorter & Zwarts (1983), who reported large numbers feeding in rice fields and estimated the wintering population at 100 000-200 000 birds in December 1982. Large-scale surveys in 1983/1984 (Altenburg & van der Kamp 1985) and 2005/2006 (Kuijper *et al.* 2006) proved that godwits were distributed mainly in the West African coastal zone from South Senegal (Casamance) to Guinea, Guinea-Bissau being a clear stronghold. In the Casamance, numbers early in the season (July-October) are much higher than in November-January when rice fields have been harvested and become desiccated (Table 23). This suggests movements to the south, related to rainfall and rice farming. Outside the core wintering area, small numbers are present in Sine Saloum and The Gambia, as well as in the rice and mangrove zone of Guinea (Table 23). Surveys in Guinea revealed their presence, unexpectedly, in the tidal zone, feeding beyond the mangroves on mudflats; in 1990 numbers here were estimated at 17 400 birds (Altenburg & van der Kamp 1991), but counts during the five consecutive winters of 1997/1998 – 2001/2002 arrived at less than 10% of the 1990 estimate (Trolliet & Fouquet 2004; Table 23).

There was no change in the distribution of godwits during the non-breeding season between the 1980s and 2000s in Guinea-Bissau and the adjacent Casamance (Kuijper et al. 2006; Fig. 190), where prime habitat comprises rice fields reclaimed from mangroves (Chapter 11). From July onwards, godwits arriving in the Casamance and Guinea-Bissau feed in fields where rice has been seeded or planted (van der Kamp et al. 2006, 2008). After the harvest (November-January), godwits take spilt rice grains from the ground (Tréca 1984, Altenburg & van der Kamp 1985).

The surface area of the rice culture in Guinea-Bissau, as measured in the early 2000s, amounts to 65 000 ha, of which about 53 000 ha consists of mangrove-swamp rice and rain-fed lowland rice, both suitable as habitat for godwits (Bos et al. 2006; Chapter 11). In 1983/1984, the area of mangrove and lowland rice amounted to >75 000 ha (Altenburg & van der Kamp 1985); the survey covered 15 590 ha (21%) and 6730 ha (9%) of less suitable inland rice fields (with only a few godwits). Using these data, and via weighted extrapolation for habitat quality, the Black-tailed Godwit population in Guinea-Bissau was estimated at 110 000-120 000 birds in December 1983. Kuijper et al. (2006) covered 22 000 ha of the 65 000 ha (41%) in 2005/2006 and arrived at an estimate of 35 000-40 000 birds, suggesting a decline of about 67%. A similar difference (60%) was found in the rice fields along the Rio Mansoa, counted in early January 1983 (17 000, L. Zwarts unpubl.) and December 2005 (3500, Kuijper et al. 2006). In other parts of the wintering range along the West African coast an equal or larger decline was noticed (Table 23).

Migration and wintering

For birds breeding in eastern and central Europe, flyway locations are largely conjectural, because recoveries are sparse, ranging from Turkey (ringed in Po delta) to the Camargue (France). Very few recoveries link East and Central European Godwits to their wintering quarters: the EURING-database contains one sub-Saharan recovery, a Polish bird recovered in the Inner Niger Delta in Mali. Two other birds ringed or controlled in southern Ukraine, and originating

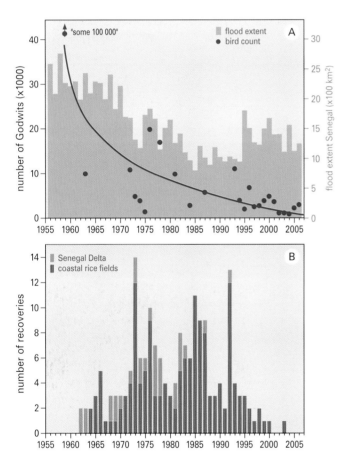

Fig 189 (A) Maximum number of Black-tailed Godwits staying in the Senegal delta in December-February 1958-2006. The exceptional number in 1958 is indicated by the arrow (Roux 1959b; see text). From sources given in Chapter 7. (B) Annual frequency of Black-tailed Godwit ring recoveries in the Senegal Delta and in the coastal rice and in the mangrove zone of West Africa. Same source as Fig. 186.

When the first Portuguese reached the West African coast in the 15th century, the local people had already perfected a system of growing rice in polders reclaimed from the mangrove zone, where salinity and acidity demand a subtle management of local hydrology to be successful in producing rice (Chapter 11). These polders were, and still are, located in tidal marshes in the zone between The Gambia and Sierra Leone. The majority of European Black-tailed Godwits are concentrated in these coastal rice fields, where they feed in shallow, open water or on bare ground. The aerial photos, taken in Guinea-Bissau in early September 2008 show the variety of rice growing systems: smaller and larger fields, water inlets, dikes (which we used during bird counts), rice planted on small ridges between July and October. Remember that almost all, if not all, work in these fields is manual, which results in even greater variety between and within plots.

Table 23 Maximum number of Black-tailed Godwits during November- December in coastal West Africa in 1983-1990 and 2001-2005. Sources: Altenburg & van der Kamp (1985, 1991), Trolliet & Fouquet (2004), Kuijper *et al.* (2006), van der Kamp *et al.* (2006) and database African Waterbird Census (Wetlands International). Year of assessment is given in brackets. Note that the 1983 census in the Sine Saloum and Casamance coincided with very dry conditions.

Region /basin	1983-1990	2000-2005
Senegal Delta	3300 (Oct. 1983)	2000 - 2500 (Dec. 2005)
Sine Saloum (Senegal)	4500 - 5000 (1983, estimate)	390 - 629 (2005)
The Gambia	5000 -10 000 (1980s, estimate)	65 (2004)
Casamance (Senegal)	1400 - 1700 (Nov. 1983 - dry)	329 (2005)
	'many thousands' (Nov. 1982)	>9000 (Sept. 2006)
Guinea-Bissau	110 000 - 120 000 (1983)	35 000 - 40 000 (2005)
Guinea (tidal mudflats)	17 400 (1990)	1480 (2001)

from Estonia and Lithuania, may be indicative of an eastern flyway (Diadicheva & Matsievskaya 2000).²

Nominate Black-tailed Godwits from West Europe may use several staging areas along the way when migrating to and from their African wintering quarters, but fewer during outward migration (*contra* Beintema & Drost 1986). These areas include the Atlantic coast of France (Vendée), Portugal (Tagus and Sado estuaries), northern Morocco and, to a lesser extent, sites in the Rhône (France)

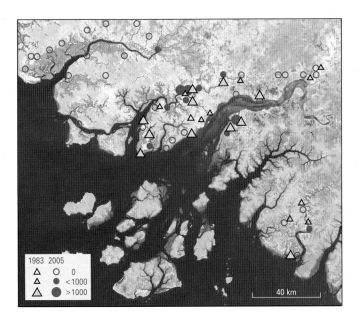

Fig. 190 Distribution of Black-tailed Godwits in December 1983 (Altenburg & van der Kamp 1985) and November – December 2004 or December 2005 – early January 2006 (Kuijper *et al.* 2006) in rice fields in Guinea-Bissau. Visited locations and number of Black-tailed Godwits are shown.

and Po (Italy) Deltas. Most Dutch Black-tailed Godwits leave the country in July, adults preceding juveniles. The timing of migration in central Europe is more or less similar (János 1996), but may last till September in the east (Dementiev *et al.* 1969).

Based on observations at communal roosts that are used prior to and after breeding (van Dijk 1980, Piersma 1983, Gerritsen 1990; Fig. 192), we have indirect evidence that the timing of post-breeding migration in The Netherlands has advanced by more than one month since the late 1960s. In the 1960s and early 1970s, numbers concentrating on post-breeding roosts peaked in mid-July.³ By the early 2000s, peak numbers at roosts were recorded in early- or mid-June (Fig. 193). This forward shift outpaces the concomitant advancement of the breeding season (Beintema *et al.* 1985), and illustrates the steep increase in the number of failed breeders. Earlier mowing, a trend started in the early 20th century and continuing unabated today, progressively affects a larger section of the breeding population by killing chicks (directly, or indirectly by removing all vegetation) before they can fledge (Kruk *et al.* 1997, Wymenga 1997, Groen & Hemerik 2002, Kleefstra 2005, Schekkerman *et al.* 2008). Failed breeders tend to leave the breeding grounds early and concentrate in the vicinity of communal roosts on favourable feeding grounds, but their numbers are now a significant proportion of all those attempting to breed. The advancement of post-migratory roosting is exemplified by the recovery on 11 May 2007 in Guinea-Bissau of a colour-ringed female, which had been sighted on the Frisian breeding ground on 11 April 2007 (J. Hooijmeijer, pers. comm.). Earlier roosting as a proxy for an advancement of post-breeding migration was unexpectedly validated by the experiences of rice farmers in the Casamance, who complained that nowadays godwits already appeared in their rice fields by early July (van der Kamp *et al.* 2008).

The flight from western Europe to West Africa may be interrupted halfway in France, Iberia and/or Morocco, as postulated by Beintema & Drost (1986). However, we are convinced that only a modest

Box 22
Godwits fatten up fast after the breeding season

Immediately after the breeding season, Black-tailed Godwits concentrate in large feeding flocks and aggregate at night in still larger groups on communal roosts. In June, just after chicks have fledged, they spend nearly 17 hours on the feeding grounds (meadows), where they consume chiefly leatherjackets, the larvae of the crane fly *Tipula paludosa*. With the naked eye, an experienced observer can see the birds getting fatter by watching the expanding abdominal profile.

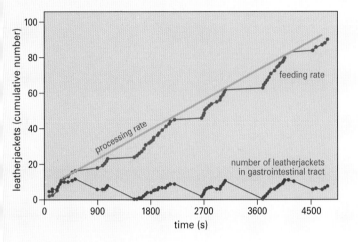

Fig. 191 Cumulative number of leatherjackets eaten by a female Black-tailed Godwit during 5000 seconds of non-stop observation. Her feeding rate cannot be higher than her average processing rate (*i.e.* defecation rate times number of leatherjackets in her droppings). The number of leatherjackets in her gullet, stomach and intestines can be derived from the difference between feeding and processing rates. The bird reaches her digestive bottleneck after swallowing about 11 leatherjackets.

To fatten up, birds must eat much more than usual, accomplishing this feat by foraging on leatherjackets between dawn and dusk, but taking frequent resting periods (covering 30-40% of the daylight period) to allow digestive pauses where their body processes can catch up with the food intake rate (Fig. 191). The daily *per capita* consumption by fattening godwits amounts to 1100-1300 leatherjackets, in any terms equivalent to 1600-2000 kJ. Given that 80% of the energy intake is assimilated, net energy intake is about twice as high as is needed to keep body weight constant (such as measured in captive godwits). This enables them to increase their body weight by about 10 g per day, a 3-4% increase relative to body weight during the breeding season. After a fortnight at this rate, they have stored enough fuel to fly to Africa non-stop. Watching the preparation of the actual intercontinental departure is exhilarating, as it appears to be for the birds themselves. Usually 2-3 h before sunset on a sunny evening the noisy birds, after several nervous false starts, leave, disappearing into the blue sky and heading straight for their destination (as described for other waders by Piersma *et al.* 1990). The Dutch meadows, which were formerly such productive godwit breeding grounds, still act as excellent feeding grounds, enabling the species to double its food intake rate for outward migration. Adults need about two weeks of foraging to attain sufficient migratory fat and although we lack this information for juveniles, we may assume that their food intake rate is slower, implying a longer period needed to build up body reserves; another hypothesis would be that they depart with smaller body reserves and fly a shorter distance. As the proportion of juveniles on the roosts steadily increases from 0% in June to almost 100% in September (Timmerman 1985, Wymenga 1997), it is likely that they do indeed take longer to fatten up than adults. Pre-migratory fattening in Africa is altogether more difficult, hinging on the exploitability of various prey types and flood height (see Box 7). (From: L.Zwarts & A-M. Blomert unpubl.).

number of juveniles (85% of the godwits shot in France in July are juveniles; Fig. 194) and even fewer adults use this option. Bird counts show that in late summer, SW European staging areas attract only very small numbers (Le Mao 1980, Díaz *et al.* 1996, Trolliet *in litt.*, R. Rufino pers. comm.). In July-September, Moroccan wetlands harbour larger numbers (3000-5000; Kersten & Smit 1984, D. Tanger pers. comm.), but these are still insubstantial compared with the size of the population.

The arrival of Black-tailed Godwits in West Africa is rather well documented. Older sources mention the presence of godwits in July-August in the Senegal Delta, where newcomers mix with summering birds (Morel & Roux 1966a, Tréca 1975, 1977, 1984). The arrival of juveniles in early August is confirmed by several recoveries of Dutch birds in the Senegal Delta and in the Casamance (Morel & Roux 1966a, 1973; Fig. 194). Recently, van der Kamp *et al.* (2006, 2008) concluded that birds in the Casamance had advanced their arrival into the first half of July, or even into late June, during the preceding 10-20 years. In the Inner Niger Delta, godwit numbers remained modest in June-August 1998-2005 (van der Kamp *et al.* 2005). June counts revealed only summering birds in winter plumage, but a flock of birds, presumed to be fresh arrivals in moulting summer plumage, was observed in late June 2004 in the *Office du Niger* zone, just west of the Inner Delta. Whether this observation is a harbinger of a general advancement of arrival dates, as in Senegal, is not yet known. First arrivals in the Chad Basin in the 1970s have been recorded for early August, considerable passage occurring in September (Newby 1979).

Black-tailed Godwits using a communal roost after the breeding season. This photo may evoke nostalgia: the time has gone when 10 000 birds roosted in wetlands in Fryslân, where the picture was made in July 1990 (Fig. 192).

Regarding the advancement of arrival dates in West Africa (of birds from western Europe), it would be very interesting to know recent arrival dates in the Chad Basin. Since godwits wintering in central Africa are likely to originate from eastern Europe, where early mowing related to modern dairy farming is not yet a large-scale practice, an advancement of arrival is not yet expected.

The finding that godwits leave their West European breeding

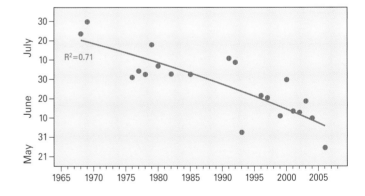

Fig. 192 Average number of Black-tailed Godwits on five major roosts in three different periods along the Frisian IJsselmeer coast, The Netherlands. From the intervening years, monthly censuses are not available. Source: van der Burg & Poutsma (2000) and database It Fryske Gea, compiled by J. Hooijmeijer.

Fig. 193 Average date on which maximum numbers of Black-tailed Godwits were present on post-breeding roosts in Fryslân, the core breeding area for godwits in The Netherlands. Based on 44 series of roost counts in May-September 1968-2006. Sources: Mulder 1972, Koopman & Bouma 1979, Kuijper et al. 1983, Kleefstra 2005, 2007, Wymenga 2005, unpubl. data E. Wymenga, It Fryske Gea and J. Hooijmeijer.

quarters prematurely (Fig. 193) is an indication of consistent reproductive failure, the wider ramifications being earlier arrival on the wintering grounds, increasing potential of crop damage in rice fields and increasing risk of birds being shot by local farmers (van der Kamp et al. 2008). Shooting is (has been?) an important mortality factor in the non-breeding season (see 'Population change').

Northward migration of Black-tailed Godwits wintering in the rice and mangrove zone of West Africa commences by late December. Observations in Guinea-Bissau show that rice fields are rapidly deserted in early January, coinciding with the end of the rice harvest and the desiccation of rice paddies (Altenburg & van der Kamp 1985, unpubl. data J. van der Kamp). Timing and presence of godwits along the flyway suggest that most birds fly nonstop to their major staging sites in Portugal and Spain, where huge concentrations accumulate from early January onwards (Fig 195 and Box 26; Leguijt et al. 1995, Kuijper et al. 2006, Lourenço & Piersma 2008). The Senegal Delta is of minor importance as a staging area, showing small influxes in January: 2000-2500 birds in December 2005 (1736 counted, Kuijper et al. 2006) and 2961 in January 2006 (P. Triplet, in litt.). Sim-

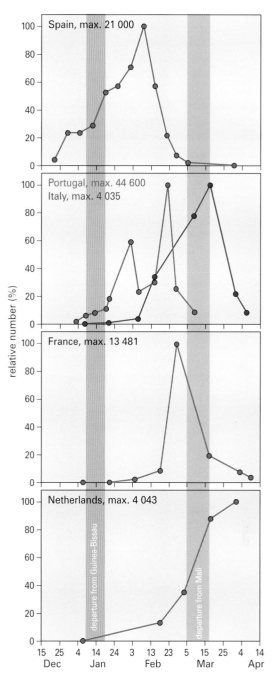

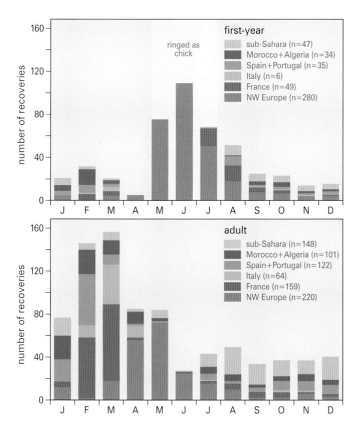

Fig. 194 Monthly recoveries of yearling (between May and the following April; upper chart) and adult (lower chart) godwits, for six zones in Europe and Africa. NW Europe = The Netherlands, Belgium, Germany and United Kingdom. Same source as Fig. 186.

Fig. 195 Relative number of Black-tailed Godwits on staging sites in Spain, Portugal, Italy, France and The Netherlands during simultaneous counts in spring 2006 (from Kuijper et al. 2006). Numbers are expressed relative to the maximum number counted. Timing of departure from Guinea-Bissau and Mali is indicated by bars derived from counts and observed departing flocks (Guinea-Bissau in 1983, 1984, 2006 and Mali in 1992-2007).

Black-tailed Godwit *Limosa limosa*

ilarly, the stopover role of Moroccan wetlands is nowadays limited, and this may have arisen through habitat deterioration of Moroccan coastal wetlands (Green 2000) coinciding with the creation of large rice complexes in Iberia, which form an attractive alternative (Kuijper *et al.* 2006). Numbers in Morocco in January 2006 did not exceed *c.* 5000 birds (Kuijper *et al.* 2006), compared to more than 10 000 birds in the 1970s and 1980s (Zwarts 1972, Kersten & Smit 1984) and an exceptional concentration of 80 000-120 000 in January 1964 in the Merja Zerga (Blondel & Blondel 1964), when the surroundings of this tidal lagoon were flooded. These observations indicate that Morocco must formerly have been an important spring staging site, and this is also illustrated by the high number of spring recoveries at that time (Fig. 194).

In the Inner Niger Delta, Black-tailed Godwits have been seen departing between late February and mid-March, with a clear peak after the first 10 days of March. Based on their food intake in the pre-migratory period, these godwits are able to undertake a non-stop flight of >4000 km in only a few days, bringing all major European staging sites within reach, including the breeding areas in The Netherlands (Box 7). By early March godwits using the Spanish, Portuguese or French staging sites have left for the breeding grounds. However, the timing of departure from the Inner Niger Delta does closely fit the spring migration pattern found in Italy (Fig 195), though spring totals in Italy are marginal compared with the wintering population in the Inner Niger Delta (Fig. 195; Serra *et al.* 1992). This is consistent with our supposition that the bulk of the birds departing from Mali fly directly to other destinations, to a range of suitable wetlands either in the Mediterranean or in Central Europe (Kube *et al.* 1998, van der Have *et al.* 1998), not forgetting the Dutch breeding areas. Recoveries of Dutch birds in Mali (Fig. 186) show there is indeed a link, but their general timing of departure from Mali, some three weeks before the first clutches are laid in The Netherlands, suggests that the majority of birds wintering in Mali must be of a Central or East European origin. This is supported by observations in the Inner Niger Delta by van der Kamp (1989) who recorded not any bird with a ring among 1000 godwits in January 1989 and 799 birds in February 2008. Hence, the fraction of Dutch godwits in the Malian population must be very low, since one in every 150 birds in the Dutch population carried a ring in the 1980s and one in every 120 in the 2000s (Beintema & Drost 1986; J. Hooymeijer pers. comm.). In eastern Europe and further east very few godwits have been ringed over the years, *e.g.* only 62 in the former Soviet Union and Russia in 1979-1999 (Gurtovaya 2002, and earlier reports of the Bird Ringing Centre of Russia).

We have not yet addressed the conundrum surrounding first-year birds, of which an unknown proportion does not undertake a return migration from Africa in their first year of life (Haverschmidt 1963, Beintema & Drost 1986). For the Senegal Delta, Morel & Roux (1966a) recorded *c.* 5000 birds on 22 May, and 1000 birds on 30 May, but much smaller numbers between 11 June and 17 July (from dozens to about 300). There is little recent information on oversummering birds in the Senegal Delta. Numbers in the Inner Niger Delta in April-June 1999-2006 were usually very small, except in June 2002 (1958 birds) and June 2005 (1770 birds; van der Kamp *et al.* 2005 and unpubl.). There is thus no evidence that first-year birds move from coastal West Africa to the Inner Niger Delta during the dry season, as previously

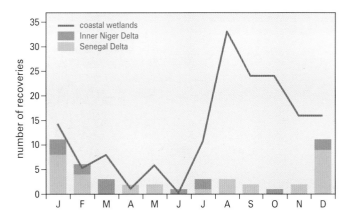

Fig. 196 Monthly recoveries from the coastal wetlands between The Gambia and Sierra Leone (red line), compared to those of the Senegal Delta and Inner Niger Delta. The difference is indicative of habitat-related periods of highest vulnerability to human predation: rice planting in August-September versus desiccation of floodplains in December-February.

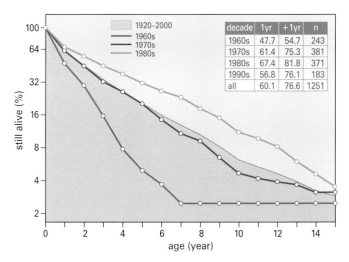

Fig. 197 Survival rates for Black-tailed Godwits ringed as nestlings in The Netherlands (blue shading). The three lines show that the survival rates have improved since the 1960s. Note that survival is plotted on a log-scale. The table gives the average annual survival rate (%) for first-year and older birds. Data from: Dutch Centre for Avian Migration & Demography.

hypothesised by Beintema & Drost (1986). Lastly, oversummering has also been reported in the Chad Basin, where Bates shot two birds on 6-7 June and saw several more (1927, in Grote 1928).

Population change

Wintering conditions The vast majority of Black-tailed Godwits recovered in the Sahel had been shot (91%, n = 213). Unlike migrants such as Ruff and Garganey, which are captured in large numbers in nets in the Inner Niger Delta (Kone et al. 2002, 2006), godwits are rarely exploited for commercial reasons. Their erratic flight performance, rushing down in near-perpendicular zigzags from high altitudes, defies any attempt at catching them on roosts or feeding sites. On being flushed, godwits also easily outmanoeuvre standing nets. Shooting is the only reasonably reliable method of obtaining godwits. In 2003, for instance, tourist shooters in the Senegal Delta were recorded as killing 214 godwits, but only 3 were reported as meeting this fate in 2002 (Baldé 2004); the actual number must have been higher.

The clear peak of December-February recoveries in the floodplains highlights the birds' vulnerability in that period when they concentrate in the last remaining vestiges of water as the dry season advances (Fig. 196); these are precisely the areas where fishermen and pastoralists will congregate as well. In coastal rice fields most birds are recovered from August to October, associated with measures to prevent crop damage during rice planting and sowing. The perceived damage is twofold: consumption of sown rice kernels in the seedbeds and trampling of newly planted rice shoots (Tréca 1977, 1984). Van der Kamp et al. (2008) showed that godwits are still being shot for the same reasons that were outlined by Bernard Tréca. During relatively dry years, two-and-a-half times as many birds are shot in rice fields as during wet years (van der Kamp et al. 2008). Just as in floodplains, birds are more vulnerable during dry conditions as they concentrate in the shrinking wet areas. Interviews with hunters in the Lower Casamance allowed us to estimate the annual hunting bag conservatively at 5% of the local godwit population in 2005-2007 (van der Kamp et al. (2008).

The impact of Sahel conditions on survival cannot be derived directly from the annual number of recoveries, because the number of ringed godwits in the population was high in the 1960s and low in the 1990s. A simple way of correcting for this discrepancy would be to calculate the proportion of recoveries from sub-Saharan Africa separately for each year. We then find that, in the nine years when rainfall was at 20% or less than the 20th century average, between 4 and 12% of the annual recoveries came from sub-Saharan Africa; in the nine years with above-average rainfall this proportion was as small as 0-3%.

Godwits can live remarkably long lives (Box 23). We used the Dutch recoveries of Black-tailed Godwits ringed as chicks to calculate survival rates. Of the yearlings, 60.1% survive to their second year. For birds 2-3 years old, annual survival rate is 76.6%, which remains essentially constant from then on (Fig. 197). Surprisingly, given the steep population decline, godwit survival rate has increased over time (Fig. 197). What other factors apply here?

We calculated the expected annual number of recoveries by determining separately for all ringing years the number of recoveries expected in the years to follow, given the total number of ringed birds and the average life expectancy (shaded blue in Fig. 197). The annual expected number of recoveries, with some delay, kept up with the number ringed annually (compare blue and red line in Fig.

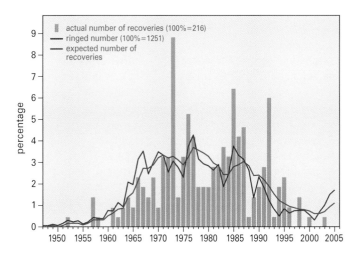

Fig. 198 Actual and expected annual number of recoveries (%) in the Sahel. The expected number is derived from the annual number of ringed birds in The Netherlands, assuming the survival rate is similar for all years. Same source as Fig. 186.

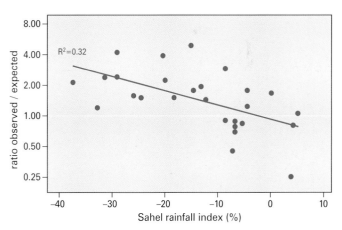

Fig. 199 Relative survival rate (given as the ratio of observed and expected annual recoveries, derived from Fig. 198) as a function of rainfall in the Sahel (from Fig. 9). The selection applies to the period 1968-1992, when the population of ringed birds was large enough to expect more than four recoveries annually (or 2% in Fig. 198).

198). The graph is based on only 216 recoveries, and so a large variation is to be expected. Nevertheless, in very dry years the number of recoveries reported in the Sahel was 2-3 times greater than expected. The negative correlation between mortality rate and Sahel rain index is significant (Fig. 199; P<0.001). The relative survival rates before and after 1980 (on average corresponding to wet and dry periods respectively) suggest that a decline in survival, at the population level, may be expected. However, the overall increase in survival rate contradicts this expectation, thus effectively removing the Sahel as a significant contributor to the recorded population decline. [4]

Because most recoveries of Black-tailed Godwits relate to birds shot, temporal changes in hunting pressure may explain part of the puzzle surrounding godwit fortunes. The available data suggest that fewer godwits are shot today than in the past, in line with the purported decrease in hunting pressure in Europe and a concomitant protection of spring staging areas (Kuijper et al. 2006). Reduced hunting along the flyway would translate into progressively fewer birds reported being shot since about 1980 (Fig 200A). This seemingly clear trend may be prone to corruption by any decline in willingness of hunters to report shot birds with rings. However, legislation has certainly improved the lot of godwits in Europe. In Italy the open season for hunting waders in spring was first advanced to 31 January, and hunting has been banished completely since 1997. Shooting waders in spring is now also prohibited in Morocco, Portugal, Spain and France, but other species such as Snipe, which occupy similar staging sites, remain legal quarry species (Kuijper et al. 2006). Shooting, legal or illegal, is no longer considered a significant threat to godwits, at least not in the countries along the western flyway (Kuijper et al. 2006). Trolliet (in litt.) estimates that the French hunting bag of godwits in 2000-2006 has declined to fewer than 200 birds annually. The diminishing impact of shooting is therefore the likely explanation for the decline in recoveries in Europe and northern Africa (Fig. 200B). Assuming that other mortality factors have not changed, the reduced shooting pressure must have had a positive impact on survival, especially after 1985, and the population decline therefore cannot be linked to hunting.

A continuous decline The nominate race shows a negative trend, particularly in western Europe where the decline has been a cause for concern for decades (Thorup 2006, Birdlife International 2004a). The Dutch population reached its likely maximum of 125 000-135 000 breeding pairs in the 1950s and 1960s (Mulder 1972), declining to 85 000-100 000 pairs in the 1980s (van Dijk 1983, Piersma 1986) and 45 000-50 000 in the mid-2000s (Thorup 2006). The downward trend shown in Fig. 201B increased from 1.5% in 1970s through 2% in the 1980s and 3.5% in the 1990s to 4% in the 2000s (Altenburg & Wymenga 2000, Teunissen & Soldaat 2006).

The German population, being about ten times smaller than the Dutch population, declined on average by 2.4% annually between 1977 and 2006, although the decline between 1976 and 1985 was as high as 3.8% annually, compared to more stable figures between 1995 and 2005 (Fig. 201A).

Despite the overall decline of the West European population, the small populations in Great Britain (Fig. 201C), Flanders (Vermeersch et al. 2004), France (B. Trolliet in litt.) and coastal Germany (Hötker et al. 2007) have been stable, or expanding slightly, in recent years. Accurate data on trends in breeding populations of central and eastern Europe are scarce. Godwits in Ukraine and European Russia are thought to be in decline (Belik 1998, Lebedeva 1998, Zubakin 2001, Birdlife International 2004a, Thorup 2006), but the Polish population appears to have been stable in 1951-2000, after increasing in the period before 1950 (Tomiałojć & Głowaciński 2006). Nikiforov & Mongin (1998, in Thorup 2006) reported a decline in Belarus from 15 000-17 000 pairs in 1990 to 6000-8500 pairs in 1998. The Estonian godwits declined by 10-50% from 1970 to 1990, then remained stable at 600-1000 pairs from 1991 to 2002 (Elts et al. 2003).

The West European breeding population of Black-tailed Godwits has halved between 1983 and 2005. Since juveniles comprised 17%

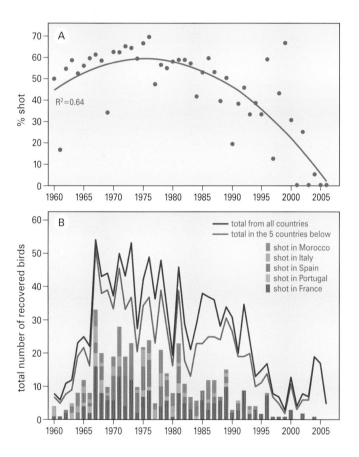

Fig. 200 (A) Percentage of shot birds in the total number of recoveries over time. (B) Number of recovered shot birds per country compared to the total number of recoveries in these countries. Same source as Fig. 186.

of the wintering population in the 1980s, but almost zero in the 2000s (Fig. 202), we would expect the winter population to have decreased from 240 000 to 100 000 birds, some 57%, in that 22-year period, but the decline found on the wintering grounds was about 70% (Table 23). [5]

Sahelian winter conditions have had no significant impact on the five trends shown in Fig. 201, on the annual changes in the populations, or on the deviations from the regression lines given in Fig. 201 (to correct for the overall decline). From this we conclude that the observed decline cannot be attributed to the wintering conditions.

Breeding conditions Since neither wintering conditions in Africa nor mortality are contingent upon population decline, reproductive failure is therefore a likely candidate. We have already alluded to this scenario in our analysis of roosting behaviour over time, and especially in the advancement of post-breeding roosting dates (Fig. 193). Even when taking into account the many factors involved in timing of post-breeding roosting, such as advanced egg-laying dates and age-specific temporal use of roosts, it remains that: (1) nowadays roosts are occupied long before any chicks could have been raised to independence, and (2) very few young godwits are now recorded.

Recovery data from ringed birds do indeed indicate a long-term

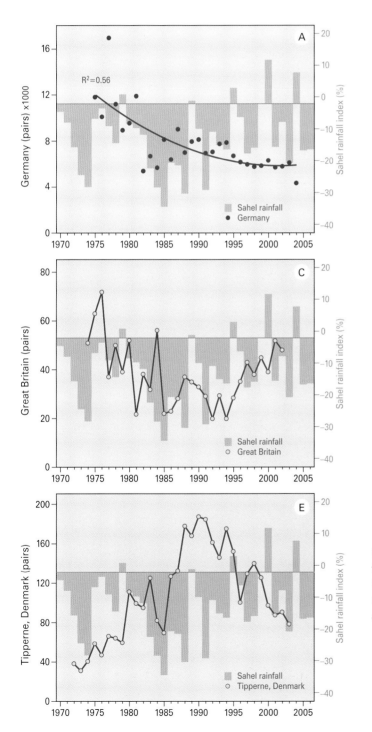
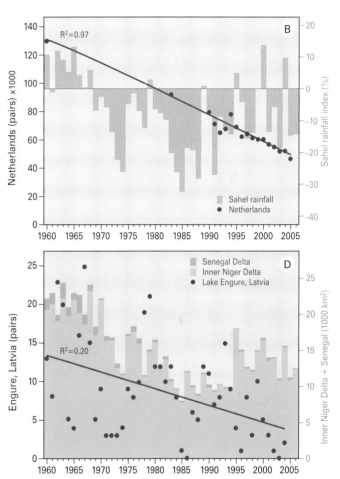

Fig. 201 Population trends of Black-tailed Godwits in parts of five countries related to Sahel rainfall or flood extent in the Inner Niger Delta. From: British Birds 97, 2004 (Great Britain), Mulder 1972, van Dijk 1983, SOVON (The Netherlands), Thorup 2004 (Tipperne, Denmark), Hötker et al. 2007 (Germany), Viksne et al. 2005 (Latvia, Lake Engure, 42 km²).

Box 23

Masking the decline: slow extinction in a long-lived bird

Farmers and birdwatchers in The Netherlands have long known that Black-tailed Godwits were in decline. Haverschmidt (1963) had reminisced about the loss of a colony on a moor at Leersumsche Veld, known to and visited by him since the 1920s. As a 13-year old boy, he tramped the moor, illegally because his request to study the site was flatly refused, up to the early 1940s when it became the hunting ground of German occupation troops. During his last visit in May 1942 he "ran into a gentleman with an outspoken Teutonic appearance in typical German hunting dress, armed with a shotgun and accompanied by a hunting dog, which boded no good in the middle of the breeding season." Returning 13 years later, in 1955, "the moor was dead and almost without bird life. The godwits were gone..." Decades later, notably since the mid-1970s, similar stories, but without Teutonic hunters, circulated among birdwatchers active in grasslands, the godwit habitat *par excellence* in the 20th century. Some of these stories were accompanied by evidence. In grassland plots of 11-125 km^2 in the province of Drenthe, intermittently censused between 1958 and 1980, the decline amounted to 25% between 1958 and 1974, and 27% in 1975-80 (van Dijk & van Os 1982). By 2000, the decline here since the late 1960s was calculated at 82%, representing an annual decrease of 6% since the late 1980s (van Dijk & Dijkstra 2000).

Black-tailed Godwits are long-lived birds, the record being a bird recorded in the EURING files as being 30 years of age, having been ringed as chick in a Dutch meadow in 1976 and recaptured on her nest in 2006 only a few kilometres away. Another bird ringed as a chick in 1969 was shot in Guinea-Bissau over 29 years later. Reproductive failure, even when occurring year after year as happens today, only shows at the population level when veteran birds start dying off. This natural lag masks the extent of any decline for a long time. Had godwits been like the short-lived Skylarks, after only a few years a greater than 90% decline would have materialised quickly over vast stretches of farmland.

decline of first-year birds in the population, especially after the 1960s (Fig. 202). This decline is not caused by a reduction in ringing effort, but is largely the consequence of fewer and fewer godwit chicks being available to be ringed nowadays. In The Netherlands, for example, in 1966-1974 annually 1006-1442 chicks were ringed, compared to 125-377 annually in 1995-2005, when a greater ringing effort was deployed. This pointer to reproductive failure is convincingly underpinned by Dutch studies on nesting success (Beintema & Müskens 1987) and chick production (Groen & Hemerik 2002, Schekkerman & Müskens 2000, Schekkerman et al. 2008). From the earliest modernisation of farming, especially since the introduction of artificial fertilizers in the early 20th century, improved living conditions were created for godwits through a boost in food supply, but further intensification led to habitat loss, lowered groundwater tables, reduced botanical and arthropod diversity and advanced and shorter mowing cycles (Beintema & Müskens 1987, Bijlsma et al. 2001, Schekkerman et al. 2008, 2009, see also Verhulst et al. 2007). These changes led to the impoverishment and fragmentation of suitable grassland habitat, in turn causing very low godwit recruitment and increased predation risks (Wymenga 1997, Schekkerman & Müskens 2000, Teunissen et al. 2005, Schekkerman et al. 2008, 2009). Apart from local improvements through innovative management (Oosterveld et al. in prep.), attempts to improve general living conditions for meadow-inhabiting birds, including Black-tailed Godwits, have so far been less than effective (Kleijn & Sutherland 2003, Verhulst et al. 2007, Schekkerman et al. 2008): the unchecked downfall of godwits in West Europe is mute testimony to that effect. Godwit populations in eastern Europe seem also to have a far from rosy future; loss of prime habitat for meadow-breeding birds is caused by: (1) the scaling-up of dairy farming, (2) increased use of

One of the many problems encountered by present-day Black-tailed Godwits breeding in the stronghold of The Netherlands is drainage of their breeding haunts. This bird clearly shows how deep its bill can (or must) penetrate the top soil to extract leatherjackets or earthworms. Nowadays foraging in the much drier meadows demands even deeper penetration of the top soil to access the deeply buried prey. In dry springs, the soil may be too dry to penetrate at all.

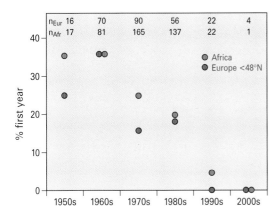

Fig. 202 Percentage of first-year godwits in the recoveries from the 1950s to the 2000s, as reported for Africa and for Europe south of the breeding area (i.e. France, Spain and Italy). Numerals indicate number of recoveries per decade. Same source as Fig. 186. From: EURING.

artificial fertilisers, (3) drainage of wetlands and meadows, (4) advancement of mowing dates in productive meadows, and (5) abandonment of those floodplain meadows and other semi-natural grasslands that are too labour-intensive to remain profitable (P. Pinchuk *in litt.*, Báldi *et al.* 2005).

Conclusion

At the present rate of decline (4% annually), the West European Black-tailed Godwit population, which still accounts for more than 50% of the entire European population, is heading for an all-time low. The situation in Central and East Europe is less bleak, but may deteriorate quickly in the wake of modernisation and agricultural changes. After an increase in the first half of the 20th century, the godwits' declining trend started in the 1960s/1970s, and accelerated in the 1990s, the major cause being reproductive failure, which relates to changes in agriculture and in farmland generally. Ironically, in such circumstances improved survival rates had led to an ageing population with poor reproductive prospects. Indeed, a considerable proportion of the breeding population cannot complete the breeding process, mainly because the earliest meadow mowing dates have advanced by a full month in the 20th century. Consequently, most godwits return ever earlier to their wintering sites in West Africa, nowadays often by June. Rice fields in Guinea-Bissau form the heart of the wintering range for western European godwits, as they have for decades, but the former important wintering site in the Senegal Delta has declined much in significance. An important factor in West Africa is the local perception that godwits damage crops, specifically rice, in July and August, which has led to many birds being shot. On average, survival in West Africa is better in 'wet' years than in 'dry' years, suggesting that the impact of shooting in dry years, when godwits are vulnerable because of reduced foraging opportunities, is significant. Such an effect in the long term has been overridden by overall improved survival rates, probably associated with stricter legislation on shooting in Europe and North Africa, leading to a decline in hunting pressure.

The wintering population in the western Sahel has declined by some 70% between 1983 and 2005, even higher than the estimated 50% decline on the West European breeding grounds in the same period. No such decline is noticeable in the Inner Niger Delta (nearly 40 000 since the 1990s) and Lake Chad; these populations largely originate from the more easterly breeding areas, where population declines have not yet reached West European proportions.

Endnotes

1. To separate Icelandic (*L. l. islandica*) from continental breeding birds (*L. l. limosa*) all godwits ringed or recovered in Iceland or along the NW Europe coast during October – February were defined as *L. l. islandica*. All other recoveries in the database were supposed to be *L. l. limosa*. Unless stated otherwise, the data presented in this chapter refer to the continental population only. More information on the migration ecology of Icelandic godwits, based on numerous sightings of colour-marked individuals, is provided by Gill *et al.* (2002).
2. Additionally, an intriguing observation of 300-400 godwits, feeding by hovering on semi-ripe standing sorghum in flooded fields in southwestern Saudi Arabia on 8-9 February 1992, suggests passage of birds on their way to the breeding quarters (Rahmani & Shobrak 1992), or even a small wintering population. Newton (1996) showed that several migratory Palearctic species regularly winter in southwest Arabia, although his list does not include Black-tailed Godwits. Large irrigation projects here may have improved wintering conditions in recent decades.
3. In 1968-1970, L. Zwarts (unpubl. data) actually witnessed the onset of migration in the N and SW Netherlands (Box 22): 13 groups left between 17 and 30 July, and single flocks on 5 July and 14 August. Similar departure dates were recorded in the northern Netherlands, *i.e.* 20 July 1978 and 3 August 1979 (van Dijk 1980), which timing coincides with the seasonal decrease in numbers visiting the roosts (Fig. 192).
4. Recently, there have been two survival analyses of the Dutch Black-tailed Godwits. Van Noordwijk & Thompson (2008) found a large variation in the annual survival. This variation (their Fig. 7) appeared not to be related to the Sahel rainfall in the preceding year. Roos Kentie (in prep.) reanalysed the same data and entered the Sahel rainfall index into her model. She found that a small part of the annual variation in adult survival may be explained by annual fluctuations in Sahelian rainfall.
5. The midwinter count in 1983/1984 produced about 150 000-170 000 birds in the Senegal Delta and coastal rice fields (Table 23). In the early 1980s, the W European population amounted to 97 000 pairs, equivalent to a winter population of 233 000 birds, given a production of 0.4 young per pair (Fig. 202). Hence, 73% of the population had been encountered in midwinter. We may assume that the missing birds mingled with East European birds wintering in the Inner Niger Delta, or were present in the unvisited rice fields of Guinea and Sierra Leone. The winter population in 2005/2006 may be estimated at 100 000 birds, of which 50 000 were in the coastal wintering areas (Table 23).

Wood Sandpiper
Tringa glareola

A few days after the snow has melted, mid-May or perhaps slightly later, the peatbogs and aapa mires of northern Finland are alive with recently-arrived Wood Sandpipers. The minerotrophic aapa mires, swamps that catch nutrients from the surrounding land via run-off, stretch from northern Finland well into Siberia. The formation of these mires started some 9000 years ago, at the end of the last ice age. At an average peat accumulation of less than 0.6 mm annually, aapa mires gradually convert into raised bogs, a process that is usually slowed by large-scale flooding in spring. Mammals, including man, find traversing broad aapa mires extremely difficult, because of flark-fen pools. In such relatively safe havens Lapland's Wood Sandpiper reached densities of up to 13-18 pairs per km^2 in the 1800s and early 1900s, but present-day figures indicate a decline, possibly associated with fen drainage. Between the breeding and wintering grounds, some 7000 km apart, drainage is also diminishing many of the European staging areas on which Wood Sandpipers depend; here they accumulate the fat reserves necessary to cross both the Mediterranean and the Sahara successfully. As with so many migratory species, life is becoming increasingly complicated.

Breeding range

The monotypic Wood Sandpiper breeds across northern Eurasia from western Fennoscandia, the Baltic States and northern Ukraine through European Russia and Siberia east to Anadyrland, Kamchatka and the northern Kuriles, and south from the southern edge of the tundra to the southern Urals, Kirgiz Steppes, Russian Altai, northern Mongolia and the Amur River.

The northwest European population is estimated at 285 000-407 000 pairs, mostly in Finland, representing an estimated 0.86-1.22 million birds. Applying Finnish densities to European Russia would produce another 0.6-0.9 million pairs (Stroud *et al.* 2004). Numbers breeding in western Siberia are unknown but presumably very large, for it is the most abundant wader in the forest-tundra (Rogacheva 1992); this population is estimated at >2 million birds (Stroud *et al.* 2004).

Migration

Wood Sandpiper is entirely migratory and winters in Africa south of the Sahara. Breeding birds from the northern and central Palearctic, west of 40° E, winter largely in sub-Saharan West Africa. The NW European birds follow a south-southwesterly course through Europe (Fig. 203). Extensive ringing has shown that in late summer Wood Sandpiper migrates southwards initially in short hops, staging at foraging sites for only a few days and at first carrying low fat loads. The observant prose stylist Sergei Timofeevich Aksakov, also hunter *extraordinaire* in the Orenburg region, in 1852 had even then recorded in his *Notes of a provincial wildfowler*, that "*fifi*", a local onomatopoeic name for Wood Sandpiper, "are never fat. They depart so early that they have no time to put on weight".

Adults migrate ahead of juveniles, and on average carry higher fat loads. The short-hop migration strategy calls for longer stopovers in southern Europe in order to accumulate sufficient energy reserves for the non-stop flight across the Mediterranean and the Sahara (Scebba & Moschetti 1996, Wichmann *et al.* 2004). One such site is the Camargue, where large numbers of Wood Sandpipers are known to assemble in autumn (Hoffmann 1957). At some sites in northern and central Europe, Wood Sandpipers accumulate sufficiently high energy reserves to cover longer distances, such as from southern Sweden (Persson 1998), Jeziorsko in western Poland (Włodarczyk *et al.* 2007) and Münster in western Germany (at a sewage farm; Anthes *et al.* 2002). Other individuals use a scattering of refuelling sites throughout Europe, putting on weight *en route* (Leuzinger & Jenni 1993, Meissner 1997). Birds with adequate fat reserves may fly uninterruptedly to stopovers in southern Europe, notably the Rhône delta, where energy reserves can be replenished. One Münster-ringed bird was shot the next day at Bouches-du-Rhône (1000 km; Anthes *et al.* 2002). Many Fennoscandian birds take a more southeasterly course via the Balkan countries, or through the Ukraine (Sivash as a stopover) and across the Azov-Black Sea area. Birds ringed in the Ukraine have been recorded in Chad and Nigeria, where they overlap with the wintering range of birds from NW Europe (Lebedeva *et al.* 1985, Nankinov 1998, Diadicheva & Matsievskaya 2000). The Wood Sandpipers from Russia and western Siberia migrate mostly to eastern and southern Africa (Lebedeva *et al.* 1985, Vandewalle 1988, Oschadleus 2002, Stroud *et al.* 2004).

Homeward migration speed in most regions is higher than on outward migration and has shorter stopovers. The low body masses measured in birds reaching southern Europe and the Middle East (Akriotis 1991, Yosef *et al.* 2002) are rectified quickly through frantic feeding during stopovers; weights gradually increase during northward migration. In southern Sweden, birds had reached a mean body mass of 71 g (range 56-86 g, n=43; Persson 1998); mean masses of 67-70 g in North and NE Poland also attest to birds fattening well (Remisiewicz & Wennerberg 2006). Even when in a hurry, Wood Sandpipers evidently are able to accumulate fat reserves along the way (Remisiewicz *et al.* 2007, Muraoka *et al.* 2009).

Distribution in Africa

Wood Sandpipers are widespread throughout West Africa in winter, occurring wherever shallow fresh, brackish, or even salt water is present for some time, but rarely assemble in large concentrations.

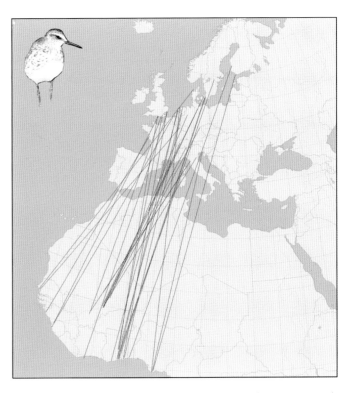

Fig. 203 European ringing locations of 69 Wood Sandpipers recovered or captured in West Africa. From: EURING.

Wood Sandpiper *Tringa glareola*

Density counts in plots of known size revealed that 300 000 Wood Sandpipers spend the northern winter in the Inner Niger Delta, and more than 100 000 in West African rice fields and the Senegal Delta. In the floodplains, densities varied considerably, depending on vegetation type and water cover. Densities were higher in *bourgou* (2.2/ha), shown here for the Inner Niger Delta (January 1994), than in wild rice (0.5/ha), but irrespective of habitat type, shallow water of <20 cm depth was highly favoured (4.5/ha) (see Chapter 6 for details).

The EURING database holds 60 recoveries and 9 captures, concentrated in Mali (n=24, of which 22 are from the Inner Niger Delta), Ghana (n=15), the Senegal Delta (n=9) and the coastal rice fields in Senegal and Guinea-Bissau (n=6) (Fig. 203). Most of the recovered birds had been ringed in Sweden (n=21) and Finland (n=15) between 14 June and 29 August, consistent with a predominantly Fennoscandian provenance of birds wintering in West Africa. Birds from elsewhere in Europe had been ringed during migration in spring (23 April-9 May, n=7) or late summer (5 August-5 September; n=17). Birds from Finland and Sweden fully overlap in their West African wintering area, presumably irrespective of the routes taken (SW or S). The apparent concentrations of ring recoveries in the Inner Niger Delta (Mali), Ghana and Senegal Delta show reporting biases related to shooting (n=22) and captures (8 in Mali, 7 in Ghana).

Density counts   Until recently, only rough estimates were available for wintering Wood Sandpiper in various parts of Africa (summarised in Stroud *et al.* 2004). The species is ubiquitous but in scattered locations throughout the region. Region-wide estimates are therefore impossible unless stratified sampling is employed (see Chapter 6 for details).

Based on plot-sampling, a total of 300 000 Wood Sandpipers was estimated for central Mali in the 1990s and early 2000s, of which 200 000 were in the floodplains of the Inner Niger Delta (2.2/ha; Chapter 6), 66 000 in the irrigated region of *Office de Niger* in the central part of the country (1.2/ha; Chapter 11) and 40 000 on floating vegetation in Lake Télé (a staggering 12.7/ha; Chapter 6).

The floodplains in the Senegal Delta, being converted into irrigated rice fields, have lost much of their attraction to Wood Sandpipers. Recent density counts resulted in an average density of 1.2/ha, varying from 0.13/ha in irrigated rice fields to the highest density in the floating vegetation, *Sporobolus* (2.4/ha), bringing the estimated total for the Senegal Delta to 36 000 (Chapter 7).

Elsewhere in West Africa, rice fields attracted large numbers of the species, for example 5100 in irrigated areas near the Selingue Dam (4.0/ha; Chapter 11) and 55 000 in coastal rice fields between The Gambia and Guinea (0.5/ha; Chapter 11). The latter estimate is of the same order of magnitude as the estimate of 25 000-50 000 in the 180 000 ha of rice fields of Guinea-Bissau in 1983 (Altenburg & van der Kamp 1986). Numbers in the Chad Basin in 1993, covering some 10 000 km^2, were calculated at slightly over 9000 birds (van Wetten & Spierenburg 1998), possibly an underestimation.

In total, more than 400 000 birds are concentrated in Sahelian floodplains and rice fields, equivalent to 25-40% of the European population of 1-1.5 million birds, but Wood Sandpipers will also inhabit small wetlands, and we assume that a large part of the European population is confined to West Africa in winter. It is one of the most frequently encountered Palearctic bird species in shallow water habitats.

Population trends

For a species so strongly associated with wetlands of all sorts, we would expect a higher recovery rate in African wintering area in dry years, which in general, was the case (Fig. 204).[1]

Most Wood Sandpipers wintering in West Africa breed in Fennoscandia and Russia.

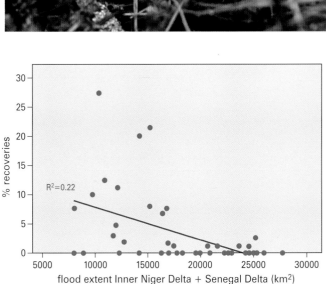

Fig. 204 During years with poor flooding, a higher proportion of Wood Sandpipers was recovered from West African floodplains (4°-20°N) than in years with good flooding, suggesting better survival in years with high floods. The recovery rate is expressed as the percentage of recoveries in West Africa based on the total number recovered from Europe and Africa (dead birds only) in 44 years with >10 recoveries between 1947 and 2001 (P<0.001). From: EURING.

In the Sahel, the total number of recoveries showed an increase in the course of the wintering period: 1 in November, 8 in December, 12 in January, and 15 in February. This increase was absent in the Sudan zone: 4 in December, 6 in January, and 3 in February. Birds wintering in the Sahel, in contrast to those in the Sudan zone, face a gradual reduction both of the extent of flooded areas and of rain-fed depressions as winter progresses, and hence become increasingly concentrated in fewer areas. At the same time fishermen and pastoralists also encroach upon these areas, increasing the risk of the Wood Sandpiper being shot or captured. This higher predation risk contributes to the difference between dry and wet years (Fig. 204). Years with low floods not only force the birds to concentrate in fewer areas, but also result in a seasonal advancement of concentrations, and so extending the period of high risk (Chapter 6, 23).

Breeding populations have been monitored in Finland since 1982 (Väisänen et al. 1998, Väisänen 2005), and for even longer, since 1975, in Sweden (Lindström & Svensson 2005). The Finnish population shows a significant decrease of 1% per year on average [2], but this

Wood Sandpipers feed in shallow water and on moist ground. Areas with high vegetation cover are avoided, as are entirely open flats. Pond-like habitats are preferred throughout Africa, the birds never reaching high densities anywhere, which suggests that feeding grounds are defended against congeners. Such habitats are used by a multitude of waterbirds, including Sacred Ibis and Little Egrets.

trend was not correlated with Sahel parameters. Drainage of Finnish wetlands, especially in the south, is thought to have caused this decline (Väisänen et al. 1998). The situation is complex, though, as the species did show an increase in Lapland (almost doubling between 1984 and 2006 in Varrio Nature Reserve; http://veli.pohjonen.org). Similarly, in Lapland Nature Reserve in NW Russia a gradual increase was recorded in the 1972-1991 period (Gilyazov 1998), but with distinct annual fluctuations. The Swedish breeding population numbers were only weakly correlated with flooding of the Inner Niger Delta and the Senegal Delta (Fig. 205).[3]

Fig. 205 Population trends of Wood Sandpipers in Sweden and Finland (TRIM indices, summer point counts) related to the flood extent of Inner Niger Delta and Senegal Delta in the preceding year. From: Lindström & Svensson 2005 (Sweden) and Väisänen 2005 (Finland).

Conclusion

The majority of European Wood Sandpipers winter in West Africa, where sites with shallow water are favoured; rice fields and the floodplain of the Inner Niger Delta are particularly important. Population trends of Fennoscandian Wood Sandpipers are not, or only weakly, related to Sahel conditions in the preceding winter, despite the observed higher winter mortality in dry Sahel years. During autumn and spring migration, Wood Sandpipers depend on a series of stopover sites between breeding and wintering sites, fattening *en route* to either goal area.

Endnotes

1 The relative number of recoveries is significantly correlated with the flood extent (Fig. 204), but not with the Sahel rainfall index ($R^2=0.06$, $P=0.10$). This was to be expected since large numbers are concentrated in floodplains.
2 The decline of the Finnish population is significant ($R^2=0.15$).
3 The correlation between flood extent and the Swedish breeding population: $R= +0.35$, $P<0.05$.

Border disputes among Wood Sandpipers, defending their feeding territories along the shores, are well-known to every birder who has spent time watching them. Black-winged Stilts, however, like Ruff and Black-tailed Godwits, feed in flocks and are not territorial on their feeding areas. The frequency and vigour of disputes among Wood Sandpipers, indicate that wintering areas can play a significant role in the regulation of their numbers, especially during droughts.

Ruff
Philomachus pugnax

5 March 1905. In a Frisian newspaper, the Leeuwarder Courant, a birdcatcher complains in a reader's letter about the new regulation on wader catching in winter, which forbids the use of clap nets after 31 March. He wonders: why limit the catching of Ruff, when they are so abundant during migration in April? In those days, about 100 000 ha of grasslands in central Fryslân, in the northern Netherlands, was flooded during winter time and well into spring. The inundated grasslands were home to a large breeding population of Ruff and huge numbers of spring migrants. They were caught by local birdcatchers, a welcome supplementary income on top of the profits reaped from the 20 000 to 40 000 Golden Plovers that were killed annually in midwinter (Jukema *et al.* 2001a). The birdcatcher's complaint was, in hindsight, a reardguard action, because not much later legal commercial trapping of Ruff ceased entirely. In the course of the 20th century the Frisian floodplains were reclaimed (Claassen 2000), and by the year 2007 a mere 2000 ha was left of the former inundation zone. Ruff disappeared as a breeding bird in the second half of the 20th century, but the Frisian meadows are still an important staging site during spring migration (Wymenga 1999).

Ruff are easy to catch, due to their habit of hugging the contours of the landscape during the approach of roosts and feeding grounds. Perhaps an effective way to avoid avian predators, but spelling doom when birdcatchers with nets are around. More than 4000 km to the south of Fryslân, local fishermen in the Inner Niger Delta have also learned that Ruff are easy prey and extra income in difficult times. Together with Garganey, Ruff are among the most frequently captured species, and many thousands are annually sold on the market in Mopti. Draining the floodplains on the breeding grounds in the past had – and still has - a negative impact on Ruff populations, as may have the present catching in Mali during the northern winter.

Breeding range

The Ruff is breeding in a wide belt across Eurasia from Scandinavia to eastern Siberia, mostly north of 60°N in wet pastures, floodplains, bogs and (sub)arctic tundra (Hagemeijer & Blair 1997, Lebedeva 1998). In the second half of the 20th century, the Ruff has vacated much of its territory in western Europe (Zöckler 2002), not unlike the southernmost breeding grounds in Hungary, Poland, Ukraine and Russia (Thorup 2006).

The core populations are centred in Scandinavia and in the Russian tundra, the latter accommodating c. 95% of the total population (Zöckler 2002). The breeding population in the tundras and peatlands of Fenno-Scandinavia are estimated at 60 000-90 000 females and the Russian population at 140 000 – 420 000 females (Birdlife International 2004a, Thorup 2006). The European breeding population has been estimated at 0.2–0.5 million females (Thorup 2006). This is equivalent to a wintering population of 0.5-1.3 million birds (given the fact that 22.7% of the birds in winter are in their first year of life and the population consists of 40-50% males). The Eurasian population of the Ruff would comprise 2.3-2.8 million birds (Zöckler 2002).

Distribution in Africa

The main wintering grounds encompass Africa and India, with more peripheral sites in West Europe (up to 2100 in the 1980s, up to 3150 in the 1990s and up to 900 in 2000-2008 in the SW Netherlands; Castelijns 1994, unpubl.), the Mediterranean Basin (probably more than 6000; van Dijk *et al.* 1986, Meininger & Atta 1994), coastal Arabia and incidentally wetlands east of India (Li Zuo Wei & Mundkur 2004). During winter it is gregarious, occupying all kinds of open, fresh to brackish wetlands: floodplains, rice fields, lake sides, riverbeds and shallow water bodies. Even the smallest temporary pools may attract Ruff. Like Black-tailed Godwits, Ruff are mainly granivorous on the wintering grounds, both in Africa (Tréca 1975, 1977) and in Europe (Castelijns *et al.* 1994). Rice (*Oryza* sp.) make up 80% of their diet, supplemented with grass seeds, millet, sorghum and various small invertebrates and chironomid larvae (Tréca 1993, Stikvoort 1994).

Ruff are common in African wetlands, and especially so in the Sahelian floodplains. Floodplains also stand out in the distribution of ringed birds recovered in Africa (EURING): 183 of the 217 recoveries

Male Ruff in breeding plumage displaying on a lek. Leks are often situated on slightly elevated spots in wet grasslands such as the border of water courses and ditches.

(84%) were from the major floodplains, *i.e.* 123 from the Inner Niger Delta, 53 from the Senegal Delta, 4 from the floodplains south of Lake Chad and 3 from the Sudd. Another 16 Ruff were recovered in coastal rice fields and only 18 came from elsewhere in West Africa.

The first attempts to count Ruff in Sahelian floodplains date from the 1960s and 1970s, and were rough indications at best. Roux (1973) estimated the wintering population in the Senegal Delta in January 1972 at 500 000, while in February 1972 one million birds were thought to have been present (reported by G. Jarry; Roux 1973). During the same survey, Francis Roux counted 110 000 Ruff in the Inner Niger Delta, but remarked that 'millions' might have been present. In the delta of the Yobe river, along the western shore of Lake Chad, Ash *et al.* (1967) used daily counts along a fixed mile of shore in one hour, and periodic observations of evening roosting movements, to arrive at 500 000 Ruff in March-April 1967. They estimated that "there may have been as many as a million within a 15-mile radius of the mouth of the River Yobe". For the Sudd, we know next to nothing, except for the observation of Nikolaus (1989) that, in 1976-1980, the species "occurs in very large numbers on open marshy grassland". These observations indicate that, at least in the 1960s and 1970s, Ruff wintering in the Sahel must have numbered in the millions. Are such numbers still present in the Sahel?

The waterbird surveys of Sahelian floodplains by the ONCFS (*e.g.* Triplet & Yésou 1998, Trolliet & Girard 2001) suggest otherwise (Fig. 206). Since 1984, totals never exceeded half a million birds for the Senegal Delta, the Inner Niger Delta and Chad Basin combined (Fig. 206). The largest concentrations were found in the Inner Niger Delta and the Chad Basin, each with several 100 000 Ruff at most. Present-day numbers in the Senegal Delta are a far cry from the million or so Ruff in the early 1970s, not least because of droughts, embankments and flood control (Chapter 7). However, up to 1998 large numbers were occasionally present (on average 135 500 birds in 1989-1998), feeding mainly in rice fields and roosting in the remaining wetlands and lake sides (Trolliet *et al.* 1992, Trolliet & Girard 2001). Since then, wintering numbers in the Senegal Delta dropped considerably (Fig. 206), even taking into account that some surveys may have been incomplete (Triplet & Yésou 1998, Trolliet *et al.* 2007).

Outside these core areas, concentrations of Ruff in the Sahel are known from the Hadejia-Nguru floodplains in Nigeria (50 000-70 000 in 1994-1998, Chapter 8), and Mare de Mahmouda (44 000 on 19 January 2001) and Lake R'kiz (22 000 on 15 January 2001) in southern Mauritania (AfWC database; Wetlands International). In eastern and southern Africa, midwinter counts indicated the presence of at least 19 000 Ruff in the lakes of the Rift Valley (1975-1985; Summers *et al.* 1987) and 50 000-500 000 in southern Africa (Underhill *et al.* 1999).

Large fluctuations in numbers on floodplains can be partly attributed to counting errors in aerial surveys (the species is notoriously difficult to detect from a plane) and to distributional shifts related to rainfall and flooding. In dry years, many Ruff leave the desiccated floodplains in favour of temporary pools, riverbeds and lake sides where they normally are all but absent. In January 1984, for example, when the floodplains held little water, 47 000 Ruff were counted at the Gaya-Kanji dam in Nigeria and 28 000 in the riverbed between Labbezanga – Niamey – Gaya (AfWC database). In the same season, in December 1983, Altenburg & van der Kamp (1986) found the Senegal Delta almost devoid of Ruff. Similarly, birds may have shifted southward in this particularly dry year, to occupy rice fields and mangroves in the Sine Saloum and Guinea-Bissau, where in December 1983, Ruff numbered 20 000-25 000 and 50 000-75 000 respectively (Altenburg & van der Kamp 1986); for the Sine Saloum much lower numbers are normally reported (Dupuy & Verschuren 1978; Chapter 11).

Sex-related differences in winter distribution   To collect representative sex ratios in the wintering quarters, sexing should take place be-

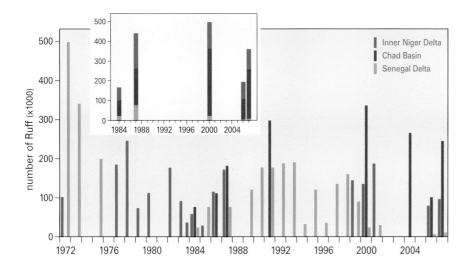

Fig. 206 Ruff numbers in the Senegal Delta, the Inner Niger Delta and the Chad Basin, based on aerial counts (including ground surveys in Senegal). Inset shows stacked totals for years when all three floodplains were counted simultaneously. From: Table 4, 5, 12, 16.

Ruff in West Africa usually feed in shallow water, often together with Black-winged Stilts and Black-tailed Godwits (Inner Niger Delta, February 1994).

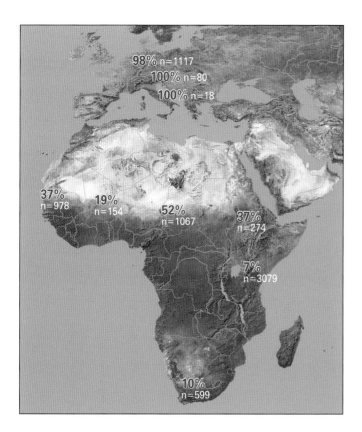

Fig. 207 Sex ratios (% male, sample size) of Ruff across Europe and Africa in midwinter (November-February, but March for Italy), based on mist-netting and field observations. From: Schmitt & Whitehouse (1976), Pearson (1981), Castelijns et al. (1988), van der Kamp (1989), OAG Münster (1991b, 1996), Wymenga (1999, 2003), Jukema et al. (2001b), Dijkstra et al. (2002), Wijmenga & Komnotougo (2005), H. Henson in litt (Italy, March 1999), J. Naber in litt. (Austria, February 2004).

tween November and February, as males commence migration earlier than females, *i.e.* by late February; OAG Münster 1989b, Wymenga 1999). Also, samples should cover all types of habitat and all hours of daylight, to account for sex-specific habitat choice and diurnal rhythm (OAG Münster 1991b, E. Wymenga unpubl.).

Males, the larger sex in Ruff, on average winter further to the north than females do (OAG Münster 1996). The small numbers wintering in Europe are almost exclusively males; in southern and eastern Africa, females overwhelmingly outnumber males (Tree 1985, Fig. 207). In the Sahel, where the majority of Eurasian Ruff winter, males on average constitute 36% of the population (despite a large variation; SD=0.20, n= 16 studies). Given the importance of the Sahel as a wintering site of Ruff, females probably outnumber males on the population level.

Migration

Origins   Ruff travel the length and width of Africa and Eurasia (Fig.208). A bird ringed in the Eastern Cape, South Africa, was recovered in the Kolyma basin (164°E), 15 392 km from its source (great circle route), reputedly the longest distance recorded for a terrestrial bird (Underhill *et al.* 1999). Ruff migration between Eurasia and Africa follows several major pathways (Fig. 208A), judging from ring recoveries and sightings of colour-ringed birds (Viksne & Michelson 1985, OAG Münster 1989a):

(1) Ruff from eastern Siberia fly west across the Lena, Yenisey and Ob basins through Kazakhstan to the Caspian and Black Seas, and from there via the Middle East into the Rift Valley southwards to wintering areas in eastern and southern Africa. Some of these birds may overshoot the Caspian and Black Seas and reach Europe; this could explain the recovery of a German-ringed Ruff in the Western Cape, South Africa (Underhill *et al.* 1999). An unknown proportion of East Siberian birds ends up in India (Fig. 208A), where they mingle with Ruff from Central Siberia. Several ring recoveries suggest an ex-

Capturing and ringing birds in the African wintering areas has contributed a lot to our knowledge of migratory birds (Inner Niger Delta, March 2001). In spring 1985, ornithologists all over Europe and North Africa watched out for Ruff captured and dyed with yellow underwings in the Senegal Delta (OAG Münster 1989b). On a Frisian roost 15 of those were seen among some ten thousands of Ruff. The yellow underwings proved to be effective markers in the clouds of incoming Ruff. The right-hand photo shows captured Ruff just before they were released, in this case with yellow-painted breasts.

change between Indian and South African wintering quarters.
(2) Part of the central Siberian breeding birds head more or less south and winter on the Indian subcontinent; another part winters in Africa where they extensively overlap with birds from Europe in the floodplains of the western Sahel (McClure 1974, Zöckler 2002). Central Siberian Ruff may fly via West Europe to and from Africa (Rogacheva 1992).
(3) Birds ringed in North and West Europa take a southwestern route and cross the western section of the Mediterranean on their flight to the West African wintering quarters, notably the Senegal Delta and Inner Niger Delta (Fig. 208A).
(4) Many Ruff from Fenno-Scandinavia and Russia follow directions between S and SE, which take them across the central and eastern Mediterranean. The latter course overlaps with that of birds from Siberia, especially in the Black Sea region and the Middle East (Fig. 208A).

The extent of overlap in the wintering quarters between eastern and western populations is highest in the Sahel, where Ruff populations originating from latitudes between 10° and 160°E can be found. Migratory connectivity is comparatively small, although Ruff wintering in Senegal on average derive from more westerly breeding grounds than those wintering in the Inner Niger Delta and near Lake Chad (Fig. 208). The western Sahel is a melting pot of European and Central Siberian birds, as demonstrated by Ruff captured during migration in Italy (Fig. 208C). Birds ringed in Sweden and Finland were reported from the entire width of the Sahel, where they extensively overlap with Ruff from Siberia (from up to the Kolyma Basin at 160°E; Viksne & Michelson 1985, Fig. 208B).

Seasonal pattern During July-September Ruff from as far east as 160°E cross Europe on their way to African winter quarters. The timing of passage shows successive, but overlapping, waves, starting with western birds and ending with the easternmost breeding birds (Fig. 209). Post-breeding movements cover more stopovers, and progress at a slower pace, than pre-breeding movements, which enables adults to perform a partial moult in Europe (Koopman 1986, OAG Münster 1989a, 1991a).

The timing of spring migration in West Europe, as elsewhere (e.g. in Belarus; Karlionova et al. 2007), is a mirror image of the one in autumn: wave upon wave of Ruff descend upon the major staging sites, to replenish body reserves in order to continue migration, starting with western breeding birds. Ruff from Siberia are among the last to pass through (Wymenga 1999; Fig. 209). East Siberian birds do not normally turn up in West Europe in spring, implicating a more easterly route during spring (via Italy or the Middle East), having a direct bearing on the breeding grounds without any detours. This may also explain the many recoveries in the Black Sea area, where various suitable wetlands can be found in spring (Kube et al. 1998). The delayed passage of Siberian Ruff, known to arrive on the breeding grounds in the last week of May in Central Siberia (Rogacheva 1992), is related to the much later phenology in Siberia. For example, Henry Seebohm during his 1877-journey had to wait till 31 May before the ice broke on the Yenisey (latitude 65°N), and during his stay migratory birds started to arrive in numbers between 31 May and 18 June (Seebohm 1901). By that time, the first chicks were already searching for insects in the Dutch grasslands.

In contrasting to the outward migration, the return migration to

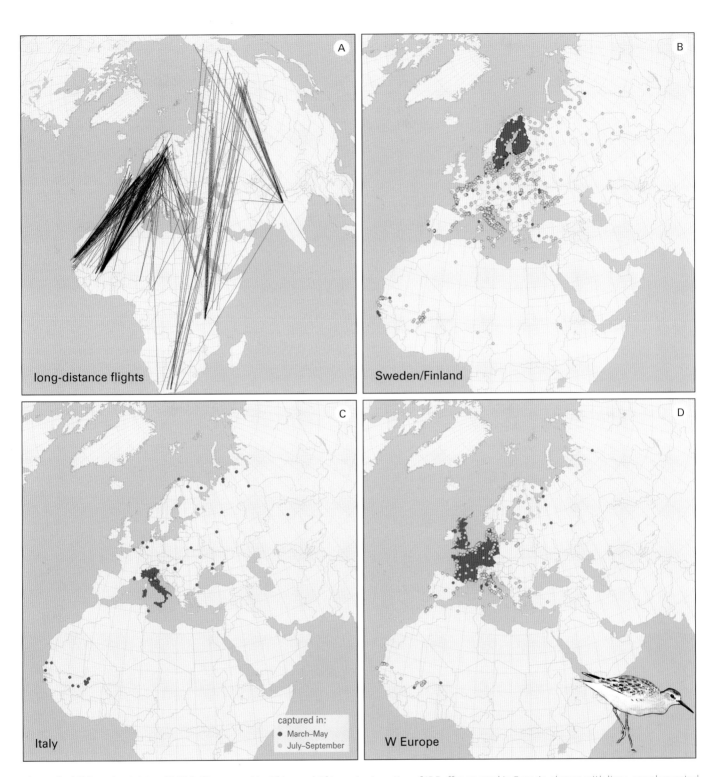

Fig. 208 (A) Eurasian origin of 243 Ruff recovered in Africa and African ringing sites of 15 Ruff recovered in Eurasia, shown with lines, supplemented with 21 long-distance recoveries of birds captured in northern India. (B) Recoveries of Ruff ringed in Sweden+Finland in March-May (n=64) and in July-September (n=857), (C) as B for Italy (92 in March-April and 3 in July-September) and (D) as B for NW Europe (n=226 in March-April and 496 in July-September). From: EURING, except for those ringed in South-Africa (Underhill et al. 1999), Kenya (Pearson 1981), India (McClure 1974) and some birds from Siberia (Viksne & Michelson 1985) and Sudan (Nikolaus 1989).

Ruff *Philomachus pugnax*

the breeding grounds is more contracted in time. From the African wintering grounds, the birds make a single flight to the European staging areas. The maximum pre-migratory weight gain of 40% in the Sahelian floodplains enables a distance to be covered of some 4000-5000 km. This brings all major transit areas in Europe within reach (OAG Münster 1989b, 1998, Zwarts et al. 1990), i.e. in Italy, Hungary, The Netherlands, Black Sea and Belarus. Of the 1988 Ruff colour-dyed in the Senegal Delta in 1985 and 1987, 27 were sighted in The Netherlands and 9 in Italy (OAG Münster 1989b); despite simultaneous observations across North Africa and Europe, none was recorded in areas in between. Some birds may even fly in one go to the Turkmen shore of the Caspian Sea, where all eight birds trapped between 24 March and 24 April were reported as "very exhausted" (Khokhlov 1995).

Sex-specific migration strategies Males winter further to the north than females, and their migration phenology is more advanced (van Rhijn 1991).[1] Male Ruff in the Senegal Delta started to increase their weight from half January onwards and left around early March; females lagged three weeks behind (OAG Münster 1989b). The early departure of males fits their advanced arrival in Europe, where across all major staging sites males predominate in the first part of spring migration. In Fryslân, northern Netherlands, among the thousands arriving around mid-March some 80% were males; this fraction remained high until mid-April, from when on females were the majority (Wymenga 1999, 2005; Fig. 210). A similar pattern was recorded in the Pripyat floodplains, Belarus (Karlionova et al. 2007), the Po Valley, Italy (Serra et al. 1990) and the sewage farms near Münster, Germany (OAG Münster 1989a).

A comparison of staging sites in Europe shows widely different sex ratios, for which several explanations have been offered. Harengerd (1982, cited in Melter 1995) found a turnover of 16 days for males and 9 days for females near Münster, Germany, accounting for a higher sex ratio. Secondly, females seem to take a more easterly route during spring migration, via the Mediterranean basin and Black Sea area, accounting for a higher sex ratio in western staging areas (Wymenga 1999).[2] Finally, sex ratios differ between breeding populations, showing a declining trend from west to east; overall, females probably outnumber males.

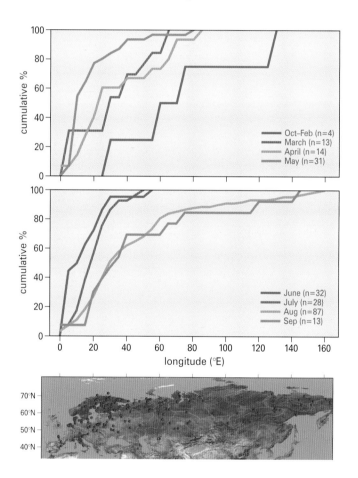

Fig. 209 Breeding origin (°E) of Ruff captured in West Europe in different months. A selection is made of birds recovered in May or June (red dots on the map); recaptures are omitted.

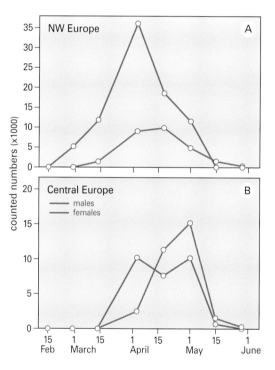

Fig. 210 Timing of spring migration of male and female Ruff in (A) NW Europe (Fryslân, The Netherlands) and (B) Central Europe (Belarus, Czech Republic and the Sivash, Ukraine), based on sex ratios and roost counts. Data from WSG Ruff Project in 1998 (Wymenga 1999).

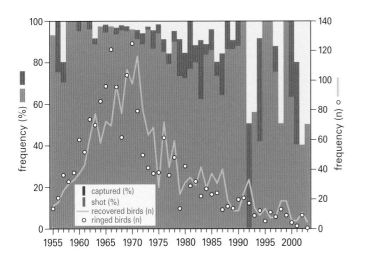

Fig. 211 Relative number of Ruff shot or captured annually (left-hand axis), and the number of ringed and recovered Ruff (right-hand axis; the annual number of ringed birds refer to the ringing year of recovered birds only). From: EURING.

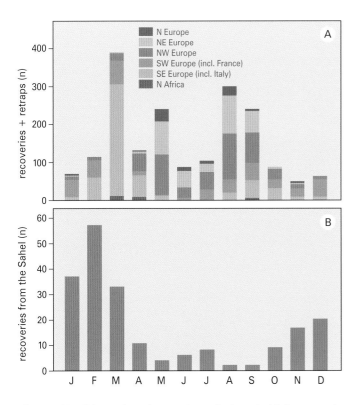

Fig. 212 Monthly number of recoveries and retraps in (A) Europe and North Africa (in total 1840) and (B) the Sahel (in total 206). From: EURING.

Population change

Mortality Of the dead birds recovered in the 1960s, 95% were recorded as shot and 2% as trapped (100%=598); these figures changed to 64% and 14% respectively after 1990 (100%=84) (Fig. 211). After the early 1970s, the number of ringed (and recovered) birds has steadily declined, partly in conjunction with the collapse of the breeding population in West Europe.

In Europe and North Africa, the recoveries peak in the pre-breeding (March-May) and post-breeding (August-September) periods, when the birds are flocking on feeding and roosting sites (Fig. 212A). In particular in Italy, large numbers were shot in spring, although less so during the post-breeding migration. Hunting on waders in spring has been prohibited here since 1992. Shooting waders, inclusive Ruff, is still rampant in NW and NE Europe (Russia). Many fewer birds were recovered from the sub-Sahara, due the presumed lower reporting rate in Africa (Fig. 212B).

Impact of wintering conditions From October to March, Ruff abound in wetlands throughout the Sahel. The recoveries show a clear peak in February, *i.e.* just before the birds start departing for their breeding grounds. This seasonal trend is strikingly similar to the ones found in other waterbird species depending on floodplains, as, for example, Garganey (Chapter 23) and Caspian Tern (Chapter 31). Towards the latter part of the northern winter, the receding flood forces birds into ever smaller vestiges of wetlands, where they become an easy prey for the local people. The risk of human predation is particularly high in the Inner Niger Delta, where during the deflooding large numbers of Ruff are captured with old fishing nets

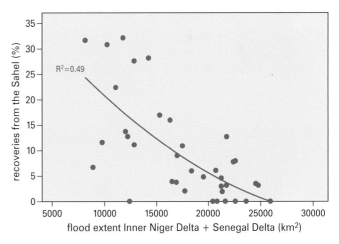

Fig. 213 Percentage of recoveries of Ruff from the Sahel relative to the annual number of recoveries as a function of the size of the floodplains (Inner Niger Delta + Senegal Delta). A selection is made for the period 1956-1988, when the number of recoveries amounted to >20 per annum. From: EURING.

(Kone et al. 2002). Bird catching on the scale as in the Inner Niger Delta is not seen elsewhere in the Sahel (Kone 2006). In the Senegal Delta, for example, recoveries are biased towards shooting (n=50) rather than catching (n=11). Hunting in the Inner Niger Delta is not commonly practiced (van der Kamp et al. 2005), and if so, mainly by city-dwellers or tourists who are more likely to report a ringed bird than a local fisherman. This explains why almost as many Ruff were reported as being shot (33, compared to 37 being captured), despite the vastly larger number of Ruff taken by catching.

Flood dynamics explain seasonal variations in recoveries in the

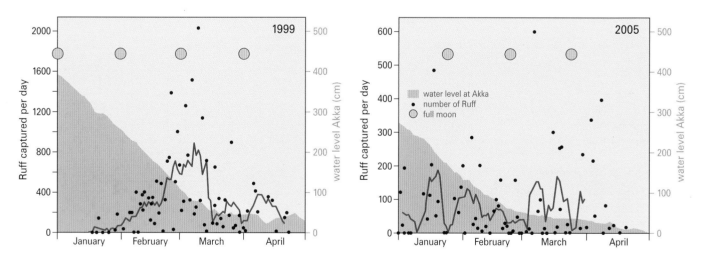

Fig 214 Daily totals of Ruff captured in the southern Inner Niger Delta in 1999 and 2005. The line shows the running mean over 7 days. The decreasing flood level is shown in blue (right axis), and the period of full moon is indicated. If the water level is less than 250-300 cm and the night is dark, many birds are captured, until most have departed by March/April. Data from: Kone et al. (2002, 2005), Kone 2006.

Many Ruff are captured in the Inner Niger Delta to be eaten or to be sold on the market. Note the fat layer on the deplumed birds (Mopti market, 5 March 2001).

Sahel, but just the same is true for annual variations. In the wet 1960s few dead ringed birds were recovered, but the dry years of 1973 and 1984 yielded 20 and 11 recoveries respectively. The absolute number of recoveries is misleading, however, since the number of ringed Ruff slowly increased until the early 1970s, then decreased to ever lower numbers till at least the early 2000s (Fig. 211). Therefore, we expressed the annual number of recoveries in the Sahel as a percentage relative to the total number of recoveries. This correction shows unequivocally that up to 30% of the recoveries originate from dry Sahel years; in wet Sahel years this is close to zero (Fig. 213). The higher mortality in dry years is, to a large extent, related to bird exploitation. We can illustrate this point by data obtained in Mali in January-April 1999, when four interviewers visited as many markets and birdcatchers in the Inner Niger Delta as possible. Their survey revealed that at least 24 482 Ruff had been captured and traded in the southern half of the delta. After a prudent correction for missing dates, we arrived at a total catch of 40 800 Ruff (see also Chapter 23: Garganey). This exercise was repeated in 2000, 2004 and 2005, when fewer Ruff were captured, *i.e.* 6133, 2463 and 6305 being registered respectively (after correction for missing dates, estimates arrived at 31 000, 9100, 13 600 respectively).

As we substantiated for the Garganey, the number of Ruff captured can be explained by flood level (more birds captured during low floods), lunar cycle (more birds captured on moonless nights) and season (fewer birds captured after late February from when on Ruff start departing). In contrasting to Garganey, the daily captures of Ruff show substantial variation but hardly any seasonal trend (Fig. 214). This may relate to the fact that many Ruff are eaten by birdcatchers and local people, whereas most Garganey, being more profitable (300-500 g, compared to 75-220 g for Ruff), are offered for sale on the markets. This local consumption was not quantified and is therefore not included in the calculations. The recorded annual catch is estimated to vary between 10 000 and 40 000 Ruff, but twice this number is possibly taken (20 000-80 000) in the Inner Niger Delta. This would amount to 15-60% of the population wintering in the area, a staggering figure which does not include natural mortality from starvation.

Catching Ruff in the Inner Niger Delta is highly biased against females. Around 10 March, very few males remain in Sahelian wetlands. Bird catching, however, continues till the last females have left in April. Given an average sex ratio in the wintering population of 19% males (Fig. 207), more than 80% of the captured Ruff must have been female in dry years (catching from late December onwards). This proportion increases to almost 100% in wet years (catching from late February onwards). Even when only 15% of the wintering birds is captured annually (a minimum), the selective killing of females must have a considerable impact on the population level.

Fattening in dry years The increased predation risk in dry years is just one problem to cope with. Another is related to food supply, and the necessity of depositing fat prior to the return migration. Most males Ruff commence fattening from 20 January onwards, females in the first half of February. The data presented by Pearson (1981) for Kenya suggest that Ruff wintering in East Africa start fattening 3-4 weeks later still, males as well as females; given the preponderance of Siberian birds, this timing does not come as a surprise. Average winter weights of Ruff captured in Guinea-Bissau (1986/1987), Senegal (1984/1985, 1986/1987), Mali (1998/1999, 2004/2005) and Cameroon (1990/1991, 2000/2001) did not differ significantly between years, for males nor females. Rates of fattening up were also very similar between sites and years. On average, males in early

March are up to 40% heavier than in midwinter, a surplus needed to fuel the flight from West Africa to Europe (as found for the other way round; Koopman 1986). Females depart at a later date, and consequently start depositing fat from late February onwards (Fig. 215). After fattening, body masses of males and females on average correspond to 246 g and 145 g respectively (Melter & Sauvage 1997).

Under average Sahel conditions, males are able to achieve a weight gain of 1% per day when fattening up; it then takes about five weeks before fat deposition is completed, *i.e.* by about mid-March. Assuming a similar rate of weight increase in females, they would be ready to leave around 1 April. In 1985, a very dry year, however, by mid-March 82% of the captured females in the Senegal Delta still had midwinter masses. Obviously, the birds had been unable to build up pre-migratory fat deposits (although they spent more time feeding in dry years: Tréca 1983). The situation was even more unfavourable in the Inner Niger Delta in 1985, where by late March thousands of reeves were recorded standing inactive on dry ground in the blazing sun (van der Kamp et al. 2002b). In view of the body masses of 27 females captured in March of this exceptional year, which were well below midwinter values (Fig. 215), the birds were emaciated and doomed to die (as many did). The toll taken under dry Sahel con-

ditions must be high, especially among females. Counts on a roost near Valle Zavelea, in the Po Delta in Italy, where in spring Ruff pass through that have wintered in the western Sahel (Fig. 208), revealed up to 2000 birds in 1985, compared to 10 000 in 1983 (Serra et al. 1990); whether the five times lower figure in the dry Sahel year really reflects winter mortality remains a question, but it is tempting to suggest.

Apart from on-the-spot mortality, we have evidence that even moderately dry Sahel years have a large impact on Ruff after they have left the floodplains. First, birds arrive with lower than average body masses on the European refuelling stations and, secondly, survival of juveniles is lower. These findings are based on Ruff caught and ringed during spring migration on major staging areas in Fryslân, northern Netherlands (Jukema et al. 1995, 2001b) and on the Pripyat floodplain in southern Belarus (Pinchuk et al. 2005, Karlionova et al. 2007). The rate of weight gain at these sites varied between years (Fig 216). In Fryslân, the variation in body weight was not related to local rainfall ($R^2=0.00$) or April temperature ($R^2=0.04$; weather data from www.knmi.nl), variables which influence the availability of food in grasslands. However, if weight gain was plotted against flood size in the Sahel in the preceding winter, a positive effect became apparent, for males as well as females, in Fryslân (Fig 216C) as well as in the Belarus (Fig 216D). The number of years is still too low to give statistically significant trends. We need more data to substantiate carry-over effects of dry Sahel years on arrival date and body mass at arrival in Europe, and to what extent such effects influence reproduction and later survival.

The proportion of first-year birds among migrants captured on European refuelling grounds is high during the outward migration (53.1% in July-October, n=538), much lower on the West African wintering grounds (22.3% between 1 November and 20 February, n=2149) and still less during return migration (13.3% between 21 February and 31 May, n=5412). This decline probably reflects a higher mortality rate among first-years compared to older birds. Even so, the proportion of first-years among captured migrants on staging sites in The Netherlands (Fryslân) in spring showed wide annual fluctuations, *i.e.* varying between 2.8% and 18.5% in 2003-2007. On average, fewer juveniles returned after dry Sahel years, but the correlation (R=+0.33), based on the small sample size of five years, was non-significant.

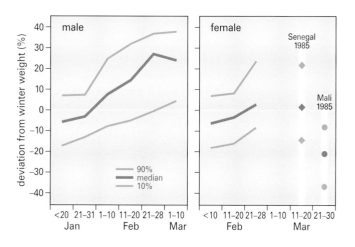

Fig. 215 Frequency distribution of body masses relative to average winter mass (set at 0) per decade for males and females. [3] Birds are captured in the Senegal Delta (Feb-Mar 1985, Nov-Dec 1986, Jan-Feb 1987; n = 1398), Inner Niger Delta, Mali (Mar 1985, Jan 1999, Mar 2001, Feb 2005; n=120), Waza Logone floodplain (Jan 2001; n=60) and Guinea-Bissau (Jan-Feb 1987; n=29). Data before 20 January and 10 February for respectively males and females have been lumped, because fattening starts at a later date. Note that for females the data only cover the start of the fattening; later catching efforts were restricted to the exceptional March 1985. From: OAG Münster (1989a-b, 1991b, 1998), A. Beintema (unpubl.), G. Gerritsen (unpubl.), B. Spaans (unpubl.), J. Wijmenga (unpubl.), own data. The weights of birds in Mali in March 1985 refer to dying birds.

Breeding conditions Tracking trends in breeding numbers is an arduous affair in Ruff; the variable mating system (monogamous or polygynous), varying operational sex ratios (excess of males in the northwest, excess of females in the east), lack of territoriality and unobtrusive breeding behaviour of females frustrate standardisation of census methods. Estimates of breeding numbers are often based on counts of breeding reeves or males on leks, and should be – at best – regarded as educated guesses. Zöckler's (2002) summary of known population estimates and trends, including many Russian and Siberian sites, paints a varying picture of declining, but also, locally, stable or even increasing populations. The vast majority breed in remote, relatively unspoilt (sub)arctic peatlands and tundras, where numbers in the long run have been more stable or in slight decline only (Zöckler 2002). Any changes here in density are more likely to reflect local weather and conditions in the wintering quarters than land changes in the tundra. Indeed, the breeding density on the Yamal peninsula showed large annual variations in synchrony with the size of the floodplains in the Sahel, some 8000 km away, in the previous winter. The very cold spring of 1983 resulted in delayed arrival on the Yamal peninsula and a lower breeding density (Fig. 217). For the tundra populations it is hypothesised that a

Unmanaged flood meadows offer important breeding habitat for Ruff. Such ecosystems still exist in eastern Europe. Timing and extent of haymaking depend on the growth stage of the grass and the accessibility of the meadows, being more protracted in wet seasons. The cut hay is dried on haycocks (see lower left) and usually transported to the haystack months later over the ice. When not cut at least once a year, meadows rapidly become overgrown with alder saplings, and turn ultimately into flood forests. (Belarus, July 1998).

range contraction may be in order when the present climate change is going to continue unabated (Zöckler 2002). Huntley *et al.* (2007) derives from the current predicted climate change that in Ruff its simulated future potential distribution is reduced by 58%, gaining a bit in the very north, but losing a lot in the southern part of its distribution area.

In the western and southern parts of the breeding range, Ruff breed in wet meadows in farmland and river valleys. These habitats underwent vast man-induced changes in the 20th century, notably via flood control, drainage and use of fertilisers, and many former breeding haunts have been abandoned (van Rhijn 1991, Hagemeijer & Blair 1997). The decline in Germany and The Netherlands during the second half of the 20th century averaged 14% per annum; by 2000, in both countries the species neared extinction (Fig. 217). In Denmark, where in the early 20th century Ruff used to be a widely distributed and common breeding bird, numbers had dropped to 1219 'pairs' in 1964-1972, 750 'pairs' in the mid-1980s, 500 in the mid-1990s and 150 females in 2000-2002 (Thorup 2004). Here, the decline was much more moderate in well-managed nature reserves than elsewhere. Changes in management also played a decisive role in the fortunes of Ruff breeding in Lake Engure, Latvia, where the termination of grazing of meadows in 1957 favoured vegetation succession with subsequent loss of breeding habitats for waders (Viksne *et al.* 2005). The habitat-related changes in farmland in Europe are so vast and continuous that any Sahel-effect on numbers is overridden.

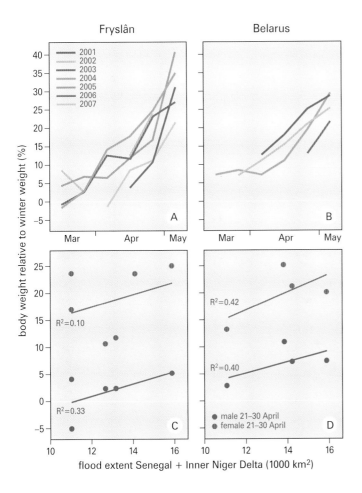

Fig. 216 Weight gain (%) of male Ruff in spring in (A) Fryslân (The Netherlands) and (B) the Pripyat Valley (Belarus) relative to their winter weight between mid-March and early May. The relative body weight of male and female in late April in (C) Fryslân and (D) Belarus in different years as a function of the flood extent of the Senegal and Inner Niger Deltas in the preceding winter. From: J. Hooijmeijer, Y. Verkuil & T. Piersma (unpubl.) for the Frisian data and Karlionova *et al.* (2007) and N. Karlionova (unpubl.) for Belarus.

Conclusion

The vast majority of Eurasian Ruff winter in Africa, notably in Sahelian floodplains. Population fluctuations on the Yamal peninsula mirrored variations in the flood extent of Sahelian floodplains in the preceding winter. Droughts in West Africa increase the vulnerability of wintering Ruff to human predation, because smaller floods enforce the birds to concentrate in fewer wetlands earlier on in winter. At the same time, pre-migratory fattening is more difficult (less food in fewer sites). Human predation on Ruff in the Inner Niger Delta varied between 10% and 60% per annum (measured in the 1990s and 2000s), with the highest predation rate in the driest years. Reeves are disproportionately affected during droughts, because males fatten up and depart earlier (from mid-February onwards) and hence are exposed to human predation for a much shorter period of time. Also, females carry the full burden of droughts, having to fatten up under extremely adverse conditions. In some years, such as 1985, this may lead to mass mortality from human predation and starvation. Carry-over effects of droughts are visible in a delayed arrival, and lower body masses, on the staging sites in Europe. Irrespective of Sahelian conditions, Ruff in western Europe have been in steep decline throughout the 20th century, mainly due to loss of breeding habitat. By the turn of the century, fewer than 10% of the West European numbers remain. Overall, the Eurasian Ruff population may be in decline, as evident from the declining numbers counted in the Sahelian floodplains since the 1970s.

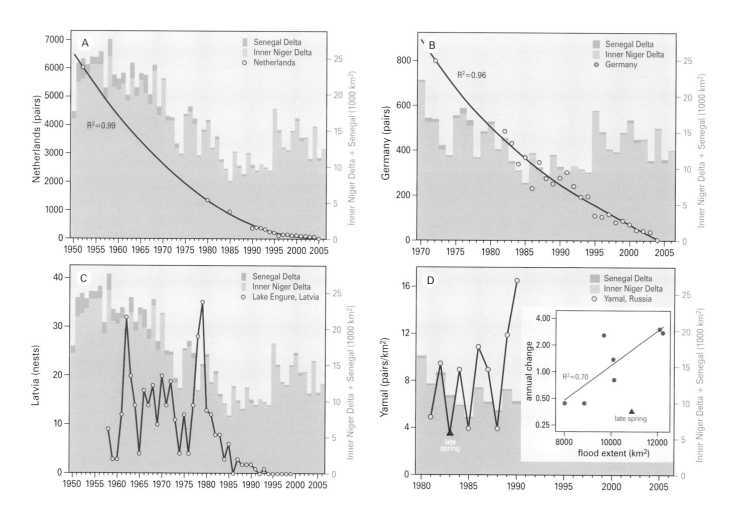

Fig. 217 Population trends of the Ruff in (A) The Netherlands, (B) Germany and (C) Latvia (Lake Engure) and (D) Yamal Peninsula (Russia) compared with the flood extent of the Senegal and the Inner Niger Deltas. The % change in density on Yamal between two successive years is positive with high Sahelian floods in the previous winter, and negative during low floods. In 1983, May temperature on Yamal was 4°C below average (measured at three meteorological stations nearby). The trend, excluding 1983, is significant (P<0.01). From: Bijlsma et al. 2001, sovon (The Netherlands), Hötker et al. 2007 (Germany), Viksne et al. 2005 (Latvia), Ryabitsev & Alekseeva 1998 (Yamal).

Many Ruff from Chukotka in the northeastern extremity of Asia (right) spend the northern winter in S and E Africa, but most Ruff from Fenno-Scandinavia (left) winter in the western Sahel.

Endnotes

1 Why are males the cock of the walk during spring migration? Several explanations have been put forward for the widespread phenomenon of protandry (e.g. Myers 1981, Cristol et al. 1999, Tøttrup & Thorup 2008). One explanation for a latitudinal segregation and early spring departure of males postulates a selective premium on being close to the breeding grounds (Myers 1981). For male Ruff, arriving early on the staging grounds, halfway the northward migration, it may be particularly important as it is here that they moult into breeding plumage before going further north to start lekking on the breeding grounds (Jukema et al. 1995, Jukema & Piersma 2000, Jukema et al. 2001b). Moulting into breeding plumage on the wintering grounds is not a serious option for males, as food is in short supply and barely sufficient to permit pre-migratory fattening. It is therefore profitable to postpone the energetically costly moult of the contour feathers till after the first extended leg of the return migration; in the temperate staging areas, food is abundant and competition for food restricted. In contrast to males, females are less in a hurry as the food on the breeding grounds will only be accessible from May onwards. On a species level, differential migration may thus reduce competition on the wintering and staging grounds.

2 The dataset of trapped birds in Fryslân (The Netherlands), Belarus (Karlionova et al. 2007), and Italy contains the following male-percentages per month: Fryslân in 2003-2007 for March, April and May resp. 97.5, 84.1 and 72.7%, Belarus in 2001-2004 resp. 100.0, 70.7 and 47.5% and Italy in March and April 1985, 1988 and 1989 61.1 and 33.9% (Serra et al. 1990).

3 To correct for size-related differences in body weight, body weight (BW, g) was regressed against wing length (W, mm). A selection was made for birds captured in West Africa from 1 November to 20 January (males; n=895, weight 175.6 g on average) or from 1 November to 10 February (females; n=225, 103.8 g on average). The equations appeared to be different for the sexes: males $BW= 0.003328 W^{2.067}$ ($R^2=.994$) and females: $BW=0.09257 W^{0.929}$ ($R^2=.929$); regressions were calculated for average body weight per mm class of wing length.

Gull-billed Tern
Gelochelidon nilotica

It was 5 October 1988. In the desert north of Iwik on the Banc d'Arguin, someone looking down from above would have seen the Mauritanian landscape take on a flickering quality, but at ground level the speckling resolved itself into a transient phenomenon that is one of Nature's finest – butterfly mass migration across a treeless landscape. Innumerable Painted Ladies *Vanessa cardui* had been passing on their southward migration in an uninterrupted stream for three hours; as far as the eye could see, butterflies galore. To get some idea of the scale of this happening we came up with the following figures: over a band 50 m wide, an average of 231 butterflies passed every 5 minutes. This tranquil scene suddenly changed, however. The wind direction swung round to the north, the sky turned inky-black and we were buffeted by gusts of wind, harbingers of a heavy thunderstorm. Once the storm had passed, first a few and then abruptly huge numbers of Desert Locusts drifted with the desert wind towards the Atlantic Coast, and beyond. On this occasion, the cliché, 'the sun was literally blanketed', was deserved. So many locusts! The beaches and tidal flats of the Banc d'Arguin were smothered by living, dying and dead insects, piles littering the tide line. It took some days for numerous Whimbrels and Turnstones, and the occasional Grey Plover, to begin to exploit this profusion of easy prey. Other wader species ignored it, sticking to their usual diet of tiny worms. But to the Gull-billed Terns, this was manna from Heaven! Without hesitation, they abandoned the pursuit of their usual prey, fiddler crabs, and began immediately to crush and swallow locusts as if these had always been their day-to-day preference! In West

Africa, Gull-billed Terns are flexible in their prey and habitat choice. In Guinea-Bissau they feed on fiddler crabs as long as the tidal flats are exposed (Stienen *et al.* 2008), but at high water they will patrol above tidal creeks in the mangroves, fishing or taking insects from the air, vegetation or bare ground. If a tern were defined as a fish-eater, the Gull-billed Tern would fail that test in its African wintering quarters and on its breeding grounds, but even such adaptable feeding behaviour cannot prevent a steady decline in numbers over much of its range.

Breeding range

The world population of the Gull-billed Tern was estimated at 55 000 pairs (del Hoyo *et al.* 1997), of which 10 000 – 12 000 pairs in Europe and Africa between 1980 and 2000 (Sánchez *et al.* 2004, correcting for the recent decline on the Banc d'Arguin, Isenmann 2006). The species almost disappeared from its northernmost breeding range in Denmark and Germany (50-60 pairs), but has shown increasing numbers in South Europe: Italy (325-450 pairs), southern France (250-300 pair) and Spain (3000-3500 pairs). These, the 180-1000 Mauritanian pairs and the much smaller numbers elsewhere in NW Africa, form the western population (4660-4850 pairs). The eastern population in Greece, Turkey and Ukraine may be estimated at 5200-6800 pairs.

Migration

Ringing recoveries and observations of visible migration both suggest that breeding birds from SW Europe winter in West Africa, and that birds wintering in East Africa originate from the Balkans and eastern Europe (Møller 1975c; Cramp & Simmons 1983). There are very few recoveries from sub-Saharan Africa to confirm such a dichotomy, namely, a recovery of a Danish bird from the Casamance (S Senegal) and another one from Mauritania. An ornithological expedition to Guinea-Bissau (Wymenga & Altenburg 1992) captured 112 Gull-billed Terns in 1986/1987, of which four had already been ringed. Three of these had been ringed as nestlings on 1 July 1986 at a single site (Lucio del Cangrejo in the Coto Doñana, Spain), their recapture being to the utter surprise of their ringers! The fourth

Gull-billed Tern with large fiddler crab, its main prey while wintering along the West African coast.

bird, however, carried a Russian (Moscow) ring and had been ringed as chick in Ukraine, on the Black Sea coast. That some eastern European birds winter in western Africa was further corroborated, and in rather remarkable circumstances, by the only recovery, from the Inner Niger Delta; this bird had also been ringed in Ukraine, but at exactly the same spot (85 km east of Odessa) and on the same day (14 June 1983) as the other 'eastern' recovery!

Non-breeding birds may leave Africa and wander around in southern Europe during the northern summer (Møller 1975c; Cramp & Simmons 1983). Counts in the African wetlands, however, suggest that many non-breeders may remain south of the Sahara. On average 503 birds (range 111-1190) were registered during eight June counts of the central lakes in the Inner Niger Delta between 1999 and 2006, which represents, on average, 21% of the numbers counted in the preceding January-March. If all wintering birds less than 3 years old remained in the area during the northern summer, we would estimate that the oversummering population would be 40% of the numbers comprising the wintering population. In two years out of eight, that proportion was actually 38% and 39%, but was lower in all other years.

Distribution in Africa [1]

Gull-billed Terns have a disjunct winter distribution in Africa, concentrated both in West Africa (mainly coastal, but with substantial numbers in the Sahelian floodplains) and in East Africa (centred around Lake Victoria) (Fig. 218).

The combined counts along the West African coast indicate a wintering population of 11 000 - 15 000 Gull-billed Terns. These birds adhere strictly to the tidal zone, possibly except those on the Langue de Barbarie, a beach 25 km south of Saint-Louis, that we assume have to feed offshore. 5000-8000 Gull-billed Terns spend the northern winter in the West African floodplains and lakes.

Gull-billed Terns winter on the Banc d'Arguin, Mauritania, in relatively small numbers, from 61 to 860 during midwinter counts (Hagemeijer et al. 2004). After the breeding season, most birds desert their colonies because the species' main summer prey, fid-dler crabs Uca tangeri, in winter emerge from burrows only at spring low tides, and so are mostly inaccessible (Zwarts 1990). In Senegal, wintering numbers counted at Langue de Barbarie showed wide fluctuations; i.e. 800 in 2001 and 4135 in 2003. In the Sine Saloum, the number of Gull-billed Terns varied from 612 in 1997, to 540 in 2000, 517 in 2003 and 233 in 2004 (the 2003 count was complete, but the status of those from other years is not known). We have not found any counts for the Casamance River and only incomplete counts for the Gambia River (maximum 86 in 1999).

The tidal flats between Guinea-Bissau and Sierra Leone attract some 7200 Gull-billed Terns, of which 4000 are in Guinea-Bissau (3 birds/km² tidal flat; Zwarts 1988, Wymenga & Altenburg 1992), 2400 in Guinea-Conakry (3.5 birds/km²; Altenburg & van der Kamp 1991) and 833 in Sierra-Leone (3.8 birds/km²; van der Winden et al. 2007). Further south and east along the Atlantic coast of West Africa, there are no extensive tidal flats, but small coastal wetlands prove equally attractive to Gull-billed Terns, such as the Vidri canal near Abidjan, Côte d'Ivoire (914 in 1999, and 1000 in 2000), Keta Lagoon, Ghana (max. 448 between 1986-1994; Piersma & Ntiamoa-Baidu 1995) and the Mono River Delta, Benin (110 in 1999, and 90 in 2000).

The floodplains and large lakes in the Sahel also are exploited by Gull-billed Terns. Midwinter counts of the Senegal Delta between 1988 and 2004 produced between 23 and 162 birds and the Hadejia-Nguru floodplains (N Nigeria) 31-174 (1994-1998). During an aerial survey of Lake Chad in 1987, 210 Gull-billed Terns were counted, probably an underestimate because 426 birds were reported for the Niger part of Lake Chad alone (NE corner) in March 2001, another 143 in a part of the Waza floodplains in 1996, and 220 along the Chari River upstream of N'Djamena in December 2003. The species is not common, however, since WIWO-expeditions in 1999 and 2001 to Waza-Logone and the southern edges of Lake Chad reported between 30 and 100 birds (Dijkstra et al. 2002; unpubl. WIWO-report). In January 1993, extrapolations from transect counts, river counts and roost counts produced an estimate of 500 birds in the Chad Basin and another 460 along the Logone River (van Wetten & van Spierenburg 1998). Hence we assume that about 1000 Gull-billed terns winter in Lake Chad and surroundings.

Winter aerial surveys of the Inner Niger Delta since 1972 counted

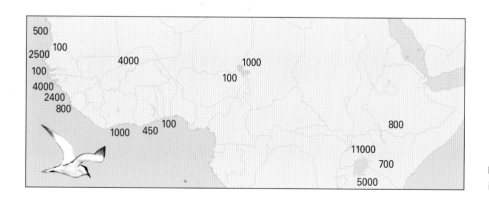

Fig. 218 Winter distribution of Gull-billed Tern in Africa.

Gull-billed Tern (top) and Whiskered Tern (below) winter in large numbers in the Inner Niger Delta.

between 102 and 3990 birds, with highest numbers in 1972 (3990) and 2001 (3725). However, we consider the early counts incomplete. Ground surveys in the central lakes in the 1990s, situated in the heart of the Delta (Chapter 6), came up with 1750 Gull-billed Terns, but since then numbers steadily have increased to 3900 birds in February 2007 (year maxima in Table 6). It is unlikely, however, that this increase is representative of the Inner Niger Delta as a whole. In 1992-1994, an estimated 6000-7000 Gull-billed Terns, including the birds counted during daylight, gathered at dusk in the central lakes. Many arrived from northern and western directions, presumably from feeding areas in the arid zone beyond the lakes. Pellets found on the roosts consisted mostly of remains of beetles, but also of grasshoppers and locusts, indicating a predominantly insectivorous diet during these dry years. In the 2000s, when conditions were wetter, such flights refer to small numbers, apparently because most Gull-billed Terns were concentrated in the central lakes where they foraged on fish. This would indicate that the population wintering in the central lakes of the Inner Niger Delta has decreased by 40% (from 6000-7000 to 3900), rather than increased as suggested by the counts in the central lakes. Even when taking account of the estimated 100-200 Gull-billed Terns present in the Inner Delta outside the central lake area in the early 2000s, the decline stands.

In East Africa, Gull-billed Terns are found mainly on Lake Victoria and on water bodies scattered along the Rift Valley. Summed midwinter counts for East Africa varied between 1705 in 1992 and 14 634 in 1998, a variation largely attributed to incomplete coverage of the wetlands involved. The peak number in 1998 hinged mainly on the 10 958 birds counted in Uganda, when five teams covered all major wetlands including the northern half of Lake Victoria. That same year 2241 Gull-billed Terns were recorded in Tanzania, 760 in Ethiopia and 675 in Kenya. The largest number in Tanzania (4992) was counted in 1995, the only complete waterbird survey there in the 1990s. Taking all this information together, we may tentatively estimate that some 17 000 Gull-billed Terns spend the winter in Ethiopia, Uganda, Kenya and Tanzania.

Of the bird species wintering in Africa, the Gull-billed Tern is one of the few whose numerical distribution is pretty well known: 11 000-15 000 along the West African coast, 4000 in the Inner Niger Delta, 1000 in Lake Chad, and 17 000 in the Rift Valley, in total 33 000-37 000 birds. This is similar to the European-African breeding population of 10 000-12 000 pairs, being equivalent to a midwinter population of 30 000-36 000 birds. [2]

The numerical distribution across Africa is another clue that Gull-billed Terns wintering in West Africa must originate partly from eastern Europe. The ratio of western (including NW Africa) versus eastern European breeding populations is 44:56, but the ratio of wintering populations in western and eastern Africa is roughly 52:48. If we assume that 100% of the western breeding population winters in West Africa, about 15% of the eastern European birds should also winter there.

Population change

The Gull-billed Tern is in a decline in most of its European breeding range (Sánchez et al. 2004). It is not clear, however, why the species in different parts of its distribution area shows conflicting trends. The Danish population, fluctuating between 300 and 500 pairs before 1950, plunged to near-extinction in the 2000s (Møller 1975b, Rasmussen & Fischer 1997, Fig. 219). A small fraction of the birds had moved from Jutland to the (German) Wadden Sea in the late 1980s. Fluctuations in breeding numbers in Denmark and Germany seem to be correlated closely with annual flooding of the Inner Niger Delta (Fig. 219A). However, at the confirmed fixed annual 4% rate of decline, the impact of flooding disappears. [3] If the Sahelian flood extent would have an impact on the population, we expect that the annual rate of decline would be large in dry years and small in wet years, but no such a difference was found. From this we conclude that the decline is not related to the winter condition in the Sahel, i.e. the flood extent of the Inner Niger Delta itself.

The trend in SW Europe is very different. Gull-billed Terns in the Camargue, France, though steeply declining in the 1950s and 1960s, proved resilient and increased again since the early 1970s (Fig. 219B). This increase continued unabated during the Great Drought in the Sahel in the 1980s, and it is therefore unlikely that fluctuations in numbers are related to wintering conditions in the Sahel. Populations in Italy and Spain, on average, showed positive trends since the 1990s, but numbers at some sites in southern Spain fluctuated steeply between 0 and 2200 (Coto Doñana) and 0 and 1600 (Fuentepiedra) (Sánchez et al. 2004). The supposed decline in Turkey, Black Sea and Sea of Azov is difficult to substantiate with hard data (BirdLife International 2004a, Sánchez et al. 2004), and may reflect merely short time-series (as for the Black Sea Biosphere Reserve: in 1989-1993; respectively 520, 340, 127, 170 and 183 pairs;

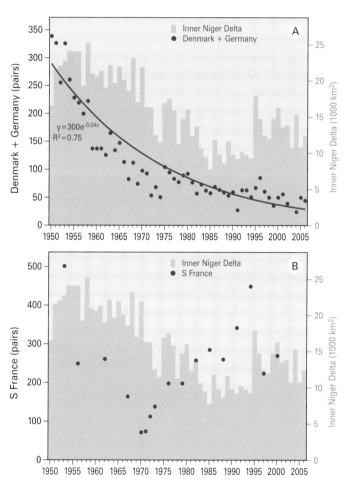

Fig. 219 The flood extent of the Inner Niger Delta related to the Gull-billed Tern breeding population in Denmark + Germany and in South France. From: Møller 1975b, Dybbro 1978, Rasmussen & Fischer 1997 (Denmark); Eskildsen & Hälterlein 2002 (Germany), K. Koffijberg (pers. comm.; both countries since 1992); Kayser et al. 2003 (South France).

Rudenko 1996) or steep fluctuations typical of this tern species' penchant for breeding in unstable habitats.

The breeding colonies on the Banc d'Arguin in Mauritania apparently have been stable for decades. Data collected by De Naurois (1959, see also 1969 for a reappraisal and field notes) indicated 1500 pairs in 1959 (not 6500, as cited by Møller 1975b in his review).[4] Subsequent surveys differed little: 1600 pairs in 1974 (Trotignon 1976) and 1180 in 1984-1985 (Campredon 1987), but since then, the breeding population did decline to 660 pairs in 1995 (Gowthorpe et al. 1996) down to 180 in 1997, although numbers recovered to 1000 pairs in 1999 (Isenmann 2006).

Long-term declines are associated mostly with a low reproduction (Rasmussen & Fischer 1997) and permanent habitat loss on the breeding grounds (Sánchez et al. 2004). There have been no large, negative habitat changes on the African wintering grounds: the Sahelian floodplains and along the coastal and littoral zones of the Atlantic Ocean. The impact of hunting in West Africa is probably smaller than in coastal *Sterna* species, although in the Inner Niger Delta, the local people successfully catch Gull-billed Terns by targeting nocturnal roosts.

Conclusion

For much of the 20th century, the European Gull-billed Tern has been in decline. This trend almost led to its loss as a breeding bird in Denmark and Germany. The wintering grounds are mainly in West Africa (Atlantic Coast and inland floodplains) and on lakes in the Rift Valley (notably Lake Victoria). An estimated 15% of the eastern European breeding population is thought to winter in West Africa. Since the 1980s there has been some recovery recorded in Spain, southern France and Italy. None of the trends could be linked to conditions in Sahelian wintering areas.

Endnotes

1. All data summarized in this section refer to counts in January, taken from the AfWC database (Wetlands International), unless stated otherwise.
2. The applied multiplier (winter numbers = 3 x pairs) is based on: (1) Fledging success of 0.8 young per pair (Cramp & Simmons 1983, Molina & Erwin 2006); (2) A mortality rate of 52% in the first and 23% in subsequent years (Møller 1975a); (3) Half the post-breeding mortality occurs before January; (4) First breeding occurs at 3 years of age (Møller 1978).
3. The size of the Danish-German population correlates well with the flood extent of the Inner Niger Delta and Senegal Delta ($R^2=0.62$; second degree polynomial; $P<0.001$). The fit is much better ($R^2=0.75$, $P<0.001$; Fig. 219A) when we assume that the population showed a constant decline of 4% between 1950 and 2007. The flood extent cannot explain the deviation from this long-term trend.
4. Møller (1975b) refers to De Naurois (1959) for his figure of 6500 pairs on the Banc d'Arguin. De Naurois (1959), however, mentions 2000-3000 nests for the island of Zira and fewer than 500 elsewhere on the Banc d'Arguin (some dozens on two islets and 100-200 on three other islets), which would give a total of 2500-3500 nests in 1959. In his detailed overview ten years later De Naurois (1969) summarised his field data for 1959-65. Since not all colonies were visited each year, it is difficult to derive annual totals, but large colonies such as on Zira in 1959 were no longer encountered. He mentions 400-500 nests as the average colony size on Zira. Moreover, he downscales the earlier estimate for Zira to 1000 nests, bringing this figure more into line with his other field data (surface area of Zira is 4 ha, of which the tern colony, according to his own map, covered only a part, with nearest-neighbour distances estimated at 2, 5 and 10 m). Consequently a figure of roughly 1500 breeding pairs on the Banc d'Arguin is more realistic for 1959, instead of 6500.

Caspian Tern
Sterna caspia

Just opposite Akka, on the sandy beach of Youvarou, a small child is sticking twigs in the sand, about 30 cm from the water's edge and about 10 m apart. How old can he be, 8 or 9? Clearly he is not playing but involved in serious business: catching birds for food. Attached to each stick is 30-50 cm of fishing line with a single hook that is baited with a small fish, about 5-10 cm long. The fish is carefully laid in shallow water, almost entirely submerged but clearly visible from above. Quite near the boy, and not in the least disturbed by his presence, large terns patrol some 5 to 25 m above the water's edge. They utter a loud, harsh and protracted squawk every now and then. Flying back and forth, red bills pointed downwards, they occasionally plunge-dive into the water or sweep the shallow margins to catch a prey by bill-dipping. Caspian Terns! In this world of few gulls, Caspian Terns may at first seem something of an anomaly in the Inner Niger Delta. In fact, this wetland is one of the most important, yet dangerous, wintering sites for Baltic and Ukrainian breeding birds. The boy laying out baited fishing lines represents a very real danger, for the Caspian Tern is particularly vulnerable to such lures.

Many islands along the West African coast are safe breeding sites for terns, which on the mainland would fall victim to predators like Jackals. Caspian Terns are mostly concentrated on the Banc d'Arguin, Mauritania (10 900 pairs in 1998), but they also breed in the Bijagos Archipelago, Guinea-Bissau (left; November 1992). Royal Terns have huge colonies in estuary of The Gambia (right; February 2004), totalling some 43 000 pairs in 1999, as well as on the Banc d'Arguin (15 000 pairs in 2004) (data from Isenmann 2006).

Breeding range

Breeding almost worldwide in geographically isolated populations, in Europe Caspian Terns also have distinctly disjunct breeding areas. The larger breeding colonies are concentrated in the north Caspian (the Volga Delta, 3000-5500 pairs in 1984-1988) and 250-800 pairs along the northern coast of the Black Sea (Heath *et al.* 2000. Apart from 2-3 small colonies in Turkey (totalling about 300 pairs; Heath *et al.* 2000, BirdLife International 2004a), the only other population of importance in Europe breeds in the Baltic. In Finland, Sweden and Estonia, some 1500-1600 pairs were present in the 1990s and early 2000s (Hario *et al.* 1987, Svensson *et al.* 1999, Hario & Stjernberg 1996, T. Stjernberg pers. comm.).

The Banc d'Arguin (Mauritania) harbours large colonies (10 900 pairs in 1998; Isenmann 2006). Small numbers are known to have bred occasionally in Tunisia, *i.e.* in the late 19th century, mid-1950s (at most, 23 nests) and mid-1990s (Isenmann *et al.* 2005). The Egyptian population, breeding on islands at the mouth of the Gulf of Suez and in the Red Sea, was estimated at 250-350 pairs in the 1980s (Goodman & Meininger 1989).

Migration

The high recovery rate of Caspian Terns in Africa, based on decades of intensive ringing in the Baltic and the Black Sea (in the 1970s, about 50% of the 1000 Swedish pairs carried a ring; Staav 1979) has allowed several detailed analyses of their whereabouts and chances of survival outside the breeding season (Shebareva 1962, Soikkeli 1970, Staav 1977, Staav 1979, Staav 2001, Kostin 1983, Kilpi & Saurola 1984). The number of recoveries has increased since then. For our analysis, we used 640 ring recoveries in sub-Saharan Africa north of the equator. The great majority of these birds were killed, only 7 were reported as captured and released.

Contrary to most other tern species, Caspian Terns routinely fly overland in Europe and in Africa on their way to and from their wintering quarters. The Sahara is clearly not a barrier, as shown by inland ringing recoveries on 24 August (Tunisia), 7 September (Mali) and 15 October (Algeria). The Baltic population, augmented by birds from the Black Sea and possibly from the Caspian (Voinstvenskiy 1986), largely concentrates in the Inner Niger Delta, where they take advantage of the seasonal boom in food supply in the inundation zone (Fig. 220). The Niger River flow reaches a peak in September, inundates the Inner Delta and temporarily converts an area of up to 20 000 km² into endless spawning grounds for fish, by which time Baltic Caspian Terns have started their southward overland migration, reaching the central Mediterranean in September and October (Staav 2001). By November most birds have crossed the Sahara to reach their wintering quarters in Mali (Kilpi & Saurola 1984, Staav 2001).

Birds from the Black Sea and Volga regions depart from the breeding grounds from mid-August until October, arrive in the eastern and central Mediterranean in October to reach the wintering grounds along the Guinean Gulf and in the Inner Niger Delta in November (Voinstvenskiy 1986, Patrikeev 2004).

The significance of Mali is just as high for Caspian Terns breeding in Sweden, Finland and the Ukraine. Of the recoveries in Africa,

78% of Finnish and 74% of Swedish birds came from Mali; even the much smaller sample from the Ukraine showed a clear preponderance of recoveries in Mali (8 of 9 birds). The winter distribution of Baltic and Black Sea populations south of the Sahara does not differ along an east-west gradient; the concentration in the floodplain of the Inner Niger Delta is overwhelming, despite the scattering of recoveries from southern Chad to as far as Senegal. We also investigated whether birds from the northern Baltic spend the winter further south than birds from the southern Baltic, but no such difference was found.

Most recoveries were from Senegal (n=20), Ghana (49), Egypt (89) and Mali (383), out of whose total 352 came from the Inner Niger Delta alone. Remarkably, of the 24 recoveries of Baltic birds before 1950, 23 were reported from Egypt and only one from Mali. The Nile Delta may have been a more important wintering area for Caspian Terns than nowadays due to the loss of feeding grounds. Support for this idea comes from the ratio of recoveries in November-February to that in March-October: 15 out of 23 recoveries (65%) before 1950 came from the Nile Delta, as opposed to 22 out of 68 (25%) after 1950. Unfortunately, very few field data exist to substantiate the ringing data, especially for the pre-1950 period. Meinertzhagen (1930) considered the Caspian Tern in Egypt uncommon during return and outward migration, and even less common in winter.

Return migration from the Inner Niger Delta is initiated from the central lakes in late March/early April, when suddenly the concentration of Caspian Terns dissolves. This phenomenon has not actually been witnessed, despite many years of observation at Lake Debo from the 1980s onwards. This suggests a nocturnal departure, presumably non-stop to Europe given the scarcity of spring observations from waters between Lake Debo and Timbuktu and the near-absence of spring observations in Morocco east of the Straits (Thévenot et al. 2003), in Algeria (Isenmann & Moali 2000) and Tunisia (Isenmann et al. 2005), as opposed to regularity of occurrence in autumn. Europe also experiences a similar scarcity of spring observations (Cramp 1985). Arrival on the breeding grounds, as summarised in Glutz von Blotzheim & Bauer (1982), averaged 10 April in southern Sweden (first on 1 April), 12 April in Estonia (range 4-22 April; Leibak et al. 1994), 17 April in S Finland (range 12-26 April), 22 April in SW Finland (Turku, range 14-29 April) and 4 May in the northern Gulf of Bothnia (Finland, Oulu, range 28 April-10 May). Black Sea breeding birds must arrive in early April on the breeding grounds, as pair-formation is confirmed immediately after arrival, usually the first eggs being produced around 20-25 April, the bulk following 5-10 days later; in early spring 1978, egg laying had already started by 10-15 April (Voinstvenskiy 1986). The time difference of only about a week between departure from Mali and arrival in the Baltic and Black Seas suggests a direct and almost non-stop overland flight to the breeding quarters.

Distribution in Africa [1]

The midwinter counts suggest the presence of 25 000 – 30 000 Caspian Terns along the Atlantic coast of West Africa. The largest con-

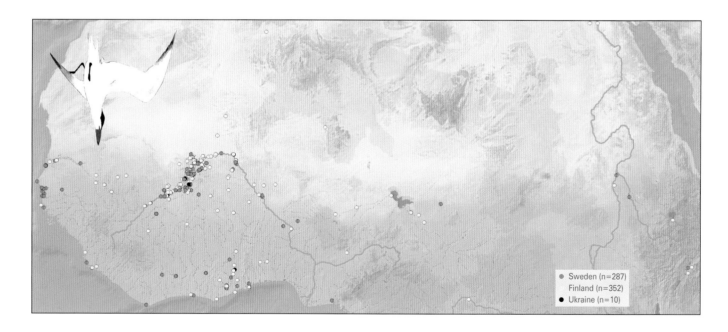

Fig. 220 European ringing locations of 518 Caspian Terns recovered or captured in Africa between 4° and 33°N. From: EURING, except for 11 birds from Mali (Wetlands International).

centrations are found in the Baie d'Arguin (1920 in 1997), Banc d'Arguin (5069 in 1997), Bell (1543 in 1999), El Ain (2400 in 2004), Langue de Barbarie (669 in 2004), Sine Saloum (2788 in 2001) and the Gambia River (8674 in 1998). 2900 Caspian Terns were counted in Guinea-Bissau in 1986, but the total number was estimated at 5000 (Wymenga & Altenburg 1992). Smaller numbers have been observed in Guinea (2500 in 1988, Altenburg & van der Kamp 1991) and Sierra Leone (44 in 2005; van der Winden et al. 2007). The two midwinter counts in Cote d'Ivoire produced almost the same numbers: 1374 in 2001 and 1290 in 2002.

Much lower numbers occur on inland waters. The Senegal Delta midwinter counts (excluding the estuarine part) revealed 100-150 birds; 76 were seen during the aerial survey of Lake Chad in December 2003 and only about 8 in the Hadejia-Nguru floodplain midwinter counts from 1994 to 1998. Aerial censuses of the Inner Niger Delta were more rewarding: 2414 in 1979, 2467 in 1994, 2956 in 1996, 2917 in 2003 and 2606 in 2004, a remarkably consistent series; most of these birds are concentrated in the central lakes, where 2500 - 3000 have been counted during ground-based surveys between 1992 and 2007 (Fig. 54).

The many recoveries from the Inner Niger Delta and the few from the Atlantic coast (Fig. 220) suggest that the birds in Mali originate from Europe and those from the Atlantic coast mainly from local breeding areas.

Population changes

Winter mortality Our database contains 640 recoveries and captures from northern Africa, 633 of which involving birds that were killed or found dead. In between being ringed as a chick and recovered dead in Africa, 16 birds were recaptured more than once on the breeding grounds. For example, a bird ringed as a nestling in Finland in 1970 was controlled as a breeding bird on its nest in Sweden in 1981, 1982, 1983, 1985 and 1988, only to be killed at 20 years of age in Ghana in January 1990. Another Finnish bird, ringed in 1968 also as nestling, was observed on its nest in 1976, 1977 and 1989, before being shot in the Inner Niger Delta in February 1992 at the age of 24. A Swedish bird, captured in the Inner Niger Delta by hookline on 1 April 1997, had been ringed 31 years earlier as a nestling in June 1966 and had been on its nest in 1977 and 1984. This last bird, if it had commuted annually between Sweden and Mali, a distance of 6000 km, would have covered at least 360 000 km on migration during its lifetime.

The many recoveries from the Inner Niger Delta permit an analysis in relation to annual variations in flooding from September through March. Most recoveries came from the dry climatic episode of 1973 to 1994, and in particular, from the extremely dry winter of 1984/85 (henceforth, '1985') (Fig. 221). Fewer recoveries were reported in the wetter years that preceded or followed this period. To calculate survival rates, we have to take age-specific mortality into account, under the assumption that mortality did not change between years. This calculation is possible for 505 Caspian Terns ringed as chicks and recovered south of the Sahara (exact age known, but it excludes first-years that died en route to wintering areas). On this basis, survival of first-year birds was calculated at 68.5%, of second-years at 76.6%, of third-years at 80.4%, and of older birds, at 82.3%; survival in the last-named category did not change further with age. Applying this calculation to all birds ringed between 1947 and 2002 in Finland and Sweden (n=64 997; see blue line in Fig. 221) and recovered in the Inner Niger Delta, the summation of all survival probabilities results in the red line in Fig. 221.[2] Between 1950 and 1970, the chances of finding a ringed Caspian Tern increased gradually, stabilised at a high level until 1990 and decreased afterwards.

Fig. 221 seems to indicate that low flood levels in the Inner Niger Delta result in proportionally higher recovery rates. To examine this tentative conclusion, the ratio of observed to expected recoveries was determined for seven categories of flooding (Fig. 223). Above-average recovery rates clearly are associated with low flood levels (<16 000 km^2), whereas 30% below-average recovery rates were found at flood levels exceeding 16 000 km^2. In other words, mortality as reflected in recovery rates is lower when flood levels are high.

Just as the annual variation in the number of recoveries is pronounced (Fig. 221), so is the seasonal variation, with a distinct peak in December-January (Fig. 224). The small number of recoveries between April and August is to be expected, since most Caspian Terns are going about their breeding business in Europe and the floodplain is not yet inundated. Further analysis is therefore restricted to the period September-March. Within this time-frame the birds face

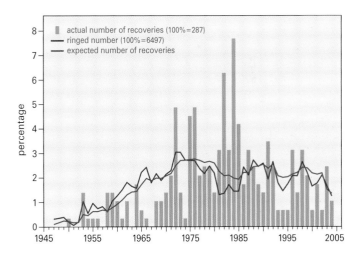

Fig. 221 Annual and expected number of recoveries (%) in the Inner Niger Delta (100% = 287 birds, for which recovery date is accurate to within 6 months). The expected number is derived from the annual number of ringed birds in the Baltic assuming an age-related survival rate (see text).

a vast annual variation in the position and size of their fishing grounds in the floodplain of the Inner Niger Delta. When the flood level is high, peak inundation is reached in December and a substantial part of the area is still covered by water when the birds leave for Europe around 1 April. In contrast, in a dry year peak flooding will have occurred in November and the floodplain will be largely dry by February, except for the Niger River, its tributaries and some permanent lakes in the centre (Fig. 37-39).

We calculated the size of the floodplain for all months from September to March for all years separately (from data given in Chapter 6). A comparison between the monthly number of recoveries and the size of the floodplain clearly shows that floodplain size is inversely correlated with the number of recoveries (Fig. 223). To investigate whether this result is spurious, due to the change in numbers of ringed Caspian Terns during the past 60 years, we repeated the analysis for 10-year periods, and although variation increased, the trend remained the same for these periods: the less flooding, the more recoveries. Put succinctly, mortality of Caspian Terns in the Inner Niger Delta is high at the end of their wintering period, especially during dry years.

If the calculations so far suggest that more Caspian Terns die at low flood levels, we would expect that annual fluctuations of the size of the wintering population would match the variation of flooding of the Inner Niger Delta. Counts of Caspian Terns scattered over the entire Inner Delta are lacking, but as soon as the water level at Akka (Central Lakes) has dropped to 200 cm or less, Caspian Terns start to concentrate in the central lakes and their presence remains more or less constant for the rest of the wintering period despite the ongoing decrease of the water level caused by the receding flood. As

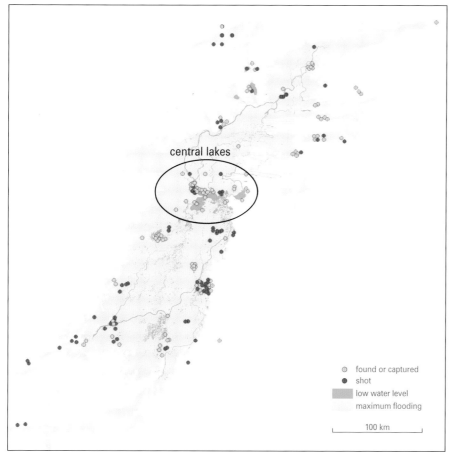

Fig. 222 Distribution in the Inner Niger Delta of Caspian Terns shot, captured or found dead. Most birds captured or found dead occurred where they are present in large numbers, while those around the outer edge of the floodplains or near cities are more likely to have been shot.

explained in Chapter 6, the central lakes become a last refuge when the surrounding floodplains dry up. For 15 different years, we have available counts of Caspian Terns in the central lakes (before 15 March) when the water level at Akka was less than 200 cm (Chapter 6). Before 1994, in the dry years when flooding was limited, wintering numbers were low at fewer than 2200 birds, but in the much improved flooding of later years, numbers consistently were higher, up to 3500 (Fig. 54C).

Because the Caspian Tern is a long-lived species, breeding for the first time at six years of age (Staav 2001), the population size is dependent on the annual winter mortality during the previous decade and not just the previous year. Hence a low flood level in a given year does not cause an immediate reduction in the breeding population. Instead, the wintering population in the Inner Niger Delta shows gradual changes. Nevertheless, the higher winter mortality in dry years, as revealed by the ringing recoveries (Fig. 223), can be confirmed, although indirectly, for first-year birds oversummering in the Inner Delta. Oversummering numbers in the 1999-2006 period varied between 54 and 282, corresponding to a 1.9 - 10.1 % range relative to numbers counted six months previously. This variation is attributable to the flood level: the higher the flood six months previously (in the local winter season), the greater number of birds that were still alive in June (Fig. 226). We therefore conclude that higher mortality in dry years (Fig. 223, 225, 226) has indeed its expected repercussions on wintering numbers in the Inner Niger Delta.

Wintering conditions The longest breeding bird survey for Caspian Terns is available for the Krunnit Islands (Finland, Gulf of Bothnia), where the population increased in the 1960s, but plunged in the early 1980s (Fig. 227A). The entire Finnish population was estimated or censused several times during the 20th century before an annual survey was instigated in 1984 (Fig. 227B). Numbers increased steeply from 200 pairs in the 1930s to 500 pairs in the 1950s and 1200-1300 in 1971 (Väisänen et al. 1998). Numbers had declined to an estimated 850 pairs in 1984; this decline continued in the late 1980s, stabilising at around 700-750 pairs in from 1990 to 97, and increasing slightly to 800-870 pairs to 2005. A similar trend was ascertained in Sweden, where 900 pairs in 1971 also represented a high, declining to 600-700 pairs in 1984 and stabilising at 400-500 pairs from 1984 at least until 1996 (Staav 1985, Staav 1988, Svensson et al. 1999). In Estonia, where up to the early 1950s a few pairs bred irregularly, the first large colony became established in 1953, followed by an increase through the 1950s and 1960s to 360 pairs by 1971. Unlike in Finland and Sweden, Estonian numbers were still high in 1984 (estimated at 400 pairs) and 1988 (360 pairs), but then declined to 300 pairs in 1990 (300) and 225-250 pairs in 1993 (Hario et al. 1987, Leibak et al. 1994).

If population size were regulated largely by the flood level in wintering areas, we would expect a high number of Caspian Terns in 1971 (just prior to the severe drought of 1973-1974), much lower numbers in the 1980s (when extremely dry winters occurred mid-decade) and some increase after 1994 (improved flooding). In general, the populations which have been monitored did indeed show fluctuations that related to flooding conditions in West Africa.

Breeding conditions Breeding in island colonies, Caspian Terns are vulnerable to changing conditions in the breeding areas. Formerly, human predation was a major factor in reducing numbers, both in the Baltic (Hario & Stjernberg 1996) and in the Black Sea (Voinstvenskiy 1986), but since legal protection has been realised, this threat effectively has been thwarted; e.g. the Caspian Tern are now in Ukraine's Red Data Book (Rudenko 1996). In the Baltic, predation by minks is of some concern (Hario & Stjernberg 1996). In the northern Black Sea, Caspian Terns breeding in the Tendra Bay, Ukraine, are exposed to pesticide pollution from the Krasnoznamenskaya ir-

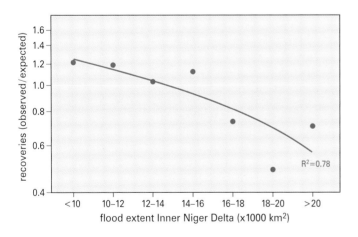

Fig. 223 Ratio of observed to expected annual numbers of recoveries (data from Fig. 221), for 7 different flood levels. Note that a log-scale is used.

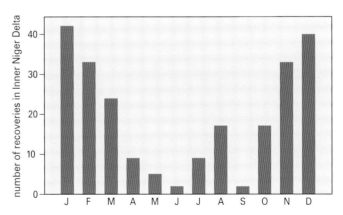

Fig. 224 Monthly numbers of recoveries in the Inner Niger Delta. Same data as in Fig. 221 and 223, but recovery dates must be accurate to within 6 weeks or less (n=233).

Fish, usually captured after a plunge-dive, are swallowed in flight (Djoudj, December 2005). Observations in autumn 1981 in the Gulf of Suez showed that out of 61 diving attempts of Caspian Terns only 25 were completed, of which 16 resulted in a visible catch (Bijlsma 1985a). Aborting dives before hitting the water surface is energetically profitable when the chances of a successful capture are small.

rigation system; breeding success of species like Slender-billed Gull and Sandwich Tern declined in the early 1990s ('many infertile eggs', Rudenko 1996).

The construction of the North-Crimean channel in 1963 temporarily resulted in a boost in numbers on the Chongarskie Islands and Lebyazh'i Islands (Fig. 227C). Likewise, the upsurge in numbers in the early 1980s is said to have resulted from the establishment of fishing farms in the vicinity of the breeding sites (Voinstvenskiy 1986).

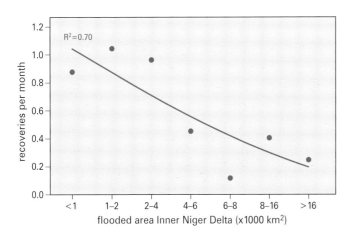

Fig. 225 Monthly numbers of recoveries in the Inner Niger Delta as a function of floodplain size in the same month, calculated for the period September through March ($p < 0.01$). Same data as in Fig. 224.

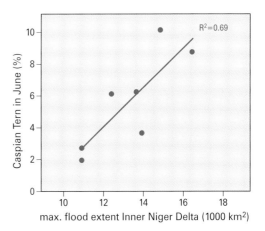

Fig. 226 Numbers of Caspian Terns counted between 1999 and 2006 in the central lakes of the Inner Niger Delta in June, given as % of the numbers counted in January or February earlier the same year, relative to the maximal flood extent the previous year ($p < 0.01$). The June 2004 count was omitted because the exceptionally low water level (4 cm at Akka) had driven all birds save 13 from the area.

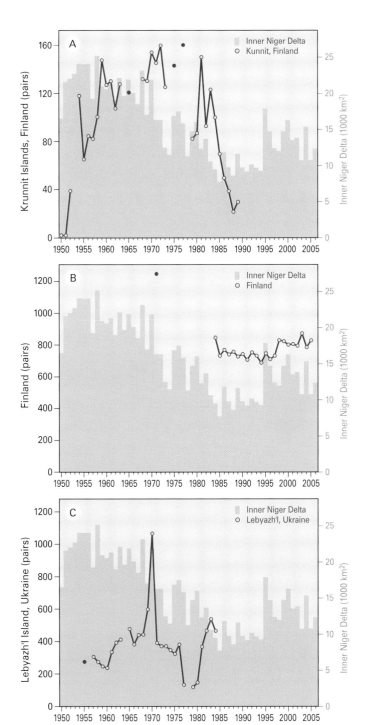

Fig. 227 Population trend of Caspian Terns on the Krunnit Islands (Finland), in Finland as a whole and on the Lebyazh'i Islands, Black Sea, Ukraine. From: Väisänen et al. 1998 (Krunnit), Hario et al. 1987, Hario & Stjernberg 1996 and T. Stjernberg pers. comm. for 1997-2005 (Finland) and Voinstvenskiy 1986 (Ukraine).

Conclusion

European Caspian Terns breed disjunctly in the Baltic, Black Sea and Volga Delta. Their numbers have increased in the 20th century, apparently peaking around 1970. Since then, numbers have declined, to improve again slightly in the late 1990s. Fluctuations in the latter half of the 20th century are related strongly to the hydrological conditions of their major wintering site, the floodplain of the Inner Niger Delta in Mali, where high floods coincide with low mortality. Caspian Terns in the Inner Delta are particularly vulnerable to human predation during low floods, when the birds are forced to concentrate in the last remaining water bodies.

Endnotes

1 Unless stated otherwise, all data refer to mid-winter counts, extracted from the African Waterbird Census database of Wetlands International.
2 There are 352 recoveries from the Inner Niger Delta, but we restricted our analysis to the recovery dates accurate to within 6 months (n=287; Fig. 221) or 6 weeks (n=233; Fig. 224) since we are interested to know the actual number of recoveries per year or per season, respectively.

European Turtle Dove
Streptopelia turtur

"Uncountable numbers, certainly more than 1 million birds, passed over the river near Kaur on 6 March. Flocks were passing overhead constantly as our boat moved upstream for more than an hour before dusk. Very large numbers were also seen passing over Jakhaly swamp the following evening, flying in the same westerly direction." Adding later: "I have never seen so many birds streaming overhead for so a long a period across a front of at least 10 km wide...". Is this a boyhood story of John Muir describing the spring arrival of Passenger Pigeons in Wisconsin in the 1800s? Far from it! It is an eye-witness report of roost flights of European Turtle Doves in The Gambia in the 1970s by Michael Gore (1982). In those earlier decades, the Sahelian zone of Africa hosted huge numbers of wintering European Turtle Doves (in the 1960s *plusieurs millions* in northern Senegal; Morel & Morel 1979). Even today, large flocks can still be found, but these are small by the standards of those days. Europe's "passenger pigeon" is in decline, and although its ecology and behaviour differ greatly from its extinct North American relative, which was a colonial breeding bird numbering billions, the present breeding status of our attractive Turtle Dove in much of Europe is but a fraction of what once was. The past abundance is but a fading memory.

Breeding range

The European Turtle Dove breeds across the temperate, Mediterranean, steppe and semi-desert climate zones of the western Palearctic, but mainly south of 55°N. Preferring dry and sunny lowlands, it is largely absent from northern Europe and from the higher altitudinal ranges of mountainous regions. In the 1990-2002 period, the population was estimated at 3.5-7.2 million pairs (BirdLife International 2004a), but much lower figures prevail nowadays due to its continued decline. This species, the only long-distance migrant among European doves, winters in the Sahel and adjacent parts of the Sudanian vegetation zone across the entire width of Africa.

Migration

From late July and early August onwards, European Turtle Dove abruptly starts assembling in flocks, even deserting recently started clutches and broods (Murton 1968, Bijlsma 1985b). Patches of land where food is abundant may attract many, sometimes hundreds, which fatten on a wide variety of weed seeds or on spilt grain (Glutz von Blotzheim & Bauer 1980, Browne & Aebischer 2003a). Turtle Doves breeding in Britain have shortened their breeding cycle by some 12 days over the 1963-2000 period, mostly as a consequence of an autumn departure date now eight days earlier (Browne & Aebischer 2003c). Because this population is peripheral, it is not known whether this phenomenon applies across the entire European population. Birds captured during autumn migration (August-September) weighed on average 126 g (juveniles) and 152 g (adults) in the Camargue and 124 g in Portugal, *i.e.* little different from midsummer weights (Glutz von Blotzheim & Bauer 1980).

Autumn migration in western Europe is rather a covert affair, involving small flocks flying low and fast, most passage possibly occurring at night. For example, in the foothills of the pre-Alps in southern Germany only 8 birds on 7 days were recorded in 30 years of systematic recording of autumn migration (Gatter 2000)! Furthermore, numbers observed in autumn during systematic migration surveys in The Netherlands and Britain normally are considerably smaller than those seen during spring watches (LWVT/SOVON 2002, Browne & Aebischer 2003c), despite winter mortality obviously having an effect on numbers in the intervening period. The perception of migration in southern Europe is quite different; "copious" (northern Spain, coming in from the Bay of Biscay on a broad front), "big passage" (over Valencia), "immense numbers" (Portugal, concentrating along the coast during easterlies), "thousands on a good morning" near Oporto in Portugal and "one of the most conspicuous migrants throughout the whole of the Mediterranean when the birds are passing to or from their winter quarters" (Tait 1924, Moreau 1953, Bannerman 1959). Note, though, that these observations were made during the first half of the 20th century, when the species still was abundant. Murton (1968) in his thorough analysis of migration through France and the Iberian Peninsula suggested that first-year birds are more prone to westward drift, which would account for the age ratio in westernmost Iberia being biased towards juveniles.

As in central and southern Europe, also further east impressive daytime migration has been recorded, but not recently: "immense flocks" on the Alfold Plain east of Tisza, Hungary (A. Keve *in* Ash 1977); sample counts by J.S. Ash (1977) on 9 September 1965 (at 13:00-15:15 local time) just west of Karçag on the Hungarian Hortobagy Plain produced some 10 450 Turtle Dove flying due south on a narrow front some 400 m wide and less than 15 m above the ground, and in Iraq, Marchant (1963) similarly described autumn passage in the early 1960s as spectacular, vast numbers flying due west near Baghdad throughout September. It is worth examining this observation in some detail. An estimated 40 birds per minute passed through a 1000 m-wide cross section of the migration front for four hours from 06:00 local time, when movement stopped or diminished. Assuming a conservative (Marchant's evaluation) estimate of the whole front being 100 km wide and that the average passage rate was 4 birds per hour (the low rates allow for days with little migration in September) he arrived at an autumn passage total in this region of three million birds. This estimate included the speculation that most passage, at least in this region, occurs at ground level in

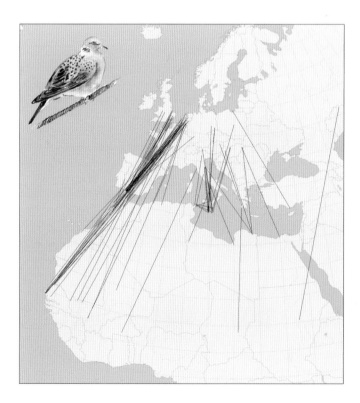

Fig. 228 Origins of 124 European Turtle Doves recovered or captured between 4° and 37°N, of which 28 south of the Sahara. From: EURING.

Turtle Doves are among the most commonly encountered migratory bird species in the desert (along with Barn Swallow, Northern Wheatear and Short-toed Lark). A trip from Bordj-Moktar to Reggane in the Algerian desert (650 km) on 17 and 18 April 1973 revealed 74 birds migrating north in flocks of 2-4 (maximum 8) in low zigzag flight. Some were found sheltering from the sun and wind in the shade of war-wrecked vehicles along the route (Haas 1974). A similar survey from Reggane to Tessalit (795 km) from 16 to 19 April 1977 came up with 394 migrant Turtle Doves and individuals that were fatigued (5), dehydrated (48), or freshly dead (9) (Haas & Beck 1979). The freshly dead birds were very fat, but their water index was quite low, indicating that the birds perished from water loss, not from lack of body reserves (Haas & Beck 1979). ¹ The Hoopoe Lark, a resident in the area, in the left picture was observed to attack the Turtle Doves.

daylight. His observations in the 1960-1962 period are unlikely to be repeatable nowadays, unless the eastern populations have fared better than those in the west – there is indirect evidence that declines locally may not have been universally as bad as in the west. For example, in Azerbaijan (86 600 km^2) the species has been described recently as very common (>100 000 pairs; Patrikeev 2004) and in Israel (Shirihai 1996) and Egypt (Goodman & Meininger 1989) it has been listed as a common passage migrant during September, peak numbers at roosts occurring in the first week of September.

European Turtle Doves cross the Sahara on a broad front, supporting evidence being the prey remains found at Lanner Falcon breeding sites in the super-arid Darb el Arba'in Desert in southwestern Egypt (Goodman & Haynes 1992). Of the 19 migrant species in the Lanner's diet, Turtle Doves comprised 34% of 108 identified prey individuals. Only Common Quail had perished more often, all other bird species being represented in much smaller proportions (0.9-3.7%). Lanners clearly have a different perception of Turtle Dove abundance than had Moreau (1961), who described "the paucity of Turtle-Doves in Tripoli and East Sahara on both passages" as "fairly well attested" (but his 1972 *magnum opus* would contain several eyewitness accounts to the contrary).

Spring migration is initiated from early March onwards. At the Damietta market in Egypt (31°26'N, 31°48'E) in 1990, offered for sale was a total of 634 Turtle Doves, the first being available on 7 March; none had been offered in the preceding winter months. Migrant birds were recorded up to 6 May at least (Meininger & Atta 1994). Prior to spring migration, the birds fatten just south of the desert's edge, increasing in body mass from a midwinter low of 100-120 g to some 200 g in March (Morel 1986, Morel & Morel 1988). This level of fat reserve enables them to cross the Sahara, North Africa, the Mediterranean Sea and much of southern Europe nonstop. The importance of water during migration seems to be corroborated by the observations near Khartoum, where northwards migrating Turtle Doves were confined strictly to the narrow belt of green vegetation along the White and Blue Niles. Observations, alternating daily between dawn-2 pm and from 10 am-sunset, produced 6180 Turtle Doves flying north from 12 February to 12 April 1961 (Mathiasson 1963). In Morocco, the bulk of the visible passage in spring follows the Atlantic coast and river valleys, the first wave of return migrants arriving mid-March (in the southwest) or early April (southeast) (Thévenot *et al*. 2003). Peak migration at the Moroccan oasis of Defilia (32°07'N, 01°15'W, 930 m asl) in spring 1963 was recorded from 2-8 May, substantial numbers also passing from 9-14 May (Smith 1968). In much of the Mediterranean area migration takes place from early April through to mid-May, peaking in late April. Usually during return migration larger numbers are seen than during autumn migration, as in Greece (Handrinos & Akriotis 1997), on Corsica (Thibault & Bonaccorsi 1999) and at the Strait of Gibraltar in southern Spain (Finlayson 1992). This must imply different migration strategies for spring and autumn, such as Turtle Doves migrat-

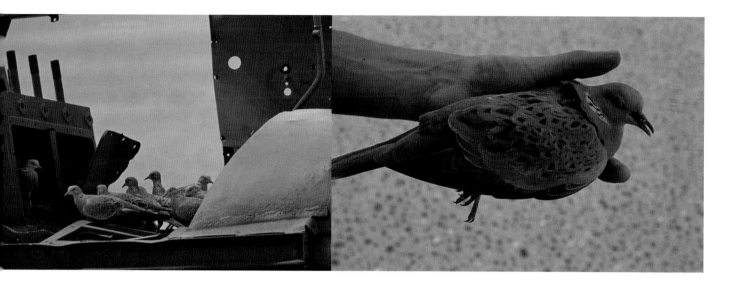

ing at lower altitudes and more often during daylight in spring, which perhaps is not surprising after having crossed the Sahara and the Mediterranean Sea. Further north, some funneling is reported from the Gironde region of SW France, where in the 1984-2004 period, from 13 500 to 46 000 were logged during counts from mid-March to mid-May (Ligue pour la Protection des Oiseaux, France).

The predominantly southwesterly to southerly flight heading for outward migration occurs over the entire width of the breeding distribution and should result in an overall longitudinal separation of eastern and western wintering populations, with some overlap in the middle. The EURING recoveries from Africa seem to corroborate this view, despite the scarcity of material from the eastern populations (Fig. 228). Birds from western Europe are spread from Senegambia through to Mali, those from central and eastern Europe presumably from Mali and Burkina Faso eastwards to Sudan and Ethiopia. That a more westerly course is undertaken by juveniles, as reported for British birds (Murton 1968, Wernham et al. 2002), finds some support from ringing recoveries (EURING): 0 out of 7 recovered birds in Mali and Burkina Faso were juvenile, compared with 3 out of 9 coastal recoveries in Morocco, Western Sahara, Mauritania and Senegal. However, between 5 and 13 March 1990, of 265 Turtle Doves captured by three different netting techniques at a roost in the Forest of Nianing (50 km SSE of Dakar, Senegal), 47.9% were in their first calendar year (Jarry & Baillon 1991).

Distribution in Africa

The European Turtle Dove in Africa typically concentrates in regions where: (1) Food supply is abundant, (2) Drinking water is available, (3) Large trees and woodlots provide cover for loafing and roosting (Jarry & Baillon 1991). Absence of one of these variables at any site ensures their presence is temporary, except on those occasions when they are preparing for the return migration in early spring at the Sahel's northernmost edge, when the birds find areas lacking cover acceptable (see below: Mali and Nigeria).

In **Senegal**, the first Turtle Doves arrive on outward migration in late July, the movement continuing until August and September (Morel & Roux 1966b). They visit rice fields to feed on grass seeds, later also scavenging spilt grains from harvested rice. In years of abundant rainfall, large numbers of Turtle Doves may aggregate in the October-December period in the usually arid Ferlo region south of the River Senegal, wherever there are rain-filled pools (Morel & Morel 1979), but most settle in wintering areas in The Gambia and further south into savanna woodland down to 9°N between Guinea and Cameroon (Morel 1985, Morel & Morel 1988). From February onwards, Turtle Doves wintering south of The Gambia gradually start drifting northwards, a process which in the 1970s involved vast numbers – e.g. the over one million birds recorded on 6 March near Kaur, heading for a roost (Gore 1982). The birds head for northern Senegal where large groups exploit spilt grain by concentrating in harvested rice fields in floodplains and near rain-fed pools. Here, at the Sahel's northernmost border, the birds complete their moult (Morel 1986, Jarry & Baillon 1991) and prepare for the leap across the Sahara by fattening (Morel & Roux 1966a, 1973). Thousands once congregated in the Senegal Delta, as in the 1960s, when at least 150 000 (but probably 3 times that number) were on 6000 ha of rice fields near Richard Toll (Morel & Roux 1966b) and 450 000 were estimated at a roost near there on 13 March 1973 (G. Jarry in Morel & Roux 1973). Conceivably, totals in the Senegal Delta may have comprised several millions in the 1970s (Morel & Morel 1979), but following the 1980s drought-related scarcity of grain from wild and cultivated rice, numbers have dropped steeply (M.-Y. Morel in Skinner 1987). The maximum count of 32 100 birds at a roost in the forest of Nianing in March 1990 (Jarry & Baillon 1991) certainly is a far cry from the hundreds of thousands in northern Senegal in the

Although seeds of grasses constitute the major source of food in early winter, wild rice (facing page) is also eaten by European Turtle Doves. However, if cultivated rice becomes available after the harvest (spilt grains on the ground), they prefer the latter. The local people vigorously defend the ripening rice against roaming weavers and other small seed-eaters, which can, unlike Turtle Doves, cling to standing rice and swallow the ripening seeds. Children have quite a job in chasing the huge flocks of weavers.

1970s. Morel (1987) issued an early alert that European Turtle Doves had been diminishing in number while wintering in Senegal since the 1970s, despite the species' main food source, rice, becoming increasingly available through increased cultivation in northern Senegal (see Chapter 7). The dam construction programme along the Senegal River had reduced the flooding, allowing the floodplain gradually to be converted into orderly cultivated rice fields, but at the same time led to the disappearance of floodplain-associated vegetation, such as wild rice and *Acacia nilotica*. The former is an important winter food item for European Turtle Doves, now superseded by cultivated rice, and the latter provided essential sites for roosting and for loafing in daylight. These circumstances were unbalanced by the 1970s and 1980s droughts, which impacted seed production of grass and rice negatively (Morel 1987). Tree-felling, as in the protected forest of Nianing-Balabougou (Jarry & Baillon 1991), only added to the problems facing the species.

In **Mali's** Inner Niger Delta, Turtle Doves are abundant in the cultivated area of the *Office du Niger* or *Delta mort*, north of Segou, from where they once would spread northwards throughout the transition zone and upper floodplain levels as the Niger flood receded. They would feed in treeless areas, either amongst grass tussocks or on bare steppe-like terrain, remaining there throughout the heat of the day (Curry & Sayer 1979). Curry (1974, cited by Morel 1987) counted 700 000 birds on roosts in the Inner Niger Delta. Lamarche (1980) mentions hundreds of thousands of Turtle Doves, mainly in the wooded dunes fringing the delta at the latitude of Mopti and Erg de Niafunké (based on Curry 1974, or his own observations in the 1970s?). The species sometimes foraged (presumably on spilt rice) alongside flocks of Ruff and resident Chestnut-bellied Sandgrouse, but rarely associated with native doves and pigeons (Curry & Sayer 1979). Since the Great Drought of the 1980s, wintering numbers in the floodplain have dwindled, first to thousands (maximum of 5000 in January 1987, but normally flocks of fewer than 100; Skinner 1987), then to sparse flocks of at most several dozens (observations from 700 field days in the northern winter in the 1992-2007 period: J. van der Kamp & L. Zwarts), representing a decline of 95-99%. Now-

Perhaps Curry & Sayer (1979) did not fully realise the privilege when they observed immense flocks of Turtle Doves in the Inner Niger Delta in the early 1970s, feeding together with Ruff, on emergent rice fields. When we studied the birdlife on these floodplains in the 1990s and 2000s, Turtle Doves had become extremely scarce.

adays, Turtle Doves are confined almost entirely to the irrigated rice cultivation of the *Office du Niger*, west of the Inner Niger Delta (estimated at 100 000; van der Kamp *et al.* 2005c) and >4000 birds near Lake Tele (February 2009). As in Senegal, shooting Turtle Doves at roost and drinking pools is common practice, facilitated by European travel agencies specialising in shooting forays that guarantee daily bags. The shooters position themselves near roosts under the flight paths of returning doves. On 15 February 2004, one such party of seven Europeans encircled a small woodlot near Kogoni. Between 16:55 and 18:05 local time, some 40 000 Turtle Doves were counted while entering the roost and 623 shots were fired, of which 101 were during the five minutes when the doves' arrival rate was at its peak (van der Kamp *et al.* 2005c). Shooting success was impossible to quantify, but the disturbance (presumably night after night) must have been tremendous. This woodlot also contained a heronry, but a year later, every tree had been lost through felling or fire; the dove roost and the heronry no longer existed. The present scarcity of woodlots makes potential roosting sites few and far between, which renders Turtle Doves more and more vulnerable to this type of indiscriminate shooting.

The Sahel zone in **Burkina Faso** is an important wintering area for European Turtle Dove. In late October 1986 some 100 000 low-flying birds were recorded near the Oursi wetland (14°30'N, 0°30'E). Between January and March, a northward displacement occurs, the birds concentrating along the Beli and to drink at the Oursi and Djibo wetlands and near clusters of *Acacia seyal* to roost. Further west, they favour the Kou valley rice fields and wetlands (January 1983; Y. Thonnerieux & J.F. Walsh *in* Morel 1987).

In contrast, the European Turtle Dove is scarce or absent in **Niger**, even in regions where suitable habitat seems available, such as along the Niger River and near Lake Chad (Giraudoux *et al.* 1988). In January 2005 and 2006 no Turtle Doves were recorded during Montagu's Harrier surveys in the southern part of the country (B. Koks pers. comm.).

In **Nigeria's** far north, Turtle Doves are fairly common winter visitors from Sokoto to Lake Chad, where flocks assemble before setting out to cross the Sahara. Birds wintering in the more arid parts just south of the desert sometimes are forced to move south as the dry season advances (Elgood *et al.* 1966). Large flocks of 1000+ have been reported for Kazaura (12°39'N, 8°23'E) in January (Elgood *et al.* 1994). As in Mali, hundreds may feed together on the open plains around noon; cover and shade are absent (Elgood *et al.* 1966). A strategy to avoid competition from local dove and pigeon species, that stay in the shade at this time of the day and year (Morel & Morel 1973), or a necessity to complete pre-migratory fattening as close to the desert's edge as possible?

Across **Cameroon** Turtles Doves were once common locally between 11° and 13°N, where apparently they resided in the Sudanian vegetation zone, concentrating near pools and watercourses (Moreau 1972). Over 60 000 were recorded in the Waza National Park in northern Cameroon in late November 1969 (Fry 1970). Holmes (*in* Pettet 1976) found Turtle Doves to be very numerous in February 1972, possibly even more so than recorded by Fry (1970). In late March 1969, Broadbent (1971) also recorded large numbers here. These data suggest that Turtle Doves pass through northern Cameroon mainly in October and November and in February and March, and occur in low numbers outside these months (Pettet 1976). Southward penetration was particularly marked in the winter of 1971/72, following a poor wet season. In years of good rains, Turtle Doves remain in northern Cameroon in higher numbers throughout the dry season. Pettet (1976) suggests that the relative scarcity of Turtle Doves in the Waza National Park during the dry season is related to shade and food; his only observation from 24-28 December 1971 was of a flock of *c.* 100 birds in *Balanites-Acacia polycantha* vegetation near a waterhole, with a few birds outside the reserve in marginal *Balanites-Acacia seyal* vegetation bordering marshland. *Balanites*, *Mitragyna* and *Anogeissus*, though not leafless, had shed many

European Turtle Doves are seed eaters which need to drink regularly. In the Sahel, they feed on drylands near drinking sites. To prevent dehydration, they drink at least three times per day, spaced evenly across the day. Apart from drinking sites, the presence of trees is another requirement of Turtle Dove habitat in the Sahel. Shade and cover are essential to be able to digest the food and loaf in the shade without disturbance. Such sites are not that common in the Sahel, hence their massed occurrence in few localities.

leaves, and shade near waterholes had become a declining asset, but grain may have been in short supply following widespread grass burning; dry-season *Sorghum* in the north of the reserve did not ripen until December in 1971 (Pettet 1976).

From the eastern end of the Sahel little information is available. In **Sudan** European Turtle Dove is considered an abundant winter visitor in *Acacia* woodland and near water between 11° and 15°N, and east of 30°E (Nikolaus 1987). Roosts of up to 100 000 birds were recorded at Sennar (13°30'N, 33°37'E, 425 m) and at the Roseires Dam (11°40'N, 34°25'E, 460 m), but that is the sum total of the information. A fascinating account from 17 April 1860 comes from Hartmann (1863), who traversed the Bayuda steppe (tellingly named Bayuda desert today) by camel from El-Debbah to Khartoum. Just south of 17°N, the desert became savanna patchily overgrown by *Andropogon* grasses and *Acacia* woodlots with some *Balanites*, *Sodada*, *Capparis* and *Ficus*. The lower parts of the terrain, well-watered during summer downpours, held a lush woodland vegetation. In one of these sites, 'millions of Turtle-Doves' (European Turtle Doves, but also Laughing Doves) flocked in dense *Acacia seyal* and *A. tortilis* woodlots near a drinking site. The site was well known to the local Bedouins, who named it 'roost or playground of Turtle-Doves'.

Food and feeding in the Sahel   Turtle Doves enter the Sahelian zone in August and September; temporary pools in Senegal usually dry out between December and February (Morel 1975). To a seed-eater *par excellence*, this change spells increasing dependency on a declining number of water sources. Turtle Doves now have to cope with an environment containing an abundance of indigenous dove species. The sounds of pigeons – muffled, plaintive, emphatic, quavering, jeering, flute-like or long-drawn-out coos, gurgles, purrs and chuckles highly characteristic of Africa – can be heard even in the heat of the day or through the night, when most other bird species are silent; the background sound stage of the incessant singing of doves creates a soporific atmosphere. All these pigeons are seed-eaters, just like the European Turtle Dove. Competition for food becomes critical for survival – the greater the drought, the greater the food shortage. However, the behaviour of Turtle Doves is quite different from that of local pigeon species, for it is a function of a gregarious species, not of the solitary residents. Resource partitioning enables the European Turtle Dove to exploit scarce food supplies

Table 24 Main foods of closely related doves in northern Senegal during wet and dry years, based on analyses of stomach and crop contents (Morel & Morel 1973, Urban et al. 1986).

Species	Wet year	Dry year
Namaqua Dove *Oena capensis*	80% *Panicum laetum*	50% *Panicum*; *Dactyloctenium aegyptium*, rhizomes of *Fimbristylis*
African Mourning Dove *Streptopelia decipiens*	50% *Panicum*, 50% *Dactyloctenium* & *Tribulus terrestris*	Almost exclusively *Tribulus*
Vinaceous Dove *Streptopelia vinacea*	Up to 80% *Panicum* & *Brachiaria*	Mainly *Zornia glochidiata* & *Alysicarpus*
African Collared Dove *Streptopelia roseogrisea*	75% *Panicum*; cereals and sedge nuts	Up to 80% *Tribulus*
Laughing Dove *Streptopelia senegalensis*	70-80% *Panicum*	70% *Gisekia pharnacoides* & *Tribulus*
European Turtle Dove *Streptopelia turtur*	Mainly *Panicum*	Mainly *Oryza*

well during droughts (see below). Its daily behavioural cycle was studied in the area of Nianking, Senegal, in mid-March 1990 (Jarry & Baillon 1991). [2] Very little is known about seasonal variations in this pattern, but the need for continued foraging throughout the day, as recorded for the Inner Niger Delta and northern Nigeria (see above) indicates that heat stress is tolerated when necessary (pre-migratory fattening, food shortage, or an absence of larger seeds). The late afternoon foraging bout is longer, possibly affording the birds the opportunity to carry a larger amount of food to the roost (a maximum of 14 g dry weight in crop and gut; Morel 1987) than it would be worthwhile collecting in the early afternoon. The large food store in the crop can be digested at leisure during the long night, and it would be interesting to determine whether retention time is also longer at night to enhance digestibility and thus achieve higher productive energy (for geese, see Prop & Vulink 1992). Such a process would greatly facilitate pre-migratory fattening.

Food choice has been investigated in both dry and wet years at Richard Toll in northern Senegal (Morel 1987). Samples were taken in the evening during the 1970s, and during the morning hours in the 1980s; sample size differed vastly between years and seasons (1-144 doves). Seeds, rhizomes (5 species of Cyperacea) and fruits (*Salvadora persica* and *Cocculus pendulus*) of at least 33 plant species, belonging to 16 different families, were taken. However, between 80% and 100% of the food consisted of three grass species (*Panicum laetum*, wild rice *Oryza barthii* and cultivated rice *O. sativa*) and Puncture Vine *Tribulus terrestris*. Wild rice is abundant on floodplains, but also in depressions with rain-fed pools ('mares') and extends widely into woodland savanna. The seeds of wild rice gradually become available during the retreating flood and as pools dry out from November onwards. The caloric value of *O. barthii* is only slightly less than *Panicum*, i.e. 12 and 14.6 kJ/g respectively (Morel 1987). *Panicum laetum* is an annual, low and branching grass typically found near seasonal wetlands and in woodland savanna on moist soil. Ground cover often reaches 100%, and the tiny seeds become available from September onwards. Seeding is particularly abundant when the rains are early and prolific; however, in dry years, *Panicum* refrains from producing seeds. A full Turtle Dove crop and stomach may contain 15 000 *Panicum* seeds (Morel 1987), which roughly equals a pecking rate of one seed per second in 4 hours of nonstop foraging. During dry years, such as the early 1970s and 1980s, seeds of *Panicum* nor wild rice are available in sufficient quantities for Turtle Doves, which then switch to cultivated rice and Puncture Vine. The latter is a prostrate annual herb forming dense mats, growing on disturbed soils enriched with nitrogen (usually cow dung). It has a short vegetative cycle and is drought tolerant. The seeds in the spiky fruit have low caloric value (7.1 kJ/g; Morel 1987) and represent a poor but readily available substitute when other seeds are scarce during periods of little rainfall such as in 1970/1971 and 1972/1973 (Morel 1987). Similarly, cultivated rice is an important food source when droughts prevail. Its high caloric value of 14.6 kJ/g and the abundance of spilt grain after the December harvest make it a valuable source of food; during the extremely dry years of 1983 and 1984, 99-100% of the Turtle Dove's food during pre-migratory fattening from early March to early May comprised cultivated rice (Morel 1987). In the *Office du Niger*, Mali, the steady increase in cultivated rice has been instrumental in making available sufficient food for Turtle Doves in a drought-stricken environment.

In years of food abundance, when rains have been plentiful and the flood high, for the Turtle Doves and the local doves and pigeons competition for food is non-existent because of the superabundance of *Panicum* seeds and cultivated cereals like *Sorghum* spp. and rice. However, drought-induced food scarcity leads birds specialising in diet selection (Table 24), Turtle Doves concentrating almost entirely on spilt rice, this switch seemingly attaining a more permanent nature after the Great Drought of the 1980s, if the behaviour of birds wintering in Mali is any yardstick. Here, the large flocks in the southern Inner Niger Delta referred to previously have vanished like snow; only a few parties of up to 30 birds are known to frequent the Mayo Dembe and Niger rivers to drink and to feed in small irrigated village rice fields. The only sizeable concentration of Turtle Doves in Mali appears on the irrigated rice fields of *Office du Niger* and comprises at least 100 000 birds (Wymenga *et al.* 2005).

Population changes

Very few long-term trends are available from European countries, the longest one from the peripheral British population. British Turtle Doves showed an increase between 1961 and 1976, but decreased between 1976 and the early 1990s; since then, low indices have prevailed (Fig. 229A). A reconstruction of Turtle Dove fortunes in Britain in the 19th and 20th centuries led Holloway (1996) to believe that this species reached an all-time high in the late 1960s and early 1970s. Between 1968 and 1998, its abundance declined by 69%, and its breeding range contracted by 25% (Browne & Aebischer 2004).

In The Netherlands, high numbers were prevalent during much of the 20th century. In the 1930s, the species was considered a 'very common and widespread breeding bird', whose song could be heard 'everywhere', even in cities (Eykman *et al.* 1941); this was no exaggeration, for just one ringer was able to capture and ring 561 nestlings and 477 adults in a small orchard in the central Netherlands from 1932 to 1937 (Eykman *et al.* 1941), and even in the 1977-1981 period, 143 nests were located on some 300 ha of coniferous woodland near Wageningen (Bijlsma 1985b). Such days are long gone. Dutch Turtle Doves showed a steep decline from the mid-1980s onwards when country-wide monitoring started, with another noticeable drop in 1991. Numbers have not recovered and the species has now become scarce in much of the country (Fig. 229B). This scenario is corroborated by regional breeding censuses, showing declines of between 30% and 97%, the steepness of the decline depending upon habitat, region and time scale (Bijlsma *et al.* 2001, Hustings *et al.* 2006). Censuses in coniferous woodland in the Netherlands indicate that Turtle Doves may have been in decline since at least the late 1970s, consistently to less than 10% of the population levels of the 1975-1979 period from the early 1990s onwards (Fig. 229C). Even more recently the declines have seemed to continue. For example, in rich deciduous and mixed woodland possessing a dense understorey in the recently reclaimed polders of the central Netherlands (clayey and loamy soils), Turtle Doves declined from densities of 8.0-18.3 territories/100 ha in 1989-1994 to 0.0-6.2 territories in 1995-2000 (van Manen 2001). Similarly, in Limburg province, local populations declined on average 40% during the 1990s and early 2000s, after having experienced a 90% decline from the 1970s through the early 1990s (Hustings *et al.* 2006). These data suggest that the published

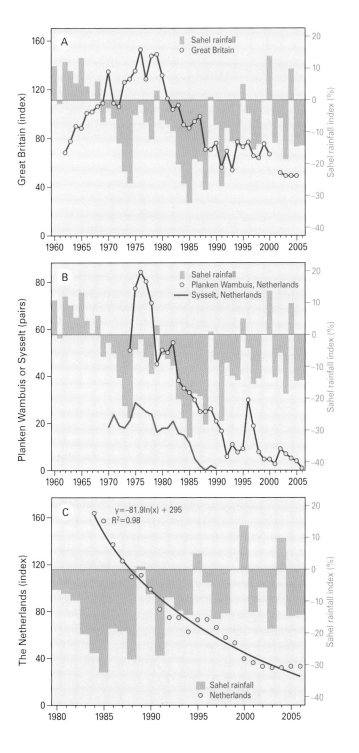

Fig. 229 Population trends for European Turtle Doves, related to rainfall in the Sahel in the preceding year. From: (A) BTO for Great Britain. (B) Bijlsma 1985, *et al.* 2001 and unpubl. for the woodlands of Planken Wambuis (2000 ha) and Sysselt (350 ha). (C) SOVON for The Netherlands as a whole.

European birders, born after 1970, may not realise how common Turtle Doves once were. Their point of reference, from the mid-1980s onwards, overlaps with a population dip from which the species never recovered. One of us, born in 1955, searched for, and found, literally hundreds of nests in the 1960s and 1970s in The Netherlands, a feat impossible to repeat in the late 2000s simply because of lack of birds. They have become scarce, and hearing the soft purring nowadays brings joy and a sense of loss at the same time. (6 August 2008, Milsbeek, The Netherlands).

decline of 75% for The Netherlands between 1973-85 and 1998-2000 actually was an underestimate. During that same overall period, the species' range contracted by 26% (SOVON 2002).

Elsewhere in Europe, monitoring scheme time series usually are shorter and show some vicissitudes: an increase in Switzerland from a low in the mid-1980s (1985-2000; Schmid *et al.* 2001), more or less stable indices for Constant Effort Sites in France from 1989 to 2001 (Boutin *et al.* 2001) and a slight decline (-10%) from point counts in France from 1989 to 2007 (Jiguet 2007), and a slight increase in Spain (point counts 1996-2005; www.seo.org). Compared with data from 1970-90, however, a long-term decline has probably occurred in much of western and southwestern Europe (Tucker & Heath 1994). Central European populations in the 1970s-1990s are reported as generally stable, but suffer erratic ups and downs (Bauer & Berthold 1996). Those in eastern Europe appear to have declined slightly (Tucker & Heath 1994). However, few quantitative data exist to substantiate the above conclusions.

The perceived steep decline is therefore based mostly upon the well-studied, but peripheral, British population, and the equally peripheral, but poorly studied, Dutch population. Annual population changes in Britain and The Netherlands are to some extent synchronised (R=0.77), suggesting one or more common causal factors. Breeding population fluctuations often run counter to rainfall patterns on the wintering grounds (Fig. 229). The decline in numbers is larger in dry years than in wet years, but overall the differences are not significant in the British or Dutch series; although there may be some influence of Sahelian rainfall on population dynamics, it cannot be held to account exclusively for the marked overall decline, in our opinion. [3]

Breeding conditions   The in-depth studies of Stephen Browne and Nicholas Aebischer in Britain have targeted several life history parameters possibly related to the decline in the past decades. Regarding breeding ecology, for the 1941-2000 period no trends were detected in nest height, first-egg date, clutch and brood size and nesting success of individual breeding attempts (Browne & Aebischer 2005). However, autumn (return) migration in recent years is on average taking place 8 days earlier than in the 1960s (an earlier departure was also recorded locally at Oxford between 1971 and 2000; Cotton 2003), and the breeding season has shortened by some 12 days compared with the mid-1960s (Browne & Aebischer 2003a). Recent fieldwork on Turtle Dove breeding ecology, 30 and 80 km from Carlton where Murton (1968) carried out his pre-crash study in the 1960s, showed that present-day Turtle Doves have a shorter breeding season and produce about half the number of clutches and young per pair compared with the 1960s (Browne & Aebischer 2004). This rate of productivity decline if consistent would result in a population decline of 17% per annum, whereas the actual decline averaged 4.6% on farmland and woodland from 1965 to 1995; Browne *et al.* 2004). An analysis of nesting habitat availability over time showed that declines in breeding density were associated with nesting habitat loss, notably hedgerows, shrubs and woodland edges (Browne *et al.* 2004). The species' feeding ecology also changed significantly between the 1960s and the late 1990s; weed seeds (mainly fumitory) comprised 95% of the adult birds' diet in the 1960s, but only 40% by 1998-2000 – for nestlings, 75% and 31% respectively. Nowadays British Turtle Doves feed mainly upon cultivated plants, *i.e.* rape and wheat seeds – in terms of caloric value, this shift might be beneficial, but the exploitability of spilt grain is poor relative to the dove's breeding season (Browne & Aebischer 2003a). In addition, the loss of weed seeds from farmland is an inevitable consequence of today's industrial farming in Europe, where decades of herbicide usage have eradicated common weed species over vast areas (Bijlsma *et al.* 2001, Newton 2004).

The fact that eastern European Turtle Doves apparently suffered less drastic declines than their western European counterparts (Tucker & Heath 1994; BirdLife International 2004a) may well be associated directly with the formerly less intensive farming system in eastern Europe and Russia and also to these regions' reduced habitat loss, not yet as devastating as in western Europe. Unfortunately, rapid changes have begun to occur here since the 1990s; see for example Poland and Latvia (Tomiałojć & Głowaciński 2005 and Aunins & Priednieks 2003, respectively). Yet another difference from western Europe may be the differential impact of climate change, although so far, the slightly earlier arrival dates for Turtle Doves in

Preferred habitat of European Turtle Doves in the Inner Niger Delta (left) includes mature trees, which offer shade whilst the birds loaf during digestive pauses, and water pools close to harvested rice fields with spilt grains. The millions of Turtle Doves which once held sway are sadly a thing of the past. Water and rice fields are still available (right), but trees have disappeared from large parts of the Inner Niger Delta, barring the few branchless trunks left behind (November 2008).

both Poland (Tryjanowski *et al.* 2005) and Moravia (Hubalek 2004) are not significant over time with respect to population size and trends, which correlates to absences of changes to arrival dates in western Europe (Browne & Aebischer 2003a).

The role of shooting   In the Mediterranean area Turtle Doves have been killed on a large scale for ages (Woldhek 1980, Magnin 1991). Despite European legislation to the contrary, spring shooting continues in Spain, Portugal, France, Malta, Italy and Greece. In much of the Balkans shooting remains legal. In Serbia, for example, the annual bag, mainly by visiting foreign hunters in the northern province of Vojvodina, is estimated at 10 000-50 000 in the five years 1996-2000 (www.dravanews.hu, 19 March 2007). In Malta, notorious for its lack of enforcement of hunting-related EC laws to which it has agreed, annual bags have been estimated at 20 000-200 000 (Woldhek 1980), 100 000-200 000 (G. Magnin *in* Fenech 1992), 160 000-480 000 (Fenech 1992, based on the numbers of shooters averaging a bag of 10-30 each) and 100 000 (Hirschfeld & Heyd 2005), mostly obtained during spring migration. In SW France, notably in the Médoc, an area once riddled with shooting platforms, spring shooting has continued despite legislation to the contrary (European Directive 79/109). Here, low-flying Turtle Doves, funnelled into the Médoc by the geography of the Gironde, run the gauntlet of guns commanding the length and breadth of their flyway. Members of *La Ligue pour la Protection des Oiseaux* have counted passing Turtle Doves at Pointe de Grave each April and May since 1984. A detailed study in 1999 revealed that 600-800 poachers killed that year 51 000 Turtle Doves out of a total of 67 500 birds passing through the Médoc, a staggering 76%. One hundred policemen patrolling the area registered only some 50 violations against the law, or 0.08% of the total number (Fournier 2001). More recently, the situation has improved somewhat: the number of local communities involved in spring shooting has declined in ten years from 43 to 8 in 2002, whereas the number of shooting platforms has fallen (literally in many cases) from 3000 to 226 (www.kjhall.org.uk/lponews2002.htm).

The total number shot in Europe each year can be only roughly estimated; hunting bags are not usually registered centrally, nor is registration always compulsory or realised reliably. The total hunting bag for the EU members has been estimated at 2-4 million birds by Boutin (2001), and at 2.36 million birds (EU plus Norway and Switzerland) by Hirschfeld & Heyd (2005). Both estimates are based on the voluntary cooperation and unvalidated data of hunting societies in the respective countries, and exclude poaching.

Whereas much attention has been focussed on illegal spring hunting in Europe, very little notice has been taken of hunting in West Africa where the resources to enforce hunting laws are much more limited than even in southern Europe, and the motivation to do so even less. Foreign shooters from Europe cause havoc among Turtle Doves by targeting congregations at roosts and drinking pools. In the Sahel, Turtle Doves face many serious problems of survival, from food shortage to scarcity of roosting sites, increased human disturbance (not just shooting), drought, scarcity of fresh water and the need they have to almost double in weight prior to crossing the Sahara in spring. The last-named can be attained only if they can forage for at least 8 hours daily and if they have ample diurnal opportunity to rest and drink. Continuous shooting, as witnessed in Mali (see above) and Senegal (see above) at roosts is likely to impact survival, not only via direct mortality from shooting, but also by depriving the birds from safe havens necessary for the pre-migratory fattening process.

Conclusion

Since the 1970s, European Turtle Doves in much of western Europe have experienced declines of 70-90%; elsewhere in Europe smaller declines have been recorded but hard data are scarce. Adverse conditions on the breeding grounds have played a major part in this outcome, notably changes in farming techniques; herbicide use, intensification, destruction of breeding habitat all lead to food shortages and loss of breeding sites. In the Sahelian wintering quarters, conditions started to deteriorate during the extensive drought in the 1980s, when food supply dwindled and acacia woodlots (necessary for roosting and loafing) disappeared rapidly. Despite slightly improved rainfall figures since the mid-1990s, Turtle Dove flocks and roosts of over 100 000 birds seem to have vanished completely. At remaining roosts, doves are frequently harassed by European shooters, disturbances that also interfere with pre-migratory fattening. Spring shooting in southern Europe is still rife, though declining somewhat in extent (including autumn shooting, annual Turtle Dove casualties for EU countries only are estimated at 2-4 million).

Endnotes

1 The birds analysed by Haas & Back (1979) had a fat load of 79.5 g and a fat-free dry body weight of 72.5 g. Their average total weight was 244 g, including 92 g water. The fat reserve (relative to the total body weight) was 32%, sufficient to fly many thousands of km. The water index was rather low, however: 62.8% relative to the fat-free weight, while it should be 65-70%.

2 The daily rhythm, as observed in mid-March, is closely attuned to sunrise and sunset (07:19 and 19:20 local time respectively on 15 March). The first birds left the single-species communal roost by 07:15, followed by the bulk around 07:25, the last stragglers leaving by 07:30. Before heading to foraging sites about 1 km away, the doves visited a drinking site to sip lightly. Arriving at the feeding sites, the birds alighted in trees before setting off and starting to forage (the latter on average by 07:30). Including a disturbance which took 40 minutes of foraging time, the birds left for daytime loafing between 10:00 and 10:45 (11:15 at the latest), stopping briefly for another drink. The first doves re-entered the feeding area by 14:00, but the majority arrived between 15:00 and 15:30 (always drinking beforehand). Feeding continued at least till 17:30, but most doves left for the roost much later, arriving at the roost between 19:10 and 19:30 (Jarry & Baillon 1991). This makes for two foraging bouts of roughly 185 and 285 minutes each, amounting to almost 8 hours of foraging within a daylight period of 12 hours. The interval in between foraging bouts is likely used for digesting the food in the crop (for granivores, this normally takes 40-100 minutes), feather maintenance and loafing during the hottest part of the day.

3 Using capture-recapture data from SW France between 1998 and 2004, Eraud *et al.* (2009) found a correlation between mean survival rate of adult Turtle Doves and the annual production of rice, millet and sorghum in Mali and Senegal. Neither rainfall nor NDVI explained a significant amount of the temporal variation in survival. However, millet production in West Africa matches the Sahel rainfall index (Fig. 31), which hints at the possibility that rainfall might have an indirect effect, via food supply, on survival. The surface area in West Africa planted with millet, sorghum and rice has more than doubled since the late 1980s (Fig. 30), but this trend did not prevent western European populations of Turtle Doves from crashing to low densities, with little evidence of recovery so far. Survival probabilities do not seem to have changed much during past decades, being estimated at 0.48 and 0.53 (depending on method of calculation) in Britain up to 1960 (Murton 1968), at 0.623 and 0.525 for adults and 0.222 and 0.185 for first-years (stable and declining trends respectively) in Britain between 1962 and 1995 (using ring recovery data; Siriwardena *et al.* 2000) and on average at 0.51 for adult Turtle Doves in SW France between 1998 and 2004 (Eraud *et al.* 2009). These data suggest that variables other than survival drive population trends.

Eurasian Wryneck
Jynx torquilla

The enigmatic Eurasian Wryneck exemplifies many of the problems associated with correlational studies. Most of the research on this species has been concentrated on the nestbox-breeding populations in the westernmost and northernmost parts of its vast breeding range, and little evidence has been obtained on trends elsewhere. Wrynecks pour into Africa after the breeding season in its millions, which seem to disappear almost completely. Only a tiny amount has been published on its habitat choice, food and behaviour in the wintering quarters, and so we must ask, 'How biased is this information?' It is perhaps surprising that the impact of Sahel rainfall should override that of the many adverse changes in its breeding quarters, for these have been quantified in several studies, and seemed convincing enough. But is this the case? To answer that question fully, we would need a number of closely-spaced years of high rainfall in the Sahel, as happened in the 1950s and early 1960s. Would the Wryneck then be restored in numbers and recolonise breeding grounds from which it has long been lost? Probably, we think.

Breeding range

The European distribution of Eurasian Wrynecks encompasses the Mediterranean and the temperate and boreal zones, but it extends into Eurasia and North Africa. Its northernmost limit lies almost at 65°N in northeastern Fennoscandia, European Russia and all the way to East Asia. In much of this range, it is a rather common breeding bird, except along its fringes in western and southeastern Europe, where Wrynecks are scarce and exhibit spasmodic population fluctuations overlying a long-term decline throughout the 20th century. In the 1990s, the European population was estimated to be in the range of 580 000-1300 000 pairs, of which more than half breed in Russia (BirdLife International 2004a).

Migration

During outward migration in August and September, European breeding birds of the nominate race move in a broad front across Europe in directions lying between SW and SE (Reichlin *et al.* 2009). Birds from Norway showed a wide scatter of directions, with long-distance recoveries predominantly indicating headings of between SW and SSW towards Iberia (Bakken *et al.* 2006). Birds preparing for departure gain weight rapidly, at about 3.0 g per day in recaptured birds (Isle of May, autumn: Langslow 1977), suggesting they were capable of flying non-stop from this island at 56°09'N to Iberia, but apparently navigating by short sea-crossings (Wernham *et al.* 2002). This rate of gain complies with findings in Ottenby (S Sweden) during autumn migration, where body mass averaged 37.5 g (n=28, range 31.2-45.0) and fat deposits on average scored 4.2 (range 0-6, highest fat loads corresponding to 38% of lean body weight). Body masses of Wryneck captured in northern Algeria, from 24 September through 1 November 1985 at Oued Fergoug (35°32'N, 0°03'E), varied between 28.8 and 48.7 g (mean 34.8 ± 5.4 g, n=15), and presumably involved only migrants (Bairlein 1988). These data suggest that Wrynecks fly nonstop to southern Europe or northern Africa in a single flight, and after refuelling there, fly nonstop across the Sahara to the Sahel and regions further south.

We know even less of the homeward migration. The first observations in spring 1966 at Tripoli, Libya, were made on 27 and 30 March, then almost daily 1-5 birds from 5-10 April and 15 birds daily from 25-27 April. Actual numbers may well have been much higher because of the species' secretive behaviour during migration. The effort of crossing the Sahara ensures that weights recorded in northern Africa are considerably lower than those for fattening birds in Nigeria just prior to crossing the desert.[1] Weights of spring migrants in the Camargue varied between 22.5 and 53.8 g (mean 33.2 g, n=135), apparently a mixture of lean arrivals and fattening birds preparing for the next leg to the breeding quarters (Glutz von Blotzheim & Bauer 1980). Pre-migratory masses as measured in Nigeria would suffice to cross both the desert and the Mediterranean. Grounded birds in northern Africa either fattened (Tripoli) or not (Morocco) probably are indicative of the prevailing pre-migratory conditions in the Sahel and variations in weather conditions *en route* north, any adverse combination being likely to require additional stopover sites.

Distribution in Africa

European and West Asian birds migrate to Africa, where they winter in the Sahel and the drier parts of the adjacent Sudan vegetation zone across the entire width of Africa. Even during wintering the Wryneck is secretive and solitary, and most observations refer to accidental encounters with ground-foraging birds or mist-netted captures. The sparse data make a balanced assessment of habitat choice in winter difficult; reported habitats cover a wide range – "open country such as bush, steppe and 'dead' trees in open savannas but also woodlands and transitional rain forests" (Curry Lindahl 1981). Within the Sahel and adjacent Sudan vegetation zone Wryneck prefers *Acacia* savanna, fragmented woodland, cultivated areas, forest clearings and 'derived savanna' (savanna-like habitats where rainforest has long been cleared) south of the Guinea savanna, up to 2500 m asl (Nicolai 1978, Fry *et al.* 1988).

A succession of ringing expeditions in the lower Senegal river valley captured 43 Wrynecks between 1987 and 1993, one of which was retrapped at the same site in three different winters (in the second winter only 400 m away) (Sauvage *et al.* 1998). At Ginak Island in The Gambia (13°34'N, 16°32'W) some birds were retrapped within the same northern winter, but numbers (n=12 initial captures) were too small to judge site fidelity (King & Hutchinson 2001). Similarly, only two birds were captured in the Comoé National Park in Ivory Coast (8°30'N-9°40'N, 3°00'W-4°30'W; northern Guinea savanna zone) from spring 1994 to spring 1997; neither was recaptured (Salewski *et al.* 2000). Near Enugu in eastern Nigeria, on three consecutive days (24-26 November 1971) Nicolai (1976) observed a ground-foraging Wryneck selecting ants or termites. In the same region, Serle (1957) recorded a single bird searching for insects in trees in the recently burnt bush savanna; a male of the nominate race captured here on 26 January 1955 was in heavy moult.

Population trends

Wintering conditions Wrynecks are well-known for their erratic fluctuations in breeding numbers; sudden outbreaks may be followed by equally abrupt declines (Glutz von Blotzheim & Bauer 1980, Bijlsma *et al.* 2001). Even so, in much of Europe the species once was quite common, even abundant throughout the 19th century. All over Europe, declines had already been noted in the late 19th century, and increasingly so in the early 20th century. The peripheral Wryneck population in Britain, having dwindled from widespread over most

Wrynecks are widespread, but nowhere common in the Sahel. Although rarely seen, they have been captured wherever ringing expeditions have been active, *e.g.* in the Senegal Delta (left) and Mauritania (right). Next to nothing is known about their winter ecology.

of lowland Britain in the late 19th century to small isolated numbers in the 1960s (Peal 1968), effectively became extinct in the late 20th century (Holloway 1996). A similar outcome materialised in The Netherlands in the early 2000s, despite marked fluctuations in the preceding century (summarised in Bijlsma *et al.* 2001, Fig. 230D). Steep declines were apparent also in Finnish breeding indices from the late 1960s onwards (Linkola 1978) and again in the early 1980s and early 1990s (Väisänen 2001, 2005; Fig. 230B). Elsewhere, the number of ringed broods in Sweden from 1962-2001 (Ryttman 2003; Fig. 230E), nestbox studies in northern and central Switzerland (1988-99; Schmidt *et al.* 2001), nestbox studies and nestlings ringed in Braunschweig (Berndt & Winkel 1979, Winkel 1992), Bayern, Baden-Württemberg, Saarland and Rheinland-Pfalz in Germany since the late 1960s (Fiedler 1998, Fig. 230A), and the breeding index for the Czech Republic (1982-2006, consistently low values since 1999; Reif *et al.* 2006, Fig. 230F) have shown similar downward trends. These circumstances are typical for northern and western Europe (BirdLife International 2004a), and also apply to the declining numbers captured at Constant Effort Sites in both spring and autumn at Falsterbø (southern Sweden; Karlsson *et al.* 2004), Denmark (Lausten & Lyngs 2004), and at Mettnau (southern Germany; Berthold & Fiedler 2005), as well as to less standardised ringing efforts on Helgoland (Fig. 230C) and along the coast of Schleswig-Holstein in northern Germany (Busche 2004).

The well-documented decline and partial demise of Wryneck in western Europe is partly reflected by its fortunes in eastern Europe. In Poland, for example, Wryneck numbers generally were thought to be stable between 1850 and 1950, but then declined consistently up to at least 2000 (Tomiałojć *et al.* 2006). In Estonia, stable numbers and locally some decline were noted from 1971-1990, but a decline of >50% followed in the 1991-2002 period (Elts *et al.* 2003). The supposedly stable figures for countries further east (BirdLife International 2004a), harbouring the majority of the population, may hint at unspoilt breeding areas, but we think that scarcity of quantitative data is more likely hiding a negative trend.

Condensing local and regional trends into 5-year European means produced high index values in the late 1950s and 1960s, then a steep and ongoing decline to more stable indices in the 1990s and early 2000s at some 30% of the starting values (Fig. 231). Because this decline has been ascertained throughout Europe, a common denominator rather than some local factor is likely to be at the root of the trend. Such a factor might be rainfall in the Sahel, a major wintering area for Wryneck. Indeed, population changes correlate quite well with the rainfall anomaly pattern for the Sahel, where positive changes are associated with high rainfall (Fig. 232). The improved rainfall in the 1990s and early 2000s presumably improved winter survival and stalled the decline, but was insufficient for Wrynecks to regain the levels of the 1950s and 1960s (Fig. 232: notice discrepancy in rainfall between both periods). The longer-term and steady deforestation in the Sahel probably does not play a role, although the possible difference in trend between the western + northern and the eastern European breeding populations may be interpreted as a within-Sahel difference in the scale of deforestation.

Breeding conditions   Light soils, sparse ground cover, open forest and high ant densities are among the many prerequisites of Wryneck habitat in the breeding season. Intensified land use and eutrophication led to habitat loss, increased ground cover and reduced ant densities. Such changes were particularly devastating in western and northern Europe in the second half of the 20th century, where farming and forestry have been industrialised. Ample evidence exists that habitat loss with its associated decline in food supply likely contributed to the decline in Wryneck, from Germany (Epple 1992) and The Netherlands (Bijlsma *et al.* 2001) through to Switzerland

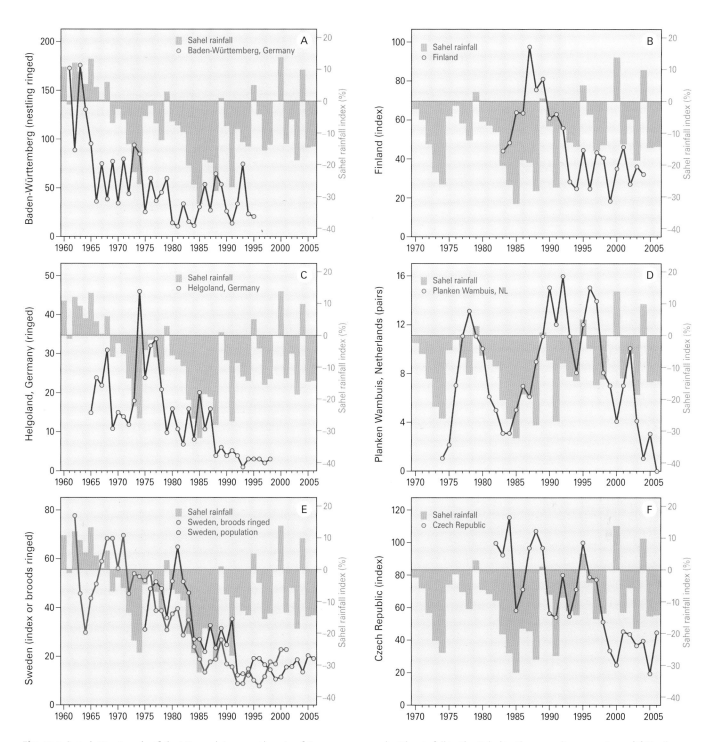

Fig. 230 Population trends of the Wryneck in several parts of Europe compared with rainfall in the Sahel in the preceding year. From: (A) Fiedler (1998) for Baden-Württemberg (Germany); (B) Väisänen et al. (2001, 2005) for Finland; (C) Busche (2004) for Helgoland (Germany); (D) Bijlsma et al. (2001; unpubl. for recent years) for Planken Wambuis (The Netherlands); (E); Ryttman (2003) and Lindström et al. (2008) for Sweden; (F) Reif et al. (2006) for the Czech Republic.

Almost all long-term Wryneck studies in Europe have been carried out on birds breeding in nestboxes. This may have, to some extent, biased the outcome of experiments and analyses of habitat choice and reproductive performance (Wesołowski & Stańska 2001, Møller 2002). However, our knowledge of Wryneck ecology in the breeding quarters is overwhelming compared to that of studies in Africa (which are non-existent), perhaps the reason why all studies unambiguously suggest that the decline is caused by factors playing in Europe.

(Weisshaupt 2007), Sweden (Ryttman 2003) and Finland (Väisänen 2001). Feeding behaviour, nestling growth and nest survival of Wrynecks breeding in the Upper Rhône Valley in Switzerland were affected by local adverse weather (high rainfall, high and low temperatures), notably when of prolonged duration (Geiser *et al.* 2008). This would make peripheral populations in N and NW Europe, where an Atlantic climate dominates, more prone to declines and range contractions, as indeed has been the case.

Conclusion

Wryneck fortunes in the 20th century are closely associated with rainfall patterns in the Sahel. The drought in the 1970s, 1980s and early 1990s led to steep declines, aggravated by loss of breeding habitat in western and northern Europe.

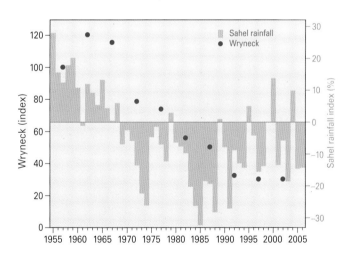

Fig. 231 Average decline per 5-year period in Europe, based on 26 series in 19 studies [2]; the long series are shown in Fig. 230. The index for 1955-1960 is based on 2 German and 1 Dutch series (Berndt & Winkel 1979, Winkel 1992, H. Stel & J. van Laar unpubl.); all other indices are averaged over 15 to 26 series, except 2000-2005 (8 series).

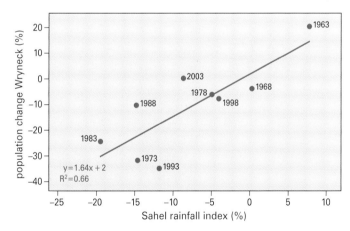

Fig. 232 Population change during nine five-year periods (data from Fig. 231) as a function of Sahel rainfall; n=9; P<0.01; figures refer to midpoints of five-year periods, thus 1973 compares the population change in 1971-1975 relative to 1966-1970.

Endnotes

1 For example, a Wryneck captured at Vom on the Jos Plateau in Central Nigeria weighed 45.5 g on 23 March 1964 and 52.5 g on 31 March (both weighed at 09:00), showing a weight increase of 0.83 g day[1] (Smith 1966). At Tripoli, however, average body masses were 31.5 g (SD=4.58, n=5) and 29.4 g (SD=3.81, n=4) for respectively 1-19 April and 25-29 April 1966. Recaptured birds increased in weight by 1.7 and 2.8 g in 24 hours, by 3.5 g in 30 hours and by 6 g in 82 hours (Erard & Larigauderie 1972). Slightly lower weights were recorded at Defilia Oasis in SE Morocco (32°7'N, 1°15'W), between 5 and 9 April 1965, *i.e.* an average 27.4 g (n=35); numbers included birds captured during a cold spell (Ash 1969). At this oasis, Wryneck captured in the morning hours (05:00-12:00) weighed slightly more than those captured in the afternoon and later (12:00-24:00), *i.e.* respectively 28.1 g (n=15) and 27.5 g (n=26). This stopover was not used as a food stop (or only marginally so), which was evident also from recaptured birds that had not gained weight (despite body masses being at the lower end of the range). Two birds captured near Shakshuk in the Fayum, Egypt, on 14-15 April 1990 weighed 27.7 and 28.5 g (Meininger & Atta 1994).

2 Fig. 231 is based on the following studies: Affre 1975, Becker & Tolkmitt 2007, Berndt & Winkel 1979, Berthold & Fiedler 2005, Busche 2004, Fiedler 1998, Hustings 1992, Karlsson *et al.* 2002, Lausten & Lyngs 2004, Østergaard 2003, Reif *et al.* 2006, Schmid *et al.* 2001, SOVON, H. Stel & J. van Laar unpubl., Winkel 1992, Linkola 1978, Väsiänen *et al.* 1998, Väsiänen 2005, Winkel & Winkel 1985.

Common Sand Martin
Riparia riparia

Centuries ago, major rivers in Europe were largely untamed. Catastrophic floods, sweeping away bridges, houses, mills or even entire villages, created new banks. In those days, the more appropriate name for Sand Martins would surely have been Bank Swallow – even today, large rivers in eastern Europe, like the Bug, Narew, Vistula, Danube and Tisza, harbour huge colonies along their banks. The soft warbling twittering of the hawking Sand Martin along rivers links species and location inseparably, but nowadays, in most of western Europe, the Sand Martin is associated with pits and quarries for sand and gravel extraction, breeding even where heavy machinery is in use daily. Ironically, economic growth has boosted Sand Martin habitat, for in this part of their range, the species rarely breeds in natural sand cliffs. This habitat change in turn has led to popular journals and manuals describing how to protect Sand Martins, how to create artificial sand cliffs, and even how to build concrete walls with prefabricated burrows. Protection in the breeding habitat, though, is not enough, for it is the rainfall in Africa that holds the key to Sand Martin survival.

Breeding range

A small hirundine, the Common Sand Martin is common and widely distributed over much of Europe, particularly so in the floodplains and lowlands of western and eastern Europe. Large wetlands are important as foraging and roosting areas. The species is multi-brooded when onset of breeding is early. In spring it is among the first insectivorous migratory passerines to return to Europe, although breeding may be postponed due to adverse weather, whether during migration or on the breeding grounds, or to late arrival following adverse wintering conditions. Breeding site fidelity is variable, and within- and between-season displacements of hundreds of km have been recorded (even when breeding sites remained intact; Leys 1987). The European population in the 1990s was estimated at 5.4-9.5 million pairs (BirdLife International 2004a). Sand Martins are entirely migratory, wintering in sub-Saharan Africa (western populations) and eastern Africa (eastern populations).

Migration

Sand Martins breeding in western, central and much of northern Europe generally migrate on headings between SW and S, but the more easterly European populations tend towards S and SE. They do not avoid either the Mediterranean Sea or the Sahara; recoveries from Algeria and Libya show that the Sahara is crossed in broad front (Fig. 233).

Given the outward migration headings of northern, central and eastern European birds (Fig. 233), the central Sahel seems to be a major wintering domain of these populations, from the Inner Niger Delta through to Lake Chad, but another such area extends into eastern and southern Africa. There have been very few recoveries from these regions to substantiate this hypothesis; a Swedish bird was recovered in Mali (Fig. 233) and birds from Norway have been recovered from Senegal (1), Mauritania (1), Nigeria (1), Congo (1) and South Africa (11) (Bakken *et al.* 2006). The recoveries of birds in the Central African Republic (one each from Finland, Sweden and Denmark) suggest that, in line with the sparse Norwegian data, many which are northern breeders continue overland to wintering areas in South Africa. Russian and Ukrainian Sand Martins presumably winter mainly in eastern and southern Africa (Cramp & Simmons 1988).

Based on ringing data, West European birds, and particularly those from Britain, end up in the westernmost part of West Africa, *i.e.* in the Senegal Delta. Of the >15 000 Sand Martin ringed in Senegal (Sauvage *et al.* 1998), 86 were recovered from Great Britain, 20 from Ireland and 11 from The Netherlands. This western European component is increased by recoveries from Portugal, Spain, France, Germany and Denmark. Cramp & Simmons (1988) and Wernham *et al.* (2002) suggested that the Sand Martin appears to be systematically nomadic, with British birds moving east from coastal sites in Senegal through the Sahel to the Inner Niger Delta in Mali during winter, from where the return migration would be initiated. This presumed movement of the wintering population is surmised on the basis of a wide distribution of ringing recoveries across northern Africa in March to April (Wernham *et al.* 2002), but is contradicted by a similar spread in autumn recoveries in Europe (SW-SE), suggesting that an unknown proportion of the British Sand Martin population flies directly to Mali to winter.

The use of discrete moulting areas in Africa, as suggested by the analysis of mineral profiles (stable-isotope ratios) in feathers (Szép *et al.* 2003b, Fox & Bearhop 2008), substantiates the use of fixed wintering sites. The African environment is richer in various metals than the European breeding grounds, resulting in a finer-tuned elemental composition in African-grown feathers, and this adds further weight to the idea of Szép *et al.* (2003a) that the birds use discrete moulting areas during winter; their analysis of trace elements in feathers showed that Spanish, Danish, British and Hungarian populations differed markedly in elemental composition, achievable only if these populations used different moulting areas in Africa. Using tail feathers from the same individual in its first and second

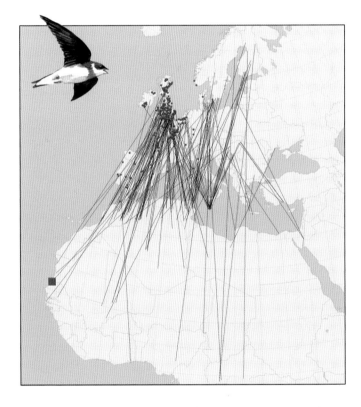

Fig. 233 European ringing locations of 870 Sand Martin recovered (n=132) or ringed (n=738) between 4° and 37°N (From: EURING). In order not to clutter the figure, lines are not shown for birds ringed or captured in the Senegal Delta; instead, red dots indicate ringing sites of birds recaptured in Senegal (n=123) or birds ringed in Senegal and recaptured later on in Europe (n=473).

Three abundant Palearctic hirundines, i.e. Barn Swallow, Sand Martin and House Martin, spend the northern winter in sub-Saharan Africa, where they occupy different niches (respectively drylands, wetlands and high air) in different zones across the continent. Sand Martins are particularly abundant near floodplains and lakes in the Sahel, here hawking insects, amidst foraging Little Stints and Common Ringed Plovers, above drying pools in the Inner Niger Delta in Mali (February 2008).

year of life, it became clear that year-on-year feather elemental composition was largely similar, strongly suggesting that wintering areas (or habitats) are the same between years. This is an interesting finding in the light of the question as to whether the birds are nomadic in winter or not. There was no significant difference in elemental composition of feathers grown in Africa between males and females from any one breeding site, which indicates that they use the same wintering area or habitat (Szép et al. 2003b).

Distribution in Africa

The wetlands in the Sahel are major wintering areas for Sand Martins. Large numbers congregate in floodplains, notably the Senegal Delta, the Inner Niger Delta in Mali and Lake Chad. During daylight, numbers are widely dispersed, most foraging in loose flocks above wet areas rich in insects, but occasionally also over drier terrain.

In the Senegal Delta, Sand Martins are very abundant from late October to April, an estimated 2 million wintering in the Djoudj National Park and its immediate surroundings. The birds roost in reedbeds and often rest on the ground during the day (Rodwell et al. 1996, Peeters 2003). Large numbers have been ringed at roosts, the location of which frequently changed by up to 20 km within the same winter (Sauvage et al. 1998).

Roosts of more than half a million birds have been recorded in early December 1999 in the central lakes of the Inner Niger Delta (van der Kamp et al. 2002a). After the birds arrive, there is a gradual movement from the northern to the southern Inner Delta, numbers accumulating in the northern delta from September to November, during which period it has not yet flooded. In the same months, only a few are present in the fully flooded southern Delta (Lamarche 1980, van der Kamp et al. 2002a).

Along the western shore of Lake Chad, 1 million Sand Martins were estimated to be present between 21 March and 13 April 1967, some 100 000-175 000 birds per hour moving north in late afternoon and evening; whether this activity involved migration or roosting flights was not determined (Ash et al. 1967). Midwinter flocks of 2000 birds are known from the Hadejia-Nguru floodplains (Elgood et al. 1994).

North of the Sahel, wintering is supposed to be a rare event, as in the Nile Delta and Nile Valley, where only small numbers have been recorded. On 10 November 1981 at Lake Manzala in the Egyptian Delta, the presence of an assembly of hundreds of thousands was thought to have included Eurasian birds and may have involved birds still en route to the sub-Saharan region (Goodman & Meininger et al. 1989).

In the Sudan and Guinea zones, the species is regarded mainly as a migrant, perhaps confirmed by the paucity of ringing recoveries in the Sudan vegetation zone. The Portuguese bird mistnetted at Keta Lagoon, coastal Ghana, on 10-11 December 1996 (among 12 captured and less than 100 observed, B. van den Brink pers. comm.) is the only one from this region; the other recoveries from the same zone further to the east relate to captures of migrants visiting Barn Swallow roosts (Fig. 233).

Population changes

Wintering conditions Long-term censuses of Sand Martins are available from many parts of Europe, varying from studies at local colonies to country-wide censuses (Fig. 234). In the long run, most trends are more or less stable, but suffer marked fluctuations (varying by factors of 3-6). Positive or negative trends are usually associated with short time series, i.e. showing that time series should

Box 24
Mixed roosting

Sand Martins and Barn Swallows in Africa are almost completely segregated, except during migration periods, when mixed flocks may occur on a large scale. Their foraging requirements usually confine Sand Martins strictly to large wetlands in the Sahel and eastern Africa; only exceptionally have they been found roosting in the company of Barn Swallows (Table 24). The latter forage widely over huge scattered areas during the day, often near human habitation and in dry habitats, and then congregate in large roosts in reedbeds, Elephant grass and sugarcane fields from the Sudan vegetation zone southwards as far as Botswana, Namibia and South Africa. However, Sand Martin roosts often overlap with those of Yellow Wagtails. For example, we recorded low-altitude mixed roosting flights involving tens of thousands of both species between 27 January and 5 February 2004 in the central lakes of the Inner Niger Delta between 06:20 and 07:10 local time and again between 17:50 and 18:10.

Table 25 Proportion of Sand Martin at various Barn Swallow roosts in Africa (n = number captured).

Country	Period	Barn Swallow (n)	Sand Martin (n)	% Sand Martin	Source
Ghana	Dec/Jan 1996/97	1828	0	0.00	van den Brink et al. 1998
Ghana	Dec 1997	983	0	0.00	Deuzeman et al. 2004
Nigeria	Feb 2001	7362	1	0.01	B. van den Brink pers. comm.
Zambia	Nov /Mar 2007/08	6358	241	3.65	B. van den Brink pers. comm.
Botswana	Dec/Jan 1992/93	5761	9	0.16	van den Brink et al. 1997
Botswana	Dec/Jan 1993/94	10 069	19	0.19	van den Brink et al. 1997
Botswana	Dec/Jan 1994/95	2594	19	0.73	van den Brink et al. 1997
Botswana	Jan-Feb 2003	4503	7	0.16	B. van den Brink pers. comm.

cover many decades before long-term trends can be distinguished reliably from short-term fluctuations (Box 27). Furthermore, for a species like the Sand Martin, occurring patchily in colonies in often unstable environments, census areas need to be large enough to allow for colony-site switching, otherwise chances are high that colonies will move outside the census plot boundaries; see for example the trend in the Czech Republic: Reif et al. 2006, www.birdlife.cz, J. Reif pers. comm.).

Cowley & Siriwardena (2005) have concluded that the Sahel rainfall index was significantly correlated with absolute abundance of Sand Martins in Nottinghamshire, England, as reflected in the total annual marked male population size (an indicator of total population size). Other studies have also shown that steep population increases and decreases can be related closely with flood level in the Inner Niger Delta and Senegal Delta, or with rainfall in the Sahel (Fig. 234). In general, low flood levels and, in particular, poor precipitation in West Africa translate into declining Sand Martin numbers.

Since in winter so many Sand Martins are concentrated in the floodplains, the expectation was that bird numbers would correlate better with flood extent than with rainfall or NDVI. This appeared not always to be the case. [1] Fig. 234 shows five trends as examples, typical of the many long-term studies available. The changes in the different populations are synchronised: all populations showed a decline in the mid-1980s, when the Sahel rainfall was more than 25% below the average and then went up in wet years in the 1990s (Table 25). Remarkably, however, that the population in Sweden and Denmark has scarcely increased after the improvement of the Sahel rain in the 1990s.

Sand Martins were hit hard by severe drought in the Sahel: the average decline in the 1981-1985 period was 69% in the regions mentioned in Table 25, ranging from 60% in Sweden to 85% in SE France. The species may recover by about 20% after a wet year in their wintering area. Lack of an immediate positive response, as shown for Central Westphalia in Germany and along the Upper Tisza in Hungary, is usually associated with locally adverse conditions. For example, the population along the entire Tisza in Hungary doubled from 10 528 pairs in 1999 to 21 365 pairs in 2000, although numbers along the Upper Tisza remained low in 1998-2000, the latter associated with catastrophic floods (Szép et al. 2003a).

The relationship between population size and Sahelian rainfall is quite straightforward. The British Sand Martin population, for ex-

Sand Martins leaving the nocturnal roosts around sunrise in the Senegal Delta (left) and Inner Niger Delta (right), where altogether millions are concentrated.

ample, starts declining when Sahelian rainfall drops 5% below the long-term rainfall average for the 20th century (Fig. 235A). Under such circumstances, the return rate of adult males was still 35%, as found in a study in Oxforshire (Holmes et al. 1987). However, when rainfall dropped by 25% below the long-term average in the Sahel (as in 1983), return rates plummeted to 14% in males and 7% in females. Hardly any juveniles returned when hit this hard, i.e. only 3.3% in 1983 (Fig. 235B). Similar results for adults were found by Szép (1993, 1995a, 1995b) in Hungary and by Cowley & Siriwardena (2005) in central England.

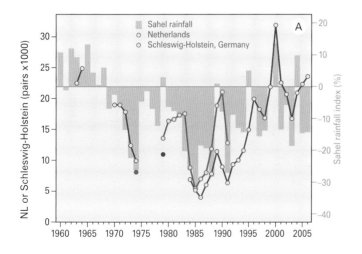
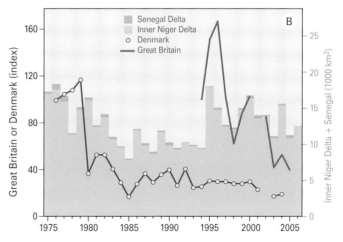

All three studies were consistent in the lower survival rate (as measured by return rate or capture-recapture rates) of females compared with males. How this sex-related difference in mortality comes about is still shrouded in mystery. First, it is not likely to be related to the sexes wintering in different regions (Szép et al. 2003b). Furthermore, the birds from the EURING files, as sexed in the wintering areas (mostly in Senegal), were in a 50:50 sex ratio, at 76 males and 77 females. It is difficult to imagine how males could outcompete females for food, given the hypothesised lack of territoriality in these aerial insectivores. To make things even more mysterious, the elegant studies of Jones (1987) and Bryant & Jones (1995) showed that smaller Sand Martins (as measured by keel-length, a fixed attribute of an individual and likely to be heritable, unlike wing length; Bryant & Jones 1995), survived better under the adverse conditions of severe drought in the Sahel. A smaller than normal size was evidently favoured during the population crash of 1983-1984, due to selective mortality of large individuals between breeding seasons (Bryant & Jones 1995), but this selection seemed to act in similar directions at any given time in either sex. Although size differences between the sexes are not significant (Cramp & Simmons 1988), females on average are slightly larger. In this respect, it is interesting to note that females are hit especially hard under severe drought conditions in sub-Saharan Africa (return rate only half of that of males; Fig. 235B). Again, how exactly body size is impacted by drought remains to be investigated. We have no evidence that a small Sand Martin is of inferior reproductive capability or perform-

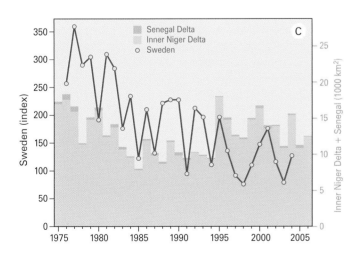

Fig. 234 Sand Martin trends in European populations, related to Sahelian flood extent or Sahel rainfall. From: (A) Berndt et al. 1994 (Germany), Leys 1987 & SOVON (The Netherlands). (B) BTO (Great Britain), Heldbjerg 2005 (Denmark). (C) Lindström & Svensson 2005 (Sweden).

Table 26 Percentage population change in European Sand Martin populations between 1981 and 1985 (1983-1985 extremely dry years in West Africa), between 1994 and 1995 (intervening winter extremely wet in West Africa compared with preceding 12 years) and between 1999 and 2000 (very wet winter in between). Rain indices expressed relative to the 20th century average.

Country	1981-1985	1994-1995	1999–2000	Source
Sahel rain index (%)	-26.36	+7.47	+1.07	
Sweden	-60.0	+75.9	+34.2	Lindström & Svensson 2005
Denmark	-67.9	+23.1	+10.7	Heldbjerg 2005
Germany, Schleswig-Holstein	-67.4	+17.2		Berndt et al. 1994
Germany Mittelwestfalen		-26.1		Loske & Laumeier 1999
Germany, Niedersachsen		+2.4	+12.9	Zang et al. 2005
Germany, Hannover		+38.5	+12.0	Zang et al. 2005
The Netherlands		+28.8	+31.2	Leys 1987, SOVON
Belgium, Flanders		+14.3	+18.3	Vermeersch et al. 2004
Great Britain		+52.0	+18.0	BTO News
France, Durance valley	-85.2			Olioso 1991
Switzerland (Freiburg, Bern, Solothurn)	-64.3	+14.6	+27.6	Weggler 2005
Hungary, Upper Tisza		-5.6	-5.1	Szép 1993, 1995a, 1999, Szép et al. 2003a
Spain			+53.8	www.ebcc.info
Average	-68.96	21.79	21.25	
SD	9.61	29.25	16.17	

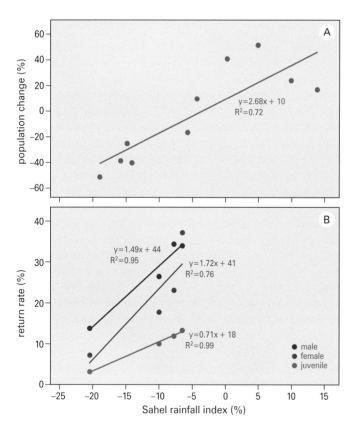

Fig. 235 (A) Annual change in the British Sand Martin population as a function of the Sahel rainfall 1995-2006; same data as Fig. 234B. (B) The annual survival for males, females and juveniles as a function of Sahel rainfall, such as determined by Holmes et al. (1987) using captures and recaptures to calculate the return rate of birds breeding in Oxfordshire between 1980 and 1983; the regression for the females is weakly significant (P=0.02), the three other highly significant (P<0.001).

ance, only that it is seemingly at less of a disadvantage than a larger individual under drought conditions in Africa. Upon return to normal winter rainfall conditions, body size tends to return to the previous average, which suggests that some other environmental pressure is dominant.

It is obvious that rainfall in the Sahel has an overriding impact on Sand Martin population dynamics, but one important question has still to be answered: do relatively more birds die if the population size is large? In other words: do proportionately more birds survive the winter when the population is small and the concomitant competition among individuals is also less? Fig. 236A shows that the latter hypothesis is the most likely. The Dutch Sand Martin population increases after a wet winter and decreases after a dry winter, but, with no change in Sahelian rainfall, relatively fewer birds return the next year if the population is large. This density-dependent effect may also be shown in another way. When population change is plotted against population size, the numbers tend to increase at a low population level and to decrease if the population is large, but the actual population change depends more on the Sahel rainfall (Fig. 236B). [2]

Breeding conditions   Wherever Sand Martins are still breeding in natural ecosystems, such as untamed rivers with steep walls, high floods can have positive as well as negative impacts on breeding success. Before the breeding season, high floods have a positive effect overall because they wash away parasites from the previous years' burrows and create new walls. Later flooding occurring during the birds' breeding cycle could have a series of adverse effects. Early-returning birds are forced to switch to secondary breeding sites, where

Sand Martins nest in river banks, but in western Europe an increasing part of the birds breed in artificial sand cliffs.

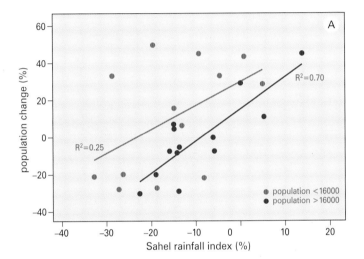

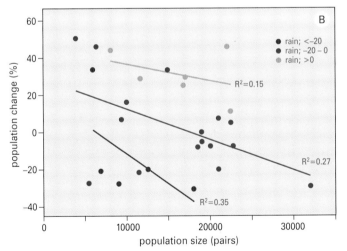

Fig. 236 Change in the Dutch Sand Martin population relative to the previous year as a function of: (A) Sahel rainfall, at two population levels, and (B) the population size in the previous year, at three levels of Sahel rainfall. Same data as Fig. 234A.

natural and human threats are greater. Floods occurring even later can destroy all first broods, as happened along the Tisza River in Hungary in 1998 in Tibor Szép's study area (Szép et al. 2003a). Although the frequency of abnormal flooding has been historically low, at least in Hungary, it has nevertheless been much higher recently, a fact possibly linked to climate change.

Apart from such natural disasters, human error can also cause great environmental damage. From January to March 2000, the entire length of the Tisza River in Hungary suffered from the collapse of reservoirs belonging to Romanian goldmines upriver, the reason being that the storage and treatment of mine tailings (acids, cyanide and heavy metals, Szép et al. 2003a) was haphazard and unsupervised. By two strokes of luck, the pollution occurred at the biologically most inactive period of the year and was partly diluted by a vigorous spring flood that did much to replenish the losses to invertebrates and fishes.

In the Sand Martin's northern breeding areas, such as the subalpine birch belt in Swedish Lapland, onset of laying was correlated with the date when 50% of the ground was snow-free, normally one month before laying started. When mean temperatures were below 5-7 °C or maximum temperatures below 10-12 °C, onset of laying was delayed. Presumably, these variables correlate with insect abundance a month later. Although clutch size declined by 0.2 eggs per week within years, early and late clutches were of about the same size, independent of the onset of laying in any particular year (Svensson 1986). These data imply that snow and temperatures affected only the start of laying, but had no consequences for reproductive output.

The well-established impact of hydrological conditions in the wintering areas (see above) does not preclude a similar impact of rainfall in the breeding quarters. A study in Nottinghamshire, England, from 1967-1992, showed that the previous summer's rainfall

Mesquite *Proposis juliflora* and *Eucalyptus*, exotic trees planted along the eastern shore of Lake Horo (Mali), offer Common Sand Martins ideal feeding circumstances during the *Harmattan*, when huge numbers of small flies are concentrated in the leeside of cover along the shore. We roughly estimated some 130 000 birds feeding along the eastern shoreline of the lake in February 2009 (13 km and 10 000/km).

was strongly and negatively correlated with survival rates, *i.e.* opposite to the effect of rainfall south of the Sahara. However, we should bear in mind that the situation may be different in central and eastern Europe, where the continental climate results in less humid summer conditions than in the marine temperate climate of Britain. Rainfall in breeding and wintering areas has different impacts on aerial insect abundance. Precipitation in the breeding areas, especially during the most stressful early breeding period, reduces the availability of airborne insects. This in itself does not normally reduce adult survival, but the added stress of raising a brood under adverse conditions may well reduce the adult's fitness for migration and overwintering. Furthermore, nestlings and fledglings may starve due to the adult's inability to provision them sufficiently (Cowley & Siriwardena 2005). In the Sahel, however, rainfall enhances vegetation growth and boosts insect abundance, leading to improved foraging conditions.

Conclusion

Sahel drought has an immediate negative impact on the population size of Sand Martins in Europe. Drought directly influences survival rate, its effects visible on breeding grounds in the declining return rates. Drought also exerts a selective mortality of large individuals, and possibly of females. Wet years in the Sahel, on the contrary, result in improved survival and recovery of losses suffered during droughts. Full recovery from severe droughts takes a couple of years of normal and abundant rainfall in sub-Saharan Africa. The population change from year to year is related to the rainfall in Africa, but is also dependent on the size of the population. Environmental factors on the breeding grounds may result in patterns deviating from the general picture, but these are either short-lived or local. Another variable that impacts population size may be the previous summer's rainfall, high rainfall resulting in a decline (as found in Britain). Whether this relationship also holds true in the climate conditions of continental Europe has not yet been investigated.

Endnotes

1 Correlations of the size of the Dutch Sand Martin population (N=31) with flood extent (R=+ 0.684) and with rainfall (R=+ 0.647) do not differ and are both highly significant (P<0.001). The size of the population in Schleswig-Holstein (N=14) was correlated with rainfall (R=+0.682, P=.007) but only slightly with flood extent (R=+0.399, P=0.157). In the 17 studies available, the correlation between population size and rain was 10 times larger, and 7 times less, than the correlation of population size with flood extent.
2 The annual population change for Dutch Sand Martins is dependent on Sahel rainfall (P<0.001) as well as on population size (R^2=0.52, P=0.003, N=27).

Barn Swallow
Hirundo rustica

"The swallows which arrived at the beginning of the cold spell in 1965 soon showed signs of fatigue and it was clear that they were not finding food. They concentrated round the Poste de Secours, the only building in the area, upon and within which they settled on various sheltered perches to rest... About 300 were present on 6 April, but these soon dwindled to 150 the next day, and there were apparently no fresh arrivals. Three main perches were used... On all these places Swallows constantly congregated in heaps, row upon row of them settling down upon each other until they were sometimes four rows deep. A distinct impression was gained that this was a deliberate action, presumably designed to conserve heat in the unusually cold weather. On being disturbed, the birds would very soon return, and could be watched from a few inches range as they shuffled down one upon another, some with their heads facing inwards and others outwards. In this position birds died, especially when roosting at night, but also during the day. On the two broader shelves the dead lay where they died, but from the narrow beam the dead fell to the floor as their places were filled by live birds." This graphic description of J.S. Ash (in Smith 1968) is one of many accounts of mass mortality of Swallows under adverse weather. His birds had just passed the Sahara, an exacting exercise in itself, and were confronted with a cold spell on the Moroccan-Algerian border. Barn Swallows, being insectivorous and always flying low, are particularly prone to weather-induced mortality. Harsh conditions on the breeding grounds, during migration and in Africa take a heavy toll. And the more so when such conditions are long-lasting, as abundantly demonstrated during droughts in southern Africa and in the Sahel. For the surviving Barn Swallows, residuals of past hardships continue to impact their lives long after the actual events.

About 200 million Barn Swallows fly from Eurasia to Africa. Large pre-migratory and pre-roost gatherings on wires are a common sight in Europe, but precisely the same scene can be seen in Africa, as pictured here at the Mafundzalo Range near Kabwe in Zambia (late November 2007). Such gatherings are bristling with activity: twittering, song, preening, panic at the approach of a predator.

Breeding range

The Barn Swallow breeds throughout low altitude Europe, reaching highest densities in the rural landscapes of western, central and eastern Europe. It is closely associated with human occupation, especially farming. Agricultural intensification, especially in the second half of the 20th century, has led to declines in western Europe and possibly elsewhere. The European population, including Russia, Ukraine and Belarus, was estimated at 16-36 million pairs in 1990-2000 (BirdLife International 2004a), amounting to a post-breeding population of 128-288 millions birds (based on 6 fledglings/pair per year; Turner 2006).

Migration

In the past decades large numbers of Barn Swallows have been ringed, both on the breeding grounds as in the areas south of the Sahara where they spend the northern winter.[1] For the early 2000s, Szép et al. (2006) showed that annually some 90 000 Barn Swallows were ringed in western and northern Europe (western flyway), and some 37 300 in eastern Europe (including Finland), constituting respectively 1.58% and 0.09% of the total population. On the receiving end of the flyway, from Ghana, Nigeria, Central African Republic and Congo-Kinshasa to Botswana and South Africa, 100 000s of Swallows have been ringed on their roosts (Oatley 2000, van den Brink et al. 2000).

For a passerine this size (17-20 g), wintering areas in Africa are relatively well known based on ring recoveries. All Barn Swallows are

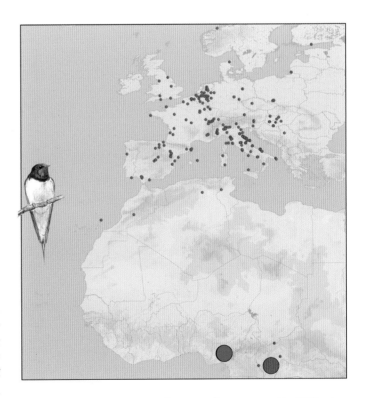

Fig. 237 Recoveries or origins of Barn Swallows captured in SE Nigeria (blue dots; n==300) or in the Central African Republic (red dots; n=117). Data from EURING and P. Micheloni (pers. comm.).

Barn Swallow *Hirundo rustica*

trans-Saharan migrants, but they differ widely in wintering areas in relation to their geographic origin. African wintering sites are headed for in more or less straight flights from the breeding grounds. The latitudinal distribution in Africa is by and large a mirror image of the one found in Eurasia: western birds in the west, eastern birds in the east. This is nicely illustrated by captures at two sites in Central Africa, only 800 km apart (Box 25): the birds wintering in or passing SE Nigeria originated from more westerly parts of Europe than those in

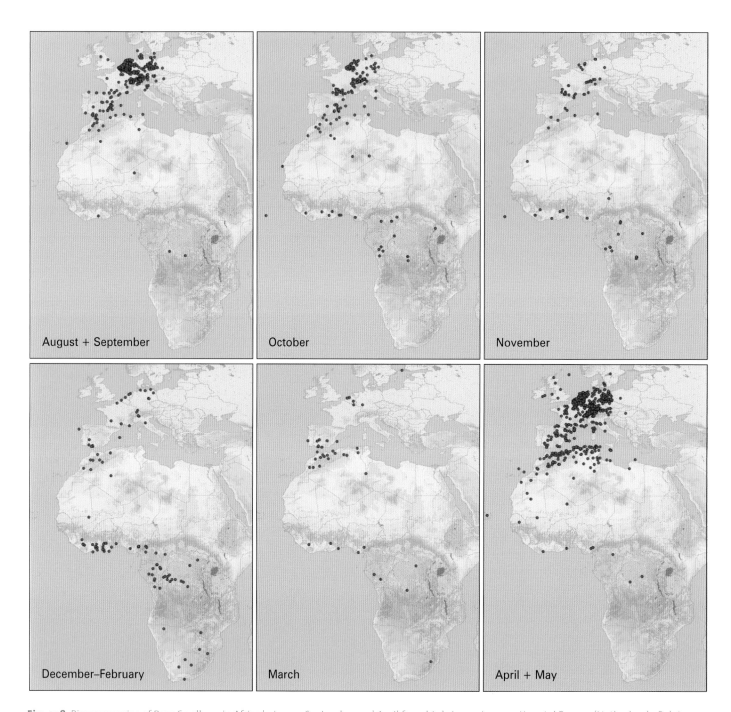

Fig. 238 Ring recoveries of Barn Swallows in Africa between September and April from birds in western continental Europe (Netherlands, Belgium, Germany, Switzerland, France, Spain, Portugal; n=451; left) or ringed in Britain (n=383; right). From: EURING.

the Central African Republic that came from a more central section of Europe (Fig. 237). Autumn migration commences on a rather broad front, without clear concentrations at small sea crossings. Consequently, fat loads of Barn Swallows leaving Italy in autumn are 1.5-2.0 g heavier than in Iberian birds, in anticipation of the longer sea-crossing for Italian Barn Swallows heading due south (400 resp. 800 km). The fat reserves probably also suffice to cross the Sahara without substantial refuelling in North Africa (Rubolini et al. 2002).

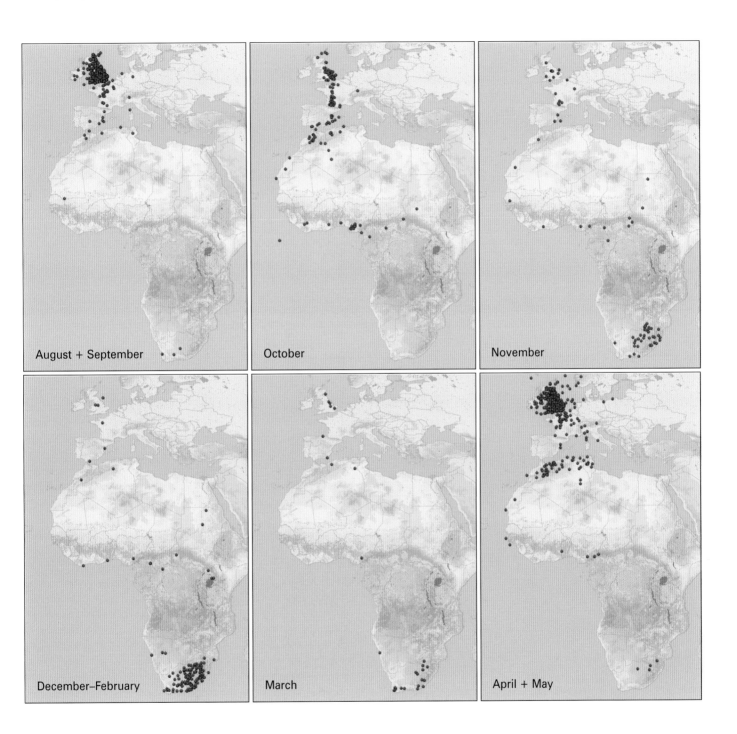

Barn Swallow *Hirundo rustica*

Outside the breeding season, Barn Swallows greatly prefer reedbeds as roosting sites, and they will fly up to 75 km to attend suitable roosting sites. In the absence of reed almost any similarly structured vegetation may do: maize, sugarcane, elephant grass, *Typha*, even trees. (Kasanka National Park, Zambia, early November 2007).

Winter distribution in Africa

Populations from northern Europe and the United Kingdom leapfrog those from central and southern Europe (Fig. 237-239), and occupy wintering areas in the southern third of Africa (mostly South Africa and Botswana, but also Zimbabwe and Zambia; Oatley 2000). Large traditional roosts of up to several millions birds are known here in reedbeds (Harrison *et al.* 1997). In the absence of reedbeds, copses of *Acacia* may also be used, as for example by 1-2 million birds at Jwaneng (24°35'S, 24°43'E) along the edge of the Kalahari in Botswana in January and February 2003, and in previous years (van den Brink *et al.* 2003). At least in the dry Karoo, SW Africa, Barn Swallows may also form many small roosts (rarely >1000 birds per roost) that are far apart and probably variable in time and space (Szép *et al.* 2006). British and northern European (including Denmark) Barn Swallows tend to mostly winter in western and southern South Africa (taking the highly biased ringing effort towards the eastern part of the country into consideration, and accounting for higher reporting rates in eastern South Africa; Oatley 2000, Szép *et al.* 2006). Those from west of the Urals (including Finland) are mostly found in Transvaal and Botswana (van den Brink *et al.* 2003), but may winter as far north as Congo-Kinshasha (de Bont 1962). The huge population from east of the Urals, with a recovery of a Transvaal-ringed bird beyond the Yenesei as far as 92°E (Rowan 1968), winters in the eastern Cape Province and KwaZulu-Natal, north into Zimbabwe,

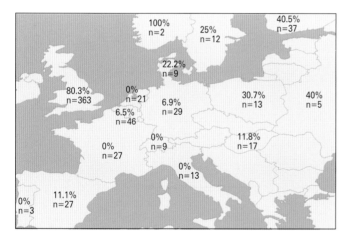

Fig. 239 Percentage of ring recoveries in southern Africa (south of 20°S) relative to the total sub-Saharan number; n= number of sub-Saharan recoveries. Data are given per country, but joined for the countries of the ex-USSR. From: EURING.

Botswana and Zambia (Harrison et al. 1997, Oatley 2000, Dowsett & Leonard 2001). One of the biggest roosts known in South Africa is situated at Mount Moreland near Durban (2-3 million birds in December, declining to 500 000 in April); and has been in use since at least the early 1990s (Piper 2007).

Central and southern European Barn Swallows on average winter furthest north in Africa, mainly in the equatorial forest zone from Guinea through Nigeria, Cameroon and Central African Republic to Congo-Kinshasha (see also Moreau 1972). The rainforest itself is largely avoided, the Barn Swallows concentrating in relatively small bands along its northern and southern edges (Fig. 238) where roosts hug the nearby savanna and/or floodplains of large rivers (4-6° on either side of the equator). Huge roosts of up to several million birds are known to occur on mountain slopes with patches of Elephant grass *Pennisetum purpureum* amidst rain forest, as for example at Boje-Ebbaken in SE Nigeria (Loske 1996, Bijlsma & van den Brink 2005). Wintering in Central Africa is confined to forested regions, where Cattail *Typha* sp., Cassava *Manihot esculanta* and Elephant grass are used for roosting (de Bont 1962; Box 25). Roost sites in West and Central Africa are used by Barn Swallows in transit to or from southern Africa (Fig. 238). Like in the Karoo (Szép et al. 2006), smaller roosts are in use throughout the forest zone, as exemplified for Ghana where in November-January 1996/1997 roosts ranged in size from tens up to 11 000 (van den Brink et al. 1998), with a single larger roost of more than 200 000 birds on 30 November 1996 near Essaman (A.P. Møller in van den Brink et al. 1998). Small roosts rarely are traditional, shifting site in accordance to local conditions.

Population change

Wintering conditions Rainfall and vegetation in the wintering quarters have a profound impact on body condition, duration of moult, timing of migration, arrival on the breeding grounds and survival. In Botswana, droughts negatively affected insect abundance as well as availability and quality of roosts. In years with abundant or normal rainfall, renewing all primaries took 120-130 days compared with 155-190 days during a drought (van den Brink et al. 2000). Body masses in juveniles and adults in late December-January was in the dry 1995 even 10% below the wet 1993 (Fig. 240). The latter period was also characterised by much smaller numbers and a higher proportion of emaciated Barn Swallows (van den Brink et al. 1997). If moult is delayed to such an extent that it overlaps with migration, it would impair migration performance and could have fitness consequences (Pérez-Tris et al. 2001).

Consistent with the above findings in the wintering quarters, winter mortality in a small Danish population, either during migration or in the African winter quarters, was shown to be correlated with precipitation patterns in South Africa (Møller 1989). The suggested impact might have been spurious, however, as Møller used data from weather stations at Pretoria and Johannesburg in the NE part of South Africa, and later research indicated that Danish Swallows presumably mostly winter in the Karoo in the SW part of South Africa (Szép et al. 2006). [2] Indeed, the long-term trend of Møller and the Danish trend, as indexed on the basis of point-counts in 1976-2004 (Heldbjerg 2005), failed to correlate with rainfall patterns in South Africa, here based upon 65 weather stations across South Africa with data from the entire 20th century (Fig. 241H). [3]

Ups and downs of Finnish Swallows (Väisänen et al. 2005; Fig. 241D), also known to mostly spend the northern winter in southern Africa, show a similar lack of correlation with rainfall in South Africa. [4] Like Robinson et al. (2003a), we are also unable to relate changes in breeding British population (Fig. 241A) to rainfall in southern Africa (the major wintering area of British Swallows). [5]

This counter-intuitive finding (see Fig. 240) demands a further analysis. The 463 British recoveries from Africa were used to investigate whether the winter mortality is related to the yearly variation in rainfall. To correct for long-term variation in ringed number of Swallows, the yearly number of African recoveries (varying between 0 and 29°S and 0-11°N of the equator) was expressed as percentage of the total number of recoveries in the same year (varying between 11 and 150). The relative numbers of African recoveries north and south of the equator were related, respectively, to the rainfall in S Africa and in the Sahel. The impact of South African rainfall on the British Swallows appears to be obvious in this data set (Fig. 242A). [6] From this we conclude that the conditions in Africa must have an impact on the population level and that the lack of a correlation in some of the trends is partly due to the less reliable data in a species difficult to monitor (Fig. 241).

Spring conditions en route The many accounts of mass mortality during migration abundantly illustrate the vulnerability of Barn Swallows to adverse weather. But even when mass mortality does not occur, mortality is likely to be high anyway, especially during

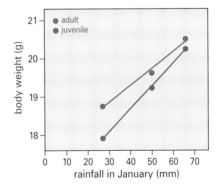

Fig. 240 Body weight of Barn Swallows captured near the Okavango Delta (Botswana) in January 1993-1995 compared to the rainfall in the same month. From: van den Brink et al. (1997). Rain was registered in the nearby station of Maun (23°25'S and 19°59'E).

Box 25
Fishing for Swallows Pierfrancesco Micheloni

Subsistence hunting is widespread in the forest zone of West and Central Africa. Bushmeat is high on the target list, but the search for protein extends well beyond mammals and covers anything from insects to fishes and birds. Swallows are a case in point. Weighing only 17 grams, it can hardly be considered a profitable prey for man. That is to say: unless swallows occur in large numbers and concentrated. In Africa various ingenious methods have been developed to capitalise on such food bonanzas, interestingly of quite different nature between geographic regions.

In SE Nigeria, near the communities of Ebbaken-Boje-Ebok (6°38'N, 9°00'E), Swallows roost in huge numbers on steep hill slopes covered with Elephant Grass amidst rainforest. In mid winter, numbers may comprise up to 1.5 million birds (Ash 1995, Bijlsma & van den Brink 2005). The total number using this roost during the northern winter must be much bigger than that because in autumn and spring birds from northern Europe and the UK pass through on their way to southern Africa, and back again to their breeding haunts (Nikolaus et al. 1995, Oatley 2000). Local villagers catch the Swallows during moonlit nights, using lime-smeared, dandelion-shaped brushes radiating from the top of 4-5 m long palmsticks. Swallows are captured upon descent and after disturbance once settled, waving the trapping device above the grass in order to entangle low-flying and fluttering birds with the sticky twigs (60-80 cm long). The average night bag amounts to 100 birds per catcher, with a record of 4425 Swallows caught by 30 people on 10 March 1995 (Nikolaus et al. 1995). The grand total captured per season (normally January-April) has been estimated at 100 000-200 000 (Ash 1995, Loske 1996).

A crude version of this capture technique is practiced near Atta (6°28'N, 11°19'E), across the Nigerian border in Cameroon, some 150 km to the east of Ebbaken (Micheloni 2000). Here, Swallows are captured at the roost in elephant grass using long sticks with glue attached to the end of the stalk, i.e. without the more effective umbrella-shaped brushes. Catches are consequently much smaller than near Ebbaken, also because in the dry season the grass is grazed by cattle rather than protected from grazing and fire to ensure the continued existence of the roost.

In the southwestern Central African Republic, swallows are also present in huge numbers but roosting sites – presumably along the river of Lobaye, marking the transition zone between the Guinea-Congolian rainforest of N'Gotto in the south and the savanna in the north - have not yet been located by the local people ('the hidden city of swallows") (Micheloni 2000). Consequently a different technique has been perfected to get hold of the birds. This technique hinges on the close of the rainy season (October- early December, but mostly November): rain showers rather than continuous rain, weak sunshine and flying termites. The latter start swarming after bursts of rainfall, especially under conditions of weak sunshine (powerful sunshine hampers swarming). After rain showers, Swallows are present in huge numbers, foraging on the ascending termites. The swallow catchers now one by one release termites that had been collected in the previous night and stored in jars, each equipped with a small hook through the abdomen and attached to a rod and thin fishing line. Strong fliers mingle with the swarming termites, where they are caught by swallows and reeled in just like a fish in the river. The hooks are self-made, and for lines nowadays thin nylon is used (but the technique was already in use before the arrival of nylon). How effective this strategy is, becomes manifest when 50 kg bags full of Swallows are being transported to the market, containing 1000s of swallows, together with many Swifts, bee-eaters, some House - and Sand Martins, and a few local swallow species. In 1999, the swallows were sold at a price of 100 CFA francs (=0.15€) on the market of Boda (Micheloni 2000). During a survey of 25 villages in the Central African Republic, centred in a region of 50x50 km along the Lobaye River near Boda (4°19'N, 17°26'E), in October-November 2000, interviews with the local people showed that between 25 and 8545 Swallows were captured per village, and between 11 and 1021 persons per village were involved in fishing for Swallows. Some villages bordering the savanna and noted for their abundance of foraging Swallows attracted hunters from more forested parts of the range. The average catch per person varied between 2 and 25 swallows per catching day. During the survey, a total of 58 754 Swallows was registered being captured, but the actual number must have been at least one order of magnitude higher. The area surveyed encompasses the heart of the swallow-fishing business; villagers living beyond this region replied negative in response of swallow-related questions.

The impact of this type of subsistence hunting on Swallow pop-

spring migration when the conditions in Sahel, Sahara and North Africa are often adverse (dry, strong head wind, bouts of cold). In spring, numbers are already at a seasonal low, and further losses may have a direct bearing on population numbers in the breeding quarters (Newton 2007). In this respect, we are particularly interested in the role played by Sahel and Sahara, a vast stretch of inhospitable habitat where not a single Barn Swallow winters but nonetheless millions of Barn Swallows must pass in spring. We checked whether population trends in the breeding quarters are linked to variation in the annual rainfall in the Sahel and in the Maghreb, the zone north of the Sahara desert (see Fig. 11).

An analysis of recoveries from North Africa shows that many more Barn Swallows are recovered in spring than in autumn, consistent with the expected higher mortality during the more adverse conditions in spring (Chapter 42; Fig. 258). The number of recoveries also showed a negative correlation with precipitation in the Sa-

As the last daylight fades, Barn Swallows enter the roost like a whirlwind. The timing (as late as possible), the extreme reluctance to venture solo into the vegetation, the synchronicity of the descent and the large numbers involved suggest that the principal selective force moulding this behaviour has been the risk of predation (Winkler 2006). (Kasanka National Park, Zambia, early November 2007).

ulations is difficult to ascertain. For the overall population of 170 million birds the capture of a million birds may seem insignificant, especially regarding the fact that most Swallows in the Central African Republic are captured in autumn and early winter (when most winter mortality has yet to occur). However, the recovery data of the 130 rings among 58 754 swallows killed in the Centra African Republic (a reporting rate of 0.22 Swallows/100 birds) showed that these birds originated from a restricted zone within Europe, *i.e.* mostly Italy, Serbia, Croatia, Slovenia and Hungary between 10 and 18°E (but with recoveries extending north into Lithuania and Finland and west into The Netherlands and UK). The impact of this hunting mortality is therefore presumably restricted to a rather small section of the European population, amounting to 3.6-6.3 million pairs (excluding the Baltic States and Finland; BirdLife International 2004a), *i.e.* 14.4-25.2 million birds. Human predation, even on a scale as in the Central African Republic, therefore seems to constitute a mortality factor of minor importance compared with drought-related mortality in the wintering quarters and losses incurred whilst crossing the Sahel and Sahara in spring.

hel in the preceding six months (Fig. 242B).[7] In other words, survival tends to decrease when the Sahel is suffering from drought. The same relation was already shown for Britain (Robinson *et al.* 2003a) and Italy (Saino *et al.* 2004).

The conditions in southern Africa, Sahel and Sahara might accumulate in Barn Swallows having spent the northern winter in southern Africa, notably those from Britain, Denmark and Fennoscandia. We thus expected Sahelian conditions to have a larger impact on these northern populations, than on western and central European birds that winter in the forests of the Sudan-Guinea and Guinea-Congolan vegetation zones in Central Africa where rainfall patterns are more stable (Fig. 11). The British population of Barn Swallows correlates quite well with rainfall in the Sahel, but the trends in Sweden, Finland, Denmark, Germany, Switzerland, Czech Republic and The Netherlands apparently fluctuated independent of rainfall in the Sahel (Fig. 241) and Maghreb.[8]

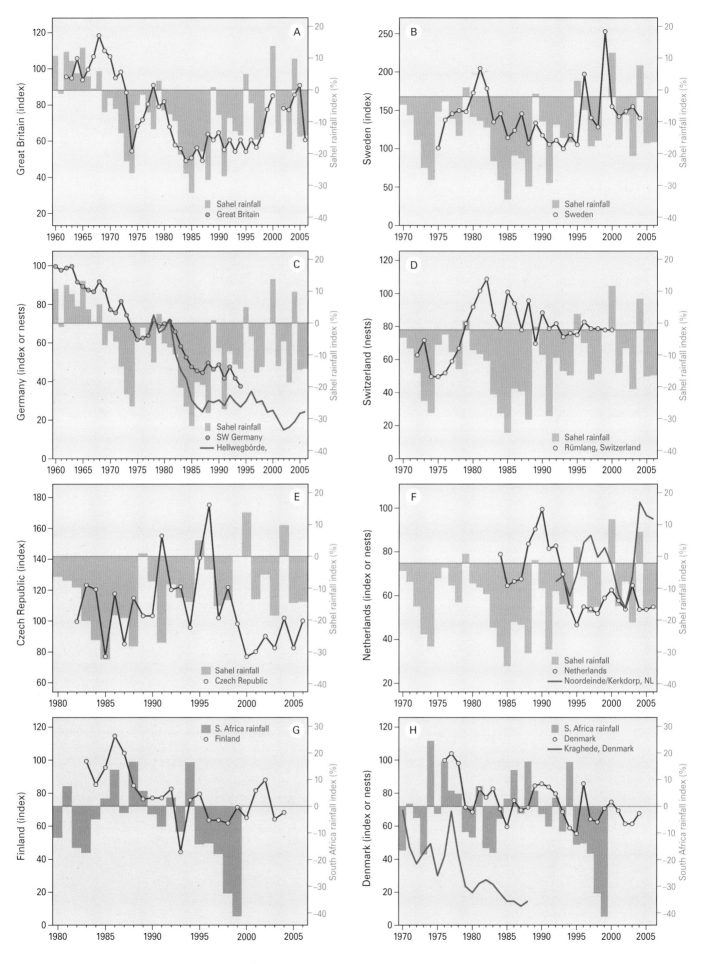

In the African wintering quarters, juveniles (as in the centre) can be easily distinguished from adults by their more brownish and spotted throat patch and light forehead; when they have finished their moult, usually in March, they resemble adults. Similarly, females (as left) differ from males (as right) by their shorter outermost tailfeathers, unless these are moulted. (Mafundzalo Range, Kabwe, Zambia, late November 2007).

Between-country comparisons showed that, except Germany and Switzerland which correlate quite well, trends in other European countries fluctuated widely on top of a long-term decline in most countries (Fig. 241).[9] The common denominator for the European-wide decline may be changes in farmland use, but asynchronous ups and downs should be of more local and regional origin (see below).

Fig. 241 Population trends of the Barn Swallow in several European countries compared with the rainfall index in the Sahel or in South Africa. Data from (A) BTO in series (Great Britain), (B) Lindström & Svensson 2005 (Sweden), (C) Hölzinger 1999 (SW Germany) and Loske 2008 (Hellwegbörde), (D) J. Muff in Schmid et al. 2001 (Rümlang, Switzerland),(E) Reif et al. 2006 (Czech Republic), (F) SOVON in series (The Netherlands), van den Brink 2006 (Noordeinde + Kerkdorp, NL), (G) Väisänen 2005 (Finland), (H) Heldbjerg 2005 (Denmark), Møller 1989 (Kraghede, Denmark). It should be noted that Barn Swallows are difficult to monitor reliably unless counting nests in fixed plots (like in Kraghede, Hellwegbörde, Noordeinde + Kerkdorp); the trends in The Netherlands, as illustrated by indexed territory mapping and nest counts, show how different the outcome can be, presumably largely caused by methodology.

Breeding conditions   At the more local level, long-term trends are often very different, as reported for England where declines in the east and southeast contrasted with increases in the west (Robinson et al. 2003b, Evans et al. 2003); such differences reflect variations in land use patterns (arable versus grassland, degree of nest-site losses, and possibly agricultural intensification). The extent of declines is masked by the lack of information from the first half of the 20th century. In 1930, the farmer's village of Labenz near Hamburg in northern Germany consisted of 62 houses with some 350 inhabitants (which were specifically described as 'im ganzen gutmütiger Gesinnung", in other words 'overall good-natured', and therefore friendly to animals). This village held 110 nests of Barn Swallows, raising 210 broods and 1024 young. A staggering 66% of the houses had one or more Swallow nests (Matthiesen 1931). Such densities are a far cry of present-day figures.[10] For example, Berthold (2003) compared bird communities in two villages in southern Germany between the mid-1950s and early 2000s (Möggingen) and between the mid-1970s and early 2000s (Billafingen); both villages showed a decline of some 75%. Regional declines of 28-78% have been compiled by Turner (2006) for various parts of western Europe, mostly since the 1960s. Monitoring programs that started in the 1970s often show fluctuations without a clear or a steep decline (see also Fig. 241), presumably because most

Barn Swallow *Hirundo rustica*

Swallow populations already had dwindled to low figures before the start of the program. Loss of habitat, loss of nesting sites and reduced insect populations in farmland have all been mentioned as major drivers of population declines and regional variations in the extent of declines (Evans & Robinson 2004, Turner 2006), factors perceptively alluded to by Von Vietinghoff-Riesch (1955) in his monograph.

Short-term variations in trends and reproductive output may be driven by local weather. Loske & Lederer (1987), for instance, noticed marked negative effects of prolonged precipitation and low temperatures on hatchability of eggs and survival of nestlings in the Hellwegbörde in western Germany. Whereas reproductive output and local trend were correlated, and annual productivity mainly hinged on the success of the first brood, weather conditions in late May and June were particularly crucial. From 1981 onwards, the incidence of adverse weather events in their study area increased for late May and

On the breeding grounds Barn Swallows show a strong commensalism with farming. Traditional farming systems must have boosted breeding numbers in the 19th and 20th century, but present-day industrial farming offers many fewer opportunities for Barn Swallows (food and nesting sites), to such an extent that large declines have occurred throughout western Europe.

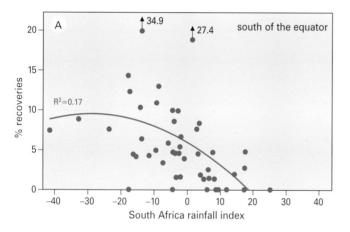

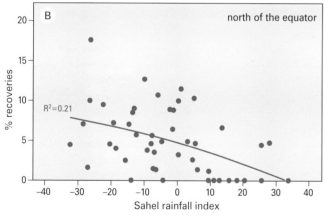

Fig. 242 The relative annual number of British recoveries of Barn Swallows from Africa, calculated separately for the land mass (A) south and (B) north of the equator, shows a relation to rainfall in either SE Africa (from Fig. 11) or the Sahel (from Fig. 9). Based on 6136 British recoveries of dead Barn Swallows (no retraps), of which 463 from Africa: 208 south and 155 north of the equator. From: EURING.

June, thought to be partly responsible for the long-term decline (Fig. 241). Such local factors are likely to interplay with mortality in, and carry-over effects from, the wintering quarters (see Chapter 43).

Conclusion

Barn Swallows in northern and eastern Europe and those from Britain leapfrog the western and central European birds, and winter in the southern third of Africa. Western, southern and central European Swallows mostly winter in the small band of 4-6°N in the transition zone of Guinea and Sudan vegetation zones. During migration, populations may mingle on communal roosts in between. British Swallows show a negative relation with rainfall patterns in South Africa (as reflected in the relative number of annually recovered birds), and with the Sahel rainfall index (but less so). Sahel nor Maghreb rainfall correlated with population fluctuations in northern and western Europe. Human predation on roosting Swallows, although massive in absolute numbers in Nigeria, Cameroon and Central African Republic, is unlikely to impact population trends. The generally negative population trend in much of Europe is probably largely caused by local factors related to farming practices, but rainfall patterns in South Africa, Sahel and Sahara may also have some bearing on trends via survival and carry-over effects.

Endnotes

1 Of course, it is odd to speak about " wintering in southern Africa", while the swallows in December-January profit from the austral summer conditions. For brevity's sake, however, " winter" will be used here to indicate 'northern winter', taking the risk to be blamed being Eurocentric.

2 The correlation between the South Africa rainfall index (Fig. 11) and individual raining stations is rather well, suggesting that the rainfall over large parts of S Africa is synchronised. The rainfall in Johannesburg and Pretoria is not related, however, with the rainfall in and around the Karoo.

3 Correlation of rainfall in S Africa with the Danish index: $R=+0.325$, $P=0.121$, $N=24$ and with Kraghede: $R=-.016$, $P=0.949$, $N=19$. Correlation of rainfall in S Africa with log(population change) for the Danish population: $R=+0.350$, $P=0.154$. $N=23$ and for Kraghede: $R=+0.051$, $P=0.817$, $N=18$.

4 Correlation of rainfall in S Africa with the Finish index: $R=+0.434$, $P=0.082$. $N=17$ and with log(population change): $R=+0.168$, $P=.534$, $N=16$.

5 Correlation of rainfall in S Africa with the British index: $R=-0.192$, $P=0.248$. $N=38$ and with log(population change): $R=-0.110$, $P=.515$, $N=37$.

6 The polynomial function is highly significant ($P=0.004$, $N=48$). The Sahel rainfall has no impact on the relative number of recoveries south of the equator ($P=0.815$). We found no explanation for the two outliers in 1968 and 1970.

7 The polynomial function is highly significant ($P=0.003$, $N=54$). The rainfall in S Africa has no impact on the relative number of recoveries north of the equator ($P=0.27$).

8 The British population is significantly correlated with the Sahel rainfall ($R^2=.580$, $P<0.001$, $n=48$), weakly correlated with the rainfall in N Africa ($R^2=0.121$, $P=0.044$, $n=33$) with a significantly linear decline ($R^2=0.312$, $P<0.001$, $n=48$). In a step-wise multiple regression analysis, in which year is entered first, the Sahel impact remains significant ($P<0.001$), but not the contribution of rainfall in N Africa (result for year+Sahel: total: $R^2=0.523$, $P<0.001$). The decline of the German population is still more pronounced as in the British population ($R^2=0.969$, $P<0.001$, $n=34$). The Sahel impact is significant too ($R^2=0.502$, $P<0.001$, $n=34$). Assuming that there is an overall linear decline, the additional impact of the Sahel remains present, although its contribution becomes weakly significant ($P=0.041$).

9 There are in total 110 correlations between the 11 series shown in Fig. 241, of which 16 significantly positive and 2 significantly negative; the other correlations are, although usually positive, non-significant. Eight of the 11 series show a clear decline, which explains already why some series are positively correlated. Hence, the population trends do not have much in common, beside the decline. Also the correlation matrix with log(population change) reveal many non-significant correlations, with some exceptions; for instance, the fluctuations in the Swiss and both German sites are highly significant, and also the Danish and Swedish trends are intercorrelated.

10 Fridtjof Ziesemer pointed out that the breeding birds of Labenz had been mapped again in 1995 (Jeremin K. 1999. Brutvögel des Dorfes Labenz in 1931 und 1995 – Wandel von Dorfstruktur und Vogelwelt. Corax 18: 88-103). The average density of breeding birds had increased from 130 territories/10 ha to 149/10 ha, mainly caused by urbanisation-related processes. However, Barn Swallows had decreased to 53 nests, compared to the 117 nests in 1931.

Yellow Wagtail
Motacilla flava

Transhumance in the Sahel has special importance for Yellow Wagtails. Thousands of Wagtail flocks attend the cattle, sheep and goats as they sweep across plains, savanna and desert, profiting from seasonal variations in food supply. The region covered by the 110 million herbivores in the western Sahel lies roughly between the zones that receive less than 100 mm of rainfall (insufficient for grass) and more than 1000 mm of rainfall (where tsetse flies reign), and within it, floodplains provide temporary food bonanzas when grasses put on growth spurts once the water has receded. From December to April, depending on the height of the flood, a paradise for fisherman is transformed into a pastoralist's idyll. The arrival of large herds in the deflooded areas heralds an influx of Yellow Wagtails, whose thin high-pitched calls are easily audible above the bellows, grunts and trampling of the cattle. Nowhere particularly abundant during daytime, Wagtails leap to prominence just before dusk, when the magnitude of their numbers becomes obvious, as a continuous stream starts heading for the roost. For as far as the eye can see along the column, it's Wagtails galore. Counting is a near-impossibility, because the bulk passes when it is almost dark. The scale of roost migration, however, does give a fair impression of the populations hosted by huge Eurasian hinterland where these wagtails breed, and to which they are preparing to return, despite often harsh conditions *en route*.

Breeding range

Yellow Wagtails breed across Europe as far north as the 10°C isotherm, covering all climatic zones between Arctic tundra and Mediterranean. Densities are highest in northern and eastern Europe. The latest estimate for Europe amounted to 8-14 million breeding pairs in the late 1990s, mostly in eastern Europe, Belarus, Russia and Turkey (BirdLife International 2004a). Several subspecies are recognised, of which the Egyptian *M.f. pygmaea* is an endemic resident and the Moroccan population of *M.f. iberiae* partly so (Thévenot *et al.* 2003). In Europe, all populations are migratory and winter throughout Africa south of the Sahara (*i.e. flavissima, flava, thunbergi, cinereocapilla, feldegg, beema* and *lutea*). The easternmost recovery of a Yellow Wagtail ringed in Africa came from the Ob in western Siberia at about 77°E (Zink 1975); this bird must have been *M.f. thunbergi*, of which a large part of the population winters in Africa (otherwise on the Indian subcontinent).

Migration

The distribution and abundance of Yellow Wagtails in Africa south of the Sahara is complex, with racial variations in distribution, latitudinal variations in sex ratio and seasonally induced fluctuations in numbers (Wood 1992, Bell 1996, 2007a, 2007b). The available evidence converges on a west to east gradient of races wintering in Africa similar to those that comprise the Eurasian breeding distribution (Zink 1975; Fig. 243). Western European Yellow Wagtails of the race *M.f. flavissima* winter mainly in Senegambia through Guinea-Bissau and Guinea to Sierra Leone. The races *flava* and *thunbergi*, originating from breeding areas between 10° and 30°E, winter largely in the West African sector between the Inner Niger Delta and Lake Chad. Birds in the eastern half of the Sahel and in eastern Africa constitute a mixture of northern and eastern European races like *thunbergi, flava, feldegg* and *lutea*; birds ringed in Kenya were recovered mainly in Europe east of 30°E in Russia (Zink 1975). Russian and Kazakh Yellow Wagtails breeding east of the Caspian Sea winter partly in the Indian subcontinent (McClure 1974), but *M.f. lutea* is also a widespread winter visitor to East Africa (Keith *et al.* 1992). Within the Sahel zone, the British birds winter, on average, 1000 km further to the west than Swedish birds, and Finnish birds 1350 km further to the east of Swedish birds (Fig. 243). This is just a crude division, and in reality there is much overlap between the wintering areas throughout the Sahel, and even in smaller geographic components like Nigeria (Wood 1975).

On top of a west-east gradient in distribution within the sub-Sahara depending on breeding origin, northernmost Yellow Wagtails are known to undertake leap-frog migration over more southerly breeding birds to winter further south (Wood 1975, 1992, Bell 1996). In Nigeria, for example, *thunbergi* (from Scandinavia) winter south of *cinereocapilla* (Italy), *feldegg* (Balkan) and *flava* (central Europe), which occupy central and northern parts of the country from Vom to Nguru and Lake Chad (Fry *et al.* 1970, Wood 1975 & 1992, Bell 2006). The evidence from ringing also indicates a latitudinal variation in sex ratio, with males on average wintering further north than females (Wood 1992).

Distribution in Africa

In sub-Saharan Africa Yellow Wagtails are among the most ubiquitous of insectivorous Palearctic migrants. The locally breeding Plain-backed Pipit *Anthus leucophrys* is one of the few insectivorous residents that uses the same niche, but it is greatly outnumbered by the millions of Yellow Wagtails. There are few African habitats totally devoid of Yellow Wagtails, mostly crowded human settlements or woodland without glades, but densities vary considerably.

Based on densities found during plot sampling in various habitats in West Africa (Chapter 11), we found Yellow Wagtails to be the commonest winter visitors in rice fields in Senegambia, Guinea-Bissau and Guinea-Conakry. The average density in the winters of 2004/2005 and 2005/2006 was 2.37 birds/ha, which results in an estimated number of 268 000 birds wintering in the 1120 km² of rice

Fig. 243 European origins of 354 Yellow Wagtails recovered between 4° and 37°N. From: EURING. To avoid cluttering the figure, lines are not shown for recoveries in southern Spain and NW Africa.

fields in these countries (Bos *et al.* 2006; Chapter 11). This density, at 0.9 birds/ha, was much lower in patches of natural vegetation within rice fields. Using also density counts, Wymenga *et al.* (2005) arrived at 320 000 Yellow Wagtails in the irrigated rice fields of *Office du Niger* in Mali (700 km^2).

Plot sampling in the Senegal Delta in the early 2000s gave rise to an estimate of the wintering population of about 500 000 Yellow Wagtails (Chapter 7). For the entire Inner Niger Delta, the wintering population is estimated at one million birds on the lower floodplains (Table 8), but since the species is found also on the surrounding emerged plains, the total might be some 1.5 million. In floodplains, it is the most common Palearctic winter visitor, which occurs in 40% of the sampling plots at an average density of 5.4 birds/ha where present. Highest densities were recorded in humid rice fields, floating *bourgou* and grasslands or stagnant waters whose water was less than 80 cm deep. In general, dry areas held low densities, unless cattle were present; densities of Yellow Wagtail associating with cattle were five times higher than in similar habitat without cattle (Chapter 6). It is possible that the doubling of the numbers of cows and tripling of the number of sheep and goats in the western Sahel between 1985 and 2005 (Chapter 5) may have created additional feeding grounds in winter for Yellow Wagtails.

The many millions of Yellow Wagtails in Africa inhabit a wide range of habitats, from very dry to wet. Water's edges, providing an abundance of insects, are preferred; birds here are often territorial, even when feeding habitats are suitable for short periods of time only, with males displacing females and juveniles.

Wintering strategies in the Sahel   Yellow Wagtails forage on the ground or by making short sallies. In the Sahel they often associate with wetlands and farmland. Commensal feeding associations with herbivorous mammals, which attract and disturb insects while grazing, are typical of the region, as during migration elsewhere (Källander 1993). The preference for insects such as Coleoptera, Hemiptera, Hymenoptera and Orthoptera is clearly borne out by the analysis of stomach contents of birds collected in central Nigeria (Vom, 9°42'N, 8°45'E; Wood 1976). In this respect, Yellow Wagtails are at a clear disadvantage to many warbler species that preferentially consume fruit, because arthropod biomass steadily declines in the course of the dry season (October-March), as soil moisture decreases. Arboreal or aerial insectivores are less likely to suffer than terrestrial insectivores, because the severity of conditions may be ameliorated by the presence of bushes and trees that, by virtue of their access to water far below the ground surface, are able to retain their leaves and harbour arthropods throughout the dry season (Morel 1968, 1973). Terrestrial insectivores depend entirely on insect availability on the ground, where the impact of drought is highest. The decline in biomass in the course of the dry season is reflected in the declining foraging success by about 25% per month as the season progresses, at the same time as arthropod biomass in foraging areas fell at 57% per month (near Vom, central Nigeria; Wood 1978). Wagtails therefore may face particularly difficult circumstances just before spring migration, when they are known to spend over 75% of daylight hours foraging actively (Wood 1992). Whereas the birds needed only 6.5 paces in between successive prey captures in late November, this had gradually increased to 22.6 paces by late March (Wood 1979), a clear indication that the birds were experiencing increasing difficulties in catching food; note, however, that this figure does not tell us anything about food intake. Yellow Wagtails may partly circumnavigate this problem by dispersal to more favourable regions further south, although the ringing evidence for this strategy is sparse. The alternative, which would be an annual *in situ* starvation-induced mortality of 68% (going by the numbers left at Kano, northern Nigeria, 12°00'N, 8°30'E; Ashford 1970), is not very likely. The 64% decline of Yellow Wagtails from the roost at Vom in central Nigeria has therefore been attributed to mortality plus the movement of birds to other roosts, possibly further south (Wood 1992).

Pre-migratory fattening and departure strategies in the Sahel   Yellow Wagtails deposit up to 40% fat relative to their live weight just before departure to the breeding grounds (Ward 1964, Smith & Ebbutt 1965). In Nigeria, birds wintering in the north fatten and depart first, despite not having experienced any rainfall whatsoever in the last part of their winter stay. This is something of a paradox, regarding the progressive decline in food availability in the course of the winter. Those wintering further south, *i.e.* the northern European populations which leapfrog those of central and southern Europe, follow sequentially, and do not leave for Europe until some weeks after the first rains of the wet season in the Sudan and Guinea vegetation zones; pre-migratory fattening occurs about 25 days later than in the northern Sahel (Wood 1992, Bell 2007b). On average, females winter more to the south than males and fatten later than males (caused by dominance-mediated access to food?). The amount of fat stored by Yellow Wagtails just suffices to cross the Sahara in a nonstop flight of some 70 hours (Wood 1982). Pre-migratory fattening commences just before the partial pre-alternate moult is completed, usually in mid-March in the northern Sahel, late March and early April in the southern Sahel and around mid-April in the Sudan zone (Bell 2007a). Yellow Wagtails wintering in the Sudan zone, *thunbergi* mostly, may profit from the rains which start in March, but still – when hopping northwards towards the edge of the Sahara before making the long jump across the Sahara – encounter withered land in the northern Sahel where rain has been lacking for more than seven months in a row. By that time, southern and central European Yellow Wagtail populations have already vacated the region, and local resources are at their lowest ebb (Wood 1979, Bell 2007b). The need to fly additional distance to that required to cross the desert, be it caused by drought or habitat degradaton, could be critical in determining how successful such birds would be in reaching their European breeding areas. This additional obstacle would particularly affect breeding birds from northern Europe (given a leap-frog migration strategy, and their timing of migration; Bell 2007a) and females (which winter further south than males). A similar problem would arise should the north-south expanse of the Sahara increase in the wake of long-term changes in rainfall (Chapter 4), an effect that may already be occurring given that so many changes in land use have taken place, as for example in the Hadejia-Nguru floodplains in northern Nigeria, where, due to a reduction of the flood dynamics, proliferating Cattail *Typha* encroaches upon huge areas and reduces the area under cultivation (Chapter 8). When farmland comprises the main foraging area, as it does near Nguru, such habitat losses may explain not only the observed delay in pre-migratory fattening and the increase in assemblage density (and hence aggression), but also the intense competition for shrinking resources (Bell 2006). These signs of stress provide circumstantial evidence for density-dependent mortality in the form of poor body condition, and consequently possibly for lowered breeding success, if delayed spring migration causes birds to lose out in competition for breeding territories with birds that wintered elsewhere. Furthermore, it should be noted that the findings near Nguru suggest that Yellow Wagtails are prisoners of their wintering sites, apparently not able to move elsewhere as a response to loss of their traditional wintering sites (Bell 2006). On a larger scale, this would imply that habitat loss in sub-Saharan wintering areas, especially just south of the Sahara (Chapters 6 and 7), has the potential to contribute to overall population declines (see Fennoscandian trends below). However, in much of Europe no such effect is discernible, and indeed habitat choice of Yellow Wagtails in Africa might be sufficiently wide to escape the immediate rigours of droughts (for example, farmland, following cattle and use of floodplains and water reservoirs) unless these are long-term and wide-ranging.

Phylogenetic analysis has revealed three clades within traditionally recognised *Motacilla flava*, each group to be considered a separate species (Pavlova *et al.* 2003). The western group encompasses Europe and SW Asia, and largely winters in Africa. Among the many subspecies wintering in Africa, *flava* (here adult male, Senegal, January 2007) and *thunbergi* are widespread and abundant. Subspecific identification is not always unambiguous, as noted by Brian Wood (1975) who intensively studied Yellow Wagtails in Nigeria in 1973-1974.

Population change

Wintering conditions There are 58 recoveries of Yellow Wagtail south of the Sahara, representing 3% of the 1910 recoveries of dead Yellow Wagtails in the EURING data bank, but many more (231 or 12.1%) were recovered from the Sahara and the Maghreb. The relative number of recoveries from Africa varies per year, being significantly higher in dry Sahel years (Fig. 244). The impact is weakly significant for the region south of the Sahara, but sample size is small (Fig. 244, upper panel). The significant Sahel effect on the number of recoveries from the Sahara and the region directly north of it (which are nearly all from spring) suggests that during dry Sahel years Yellow Wagtails are unable to build up sufficient body reserves to cross the desert (Fig. 244, lower panel). This is discussed further in Chapter 42.

Of the many local, regional and country-wide trends published for eight European countries, those of Britain (Fig. 245A), The Netherlands (Fig. 245B), Finland (Fig. 245E), Sweden (but not Lapland, where a steady increase has been noted since the late 1960s; Enemar *et al.* 2004; Fig. 245C), Germany (Fig.245D) and Poland have shown a decline in the late 20th century, despite temporary respites. British, Dutch and Fennoscandian series seem to fluctuate in synchrony.[1]

To what degree are these fluctuations in population size related to wintering conditions in sub-Saharan Africa? And precisely which wintering conditions matter most? To answer these questions, we have to take into account that Yellow Wagtails are widely distributed

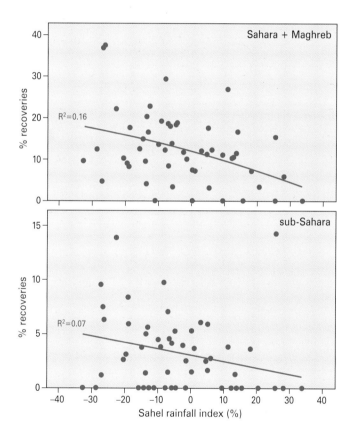

Fig. 244 Annual number of Yellow Wagtail recoveries from Sahara + Maghreb (above 20°N) and sub-Sahara (4°-20°N) as a percentage of the total number recovered across Europe and Africa (dead birds only) between 1949 and 2002 (respectively P<0.001 and P<0.05; selection for 54 years with >10 recoveries). From: EURING.

During deflooding, the Inner Niger Delta in Mali temporarily hosts millions of cattle, goats and sheep. The herds are accompanied by huge numbers of Cattle Egrets and Yellow Wagtails. This picture was taken from Gourao, and shows the only other elevation, Soroba, in the otherwise flat floodplain in the background. (January 2007).

across many different habitat types, varying from wet to dry, from floodplains to woody savanna, with or without cattle, in Sahelian, Sudan and coastal vegetation zones. Because large numbers are concentrated in the floodplains (Chapter 6), we should expect some relationship between the size of the floodplains and population size. Conversely, large numbers also roam across derived savanna associating with herbivores, and so we would expect a correlation with rainfall and vegetation cover (Normalized Difference Vegetation Index, NDVI).

Because northernmost populations migrate furthest south, and are the latest to depart from their wintering haunts, on their return migration they experience the harshest conditions on their take-off point just south of the Sahara (at the end of the 7-months long dry season). Droughts would therefore hit Yellow Wagtails breeding in Fennoscandia hardest (Bell 2007b, Fig. 245E). However, after correcting for a long-term linear population change the correlation between population size and Sahel parameters became smaller, except in two studies. First, the standardised ringing totals of Sweden (Österlof & Stolt 1982) increased significantly between 1960 and 1979, but relatively more Yellow Wagtails were captured after a winter when a larger extent of Sahelian floodplain had been inundated. Secondly, the trend in a small study plot in Niedersachsen, Germany, was stable in the long run (Zang & Heckenroth 2001), but also fluctuated in synchrony with the flood extent in Sahelian floodplains in the preceding winter (Fig. 245C). [2]

The lack of an unambiguous impact of flooding and rainfall on

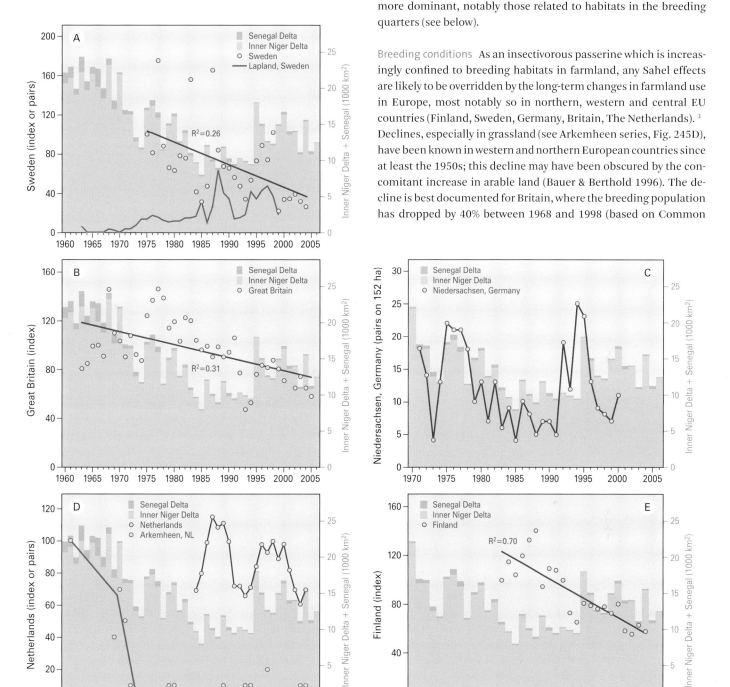

Fig. 245 Population trends of Yellow Wagtails in several parts of Europe in relation to the flood extent of the Inner Niger Delta and the Senegal Delta in the preceding winter. The number of pairs in Arkemheen has been exaggerated tenfold to make the series visible. From: (A) Lindström & Svensson 2005 (Sweden, index), Enemar et al. 2004 (Lapland, pairs), (B) BTO (Great Britain), (C) Zang & Heckenroth 2001 (Germany, Niedersachsen, SE Hitzäcker), (D) SOVON (The Netherlands), van den Bergh et al. 1992, van Manen 2008 (The Netherlands, Arkemheen) and (E) Väisänen 2005 (Finland).

population size in most studies suggests that other factors might be more dominant, notably those related to habitats in the breeding quarters (see below).

Breeding conditions As an insectivorous passerine which is increasingly confined to breeding habitats in farmland, any Sahel effects are likely to be overridden by the long-term changes in farmland use in Europe, most notably so in northern, western and central EU countries (Finland, Sweden, Germany, Britain, The Netherlands).[3] Declines, especially in grassland (see Arkemheen series, Fig. 245D), have been known in western and northern European countries since at least the 1950s; this decline may have been obscured by the concomitant increase in arable land (Bauer & Berthold 1996). The decline is best documented for Britain, where the breeding population has dropped by 40% between 1968 and 1998 (based on Common

Table 27 Contraction of breeding range in various European countries since the 1960s and 1970s.

Country	Squares occupied	% change	Source
Finland			
1974-79	2867		
1980-89	2668	-6.9	Väisänen et al. 1998
Denmark			
1971-74	770		
1993-96	521	-32.2	Grell 1998
Switzerland			
1972-76	57		
1993-96	31	-45.6	Schmid et al. 2001
Czech Republic			
1973-77	279		
2001-2003	217	-22.2	Šťastný et al. 2005
The Netherlands			
1973-77	1416		
1998-2000	1287	-9.1	Hustings & Vergeer 2002
Britain			
1968-72	1155		
1988-91	1047	-9.4	Gibbons et al. 1993

The enormous changes in the western European agricultural landscape in past decades, here illustrated by grasslands used for dairy farming in Fryslân, The Netherlands, have an overriding impact on population trends of many birds, including long-distance migrants wintering in the Sahel (May 1992).

Bird Census data; BTO), and by 14% between 1994 and 2002 (Breeding Bird Survey data; BTO). At the same time, the breeding range contracted by almost 10% between 1968-1972 and 1988-1991 (Table 27). Simultaneously with these declines, a shift from wet grasslands to arable farmland with spring-sown crops took place (Chamberlain & Fuller 2000); declines in wet grasslands in lowland England and Wales averaged 65% between 1982 and 2002 (Wilson & Vickery 2005). Although the mechanisms behind these changes are not fully understood, changes in farming practices such as drainage, application of fertilizers, mowing regimes and use of herbicides are likely to have reduced the suitability and extent of foraging and breeding habitats (Vickery et al. 2001, Newton 2004).

The British experience is not unlike the ones across much of Europe, where farming has intensified to a similar or even greater degree (Bauer & Bertold 1996, Pain & Pienkowski 1997). Range contractions, assessed from repeated atlas surveys showing presence or absence in fixed squares, have been recorded for Finland, Denmark, Switzerland, Czech Republic and The Netherlands (Table 27). Apparent increases in presence between first and second surveys, as recorded for France (1970-1975 versus 1985-1989; Yeatman 1976, Yeatman-Berthelot & Jarry 1994) and Slovakia (1973-1977 versus 1985-1989; Šťastný et al. 1987, Danko et al. 2002), are attributable to the better coverage obtained during the second survey. The shift from wet meadows and grasslands to arable land after the mid-20th century is detailed explicitly for Switzerland (Schmid et al. 2001), The Netherlands (Hustings & Vergeer 2002), Flanders (Vermeersch et al. 2004) and France (Yeatman-Berthelot & Jarry 1994), with commensurate declines in grassland areas and increases in arable land. However, intensification of farming in arable fields also leads to declining numbers, as recorded, for example, in Poland (Tryjanowski & Bajczek 1999) and Central Europe (Bauer & Berthold 1996).

It should be noted that the vast majority of European Yellow Wagtail breed in the eastern half, notably in Belarus, Russia and Ukraine, which hold more than 60% of European numbers (BirdLife International 2004a). Very little information is available on the fortunes of populations east of Poland, except that the species remains one of the commonest breeding birds, as it was in the early 20th century (Zedlitz 1921, Tomiałojć & Głowaciński 2006), in comparatively unspoiled breeding habitats.

Conclusion

The overall picture for western, central and northern Europe is unambiguous: steep declines in numbers and considerable range contractions, mostly in association with farming intensification after the mid-20th century. Fortunes of eastern European birds are unknown, both in East Africa, where many winter, as on the breeding grounds, where any changes probably are smaller than in the rest of Europe. We assume that Sahelian rainfall and flooding affect the pre-migratory fattening process in spring, because fewer birds are able to cross the Sahara desert successfully in dry years. The latter results in a southwards contraction of the Sahel in West Africa, forcing Yellow Wagtail to make longer nonstop flights than desirable in spring. The overall impact of the Sahel on the population trends is limited, however, and overridden by the changing breeding conditions in Europe.

Endnotes

1 The British index, being the longest series, varies significantly in concordance with those of Sweden (R=+0.673, P=0.031), Finland (R=+0.714, P<0.001) and The Netherlands (R=+0.442, P=0.051). However, when annual changes are correlated, the correlations are no longer significant, except for Britain vs. Finland (R=+0.532, P=0.016, N=20).

2 A multiple regression on the standardized ringing totals of Sweden for 1960-1979 (Österlof & Stolt 1982) shows a significant relation with year (P<0.001) and with the extent of Sahelian floodplains (P=0.029); N=19; R^2=0.512. In the German data set for 1971-2000 (Fig. 245D), only flood extent correlated significantly (P=0.008; year: P=0.732); N=29; R^2=0.260. The explained variance was less, when instead of flood extent, rain or NDVI was entered into the analysis.

3 The decline of Yellow Wagtails is highly significant in Finland, Sweden and Britain, and non-significant in The Netherlands (Fig. 245). The trend is also significantly downward in western Poland (R=-0.892, P<0.001, N=11; Tryjanowski & Bajczyk 1999). The long-term trends for Germany (Flade & Schwarz 1996); Germany, Niedersachsen, SE Hitzacker (Zang & Heckenroth 2001), and Switzerland, Gampel and Grossen Moos (Schmid et al. 2001), Switzerland, Zürcher Weinland (Schümperlin 1994) are non-significant. Two trends show a significant increase: Switzerland, Thurgau & Zürcher Weinland between 1989 and 2001 (R=+.589, n=13, P<0.01; Schmid et al. 2001) and France between 1983 and 2000 (R=0.872, N=13, P<0.001; www.mnhn.fr/mnhn/crbpo/fiches_especes/caille.htm).

Common Redstart
Phoenicurus phoenicurus

When Gerrit Wolda, a maths teacher in The Netherlands, started his research on "physiological phenomena in some bird species" in the winter of 1909, by providing 100 nestboxes in the woodland surrounding the sanatorium "Oranje Nassau's Oord" near Wageningen, he could not have dreamed of the far-reaching impact that his initiative would have on ornithology and ecology (Bijlsma 2006a). By "physiological phenomena" he alluded to site fidelity, mate choice and reproductive strategy (more precisely: number of broods per annum; Wolda 1918). His nestbox approach was expanded to the Hoge Veluwe in 1921 by H.N. Kluijver (Kluyver 1951); a study that is still flourishing (Both *et al.* 2006), and since copied all over Europe. An unforeseen side-effect of his study was the occupation of nestboxes by Redstarts and Wrynecks. The latter were considered such a threat to other nestbox breeders (by their habit of removing the nests and clutches of tits and redstarts) that the advice given was to remove Wrynecks from nestboxes (as was to happen to Tree Sparrows half a century later - like Wryneck). As it happens, since he was above all a curious naturalist, Wolda did no such thing, but instead studied Wrynecks as the opportunity arose, unwittingly providing a wonderful calibration point for studies on this species during their declining phase in the late 20th century (Bijlsma *et al.* 2001). Returning to the Redstarts in Wolda's days, and even much later when Ruiter (1941) did his studies on Oranje Nassau's Oord in the mid-1930s, this species was second in abundance only to Great Tits. How much the world has changed since then.

Breeding range

Redstarts occupy the boreal and warm temperate zones of Eurasia, essentially between the July-isotherms of 10°C and 24°C (Hagemeijer & Blair 1997). Densities increase from south to north. The European population was estimated at 6.8-16 million pairs in 1990-2002 (BirdLife International 2004a).

Migration

Most African ringing recoveries are from Morocco and the Mediterranean coast of Algeria and Tunisia (Fig. 246). In total, 291 birds were reported from the Atlas mountains and the relatively humid zone north of it, of which 16.5% occurred during the spring migration (21 March-31 May), 65.7% during the autumn migration (21 August-20 November) and only 7.4% during the 13 intervening winter weeks. A surprisingly high number (n=116) were recovered from the Sahara desert, birds which were also mainly migrants: 74.7% in spring, 14.1% in autumn and 9% from the winter period.

From the wintering areas proper, ring recoveries are scarce. Two British-ringed birds in the Senegambia were captured on 23 October and 26 November. A Redstart in Chad was recaptured at the same locality one month later in November. A bird ringed on the north coast of the Black Sea on 1 October 1991 was recovered on a small island in the Red Sea on 25 April 1996.

Recoveries in Europe show that Redstarts generally follow a southwesterly migration direction in autumn (an arc of WSW-SSW), except for British birds, which first seem to take a more southerly course until they reach Iberia (Zink 1981, Wernham et al. 2002). The Mediterranean is crossed on a broad front, as attested by abundant catches in northern Egypt (60 near Alexandria from 8 September to 13 October 1965, body mass 11.5-20.7 g, mean 16.9 g; Moreau & Dolp 1970) and in northwestern Algeria (52 from 18 September to 19 November 1985, body mass 11.9-19.0 g. mean 14.7 g; Bairlein 1988). The latter study showed that 40% of the birds captured for the first time had no fat; those recaptured 2-23 days later had increased their body mass on average from 13.8 to 17.4 g. Of Redstarts captured at oases between northern Africa and the wintering quarters, the average body masses and fat reserves were well above pre-migratory levels as measured in SW Germany (Bairlein 1992), enabling them to continue their migration without refuelling at such stopovers.

Distribution in West Africa

The winter distribution in the Sahel presumably follows more or less the same east-west gradient as in the breeding quarters (Fig. 246). In the westernmost part of the Sahel, *i.e.* in Senegambia, two British birds were reported, but this region may be a melting pot of birds from western Europe and southern Fennoscandia (Fig. 246; Bakken et al. 2006). Although not backed by ringing data, it is likely that central and eastern European birds winter further to the east in the Sahel (Glutz von Blotzheim & Bauer 1988).

The first Redstarts arrive in their wintering quarters in Senegal in the first half of September, but it takes another month before the species is really abundant (Morel & Roux 1966b). This schedule fits the dates of passage as noted in NW Algeria (mid-September through mid-October, peaking early October; Bairlein 1988). They are eclectic in their habitat choice, ranging from very dry scrub savanna with *Acacia* species through more humid and lush *Acacia nilotica* thickets, possessing an inextricable understorey, to park-like forests in the Sudan vegetation zone. Redstarts are absent when woody vegetation is lacking (Morel & Roux 1966b).

Morel & Roux (1966b) suggested that Redstarts exhibited high site fidelity from mid-October onwards, because the same numbers of birds of the same age-composition were recorded week after week at a site in Senegal, noting that the same site was used by Redstarts in successive winters. However, of 136 Redstarts ringed in Senegal in 1957-1977, only 2 were recaptured and even fewer for Redstarts ringed in 1985-1993 in the Djoudj: none from 140 (Sauvage et al. 1998). On the coastal island of Ginak, The Gambia, site fidelity was highly significant within each winter, but not between winters. The recurrence rate – retraps within the whole study area of *c.* 1 km diameter – was moderate, but 4 of the 17 recurrent birds had been

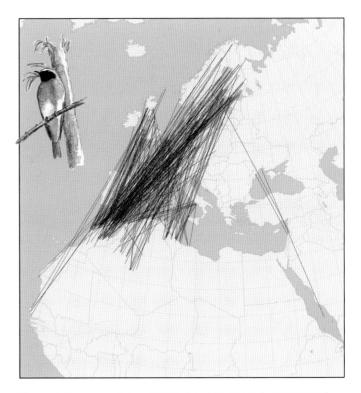

Fig. 246 European origins of 502 Redstarts recovered or recaptured between 4° and 37°N From: EURING.

Acacia flowers are rich in nectar and attract many insects. In sub-Saharan Africa, Redstarts and other insectivorous passerines profit from the successive flowering phenology of indigenous trees. *Acacia nilotica* (left) and *A.(=Faidherba) albida* (above) flower early in the dry season (October-November), whereas *Acacia seyal* (centre and right) flower in the second half of the dry season (January-March). The *albida* shown is a solitary tree along the River Niger; this particular tree has been in use as a Cattle Egret roost since at least the early 1990s, as visible from the whitewash. (Levees in the Inner Niger Delta, Mali, mid-February 2007).

ringed and retrapped in spring, suggesting passage visits (King & Hutchinson 2001). A low recurrence rate was noted for Kano, northern Nigeria: 2 out of 90 captured birds (Moreau 1969, based on data from R.E. Sharland).

In the Djoudj, Senegal, faecal analyses revealed that in March 1993 Redstarts still foraged mostly on arthropods (85% by weight), but less often on fruit (15%); the latter is considered an important food source for Palearctic warblers whilst fattening prior to spring migration (Stoate & Moreby 1995). The mean fat score of 0.72 was much lower than that of four species of *Sylvia* warblers present (4.00-5.84). Either Redstarts start pre-migratory fattening later, or elsewhere, or migrate northwards in shorter hops than do *Sylvia* warblers (Stoate & Moreby 1995). This observation is an echo of Fry's (1971) finding that departing Redstarts at Zaria (11°01'N, 7°44'E) in Nigeria did not fatten in spring, apparently doing so further north, where at that stage of the dry season only *Salvadora* berries may have been available (Moreau 1972). However, at Vom in central Nigeria (9°50'N, 8°50'E), three males first caught between 12 March and 2 April in the mid-1960s showed increases in weight after recapture a few days later (Smith 1966). Two birds collected on 11 March 1962 around Maiduguri in NE Nigeria, close to the southern end of Lake Chad and some 250 km south of the desert's edge, weighed 15.5 and 17.5 g and contained 14% and 20% fat respectively (as a percentage of live weight; Ward 1963). Spring arrival in Morocco on the Algerian border (Defilia oasis, 32°7'N, 1°15'E) peaked during late March and early April (Smith 1968); in spring of the years 1963-1966, birds weighed between 10.6 and 17.1 g (mean of 66 males 13.2 g, of 44 females 13.0 g; Ash 1969). *En route* from the northern Sahel to northern Africa, low departure weights may mean that many perish crossing the Sahara. Of five birds encountered along the 975 km long desert track of Reggane-Tessalit (Algeria-Mali) during 16-19 April 1973, 16-19 March 1977 and 16-21 April 1977, three were dead (fresh weight of one bird: 9.74 g), but the other two were apparently in good condition. Moreover, birds lured to the ground with sound, in a wadi 500 km inland in Mauritania (20°59'N, 11°40'E), and retrapped later in spring 2003, were able to gain weight by 0.25 g per day on average (Herremans 2003). Birds captured at the oasis of El Golea in Algeria (30°35'N, 2°51'E) weighed on average 11.37 g (Haas & Beck 1979, see also Haas 1974). Evidently, birds trying to cross the Sahara northwards in short hops need to refuel along the way, and apparently do so when conditions permit.

Population change

Wintering conditions Together with Whitethroats and Sand Martins, Redstarts are often regarded as a reliable indicator for tracking regime shifts in the Sahel (Berthold 1973, 1974). Contemporaneous monitoring schemes showed unequivocal crashes during severe droughts, amounting to declines of 68-88% between 1968 and 1973 in Germany (Fig. 247C) and Britain (Fig. 247D). This turning point in the rainfall pattern of the Sahel region was also reflected in Redstart numbers in Switzerland (Fig. 247C), Sweden (Fig. 247B) and The Netherlands (Fig. 247A), which crashed in the late 1960s and early 1970s and have remained at consistently low levels since.

In other countries, a moderate increase has been noted since the mid-1970s (Britain; Fig. 247D) or 1980s: Denmark (Heldbjerg 2005), Finland (Fig. 247F) and the Czech Republic (Reif *et al.* 2006). In the case of Britain, the present population fluctuates around the pre-crash levels of the early 1960s. Contrary to expectations, this apparent recovery from the crash in the 1970s did not decelerate during the Great Drought in the Sahel of the 1980s, but levelled off when rainfall figures slightly improved on the wintering grounds in the 1990s (Fig. 247D). The increases in Denmark (Heldbjerg 2005), Finland (Fig. 247F) and the Czech Republic (Reif *et al.* 2006) are possibly recoveries from losses incurred during the 1969-1973 crash, because the negative Sahel rainfall anomalies in the 1980s and early 1990s were later reflected in declines or consistently low levels, before indices started to improve in conjunction with slightly better rainfall figures in the Sahel.

Nevertheless, the above-mentioned recoveries stand in stark contrast to the decline in much of Europe (Fig. 247, Bauer & Berthold 1996, Sanderson *et al.* 2006). Huge depletions have been noted: for example, in Niedersachsen (based on mapping, NW Germany: -63% in the 1970s compared with the 1960s, -39% in the 1980s, -54% in 1991-2002; Zang *et al.* 2005; Fig. 247E), in a nestbox population at Steckby in Sachsen-Anhalt (steep crash in 1968-1970, no recovery; Dornbusch *et al.* 2004) (Fig. 247E), at the Constant Effort Site Mettnau in southern Germany (very few captures from 1975 onwards; Berthold & Fiedler 2005; Fig. 247G), in the federal states of Saarland, Rheinland-Pfalz, Baden-Württemberg and Bayern in Germany (an ongoing decline in the annual number of nestlings and fully grown fledglings and adults ringed, especially from the 1970s onwards; Fiedler 1998) and in Switzerland (-58% in 1990-2004; Zbinden *et al.* 2005; Fig. 247C). Monitoring in subalpine birch forest in Swedish Lapland showed that the species reached high densities in the mid-1960s, experienced 28-33% lower densities in the 1970s and 1980s (with particularly steep drops in 1969-1970 and 1983-1985, severe drought years in the Sahel) and increased somewhat during the 1990s (to only 15% less than in the 1960s; Enemar *et al.* 2004; Fig. 247B).

The extent of the decline, and the time scale over which it has occurred, are particularly manifest in nestbox studies. In The Netherlands, for example, several such studies showed high occupation rates in the 1910s and 1930s, a steep decline in the late 1930s and 1940s and an increase in the 1950s and 1960s (but never reached the high occupation rate of the early 1900s). Since then, the species has virtually disappeared as a breeding bird in nestboxes in coniferous woodland, a trend not related to a change in nestbox design (H. Stel pers. com.). Fig. 247A shows the decline on the Veluwe, but the same trends were found elsewhere in The Netherlands (Jonkers & Maréchal 1990), although a sparse breeding population using natural cavities still persists into the 2000s (with signs of a slight recovery in the late 1990s in coniferous woodland, but still almost completely absent from farmland and deciduous woodland; Bijlsma *et al.* 2001). Essentially the same disappearance from nestboxes has been described for Switzerland (Bruderer & Hirsch 1984) and for Hessen in Germany (500 nestboxes near Oberursel annually held up to 4 Redstarts in the 1950s and 1960s, but none since 1965; Mohr *in* Gottschalk 1995). Although the disappearance of Redstarts from nestboxes coincided with the temporary boom of nestbox-using

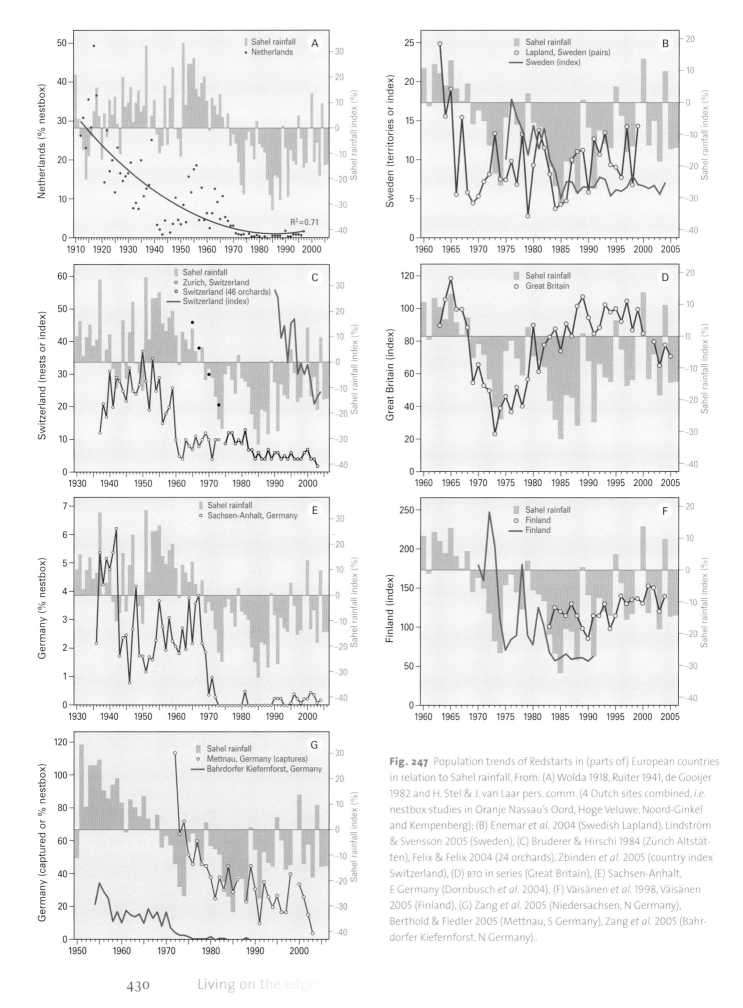

Fig. 247 Population trends of Redstarts in (parts of) European countries in relation to Sahel rainfall. From: (A) Wolda 1918, Ruiter 1941, de Gooijer 1982 and H. Stel & J. van Laar pers. comm. (4 Dutch sites combined, i.e. nestbox studies in Oranje Nassau's Oord, Hoge Veluwe, Noord-Ginkel and Kempenberg); (B) Enemar et al. 2004 (Swedish Lapland), Lindström & Svensson 2005 (Sweden), (C) Bruderer & Hirschi 1984 (Zürich Altstätten), Felix & Felix 2004 (24 orchards), Zbinden et al. 2005 (country index Switzerland), (D) BTO in series (Great Britain), (E) Sachsen-Anhalt, E Germany (Dornbusch et al. 2004), (F) Väisänen et al. 1998, Väisänen 2005 (Finland), (G) Zang et al. 2005 (Niedersachsen, N Germany), Berthold & Fiedler 2005 (Mettnau, S Germany), Zang et al. 2005 (Bahrdorfer Kiefernforst, N Germany)..

In the 20th century, a revolution took place in the cultivation of fruit in Europe. The standard apple trees in orchards were gradually replaced by semi-dwarf trees (from 1940s onwards), and then again by dwarf trees (from the 1960s). The stately old trees, at a density of 75-100 per ha, disappeared from the agricultural landscape, to be replaced by endless rows of stick-like dwarfs at a density of 1000-3300 per ha. This conversion eliminated an important breeding habitat of Redstarts, and - at least in Switzerland where the species is largely restricted to breeding in orchards - effectively prevents any recovery from declines caused by droughts in the Sahel (Felix & Felix 2004). In The Netherlands, where the decline in the surface area of standard fruit trees amounted to 74% in 1967-1977 (Schimmel & Molenaar 1982) and has continued unabated, Redstarts have almost entirely disappeared from farmland.

Tree Sparrows in the 1970s, both in Switzerland (Bruderer & Hirsch 1984) and in The Netherlands (Both *et al*. 2002), a causal relationship seems unlikely, if only because the equally sudden disappearance of Tree Sparrows in the 1980s did not result in a recovery for Redstarts. The marked increase of nestbox-using Pied Flycatchers in western and central Europe is another unlikely explanation for the demise of Redstarts (via competition for nest sites), as both species occurred alongside one another, using the same nestboxes, be-

Fig. 248 Average decline of Redstarts in three Swiss, three German and four Dutch sites between 1940 and 2003, based on the data presented in Figs. 247A, C, E and G. To make the series comparable, all figures were converted to percentages relative to the 1940 number equivalent set at 100. The lines give the running means over 5 years. Note that a log-scale is used.

In the mid 1980s, rice farmers removed nearly all flooded *Acacia kirkii* forests in the Inner Niger Delta to grow rice on the lower floodplains (Chapter 6), reducing the surface area of these forests from many hundreds of square kilometres to less than 20 km². The Akkagoun forest came into existence in the same period, when local people, supported by the IUCN, planted and seeded *A. kirkii*. When the picture was made (November 1999), Akkagoun and the surrounding bourgou fields was flooded by 2-3 m; note the large heronry (white trees in the middle). The *A.kirkii* forests attract many warblers, among which Chiffchaff, Olivaceous and Subalpine Warblers, but also Redstarts; see Fig. 34.

fore Redstarts began to decline (Bruderer & Hirsch 1984, de Gooijer 1982).

These data suggest: (1) An overall long-term decline of the continental population of 4.7% annually, on average (Fig. 248); (2) a large decline in dry Sahel years and a stable population in wet Sahel years; (3) various starting points of European declines, some of which were already apparent in the 1950s up to the mid-1960s; (4) a synchronous steep drop in the late 1960s and early 1970s; (5) some recovery in parts of Europe in the late 1990s.

The British trend is anomalous for its recovery during severe drought in the Sahel in the 1980s (associated with improved breeding performance; Baillie *et al.* 2007, see also below), when all other trends show low or declining numbers. This raises the question whether British birds, probably largely wintering in westernmost West Africa, encounter fewer bottlenecks there than continental Redstarts, which run into a multitude of problems across the enitre Sahel. Combining Swiss, German and Dutch long-term trends, we may conclude that these populations were about 20 times larger in

the 1950s than they were in the 2000s (Fig. 248), a circumstance probably applicable to much of western and central Europe (Sanderson et al. 2006), and possibly to northern Europe (Fig. 247). However, most monitoring programmes started too late to have been able fully to grasp the extent of the decline in the 20th century.

As a typical inhabitant of Sahelian woodland and scrub, Redstart fortunes should be closely linked with conditions in the Sahel. Indeed, serious droughts like the ones in 1969-1974 and in the 1980s are, with few exceptions, reflected in declines of European populations. Conversely, periods with above-average rainfall in the Sahel, like the 1920s, 1930s and 1950s, on average reflected somewhat higher densities of Redstarts. The impact of the Sahel rainfall is small, however, compared to the long-term downward trend. [1] This poses the question whether some other factor might be responsible for the large and persistent decline of Redstarts in Europe since the early 1970s (see below for local factors on the breeding grounds, which can be disregarded as being decisive in impact).

Redstarts are a species of park-like savanna, but while most Eurasian passerines bound to this habitat are wintering in the Sudan vegetation zone, Redstarts are typical Sahel dwellers. We believe that this offers the explanation for the large decline of the species. Deforestation continues steadily in West Africa (Chapter 5), but the rates at which this occurs differ – it is faster in the northern Sahel than in the region south of it. Many trees died in the northern Sahel during the Great Drought in the mid-1980s. Also, most of the riverine *Acacia nilotica* forests along the Senegal River have been removed, declining by 90% between 1954 and 1986 (Morel & Morel 1992) and by 77% between 1965 and 1992 (from 394 km^2 to 91 km^2; Tappan et al. 2004). Before the 1980s the *Acacia seyal* forests surrounding the Inner Niger Delta probably covered more than 1000 km^2, of which only tiny fragments remained in the 1990s. Entire for-

We observed Common Redstarts in Mali in inundated *Acacia kirkii* forests, as well as in dense *Acacia tortilis raddiana* forests further north (left) and in scattered *Faidherbia alba* trees, but most individuals were observed in gardens, either watered (right) or not (facing page left). In these sites, Redstarts often feed on the ground as well as low in the trees, where dead branches are used as perches; all three pictures were taken near Goundam, Mali, where Redstarts stayed in February 2009. The Sahelian gardens resemble the habitat where the birds still breed in numbers (although much more abundantly so in the past): small-scale *bocage*, an enclosed landscape with mature hedgerows (eastern Fryslân, The Netherlands, June 2007; facing page right).

ests died in the northern Delta during the low floods of the mid-1980s (Chapter 6). In the 1980s in the Hadejia-Nguru wetlands, people started to cut down the trees in the Baturiya forest (340 km²), many of which were already dead from the drought (Chapter 8). Everywhere floodplains have been turned into essentially treeless plains.

Redstarts have lost a large part of their wintering habitat due to deforestation, but further pressure on the species comes from the reduction in tree density and tree diversity in the remaining woodland, which is now severely degraded. Cresswell *et al.* (2007) compared tree density and bird density in the Watucal Forest (N Nigeria) for 1993 and 2001. Within that short period, tree density has decreased by 82%, but the commonest tree species in 1993, *Piliostigma reticulatum*, has disappeared almost completely. In 1993, there were six tree species occurring at a density of more than 20 trees/ha, but in 2001 there was but one, *Balanites aegyptiaca*. This degradation occurred despite Watucal being a protected Forest Reserve. Redstart density declined by 67% in Watucal during these 8 years (Cresswell *et al.* 2007), and declines experienced in unprotected woodlands in the Sahel must have been greater. The large-scale planting of non-native trees does not offer a viable alternative to Redstarts for the loss of indigenous trees (Box 3).

Breeding conditions   In Switzerland significant declines had been noted from the mid-1950s onwards, when Redstarts still thrived elsewhere. Most affected were orchards used simultaneously as meadows, where 50% declines were commonplace between the 1950s and the early 1970s (Bruderer & Hirsch 1984); yet even in this

Common Redstarts breed in parks, gardens, orchards, open deciduous woodland, but also in coniferous forests (Poland May 2007).

impoverished population the impact of the 1969 crash was still noticeable. Around the same time, the mid-1960s, Redstarts declined significantly in farmland until a very low-level stability was apparent in the 1980s, a trend since maintained (Felix & Felix 2004, Schmid *et al*. 2001). In other regions, the species continued to decline, as for example in the Zürich district (17 in 1984-88, 4 in 1999), despite the introduction of "ecological compensation areas" since 1990 (Weggler & Widmer 2000).

In western Europe, changes in farming practices included EU-subsidised orchard conversion from standard to half-standard trees (effectively removing many old trees with cavities from farmland), hedgerow removal, massive use of pesticides and herbicides and never-ending rationalisation of farming practices, which have led to widespread wildlife extinctions and depleted populations in farmland (Bezzel 1982, O'Connor & Shrubb 1986, Pain & Pienkowski 1997, Shrubb 2003).

Interestingly, British Redstarts showed improved breeding performance (clutch and brood size) and progressively earlier laying dates (partly explained by recent climate change) since the mid-1970s, in parallel with population recovery (Baillie *et al*. 2007). An advancement of laying date, by 15 days in the period 1986-2004, was also apparent in The Netherlands, but nest success here has not changed since 1962 apart from annual fluctuations (van Dijk *et al*. 2007). In Czech pine forests, a nestbox study in 1983-2002 failed to reveal changes in start of laying and nest success (Porkert & Zajíc 2005), and the reproductive output in a nestbox study in Braunschweig, North Germany, also did not show significant changes between 1954-1961 (5.24 young/nest), 1962-1969 (5.19) and 1970-1977 (5.23) (Berndt & Winkel 1979). Variations in breeding performance are therefore unlikely to have had an overriding impact on the negative population trend in Europe.

Conclusion

Of the widespread and common breeding birds in Europe, Redstarts have suffered one of the most dramatic and relentless declines. In much of Europe present-day numbers are a far cry from those of the 1950s and earlier. Although local factors on the breeding grounds may have contributed to this decline, the underlying cause must be sought in the Sahel, where long-lasting and serious droughts in 1969-1974 and the 1980s triggered catastrophic losses, from which few populations have recovered. A full recovery is unlikely because the birds have now permanently lost a huge proportion of their wintering habitat, the Sahelian woodland, and any remaining forests in the Sahel are rapidly degrading.

Endnotes

1 The exponential decline (annually 4.7%) between 1940 and 2003 (shown in Fig. 248) explains 81% of the variance. In a multiple regression analysis with rainfall and year, the explained variance increases to 88%, of which 7% may be attributed to rainfall, 38% to year and 43% to year or rainfall; year and rainfall are both highly significant ($P<0.001$). Correcting for the impact of rainfall, the annual population change declines from -4.7% to -3.6%.

Sedge Warbler
Acrocephalus schoenobaenus

Until 1961, the year in which Moreau published his ground-breaking review of the problems surrounding Mediterranean-Saharan bird migration, very little evidence had been available for the presence of large numbers of Palearctic migrants in West Africa. Sedge Warblers were known to winter east of Chad, and only because they were seasonally well distributed in the western Mediterranean was it considered 'pretty certain' that they wintered across the whole of Africa south of the Sahara. Large-scale ringing on the breeding grounds, especially after the introduction of mist nets in the 1970s, began to change this monumental gap in our knowledge, albeit fitfully, because even ringing more than a million Sedge Warblers, over 600 000 of which had been ringed in Britain alone (Wernham *et al.* 2002), produced few recoveries from West Africa (Zink 1973, Glutz von Blotzheim & Bauer 1991). In the 1960s, the advent of ringing expeditions to wintering grounds, so that species could be studied there, improved the chances of recovery substantially. Furthermore, they provided a wealth of data on habitat use, stopover strategies, timing of fattening and arrival and departure dates. These expeditions continue to this day. The recent introduction of stable isotope ratio techniques to establish the proportions of trace elements in feathers, thus providing clues to individual birds' breeding grounds, staging posts and wintering grounds, is an invaluable step in defining population distributional patterns and changes (Hobson *et al.* 2004). Such techniques have the potential to revolutionise our understanding of the peregrinations of migrant birds, especially those with small chances of recovery.

Breeding range

The breeding range of Sedge Warblers extends from Great Britain to the Yenisey River in West Siberia, between 45° and 65°N for the most part. Breeding limits lie between the July isotherms of 12° and 30°C in the boreal, temperate, Mediterranean and steppe climate zones. In the early 1990s, the European population was estimated at 4.4-7.4 million pairs, biased towards northern and eastern countries, including Russia (BirdLife International 2004a). The entire population winters in Africa south of the Sahara.

Migration

The essentially single-brooded Sedge Warblers are, from the moment the young have reached independence, in a hurry. They start leaving the breeding sites from late July and early August onwards (Finnish and Swedish breeding sites; Koskimies & Saurola 1985, Hall 1996), or slightly later (mid-August, Central Europe; Procházka & Reif 2002). On average, adult peak migration is between 4-7 days (in some years up to 22) ahead of that of juveniles, a feature typical throughout Europe (Insley & Boswell 1978, Bibby & Green 1981, Røstad 1986, Literák *et al.* 1994, Gyurácz & Csörgö 1994, Basciutti *et al.* 1997, Bermejo & de la Puente 2002, Zakala *et al.* 2004). Juveniles may at first disperse in any direction, including north, before heading towards the wintering grounds (Gyurácz & Csörgö 1994, Basciutti *et al.* 1997, Procházka & Reif 2002). The adults, whose body masses remain consistently higher than those of juveniles, head directly for their wintering quarters and are more adept in finding food bonanzas *en route*. One of the main prey, aphids, show steep seasonal declines, to the point that aphid abundance in southern Europe has already dropped below levels that are worth exploiting by the time Sedge Warblers begin to migrate (Bibby & Green 1981). Furthermore, aphid abundance differs substantially between areas and years, from almost non-existent to superabundant (Bibby & Green 1981, Bargain *et al.* 2002). Adult experience in locating and exploiting patchy food supplies may enhance migration speed and efficiency of fat loading. However, it is an oversimplification to consider aphids as the sole diet of migrating Sedge Warblers. At Bouche d'Ognon near Lac de Grand Lieu in France, Bibby & Green (1983) considered Sedge Warblers fattening on mayflies as an interesting exception, but the extensive research of Chernetsov & Manukyan (2000) on the Baltic's Courish Spit showed unambiguously that this species actually forages on a wide range of small invertebrates associated with aquatic or moist habitats and reeds, *i.e.* on Chironomidae, beetles, spiders, aphids and aphid consumers (mostly small parasites and predators). Depending on seasonal variations in abundance, the species easily switched from one group of abundant invertebrates to another. Sedge Warblers are clearly efficient in finding unpredictable patches possessing a high variety of arthropods, and in tracking the phenology of these ecological complexes (Chernetsov & Titov 2001). This should translate into low stopover site fidelity, as suggested by Bibby (1978), but exceptions do occur: in a reedbed near Madrid in central Spain, individual birds were recaptured frequently during autumn (outward) and spring (return) migration, indicating site fidelity (Bermejo & de la Puente 2002). The higher speed of migration compared with Reed Warblers (55 km versus 39 km per day) and the lower fat loads of Sedge Warblers at departure (southern Sweden, Bensch & Nielsen 1999) indicate that the latter must accumulate energy reserves on their way south, as is known in western France (Bibby & Green 1981, Bargain *et al.* 2002) and is thought to occur in the Pannonian Lowlands in Central Europe (Gyurácz & Csörgö 1994, Procházka & Reif 2002). Fat reserves accumulated by juveniles during migration in southern Hungary theoretically sufficed to cover some 1400 km, enabling nonstop flight across the Mediterranean (Gyurácz & Bank 1996). However, Scandinavian birds captured in northern Italy carried only moderate or low amounts of fat (Basciutti *et al.* 1997). Minor daily increases in body mass were recorded for this region, as were very short stopovers, and so there can be no doubt that these birds must use fattening areas further south in Italy or in northern Africa. This

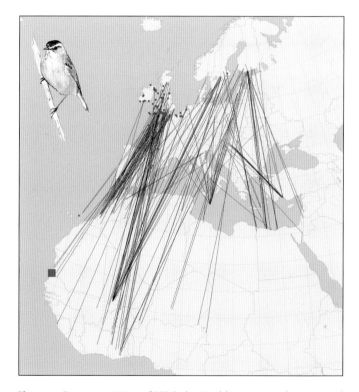

Fig. 249 European origins of 372 Sedge Warblers recovered or captured between 4° and 37°N. To avoid cluttering the figure, lines for birds ringed or captured in the Senegal Delta (red square) are omitted; instead, red dots indicate ringing sites of 33 birds recaptured in Senegal or 107 birds ringed in Senegal and recaptured subsequently in Europe. From: EURING.

Large numbers of Sedge Warblers spend the winter in the northern Inner Niger Delta, including Lake Tele, which is covered by floating vegetation (left), and Lake Horo. The dense vegetation of *Aeschynomene nilotica*, locally known as *poro*, *foro* or *gadal* (right) also holds high densities of Sedge Warblers (20-40 birds/ha in flooded fields, but less than 2/ha in dry *poro*; February 2009). These *poro* fields cover a large part of the floodplains in the northern Inner Niger Delta.

short-hop strategy may well be typical of Sedge Warblers on the eastern migration route (Schaub & Jenni 2000a & b). The crossing of northern Africa and the Sahara is rather an enigma, because body masses of migrating Sedge Warblers in southern Europe are only slightly higher than those of birds caught further north (Schaub & Jenni 2000b). Either these warblers represent a cohort that has not yet attained sufficient fat to make the crossing and may be predisposed to failure, or they have to refuel in northern Africa before setting off. So far, there is very little evidence which option is representative (Bairlein 1988, Biebach *et al.* 1991), but the latter certainly is problematic, because the vegetated belt in northern Africa is so narrow and decreases in width from west to east, particularly beyond Tunisia. Furthermore, this area normally does not receive rainfall until October, and so during autumn migration it is in its most desiccated state (Moreau 1961) and unlikely to provide much food.

Return migration is preceded by the birds fattening in their sub-Saharan wintering grounds from mid-March, but continuing as late as the end of May near Lake Chad (Fry *et al.* 1970, Aidley & Wilkinson 1987). A similar spread of dates applies to East African wintering birds; in Kenya thousands of birds were present during late April and early May 1972, staying to fatten for 1-3 weeks, but lean birds continued to arrive from further south well into May. Fattening rates amounted to 0.31 and 0.64 g per day at Lake Nakuru and the Athi River respectively, resulting in departure weights of 16-21 g, which enabled a nonstop flight to the Middle East (Pearson 1979). Body masses of Sedge Warblers captured during spring migration at Eilat in Israel showed considerable annual variations between 1984 and 2001, from an average of 9.51 g in 1984 to 13.1 g in 1998 (Yosef & Chernetsov 2004). Based on the temporal distribution of captured birds at Eilat, it is quite possible that in spring two discrete populations pass successively through the region, each refuelling before continuing onwards (Yosef & Chernetsov 2004). Several waves of migrants, possibly representing different populations, may also pass through the sub-Saharan region where Fry *et al.* (1970) recorded a bimodal weight distribution (peak weights in late March-early April and mid-late May). At each stopover, Sedge Warblers presumably try to refuel opportunistically, and so it is hardly surprising that birds captured in late April and May at Rybachy along the Courish Spit weigh 13-14 g at departure (Bolshakov *et al.* 2001), a reserve margin that allows for an additional unrefuelled range of at least 500 km to high latitude breeding sites, where feeding conditions are unpredictable so early in the season (Chernetsov 1996); excess fuel stores may represent the means of surviving at such critical times (Bolshakov *et al.* 2001).

Intensive ringing expeditions took place to Nigeria (from 1967 onwards), Mali (1979-82), Ghana (1987), Senegal (1984-1993, with numerous teams ringing a total of 82 468 birds, of which 1460 in 31 Palearctic species were recaptured in at least one subsequent winter or passage period; Sauvage *et al.* 1998), Djoudj, Senegal (18 January-10 February 2007: 1963 Palearctic birds in 22 species captured; Flade 2008), Niger (2002-2004) and Mauritania (spring and autumn 2003: 9467 migrants in 55 species; Herremans 2003) have produced a multitude of recaptures and at least 41 recoveries from Europe (Fig. 249). The ringing recoveries clearly show that the western European Sedge Warbler population winters in the western Sahel (all 20 Irish and 6 Belgian birds came from Senegal, as did 24 out of 28 French birds, for example). Southern Scandinavian, western and central European birds occurred mainly in the central section of the Sahel (9 out of 10 Swedish birds from Mali and Niger), and birds

In Lake Horo, one of the lakes in the northern part of the Inner Niger Delta, Sedge Warblers reached an extremely high density of 103 birds/ha in floating vegetation of *Polygonum senegalense* (left, March 2003 and 2004), locally known as *kouma*. For this lake alone, we estimated the number of wintering Sedge warblers at 390 000 – 750 000. Much lower densities, but still considerable, were found in *Mimosa pigra* scrub (middle) and *bourgou* floating grass fields (right) in the Inner Niger Delta, *i.e.* on average 6.5 and 5.8 birds per ha, respectively.

from northernmost and eastern Europe (including Russia) appeared from the Inner Niger Delta eastwards. The ringing data suggest a high migratory connectivity, with a migratory divide in central Europe (see below).

Birds from Britain and Ireland migrate along the Atlantic coast to the wintering grounds in Senegambia (and possibly further south). Recaptures of ringed birds in Senegal originated mainly from Britain and Ireland (30), with single birds from France, Belgium and The Netherlands (Fig. 249). However, British Sedge Warblers have been recovered as far east in the Sahel as Mali and Burkina Faso, and as far south as Liberia, Sierra Leone and Ghana, indicating that they occur in a wider swathe across West Africa than in just the coastal zone. Birds from the Low Countries show departure directions between SW and SSE and so are more likely to reach Africa via Iberia and Italy (Zink 1973). Central European birds show an equally wide range of departure directions, from SW through SE, with an increasing tendency to migrate southeast for Sedge Warbler breeding populations from the eastern section of this region (Literák *et al.* 1994, Trocińska *et al.* 2001, Procházka & Reif 2002, Zehtindjiev *et al.* 2003). These birds will end up mostly in the central part of the Sahel (and Sudan zone), from the Inner Niger Delta to the Chad Basin and possibly even in eastern Africa.

Birds from southern Scandinavia and the Baltic States migrate on headings between SW and S, wintering mainly between the Inner Niger Delta and the Chad Basin, and also south of this region (Fig. 249). Sedge Warbler populations from northern Scandinavia, eastern Europe and Russia follow a southerly to southeasterly course and most winter from the Inner Niger Delta eastwards and southwards, up to and including East and southern Africa (Koskimies & Saurola 1985, Dowsett *et al.* 1988, Yosef & Chernetsov 2004).

Distribution in Africa

Sedge Warblers seldom winter in Africa north of the Sahara (for example, they are irregular in Egypt; Goodman & Meininger 1989), but millions disperse over the Sahel and further south, concentrating mostly in well-vegetated wetlands and swamps.

In the lower floodplains of the Inner Niger Delta in Mali, some 250 000 are estimated to winter, reaching high densities in *bourgou* (5.8/ha) and *Mimosa* scrub (6.5/ha). For the entire Inner Niger Delta, the wintering population must be well over one million birds (Chapter 6).

Sedge Warblers are rare in, or absent from, irrigated rice fields. In the coastal rice fields between The Gambia and Guinea, density averaged only 0.08/ha; in this huge rice-growing area, fewer than 10 000 birds winter (Bos *et al.* 2006). The species was not recorded during plot sampling in several large irrigated rice-growing regions in Mali, *e.g. Office du Niger* (Wymenga *et al.* 2005) and Selingue (van der Kamp *et al.* 2005), nor in the Senegal Delta (Chapters 6 and 7). However, Sedge Warblers have frequently been observed in any bushy vegetation of *Typha, Cyperus, Phragmites* or *Scirpus* within irrigated rice fields.

Large numbers winter south of the Sahel in the Sudan vegetation zone, where any stretch of water bordered by lush vegetation holds, at separation distances of 10-50 m, territorial Sedge Warblers which can be heard singing in early morning (as recorded in *Typha* vegetation in Lake Bosomtwi, 6°38'N, 1°25'W, 17 December 1996-2 January 1997; van den Brink *et al.* 1998). For example, Ghana's Lake Volta, created in 1966 and covering some 8500 km², has well-developed macrophytic vegetation that provides extensive habitat for Sedge Warblers (Walsh & Grimes 1981). Similarly, the construction

of thousands of small dams and water reservoirs all over sub-Saharan Africa since the 1960s has created much new habitat for waterbirds (Claffey 1999) and warblers alike. It is doubtful if this habitat gain outweighs the dam-induced loss of wetland habitats downstream.

Population change

Wintering conditions   Although they are to be encountered all over West Africa during the northern winter, Sedge Warblers are nevertheless largely confined to well-vegetated wet or moist habitats. This translates into a concentrated occurrence in the dry Sahelian zone (mostly in floodplains, where a quarter of all recoveries have occurred), and a more widely scattered distribution in the Sudan zone.

Table 28 Percentage population change between the 1970s and 1985, after a series of extremely dry years in West Africa, and between 1994 and 1995, when the intervening winter in Africa was extremely wet compared to the preceding 14 years. All studies refer to counts in breeding areas, except France, southern Germany and a Swedish study which have used mist-netting on Constant Effort Sites.

Country	1985 vs. 1970-80	1995 vs. 1994	Source
Sweden	-79.2	+50.9	Lindström & Svensson 2005
Finland	-71.2	+12.6	Väisänen 2005
Sweden (migrants)	-51.2		Karlsson et al. 2002
Denmark		+17.3	Heldbjerg 2005
The Netherlands		+45.5	SOVON
Great Britain, farmland	-59.1	+27.0	BTO
Germany, Niedersachsen	-63.2	+30.2	Zang et al. 2005
S Germany (breeding+migration)	-75.0	+46.6	Berthold & Fiedler 2005
France (breeding+migration)		+90.7	www.mnhn.fr/mnhn/crbpo/fiches_especes/caille.htm
Average ± SD	-66.5 ± 10.5	+40.1 ± 24.7	

Rainfall in West Africa has been established as the main factor determining breeding numbers in Britain (Peach *et al.* 1991) and The Netherlands (Foppen *et al.* 1999); the entire breeding population is likely to be affected when hydrological conditions in West Africa are particularly severe. In 1972, low rainfall and poor flood levels in the Sahel suggested poor prospects for Sedge Warblers; captures at Constant Effort Sites in southern Germany dropped by 23% (Berthold & Fiedler 2005) and the standardized ringing total in Sweden showed a decline of 58% (Österlöf & Stolt 1982). Adult survival rates of birds in Britain had dropped from 35-41% in 1970-1972 to 13-28% in 1973-1975 (Green 1976, Bibby 1978). The Great Drought in 1983-1984 brought even greater troubles to wintering populations than had the earlier droughts of the 1970s, causing declines of 51-79% compared with previous wet years (Table 28). The survival rate of adults reaching southern England was estimated to be lower than 4% (0.6-22.3% asymmetric 95% confidence limits) between 1983 and 1984 (Peach *et al.* 1991). Even considering Sedge Warblers' generally low site fidelity (Bibby 1978), which could bias rates downwards, the mortality caused by the devastating drought in the Sahel in 1983 must have been enormous. Mortality of first-year birds is very difficult to assess, but an attempt by Staffan Bensch for about 16 000 birds ringed as juveniles at Kvismare in southern Sweden since 1961 came up with an average survival rate of some 10% (*in* Glutz von Blotzheim & Bauer 1991: 331). Imagine how this must have been affected as a consequence of the drought of 1984...

Whereas droughts have a drastic culling effect on Sedge Warbler populations, wet years in West Africa enable rapid recoveries. After hydrological conditions improved in the wet year of 1974, populations bounced back immediately by +42% in Sweden, +103% in southern Germany (captures; including migration period) and +116% in Britain. In 1994, following a long series of dry winters in West Africa, rainfall was again prolific and subsequent recoveries in population size varied between +10% and +91.7% (Table 28).

The pivotal role of wintering conditions in determining breeding

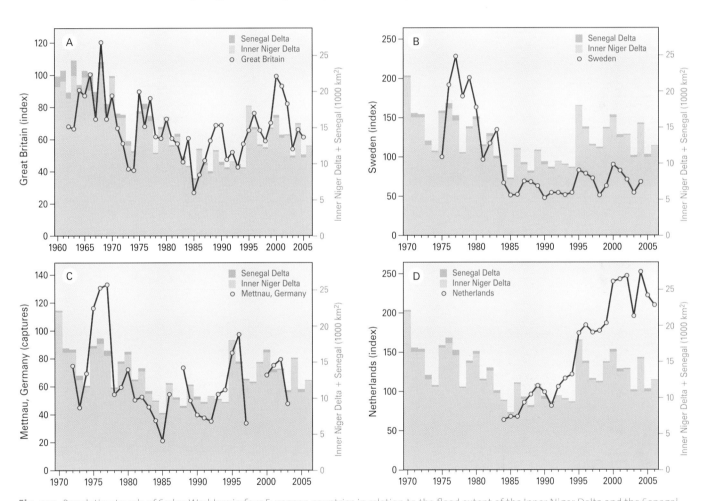

Fig. 250 Population trends of Sedge Warblers in four European countries in relation to the flood extent of the Inner Niger Delta and the Senegal Delta in the preceding winter. From: (A) BTO (Great Britain), (B) Lindström & Svensson 2005 (Sweden), (C) Berthold & Fiedler 2005 (Germany, Mettnau), (D) SOVON (The Netherlands).

Sedge Warbler on Yellow Iris

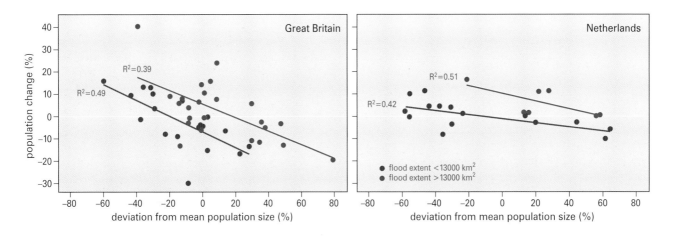

Fig. 251 Population change between consecutive years as a function of population size in the preceding year, split for winters with low and high flood extents of Sahelian floodplains, for the Dutch (top) and British (below) population. To allow both sets of figures to be comparable, the average indices (67 in Britain 1962-2006 and 155 for The Netherlands 1984-2006) were equalised to 100, all other indices being expressed as relative deviation from this mean.

numbers can be put to the test by looking at results from long-term monitoring schemes throughout Europe (Fig. 250). First, we expect that the different series from the various countries will show synchronic fluctuations, which indeed was the case: all long time series are well correlated. [1] Secondly, was the wintering population regulated primarily by the Sahelian rainfall (as implicitly assumed by Peach et al. 1991 and Foppen et al. 1999) or by the Sahelian flood extent? The statistical analysis shows that flood extent is the best predictor. [2]

The annual change in breeding numbers was highly correlated with the flood extent, notwithstanding a large variation, which could be attributed partly to the size of the population: at a high population level, populations tend to decrease, while at a low level, they tend to increase. The state of equilibrium in British (Fig. 250A) and Dutch populations (Fig. 250D), when the annual change is zero (Fig. 251), is found at a higher level if the flood extent in the preceding winter in the Sahel has been high. [3]

Breeding conditions   For a species as dependent on hydrological conditions in sub-Saharan Africa as the Sedge Warbler, synchronous fluctuations would be implicit. Even so, independent of flood extent, the Dutch population increased (Fig. 250D), but the Swedish population (Fig. 250B) and also, to a much lesser degree, the British population (Fig. 250A), decreased. The series from South Germany showed no long-term change except for fluctuations relating to Sahel conditions (Fig. 250C). [4] The Danish trend is also more or less stable (Heldbjerg 2005). As these sub-populations have overlapping wintering grounds, local conditions on the breeding grounds should play an additional role.

Land reclamation, agricultural intensification and water management all claimed substantial proportions of prime breeding habitat throughout Europe, especially in the first half of the 20th century (Bauer & Berthold 1996). In Niedersachsen, for example, the present Sedge Warbler density is thought to be a mere shadow of past numbers, notably in comparison with the late 19th century (Zang et al. 2005). In western Europe, remaining wetlands are better protected nowadays, although highly dynamic water regimes, formerly an intrinsic part of the system, have become exceedingly rare. Many marshlands therefore rapidly become converted into marshy woodland and scrubland, a process accelerated by eutrophication and the concomitant need for stepped-up management. It is most likely to be this process, habitat change through degradation, that caused Sedge Warblers to desert many peripheral breeding sites across Europe, rather than habitat fragmentation *per se* (*contra* Foppen et al. 1999). In this respect, future development in eastern Europe will serve as an interesting test case, especially in those countries that have recently joined the EU and become subject to the same kinds of environmental issues brought about by changing land-use as experienced by western Europe since the 1960s.

Conclusion

Sahel drought causes high mortality among Sedge Warbler populations, resulting in substantial Europe-wide declines. Populations recovered from droughts after wet winters in Africa. Annual changes in breeding numbers were better explained by the extent of flooding of Sahelian floodplains in the preceding year, than by rainfall. Flood extent also influences the population level at which equilibrium is reached (here defined as zero change from one year to the next), being higher after winters with a high flood extent in Sahelian floodplains in the preceding winter. Local factors explain why trends for different countries and regions are not fully synchronised.

Endnotes

1   The correlations between country indices are highly significant: Great Britain – Netherlands: R=+0.71 (N=22), Great Britain – S Germany (R=+0.588, N=29). Correlations with Finland and Sweden are also exceeding R=+0.50.

2   The correlation between population indices and Sahel parameters are significant, except for Sweden for rain and NDVI. The correlations are highest for flood extent and lowest for NDVI: Great Britain (R=+0.68 for rain, +0.72 for NDVI, +0.72 for flood; N=44, but N=24 for NDVI); S Germany (R=+0.45 for rain, +0.62 for NDVI, +0.75 for flood; N=29, but N=19 for NDVI), The Netherlands (R=+0.60 for rain, +0.68 for NDVI, +0.73 for flood; N=23, but N=22 for NDVI), Sweden (R=+0.29 for rain, +0.22 for NDVI, +0.55 for flood; N=30, but N=23 for NDVI).

3   The annual population change for Sedge Warblers in The Netherlands depends on flood extent (P<0.001) as well as on population size (P=0.003) (N=27, R^2=0.52). The same figures for the British series are: flood (ß=+0.600, P<0.001), population size (ß=-0.769, P=0.001), R^2=0.532; for S German series: flood (ß=+0.716, P<0.001), population size (ß=-0.789, P<0.001), R^2=0.591; for Swedish series: flood (ß=+0.600, P<0.001), population size (ß=-0.614, P=0.005), R^2=0.472.

4   In a multiple regression of population size with flood extent and year, flood extent was always highly significant, but the significance of the long-term trend varies and may be positive as well as negative: Great Britain (ß=+0.400, P=0.003), S Germany (ß=-0.066, P=0.622), Sweden (ß=+0.837, P=0.018), The Netherlands (ß=+0.772, P<0.001).

European Reed Warbler
Acrocephalus scirpaceus

European ornithologists know they are far from home when they see a Nut Palm Vulture taking fiddler crabs, a solitary Goliath Heron fishing in a tidal creek and exotic kingfishers dashing through mangroves. However, after a restless night, spent well aware of the presence of venomous snakes which abound in this habitat, our Europeans wake up to the calls and snatches of song of a host of Reed and Melodious Warblers, Blackcaps and other bird species typical of Europan breeding grounds; the surrounding soundscape gives the sensation of being back home. Beneath their feet anaerobic bacteria steadily produce nitrogen gas, soluble iron compounds, inorganic phosphate and methane, contributing hugely to the particularly pungent odour that rises from waterlogged systems - another reminder of home, or rather of wading through reedbeds in search of Reed Warbler nests, an experience unrelated to, in the words of Brown & Davies (1949), "nostalgic longings for the twilight zephyrs which may or may not rise from the bosom of Mother Nature (for in a reed-bed there is nothing mysterious about what rises up from that bosom...it is a persistent stink of sulphuretted hydrogen)."

Breeding range

European Reed Warblers occur in the mid-latitude lowlands of the western and central Palearctic, showing a strong association with reedbeds. The northern range limit extends into southern Scandinavia. BirdLife International (2004) estimated the European population at 2.7-5 million pairs, equivalent to a wintering population of 8-14 million birds, given the proportion of 42% first year birds captured in the wintering grounds south of the Sahara (see below).

Migration

In contrasting to many other trans-Saharan migrants, Reed Warblers generally use shorter migratory stages, both during autumn (outward) (Bibby & Green 1983) and spring (return) migration (Bolshakov et al. 2003). In the Mediterranean region Reed Warblers increase their stopover duration in the course of the post-breeding migration, from 6.1 days at the end of July to 11.1 days at the end of October. Fuel deposition rate also increases, from 0.29 g per day during peak migration (late September) to 0.40 g per day in late October (Balança & Schaub 2005), suggesting that later in the season stopover sites are farther apart and birds have to undertake longer journeys between stopover sites. In Eilat, Israel, the average rate of weight gain was 0.157 g per day, both in autumn and spring. In spring, and especially later in the season, birds are in a hurry and deposit fat more quickly (Yosef & Chernetsov 2005). These later migrants may be heading for more distant breeding grounds. Interestingly, more than half of birds arriving in spring at Rybachy, in the Baltic on the Courish Spit, had energy stores sufficient to continue migration for at least another night (Bolshakov et al. 2003). Apparently, in spring Reed Warblers initiate mass nocturnal migration when warm air of Mediterranean origin is moving slowly northwards, providing favourable flight conditions of light following winds and rising air temperatures, conducive to improved feeding conditions *en route*. Using a succession of short migratory flights of 4-6 h each, the warblers stage through Europe, whilst maintaining a positive energy balance during long pauses at stopovers (Bolshakov et al. 2003).

Distribution in Africa

The recoveries and captures of Reed Warblers in Africa north of 4°N (Fig. 252) suggest a wide distribution over the Sahel and Sudan zone, without concentrations in the Sahelian floodplains. Indeed, the species was singularly absent during the 1630 density counts in the lower floodplains of the Inner Niger Delta in Mali (Chapter 6) and none were observed during the 1365 density counts in irrigated rice fields elsewhere in Mali (Chapter 11). During the northern winter, the species must be absent or rare in these wet habitats, since the observers were keen to identify any species that was not a Sedge Warbler, a species which abounds in floodplains, if less so in rice fields. Reed Warblers were also quite scarce (9 birds/km²) in the coastal rice fields between The Gambia and Guinea (Chapter 11).

Whereas Reed Warblers are almost an oddity in Sahelian wetlands, they are very common in drier habitats further south and in the saline mangroves of Guinea Bissau (Altenburg & van Spanje 1989). If the minimum density of 5.7 birds/ha estimated for mangroves holds for the entire mangrove zone between Senegal and Sierra Leone (8 000 km²), some 4.5 million Reed Warblers might spend the northern winter here, equivalent to a substantial part of the estimated European population. The importance of mangroves as a wintering habitat is apparently not substantiated by the 438 ringed Reed Warblers found or recaptured in the Sahel zone (Fig. 252), for only 16 came from the mangrove zone, and, if recaptured birds are omitted, mangroves account for a mere 10% of all captures. However, there is a counterintuitive argument to support the posited concentration in mangroves, for the mangrove forests are dense and inaccessible on foot, making the ringing reporting rate for birds in

Fig. 252 European origins of 1167 Reed Warblers recovered or captured between 4° and 37°N. To avoid cluttering the figure, lines for birds ringed or captured in Morocco and northern Algeria and for birds ringed in the Senegal Delta (red square) are omitted; instead, red dots indicate ringing sites of 84 birds recaptured in Senegal or 56 birds ringed in Senegal and recaptured subsequently in Europe. From: EURING, except 18 from Mullié et al. (1989), Meininger et al. (1994).

The extensive mangrove forests along the West African coast, here shown for Guinea-Bissau, are almost impenetrable on foot, and hence received little attention from ornithologists. Using the MS Knud W as a basis from December 1986 to February 1987, a Dutch team studied the mangrove bird community in Guinea-Bissau by mapping singing birds and using mist-nets. The number of Palearctic bird species was small, but densities of Reed warbler, Willow Warbler and Melodious Warbler were high (Altenburg & van Spanje 1989). The significance of mangroves as wintering site for Reed Warblers and Melodious Warblers is undervalued, and in need of a reappraisal.

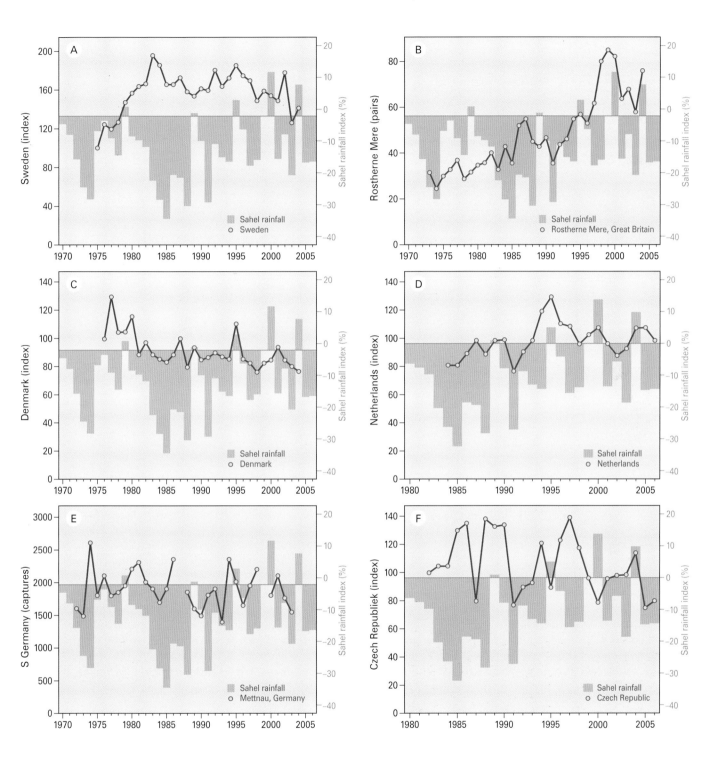

Fig. 253 Population trends of Reed Warblers in (parts of) six European countries (red lines, left-hand axis) related to the rainfall anomaly (columns). From: (A) Lindström & Svensson 2005 (Sweden), (B) Calvert 2005 (Rostherne Mere, Great Britain), (C) Heldbjerg 2005 (Denmark), (D) SOVON (The Netherlands), (E) Berthold & Fiedler 2005 (Germany, Mettnau), (F) Reif *et al.* 2006 (Czech Republic).

In a 16-day period during a 3-month stay in the mangrove zone of Guinea-Bissau in 1986/1987, Altenburg & van Spanje (1987) captured 50 European Reed Warblers, 23 Melodious Warblers, 22 Yellow Wagtails, 12 Subalpine Warblers, 6 Willow Warblers, 2 Chiffchaffs and 2 Sedge Warblers. Density-counts in six plots each of 5 ha suggested the same species distribution with the following average (minimum!) numbers per ha: 5.7 Reed Warblers, 4.3 Melodious Warblers, 2.6 Willow Warblers or Chiffchaffs, 1.0 Sylvia sp. and 1.0 Yellow Wagtail. Multiply this number by 880 000 (being the surface area in ha of the mangrove zone between Senegal and Sierra Leone (Bos et al. 2006) and one can easily envisage the huge numbers of temperate zone birds which concentrate here.

mangroves a fraction of that on dry land. During post-breeding migration, Reed Warblers start increasingly to inhabitat drier areas containing mixed grasses and scrub, and this change in habitat choice becomes quite pronounced in the wintering areas. Their near-absence in freshwater marshes may be due to competition avoidance with their congeners, African Reed Warblers (Dowsett-Lemaire & Dowsett 1987).

Those Reed Warblers wintering along the Atlantic coast of Senegambia, west of 12°W, breed mostly in northwestern Europe and migrate via the Iberian Peninsula (Fig. 252); 82% of 87 recoveries in Senegal originated from Great Britain, the remaining being from Spain (n=7), France (n=6), The Netherlands (n=1) and, the only outlier, from Italy (n=1). In contrast, between 12°W and 10°E, only one of the recoveries was British. Reed Warblers recovered in Mali had been ringed in Germany (n=7), France (n=5), Belgium (n=3) and Sweden (n=3), confirming the high migratory connectivity even within the SW-migrating population (Procházka et al. 2008). Northern European birds, from as far east as the Kaliningrad Region in Russia, follow a course between SW and SSW, with a gradual change in direction from West to East: recoveries in the same year, and at distances of 400-4500 km from the ringing sites, showed an average direction in autumn of 195° in Norway (n=60), 216° in Sweden (n=669) and 226° in Finland (n=110) (Fransson & Stolt 2005). Upon reaching Africa, a change in the mean direction of orientation is effectuated, presumably controlled endogenously (as in Garden Warblers; Gwinner & Wiltschko 1978). The resulting shorter crossing of the Sahara brings the birds mostly to wintering areas from Mali south into the Sudan vegetation zone of West Africa, while generally maintaining the west-east distribution as found on the breeding grounds (Fransson & Stolt 2005, Bakken et al. 2006, Procházka et al. 2008). Tape-luring in the western Sahara and along the coast in Mauritania seems to indicate that adults and juveniles may employ different strategies to cross the Sahara in autumn: adults migrated ahead of juveniles and travelled over the interior of the Sahara, whereas juveniles were concentrated along the coast. Few Reed Warblers were captured inland during spring migration, probably because the birds are then migrating with tail-winds at great altitudes above the wind-shear (Herremans 2003).

The research of Dowsett-Lemaire & Dowsett (1987), Schlenker (1988) and Yosef & Chernetsov (2005) indicated that the disparity in migratory directions of Central European breeders, either heading SW or SE, is evidence of a migratory divide. Birds from Serbia, Hungary, eastern Austria and the Czech Republic migrate to the southeast, a course which is almost a mirror image of the flight direction followed by northern and western European populations. These birds pass through Greece, Cyprus, Turkey and the Levant to continue south into eastern Africa. Some, or perhaps many, birds may divert from this course to end up in the central Sahelian and Sudan zone of Nigeria, Cameroon and Chad, presumably to winter. An alternative explanation could be that these birds have headed due south from Central Europe, with some support for this idea coming from a few recoveries in Italy (Fig. 252, see also Procházka *et al.* 2008); the question is as yet unresolved. The seven birds recovered in the vicinity of Lake Chad originated from the Central European region, but two French birds in Chad show that there is some overlap with birds from western Europe.

Reed Warblers from further east in Europe belong to the race *fuscus* and probably cross Saudi Arabia towards wintering areas in eastern Africa (Dowsett-Lemaire & Dowsett 1987). Their passage through Sudan in September to October is somewhat later than nominate *scirpaceus* in August (G. Nikolaus *in* Dowsett-Lemaire & Dowsett 1987).

Population changes

Effect of wintering conditions The distribution of Reed Warblers in West Africa contrasts strongly with that of Sedge Warblers. Where the latter are confined to wetlands, Reed Warblers either inhabit drier habitats or intertidal mangroves. The adverse effects of drought in West Africa are therefore expected to be smaller in Reed Warblers, and especially so in the NW European population, which winters partly in intertidal mangroves. Indeed, none of the long-term trends from Europe revealed any relationship with rainfall figures in West Africa, except the study in 1973-2004 by Malcolm Calvert (2005) in Rostherne Mere near Manchester, England: Sahel rainhall in the winter correlated here with the number of pairs the following breeding season ($R^2=0.157$, $p<0.01$), and with the percentage change between years ($R^2=0.113$, $p<0.02$). The extremely dry winter of 1973, however, did not affect the normal level of birds caught the following year (1974) in Sweden (Österlöf & Stolt 1982) or Belgium (Nef *et al.* 1988). Breeding numbers in two North German plots (Zang *et al.* 2005) remained stable in 1971-2003. In southern Germany, Reed Warbler captures peaked at Constant Effort Sites in 1974 (Berthold & Fiedler 2005). The Great Drought in the early 1980s also failed to cause any decreases in populations in Germany (Berthold & Fiedler 2005, Zang *et al.* 2005), The Netherlands (SOVON), Switzerland (Weggler 2005) or the Czech Republic (Reif *et al.* 2006). The long-term trends in Sweden showed an increase during the 1970s and stability since then. The trends in other European countries were more or less stable, but with locally induced fluctuations (Fig. 253). In two British populations at Gosforth Park and Wicken Fen, variations in Sahel rainfall were correlated neither with variations in return rate (a combination of true annual survival and permanent emigration) nor with adult Reed Warbler abundance in the UK at Constant Effort Sites (Thaxter *et al.* 2006).

Breeding conditions The lack of correlation with Sahel rainfall indicates that local productivity and climate variations during the breeding season are better predictors of variations in numbers. Good-quality habitat is important for survival and retention of breeding birds (Schulze-Hagen 1993). Reedbeds cut on a rotation are known to be favoured by Reed Warblers, especially in combination with the removal of dead litter; vigorous regrowth then provides better food resources and stronger reeds for nesting (Poulin *et al.* 2002, Thaxter *et al.* 2006). Population growth in a small population with 15-20 pairs at Gosforth Park, England, may have been more influenced by the return rate of adult Reed Warblers than by new recruits from immigration and natural recruitment, but the relationship between return rate and Sahel rainfall was not significant (Thaxter *et al.* 2006). Short-term variations in numbers may be caused by poor reproduction, in its turn largely dependent upon habitat quality (Calvert 2005, Thaxter *et al.* 2006). Differences in habitat quality across Europe are then the likely determinant factor for the various trends found.

Between 1970 and 2006, Reed Warblers in SW Poland significantly advanced their laying date by 18 days (median first egg) in conjunction with increasing April-May temperatures, but this shift still lags behind the advancement in reed phenology (which may be advantageous, as early nests are then built in a mixture of old and new reeds, providing better cover). In contrast, the end of laying did not change significantly. The overall change was one of the laying period becoming significantly longer, enabling a larger proportion of pairs to initiate a second brood; in the 1970s and 1980s only 0-15% of the individuals produced a second clutch, but this had increased up to 35% between 1994 and 2006 (Halupka *et al.* 2008). The increased frequency of renesting and better cover for early nests should positively affect the total number of nestlings produced, and hence lead to improved fitness.

Conclusion

European Reed Warblers mostly spend the northern winter in the Sudan zone and intertidal mangroves, and any population changes are therefore unlikely to be governed by Sahel conditions. Indeed, almost all trends from European countries failed to show a relationship between Sahel rainfall and fluctuations in numbers on the breeding grounds. The latter are probably related to habitat quality. Spring phenology has advanced since the 1970s, resulting in earlier laying, a higher frequency of second broods and an improved reproductive output per pair in Poland, for example. Improved fitness may lead to population increases, such as those which are indeed found in several European countries.

Lesser Whitethroat
Sylvia curruca

In his correspondence with H. Freiherr Geyr von Schweppenburg (1930), Baron Snouckaert van Schauburg stated his belief that Lesser Whitethroats in The Netherlands were not common breeding birds: "In 1887-1896 I lived on an estate between Haarlem and Leiden, the same where the old Temminck had lived. As a young man I industriously collected birds there, and I made quite a nice collection of local birds. I saw curruca for the first time in 1895, *i.e.* after a stay of eight years. That's when I found a dead female on the ground on 18 May, and I shot the matching male on 23 May. That is all!" His only observation on an estate near Doorn in 1903 to 1912, where the species nowadays is a widespread breeding bird, was of two birds on 3 August 1905: "and, apparently on migration, which I both shot." By that time, it was already common knowledge that European Lesser Whitethroats wintered in eastern Africa, where, for example, Robert Hartmann (1863) encountered many in "impenetrable Nile acacias" just north of Dongola on 25 March 1860. Later studies showed the species to be particularly abundant in Acacia-dominated semi-open wood- and scrubland across Sudan and Ethiopia; in fact, the only autecological study on the species in its wintering habitat, by Sven Mathiasson (1971) in January-February 1961 and 1964, comes from this vegetation zone. The past decades, however, have brought tremendous changes in the main wintering range of Lesser Whitethroats, particularly in Darfur in western Sudan. Although Schlesinger & Gramenopoulos (1996) found no evidence for climatic-induced change in the woody vegetation in Sudan in 1943-1994, the persistent drought in the 1980s did bring desertification and ecological problems. These were the triggers for conflict, which eventually led to war in Darfur in 2003 (turning into genocide and depopulation), disruption of traditional land use patterns and huge loss of livestock. The latter resulted in a rebounding vegetation in Darfur (most evident in 2007), but the adverse effects of the displacement of people on vegetation are now clearly visible in eastern Chad (Schimmer 2008). Today's Darfur is very different from the Darfur visited by Gerhard Nikolaus (1987, observations in 1976-84), R. Trevor Wilson (1982, observations in the 1970s), P. Hogg *et al.* (1984, mostly observing birds between 1930 and 1954), and Admiral Lynes (1924-1925, fieldwork in 1920). Sufficiently different to affect Lesser Whitethroats? We do not yet know.

Breeding range

The vast and continuous distributions of the Lesser Whitethroat subspecies which winter in sub-Saharan Africa extend irregularly southwards down to 37°-43°N and eastwards into Russia and beyond. Further east still, other subspecies occur. Highest breeding densities prevail in western, central and eastern Europe, as far north as southern Finland. The European population is estimated at 1.8-4.4 million pairs (BirdLife International 2004a).

Migration

The entire European population migrates south or southeast to converge in the Middle East and continues towards the wintering grounds in central West Africa and in East Africa (Wernham *et al.* 2002, Fransson *et al.* 2005; Fig. 254). Some birds straggle as far west as Senegal, where Rodwell *et al.* (1996) mention 30 records between 30 October and 22 March 1984-1994 for the Djoudj. A remarkable recovery of a Danish bird (ringed 24 April 1994) came from Guinea, where it was recaptured on 16 October 1994 (EURING; Fig. 254). The scarcity of Lesser Whitethroats in West Africa is corroborated by the few observations in Morocco, where only 4 autumn and 30 spring observations have been reliably recorded (Thévenot *et al.* 2003), and two birds in Mauritania in 2003 (16 April and 14 October; Salewski *et al.* 2005).

Post-juvenile dispersal of British birds normally starts when birds are 30-38 days old, and still in active moult of wing coverts (Norman 1992). Fat loads prior to departure from the breeding grounds are usually less than 10% of total body mass, as reported for southern Gotland, Sweden (Ellegren & Fransson 1992), enabling a non-stop flight of slightly more than 300 km. Adults had significantly higher fat loads than juveniles. Furthermore, fat loads showed a positive correlation with the progress of autumn. The fattest birds, presumably those which are about to depart, are capable of flying through two successive nights, representing a nonstop flight of some 770 km, although a maximum of 1270 km is just possible (Ellegren & Fransson 1992). If British-ringed birds follow a similar strategy on autumn migration, they should reach northern Italy, which is where a concentration of recoveries has indeed occurred (Wernham *et al.* 2002). Another similar leg on their journey would carry the birds to Greece, from where the species crosses the eastern Mediterranean on a relatively broad front, given the many recoveries in coastal Libya, the Egyptian delta and Cyprus (Fig. 254, Fransson *et al.* 2005). Obviously, many birds use this region for refuelling, as all recaptured birds caught in northwestern Jordan in autumn showed an increase in body mass, at an average of 4.45% per day (Khoury 2004). Ringing activity from 8 September-13 October 1965 near Bahig, 32 km southwest of Alexandria, Egypt, and 8 km from the Mediterranean coast, did not reveal obvious weaklings, despite a large indi-

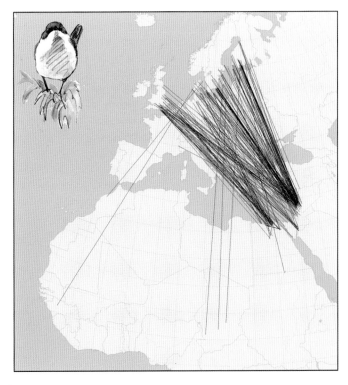

Fig. 254 European origins of 306 Lesser Whitethroats recovered between 4° and 37°N. From: EURING, Mullié *et al.* (1989) and Meininger *et al.* (1994).

vidual variation in body mass, from 9.5 to 15.5 g (Moreau & Dolp 1970). Deeper into Africa, in Egypt's Western Desert, 170 km south of the Mediterranean coast and 150 km west of the Nile, Lesser Whitethroats were, after Willow Warblers, the commonest of 28 migratory species to be captured in September-October 1983 and in September 1985, comprising 11.4% of 1028 birds (Biebach et al. 1991). Such abundance makes one wonder how many birds cross the Sahara west of the Nile. Migratory pathways in eastern Africa are not well known, but it is likely that the majority of birds continue south to reach their wintering grounds in Sudan, Eritrea and northern and central Ethiopia. Records of wintering Lesser Whitethroats in the Sahel further west, as far as northern Nigeria (Urban et al. 1997, Wilson & Cresswell 2006), indicate that a continuation of migration along the NE-SW axis may be carried out by the nominate subspecies from its eastern breeding grounds around the Caspian Sea (Zink 1973). However, it is also possible that the birds are European migrants which undertake a change of direction, from their main south-southeasterly course to southwest, after reaching Israel, Egypt or Libya, or perhaps a partial westward dispersal while wintering, to travel along the Sahel axis after reaching its easternmost limit in Sudan and Ethiopia (see recoveries in Chad, Fig. 254).

Spring migration in the Middle East occurs further east than autumn migration (Zink 1973, Glutz von Blotzheim & Bauer 1991, Wernham et al. 2002), apparently circumventing the Mediterranean Sea via Egypt (mostly east of the Nile), Israel, Lebanon and Turkey. The number of recoveries in the Middle East in spring does indeed far outnumber those in autumn: February (8), March (49), April (9) and May (1), contrasting with September (3), October (2), November (1) and December (3). Recaptured Lesser Whitethroats in Jordan in spring 2002 had short stopovers (median of 2 days) and increased only on average 1.24% in weight per day. Slightly more than 50% of the recaptured birds were found to have increased in weight; most birds carried intermediate fat loads and presumably rested without replenishing fat reserves (Khoury 2004). The average date of passage at Eilat, Israel, from 1984 to 2003, was 2 April (N=13 691; Markovets & Yosef 2004). The timing of spring migration on Christiansø in the Baltic Sea, as indicated by standardised ringing, showed a non-significant advancement over 1976-1997 (Tøttrup et al. 2006).

Distribution in Africa

The main wintering grounds comprise of dry savanna and semi-open woodland at low altitudes in Chad, Sudan, Eritrea and northern and central Ethiopia, north of 8°N (Zink 1973, Urban et al. 1997). The EURING files contain only four recoveries south of the Sahara, three in southern Chad and an outlier in Guinea (Fig. 254). The Chad recoveries refer to birds from Sweden and Poland (mid-January) and Finland (9 April).

In the core wintering areas in eastern Africa, Lesser Whitethroats are abundant in woodland savanna. A study along the Nile, some 17 km south of Wadi Halfa (21°55'N, 31°20'E), from 10 February to 10 April 1964, revealed that the species was particularly abundant (on average 8.0-12.9 birds/day), along the desert's edge, where Tamarix bushes and grassland were mixed with Acacia trees in a 1 km transect containing five different habitat types. The species was absent in shoreline reedbeds, almost absent in pure desert (0.0-0.3 birds/day) and scarce in scrub (0.9-1.5 birds/day), but rather common in a garden with some palms, Acacia and Tamarix bushes (1.3-5.8 birds/day). Only 4 out of 167 Lesser Whitethroats were recaptured here at the ringing site, demonstrating the species' low site fidelity during winter. Intraspecific aggression was often recorded, especially so when birds were close together or reached a high density (Mathiasson 1971, S. Mathiasson pers. comm.). The species was particularly numerous in lowland savanna woodland along the Ethiopian border between the rivers Rahad and Dinder, but much less so at the same latitude near Asmara, Eritrea (Mathiasson 1971). In the Sahel between 14° and 17°N, west of the Nile, where the vegetation comprises a thin and patchy cover of short annual and perennial grasses and herbs, the scattered areas of thorny scrub along wadis and other drainage lines abounded with Lesser Whitethroats (Hogg et al. 1984). Capparis aphylla was especially favoured (Hogg et al. 1984), a thorny shrub whose seeds are unusually rich in oleic acid (Sen Gupta & Chakrabarty 1964).

Point counts at 16 sites in northern Nigeria between Alagarno (13°19'N, 13°33'E) and Nguru (12°52'N, 10°27'E) from October to December 2001 and November to December 2002 revealed that Lesser Whitethroat abundance was greatest at sites having low to intermediate Acacia densities and high tree density (Wilson & Cresswell 2006). The species occupied sites of taller Balanites trees than those frequented by Subalpine Warblers (which use more or less the same microhabitat; Wilson & Cresswell 2007), but was less abundant when Salvadora persica reached medium to high densities (unlike Subalpine Warblers and Common Whitethroats). Lesser Whitethroats depart in spring prior to the peak of Salvadora fruiting and so pre-migratory fattening presumably depends upon the abundance of invertebrates, which decline over autumn and winter in Acacia, but remain consistent in evergreen species like Balanites and Salvadora (Wilson & Cresswell 2006). In line with these findings, the studies in Sudan (Mathiasson 1971) and in northern Nigeria in December and January 1993/1994 (Jones et al. 1996) also found highest densities in non-degraded (or partly degraded) dense Sahelian woodland, but low densities in, or absence from, degraded sites, in areas of only low bush cover and in desert-like habitats. In Sudan, Lesser Whitethroats were observed foraging extensively on insects (once even a 5 cm-long dragonfly, 21 March), in addition to nectar-feeding in flowering Acacias; the last-named is also explicitly referred to by Hogg et al. (1984) near the Nile. In February and March, tamarind Tamarindus indica fruit was taken regularly in northern Sudan (Mathiasson 1971). Habitat loss and habitat degradation, especially in the northern Sahel, where birds fatten prior to spring departure, may force birds to depart from further south. If this is the case, it may herald a

cascade of negative consequences, such as widening of the Sahara barrier, increased mortality during migration, delayed arrival on the breeding grounds (for which there is no evidence; 7 out of 13 European studies showed the species had a tendency to arrive earlier in the late 20th century; in 3 studies, none of which showed any degree of significance, the opposite conclusion was drawn; Lehikoinen et al. 2006) and poorer condition of individuals on arrival at the breeding grounds (Wilson & Cresswell 2006).

Population change

Wintering conditions   Trends in Europe show wide variations (Fig. 255), from stable (Sweden) to significant increase (Finland, Czech Republic) and significant decline (Great Britain, Denmark, The Netherlands, South Germany). None of these trends showed a correlation with rainfall in the Sahel, except the one based on birds captured during autumn migration in South Germany (Fig. 255C). This lack of correlation does not change much when the trends are compared to the rainfall index of Ethiopia, except for the Swedish trend which becomes significant. [1]

When compared with Common Whitethroats, another *Sylvia*-warbler wintering for the most part in the Sahel, Lesser Whitethroats show the same trends in two German studies (Wangeroog and South Germany) and in Finland, but the trends of both species are the opposite to those in The Netherlands (Fig. 255). [2] Alignment of trends is to be expected if species' populations are regulated by common mechanisms.

Whether the crux of the mechanisms controlling their population dynamics lies in the Sahel or elsewhere remains difficult to detect from population trends, although the impact of rain in the wintering area is evident in the return rate of British Lesser Whitethroats, varying for adults between 6.7% in 1985 (after the driest Sahel winter of the 20th century) to about 20% in normal years (Fig. 256). [3] The return rate for juveniles is always low and barely related to the rainfall in the Sahel.

The study in northern Nigeria by Wilson & Cresswell (2006, 2007) seems to point to Lesser Whitethroats preferring to forage at low to intermediate heights in large trees occurring in high density and diversity, a winter habitat extensively overlapping with that of Subalpine Warblers (but information on food intake and interspecific competition is lacking). They appear to be less resilient to habitat degradation than Common Whitethroats, and so would be at greater risk from gradual degradation of woodland savanna in the Sahel (Chapter 5). There is scant evidence that plans to recoup lost or dwindling tree cover by planting exotic tree species or cultivars of higher productivity in terms of fruit and wood (Augusseau et al. 2006, Ouédraogo et al. 2006) will not be beneficial to Lesser Whitethroats, which depend on insects whose niche is restricted to indigenous trees.

Breeding conditions   Long-term studies of Lesser Whitethroat in the breeding quarters are scarce. The species prefers tall, dense scrub; vegetation succession is therefore likely to affect numbers at the local scale. Interspecific interference competition with Common Whitethroats, although reported by Cody (1985), is unlikely to have a significant impact on numbers, because these two species use the same type of habitat sequentially, rather than simultaneously. In fact, a study in Lincolnshire showed that the number of territories of both species was positively correlated (Boddy 1994), although the overall trend for Great Britain suggests differently (Fig. 255A).

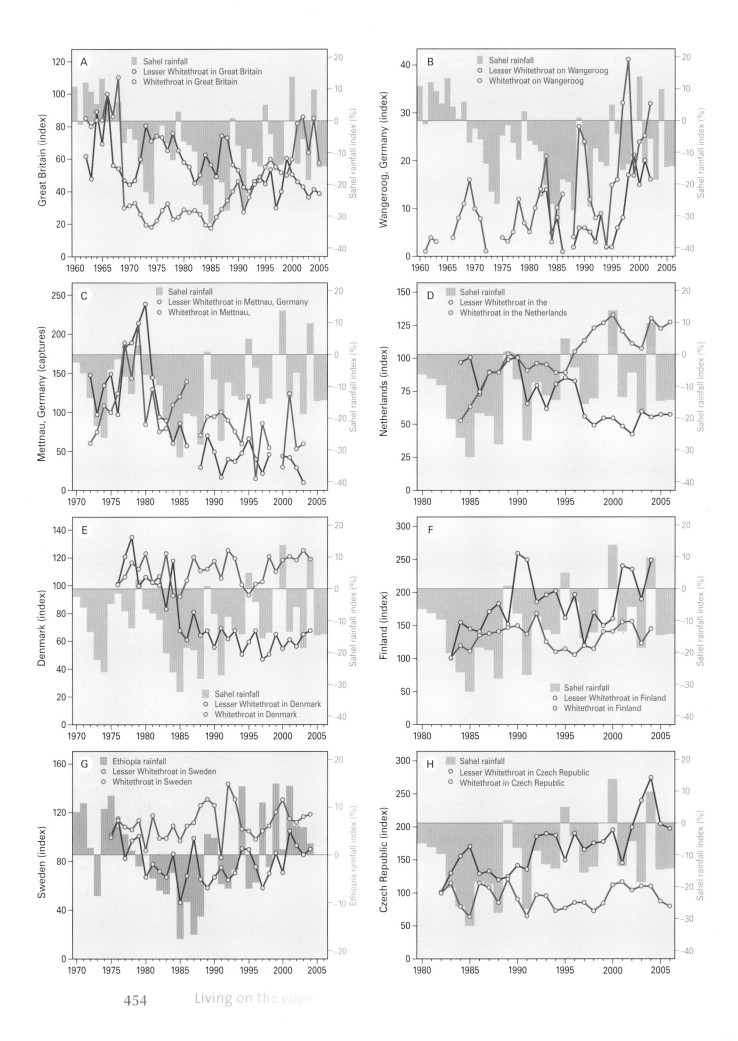

The Desert Date *Balanites aegyptiaca* (left) and Tamarind Tree *Tamarindus indica* (right) are tree species being preferred by Lesser Whitethroats. Note that the branches of the *Balanites* have been cut, either for fuel or as fodder for goats, sheep or cattle. (Inner Niger Delta, January/February 2006).

Conclusion

Lesser Whitethroats experienced distinct trend directions in various European countries, although these were not synchronised and some were even contradictory, which would imply that the conditions in the eastern Sahel, their main wintering ground, have no overriding impact on population size. Nevertheless, the only study on annual survival revealed a close relationship between rainfall in the eastern Sahel and return rate on the breeding grounds. The species' dependence upon insect life in the taller indigenous trees in the Sahel, and also during pre-migratory fattening in spring (unlike Common Whitethroats, which depart at a later date and then depend upon *Salvadora persica* fruit), may indicate that maladaptation to past changes have created a behavioural bottleneck. Much more research is necessary before we can understand the subtleties of the population dynamics of this extraordinary *Sylvia* species, the only one of its genus in Europe which migrates southeast.

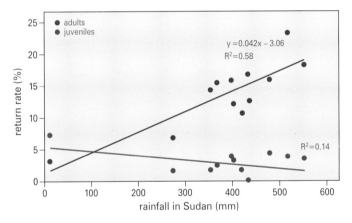

Fig. 256 Annual survival of adult and juvenile Lesser Whitethroats as a function of rainfall in Sudan, based on Boddy (1994) who used captures and recaptures to calculate the return rate of birds breeding in Lincolnshire between 1982 and 1992. The regression is highly significant (P<0.005) for adults, but non-significant for juveniles.

Fig. 255 (facing page) Population trends for Lesser Whitethroats in parts of seven European countries related to the population trends of Common Whitethroats and the Sahel rain index or Ethiopian rainfall (G). Note: Common Whitethroat numbers in South Germany have been scaled up by a factor of 5 to enable visible comparison with Lesser Whitethroats. From: (A) BTO (Great Britain), (B) Zang et al. 2005 (Germany, Wangeroog), (C) Berthold & Fiedler 2005 (South Germany), (D) SOVON (The Netherlands), (E) Heldbjerg 2005 (Denmark), (F) Väisänen 2005 (Finland), (G) Lindström & Svensson 2005 (Sweden), (H) Reif et al. 2006 (Czech Republic).

Endnotes

1. The rainfall index of the Sahel is highly correlated with the rainfall in Sudan (R=+0.89) and in Chad (R=+0.84), but less so with the rain in NW Ethiopia (Conway 2005; R=+0.51). Because the birds winter in all three countries, the population trends were regressed against the different rainfall indices. The fit did not improve in most studies when the indices were plotted against rainfall in Chad, Sudan or Ethiopia, instead of against the overall Sahel index.

2. The correlations between numbers of Lesser Whitethroat and Common Whitethroat vary between R=+.059 at Wangeroog (P<0.01, n=19), +0.56 in South Germany (P<0.001, N=30), +0.48 in Finland (P<0.02, N=22) and -0.74 in The Netherlands (P<0.001, N=23), being non-significant in the 4 other studies shown in Fig. 255.

3. The return rate was significantly correlated with the Sahel rainfall index (R=+0.64) and with the rainfall in Chad (R=+0.75) and NW Ethiopia (R=+0.56), but the correlation was even higher with the rainfall in Sudan (R=+0.87) (for rainfall data see Fig. 7). Although the rainfall in Sudan varied in 1982-1993 synchronously with the Sahel rainfall index (R=+0.92), the small difference is apparently large enough to have affected the annual survival of Lesser Whitethroats wintering in Sudan.

Common Whitethroat
Sylvia communis

When the impact of Sahelian drought on European birds made headlines in the late 1960s, the Whitethroat was its flagship. *Where have all the Whitethroats gone*, asked Derek Winstanley and co-authors rhetorically (1974) - a question that clearly had wider implications than just for Whitethroats and the Sahel. Was this a once-in-a-lifetime experience, a freak incident? On what scale did it happen (remember that monitoring was still in its infancy, no controls available), and how did the crash come about? Several decades and many gigabytes later, we know a lot more. For a start, we know that many long-distance migrants are in trouble, not only those breeding in Europe, but also – albeit partly for different reasons - many North American birds which winter in the Neotropics. The title of John Terborgh's stunning book *Where have all the birds gone?*, published in 1989, was an extended echo of the above-mentioned paper in Bird Study. His book was also another lesson learnt: with each generation of naturalists in the past century the time span of continent-wide changes has halved. Where it formerly took about a full lifetime to notice major environmental changes, in present times a few decades can suffice. For a Whitethroat struggling to find berry-bearing *Salvadora* bushes in the northern Sahel in April, the outcome may be the same: a paradise lost, with little hope of a paradise regained?

Breeding range

From the Arctic Circle to Morocco, and from Ireland eastwards to Central Siberia, Common Whitethroats live up to their name wherever open scrubland and farmland abound. In the 1990s, the European populaton was estimated at 14-25 million pairs, with particularly high densities from Denmark eastwards through eastern Europe, the Baltic and southern Finland into Belarus and Russia (BirdLife International 2004a). The entire population winters in sub-Saharan Africa.

Migration

Between arrival on the breeding grounds and the start of laying, an interval of 2 to 40 days may elapse, with delays occurring under the influence of adverse weather (and poor insect availability). Second layings, or any repeat layings for that matter, do not occur very often, and, when they do, they are usually by early starters (da Prato & da Prato 1983). Departure can be as early as late June, but continues into September in northern Europe and into October in southern Europe (Hall & Fransson 2001, da Prato & da Prato 1983). Late-departing Whitethroats from northern breeding grounds have little time to complete their moult, and instead interrupt the moult in either the primaries or the secondaries. In SE Sweden 77% of the birds showed interrupted moult (Hall & Fransson 2001), and in SE-Scotland all local breeding adults left the area before finishing moult (da Prato & da Prato 1983). Birds with incompleted moult may extend moult into the first leg of migration, as shown by the night migrants at Col de Bretolet in Switzerland (17% in moult, N=35; Schaub & Jenni 2000a). Within Europe, the proportion of migrating birds in moult decreases from north to south, and the observed mass increase towards the south is mainly attributed to the concordant decline in the proportion of moulting birds (Schaub & Jenni 2000a). In more southerly located breeding areas, the slightly earlier breeding schedule permits a full moult cycle before departure (Boddy 1992). Moult interruption facilitates an early start of autumn migration and enables the birds to cross the Sahara before the dry season in the Sahel has fully advanced. Despite leaving the breeding areas some three weeks after the juveniles, adult Whitethroats from eastern Fennoscandia and western Russia arrived on average significantly earlier at Malamfatori in northern Nigeria (Ottosson et al. 2001, Waldenström & Ottosson 2002, Ottosson et al. 2002).

The available data from ringing sites across Europe suggest that Whitethroats migrating south accumulate just enough reserves to hop from one stopover to the next; on Gotland, SE Sweden, mean flight ranges in autumn were estimated at 340 km, just sufficient to cross the Baltic Sea (250 km) in one flight to Poland (Ellgeren & Fransson 1992). The slight mass increase along the way is attributed to birds suspending moult (see above). In southern Europe or N Africa, the birds deposit sufficient energy reserves to cross the Sahara in a single stage (Schaub & Jenni 2000a).

In spring, Whitethroats in the Djoudj (northern Senegal) start pre-migratory fattening in March and especially in April, almost doubling their mass before departure (as in The Gambia, 300 km south of the Djoudj; Hjort et al. 1996). The accumulated fat enables flight ranges of 1500 km or more in still air. The heaviest birds may well fly across the Sahara and the Mediterranean without refuelling (Ottosson et al. 2001); Whitethroats grounded in inland Mauritania only gained 0.14 g per day (i.e 1% of body mass, N=24) in spring (Herremans 2003).

More to the east, at Malamfatori in northern Nigeria, Whitethroats generally follow the same strategy, but on average one month later. Birds arriving here in spring have low (often even decreasing) body masses, suggesting wintering areas further south, from where they move northwards towards the northern Sahel (see also Fry et al. 1970, for data from Zaria, about 250 km to the south, where Whitethroats had weights of 16 g at most). Fattening starts at the beginning of April, and intensifies from mid-April. Highest masses are reached in late April and early May, although a comparison of data obtained in 1967 (Fry et al. 1970) and 2000 (Ottosson et al. 2002) indicates that in recent years fattening takes place some three weeks earlier than

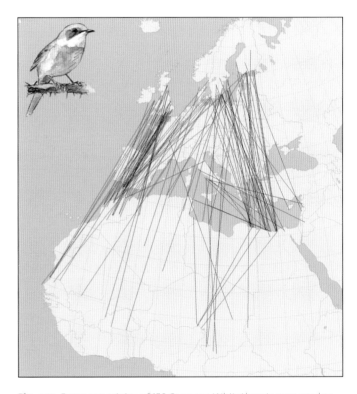

Fig. 257 European origins of 150 Common Whitethroats recovered or captured between 4° and 37°N. In addition, 49 birds were captured and recaptured on the same two sites in Niger. From: EURING, except 15 from Mullié et al. (1989), Meininger et al. (1994).

Many Palearctic passerines, like this Common Whitethroat, can be heard singing in their African wintering quarters, which is not always a sign of territoriality. The frequency of song increases prior to the return migration. Is this a measure of fitness, and if so, is singing less frequent in dry years? (Senegal, 28 February 2007).

in 1967, *i.e.* at about the same time as in The Gambia and Senegal (Ottosson *et al.* 2002).

Does this indicate that eastern European Whitethroats depart earlier these days (recoveries of Malamfatori-ringed birds came from Tunisia, Libya, Egypt and Poland; Ottosson *et al.* 2001)? An analysis of long-term phenological series from Europe did not list the Whitethroat as a species which had advanced its arrival date significantly during the late 20th century (only 2 out of a series of 15 were significantly negative, although another 6 were negative, 2 stable and 5 positive, of which 1 significantly; Lehikoinen *et al.* 2006). Many of these studies were from western Europe. From eastern Europe, where the birds wintering near Lake Chad are presumably heading, results are ambivalent, with no change recorded in phenology in western Poland in 1983-2003 (Tryjanowski *et al.* 2005; perhaps too short a series?) and a significantly earlier arrival in eastern Lithuania during 1971-2004 (Zalakevicius *et al.* 2006). This is therefore an unresolved question.

Alternatively, do the birds start the Sahara-crossing from a more southerly location than before? A comparison of the two data sets hints at such a possibility (Ottosson *et al.* 2002). In 1967, after 15 years of abundant rainfall (Chapter 2), it was noted that many birds departed before they had started fattening at Malamfatori, apparently intending to refuel north of this site. Although this was still the case to some extent in 2000, fattening conditions just north of Nigeria were thought to have been less favourable than during the BOU expedition in 1967. A departure from more southerly latitudes, thus leading to a longer Sahara-crossing, is therefore a more likely option.

Distribution in Africa

The Common Whitethroats breeding in Europe are distributed across the entire Sahel and adjacent Sudan zone during the northern winter, with a longitudinal gradient more or less similar to the distribution in Europe. Twelve Finnish birds were recovered in Sudan and the eastern Mediterranean. Swedish birds were captured or found during migration in Lebanon (1), Egypt (16), Libya (5), Tunisia (2), Algeria (2), Morocco (1), Spain (1) and Chad (2) (Fig. 257). The central Sahel, *i.e.* Mali, Burkina Faso, Niger, Chad and northern Nigeria, harbours birds from central and eastern Europe (Ottosson *et al.* 2001), and probably from northern Europe, given the general direction of migration within Europe (Bakken *et al.* 2006). In contrast, British birds were captured or recovered in Spain (8), Morocco (16), Senegal (5), but also 2 birds from Burkina Faso. Recoveries from regions south of the Sahara were largely confined to the period between November and April, in agreement with the main wintering period. The 132 birds recovered in the Sahara and north of it were on migration: 31 birds were from April and May, 76 from September and October (EURING).

Several in depth studies are available of Whitethroats in their wintering habitat in the Sahel and adjacent Sudan zone. It appears that the species is particularly common in dry shrub- and farmland with a variety of indigenous tree species and a profusion of small bushes of *Salvadora persica*. In the Djoudj National Park, northern Senegal, the low-lying wetter parts support a healthy population of *Salvadora*. In the absence of browsing domestic livestock, the proportion of small young bushes is high, and this is important because such bushes provide ripe fruit of the correct size (*i.e.* berries that can be swallowed) just prior to migration. Where livestock abounds, browsing reduces the availability of small bushes (Stoate & Moreby 1995). However, the main dietary component throughout the winter is insects, such as caterpillars (at Ngura in northern Nigeria, Jones *et al.* 1996), and ants, beetles, plant bugs, parasitic Hymenoptera, Diptera and Lepidoptera larvae (Djoudj in Senegal, Stoate & Moreby 1995). Body masses remain low throughout this period (Fry *et al.* 1970). In the Sahel zone of northern Nigeria, highest insect abundance in March and April 1995 was measured in mixed woodland, especially in *Balanites aegyptiaca* (Vickery *et al.* 1999). Distribution and abundance of Whitethroats was correlated with tree diversity, rather than tree density, foliage volume and canopy volume (but all these variables were interrelated). Degraded woodland and semi-desert held much lower densities of Palearctic migrants than dense Sahelian woodland, where Whitethroats were most abundant in sites rich in *Piliostigma reticulatum*, with up to 0.7 birds/ha (Jones *et al.* 1996). The latter tree, together with *Balanites aegyptiaca* and *Ziziphus* sp. had the highest densities of invertebrates (12.7-15.9 invertebrates per 0.5 m^3, excluding ants; Vickery *et al.* 1999). Whitethroat abundance in open farmland savanna on the border of The Gambia and Senegal near Fass (13°35'N, 16°27'W) showed a positive relationship with total shrub area, and with the area of the shrub species *Guiera senegalensis* (Stoate *et al.* 2001). The species was absent from farmland with low shrub cover (<399m^2/ha), but reached a density of 1.02 birds/ha on farmland with higher cover. Caterpillars and spiders were significantly more numerous in *Guiera senegalensis* than in any other tree species, but Miridae and Homoptera were more abundant in the non-indigenous Mango *Mangifera indica* and Cashew *Anacardium occidentale*, respectively.

Winter diet changes abruptly the moment Whitethroats start preparing for the flight back to the breeding quarters. For fattening, they need berries - lots of them, and of the correct size. In Senegal, *Salvadora persica* was the major berry-bearing species, comprising 79% of the shrub area. Total *Salvadora* berry abundance averaged 130/m^3; those exceeding a diameter of 4 mm occurred at a rate of 43/m^3. Whitethroats had a preference for the larger berries (mean 6.3 mm diameter), presumably because fruit volume increases dispro-

Whitethroats rely on berries of *Salvadora persica* when depositing fat reserves prior to migration. This shrub is known as the Toothbrush Tree, since its twigs are widely used by local people to clean their teeth.

portionally with diameter. They regurgitated approximately 70% of the seeds, an effective strategy to reduce the amount of indigestible material in the gut (Stoate & Moreby 1995). The importance of *Salvadora* for fattening of Whitethroats is also put forward for northern Nigeria, despite the fragmented occurrence of this tree species in the area (Wilson & Cresswell 2006). Boyi *et al.* (2007) specifically mention the fruit of *Canthium* and *Heranguina* spp. as the food of Whitethroats in the Amurum Forest in northern Nigeria prior to migration. During midwinter, the diet here consisted of insects which were gleaned from low bushes and trees; taking nectar of *Parkia biglobosa* was occasionally observed.

Droughts and human population growth have primed an avalanche of negative changes in the Sahel (Chapter 5), especially a decline in tree cover, reduced tree diversity, introduction of non-native tree species like Neem *Azadirachta indica* and *Eucalyptus* sp. (Box 3) and browsing of *Salvadora* and other berry-carrying shrubs by domestic livestock (leading to reduced fruiting and, eventually, removal). Whitethroats can survive in seriously degraded habitats in the

Year		1979-85	1988-89	1990-91	1994-95
Sahel rainfall index (%)		-17.4%	+1.0%	-27.1%	+5.0%
Väisänen 2005	Finland		5.7	-10.0	4.5
Hustings 1992	Finland	12.5	13.2	-12.3	
Kuresoo & Mänd 1991	Estonia		6.3		
Lindström & Svensson 2005	Sweden	-15.0	5.6	-34.4	-1.9
Pedersen 1998	Denmark	-52.9	12.8	-29.4	-17.1
Heldbjerg 2005	Denmark	-9.0	0.9	-11.1	-6.1
SOVON	The Netherlands		14.4	-11.0	6.3
BTO	UK, woodland	-42.2	21.9	-28.2	33.3
BTO	UK, farmland	-29.2	17.6	-41.3	17.0
www.mnhn.fr/mnhn/crbpo	France			-11.1	-11.4
Berthold & Fiedler 2005	South Germany	-44.2	35.7	5.2	100.0
Schmid *et al.* 2001	Switzerland		17.3	-10.0	24.1
Schmid *et al.* 2001	Switzerland, Genf				45.8
Hustings 1992	Czechoslovakia		33.3		
Reif *et al.* 2006	Czech Republic		43.5	-28.2	8.2
Average		-25.7	17.6	-18.5	16.9
SD		23.2	12.9	13.5	32.0

Table 29 Population change (%) after wet and dry episodes in the Sahel. The population in 1985 is compared to 1979 after six dry years; 1988 was the first year with above-average rainfall since 1979.

Whitethroats are most abundant in sites rich in *Piliostigma reticulatum* (facing page) and *Balanites aegyptiaca* (above).

Sahel (Wilson & Cresswell 2006), but only up to a certain point. For example, at Watucal Forest Reserve in northern Nigeria (12°52'N, 10°30'E), tree density decreased significantly by 82% between 1993-1994 and 2001-2002, notably of native species such as *Ziziphus* sp., *Cassia sieberiana*, *Acacia nilotica*, *A. senegal* and *A. seyal*. The number of Whitethroats counted per 10 points showed a concomitant decline from 5.3 to 0.4 (Cresswell *et al.* 2007).

Population change

Wintering conditions   For birds as dependent on dry scrub and woodland in the Sahel as Whitethroats, numbers are likely to fluctuate in synchrony with rainfall, the ultimate driver of vegetation, and, hence, of insect abundance and fruiting. We also expected a downward trend in breeding numbers due to the above-mentioned long-term changes in the Sahel. Most trend series were positively correlated with one another, especially those which show the highest migratory connectivity, such as the Dutch and British indices (western Sahel) and the Finnish and Swedish indices (central Sahel). [1] The steady decline in S Germany (based on standardised mist-netting in late summer and autumn at Mettnau) contrasts with stable (Denmark, Sweden, Finland, Czech Republic; Fig. 255E, F, G & H, Reif *et al.* 2006) and increasing trends (Great Britain since the 1970s, The Netherlands, Switzerland; Fig. 255A & D, Schmid *et al.* 2001). [2] The majority of the European monitoring schemes, if not all, are inadequate fully to comprehend the downfall of Whitethroat numbers in the 20th century. For that to have shown up, monitoring should have started in the 1950s; in fact, only the British series shows a glimpse of the heyday of Whitethroat fortunes, *i.e.* in the second half of the 1960s (Baillie *et al.* 2007). Numbers in the 1950s and 1960s were presumably very much higher than at any time since then. Observations near Wageningen, in the central Netherlands, since 1953, for example, averaged 51 per year between 1953 and 1958, dropped to an all-time low of 2 per annum between 1974 and 1978, then increased again to 10 per annum between 1984 and 1988 (still only 20% of the numbers in the 1950s; Leys *et al.* 1993). Recent trend data from Niedersachsen, northern Germany, show the same negative tendency when compared with censuses in the same regions in the 1950s and earlier (Zang *et al.* 2005); shorter time series starting in the 1970s and 1980s were almost invariably positive, presumably indicating (incomplete) recoveries from drought-related losses incurred in the Sahel between 1969 and the mid-1980s. Unfortunately, reliable quantitative data from the 1960s and earlier are as scarce as the proverbial hen's teeth (see also Common Redstart, Chapter 37).

Whitethroat trends on the breeding grounds were found to correlate with the Normalized Difference Vegetation Index (NDVI) and with rainfall in the Sahel in the preceding winter, but, except for the British and Dutch series, none significantly so. [3] The impact of rainfall was most obvious after a year or a period with particularly poor or abundant rainfall, usually resulting in declines and increases respectively (Table 29). Populations increased more strongly following high rainfall in the Sahel, but the effect was significant only in the British, Danish and Dutch populations (Table 30). Irrespective of rainfall, significant declines also occurred more often when population levels were high to begin with, or, conversely, increases occurred when population levels were at a low ebb, an indication of density-dependent adjustment to resources in the wintering areas, such as that also found in Black-crowned Night Herons (Chapter 17) and Sand Martins (Chapter 34). This is consistent with the British data, which suggest that fluctuations in Whitethroat populations result mainly from losses among full-grown birds, which are themselves correlated with Sahel rainfall (setting the carrying capacity of

Table 30 Population change between successive years ($\ln(pop_0/pop_{-1})$) as a function of rainfall in the Sahel (R) and the size of the population in the previous year (P); same data as in Fig. 255 and Table 29.

Country	Function	N	Population		Rain	
			R^2	p	R^2	p
Sweden	1.045 − 0.901P + 0.004R	28	.487	.000	.037	.089
Denmark	0.543 − 0.001P + 0.009R	18	.279	.011	.176	.012
Great Britain	0.459 − 0.007P + 0.013R	42	.310	.000	.066	.000
Switzerland	1.223 − 0.042P + 0.010R	14	.554	.002	.002	.144
Finland	0.651 − 0.466P + 0.002R	20	.360	.005	.017	.413
The Netherlands	0.417 − 0.003P + 0.003R	19	.498	.001	.003	.033
Czech Republic	0.832 − 0.008P + 0.006R	23	.311	.062	.168	.065
South Germany	1.045 − 0.051P + 0.000R	26	.370	.001	.008	.978

the wintering grounds; Baillie & Peach 1992). In Germany, the crash in 1969 was accompanied by lower than normal hatching and fledging rates, and, interestingly, in a high proportion of unpaired males. In 1973, only 4 out of 20 territorial males were paired (Conrad 1974), suggesting differential mortality among the sexes during droughts in the Sahel.

To check whether Whitethroats were in decline, independent of variations in rainfall, three periods were selected with rainfall of about 6% below the long-term average: 1969-1972, 1975-1982 and 1992-1998. By comparing population sizes in 1972, 1982 and 1998, and having equalised the effect of rainfall, the impact of habitat degradation in the Sahel in the intervening period might then be easier to detect. The expected negative trends were not found, however. In fact, the trends were inconsistent. For example, between 1972 and 1982, German and British populations went up by 25% and 8% respectively. In 1998, German and Czech populations were 37-38% lower than in 1982, but there was an increase of 10% in Sweden, 168% in Britain and 21% in Denmark (Fig. 255).

Breeding conditions Agricultural changes in Europe, including intensification, use of herbicides and pesticides and hedge removal, have seriously affected populations of farmland birds (Pain & Pienkowski 1997, Shrubb 2003, Newton 2004). Whitethroats captured during spring migration in Italy had very low concentrations of organochlorine residues in their breast muscles compared to those captured at Ottenby, Sweden. This discrepancy indicates that Swedish Whitethroats accumulated organochlorines during their passage through Europe, rather than in their African wintering areas (Persson 1974). Contamination of nestling Whitethroats with DDT, or its metabolites, and PCBs in Sweden (1965-1970) did not appear to affect breeding success negatively (Persson 1971).

Recovery from periodic droughts in the Sahel may be more difficult in depleted farmland populations. For example, the population recovery from the Sahel-caused crash in 1984 in an area of scrubland in Britain, which had not changed except for natural plant growth and succession, was very rapid: three times as many adults were captured in 1987 as in 1984, and the estimate of breeding pairs was 2.5 times higher (Boddy 1993). At the same time the British farmland population showed very little evidence of recovery (Fig. 255, Baillie et al. 2007). Some variation in trends within Europe following Sahel-induced declines may therefore partly mirror various stages of intensification of farming practices.

Conclusion

Conditions in the Sahel have a strong bearing on fluctuations in Whitethroat populations, especially via mortality of full-grown birds (probably with a clear bias against females). Droughts bring steep crashes, but recoveries are almost equally rapid after one or more successive wet years. Population regulation is density-dependent. The species is, to a certain extent, able to cope with habitat degradation in the Sahel. However, the scarce information on breeding numbers in the 1960s and earlier decades suggests that, despite recoveries from drought-induced declines between 1969 and the mid-1980s, present-day figures are still well below those in the first part of the 20th century. Habitat loss in European breeding quarters, notably in farmland, may lie at the heart of this decline, possibly increasingly exacerbated by similar losses in the Sahel.

Ziziphus trees (here *Z. spina-christi*, a buckthorn growing in West African inundation zones) attract many insects, either large (such as Banded Groundlings *Brachythemis leucosticta*, a common dragonfly in West Africa) or small. *Ziziphus* trees are therefore attractive foraging sites to Common Whitethroats, Common Chiffchaffs and other warblers (Inner Niger Delta, November 2008).

Acacia seyal is one of the Sahelian trees flowering in the late dry season (January-April). The sweet smell is overwhelming and the insect buzz can be heard from a distance. The conspicuous African Monarch *Danaus chrysippus* feeds on the nectar of *A. seyal* (near Timbuktu in February 2009). Nectar is an important food resource for *Sylvia* warblers, probably especially so after long-distance flights have depleted energy stores and reduced the digestive tract (Schwilch *et al.* 2001). Nectar is easy to absorb and the frequency with which the heads of *Sylvia* warblers are adorned with pollen traces testifies to the importance of this type of food. Forests of *A. seyal* attract many warblers, including Bonelli's and Subalpine Warblers, but in the *seyal* forests of the Inner Niger Delta we saw only a few Common Whitethroats.

Endnotes

1 The correlation between the Dutch and British series is R=+0.77 (N=22, P<0.001) and between the Finnish and Swedish indices R=+0.64 (N=22, P<0.01).

2 The correlation between year and the index for S Germany: R=-0.46, N= 30, P<.0.01, and for other countries: The Netherlands +0.87, Sweden: +0.65, Denmark: +0.30, Finland: +0.25 and Great Britain: -0.03.

3 When the 24 trend series available for Common Whitethroats were plotted against rainfall and NDVI (after a selection was made for the data from 1982 onwards), the correlation with rainfall was 8 times larger than the correlation with NDVI, but 16 times it was the other way round. Hence NDVI seems to be a better predictor of the Sahel circumstances relevant to Common Whitethroats. Nevertheless we used the rainfall data, since the differences are small and monitoring in several countries started before 1982. The correlation between the indices and rainfall is highly significant (P < 0.001) in Great Britain (R=+0.63) and The Netherlands (R=+0.61), but non-significant in the other countries: Czech Republic (R=+0.37), Sweden (R=+0.33), Finland (R=+0.31), Denmark (R=0.17) and S Germany (R=0.03).

Crossing the desert

Moreau (1972) described the Sahara desert for migrant birds as "a tremendous all-or-nothing hazard with practically no food, no water, no shade, in fact no chance of rehabilitating rest or of turning back. (...). In such circumstances for a bird to come to ground is worse than useless; it has no hope of food and its water is unduly depleted by the greater heat at the surface of the ground, even though it seeks shade". His conclusion was that birds have to cross the Sahara in one go, the advantage of a quick and non-stop crossing seemingly overwhelming. Neither systematic radar observations nor satellite-tracked birds were as yet available to substantiate his claim. However, Moreau underpinned the validity of his idea with compelling circumstantial evidence. First, if Wheatears are able to fly without rest from Greenland to Europe (in fact, as we now know, directly to Africa), why should they not be able to cover a similar distance across the Sahara? Secondly, birds crossing the Sahara have stored sufficient body reserves to fly for 40 hours and to cover the required 1500-2000 km without refuelling. Thirdly, the oases scattered across the Sahara harbour only small numbers of migrants, compared to the huge numbers crossing the desert (estimated by Moreau at 5000 million in autumn and half this number in spring). A watertight case, therefore? Put to the test, and using increasingly sophisticated techniques, the facts show a more complex picture.

Non-stop or intermittent flight?

Accounts of emaciated, dead and dying birds in the Sahara profusely litter the ornithological literature (for example, Zedlitz 1910, Geyr von Schweppenburg 1917, Moreau 1961, Smith 1968, Haas & Beck 1979). The Sahara is a formidable ecological barrier. Even from a height of 10 km, looking down from an airplane's window, it does not take the perceptive observer much imagination to understand that only well prepared birds have a chance of surviving such a hazardous environment. How do birds cross this vast expanse of sand and rock?

The available evidence points to a number of strategies, and more are likely to come to light in the near future. Lesser Black-backed Gulls, for example, wintering along the Atlantic coast of Senegal and thought to follow coast line or rivers between breeding and wintering sites, were found 500 km inland in Mauritania, following a straight SW-NE flight and apparently flying non-stop to the Mediterranean. Flocks of gulls, followed by the radar "Superfledermaus", flew at an altitude of 3.5 km, on average, where favourable tailwinds enabled them to reach a speed of 90 km/h (Schmaljohann et al. 2008). We assume that Garganey, after having first gorged on water lily seeds in floodplains in the Sahel, also fly non-stop across the desert when heading back to their breeding quarters. The Garganey ringed in Senegal and shot five days later in Italy (Roux & Morel 1973) must have covered this distance in 3.5 days when flying for 24 h a day (Box 20), probably nothing out of the ordinary for a Garganey. Ospreys use another strategy: they interrupt their desert crossing at night for roosting, but otherwise utilise the entire time between 09:00 and 17:00 to fly (contrary to their flight in Europe, which is interrupted frequently for opportunistic feeding bouts; Strandberg & Alerstam 2007). Satellite-tracked Ospreys covered on average about 220 km on travelling days in the Sahara (Klaassen et al. 2008). They did not substantially increase their speed while crossing the Sahara (only 14% higher in the Sahara compared to Europe, for high-altitude flights of >100 m), and it took them at least four or five days to complete the desert crossing, spending the night wherever the daily flight had taken them (Alerstam et al. 2006, Klaassen et al. 2008, Dennis 2008).

Long before the advent of satellite-tracking, Biebach (1986) discovered that many passerines crossing the Sahara come to ground, with fat birds staying in the shade (if present) and lean birds searching for food (if available). The birds were not exhausted, but had interrupted their migration during the day to resume their flight at night. Carmi et al. (1992) hypothesised that in small bird species (<23 g) dehydration, rather than migratory fat, limited flight duration; water loss whilst crossing the Sahara would prevent non-stop flights exceeding 30-40 h. Dehydration would be minimised by flying during the (cooler) night and by resting during the day. Following the same reasoning, they predicted that small birds would not fly at altitudes higher than 1000 m. An extensive study with a tracking radar did indeed show that most songbirds crossing the Sahara

The Sahara desert in Mauritania, the scene where intensive radar studies by the Swiss Ornithological Institute in the early 2000s unravelled several questions regarding desert crossings by songbirds (Schmaljohann et al. 2006). Contrary to earlier belief, passerines often use an intermittent flight strategy, resting by day (when even small bushes and scrub, as shown here, are important for their shade), and flying by night.

are grounded by day and fly during the night (Schmaljohann et al. 2007). But contrary to the prediction of Carmi et al. (1992), 90% of the spring migration birds were tracked at 1000 m above the Sahara, 50% even exceeding an altitude of 2500 m (Liechti & Schmaljohann 2007). By doing so, they avoided head winds prevalent at ground-level (the steady *harmattan*, coming from the NE) and profited from tailwinds at higher altitudes. This confirms that wind profiles, and thus energy expenditure, govern the altitudinal distribution of nocturnal migrants (Liechti et al. 2000). Water constraint contributes to the selection of flight altitudes, as presumed by Carmi et al. (1992) and confirmed during spring migration in Mauritania in 2003; the effect of water constraint increases with increasing temperatures during the season (Liechti & Schmaljohann 2007). Earlier, Biebach (1992) had already suggested from fat reserves in Willow Warblers and Garden Warblers measured prior to migration, that a successful crossing of the Sahara was only possible with tailwinds.

Whether birds interrupt their flight over the Sahara (warblers, Ospreys) or not (gulls, ducks, waders), body reserves are needed to cross the desert without refuelling. In February-April, when birds wintering south of the Sahara prepare for the return migration, the scarce, and rapidly dwindling, resources in the Sahel offer the last refuelling option. When conditions in the Sahel are drier than usual, gaining weight prior to migration is difficult. In such circumstances there is an even higher on-the-spot mortality (for example Garganey and Ruff; Chapters 23 and 29) and a reduced survival during spring migration (White Storks and Yellow Wagtails; Chapters 20 and 36).

Many other Palearctic migrants face the same predicament, albeit to varying degrees. However, without body reserves crossing the desert – by night or by day – is folly. But are the odds worse in spring, when conditions for pre-migratory fattening south of the Sahara are particularly poor after many months without rain?

Seasonal and geographical variation in mortality

To investigate seasonal variations in mortality, we analysed ring recoveries (ringed birds found dead only) for three regions in northern Africa: (1) the area north of the Atlas mountains in Morocco and Algeria where the annual rainfall amounts to 300-600 mm, (2) the Sahara desert south of these mountains up to 14°N and (3) the sub-Sahara (see inset map in Fig. 258). As expected, the number of ringed birds found dead in the Sahara peaked in the migration periods of September-October and April-May, except for Blackcaps and Common Chiffchaffs, which mostly winter north of the Sahara and, if wintering in the sub-Sahara, are the first to return to the breeding quarters (February) (Fig. 258). As far as the other species were concerned, with the exception of Willow Warblers and Sedge Warblers, ringed birds found dead in the Sahara in January-June outnumbered those in July-December (Fig. 259). This difference is the more remarkable given the mortality during the northern winter; numbers during the outward migration must have been at least twice as high as during the homeward journey.

For most passerine species, fewer ringed birds are found dead in the sub-Saharan wintering grounds than in the Sahara (Fig. 259); this is most pronounced in Pied Flycatchers (9 and 162 birds respectively) and Common Redstarts (2 and 99 respectively). Taking into account the time spent in the sub-Saharan area and the Sahara proper, *i.e.* 6-7 months (October-March/April) and twice a week at most, we would expect about twenty times fewer dead ringed birds from the Sahara if daily survival rates and reporting rates were random. However, reporting rates for the unpopulated Sahara desert should be much lower than for the populated Sahel-Sudan zone, thus biassing the reporting rate even further in favour of the sub-Saharan area. In fact, it is the other way around, suggesting high mortality among migrants crossing the Sahara desert.

Formidable as the Sahara is, and representing as it does a serious drain on body reserves, we expected the number of ringed dead birds to increase with the distance flown, *i.e.* from north to south in autumn and *vice versa* in spring. Most birds are indeed from the northern Sahara in spring, but – contrary to our prediction – the ma-

Purple and Grey Herons, Sand Martin and Glossy Ibis, found dead by Wilfried Haas during one of his trips in spring across the Algerian Sahara in 1966-1977. Wherever shade can be found, such as underneath a piece of scrap metal, there is a good chance of finding a hidden bird, whether it is dead, dying or biding its time during the hottest part of the day.

jority of the birds in autumn are also from the northern half of the Sahara (Fig. 260). This counterintuitive finding can be explained by the higher density of the human population (Fig. 35B), the higher education level and the higher level of prosperity in the northern Sahara in comparison with countries in the sub-Saharan area, all of which contribute to higher reporting rates in the northern Sahara.

For the higher reporting rate in the Sahara in spring, several explanations spring to mind. First, the trade wind from the north or

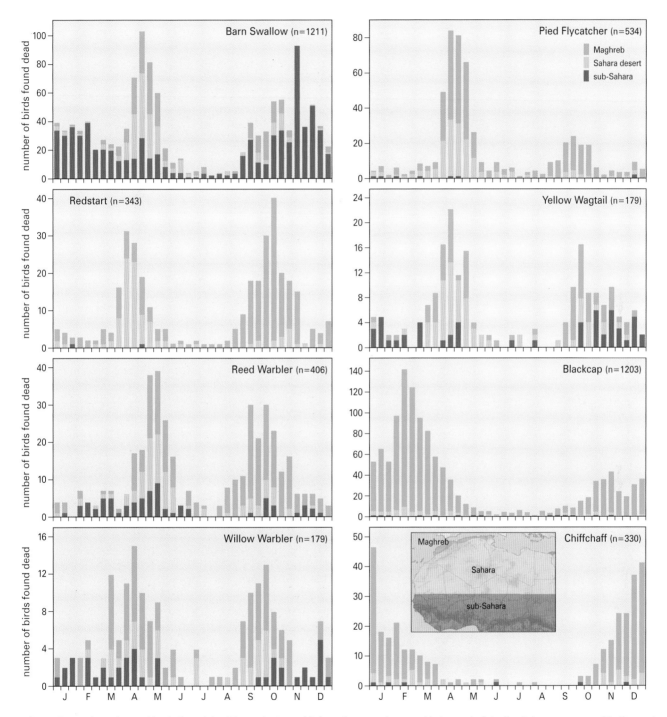

Fig. 258 Number of ringed birds found dead in a selection of Palearctic passerines per 10-day periods in the Sahara proper, and in the regions north (Maghreb) and south of the desert ('sub-Sahara'). From: EURING.

In the Sahara grounded migratory birds, such as Woodchat Shrike, Northern Wheatear, Common Quail and Stone Curlew, are desperately in search of shade and protection from the wind. When neither can be found at the same time, an on-the-spot decision is made as to whichever is most needed (shade during the hottest part of the day, wind protection in the early morning). In the bare desert, any obstacle is used, from road markers, such as drums and concrete blocks, to cars that have stopped for a while.

northeast constitutes a steady and reliable tailwind for migratory birds in autumn, but this same wind is head-on in spring and reaches its maximum speed in March and April, coinciding with the period of peak migration. Migrants may escape the rigours of a strong headwind by flying at altitudes of 1500-2000 m above ground level or more, *i.e* above the wind shear, where the wind direction is reversed and favourable tailwinds occur frequently (Liechti & Schmaljohann 2007). However, getting up there is not without cost. For a Willow Warbler, weighing 10 g and climbing at one metre per second, it takes about half an hour (and a corresponding flight distance of 7 km) to reach an altitude of 2000 m. A tailwind of 1 m/s would compensate for this energetic cost within a flight of about two hours. Because the energetic costs of climbing increase with size, for larger birds the difference in wind assistance between altitudes should be higher if climbing to altitudes with more favourable winds is to be worthwhile (Liechti *et al.* 2000). Radar technology today can distinguish between waders and waterfowl (continuous flapping), songbirds (regular intermittent flapping), swifts (intermittent flapping with irregular long flapping and pause phases) and swallows (irregular flapping) (Schmaljohann *et al.* 2007b). This is still a far cry from species-specific identification, and hence from studying differential mortality among migratory birds associated with flying height during spring migration. Data on visible mortality in the Sahara in spring seems to indicate that some species, such as Turtle Doves and Barn Swallows (both of which are known to fly at low altitudes during spring migration), are found dead much more frequently than other, equally common, species (Haas & Beck 1979).

Secondly, the conditions for pre-migratory fattening in autumn and spring are widely different. Migrants heading for the sub-Saharan zone find plenty of feeding opportunities in Europe and – if the jump across the Mediterranean and the Sahara falls short – in North Africa. Insect availability in Europe may be reduced when the weather is poor (*i.e.* in periods of low temperatures and high rainfall), but weather conditions here are hardly ever consistently adverse over long periods of time (although, when they are, they may cause mass mortality; Newton 2007). Variations in fattening rate are therefore low and mostly explained by body mass *sec* (Schaub & Jenni 2000a & b). For Palearctic birds wintering in the sub-Saharan area, conditions in spring are radically different from those encountered during autumn migration. Since their arrival in the wintering quarters, from August to November, little if any rain has fallen in the Sahel. The local vegetation has withered and insect life is at a low ebb. Palearctic passerines preparing for the return flight encounter the harshest conditions in the Sahel when fattening is of highest priority, unless the birds change their diet from insects to fruits. The abundance of lipid-rich fruits in spring, for example *Salvadora persica* and *Canthium* and *Heranguina* sp., favours frugivory and plays a key role in fattening strategies. Eating low-fat diets, such as fruit, is associacted with a significant increase in fat assimilation efficiency in Garden Warblers (Bairlein 2002). However, not all migratory species are seasonal frugivores. Instead, for the accumulation of body reserves, non-frugivores depend on seriously depleted stocks of seeds or insects. Frugivores face a similar bottleneck of food shortage when droughts and over-grazing reduce the availability of berries. Climate change and human-caused habitat change have a longer-lasting, and possibly irreversible, impact on food resources,

a cascading effect annually culminating in spring when the food demand of migratory birds is highest. We therefore expect more dead ringed birds from the Sahara in spring than in autumn. The number of spring recoveries of dead birds should be highest when rainfall has been poor in the preceding rainy period.

The hazards of migration take a heavy toll, especially among juveniles. In Barn Swallows, about 80% of all mortality, as reflected in ring recoveries (Fig. 260), concerns juveniles (Fig. 261). Given the same fraction of juveniles in the population at that time of year, juvenile mortality still does not differ from random mortality among age classes. During spring migration, when the proportion of juveniles in the population has dropped considerably, mortality among juveniles crossing the Sahara is disproprotionally heavy (up to 50-60% of all recoveries; Fig. 261). Differential mortality among age classes is well-known in Barn Swallows, with lowest survival rates found among first-year birds (*c.* 27%) and among five-year old birds and older (*c.* 17%); survival increased up to and including the third year of life, then declined (Møller & de Lope 1999). The heavy toll taken by juveniles during spring migration probably also relates to their timing, *i.e.* later than adults. The trade winds across the Sahara increase in strength during spring, and low-flying birds like Barn Swallows are likely to encounter stronger head-winds in April and May than in March. In many long-distance migratory passerines, adults arrive earlier on the breeding sites in spring than juveniles (overview in Newton 2008), as found in Barn Swallows in northern Italy, where the difference amounted to more than three weeks (Saino *et al.* 2004).

Annual variation in mortality

For birds flying close to the ground during spring migration, the *harmattan* is a real adversary whose negative impact, revealed in an increase in the number of dead ringed birds, would be highest in years with many days of strong head-on winds. To estimate the annual mortality during migration, it is necessary to correct for the size of the ringed population. For instance, the number of dead ringed birds amongst Common Redstarts declined between 1961 and 2005, in synchrony with the collapse of the population (Chapter 37). Several other species showed an increase in the ringed totals, for example, Barn Swallows; in their case, the ringing effort was hugely expanded by the instigation of the EURING Swallow Project

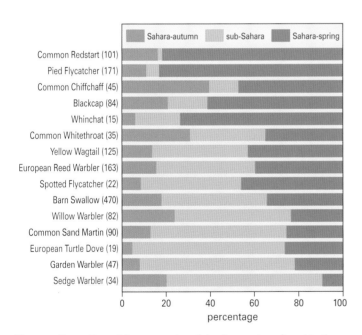

Fig. 259 Proportion of ring recoveries of dead passerines found in the Sahara during spring (n=666) and autumn (n=262), and in the sub-Saharan region during the northern winter (n=575). Same data as Fig. 258.

Table 31 The % ringed birds found dead in the Sahara in January-June relative to the numbers found dead between 4°N and 36°N and between 1 July and the preceding 1 July; total n is given between brackets. Calculated for the 9 (or 20%) most dry and most wet (preceding) years between 1961 and 2005 (nine driest years: 1972-1973, 1982-1984, 1986-87, 1990, 2003; nine wettest years: 1961-1965, 1967, 1994, 1999, 2003).

Species	dry (%)	wet (%)
Barn Swallow (n=433)	33.5	8.3
Yellow Wagtail (n=144)	45.0	15.4
Common Redstart (n=258)	26.7	22.5
European Reed Warbler (n=325)	24.8	2.4
Blackcap (n=506)	10.8	5.6
Common Chiffchaff (n=222)	15.5	0.0
Willow Warbler (n=203)	6.7	4.0
Pied Flycatcher (n=194)	47.2	14.3

(>500 000 ringed in Europe in 1997-2000; Spina 2001) and targeted ringing at roosts in sub-Saharan Africa (Chapter 35). To correct for this bias, we added up the annual number of recoveries of dead birds between 1 July and 1 July of the preceding year in southern Europe and northern Africa, between 4°N and 36°N; from this figure, we calculated the percentage of birds found in the Sahara between 1 January and 1 July.

The eight species with the highest number of recoveries showed a negative relationship between rainfall in the Sahel and mortality in the Sahara during spring migration (as reflected in the number of ringed birds found dead). Relative mortality during spring migration was on average four times higher in dry years than in wet years (Table 31).

This outcome suggests that the ecological conditions in the Sahel, as determined by rainfall in the preceding rainy period, have an impact on mortality while crossing the Sahara desert during the return migration. An alternative explanation might be that the force of the *harmattan* wind varies between years and that this has a direct bearing on mortality among migratory birds. To test the latter assumption, average monthly wind speed data were extracted from the FAO-database (Chapter 2) for the period between 1961 and 1986 for 27 meteorological stations in the southern Sahara and in the Sahel (15°-26°N); unfortunately, recent data were not available. The average wind speed, calculated for these stations, varied in April and May between 1.58 and 3.41 m/s. The wind forces in April and May are highly correlated within years [1], suggesting that wind speed does indeed vary from year to year. On average, wind speed was low in the Sahel and high in the southern Sahara. The highest wind speeds were always measured along the Atlantic coast of Mauritania in spring - up to, on average, 10m/s in May.

Variations in the average wind speed in April or May were not reflected in similar changes in the percentage of recoveries of the seven bird species in Table 31. [2] Given also the lack of a relationship between rainfall in the Sahel and wind speed in spring in the Sahara [3], the discrepancy between the number of birds found dead in dry and wet years (Table 31) cannot be attributed to the prevailing wind speed in April or May. Between-year variations in the strength of the *harmattan* fail to explain annual variations in recovery rates among migratory passerines in the Sahara.

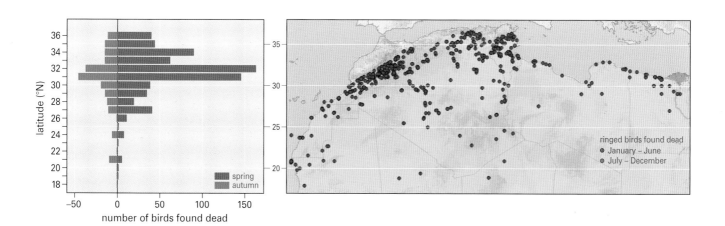

Fig. 260 Recoveries of passerines (dead birds only) from the Sahara (indicated yellow on the map) during spring (n=715) and autumn (n=229). The graph shows the numbers split per latitude. Same data as Fig. 258.

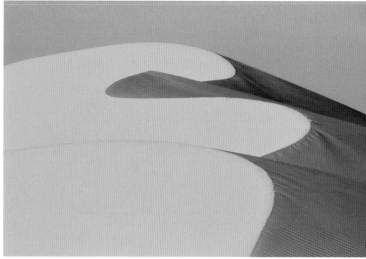

No food, no water, hardly any shade... The southern edge of the Sahara desert in northern Niger, October 2005 (left) and sand dune (with Greater Hoopoe Lark) in Mauritania (right).

Conclusion

The Sahara is an ecological barrier for migratory birds, which they nevertheless have to face twice a year. It is most likely that many more songbirds die during the few weeks of crossing of the Sahara than during the six months they spend in the sub-Saharan region. More birds are found dead in the Sahara in spring than in autumn, despite the fact that mortality has eliminated a substantial part of the total population in the intervening period. Annual mortality during spring migration is not related to the force of the prevailing headwind above the Sahara, but rather to the amount of rainfall in the Sahel in the preceding six months. Storing the required body reserves to be able to cross the desert in spring is apparently difficult, and often insufficient at times when the Sahel is drought-stricken.

Endnotes

1. Wind force in April and May are highly correlated: R=+0.93, n=26, P<0.001
2. The possible impact of the wind force on the number of birds found dead was analysed in different ways, directly and within subcategories with similar rainfall (*i.e.* categories as shown in Table 31); none revealed any effect of wind force.
3. Average wind force in April+May and Sahel rainfall in preceding year are not correlated: R=-0.08, n=26, N.S.

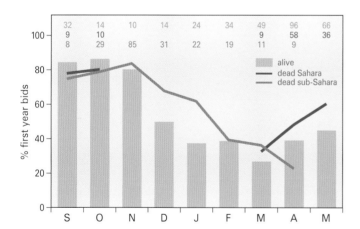

Fig. 261 The monthly percentage of first-year Barn Swallows found dead in the Sahara or sub-Saharan region compared to the age distribution of live retraps in Africa north of 4°N; total number of birds with known age per month is given (row 1: alive; row 2: dead Sahara; row 3: dead sub-Sahara). No value given if number of cases was less than 4. Same data as Fig. 258.

Carry-over effects of Sahel drought on reproduction

Ospreys in Scotland experienced a poor breeding season in 2007. Several nests remained unoccupied because adults arrived very late or did not return at all. Roy Dennis, who had been following some of his birds equipped with satellite transmitters on an hourly basis, knew why: "It was a very difficult spring migration" due to "very poor weather in April in northern Africa and Spain" (www.roydennis.org). In some years White Storks also arrive too late in Europe to start breeding; these years, known in the literature as *Störungsjahre* (disaster years), coincide with dry conditions in the Sahel (Chapter 20). When rainfall in the Sahel has been above the long-term average, only some 20% of the returning White Stork pairs in Schleswig-Holstein in northern Germany do not breed, but about 50% fall into this category after years with poor rainfall in the Sahel. Conditions in Africa not only can impact survival during the wintering period (Chapter 15-41) and mortality during the homeward flight (Chapter 42), but also reproduction in the following breeding season. The medium-term consequences of conditions experienced in Africa, or carry-over effects, can, when properly looked for, probably be found among many Palearctic migratory species, shaping their lives long after the actual moulding events have occurred.

A chain of causal relationships

In the migratory European population of White Storks, the proportion of unsuccessful pairs (but not the number of fledglings per successful pair) depends on Sahel rainfall (Fig. 148). In the well-studied Squacco Herons in the Camargue, annual variations in clutch size were correlated with the flood extent of the Inner Niger Delta, the species' main wintering area (Fig. 137), and independent of local factors (Hafner et al. 2001). A third example is the Barn Swallow, studied by Karl-Heinz Loske (1989) in western Germany. Inclement local weather in late May (hatching time) and June (nestling stage), notably low temperature and prolific rainfall, were found to influence hatching rate and nestling mortality in first broods (Loske 1994), but annual variations in breeding success did show a good correlation with rainfall in the Sahel (Fig. 262). Though not a wintering site for Barn Swallows, the Sahel and Sahara have a large impact on Barn Swallows during their northbound migration (Chapter 42). Anders Pape Møller (1989), in his study of Danish Barn Swallows, also suggested carry-over effects of precipitation in South Africa on mortality and perhaps reproductive output (Chapter 35).

Each study is suggestive in itself, but not strong enough evidence for proving a causal relationship between breeding performance in Europe and conditions in the Sahel during the preceding months. Several factors, some in concert, provide additional circumstantial evidence for wintering conditions in the Sahel to have carry-over effects on reproductive performance in subsequent breeding seasons.

1. Body condition in winter The Sahel is not a sufficiently stable environment for *a priori* guarantees of food for migrants throughout the northern winter. Indeed, many Palearctic migrants are prisoners of their own food supply, although we found some evidence that Purple Herons, White Stork, Black-tailed Godwits and Ruff may move from Sahelian floodplain to wetlands further south if necessary (Chapters 16, 20, 27 and 29). A similar strategy has been suggested for Montagu's Harriers (albeit for hundreds of km only, not moving farther than the northern Sudan zone; Chapter 26), passerines and intra-African migrants shifting from the Sahel to the Sudan or Guinea zones during the dry season (Jones 1996). Of migrants staying in the Sahel throughout the northern winter, many die from starvation in dry years; of the survivors, many will be in poor condition by the time of departure towards the breeding grounds, and so are prone to enhanced mortality rates during migration or are liable to be in poor condition on arrival on the breeding grounds. Van den Brink et al. (1997) did indeed find that Barn Swallows wintering near the Okavango Delta (Botswana) were very lean in a dry year but fatter in a wet year (Fig. 240). In a dry year it took the average Barn Swallow an extra 35-60 days to complete the moult of its primaries (van den Brink et al. 2000; Chapter 35).

2. Rate of fattening When impoverished feeding circumstances during droughts in the Sahel increase mortality, it may be out of the question for birds to achieve pre-migratory fattening. In such circumstances, the crossing of the Sahara takes a heavy toll (*e.g.* Fry et al. 1970, Dowsett & Fry 1971). The fattening up process requires increasing daily food consumption. For example, Whimbrels in Mauritania increased their daily food intake by 50% for about a month, thus depositing body reserves equivalent to 35-40% of their body mass (Zwarts 1990). The conditions in Mali permit Black-tailed Godwits to raise their daily food intake by only 1% per day in February; it takes about a month before they are ready to depart on their return flight to the breeding grounds (Box 7). A similar slow rate of

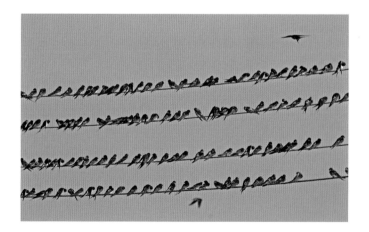

Fig. 262 Breeding success of first and second broods of Barn Swallows in the Soest district, Germany (1977-1987), in relation to the Sahel rainfall. Both regressions are significant (1st brood: p<0.001, 2nd brood: p=0.03). Data from: Loske (1989).

Box 26
Do Black-tailed Godwits return earlier from the wintering grounds in wet Sahel years?

Black-tailed Godwits wintering in Guinea-Bissau feed on rice. The feeding conditions are ideal during and after the harvest in December, when an abundance of culms and ripe rice covers the ground. However, in contrast to Ruff, Black-tailed Godwits rarely forage on rice grains on dry ground. Around harvest time, they tend to concentrate on the remaining wet rice fields. Most Black-tailed Godwits have left Guinea-Bissau by the end of January, but the timing of departure probably varies in conjunction with local rainfall, which in its turn influences foraging conditions in the ricefields. There are two observations of departing flocks in Guinea-Bissau in 2006 (3 and 8 January; J. van der Kamp, unpubl.), and for some years we know the arrival date in Portugal, a stopover on their way to The Netherlands (Fig. 263). In 1992, for example, 20 000 bird had already arrived in Portugal by the turn of the year, but in 1991 and 1993 we noticed hardly any Godwits before 10 January. The ricefields in Guinea-Bissau were relatively wet in 1992, but dry in the two other years. In the still drier year of 1984, the majority of birds had not arrived in Portugal by late January. These observations suggest that poor wintering conditions result in late arrivals in Portugal; early arrivals coincide with wet conditions on West African ricefields, but what we do not know is, whether the birds leave Guinea-Bissau later in dry years or spend additional time in staging areas between Guinea-Bissau and Portugal (perhaps in Morocco).

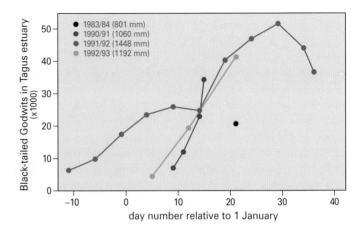

Fig. 263 Number of Black-tailed Godwits feeding in rice fields in the Tagus estuary (Portugal), based on counts of birds flying from the nocturnal roost to the feeding areas. The legend gives the annual rainfall in Guinea-Bissau in the preceding year. Data from: A-M. Blomert, R.G. Bijlsma, L. Zwarts (unpubl.).

Facing page: Feeding on the sheltered Portuguese rice fields is not without danger; predators like Peregrines, Booted Eagles and man take their toll and may temporarily disrupt daily routines. (Samora, January 2004).

weight gain has been found for Ruff (Fig. 215) and Garganey (Box 20, page 284), and is indicative of the generally poorer refuelling conditions in the wintering areas than in the breeding grounds (Box 22, page 335). Data on inter-annual variations in rate of fattening are lacking, but there is no doubt that migratory fattening is impossible under extreme drought conditions, as shown by Ruff in March 1985 (Fig. 215).

3. Date of departure Poor feeding conditions may influence departure in several ways. The birds may attempt to start fattening up earlier in order to leave the area as soon as possible, but it is more likely that they cannot find enough food to gain sufficient fat and may attempt to fatten up later, at a slower rate or a combination of both. The chances are that departure weight will also be lower than normal, making a successful crossing of the Sahara less likely. Chapter 42 gives some evidence that fewer birds are able to cross the Sahara after they have experienced a dry Sahel winter. A slowed rate of fattening in dry years may be countered by postponing the date of departure. Individuals of many large species, such as White and Black Storks, Ospreys and Lesser Spotted Eagles, have been fitted with satellite transmitters, which offer splendid opportunities to compare timing and speed of migration in relation to weather conditions in the wintering quarters and along the migration routes. For Lesser Spotted Eagle, Meyburg et al. (2007) have suggested an increasingly delayed departure since 2000 from its southern African wintering grounds due to drought and habitat destruction. In turn, this resulted in late arrival on the breeding grounds and failure to breed. (For the alternative strategy, of remaining in Africa when droughts prevail, we were unable to find conclusive evidence among ducks and waders, the groups for which data are available on numbers in Africa in the northern-hemisphere's summer.)

4. Replenishment en route Waders and ducks wintering in the sub-Saharan areas and breeding in N or NE Europe, and particularly those breeding in Asia, need to refuel in Europe to prepare for the second leg of their long-distance return migration. After a dry Sahel year, Ruff arriving at refuelling stations tended to have lower body masses (Fig. 216).

5. Date of arrival Dates of arrival in Europe can be used as a predictor of actual departure times from the Sahelian wintering quarters (which are often unknown) for many species. Arrival of Garganey in

the Camargue was retarded in dry Sahel years but annual variations were not significant, possibly because timing of arrival was based on captures (which may depend on several other factors) (Chapter 23). A better measure would be regular counts of birds present, and this approach was used for Black-tailed Godwits using a stopover site in Portugal (Box 26). As for Garganey, Black-tailed Godwits arrived later in dry Sahel years. A similar negative effect of Sahelian rainfall on arrival date was found for Barn Swallows in northern Italy (Saino et al. 2004) and in Spain (Gordo et al. 2005, Gordo & Sanz 2006, Gordo 2007); the data of Gordo & Sanz (2006) are reproduced in Fig. 264.

Spring migration, weather and climate change

The date of arrival on the breeding grounds depends not only on the date of departure from the sub-Saharan wintering areas, but just as much on the time the birds spend *en route*. Weather conditions affect the progress being made by the migrants during the homeward migration (Huin & Sparks 1998, Both et al. 2005, Saino et al. 2007), as exemplified by the timing of spring passage of migratory birds on the isle of Heligoland, in the North Sea north of Germany, which was related to the North Atlantic Oscillation (NAO [1]; Hüppop & Hüppop 2003). The spring migration of Common Whitethroat, for instance, was delayed by 10 days after 'continental' winters (high NAO) compared to more 'Atlantic' winters (low NAO). Hüppop & Hüppop (32) considered the NAO to be a reliable measure of spring conditions in Europe (including a higher probability of experiencing favourable tail winds in spring). However, since a low NAO often coincides with dry Sahel conditions [2], the delay in spring migration of long-distance migrants cannot be decoupled from the effects of dry Sahel conditions. It is nonetheless remarkable that the impact of NAO is at its maximal on species typically wintering in the Sahel (Common Redstart, Lesser Whitethroat, Common Whitethroat, Sedge Warbler, Common Chiffchaff; average variation in timing 13 days), but has minimal effect on species wintering further south (Icterine Warbler, European Reed Warbler, Garden Warbler, Willow Warbler, Spotted Flycatcher, Pied Flycatcher; annual variation in timing 8 days).

Climate change is often cited as being responsible for gradual long-term shifts in the timing of spring migration, although the effects vary according to migratory strategy (long- versus short-distance migrants, geography and to duration and type of studies; reviews in

Climate change, but also Sahel rain, have an impact on egg-laying dates of Common Whitethroat (left) and European Reed Warbler (right).

Møller et al. 2007, Gordo 2007). The study on Heligoland concluded that 21 migratory bird species showed a systematic advancement of passage date between 1960 and 2000, an effect most pronounced in Reed Warbler, Garden Warbler and Spotted Flycatcher (Hüppop & Hüppop 2003). Our re-analysis of the data for these three species indicated that year-to-year variations in passage date were not correlated with Sahel rainfall. This is hardly surprising since their main wintering grounds are situated south of the Sahel (Ottosson et al. 2005).

The number of studies showing an impact of climate change on spring migration is increasing sharply, but there are many provisos to consider (Gordo 2007). For example, calculated over the 1983-2004 period, Barn Swallows advanced their arrival date in Spain by 0.52 days per annum (Gordo & Sanz 2006; Fig. 264). This result by itself might be interpreted as an effect of ongoing climate change,

Fig. 264 Date of arrival (given as deviation from the long-term average, corrected for regional variation) of Barn Swallows in Spain since 1945 (left-hand axis), compared to the Sahel rain index and the temperature in Spain in March-April (given as % deviation from the average in 1945-2003) (right-hand axis). Date of arrival and temperature from: Gordo & Sanz (2006).

Table 31. Impact of Sahel rainfall ("rain" = daily change per % Sahel rain index) and climate change ("year" = daily advancement per year) on mean laying date of long-distance migrants in Britain in 1966-2006. The table shows the unstandardised regression coefficients, such as determined in a multiple regression analysis, explained variance (R^2) and level of significance (* $P<0.5$, ** $P<0.1$, *** $P<0.01$). From: BTO-data given in Baillie et al. (2007).

Species	rain	year	R^2
Barn Swallow	-.143*	-.223***	.375***
Common Redstart	-.147**	-.285***	.570***
Whinchat	-.073*	-.092*	.126
Northern Wheatear	-.314**	-.120	.446*
Sedge Warbler	-.119	-.107	.131
European Reed Warbler	-.201**	-.165**	.315***
Blackcap	-.189**	-.215**	.336***
Garden Warbler	-.080	-.203	.280**
Common Whitethroat	-.196*	-.260**	.327***
Common Chiffchaff	-.046	-.383***	.503***
Willow Warbler	-.044	-.159***	.378***

but data from a longer time-period would be needed to see whether there is any discernable link to climate change. The date of arrival appears to depend primarily on rainfall in the Sahel, and to a lesser degree on spring temperature in Spain. These data do not suggest, independent of year-to-year variations, a long-term trend towards earlier arrival. [3]

However, the opposite is the case when considering the date by which 95% of the Redstarts arrive on Christiansø (an island off Denmark in the Baltic Sea; 55°11'N, 15°11'E). The advancement of 0.41 days per year between 1976 and 1997 (Tøttrup et al. 2006) became stronger when Sahel rainfall was entered into the analysis. [4]

Gordo & Sanz (2006) and Gordo (2007) sensibly caution against automatically interpreting forward shifts in timing of spring migration as evidence of climate-driven evolutionary change (and hence a change in endogenous control; Jonzén et al. 2006). Changes often may be explained by short-term, year-to-year variations in weather conditions in wintering, staging and breeding areas.

Impact of climate change and Sahel droughts on egg-laying dates

Carry-over effects of wintering conditions may relate to variations in egg-laying dates. The data collected in Britain since 1966 by volunteers of the British Trust for Ornithology (BTO) show a long-term advancement in egg-laying dates for many species (Baillie et al. 2007). The laying date of British Redstarts advanced by two weeks during this period, for which Sahel conditions may be partly responsible (Fig. 265: see scattering around the trend line). In 5 of the 7 dry Sahel

The BTO data suggest that the Sahel rainfall has an impact on the productivity of Sand Martins, but the sample size is still small.

years, the birds arrived later than on average; in wet years, this occurred in only 2 out of 8 years.

A re-analysis of the BTO data reveals that long-distance migrants advanced their mean laying date between 1966 and 2006 by 4-15 days (Table 31). This long-term trend is superimposed on, and independent of, Sahel rainfall, which is responsible for short-lived delays during dry Sahel years. The NAO index appeared to have no impact on laying date, except for Garden Warbler.

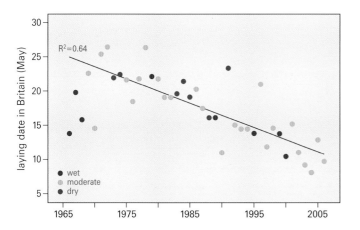

Fig. 265 The significant, forward shift of laying date in Common Redstarts in Britain between 1965 and 2006, grouped for three levels of rainfall in the Sahel in the preceding year (less than 20% below average, above average and the category in between). The regression line refers to the moderate years and excludes the wet and dry Sahel years. Data from: BTO (Baillie et al. (2007).

Table 32 Correlations between productivity (ratio juveniles:adults on British Constant Effort Sites) and Sahel rainfall in the preceding year (R-simple); R-partial corrects for the local breeding conditions; explanation in text. Data from: BTO (Baillie et al. 2007) *= $P<0.05$; N=21.

Species	R-simple	R-partial
Sedge Warbler	0.38	0.53*
European Reed Warbler	0.24	0.25
Blackcap	-0.07	0.32
Garden Warbler	0.24	0.25
Lesser Whitethroat	0.33	0.45*
Common Whitethroat	0.10	0.21
Common Chiffchaff	-0.06	0.17
Willow Warbler	-0.17	0.00

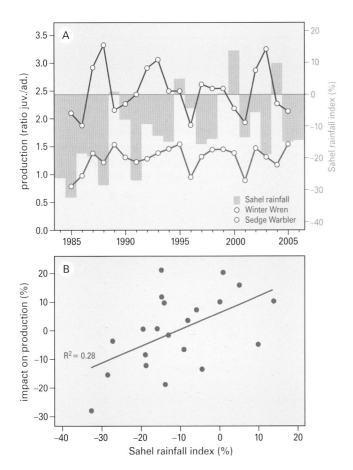

Fig. 266 (A) Average productivity (number of juveniles relative to adults captured on British Constant Effort Sites) of Sedge Warblers and Winter Wrens compared to Sahel rainfall. (B) Productivity of Sedge Warblers as a function of Sahel rainfall; same data as chart (A), but correcting for local breeding conditions as exemplified by Wren productivity; explanation in text. Data from: BTO (Baillie et al. 2007)

Carry-over effects on breeding performance

Do breeding birds produce fewer young after they have spent the preceding winter in a dry Sahel? To answer this question, we re-analysed the capture data from Constant Effort Sites (CES) in Britain (BTO, 1981-2006) and The Netherlands (SOVON, 1994-2005) and used the ratio of juvenile and adult birds as a measure of productivity. Productivity of long-distance migrants revealed positive correlations with Sahel rainfall in the preceding year, but none were significant. The problem with this analysis is that the possible Sahel impact is overruled by local factors. For instance, most bird species produced fewer young during a cold and rainy summer. This explains why the productivity of Dutch Reed Buntings (a resident), Marsh Warblers (wintering in East Africa) and Reed Warblers (wintering in West Africa) are highly intercorrelated (R varying between +0.86 and +0.94). To investigate carry-over effects of Sahel conditions on breeding performance, the conditions playing in the breeding area have to be statistically eliminated. The Dutch CES series (12 years) is still too short for such an analysis, but the British series spans 26 years. The Winter Wren appeared to be a species whose productivity correlated well with those of many other species in the BTO data set and so we used this species to keep the local breeding conditions statistically constant.

When comparing the productivity of Sedge Warblers and Winter Wrens in relation to Sahel rainfall in the preceding year, it is obvious that for both species 1985, 1986, 1996 and 2001 were poor breeding years (Fig. 266A). Sedge Warblers produced relatively more young after a wet Sahel year, but the correlation is poor. However, taking local breeding conditions into account (by entering the productivity of Wrens in a multiple regression analysis), the Sahel impact became significant (Fig. 266B). A similar analysis, done for another seven long-distance migrants (Table 32), showed that there was a similar significant impact upon the Lesser Whitethroat, but not upon the other species, although for all species Sahel rainfall has a positive effect. It shows that, no matter in how good a condition a migrant may be on return from the wintering grounds, when local conditions are poor in the wrong part of the breeding cycle, reproduction (and possibly survival) will plummet. A similar case has been described for British Sand Martins, in which rainfall during the breeding period adversely affected apparent over-winter survival (and possibly productivity), although there was a positive relationship between annual survival and Sahel rainfall (Cowley & Siriwardena 2005). These relationships are further complicated by variations in immigration and emigration rates over time, which may override other factors influencing population dynamics (Szép 1995a, 1995b).

In November 2008, very high grasshopper densities over large areas of central Senegal attracted 3500 White Storks (see Chapter 14). This is the highest concentration observed in Senegal since exactly 50 years ago, in November 1958, Morel & Roux (1966) observed 4000 individuals in the Senegal Delta. It reflects the increase of the Western population during the last decennium.

Conclusion

Conditions on the African wintering grounds, and *en route* between wintering and breeding areas, affect pre-migratory fattening, departure date, refuelling rate and migration timing. Adverse conditions during any of these stages may have a negative impact on breeding performance, as for example found for White Storks, Ospreys and Barn Swallows. We may expect similar effects for other long-distance migrants, but few data exist to substantiate carry-over effects of wintering conditions. British (BTO) and Dutch (SOVON) data on productivity, as measured on Constant Effort Sites, indicate that conditions in the Sahel in the preceding winter are – to some extent – interlinked with breeding performance, but that the consequences of local breeding conditions are often superimposed upon those of the sub-Saharan region.

Endnotes

1 NAO stands for North Atlantic Oscillation and is defined as the difference between the normalised sea-level pressures at the Azores and Iceland, averaged for the period December-March. The winters in Europe are 'Atlantic' (warm and wet) when the average NAO is positive and 'Continental' (cold and dry) when the NAO is negative.

2 The NAO index is related to many different meteorological variables, including the Sahel rain index: years with low rainfall in the Sahel are followed more often by 'Atlantic' winters in W Europe. Hence there is a negative correlation between the NAO (y) and Sahel rainfall (x) in the preceding six months: $y = -0.044x + 0.113$; $R = -0.30$, $N = 106$, $P < 0.01$).

3 A multiple regression performed on the data shown in Fig. 265 reveals that Sahel rainfall explains 38.3% of the variance ($P<0.001$); spring temperature adds 21.7% ($P<0.001$) and year (as a linear function) 0.1% (N.S.). The arrival is delayed with 0.193 day for every 1% less rain in the Sahel, and this remains the same (-0.190) after the other variables were entered into the equation.

4 We analysed the date at which 95% of the Common Redstarts had arrived, using a regression analysis (data from Jonzén et al. 2006: their Fig. 1b). The advancement was 0.41 day per year ($P=0.027$) in a simple regression, but in a multiple regression this became 0.44 day/year ($P=0.08$) with Sahel rainfall causing an advancement of 0.236 day per 1% higher Sahel index ($P=0.012$); $R^2=0.45$, $N=21$.

Impact of the Sahel on Eurasian bird population trends

The world, as birds know it, is in turmoil and has been for some time. Many bird species in Europe in the 19th century were subjected to campaigns of destruction, either on the grounds that they were harmful to man's interests, or that they were perceived as a source of income. No matter how destructive the persecution, populations usually bounced back as soon as the cause of decline was eliminated. However, in the 20th century, habitat loss and degradation took place on a scale never before witnessed. Vast wetlands disappeared, industries and ports were built on infilled estuaries and intertidal flats, chemical-based farming became the norm, urban sprawl and roads claimed ever more territory and sea life was plundered. Such large-scale habitat changes are – unlike the causes of the earlier near-exterminations – probably irreversible, given the inability of governments to act appropriately. Birds are on the receiving end of this train of events. In The Netherlands, for example, a number of breeding species that are long-distance migrants have effectively disappeared in only half a century. Species affected include Stone-curlew (mid-1950s), Woodchat Shrike (mid-1950s), Gull-billed Tern (late 1950s), Eurasian Hoopoe (late 1960s), Ortolan Bunting (mid-1990s), Tawny Pipit (early 2000s) and Eurasian Wryneck (mid-2000s). Similar stories can be told for other European countries, sometimes concerning different species. Is there a common denominator underlying these losses, like loss and degradation of habitats on the breeding grounds? And is it a coincidence that many of the steepest declines are to be found among long-distance migrants, especially those wintering in the Sahel? The first suggestion is supported by a huge body of evidence; to answer the second has been our major line of inquiry throughout this book. But which region has been the more important in determining population fluctuations, or are both vital?

The long-distance migrants involved

The first attempts to present an overview of Palearctic birds in Africa were biased towards northeastern and eastern Africa (Grote 1930, 1937; Moreau 1972). Thanks to the pioneering work of Gérard and Marie-Yvonne Morel and their collaborators, the importance of West Africa for Palearctic birds has become firmly established (Morel & Bourlière 1962, Morel & Roux 1966a & 1966b, Morel 1968, Morel & Morel 1972, 1974, 1978). Since then, great strides have been made in improving our knowledge of Palearctic birds in Africa. Atlases or avifaunas are available or forthcoming for the majority of African countries, the seven volumes of *The Birds of Africa* are an encyclopedic source of information, and there is a constant outpouring of papers and reports from across the continent. Detailed studies in the Sahel have focussed on migration strategies, site fidelity, recurrence and itinerancy, habitat use and foraging ecology, and interspecific relationships with Afrotropical residents (*e.g.* Morel & Morel 1992, Wilson 2004, Salewski & Jones 2006), but despite all these great efforts, woefully little is known about the life histories of the majority of trans-Saharan migrants. However, one aspect has become abundantly clear: long-distance migrants have markedly declined since the 1970s.

The Sahel and adjoining vegetation zones shape the lives of billions of Eurasian birds, including those on passage across the region. Rainfall, the extent of flooding of floodplains and the 'greenness' of the vegetation are the main drivers of life in the Sahel, and are potential key factors in the rise and fall of migratory bird populations. In the preceding chapters, the vicissitudes of a number of Eurasian bird species have been reviewed against the background of the sub-Saharan wintering areas and the breeding quarters. Our selection encompassed an array of species with widely different migratory and wintering strategies; some species are entirely dependent on what the Sahel has to offer during the northern winter, others less, or hardly, so.

The Palearctic hordes wintering in Africa comprise mostly waterbirds and insectivorous passerines. Their stay in sub-Saharan Africa lasts about six months, from just after the end of the rainy season when local conditions are still benign, the vegetation green and insects in abundance (Jones 1998). As the dry season progresses, the Sahel changes from green to brownish-yellow in only a few months. The vegetation in the bushy savanna withers, temporary pools dry out, many trees shed their leaves, insects dwindle in number or go into hiding. However, as pointed out by Morel (1973), the Sahel is less impoverished during the dry season than superficially appears to be the case. During the period when migrants are wintering, some evergreen or semi-evergreen tree species flower, produce nectar (which attracts large numbers of insects) or carry fruit that passerines will eat, *e.g. Salvadora*, *Ziziphus*, *Boscia*, *Grewia*, *Balanites* and *Maerua* (Fig. 267). Even when conditions seem really severe in February and March, many migrants can survive and fatten up in wooded savannas, wadis and river valleys.

Fig. 267 The phenology of some tree species in the Sahel. Yellow lines show the monthly variation in leaf biomass (maximum set at 100; Hiernaux *et al.* 1994) and white bars the flowering period (From: Arbonnier 2000). Pictures taken in Mali in November, except *Acacia* (=*Faidherbia*) *albida* (August) and *Balanites* (February 2008).

Box 27
Monitoring, trends and some pitfalls

Monitoring common bird species in Europe, using standardised methods, is a rather recent technique (Table 34). The first scheme was begun in the 1960s (United Kingdom), with two more initiated in the 1970s (Denmark, Sweden). Enthusiasm for monitoring gained momentum in the 1980s, nine more countries adopting the technique, followed by a further eight countries in the 1990s and 13 in the 2000s. Only nine countries have yet to start a monitoring scheme; the most important of these is Russia, which, due to its size (37% of the European territory), has a significant effect on any measure of overall European trends. Altogether, over 60% of the European territory is not yet covered by long-term censuses of breeding birds, especially in eastern and southern Europe. The longest-running schemes are strongly biased in favour of western Europe and Scandinavia, *i.e.* regions where human impact on the landscape has been the greatest.

Monitoring breeding birds demands dedicated observers, willing to employ a rigid methodology and to walk the same route again and again, year after year. The reward is that the participants find joy and solace in their intimate knowledge of their plots and the inhabitants. Meanwhile they may wonder about the changes that take place as they, imperceptibly, get older and balder along the well-worn way. (cartoonist: Fred Hustings).

Most time series available are insufficiently long to document trends adequately. The meaning of the expression 'long-term' has altered with time as understanding of the nature of trends has developed: in the early 1980s, a time series spanning ten years was considered long-term, but now is rated as short-term. But why would a 20-year or longer time series be long-term? After all, it often only represents an individual observer's working life. The answer is to be found in the study of the rainfall regime in the Sahel over the past centuries. We have clearly shown in previous chapters that the subtleties of the controlling biological processes in the Sahel cannot be identified with certainty over a shorter period; the likely corollary is that a time series cannot ever be long enough. Correlations with climate based on short series (of less than 20 years; Wang 2003) can be spurious and potentially misleading.

Other serious problems of existing monitoring schemes are sample size (number of plots, points or transects) and lack of randomisation (few schemes use random samples). Some countries are lucky to have hundreds, or thousands, of qualified volunteers (*e.g.* Denmark, The Netherlands, the United Kingdom), but others have to make do with many fewer (only 7 fieldworkers for Ukraine, a country 14 times the size of Denmark or The Netherlands). No matter how many statistics are used to counter the effects of imperfect design or small sample size (Kéry 2008), the outcome should always be viewed with reservations and preferably be tested against other schemes running in parallel.

Table 34 Monitoring schemes for breeding birds in Europe (surface area plus waterbodies, start of schemes and methods used). From: European Bird Census Council, www.ebcc.info (accessed 30 November 2008). Schemes used in this book are in italics; they refer to the longest time series with the largest sample sizes. Point = point counts, line = line transects, map = territory mapping.

Country	1000 km²	Start	Method(s)
Albania	29	-	-
Austria	84	1998	point
Belarus	208	2007	point
Belgium	31	2007	point
Bosnia-Herzegovina	51	-	-
Bulgaria	111	2004	point
Croatia	57	-	-
Cyprus	9	2005	various
Czech Republic	*79*	*1981*	*point*
Denmark	*43*	*1976*	*point*
Estonia	45	1983	point
Finland	*337*	*1981*	*point*
France	547	1989	point
Georgia	70	-	-
Germany	357	1989	point, line, map
Greece	132	2006	point
Hungary	93	1999	point
Iceland	103	?	various
Ireland	70	1998	line
Italy	301	2000	point
Kosovo	11	-	-
Latvia	65	1983	point, line
Liechtenstein	0.2	1981	map
Lithuania	65	1991	point
Luxembourg	3	2002	point, map
Macedonia	25	2007	line
Moldova	34	-	-
Montenegro	14	-	-
Netherlands	*42*	*1984*	*map*
Norway	324	1995	point
Poland	313	2000	line
Portugal	92	2004	point
Romania	238	2006	point
Russia (European)	3960	-	-
Serbia	88	-	-
Slovakia	49	1994	point
Slovenia	20	2007	line
Spain	505	1996	point
Sweden	*450*	*1975*	*point*
Switzerland	41	1999	map
Turkey	784	2007	line
United Kingdom	*245*	*1962*	*map, line*
Ukraine	604	1980	point

Measurement errors (often larger than 20%) are an incentive to use a variety of survey methods in parallel, which approach would provide independent estimates. Even so, the likelihood of finding significant evidence for the existence of a temporal trend could be vanishingly small, unless the time series spans at least 15 years, or the trend is truly massive (Hovestadt & Nowicki 2008). European trends of long-distance migrants illustrate the importance of long time series. Countries that initiated their monitoring programs in the 1970s and 1980s observed just such major trend changes that were consequent on a weather-regime shift in the Sahel (the start, in 1969, of a 30-year drought; Chapter 2). Populations of long-distance migrants were seriously depleted by this sustained drought. Any ensuing trend would then be likely to stabilise at a low population density as long as adverse conditions persisted, or would experience an upward shift as soon as conditions ameliorated.

This predicted pattern of events is precisely what happened. In the few cases for which we were able to reconstruct time series for the decade or decades preceding 1970, the outcome was rather dramatic. Much higher pre-1970s population levels were found for Ruff (Chapter 29), Gull-billed Tern (Chapter 30), Eurasian Wryneck (Chapter 33), Common Redstart (Chapter 37) and Common Whitethroat (Chapter 41); these levels were very much higher than any population level attained since then. What is described as a significant increase for Common Redstart in Denmark in 1976-2005 (Heldbjerg & Fox 2008), would therefore be an incomplete recovery from huge losses, had the time series included the 1950s and 1960s, as was recognised as a possibility for Common Whitethroat in the same paper. The impact of temporal scale on trend analyses has also been identified as a serious problem in Nearctic-Neotropical forest-dwelling migrant birds: population declines in the eastern and central United States are not evident in trends after 1960, because the bulk of declines had already occurred by that time (King *et al.* 2006).

The problems surrounding reliable monitoring are particularly acute when transposed on a European scale, even when the many provisos surrounding trend assessments are countered by using weighting processes and broad categories (Tucker & Heath 1994, BirdLife International 2004a). We use these data therefore in the spirit of Moreau (Box 16, Chapter 13): the last thing we too would wish for is to be regarded as having all the answers, although it does not prevent us trying to make sense of the numbers.

Once south of the Sahara, the Palearctic migrants disperse across the African continent. The entire range of available habitats is exploited during the northern winter, but to highly varying degrees. Tropical rainforests are almost completely avoided, whereas the Sahel receives a disproportionately large number of Eurasian birds (Fig. 104). Several attempts have been made to categorise Palearctic long-distance migrants according to their migratory strategy and habitat choice in Africa (see also Fig. 105 and Table 35, adapted from Morel 1966b, Morel & Morel 1992, Pearson & Lack 1992, Jones 1998, Yohannes et al. 2009):

1. Occupying winter quarters in equatorial and southern Africa (e.g. European Honey Buzzard, Eurasian Hobby, Barn Swallow).
2. Wintering largely north of the Sahara, but with substantial numbers crossing the desert to winter in the northern and southern tropics, including the Sahel (e.g. Grey Heron, Common Chiffchaff).
3. Wintering largely south of the Sahara, often widely dispersed across the continent and including the Sahel and coastal zone (e.g. Osprey, Common Greenshank).
4. Wintering south of the Sahara and moving gradually south of the Sahel to destinations in the Sudan and Guinea zone between 8° and 14°N (e.g. Pied Flycatcher); in East Africa this zone extends further south up to 5°N, and includes the northern equatorial tropics (e.g. Barred Warbler).
5. Staying in the Sahel and Sudan zone until November (11°-16°N), often (partially) moulting and fattening, before embarking on the second southward leg of migration into wintering areas in the Guinea vegetation zone (8°-10°N), or – in East Africa – even south of 5°S (e.g. Thrush Nightingale, Great Reed Warbler and Garden Warbler; Pearson & Lack 1992, Hedenström et al. 1993, Ottosson et al. 2005, Yohannes et al. 2009).
6. Wintering in the Sahel (12°-18°N) in West Africa, but extending south of 5°S in East Africa (also involving birds from eastern Europe and Asia, e.g. Montagu's Harrier, Ruff, Common Sand Martin). This category includes species whose northern European populations leapfrog those from southern Europe (e.g. Eurasian Marsh Harrier, Yellow Wagtail).
7. Staying in the northern tropical savannas throughout the northern winter, with an emphasis on the Sahel (12°-18°N) or in the adjacent Sudan and northern Guinea zones (8°-12°N). This category includes waterbirds largely restricted to floodplains (e.g. Glossy Ibis, ducks), and typical savanna dwellers (e.g. European Turtle Dove, Eurasian Wryneck, Greater Short-toed Lark).

Taking stock: negative trends predominate among Afro-Palearctic migrants

Of the 200 or so Palearctic-African migratory bird species, we restrict our analysis to 127 species which winter predominantly, or in significant numbers, south of the Sahara and for which population trends are available. [1] We reconstructed trends for long-distance migrants for 1970-2005, based on trend directions in 1970-1990 and 1990-2000 but with some adjustments. [2] To be able to discern and understand trend changes, we classified each species according to their dependence on the Sahel during the northern winter, and their habitat choice in Africa and Europe (Table 35).

The number of trans-Saharan migrants in decline increased from 39% in 1970-1990 to 55% in 1990-2000 (Table 35), a trend significantly more negative than those of residents and short-distance migrants (Sanderson et al. 2006). For the period 1970 to 2005, 75 out of 127 species were in decline, a staggering 59% of all trans-Saharan migrants under consideration, with little change in the trajectory of trends of most migrant species' populations, despite some improvement of rainfall in the Sahel from the mid-1990s onwards. Using the Danish time series, from 1975 to 2005, Heldbjerg & Fox (2008) calculated an average decline of 1.3% per annum for trans-Saharan migrants, a figure which would have been higher had monitoring started in the 1960s with high rainfall in the Sahel. The Sahel-Sudan zone, where the highest numbers of Palearctic migrants concentrate, clearly stands out as a trouble zone. Whereas most European monitoring schemes were initiated in the 1970s and later, i.e. coinciding with a prolonged period of severe drought in the Sahel, declines and low population levels of species wintering in the Sahel-Sudan zone were to be expected. For a number of species, we found a positive correlation between population size or survival and the size of Sahelian floodplains (Purple Heron, Black-crowned Night Heron, Squacco Heron, Little Egret, Glossy Ibis, Garganey, Black-tailed Godwit, Caspian Tern, Sedge Warbler), or with rainfall in the Sahel (European Turtle Dove, Eurasian Wryneck, Common Sand Martin, Yellow Wag-

The decline of Stone-curlews is mainly due to changes in their breeding habitat. The three birds shown here remained at the southernmost edge of their winter distribution (Iwik, Mauritania, January 2005).

Table 35 Population trends in 127 Afro-Palearctic bird species in 1970-1990 (Tucker & Heath 1994; trend range is from -2 = large decline to +2=large increase), 1990-2000 (BirdLife International 2004a; trend range is from -3 = large decline to +3 = large increase, no trend available for 11 species) and 1970-2005 (same scale); blanks = insufficient data available. The importance of the Sahel for Palearctic birds during November up to and including February is labelled from insignificant (1) to large (7); see text for further explanation. Habitats in Europe and Africa are simplified from Fig. 105 and BirdLife International (2004a).

Species	Long-term trend			Sahel dependence	Habitat	
	1970-1990	1990-2000	1970-2005		Africa	Europe
Grey Heron	2	2	2	2	wetland	wetland
Purple Heron	-2	-2	-2	6	wetland	wetland
Black-crowned Night Heron	-1	0	-1	6	wetland	wetland
Squacco Heron	-2	-2	-2	6	wetland	wetland
Little Egret	2	1	2	3	wetland	wetland
Little Bittern	-2	0	-2	6	wetland	wetland
Great Bittern	-2	0	-2	2	wetland	wetland
White Stork	-2	2	0	6	savanna	farmland, steppe
Black Stork	0	0	0	7	savanna	woodland
Eurasian Spoonbill	-2	0	1	3	wetland	wetland
Glossy Ibis	-1	-2	-3	7	wetland	wetland
Northern Shoveler	0	-2	-1	7	wetland	wetland
Northern Pintail	-2	-2	-2	7	wetland	tundra, mire, moor
Garganey	-2	-2	-2	7	wetland	tundra, mire, moor
Ferruginous Duck	-2	-3	-3	7	wetland	wetland
Osprey	1	2	2	3	wetland	wetland
Egyptian Vulture	-2	-3	-3	7	savanna	Mediterranean
Black Kite	-2	-3	-3	6	wooded savanna	wetland
Eurasian Marsh Harrier	2	2	2	6	wetland	wetland
Pallid Harrier	-2	-3	-3	7	savanna	farmland, steppe
Montagu's Harrier	2	2	2	7	savanna	farmland, steppe
Short-toed Eagle	0	-1	-1	7	savanna	woodland
Levant Sparrowhawk	0	-2	-1	1	woodland	woodland
European Honey Buzzard	0	0	0	1	woodland	woodland
Steppe Eagle	-2	-3	-3	1	savanna	farmland, steppe
Lesser Spotted Eagle	0	-2	-1	1	savanna	farmland, steppe
Booted Eagle	0	0	0	4	wooded savanna	woodland
Lesser Kestrel	-2	-1	-2	6	savanna	farmland, steppe
Red-footed Falcon	-2	-3	-3	1	savanna	farmland, steppe
Saker Falcon	-2	-3	-3	7	savanna	farmland, steppe
Eurasian Hobby	0	0	0	1	wooded savanna	woodland
Common Quail	-2	0	-2	7	savanna	farmland, steppe
Corn Crake	-2	0	-2	1	savanna	farmland, steppe
Demoiselle Crane	2	3	3	7	wetland	wetland
Stone-curlew	-2	-3	-3	2	savanna	farmland, steppe
Black-winged Stilt	0	0	0	3	wetland	wetland
Collared Pratincole	-2	-2	-2	7	savanna	farmland, steppe
Black-winged Pratincole	0	-3	-2	1	savanna	farmland, steppe
Great Snipe	-2	-2	-2	6	wetland	tundra, mire, moor

Species	Long-term trend			Sahel dependence	Habitat	
	1970-1990	1990-2000	1970-2005		Africa	Europe
Common Snipe	0	-2	-1	2	wetland	tundra, mire, moor
Little Ringed Plover	1	-1	0	2	wetland	wetland
Black-tailed Godwit	-2	-3	-3	7	wetland	farmland, steppe
Little Stint	0	0	0	3	wetland	tundra, mire, moor
Temminck's Stint	0	0	0	3	wetland	tundra, mire, moor
Green Sandpiper	0	0	0	3	wetland	tundra, mire, moor
Wood Sandpiper	-1	0	-1	3	wetland	wetland
Common Sandpiper	0	-2	-1	3	wetland	wetland
Ruff	0	-2	-2	6	wetland	tundra, mire, moor
Spotted Redshank	0	-2	-1	3	wetland	tundra, mire, moor
Common Greenshank	0	0	0	3	wetland	tundra, mire, moor
Marsh Sandpiper	2	-2	0	6	wetland	wetland
Gull-billed Tern	-2	-2	-2	6	wetland	wetland
Caspian Tern	-2	3	-1	6	wetland	wetland
Whiskered Tern	-1	0	-1	6	wetland	wetland
White-winged Tern	1	0	1	6	wetland	wetland
European Turtle Dove	-1	-2	-2	7	savanna	woodland
Great Spotted Cuckoo	2		2	7	wooded savanna	woodland
Common Cuckoo	0	-1	-1	1	wooded savanna	woodland
European Scops Owl	-1		-1	7	wooded savanna	woodland
European Nightjar	0	-1	-1	6	wooded savanna	woodland
Red-necked Nightjar	0		0	7	wooded savanna	Mediterranean
Pallid Swift	1		1	6	aerial	
Common Swift	0	-1	-1	1	aerial	
Alpine Swift	0	1	1	1	aerial	
Blue-cheeked Bee-eater	0	2	1	6	savanna	farmland, steppe
European Bee-eater	-1	2	0	1	wooded savanna	farmland, steppe
European Roller	-1	-3	-3	1	wooded savanna	farmland, steppe
Eurasian Hoopoe	0	-2	-1	6	wooded savanna	farmland, steppe
Eurasian Wryneck	-1	-2	-2	7	wooded savanna	woodland
Greater Short-toed Lark	-2	-2	-2	7	savanna	farmland, steppe
Common Sand Martin	-1		-1	6	wetland	farmland, steppe
Barn Swallow	-1	-1	-1	1	wooded savanna	farmland, steppe
Red-rumped Swallow	1	0	1	6	savanna	Mediterranean
Common House Martin	0	-2	-1	3	aerial	farmland, steppe
Tree Pipit	0	-1	-1	6	wooded savanna	woodland
Red-throated Pipit	0		0	6	savanna	tundra, mire, moor
Tawny Pipit	-2	-1	-2	7	savanna	farmland, steppe
White Wagtail	0	0	0	2	savanna	farmland, steppe
Yellow Wagtail	0	-1	-1	6	savanna	farmland, steppe
Common Rock Thrush	-1	-1	-1	6	savanna	Mediterranean
Bluethroat	0	0	0	7	wetland	tundra, mire, moor
Common Redstart	-2	0	-3	6	wooded savanna	woodland
Thrush Nightingale	0	0	0	5	wooded savanna	woodland

Species	Long-term trend			Sahel dependence	Habitat	
	1970-1990	1990-2000	1970-2005		Africa	Europe
Common Nightingale	0	0	0	5	wooded savanna	woodland
Rufous Scrub Robin	0	-3	-2	6	wooded savanna	Mediterranean
Northern Wheatear	0	-2	-1	6	savanna	farmland, steppe
Isabelline Wheatear	0	0	0	7	savanna	farmland, steppe
Black-eared Wheatear	-2	-1	-2	7	savanna	Mediterranean
Cyprus Wheatear	0	0	0	7	savanna	Mediterranean
Pied Wheatear	0	0	0	6	savanna	farmland, steppe
Whinchat	0	-1	-1	6	wooded savanna	farmland, steppe
Grasshopper Warbler	0	-1	-1	6	wetland	farmland, steppe
River Warbler	0	0	0	1	wetland	farmland, steppe
Savi's Warbler	0	0	0	7	wetland	wetland
Sedge Warbler	0	0	0	6	wetland	wetland
Aquatic Warbler	-2	-2	-2	7	wetland	wetland
Great Reed Warbler	0	-1	-1	5	wooded savanna	wetland
Eurasian Reed Warbler	0	0	0	5	wetland	wetland
Marsh Warbler	0	0	0	5	wetland	farmland, steppe
Olivaceous Warbler	-2	0	-2	7	wooded savanna	farmland, steppe
Melodious Warbler	0	0	0	4	wooded savanna	farmland, steppe
Olive-tree Warbler	0	0	0	1	wooded savanna	Mediterranean
Icterine Warbler	0	-1	-1	1	wooded savanna	farmland, steppe
Garden Warbler	0	0	0	5	wooded savanna	farmland, steppe
Barred Warbler	0	0	0	4	wooded savanna	farmland, steppe
Lesser Whitethroat	0	0	0	7	savanna	farmland, steppe
Common Whitethroat	0	1	0	6	savanna	farmland, steppe
Blackcap	0	1	1	2	wooded savanna	woodland
Orphean Warbler	-2	-1	-2	7	savanna	Mediterranean
Rüppell's Warbler	0	-1	-1	7	savanna	Mediterranean
Ménétries's Warbler	0	0	0	7	savanna	Mediterranean
Subalpine Warbler	0		0	7	savanna	Mediterranean
Wood Warbler	0	-2	-1	4	woodland	woodland
Willow Warbler	0	-1	-1	3	wooded savanna	woodland
Western Bonelli's Warbler	0	-2	-1	7	savanna	woodland
Chiffchaff	0	0	0	6	wooded savanna	woodland
Iberian Chiffchaff	0	1	1	7	wooded savanna	woodland
Spotted Flycatcher	-1	-1	-1	3	woodland	woodland
Pied Flycatcher	0	-1	-1	4	woodland	woodland
Collared Flycatcher	0	1	1	1	woodland	woodland
Red-backed Shrike	-1	-1	-1	1	wooded savanna	farmland, steppe
Woodchat Shrike	-2	-2	-2	6	wooded savanna	farmland, steppe
Masked Shrike	-2	-2	-2	7	wooded savanna	farmland, steppe
Lesser Grey Shrike	-1	-2	-2	1	wooded savanna	farmland, steppe
Eurasian Golden Oriole	0	-1	-1	1	woodland	woodland
Ortolan Bunting	-2	-1	-2	7	savanna	farmland, steppe
Cretzschmar's Bunting	0	-1	-1	7	savanna	Mediterranean

The population fluctuations of Black-crowned Night Herons breeding in Europe are largely determined by the annual variation in rainfall and flooding in the Sahel. (Senegal Delta, February 2008).

The Short-toed Eagle, one of few Eurasian raptor species wintering in the Sahel, shows a steady decline (central Senegal, January 2009).

tail, Common Redstart, Lesser Whitethroat, Common Whitethroat). When rainfall increased somewhat in the course of the 1990s, increases in the numbers of these and other Sahel-dwellers occurred, and indeed some species bounced back to levels from before the Great Drought (White Stork, Sedge Warbler, Common Whitethroat). Other species, although showing signs of recovery from depletion, have not yet reached population levels which were typically attained at similar rainfall indices and flood extents before the 1970s (Purple Heron, European Turtle Dove, Eurasian Wryneck and Common Redstart, to name but a few). As it is, the ongoing declines of many long-distance migrants testify to the fact that many species experience a wider range of constraints than just flood levels or rainfall. The uneven distribution of declining species across African habitats and regions, with those wintering in the savannas and wooded savannas of the Sahel-Sudan zone especially afflicted (see below), clearly hints at habitat loss as a cause of sustained declining trends.

Twelve species listed in Table 35 underwent a large decline, of which five are typical Sahel-dwellers (Glossy Ibis, Ferruginous Duck, Egyptian Vulture, Pallid Harrier and Saker Falcon) and three that use the wider Sahel-Sudan zone (Black-tailed Godwit, Black Kite and Common Redstart). Three other species in marked decline were found to winter in southern Africa, *i.e.* Steppe Eagle, Red-footed Falcon and European Roller. The only other species that declined so badly in 1960-2005 was the Stone-curlew, which winters in the Sahel, North Africa and southern Europe.

Only 15 of 127 Afro-Palearctic migrants were recorded as increasing, with another 37 species being considered as stable (Table 35, Fig. 266A). Species wintering in the Sahel-Sudan zone were disproportionally hit by declines, *i.e.* 49 out of 73 species (67%; labelled 6&7 under Sahel; Table 35), as were species wintering in southern Africa (14 of 21: 67%; labelled 1). Species wintering from the Guinea zone south of 5°S in East Africa, or more dispersed, fared comparatively better (9 of 25 in decline, or 36%; labelled 3, 4 and 5), as did the species largely remaining north of the Sahara (3 of 8 in decline, 38%; labelled 2).

Considered in terms of African habitats, the conclusion must be that birds wintering in savannas suffered the strongest declines. The few bird species wintering in woodland were affected least (Fig. 268B). The relation between habitat choice on the wintering grounds and trend direction is complicated by the species' vicissitudes in the European breeding areas. Declines, and especially steep declines, were most commonly found among Afro-Palearctic migrants breeding in farmland and steppe, less so among Mediterranean, arctic and subarctic species and least of all in wetland and woodland species (Fig. 268C). More than half of the long-distance migrants wintering in savanna and wooded savanna, i.e. 38 out of 73, are classified as birds of farmland and steppe in Eurasia. The proportion of losers in this category is higher (20 out of 38, or 47%, showing a large or moderate decline) than among the 35 savanna species which do not breed in farmland and steppe (8 out of 38; 23%). This might indicate that breeding birds of farmland and steppe are hit by a double-edged sword when wintering in African (wooded) savannas.

Declines of Palearctic birds spending the non-breeding season in Africa were especially large among birds using wetlands in the northern tropics (Fig. 269A), where their dependence on the few floodplains in the Sahel makes them highly vulnerable to droughts and human action. Elsewhere in Africa, herons, ducks and waders have a wider choice of wetlands across various climate zones, reducing their vulnerability to adverse conditions. Similarly, 7 out of the

15 species wintering in the wooded savannas of the Sahel-Sudan zone were in moderate or large decline, compared to none of the ten species heading for wooded savannas in southern Africa (Fig. 269C). As only two Palearctic species (one of which is in steep decline) spend the northern winter in savannas in southern Africa, a comparison with the species wintering in the savannas of the northern tropics (12 out of 32 in moderate or marked decline; Fig. 269B) is not in order.

Escaping the general pattern: Afro-Palearctic migrant populations on the increase or stable

Although the majority of Palearctic long-distance migrants are in steep decline, there are some notable exceptions. Several factors, not mutually exclusive, are suggested as reasons why these species escape the general trend. However, none of these increases was triggered by events occurring in the Sahel, but rather by an improvement of conditions elsewhere.

(1) Wintering throughout West Africa and along the coast The European Osprey population has doubled since the mid-1980s (Fig. 74), but the number of Ospreys staging in Sahelian wetlands did not increase in proportion, but instead declined (Chapter 24). In West Af-

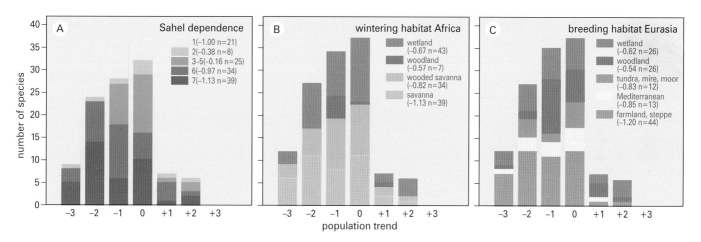

Fig. 268 Population trends of 127 Afro-Palearctic species in 1970-2005, listed in Table 35, in three categories: (A) the degree to which they depend on the Sahel as a wintering area (1 = southern Africa, 2 = Sahel, northern Africa and southern Europe, 3-5 = south of the Sudan zone, 6 = Sahel-Sudan, 7 = Sahel); see text for a full explanation, (B) their occurrence in African and (C) in Eurasian habitats. Trends are scored from - = decline, 0=stable, + = increase; 1 = small, 2 = moderate and 3 = large. The average trend score and number of species are given in the legend.

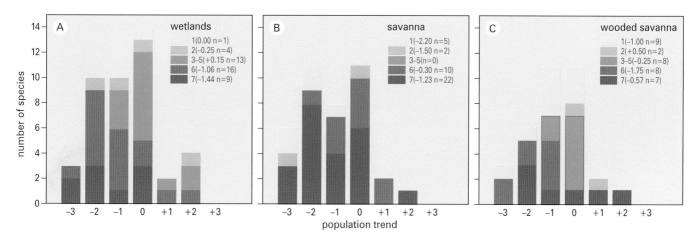

Fig. 269 As for Fig. 268A, but showing the trends for migrant birds which spend the northern winter in (A) wetland, (B) savanna or (C) wooded savanna respectively.

Box 28
The remarkable recovery of Eurasian Spoonbills
Otto Overdijk & Patrick Triplet

The Spoonbill population breeding in NW Europe and wintering in West Africa had steadily dwindled by 1968 to 151 pairs, confined to a handful of small colonies in The Netherlands (Bijlsma et al. 2001). Subsequently, this population has recovered and increased at an average rate of 6.6% per annum, reaching 2515 pairs in 2008, dispersed across 53 colonies between France and Denmark (including Britain). This development was totally unexpected and unprecedented (e.g. Bauchau et al. 1998). Is there any reason to implicate changing conditions in the African wintering areas as a trigger for these events?

Most 'Atlantic' Spoonbills (from Spain and NW Europe) spend the winter on the Banc d'Arguin in Mauritania and, 350 km further south, in the Senegal Delta. They are separated from 'continental' Spoonbills (from central and SE Europe), which winter east of Morocco and Senegal (Smart et al. 2007). The numbers of 'Atlantic' Spoonbills wintering in the Senegal Delta varied between years, with higher numbers in years with more flooding [3], but numbers remained more or less stable during the 19 years for which counts are available. In contrast, numbers on the Banc d'Arguin have increased from 650 birds in 1980 to 5000 birds in 2005; the annual increase between 1997 and 2005 averaged 20% (Fig. 270). At the same time, Spoonbills started wintering in SW Europe, where the relative increase in wintering numbers has been even larger than on the Banc d'Arguin (although totals are still low).

The rapid increase of the population wintering in SW Europe coincided both with the trend of winters becoming warmer over much of Europe since about 1990 and with improved protection. But why have numbers increased so much on the Banc d'Arguin, and not in the Sen-

During low tide, Eurasian Spoonbills on the Banc d'Arguin, Mauritania, feed in pools and creeks within the emerging eelgrass *Zostera*.

egal Delta? The carrying capacity of the Banc d'Arguin for Spoonbills seems to be about 8000 birds (Fig. 270B). In the 1980s, these consisted almost entirely of the local resident *P.l. balsaci*. In subsequent years, *balsaci* declined from 8250 in 1980 to 1500 in 2007. Is this decline the result of increased competition with birds from Europe? Perhaps it is the other way round - the decline of *balsaci*, for reasons unknown, offering an opportunity for its European congeners to occupy a vacant niche? Or are the contrasting population trends unrelated? Whatever

rica, Ospreys winter mostly along the coast and on reservoirs, where they are not affected by droughts in the Sahel. Another predominantly coastal species when wintering in Africa, Eurasian Spoonbill (from the western population), shows a similar increase in numbers (Box 28). Stable numbers were reported for several wader species, like Little and Temminck's Stints and Green Sandpiper, which winter widely dispersed across African wetlands including Sahelian floodplains. Interestingly, a similar strategy used by some passerines, of not being restricted to the Sahel-Sudan zone but also occurring throughout much of Africa, apparently has been inadequate to prevent a decline in species like Common House Martin, Willow Warbler and Spotted Flycatcher. Unlike wetland habitats, which display much resilience after temporary setbacks like droughts, woodlands and wooded savannas have suffered damage that exceeds the capability of natural restoration, with dire consequences even for passerines widely dispersed across Africa.

(2) Recoveries from human-induced declines Persecution and the use of persistent pesticides in farmland suppressed breeding numbers of many top predators in Europe to very low levels during much of the 20th century. The upsurges of species such as Little Egret (Chapter 19), Osprey (Chapter 24), Eurasian Marsh Harrier (Chapter 25) and Montagu's Harrier (Chapter 26) since about the 1970s are in fact recoveries from depletion, made possible through effective protection and bans on the use of organochlorines in agriculture. The effects of flood extent in the West African inundation zones on Eurasian Marsh Harriers breeding in western Europe became visible only after populations had bounced back to saturation levels set by the breeding grounds.

(3) Shifting the centre of gravity of wintering areas northwards Many Palearctic migrants have wintering grounds scattered from Europe through North Africa and into sub-Saharan regions. The evidence since the early 1990s indicates that the relative importance of the

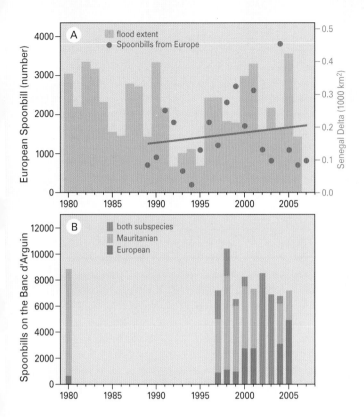

Just like waders, Eurasian Spoonbills on the Banc d'Arguin roost at high water. The largest roost is on the island of Arel, where thousands may be concentrated in one flock.

Fig. 270 (A) Number of Eurasian Spoonbills wintering in the Senegal Delta between 1989 and 2007, compared to the flood extent; numbers in 2002 and 2003 were underestimated. From: Triplet et al. (2008).
(B) Number of Eurasian Spoonbills at Banc d'Arguin (Mauritania) in January, showing the subspecies *Platalea l. leucorodia* (Europe) and *P.l. balsaci* (local residents). Recent counts of only *balsaci* reveal a further decline: 2300 in 2006 and 1500 in 2007. From: Altenburg et al. (1990), O. Overdijk (unpubl.).

sub-Sahara as winter quarters has declined for several bird species in favour of wintering areas in North Africa and in Europe. This shift coincided with a population increase, perhaps suggesting that the carrying capacity of Sahelian winter quarters had been reached (a buffer effect, necessitating the use of other wintering areas) or that a climate-induced shift had occurred, favouring wintering nearer the breeding range, triggered by more relaxed winter weather regimes in Europe after the mid-1980s. This behavioural change has been quite pronounced in Little Egret (Chapter 19), White Stork (Chapter 20), Eurasian Spoonbill (Box 28), and Eurasian Marsh Harrier (Chapter 25). Migratory flexibility enables birds to adapt to new migration routes and wintering areas arising from environmental change (Sutherland 1998). A stunning example is demonstrated by White Storks, which started to winter in their hundreds (perhaps thousands) in the Libyan desert near Marknusa, Al Kufra and Jalu, regions where fossil water is used to irrigate large tracts of desert

the intricacies affecting both subspecies on the Banc d'Arguin, in the Senegal Delta, wintering numbers had remained essentially stable, another indication that the change in Spoonbill fortunes was not triggered by events in this Sahelian floodplain.

(Hering 2008). An even more dramatic shift in wintering quarters has been documented for Blackcaps breeding in central Europe, which partly changed their migratory behaviour from wintering in southwestern Europe to Britain (Berthold et al. 1992).

Within-Sahel variations in trends

Coupling Sahel conditions with population changes on the breeding grounds is rarely a straightforward affair (Chapters 15-41). Circumstantial evidence for the existence of a Sahel effect on numbers might be gained from a comparison between species which winter in either the western Sahel (where habitat loss is mounting) or the eastern Sahel (less so), or, conversely, by a comparison of species which winter partly in the Sahel, and partly elsewhere. When it is known which part of the population depends on the Sahel during the northern winter (as in Common Chiffchaff and Willow Warbler), the chances are that different trends will be found when wintering conditions play a significant role in population dynamics.

Longitudinal differences across the Sahel Variations in annual rainfall are closely correlated between eastern and western Sahel regions (Fig. 7). Nevertheless, Palearctic birds wintering in the western Sahel (largely originating from western, central and northern Europe) encounter conditions during the northern winter that differ from those affecting the eastern Sahel, whose wintering populations derive mostly from the Asian and eastern European breeding grounds. A good example was presented in the 1970s and 1980s, when the flood extent of the Sudd was great but Lake Chad and the Inner Niger Delta were much diminished in size and the Senegal Delta even lost most of its floodplains. Secondly, the density of the human population (Fig. 35B) and its concomitant impact on the environment are much greater in the western Sahel than in the eastern Sahel. Thirdly, most birds in the western Sahel-Sudan zone are confined to a band between the Sahara in the north and the rainforest to the south, a region that spans only about 1000 km at its widest. They have to make do with what the band's vegetation zones offer during the northern winter, but when conditions in the eastern Sahel deteriorate, birds wintering there have the additional option of continuing their flight to equatorial East Africa (where the second rainy season has just started) or southern Africa, for there is no intervening barrier of rainforest or mountains. The penalty to be paid is the burden of having to fly up to 4000 km further (and back), an imposition that nevertheless poses less risk than remaining exposed to the rigours of drought.

These differences between eastern and western Sahel should reflect themselves in differential mortality under conditions of drought, *i.e.* lower mortality rates for birds wintering in the eastern part of Africa after 1969. If this were found to be true, then we should expect smaller or no declines for a larger number of long-distance migrants breeding in eastern Europe. To test this idea, we collated long-term trend data for long-distance migrants from across Europe (Fig. 271; average trend length 30 years, range 20-50 years, N=20).[4] It should be noted that most trends from Europe used here are too short to adequately cover the regime shift in rainfall patterns in the Sahel; unfortunately, longer trends are not available (Box 27).

In the western part of the breeding range, on average, declines affected a higher proportion of long-distance migratory bird species than in the eastern part. This trend was manifest in breeding bird surveys as well as in captures of migrants at Constant Effort Sites (Fig. 271), possibly attesting to the reliability of the outcome.

The fewer and smaller declines of long-distance migrants breeding in eastern Europe are not necessarily associated with conditions in wintering quarters, because the impact of man on the landscape in Europe exhibits a gradient from intensive in the west to more extensive in the east (but keep Box 27 in mind). This alone would suffice to explain differences in density and trends among long-distance migrants, as suggested for woodland birds (Gregory *et al.* 2007) and farmland birds (Skibbe 2008). For example, Ortolan Bunting numbers in eastern Europe have long been stable where farming has been less intensive, but it has been in decline in western European farmland for decades (BirdLife International 2004a).[5] The unification of East and West Germany in 1990 was another reminder of this dichotomy. Eastern Germany rapidly transformed its farming system along the lines of the European Community, causing many bird species, already at depressed levels in the west, to decline in the east as well. However, this is not yet the case for Ortolan Buntings, which have profited from the conversion of grassland into farmland and set-aside in eastern Germany (George 2004), much as found locally in Niedersachsen, NW Germany (Deutsch 2007). Nevertheless, the fact that Ortolan Buntings continued to decline in western European regions, even when farmland still offered favourable or unchanged breeding habitat (as in Sweden and northern Bavaria;

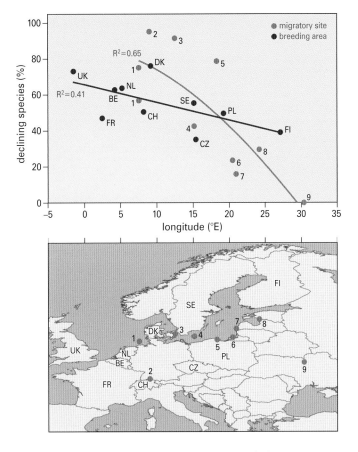

Fig. 271 Proportion of long-distance migrants in decline across Europe (see map) between 1950 and 2007, classified along a longitudinal gradient (for countries, the central longitude is taken). Notice that data from Constant Effort Sites (where migratory birds are captured via standardised procedures, allowing trend analyses) refer to populations breeding to the N and NE of the site of capture.

Pristine land in Europe is no more to be found. Even remote taigas and tundras have felt the stamp of humankind. However, the impact of human actions differs greatly across Europe, having been most severe in the most densely populated countries. In western Europe, for example, every single square metre of soil has been tilled several times in the past century. In northern and eastern Europe, the intensity of land-use has been much lower and large regions may still slumber in a semi-natural state, like these highlands in Norway (Jotunheimen, June 2007).

Stolt 1993, Lang 2007), strongly suggests that factors in the wintering areas may play a significant additional role. The sudden crash of the Ortolan Bunting population in southern Finland from around 1990 onwards, presents an interesting test case of an impact of wintering conditions on numbers, as it was found to be only partly associated with changes on the breeding grounds (Vepsäläinen et al. 2005). However, the crash coincided with an improvement after several decades of below-average rainfall in the Sahel, including the main wintering grounds in the Sudan and Ethiopia (Fig. 7), which would make a winter effect on population size less likely. Sahel conditions should be visible in local survival rates, but few data to that effect have been published. Curiously, a study in Switzerland in 1982-1988, i.e. during the Great Sahelian Drought, showed very high average return rates: 77% in adult males and 41% in females (females being more dispersive, hence the lower value; Bruderer & Salewski 2009). High survival rates are a prerequisite for numbers to remain stable in this single-brooded species, but little is known of survival rates, and changes therein, in the long run.

These examples seem to indicate that longitudinal variations in breeding trends are more likely to be caused by changes on the breeding grounds than by conditions in the wintering areas.

Latitudinal differences within the Sahel-Sudan-Guinea zones Many Palearctic passerines, upon reaching the Sahel after the Sahara crossing, escape from the rigours of the progressing dry season by trickling southwards into the Sudan and Guinea zones, or beyond (Table 35). So far, several strategies by which species cope with the progressing dry season and the resources available in the various climate zones have been unravelled. There can be little doubt that future research will reveal other, more complicated, strategies (as suggested for eastern Africa; Yohannes et al. 2009).

A study in the northern Guinea savanna of NE Ivory Coast in 1994-

1998 showed bimodal peaks of passage of Willow Warblers and Melodious Warblers in November-December and February-March (Salewski *et al.* 2002a). The low numbers of Willow Warblers in the intervening 4-6 weeks in January and February have been interpreted as itinerancy, perhaps related to the specialised foraging technique of gleaning insects from leaves, which necessitates a continuation of the southward flight in conjunction with the progress of the dry season (Salewski *et al.* 2002b). In March the returning birds have only slightly higher body masses and fat scores than during midwinter, probably because pre-migratory fattening does not commence until they are in the Sahel, *i.e.* well to the north of the Guinea vegetation zone (Dowsett & Fry 1971, Salewski *et al.* 2002a). The same strategy is found in Melodious Warblers, but this species differs in having two periods of intensive moult, one in autumn (remiges, body feathers) and the other from the second half of February onwards (body feathers, some tertials and secondaries). Willow Warblers start moulting in early winter, but essentially have a single moult cycle in the latter half of the northern winter (Salewski *et al.* 2004).

Another set of Palearctic migrants uses a two-step strategy on the journey to wintering areas in West Africa south of the Sahel, but in a different manner, as exemplified by Great Reed Warblers (Hedenström *et al.* 1993) and Garden Warblers (Ottosson *et al.* 2005). The first step occurs after arriving south of the Sahara; the birds remain in the Sahel, Sudan or Guinea zones for two months, using the time to rest, moult and fatten up. Moulting here is unusually rapid, probably being an evolutionary adaptation to cope with the advancing dry season and its concomitant decline in insect food (Bensch *et al.* 1991). Garden Warblers captured on the Jos Plateau in Nigeria in December carried fuel stores of up to 70% of lean body mass, strongly suggesting that the birds had prepared for the second leg of migration which may well take them beyond the Congo Basin (2500 km) in midwinter (Ottosson *et al.* 2005). Refuelling for the return flight takes place in the northern Guinea zone, hence avoiding the Sahel in spring and profiting from the northward moving rainbelt of the Intertropical Convergence Zone (Fig. 4). The high between-year repeatability of stable isotope profiles of individual Great Reed Warblers might indicate a preference for certain Afrotropical habitats (or the same sites?), apparently consistently providing food and shelter over the years (Yohannes *et al.* 2008).

Pied Flycatchers demonstrate yet another wintering strategy. This species flies without prolonged stopovers in the Sahel or Sudan zone to its wintering quarters in the Guinea zone, where it arrives in September and stays throughout the northern winter (Salewski *et al.* 2002a). Having a less specialised foraging behaviour, *i.e.* pouncing, hovering, jump-flights and aerial pursuits to capture insects on the ground, on leaves or in flight, Pied Flycatchers are able to defend a territory throughout the northern winter. Such a strategy allows not only for weight gain prior to the return migration but also the completion of a full moult in March (Salewski *et al.* 2004). High recurrence rates suggest that birds are faithful to wintering sites (22 of 94 birds in NE Ivory Coast; Salewski *et al.* 2000).

These strategies have in common that the driest part of the dry season is spent in regions with a more benign climate, although many birds may still be affected by adverse conditions during the return migration, during the fattening period in the Sahel or when crossing the Sahara (Chapters 20, 35, 36, 42: White Stork, Yellow

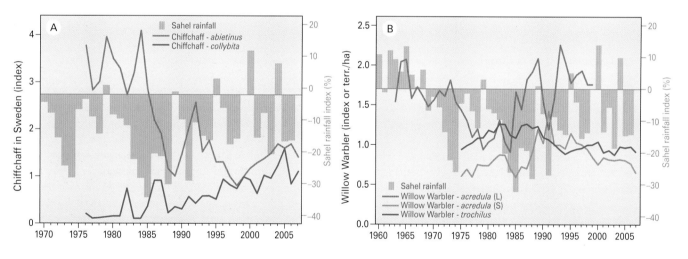

Fig. 272 (A) Trends of Chiffchaffs in Sweden, separately for *P.c. collybita* (south of 60°N) and for *P.c. abietinus* (above 60°N), showing opposing trend directions. The nominate winters in southern Europe (possibly extending somewhat into West Africa) and *abietinus* probably in the eastern Sahel and Sudan zone. Data from: Lindström *et al.* 2007, 2008). (B) Trends of Willow Warblers *Phylloscopus trochilus acredula* in Lapland (L) and Sweden north of 62°N (S) differ from the trend for *P.t. trochilus* in Sweden south of 62°N: *acredula* winters in eastern and southern Africa south of the Sahel and *trochilus* in the northern tropics of West Africa. The between-year population change in *trochilus* is related to the Sahel rainfall. Data from: Enemar *et al.* 2004 and Lindström *et al.* 2008.

Wagtail, Barn Swallow). Long-distance migrants wintering in West Africa south of the Sudan zone, namely those subjected to Sahel conditions for only a short period (Fig. 268A, codes 3-5), showed on average the smallest population declines, probably attesting to the profitability of spending the northern winter south of the Sudan zone. This idea could be tested by comparing trends of the same species breeding across a wide latitudinal range, either resulting in a migratory divide or in northern populations leapfrogging the wintering areas of their more southerly breeding congeners. Unfortunately, very few recorded trends have been differentiated according to subspecies or migratory strategy.

However, there are recorded trends at the subspecific level for Swedish Chiffchaffs. North of 60°N, birds are assigned to the subspecies *Phylloscopus collybita abietinus* and below this latitude to the nominate, *collybita*, which recently has colonised southern Sweden (Lindström *et al.* 2007). The scarcity of ring recoveries frustrates the identification of subspecies-specific wintering quarters, although those few which do exist, hint at a migratory divide (*abietinus* following a more easterly course, *collybita* a more westerly). Perhaps the wintering grounds of *collybita* do not extend as far south as those of *abietinus*, because many winter recoveries of Chiffchaffs ringed in Norway were from southern Europe and northern Africa, with only a few coming from the Sahel (Bakken *et al.* 2006). The few birds ringed in Sweden and recovered south of the Sahara were probably *abietinus* (Lindström *et al.* 2007). The distinction between both subspecies and their respective wintering areas is highly relevant, given their contrasting population trends (Fig. 272A); *collybita* showed consistently low indices between 1975 and 1985, then started to increase (the population in 2006 was estimated to be 15 times larger than in 1975, with little range expansion), but *abietinus* scored high index values up to the early 1980s, then starting to decline by about 75% between 1983 and 2006. Their breeding season habitat preferences differ, too: spruce for *abietinus* and deciduous woodland for *collybita*, but local habitat changes could not explain the differing fortunes of the two populations (Lindström *et al.* 2007). In Finland, *abietinus* had also declined (by 75% between 1983 and 1998), but then showed an increase by 2005 to about 50% of 1983 levels; Väisänen 2006). We posit that changing conditions along the migratory pathway or in the (eastern) Sahel-Sudan zone may be responsible. In Swedish *abietinus*, the proportional between-year changes in population size were weakly correlated with Sahel rainfall; no such effect was found in *collybita*. [6] The pattern exhibited by *collybita* in Sweden aligns with those recorded across western Europe. This subspecies winters predominantly in the Mediterranean Basin, rather than in the Sahel in West Africa.

All Chiffchaffs encountered singing in Mauritania, Senegal and Mali since the 1980s were Iberian Chiffchaffs *P. ibericus* (J. van der Kamp unpubl.), which is supported by recent research showing that this species is a long-distance migrant heading for West Africa (Pérez-Tris *et al.* 2003, Catry *et al.* 2005). Reliable trend data are not yet available for this species (Escandell 2006), but given its wintering areas in the Sahel-Sudan zone, a trend in line with *abietinus* is plausible. The increase of *collybita*, the subspecies wintering largely in southern Europe and northern Africa, coincides with the predominance of mild winters in Europe since the early 1990s. If *abietinus* indeed winters in sub-Saharan Africa, then most likely it would be in the wooded savannas of the northern tropics, where many more Eurasian long-distance migrants have been found to be in trouble (Fig. 268C).

Willow Warblers in Sweden provide a well-documented case study of a migratory divide with possible consequences for population dynamics: birds breeding north of 63°N are attributed to *Phylloscopus trochilus acredula*, those south of 61°N to *P.t. trochilus* (the intervening zone largely occupied by hybrids). Ring recoveries, backed up by isotope signatures from feathers collected in various parts of Africa, revealed that *acredula* winters largely in southern Africa and *trochilus* mostly in West Africa (Bensch *et al.* 2006). The trends between the two subspecies differ; *acredula* shows an increase from 1975 to 2007, *trochilus* a decline (Lindström *et al.* 2008). A separate trend for *acredula*, obtained in Lapland, northern Sweden, confirms the overall increase for this subspecies (Fig. 272B). Given that *trochilus* winters in wooded savannas in West Africa, that could explain its decline, whereas the more southerly wintering areas of *acredula* in eastern and southern Africa may be the central reason that this subspecies escaped any decline. The Sahel effect on annual population changes in *trochilus* is obvious: the proportional between-year changes in population size were strongly correlated with Sahel rainfall. As expected from its wider wintering range, population changes in *acredula* were not related to rainfall in the Sahel. [7]

The perfect species against which to test whether wintering at different latitudes in Africa might have a bearing on population trends is the Yellow Wagtail, whose subspecies can be identified in the field (males only). The northern subspecies *thunbergi* leapfrogs the central and southern European subspecies (*cinereocapilla*, *flava* and *feldegg*) to regions in West Africa that are generally less desiccated and have a shorter dry season (first rains by March) than the Sahel (Chapter 36). However, any advantage to *thunbergi* posed by these conditions is off-set by the fact that it moves gradually northwards in spring to make its trans-Saharan flight from the northern Sahel when conditions there are least favourable. Data on population trends from a sufficiently wide range across Europe are lacking, preventing a comparison of trends between the various subspecies. Moreover, although Sahel effects on trends are present, they are overridden by the impact of changes in the breeding areas (Chapter 36).

These examples show the immense complexity of migratory strategies. At present, lack of detailed information on wintering areas for the various subpopulations is one of many gaps in our understanding of the influence of sub-Saharan conditions on populations. In the case of Chiffchaffs and Willow Warblers, subspecific differences in wintering areas may be crucial in explaining their various vicissitudes on the breeding grounds.

The transition from forest to farmland takes decades. Low-lying clearings in woodland (left) may gradually lose their trees through felling or cutting back, until eventually few if any trees are left, the cleared area being converted to ricefields (right) (Casamance, September 2007).

Changing Sahel conditions: their impact on migratory birds

The effect of Sahelian drought on bird populations is epitomised by the population crash of Common Whitethroats in 1969 (Fig. 255). Common Whitethroats show high recurrence rates within winters (King & Hutchinson 2001), and are territorial on the wintering grounds (Boyi et al. 2007). Any drought effect on survival would then be highest when Common Whitethroat populations are at a high (in other words, pre-1969 levels). The negative effects of reduced food resources during the ensuing drought must have been smaller than in 1969 because populations were seriously depleted in the 1970s and 1980s and so intraspecific competition must have been less intense, improving the prospects of survival even under drought conditions.

Being territorial has clear advantages when food supply is predictable throughout the northern winter. Concomitant high recurrence rates might be expected through habitual return to areas occupied in previous winters. Recurrence is known to occur amongst many Palearctic migrant species wintering in or migrating through Senegal, The Gambia, Ivory Coast and northern Nigeria (Moreau 1969, Sauvage et al. 1998, Salewski et al. 2000, King & Hutchinson 2001), notably in flycatchers and *Hippolais* warblers. No attempts have as yet been made to differentiate recurrence rates between dry and wet years. In the Sahel and northern Sudan zone, insect and seed availability decline steeply in the course of the dry season (Morel 1968, Bille 1976); in years of poor rainfall, food availability is already low as the dry season begins. When food is scarce, erratic in availability or restricted in time and space, species that benefit are those that pursue an itinerant lifestyle, and can exploit transient food supplies (Jones 1999) or capitalise on temporary and local abundances of food consequent upon bush fires and rain fronts (e.g. White Stork and Montagu's Harrier; Chapters 20 and 26).

So many Palearctic long-distance migrants have been in decline since at least the 1970s, particularly those depending on the Sahel, that conditions in the wintering areas must have had a long-lasting and significant influence on numbers; evidence to that effect has accumulated steadily. The principal causes that have triggered the declines are climate change and human action (Chapters 2 and 4):

Climate change A reconstruction of rainfall patterns in past centuries showed long periods of high rainfall alternating with periods of drought. From about 1700 the overall rainfall trend, up to and including the 20th century, showed a steady decline (Fig. 10). The onset of a 30-year drought in 1969 (thought to be the first climate regime shift since the one about 5500 years ago that brought about the 'desert Sahara' from the 'green Sahara'; Foley et al. 2003) may have been triggered by rising sea-surface temperatures in the tropical Atlantic (Chapter 2), increasing solar radiation across the region and land degradation (see below). The persistent decline in rainfall led to an expansion of the Sahara southwards, especially in the 20th century when the periods of abundant rainfall were of much shorter duration than in previous centuries. The Sahel was particularly badly affected by this change. Furthermore, a small decline in rainfall has a large impact on the discharge of rivers which run across dry lands (Chapter 3), in its turn affecting the flood extent of Sahelian inundation areas (Chapter 6-10).

Rainfall and flooding are the life-giving instruments of the Sahelian ecosystem: abundant rainfall brings lush vegetation, promotes the leafing, fruiting, seeding and flowering of trees, shrubs and grasses, and leaves behind temporary pools that will hold water well into the dry season. The biomass and diversity of insects are correlated with the biomass of grasses and leaves, whereas temporary pools are important as refuges for insects and as vital egg-laying sites for phytophagous insects (Gillon & Gillon 1973). Even when insect abundance declines in the course of the dry season (Gillon & Gillon 1974), the total biomass of insects remains high and is available for a long period, unlike in years when the preceding seasonal rainfall has been low. This variation in weather pattern has a profound impact on the presence and abundance of Afrotropical and Palearctic birds during the dry season (Morel & Morel 1978).

Deforestation The steep human population growth in Sahelian countries, particularly evident from the mid-20th century onwards, triggered a long-term and ongoing loss of woodland and decline of woody cover in the savanna (Chapter 5), a trend exacerbated by droughts since 1969. Deforestation usually starts with cutting branches and selective felling, eventually reducing the number of tree species (Cresswell et al. 2007). Sahelian tree species differ in their phenology of leafing, fruiting and flowering (Arbonnier 2002; Fig. 267). An intact suite of indigenous trees offers Palearctic birds a sequence of food resources from September to April, which is the main reason why unspoilt woodland in the Sahel attracts more birds than degenerate woodland (Cresswell et al. 2007). Planting non-indigenous trees does not provide compensation for the loss of indigenous trees in terms of biological equivalence. The former are poor substitutes in terms of insect biomass and diversity (Box 3). Whereas deforestation means the loss of habitat for bird species utilising shrubs and woodland, it creates new habitat for species that can exploit savanna, whether original or derived. Examples of such beneficiaries are Palearctic wagtails, pipits and wheatears, Whinchats and Rufous Scrub Robins (Gatter & Mattes 1987, Vickery et al. 1999, Cresswell et al. 2007, Hulme 2007). It is worth noting that the derived savanna of Liberia is extensively used by Palearctic wagtails and pipits, but has never attracted their Afrotropical counterparts to breed (Gatter 1987). Deforestation may also clear the way for grasshoppers and their predators such as Cattle Egrets, African Swallow-tailed Kites, Montagu's Harriers and Lesser Kestrels (Chapter 14).

The steady loss of woody cover in the Sahel, Sudan and Guinea zones has triggered a trend towards deforestation of the Sahel, *sahelisation* of the Sudan and *sudanisation* of the Guinea zone. The steeper than average declines of Palearctic birds such as Common Redstarts, Western Bonelli's, Olivaceous and Subalpine Warblers is not surprising considering that they are typical winterers of northern *Acacia* woodlands. The few Palearctic birds wintering in gallery forests and rainforests, notably flycatchers, European Honey Buzzards and Eurasian Golden Orioles, seem to have escaped large declines (Table 35).

Agriculture The food demands of the growing human population resulted in a concomitant expansion of cropland at the expanse of woodland and wetlands. The clearance of trees and shrubs from farmland often brought other changes in farming systems, such as shortening of fallow cycles, conversion of non-cropland to agriculture, increased grazing pressure, greater harvesting of trees for firewood and lumber and increased encroachment of humans into parks and reserves. In degraded habitats in northern Nigeria, Afrotropical open-country birds preferred the more intensively farmed sites, but birds normally associated with savanna woodland became associated with lower-intensity farming. Where farming was less intensive, more bird species were observed (Hulme 2007). Whether Palearctic birds that usually occupy the more open habitats in their African wintering quarters are less susceptible to tree loss than Afrotropical bird species (as suggested by Cresswell et al. 2007) remains to be seen. The disproportionate declines of Palearctic migrants associated with Sahelian wintering quarters certainly suggest that Palearctic birds are under pressure.

In Africa, shifting cultivation and a variety of fallow systems traditionally have been applied to restore soil fertility on cropped land. Avian species richness is high on fallow land, but gradually decreases with time through disturbance (Söderström et al. 2003). Reduction of fallow land and intensification of land-use adversely affect bird communities, especially when removal of trees and shrubs also occurs. However, Whinchats profited from tree removal and the concomitant increase of herbaceous vegetation growth (Hulme 2007). In West Africa, fields of stunted rice remain unharvested, in some years providing high food abundance for birds over large areas long after harvest, similar to the effects of short-cycle fallow farming. Many species, including Fulvous Whistling Ducks, capitalise on such food bonanzas (Fig. 273).

Grazing The impact of the mounting grazing pressure in the Sahel is manifold. Some bird species, like Yellow Wagtails and Cattle Egrets, may have profited, but the levels of increased disturbance,

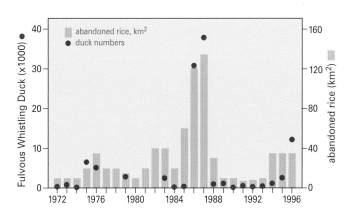

Fig. 273 Number of Fulvous Whistling Ducks in the Senegal Delta counted in January between 1972 and 1996, related to the surface area of abandoned rice fields. From: Tréca 1999.

An exclosure near Lake Do, north of Douentza in Mali, clearly shows the impact of grazing: within the exclosure (right) the ground is covered by a dense grassy vegetation, but the ground is bare in the grazed field (February 2009). Grazing has not so far affected the woody vegetation, however, with *Acacia tortillis raddiana* as the dominant tree species on both sides of the gate, and a scatter of *Leptadenia pyrotechnica* scrub and a few *Balanites aegyptiaca* trees.

wood-cutting, burning and predator eradication associated with livestock have altered habitats and destroyed wildlife. The decline of vultures, as noted by Thiollay (2006a, b, c) in the Sahel and Sudan zones of West Africa between the 1970s and early 2000s, had started in the 1940s and 1950s when the indiscriminate use of strychnine to reduce livestock predation poisoned Hyenas *Hyaena hyaena* and Jackals *Canis aureus* in southern Darfur and Mali and, as a side-effect, had wiped out vultures over vast areas (Wilson 1982).

Grazing pressure in semi-arid West Africa, which historically was always at its highest at the beginning of the rainy and dry seasons, has become more persistent across seasons. Since the droughts in the early 1970s, many pastoralists have settled to farm, expanding the cultivated area to the detriment of rangeland and leading to spatial dispersion of livestock and a higher grazing pressure relative to forage availability (Turner 2005). Although the productivity of the herbaceous vegetation is influenced mainly by soil conditions and the amount and distribution of rainfall, the negative impact of grazing is increasing. This impact varies: floodplains are more resilient to grazing than the drylands, as each flooding produces a new outburst of vegetation growth. Persistent grazing of drylands, especially under conditions of drought, eventually leads to loss of tree and shrub cover and prevents grasses from producing seeds and the remaining shrubs and small trees from fruiting. The ill-effects of grazing (and drought) are hardest felt in the northern Sahel, the region from which many Palearctic migrants depart after having fattened on fruits in February and March (*e.g.* Chapter 41, Common Whitethroat).

Loss of wetlands   The climate-driven long-term decline of the flood extent in river basins has been exacerbated by embankments, dams, the creation of water reservoirs and the increase in irrigation of fields upstream (Chapter 12). Many species that concentrate in floodplains during the northern winter have gradually lost part of their habitat, a potential risk to population sizes (Chapter 13).

The embankment of the Senegal Delta effectively converted a dynamic floodplain (one of the few in the Sahel) into static farmland, destroying a major wintering area for western European breeding birds like Purple Herons, Ruff and Black-tailed Godwits (Chapter 7). The recovery of Purple Herons in The Netherlands from the 1990s onwards, in line with improved rainfall in the Sahel, lagged behind

what could have been expected from the population levels when similar rainfall occurred in the 1970s; it is tempting to suggest the loss of the floodplain of the Senegal Delta, a major wintering area, as a causal factor. The impact on bird numbers of a loss of wintering grounds is difficult to assess, because factors affecting the breeding grounds may be dominant (Black-tailed Godwits). Indeed, newly-created habitats or adaptation to alternative habitats may dilute the immediate consequences of such losses. The embankment, and subsequent partial rehabilitation, of the Waza Logone in northern Cameroun is a telling example (Scholte 2006). In the case of floodplains, few as they are and being essential lifelines to birds in the Sahel, any reduction in their size has a negative effect. For example, population changes experienced by Purple Herons and Sedge Warblers were found to be closely related to flood extent, and only weakly to rainfall. Further permanent reductions in floodplain size must be expected, given the plans to extend the irrigated areas along the middle and lower Senegal and to increase the number of dams in the Niger and Bani Rivers (hydropower, irrigation schemes; Chapter 12). The reduction of floodplain size by dams and reservoirs also increases the vulnerability of birds during periods of low rainfall (Chapter 6-10).

In the wake of the expanding human population and droughts, human encroachment on wetlands has increased through cultivation and often over-exploitation of natural resources. Protection of wetlands may prevent or reduce many adverse human-related changes (as in the Senegal Delta; Chapter 7, Box 9), but only when it is applied effectively (Box 12; Caro & Scholte 2007). Smaller wetlands have been hit disproportionately by human-induced change, often to the extent that the natural habitat has disappeared completely. Natural riparian and marshland vegetations have been converted

The simplest of devices, such as snares, can be remarkably effective in catching large birds when their use is combined with knowledge of the behaviour and habitat choice of the intended victims. Snaring egrets and herons is common practice in wetland vegetation. Single visits to the local bird trader in Goundam, Mali, in March 2003 and February 2009, revealed that about 250 waterbirds had been delivered by the bird catchers from Lake Tele (and to a lesser degree from Lake Horo) each day. Most victims were Garganey, but we also found several Black-crowned Night Herons, Great Bitterns, Purple Herons, Common Moorhens and Purple Swamphens. Eurasian migrants are captured from January-March, so we may estimate that some 25 000 water birds find their way to the local market each year. The feathers and bird remnants along the water's edge near a fishing village in Lake Horo (again containing remarkably many Great Bitterns) bear witness that even more birds are captured by the local fishermen for their own consumption. The dense vegetation makes it difficult to estimate the size of the local bird population, but plot counts from March 2003 and 2004 suggest the presence of, for instance, 1200 Purple Herons and 3600 Purple Swamphens (Table 7). The human predation on birds in both lakes is sufficiently high to create a sink for several species.

Box 29
Swimming against the stream?

While writing this book, one of us was invited to attend a meeting of wetland wardens, to tell them something about the whereabouts of 'their' breeding birds in Africa during winter. Following the presentation, the discussion focussed on the question whether populations of migratory birds were limited by factors operating on the breeding grounds or wintering grounds. Ultimately, how did that relate to conservation efforts undertaken in the breeding areas? As one of the wardens wondered, "How can we explain our efforts and expenditure to the public and politicians, if population size is determined more than 4000 km away?" This question was relevant, for despite all their efforts to safeguard the breeding birds, they had witnessed the ongoing decline of several migratory birds that bred in their nature reserves. In other words, does it make sense to swim against the stream, given the scale of wetland loss in the Sahel?

In part-answer, the results obtained by their Danish colleagues in the Vejlerne wetland reserve are particularly illuminating, and provide some hope. Garganey have declined all over western Europe, but still thrive in Vejlerne, increasing from 4 pairs in 1985 to 184 in 2000 (Fig. 272). This increase is clearly associated with improved management, notably the raising of water levels in May. However, a statistical analysis shows that annual population changes can also be explained by conditions in the Sahel during the northern winter. The interaction between year, water level in May and flood extent in the Sahel explained much of the population trend, with each variable being significant.[8] This finding implies that bird populations may recover more quickly when breeding conditions are optimal and that annual variations in numbers depend on the conditions in both breeding and wintering areas. As we have shown, the improved flood extent in the Sahel since 1994 coincided with a gradual increase of migratory wetland species. Thus, the swimming-against-the-stream view is over-pessimistic, at least for most species wintering in Sahelian wetlands. The population development of Garganey in Vejlerne, in itself no more than an anecdote, clearly shows that conservation efforts in the breeding areas do make a difference.

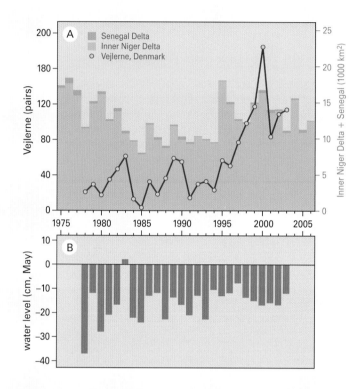

Fig. 274 (A) Breeding population of Garganey in the Veljerne wetland reserve (55 km²) in northern Jutland, related to the flood extent of the Inner Niger Delta and the Senegal Delta.
(B) Water level in May in Bygholmengen, the area within Vejlerne where most Garganey breed; water level has gradually risen some 10-20 cm from -35 cm in 1978. From: Kjeldsen 2008.

into farmland, catchment areas of rainfall and depressions into irrigated areas (Chapter 11). Although some of these schemes have attracted high numbers of ducks and weavers, the overall effect on avian diversity is negative. The newly-created reservoirs and irrigation schemes do not compensate for the overall effects of habitat loss caused by the disappearance or conversion of wetlands.

Bird exploitation   Shooting on the scale witnessed in the Mediterranean is not widespread in sub-Saharan Africa, except locally by European shooting parties (*e.g.* Turtle Dove, Chapter 32). However,

much cheaper indigenous methods have been honed to perfection to enable the capture or killing of large numbers of birds for consumption (Chapter 13 and Box 25). The toll taken by the local populace varies from insignificant to highly significant, and is location-dependent. Where birds are concentrated in few areas, even simple catching methods can cause a high death toll, the Inner Niger Delta in Mali being such a spot. At the end of the northern winter, and well into the local dry season, waterbirds are forced into remaining pockets of water, where they are easy prey to those who are adept in the ways of the quarry. Gargeney (Chapter 23) and Ruff (Chapter 29)

Eurasian long-distance migrants have been found to be in trouble in the Sahel. This is even more the case for African breeding birds, especially the larger species which are targeted as food. Many waterbird species which have become scarce in the Sahel are still quite common in eastern and southern Africa. The Sacred Ibis is one of the exceptions, as it remains widespread throughout African wetlands, including those in the Sahel.

are among the species affected most, especially in dry years when the catching period stretches to three months. Persistent droughts, as from 1969 onwards, are likely to have a depressing effect on the population levels of these species, because the survivors of autumn migration and adverse winter conditions, normally those who comprise the core breeding population, can be badly hit. Pre-migratory fattening is difficult enough to achieve, without being captured in the drought-stricken, desiccated landscape amidst intensive bird-catching in the few vestigial pools. Human-induced mortality then adds to the death toll. In the same vein, the impact of man on colonial breeding birds is devastating. Without any real protection of colonies or breeding haunts, populations of cormorants, darters, pelicans, herons, storks, ibises, spoonbills, flamingos, cranes and large bustards will become depleted in the long run (Box 9 and 12). Many species have reached that status already (Thiollay 2000b, c).

A large variety of birds is captured in sub-Saharan Africa. Although large bird species are particularly targeted, passerines and near-passerines are also taken in numbers, which must have increased over the past decades, even if there is no documentation available, for the following reasons: the human populace has multiplied, encroaching upon natural habitats (including protected sites) to an increasing degree; the standard of living has deteriorated under the present drier conditions and so income is sought from whatever source (such as birds) that can be exploited, just at the time when inexpensive means of trapping birds (nylon nets and hooks) became available across the region, thus facilitating large, marketable catches.

Some thoughts on the future

There is no denying that Eurasian birds wintering in the northern savannas, especially those in the Sahel, are in trouble. For some species, sufficient evidence has accumulated to consider a causal relationship between Sahel conditions and migrant numbers. Indeed, population fluctuations in about half the species dealt with at length in this book are thought to have been largely determined by rainfall or flood extent in the Sahel. Such declines are believed to be short-lived when related to short-lived droughts, but the regime shift in 1969 may herald a prolonged period of reduced rainfall, a circumstance for which increasing evidence is amassing. Despite some improvements in rainfall in the 1990s and 2000s, the rainfall index of the Sahel remains just below the long-term average for the 20th century. The ongoing drought may have caused a structural decline at the population level for a large number of Palearctic species dependent on the Sahel and adjacent vegetation zones during the northern winter. Furthermore, the effects of drought are magnified by negative changes in the Sahel associated with human population growth (see above section), which seems irreversible. What may have started through climate change has become a knot of inter-related circumstances whose adverse developments reinforce each other.

Is this scenario for doom inevitable, therefore? Not necessarily. First of all, the northern savannas are relatively thinly populated. We are well aware that even low densities of humans can have far-reaching consequences for the environment, but matters would have been even worse if densities had been high in the first place. Secondly, habitat changes that negatively affect woodland species can turn out to be positive for inhabitants of savannas. Loss of indigenous trees in the woody savanna, for example, adversely affects Eurasian Wrynecks and Common Redstarts, but – for the moment – may be positive for Montagu's Harriers and wheatears. Thirdly, the destruction of evergreen forest, the least acceptable habitat for Palearctic migrants but of outstanding importance to African bird species, increases the area of seasonally bare ground possessing early stages of regenerating 'bush', as was noted by Moreau as early as 1952. Fourthly, many man-made changes are, at present, small-scale and non-chemical (usage of herbicides and pesticides is virtually non-existent in large parts of Africa), often promoting avian diversity. Finally, the floodplains in the Sahel are highly resilient ecosystems. Although droughts may seem to have dramatic consequences, floodplains do retain water from previous floods for some years (thus delaying immediate negative effects), and bounce back to full life as soon as a high flood occurs. To some extent, the same can be said for drylands, except that droughts here are often accompanied by loss of trees, which take longer to be replaced. Such positive effects are inherently fragile and cannot be maintained if present trends in climate, human population growth and land-use continue unabated.

Whatever the impact of Sahel conditions on migrant numbers, the fact remains that changes in the breeding quarters and on stopover sites along migration routes may have malign influences on the same scale. The preceding chapters have shown that land changes in Europe often had an overriding influence on population trends, indirectly visible from the progressively smaller declines recorded across Europe from west to east (Fig. 271, see also Gregory et al. 2007). This cline coincides with a decline in the density of the human population, a declining intensity of farming and forestry, and an increase in woodland cover. In contrast to sub-Saharan Africa, the changes in Europe have left but few spots of natural habitat and long-term changes appear irreversible. The composition and scale of the present European bird life are quite different from those of the 1950s and even more so from those of the 19th century. In the words of Moreau (1952): "It cannot be doubted that at the present time Africa has to accommodate in the winter far fewer marsh birds, wheatears, quails, raptors and probably swallows and martins now than a hundred years ago." The list today would be a little different, but certainly much longer.

Is this then the future for sub-Saharan Africa, analogous to the earlier experience of Europe: a drastically changed landscape, a completely different composition of the Afrotropical fauna and greatly diminished numbers of Palearctic migrants? We take some comfort in the thought that ecologists have sometimes been able to explain why bird species have declined or increased in the past, but have been singularly inept at predicting future trends correctly. We hope, against our better judgment, that our predictions will also prove incorrect and that Afro-Palearctic birds will be sufficiently resilient to cope with the forthcoming changes, whether numbering 5000 million or not.

Endnotes

1 The list of 127 species comprises 84 species occurring in the western Sahel (Fig. 105). We added 34 species wintering south of the Sahel zone (European Honey Buzzard, Lesser Spotted Eagle, Steppe Eagle, Red-footed Falcon, Eurasian Hobby, Corn Crake, Black-winged Pratincole, Little Ringed Plover, Temminck's Stint, Common Snipe, Green Sandpiper, Common Sandpiper, Common Cuckoo, Common Swift, Alpine Swift, European Bee-eater, European Roller, Common House Martin, Thrush Nightingale, Eurasian River Warbler, Marsh Warbler, Olive-tree Warbler, Icterine Warbler, Garden Warbler, Barred Warbler, Blackcap, Wood Warbler, Willow Warbler, Spotted Flycatcher, European Pied Flycatcher, Collared Flycatcher, Red-backed Shrike, Lesser Grey Shrike, Eurasian Golden Oriole), 8 species occurring in the eastern Sahel (Levant Sparrowhawk, Demoiselle Crane, Cyprus Wheatear, Pied Wheatear, Rüppell's Warbler, Ménétries' Warbler, Masked Shrike, Cretzschmar's Bunting), and the recently split Iberian Chiffchaff (for which we used the trend given in Escandell [2006] for 1996-2005). We excluded Steppe Buzzard (sp. *vulpinus*), Spotted Crake, Little Crake, Baillon's Crake and Isabelline Shrike for which reliable population trends are not available.

2 In total, ten species had the combined trend score of -3 for the period 1970-2005: nine with -2 in 1970-1990 and -3 in 1990-2000 and one with -3 in 1970-1990 and -1 in 1990-2000. In total, 29 species had the combined score of -2: 23 species had -2 in 1970-1990 (of which 11x -2, 5x -1 and 7x 0 in 1990-2000), 4 species had -1 in 1970-1990 and -1 in 1990-2000 and 2 species had 0 in 1970-1990 and -3 in 1990-2000. In total, 26 species had an overall score of 0, being stable in 1970-1990 and 1990-2000, as did three other species showing contrasting trends. Other combinations also had combined trend scores of -1. However, after a comparison of the combined trends with the data included in Chapter 15-41, and with similar monitoring data for other Afro-Palearctic species, we decided to adjust the com-

bined trend for five species as follows:

Spoonbill -2 → -1; the large decline of the Russian population between 1970 and 1990 is almost nullified by the large and consistent increase of the Atlantic population since the 1970s.

Glossy Ibis: -2 → -3, based on the large decline observed in the Sahel (Chapter 21).

Ruff: -1 → -2, due to the large decline of the western population, but also in the Sahel (Chapter 29).

Common Redstart: -2 → -3, based on the large decline described in Chapter 37.

Common Whitethroat: 1 → 0. The population at the end of the last century was indeed larger than in 1970, but taking into account the decline in 1969, the population reached the late 1960s level in the 2000s (Chapter 41).

Demoiselle Crane: +3 → 0. The evidence for a strong increase is rather thin. In fact, the sub-populations of Demoiselle Cranes breeding in the Black Sea area and to the west have been in decline for decades, disappearing from Morocco in the mid-1980s (Thévenot et al. 2003) and nearly so from Turkey (just 10-20 pairs left in the northeast; Kirwan et al. 2008). The Black Sea population has been estimated at fewer than 250 pairs and declining, the Kalmykia population at 30 000-35 000 pairs and stable (www.npwrc.usgs.org/resource/birds/cranes/anthvirg.htm, accessed 15 December 2008). These birds winter mostly in the Sahel across Ethiopia and Sudan, westwards up to Chad and northern Cameroon.

3 Counts in West Africa suggest an exchange of Eurasian Spoonbills between the Senegal Delta and the Mauretanian Banc d'Arguin. First, the number of Spoonbills in the Senegal Delta is positively correlated to the flood extent of the Senegal River (R=+0.43, n=18, P=0.05). Secondly, if the number in the Senegal Delta is expressed as a percentage of the number on the Banc d'Arguin and in the Senegal Delta combined, relatively more birds spend the winter in the Senegal Delta in years with a more extensive flooding (R=+0.39, n=8, NS).

4 Data extracted from (N= number of migratory species covered): RSPB 2008 (Britain, 1970-2006, breeding birds, N=15), Bijlsma et al. 2001 (The Netherlands, 1970-2000, breeding birds, N=50), Vermeersch et al. 2004 (Flanders, 1974-2002, breeding birds, N=46), Yeatman-Berthelot & Jarry 1994 (France, 1970-1989, breeding birds, N=51), Moritz 1993 (Helgoland=1, 1960-1991, autumn migration, N=12, spring migration, N=13), Berthold & Fiedler 2005 (Mettnau=2, 1972-2003, autumn migration N=21), Schmid et al. 1998 (Switzerland, 1972-1996, breeding birds, N=53), Heldbjerg & Fox 2008 (Denmark, 1975-2005, breeding birds, N=21), Lausten & Lyngs 2004 (Christiansø=3, 1976-2001, autumn migration, N=21), Karlsson 2007 (Falsterbo=4, 1980-2006, autumn migration, N=24), Lindström et al. 2008 (Sweden, 1975-2007, breeding birds, N=33), Koskimies 2005 (Finland, 1980-2004, breeding birds, N=59), Busse 1994a, (Baltic coast=5, 1961-1990, autumn migration, N=14), Tomiałojć & Głowaciński 2006 (Poland, 1951-2000, breeding birds, N=61), Reif et al. 2006 (Czech Republic, 1981-2006, breeding birds, N=34), Sokolov et al. 2001 & Payevsky 2006 (Pape=6, 1967-1994, autumn migration, N=13), Sokolov et al. 2001 & Payevsky 2006 (Baltic coast=7, 1971-2000, autumn migration, N=17), Sokolov et al. 2001 & Payevsky 2006 (Rybachy=8, 1957-2000, autumn migration, N=17), Sokolov et al. 2001 & Payevsky 2006 (Kiev region=9, 1976-1998, autumn migration, N=9).

The data vary considerably in their explanatory power, with breeding bird data (distribution or numbers) and the longest trend series generally being more reliable. Data from Constant Effort Sites, here mostly referring to catches made dur-

Roost flight of egrets and herons. (Inner Niger Delta, November 2008).

ing the post-breeding migration, refer to a wider geographical region to the N or NE of the catching site. Fluctuations in populations may also be partly obscured in migration data by annual variations in productivity, but consistent trends or steep declines will normally show up in the catching totals.

The trends shown in Fig. 269 are significant: (countries: P<0.01, CES: P<0.05).

5 Trend data for Ortolan Buntings are in fact not available for the eastern European populations. The population estimates for Romania and Russia for 1970-1990 (Tucker & Heath 1994) and for 1990-2000 (BirdLife International 2004a) show the point. Respective population sizes in 1990-2000 had been raised by one or two orders of magnitude in comparison to 1970-1990, indicating a more realistic estimation of numbers. Assessing a population trend for a third of the European surface area, when pair numbers run into the millions, is impossible without a well-designed monitoring effort. The more stable trend in eastern Europe, although intuitively a likely scenario, should therefore be viewed with some reservation.

6 The log between-year population change was positively, but non-significantly, correlated with Sahel rainfall (R=+0.19, P=0.34, N=31) in the Swedish *abietinus*; no effect was found in *collybita* (R=+0.10, P=0.60, N=31).

7 The log population change between successive years was positively correlated with Sahel rainfall (R=+0.40, P=0.03, N=32) in the Swedish *trochilus*, but no such effect was found in *acredula* (R=-0.06, P=0.76, N=32).

8 Log population change between successive years was related to the water level in the breeding area (R=+0.34, P=0.089) and the Sahelian flood extent (R=+0.31, P=0.13). The population change was dependent mainly on the log population size the year before (R=-0.44, P=0.024), which may be interpreted as a density-dependent effect, but the counts were not precise, so a statistical artifact cannot be excluded. A multiple regression revealed that the three variables combined explained more than half of the variance (R^2=0.521); each was significant (population size: P=0.001, Sahel flood extent: P=0.013, water level in May: P=0.021).

References[1]

A

Abdou Nalam, I. 2004. Chasse aux oiseaux d'eau du Niger. Sévaré, Mali: Wetlands International.

Abou-Seedo, F., D.A. Clayton & J.M. Wright. 1990. Tidal and turbidity effects on the shallow-water fish assemblage of Kuwait Bay. Mar. Ecol. Prog. Ser. 65: 213-223.

Acharya, G. & E.B. Barbier. 2000. Valuing groundwater recharge through agricultural production in the Hadejia-Nguru wetlands in northern Nigeria. Agricult. Econ. 22: 247-259.

Adams, A. 1999. Social impacts of an African dam: equity and distributional issues in the Senegal River Valley. Cape Town: World Commission on Dams (WCD).

Adams, W.M. 1993. The wetlands and conservation. In: G.E. Hollis, W.M. Adams & M. Aminu-Kano (eds). The Hadejia-Nguru Wetlands: Environment, Economy and Sustainable Development of a Sahelian Floodplain Wetland: 211-214. Gland: IUCN.

Adite, A. & K.O. Winemiller. 1997. Trophic ecology and ecomorphology of fish assemblages in coastal lakes of Benin, West Africa. Ecoscience 4: 6-23.

Affre, G. 1975. Estimation de l'évolution quantitative des populations aviennes dans une région du Midi de la France au cours de la dernière décennie (1963-1972). L'Oiseau et RFO 45: 165-187.

Agostini, N. & C. Coleiro. 2003. Autumn migration of Marsh Harriers (Circus aeruginosus) across the central Mediterranean in 2002. Ring 25: 47-52.

Aidley, D.J. & R. Wilkinson. 1987. The annual cycle of six Acrocephalus warblers in a Nigerian reed-bed. Bird Study 34: 226-234.

Akinsola, O.A., A.U. Ezealor & G. Polet. 2000. Conservation of waterbirds in the Hadejia-Nguru Wetlands, Nigeria: current efforts and problems. Ostrich 71: 118-121.

Akriotis, T. 1991. Weight changes in the Wood Sandpiper Tringa glareola in south-eastern Greece during the spring migration. Ring. & Migr. 12: 61-66.

Aksakov, S.T. 1998. Notes of a provincial wildfowler. Evanston, Illinois: Northwestern University Press.

Alerstam, T. 1990. Bird migration. Cambridge: Cambridge University Press.

Alerstam, T., M. Hake & N. Kjellén. 2006. Temporal and spatial patterns of repeated migratory journeys by ospreys. Anim. Behav. 71: 555-566.

Altenburg, W., M. Engelmoer, R. Mes & T. Piersma. 1982. Wintering waders on the Banc d'Arguin. Leiden: Wadden Sea Working Group.

Altenburg, W. & J. van der Kamp. 1985. Importance des zones humides de la Mauritanie du Sud, du Sénégal, de la Gambie et de la Guinée-Bissau pour la Barge à queue noire (Limosa l. limosa). Leersum: Research Institute for Nature Management.

Altenburg, W. & J. van der Kamp. 1986. Oiseaux d'eau dans les zones humides de la Mauritanie du Sud, du Sénégal et de la Guinée-Bissau; octobre-décembre 1983. Leersum: Research Institute for Nature Management.

Altenburg, W., A.J. Beintema & J. van der Kamp. 1986. Observations ornithologiques dans le delta intérieur du Niger au Mali pendant les mois de mars et août 1985 et janvier 1986. Leersum: Research Institute for Nature Management..

Altenburg, W. & T.M. van Spanje. 1989. Utilization of mangroves by birds in Guinea-Bissau. Ardea 77: 57-70.

Altenburg, W. & J. van der Kamp. 1991. Ornithological importance of coastal wetlands in Guinea. Cambridge: International Council for Bird Preservation.

Altenburg, W. & E. Wymenga. 2000. Help, de Grutto verdwijnt! De Levende Natuur 101: 62-64.

Amat, J.A. 2006. Should females of migratory dabbling ducks switch mates between wintering and breeding sites? J. Ethol. 24: 297-300.

Amatobi, C., S. Apeji & O. Oyidi. 1987. Effect of some insectivorous birds on populations of grasshoppers (Orthoptera) in Kano State, Nigeria. Samaru J. Agric. Res. 5: 43-50.

Ambagis, J., J. Brouwer & C. Jameson. 2003. Seasonal waterbird and raptor fluctuations on the Niger and Mékrou Rivers in Niger. Malimbus 25: 39-51.

Anthes, N., I. Harry, K. Mantel, A. Müller, H. Schielzeth & J. Wahl. 2002. Notes on migration dynamics and biometry of the Wood Sandpiper (Tringa glareola) at the sewage farm of Münster (NW Germany). Ring 24: 41-56.

Anyamba, A. & C.J. Tucker. 2005. Analysis of Sahelian vegetation dynamics using NOAA-AVHRR NDVI data from 1981-2003. J. Arid Environ. 63: 596-614.

Arbonnier, M. 2002. Arbres, arbustes et lianes des zones sèches d'Afrique de l'Ouest. Paris: CIRAD-MNHN.

Archaux, F., P.-Y. Henry & G. Balança. 2008. High turnover and moderate fidelity of White Storks Ciconia ciconia at a European wintering site. Ibis 150: 421-424.

Arroyo, B., J.T. Garcia & V. Bretagnolle. 2002. Conservation of the Montagu's harrier (Circus pygargus) in agricultural areas. Anim. Conserv. 5: 283-290.

Arroyo, B.E. & J.R. King. 1995. Observations on the ecology of Montagu's and Marsh Harriers wintering in north-west Senegal. Ostrich 66: 37-40.

Arroyo, B.E. 1997. Diet of Montagu's Harrier Circus pygargus in central Spain: analysis of temporal and geographic variation. Ibis 139: 664-672.

Arroyo, B.E. & V. Bretagnolle. 2004. Circus pygargus Montagu's Harrier. BWP Update 6: 41-55.

Arroyo, B.E. & V. Bretagnolle. 2007. Interactive effects of food and age on breeding in the Montagu's Harrier Circus pygargus. Ibis 149: 806-813.

Ash, J. 1995. An immense Swallow roost in Nigeria. BTO News 200: 8-9.

Ash, J.S., I.J. Ferguson-Lees & C.H. Fry. 1967. B.O.U. expedition to Lake Chad, northern Nigeria, March-April 1967. Preliminary report. Ibis 109: 478-486.

Ash, J.S. 1969. Spring weights of trans-Saharan migrants in Morocco. Ibis 111: 1-10.

Ash, J.S. 1977. Turtle Dove migration in southern Europe, the Middle East and North Africa. British Birds 70: 504-506.

Ash, J.S. 1990. Additions to the avifauna of Nigeria, with notes on distributional changes and breeding. Malimbus 11: 104-116.

Ashall, C. & P.E. Ellis. 1962. Studies on numbers and mortality in field populations of the Desert Locust. Anti-Locust Bull. 38.

Ashford, R.W. 1970. Yellow Wagtails Motacilla flava at a Nigerian winter roost: analysis of ringing data. Bull. Nigerian Orn. Soc. 7: 24-26.

Augusseau, X., P. Nikiéma & E. Torquebiau. 2006. Tree biodiversity, land dynamics and farmers strategies on the agricultural frontier of southwestern Burkina Faso. Biodivers. Conserv. 15: 613-630.

Aunins, A. & J. Priednieks. 2003. Bird population changes in Latvian farmland, 1995-2000: responses to different scenarios of rural development. Ornis Hungarica 12-13: 41-50.

Axelsen, J., B. Petersen, I. Maiga, A. Niassy, K. Badji, Z. Ouambama, M. Sønderskov & C. Kooyman. 2008. Simulation studies of the Senegalese grasshopper ecosystem interactions II : the role of egg pod predators and birds. Intern. J. Pest Management: in press.

B

Baddeley, J. 1940. The rugged flanks of Caucasus. Vol. 1 and 2. London: Oxford University Press.

Bader, J. & M. Latif. 2003. The impact of decadal-scale Indian Ocean sea surface temperature anomalies on Sahelian rainfall and the North Atlantic Oscillation. Geophysical Research Letters 30: 2169.

Baillie, S.R. & W.J. Peach. 1992. Population limitation in Palearctic-African migrant passerines. Ibis 134 suppl.1: 120-132.

Baillie, S.R., J.H. Marchant, H.Q.P. Crick, D.G. Noble, D.E. Balmer, C. Barimore, R.H. Coombes, I.S. Downie, S.N. Freeman, A.C. Joys, D.I. Leech, M.J. Raven, R.A. Robinson & R.M. Thewlis. 2007. Breeding Birds in the Wider Countryside: their conservation status 2007. (http://www.bto.org/birdtrends). Thetford: BTO.

Baillon, F. & J.P. Cormier. 1993. Variations d'abondance de Circus pygargus (L.) dans quelques sites du Sénégal entre les hivers 1988-1989 et 1989-1990. L'Oiseau et RFO 63: 66-70.

Bairlein, F. 1981. Analyse der Ringfunde von Weiszstörchen (Ciconia ciconia) aus Mitteleuropas westlich der Zugscheide: Zug, Winterquartier, Sommerverbreitung vor der Brutreife. Vogelwarte 31: 33-44.

Bairlein, F. 1988. Herbstlicher Durchzug, Körpergewicht und Fettdeposition von Zugvögeln in einem Rastgebiet in N-Algerien. Vogelwarte 34: 237-248.

Bairlein, F. 1991. Population studies of White Storks (Ciconia ciconia) in Europe. In: C.M. Perrins, J.-D. Lebreton & G.J.M. Hirons (eds). Bird population studies: relevance to conservation and management: 207-229. Oxford: Oxford University Press.

Bairlein, F. 1992. Recent prospects on trans-Saharan migration of songbirds. Ibis 134 suppl. 1: 41-46.

Bairlein, F. 2002. How to get fat: nutritional mechanisms of seasonal fat accumulation in migratory songbirds. Naturwissenschaften 89: 1-10.

Bakken, V., O. Runde & E. Tjørve. 2006. Norwegian Bird Ringing Atlas, Vol. 2. Stavanger: Stavanger Museum.

Balança, G. & M.-N. De Visscher. 1990. Le peuplement des sauteriaux, II. In: J. Everts (ed.). Environmental effects of chemical locust and grasshopper control. A pilot study. Rome: FAO.

Balança, G. & M. Schaub. 2005. Post-breeding migration ecology of Reed Acrocephalus scirpaceus, Moustached A. melanopogon and Cetti's Warblers Cettia cetti at a Mediterranean stopover site. Ardea 93: 245-257.

Balas, N., S.E. Nicholson & D. Klotter. 2007. The relationship of rainfall variability in West Central Africa to sea-surface temperature fluctuations. Intern. J. Clim. 27: 1335-1349.

Baldé, O. 2004. Evaluation de l'exploitation des oiseaux d'eau dans le Delta du fleuve Sénégal. Bambey, Senegal: ENCR.

Báldi, A., P. Batáry & S.Erdös. 2005. Effects of grazing on bird assemblages and populations of Hungarian grasslands. Agriculture Ecosystems & Environment 108: 251-263.

Bannerman, D.A. 1958. The Birds of the British Isles, Vol. 7. Edinburgh: Oliver and Boyd.

Bannerman, D.A. 1959. The Birds of the British Isles, Vol. 8. Edinburgh: Oliver and Boyd.

Bannerman, D.A. 1960. The Birds of the British Isles, Vol. 9. Edinburgh: Oliver and Boyd.

Barbier, E.B. & J.R. Thompson. 1998. The value of water: Floodplain versus large-scale irrigation benefits in northern Nigeria. Ambio 27: 434-440.

Barbier, E.B. 2003. Upstream dams and downstream water allocation: The case of the Hadejia-Jama'are floodplain, northern Nigeria. Water Resources Research 39.11: 1311-1319.

Barbosa, A. 2001. Hunting impact on waders in Spain: effects of species protection measures. Biodivers. Conserv. 10: 1703-1709.

Barbraud, C., J.C. Barbraud & M. Barbraud. 1999. Population dynamics of the White Stork Ciconia ciconia in western France. Ibis 141: 469-479.

Barbraud, C. & H. Hafner. 2001. Variation des effectifs nicheurs de hérons pourprés Ardea purpurea sur le littoral méditerranéen français en relation avec la pluviométrie sur les quartiers d'hivernage. Alauda 69: 373-380.

Bargain, B., C. Vansteenwegen & J. Henry. 2002. Importance des marais de la baie d'Audierne (Bretagne) pour la migration du Phragmite des joncs Acrocephalus schoenobaenus. Alauda 70: 37-55.

Basciutti, P., O. Negra & F. Spina. 1997. Autumn migration strategies of the Sedge Warbler Acrocephalus schoenobaenus in northern Italy. Ring. & Migr. 18: 59-67.

Bashir, M., A. Hassanali & C. Kooyman. 2007. Southern Red Sea trials report. March-April 2007. Unpublished report. Nairobi: ICIPE.

Bates, C.L. 1933. Birds of the southern Sahara and adjoining countries in French West Africa. Ibis 13-14: 61-79; 213-239; 439-466; 685-717.

Bauchau, V., H. Horn & O. Overdijk. 1998. Survival of Spoonbills on Wadden Sea islands. J. Avian Biol. 29: 177-182.

Bauer, H.-G. & P. Berthold. 1996. Die Brutvögel Mitteleuropas: Bestand und Gefährdung. Wiesbaden: AULA-Verlag.

Bauer, H.-G. & G. Heine. 2008. Die Entwicklung der Brutvogelbestände am Bodensee: Vergleich halbquantitativer Rasterkartierungen 1980/81 und 1990/91. J. Ornithol. 133: 1-22.

Bauer, K.M. & U. N. Glutz von Blotzheim. 1966. Handbuch der Vögel Mitteleuropas, Band 1. Wiesbaden: Akademische Verlagsgesellschaft.

Bavoux, C., G. Burneleau, P. Nicolau-Guillamet & M. Picard. 1992. Le Busard roseaux *Circus a. aeruginosus* en Charente-Maritime (France). V – Déplacements et activité des juvéniles en hiver. Alauda 60: 149-158.

Beck, Ch., J. Grieser & B. Rudolf. 2004. A new monthly precipitation climatology for the global land areas for the period 1951 to 2000. Klimastatusbericht DWD 2004: 181-190.

Beecroft, R. 1991. Ringing in Senegal; an expedition update. BTO News 173: 5.

Beeren, W., J. Bredenbeek, B. Dijkstra & R. Messemaker. 2001. Ornithological studies in the Lake Chad Basin Area, West Africa. Preliminary report Waza-Logone project, Cameroon. Zeist: WIWO.

Beilfuss, R.D., T. Dodman & E.K. Urban. 2007. The status of cranes in Africa in 2005. Ostrich 78: 175-184.

Beintema, A.J. 1983. Meadow birds as indicators. Environmental Monitoring and Assessment 3: 391-398.

Beintema, A.J., R.J. Beintema-Hietbrink & G. Müskens. 1985. A shift in the timing of breeding in meadow birds. Ardea 73: 83-89.

Beintema, A.J. & N. Drost. 1986. Migration of the Black-tailed Godwit. Gerfaut 76: 37-62.

Beintema, A.J. 1986. Man-made polders in the Netherlands, a traditional habitat for shorebirds. Colonial Waterbirds 9: 196-202.

Beintema, A.J. & G. Müskens. 1987. Nesting success of birds breeding in Dutch agricultural grasslands. J. appl. Ecol. 24: 743-758.

Beintema, A.J. 1989. Vers un plan d'aménagement en faveur du Parc National des Oiseaux du Djoudj, Sénégal. WWF-International, Suisse / Direction des Parcs Nationaux du Sénégal, Dakar / Research Institute for Nature Management, Arnhem, Pays-Bas.

Beintema, A.J. 1991. Management of the Djoudj National Park in Senegal. Landscape and Urban Planning 20: 81-84.

Beintema, A.J., M. Diallo & M. Maiga. 2002. Monitoring, exploitation and marketing of waterbirds in the Inner Niger Delta, Mali. Sévaré, Mali: Wetlands International / Alterra Green World Research.

Beintema, A.J., J. van der Kamp & B. Kone. 2007. Les forêts inondées: trésors du Delta Intérieur du Niger au Mali. Wageningen: Wetlands International.

Belik, V. & I. Mihalevich. 1994. The pesticides use in the European steppes and its effects on birds. J. Ornithol. 135, Sonderheft: 233.

Belik, V.P. 1998. Current population status of rare and protected waders in south Russia. International Wader Studies 10: 273-280.

Bell, C.P. 1996. Seasonality and time allocation as causes of leap-frog migration in the Yellow Wagtail *Motacilla flava*. J. Avian Biol. 27: 334-342.

Bell, C.P. 2006. Social interactions, moult and pre-migratory fattening among Yellow Wagtails *Motacilla flava* in the Nigerian Sahel. Malimbus 28: 69-82.

Bell, C.P. 2007a. Timing of pre-nuptial migration and leap-frog patterns in Yellow Wagtail (*Motacilla flava*). Ostrich 78: 327-331.

Bell, C.P. 2007b. Climate change and spring migration in the Yellow Wagtail *Motacilla flava*: an Afrotropical perspective. J. Ornithol. 148 (Suppl. 2): S495-S499.

Belovski, G., J. Slade & B. Stockoff. 1990. Susceptibility to predation for different grasshoppers: an experimental study. Ecology 71: 624-634.

Belovski, G. & J. Slade. 1993. The role of vertebrate and invertebrate predators in a grasshopper community. Oikos 68: 193-201.

Ben Mohamed, A., N. van Duivenbooden & S. Abdousallam. 2002. Impact of climate change on agriculture production in the Sahel - Part 1. Methodological approach and case study for millet in Niger. Climate Change 54: 327-348.

Benjaminsen, T.A. 2008. Fuelwood and desertification: Sahel orthodoxies discussed on the basis of field data from the Gourma region in Mali. Geoforum 24: 397-409.

Bensch, S., D. Hasselquist, A. Hedenström & U. Ottosson. 1991. Rapid moult among Palearctic passerines in West Africa - an adaptation to the oncoming dry season? Ibis 133: 47-52.

Bensch, S. & B. Nielsen. 1999. Autumn migration speed of juvenile Reed and Sedge Warblers in relation to date and fat loads. Condor 101: 153-156.

Bensch, S., G. Bengtsson & S. Åkesson. 2006. Patterns of stable isotope signatures in willow warbler *Phylloscopus trochilus* feathers collected in Africa. J. Avian Biol. 37: 323-330.

van den Bergh, L.M.J., D.A. Jonkers & T.H. Slagboom. 1992. De broedvogels van de Polder Arkemheen. *In*: G.M. Dirkse & V. van Laar (eds). Arkemheen te velde: 138-151. Utrecht: KNNV Publishers.

Bermejo, A. & J. De La Puente. 2002. Stopover characteristics of Sedge Warbler (*Acrocephalus schoenobaenus*) in central Iberia. Vogelwarte 41: 181-189.

Bernard, C. 1992. Les aménagements du bassin du fleuve Sénégal pendant la Colonisation (1850-1960). Thèse. Paris: Université Paris VII.

Berndt, R. & W. Winkel. 1979. Zur Populationsentwicklung von Blaumeise (*Parus caeruleus*), Kleiber (*Sitta europaea*), Gartenrotschwanz (*Phoenicurus phoenicurus*) und Wendehals (*Jynx torquilla*) in mitteleuropäischen Untersuchungsgebieten von 1927 bis 1978. Vogelwelt 100: 55-69.

Berndt, R.K., K. Hein & T. Gall. 1994. Stabile Brutbestände der Uferschwalbe *Riparia riparia* in Schleswig-Holstein zwischen 1979 und 1991. Vogelwelt 115: 29-37.

Berndt, R.K. & J. Kauppinen. 1997. Pintail. *In*: W.J.M. Hagemeijer & M.J. Blair. (eds). The EBCC Atlas of European Breeding Birds: their Distribution and Abundance: 94-95. London: Poyser.

Berthold, P. 1973. Über starken Rückgang der Dorngrasmücke *Sylvia communis* und anderer Singvogelarten im westlichen Europa. J. Ornithol. 114: 348-360.

Berthold, P. 1974. Die gegenwärtige Bestandsentwicklung der Dorngrasmücke (*Sylvia communis*) und anderer Singvogelarten im westlichen Europa bis 1973. Vogelwelt 95: 170-183.

Berthold, P., A.J. Helbig, G. Mohr & U. Querner. 1992. Rapid microevolution of migratory behaviour in a wild bird species. Nature 360: 668-670.

Berthold, P., W. Fiedler, R. Schlenker & U. Querner. 1998. 25-Year study of the population development of Central European songbirds: a general decline, most evident in long-distance migrants. Naturwissenschaften 85: 350-353.

Berthold, P., W. van den Bossche, W. Fiedler, C. Kaatz, M. Kaatz, Y. Leshem, E. Nowak & U. Querner. 2001a. Detection of a new important staging and wintering area of the White Stork *Ciconia ciconia* by satellite tracking. Ibis 143: 450-455.

Berthold, P., W. van den Bossche, W. Fiedler, E. Gorney, M. Kaatz, Y. Leshem, E. Nowak & U. Querner. 2001b. The migration of the White Stork (*Ciconia ciconia*): a special case according to new data. J. Ornithol. 142: 73-92.

Berthold, P., W. van den Bossche, Z. Jakubiec, C. Kaatz & U. Querner. 2002. Long-term satellite tracking sheds light upon variable migration strategies of White Storks (*Ciconia ciconia*). J. Ornithol. 143: 489-495.

Berthold, P. 2003. Die Veränderungen der Brutvogelfauna in zwei süddeutschen Dorfgemeindebereichen in den letzten fünf bzw. drei Jahrzehnten oder: verlorene Paradiese? J. Ornithol. 144: 385-410.

Berthold, P., M. Kaatz & U. Querner. 2004. Long-term satellite tracking of white stork (*Ciconia ciconia*) migration: constancy versus variability. J. Ornithol. 145: 356-359.

Berthold, P. & W. Fiedler. 2005. 32-jährige Untersuchung der Bestandsentwicklung mitteleuropäischer Kleinvögel mit Hilfe von Fangzahlen: überwiegend Bestandsabnahmen. Vogelwarte 43: 97-102.

Bevanger, K. 1998. Biological and conservation aspects of bird mortality caused by electricity power lines: a review. Biol. Conserv. 86: 67-76.

Bezzel, E. 1982. Vögel der Kulturlandschaft. Stuttgart: Eugen Ulmer.

Bezzel, E. 1995. Werden neuerdings aus Italien keine Wiederfunde beringter Vögel mehr gemeldet? Vogelwarte 39: 106-107.

Bénech, V. & D.F. Dansoko. 1994. Reproduction des espèces d'intérêt halieutique. *In*: J. Quensière (ed.). La Pêche dans le Delta Central du Niger: 213-227. Paris: Karthala.

Bibby, C.J. 1978. Some breeding statistics of Reed and Sedge Warblers. Bird Study 25: 207-222.

Bibby, C.J. & R.E. Green. 1981. Autumn migration strategies of Reed and Sedge Warblers. Ornis Scand. 12: 1-12.

Bibby, C.J. & R.E. Green. 1983. Food and fattening of migrating warblers in some French marshlands. Ring. & Migr. 4: 175-184.

Biebach, H., W. Friedrich & G. Heine. 1986. Interaction of body-mass, fat, foraging and stopover period in trans-Sahara migrating passerine birds. Oecologia 69: 370-379.

Biebach, H., W. Friedrich, G. Heine, L. Jenni, S. Jenni-Eiermann & D. Schmidl. 1991. The daily pattern of autumn migration in the northern Sahara. Ibis 133: 414-422.

Biebach, H. 1992. Flight-range estimates for small trans-Saharan migrants. Ibis 134, suppl. 1: 47-54.

Biebach, H. 1996. Energetics of winter and migratory fattening. *In*: C. Carey (ed.). Avian energetics and nutritional ecology: 280-323. New York: Chapman & Hall.

Bijleveld, M. 1974. Birds of prey in Europe. London: MacMillan Press Ltd.

Bijlsma, R.G. 1985a. Foraging and hunting efficiency of Caspian Terns. British Birds 78: 146-147.

Bijlsma, R.G. 1985b. De broedbiologie van de Tortelduif *Streptopelia turtur*. Vogeljaar 33: 225-232.

Bijlsma, R.G. 1987. Bottleneck areas for migratory birds in the Mediterranean region. Study Report No. 18. Cambridge: International Council for Bird Preservation.

Bijlsma, R.G. 1993. Ecologische atlas van de Nederlandse roofvogels. Haarlem: Schuyt & Co.

Bijlsma, R.G. 2001. Waarnemingen van roofvogels op de grens van primair regenwoud in Zuidoost-Nigeria. De Takkeling 9: 235-262.

Bijlsma, R.G., F. Hustings & C.J. Camphuysen. 2001. Algemene en schaarse vogels van Nederland (Avifauna van Nederland 2). Haarlem/Utrecht: GMB Uitgeverij/KNNV Uitgeverij.

Bijlsma, R.G., W. van Manen & J. van der Kamp. 2005. Notes on breeding and food of Yellow-billed Kite *Milvus migrans parasitus* in Mali. Bull. African Bird Club 12: 125-133.

Bijlsma, R.G. & B. van den Brink. 2005. A Barn Swallow *Hirundo rustica* roost under attack: timing and risks in the presence of African Hobbies *Falco cuvieri*. Ardea 93: 37-48.

Bijlsma, R.G. 2006a. Ornithology from the tree tops. Ardea 94: 1-2.

Bijlsma, R.G. 2006b. Trends en broedresultaten van roofvogels in Nederland in 2005. De Takkeling 14: 6-53.

Bijlsma, R.G. 2006c. Winterstilte: een onderzoek naar vogels op de zandgronden van de Veluwe. Limosa 79: 72-73.

Bille, J.C. 1976. Étude de la production primaire nette d'un écosystème sahélien. Paris: ORSTOM.

BirdLife International. 2004a. Birds in Europe: population estimates, trends and conservaton status. Cambridge: BirdLife International.

BirdLife International. 2004b. Birds in the European Union: a status assessment. Cambridge: BirdLife International.

Birkett, C., R. Murtugudde & T. Allan. 1999. Indian Ocean climate event brings floods to East Africa's lakes and the Sudd marsh. Geophysical Research Letters 26: 1031-1034.

Blake, S., P. Bouché, H. Rasmussen, A. Orlando & I. Douglas-Hamilton. 2003. The last Sahelian Elephants: ranging behaviour, population status and recent history of the desert Elephants of Mali: 1-48. Nairobi: Save the Elephants.

Blanc, J.J., C.R. Thouless, J.A. Hart, H.T. Dublin, I. Douglas-Hamilton, G.C. Craig & R.F.W. Barnes. 2003. African Elephant Status Report 2002: An update from the African Elephant Database. Gland/Cambridge: IUCN.

Blondel, J. & C. Blondel. 1964. Remarques sur l'hivernage des limicoles et

autres oiseaux aquatiques au Maroc. Alauda 32: 250-279.

Bobek, M., R. Hampl, L. Peške, F. Pojer, J. Šimek & S. Bureš. 2006. African Odyssey projecyt – satellite tracking of black storks *Ciconia nigra* breeding at a migratory divide. J. Avian Biol. 39: 500-506.

Bock, C., J. Bock & M. Grant. 1992. Effects of bird predation on grasshopper densities in an Arizona grassland. Ecology 73: 1706-1717.

Boddy, M. 1992. Timing of Whitethroat *Sylvia communis* arrival, breeding and moult at a coastal site in Lincolnshire. Ring. & Migr. 13: 65-72.

Boddy, M. 1993. Whitethroat *Sylvia communis* population studies during 1981-91 at a breeding site on the Lincolnshire coast. Ring. & Migr. 14: 73-83.

Boddy, M. 1994. Survival/return rates and juvenile dispersal in an increasing population of Lesser Whitethroats *Sylvia curruca*. Ring. & Migr. 15: 65-78.

Bolshakov, C., V. Bulyuk & N. Chernetsov. 2003. Spring nocturnal migration of Reed Warblers *Acrocephalus scirpaceus*: departure, landing and body condition. Ibis 145: 105-112.

Bolshakov, C.V., A.P. Shapoval & N.P. Zelenova. 2001. Results of bird ringing by the Biological Station "Rybachy" on the Courish Spit: long-distance recoveries of birds ringed in 1956-1997. Part 1: Non-passeriformes. Passeres (Alaudidae, Hirundidae, Motacillidae, Bombycillidae, Troglodytidae, Prunellidae, Turdidae, Sylviidae, Regulidae, Muscicapidae, Aegithalidae). Avian Ecol. Behav. Suppl. 1: 1-126.

Bolshakov, C.V., V.N. Bulyuk, A. Mukhin & N. Chernetsov. 2003. Body mass and fat reserves of Sedge Warblers during vernal nocturnal migration: departure versus arrival. J. Field Ornithol. 74: 81-89.

Bonneval, P., M. Kuper & J.-P. Tonneau. 2002. L'Office du Niger, grenier à riz du Mali: Succès économiques, transitions culturelles et politiques de développement. Paris: Karthala.

de Bont, A. F. 1962. Composition des bandes d'hirondelles de cheminée *Hirundo rustica rustica* L. hivernant au Katanga et analyse de la mue des rémiges primaires. Gerfaut 52: 298-343.

Borrow, N. & R. Demey. 2004. Birds of Western Africa. London: Christopher Helm.

Bos, D., I. Grigoras & A. Ndiaye. 2006. Land cover and avian biodiversity in rice fields and mangroves of West Africa. Veenwouden: A&W/Wetlands International.

Bosshard, P. 1999. A case study on the Manantali dam project (Mali, Mauritania, Senegal). Press Release, Berne Declaration. Provided by International Rivers Network, http://www.irn.org/programs/safrica/index.php?id=bosshard.study.html.

Both, C., M.E. Visser & H. van Balen. 2002. De opkomst en ondergang van een populatie Ringmussen *Passer montanus*. Limosa 75: 41-50.

Both, C., R.G. Bijlsma & M.E. Visser. 2005. Climatic effects on timing of spring migration and breeding in a long-distance migrant, the pied flycatcher *Ficedula hypoleuca*. J. Avian Biol. 36: 368-373.

Both, C., S. Bouwhuis, C.M. Lessells & M.E. Visser. 2006. Climate change and population decline in a long-distance migratory bird. Nature 441: 81-83.

ould Boubouth, A., Y. Diawara, M. El Hacen Mint, S. Kloff & C. Vaufrey. 1999. Etouffés sous le tapis vert. Nouakchott, Mauritanie: IUCN.

Bousso, T. 1997. The estuary of the Senegal River: the impact of environmental changes and the Diama dam on resource status and fishery conditions. In: K. Remane (ed.). Africa inland fisheries, aquaculture and the environment: 45-65. Oxford: Fishing News Books.

Boutin, J.-M. 2001. Elements for a turtle dove (*Streptopelia turtur*) management plan. Game Wildl. Sci. 18: 87-112.

Boutin, J.-M., L. Barbier & D. Roux. 2001. Suivi des effectifs nicheurs d'Alaudidés, Colombidés et Turdidés en France : Le programme ACT. Alauda 69: 53-61.

Boyi, M. G., U. Ottosson & R. Ottvall. 2007. Winter ecology of common whitethroat (*Sylvia communis*) in the Amurum Forest, northern Nigeria. Ostrich 78: 369.

Böhning-Gaese, K. 1992. Zur Nahrungsökologie des Weißstorchs (*Ciconia ciconia*) in Oberschwaben: Beobachtungen an zwei Paaren. J. Ornithol. 133: 61-71.

Brader, L., H. Djibo, F.G. Faye, S. Ghaout, M. Lazar, P.N. Luzietoso & M.A. Ould Babah. 2006. Apporter une réponse plus efficace aux problèmes posés par les criquets pèlerins et à leurs conséquences sur la sécurité alimentaire, les moyens d'existence et la pauvreté. Évaluation multilatérale de la campagne 2003-05 contre le criquet pèlerin. Rome: FAO.

Breman, H. & A. M. Cissé. 1977. Dynamics of Sahelian pastures in relation to drought and grazing. Oecologia 28: 301-315.

Breman, H. & C.T. de Wit. 1983. Rangeland productivity and exploitation in the Sahel. Science 221: 1341-1347.

Breman, H. & N. de Ridder. 1991. Manuel sur les pâturages des pays sahéliens. Paris: Karthala.

Breman, H., J.J.R. Groot & H. van Keulen. 2001. Resource limitations in Sahelian agriculture. Global Environmental Change 11: 59-68.

Brichetti, P. & G. Fracasso. 2003. Ornitologia Italiana. Vol. 1. Bologna: Alberto Perdisa Editore.

Bricquet, J.P., G. Mahé, M. Toure & J.C. Olivry. 1997. Évolution récente des ressources en eau de l'Afrique atlantique. Rev. Sci. Eau 10: 321-337.

van den Brink, B., R.G. Bijlsma & T. van der Have. 1997. European Swallows *Hirundo rustica* in Botswana. WIWO-report No. 56. Zeist: WIWO.

van den Brink, B., R.G. Bijlsma & T. van der Have. 1998. European songbirds and Barn Swallows *Hirundo rustica* in Ghana: a quest for Constant Effort Sites and Swallow roosts in December/January 1996/97. WIWO-report 58. Zeist: WIWO.

van den Brink, B., R.G. Bijlsma & T. van der Have. 2000. European Swallows *Hirundo rustica* in Botswana during three non-breeding seasons: the effect of rainfall on moult. Ostrich 71: 198-204.

van den Brink, B., A. van den Berg & S. Deuzeman. 2003. Trapping Barn Swallows *Hirundo rustica* in Botswana in 2003. Babbler 43: 6-14.

van den Brink, B. 2006. Euring Swallow project Nederland, regioverslag Noord-Veluwe. Noordeinde: Published privately.

Broadbent, J. 1971. Additions to the avifauna of Waza (Cameroun) and Lake Natu (Sokoto). Bull. Nigerian Orn. Soc. 8: 58-61.

Brooks, N. & M. Legrand. 2000. Dust variability over northern Africa and rainfall in the Sahel. *In*: S. J. McLaren & D. Kniverton (eds). Linking land surface change to climate change: 1-23. Dordrecht: Kluwer.

Brouwer, J. & W.C. Mullié. 1992. Range extensions of two nightjar species in Niger, with a note on prey. Malimbus 14: 11-14.

Brouwer, J. & W.C. Mullié. 2000. Description of eggs and young of the Fox Kestrel *Falco alopex* in Niger. Bull. Brit. Orn. Club. 120: 196-198.

Brouwer, J. & W.C. Mullié. 2001. A method for making whole country waterbird population estimates, applied to annual waterbird census data from Niger. Ostrich Suppl. No. 15: 73-82.

Brouwer, J., W.C. Mullié & P. Scholte. 2003. White Storks *Ciconia ciconia* wintering in Chad, northern Cameroon and Niger: a comment on Berthold *et al*. (2001). Ibis 145: 499-501.

Brown, L. 1970. African birds of prey. Boston: Houghton Mifflin Company.

Brown, L. 1980. The African Fish Eagle. Folkestone: Bailey Bros. and Swinfen Ltd.

Brown, L.H., E.K. Urban & K. Newman. 1982. The Birds of Africa, Vol. 1. London: Academic Press.

Brown, P. E. & M.G. Davies. 1949. Reed-warblers. East Molesey: Foy Publications.

Browne, S.J. & N.J. Aebischer. 2003a. Habitat use, foraging ecology and diet of Turtle Doves *Streptopelia turtur* in Britain. Ibis 145: 572-582.

Browne, S.J. & N.J. Aebischer. 2003b. Temporal changes in the migration phenology of turtle doves *Streptopelia turtur* in Britain, based on sightings from coastal bird observatories. J. Avian Biol. 34: 65-71.

Browne, S.J. & N.J. Aebischer. 2003c. Temporal variation in the biometrics of Turtle Doves *Streptopelia turtur* caught in Britain between 1956 and 2000. Ring. & Migr. 21: 203-208.

Browne, S.J. & N.J. Aebischer. 2004. Temporal changes in the breeding ecology of European Turtle Doves *Streptopelia turtur* in Britain & implications for conservation. Ibis 146: 125-137.

Browne, S.J., N.J. Aebischer, G. Yfantis & J.H. Marchant. 2004. Habitat availability and use by Turtle Doves *Streptopelia turtur* between 1965 and 1995: an analysis of Common Birds Census data. Bird Study 51: 1-11.

Browne, S.J., N.J. Aebischer & H.Q.P. Crick. 2005. Breeding ecology of Turtle Doves *Streptopelia turtur* in Britain during the period 1941-2000: An analysis of BTO nest record cards. Bird Study 52: 1-9.

Browne, S.J. & N.J. Aebischer. 2005. Studies of West Palearctic birds: Turtle Dove. British Birds 98: 58-72.

Broyer, J., P. Varagnat, G. Constant & P. Caron. 1998. Habitat du Héron pourpré *Ardea purpurea* sur les étangs de pisciculture en France. Alauda 66: 221-228.

Bruderer, B. & W. Hirschi. 1984. Langfristige Bestandsentwicklung von Gartenrötel *Phoenicurus phoenicurus* und Trauerschnäpper *Ficedula hypoleuca*. Ornithol. Beob. 81: 285-302.

Bruderer, B. & A. Boldt. 2001. Flight characteristics of birds: I. Radar measurements of speeds. Ibis 143: 178-204.

Bruderer, B. & V. Salewski. 2009. Lower annual fecundity in long-distance migrants than in less migratory birds of temperate Europe. J. Ornithol. 150: 281-286.

Bruinzeel, L.W., J. van der Kamp, M.D. Diop Ndiaye & E. Wymenga. 2006. Avian biodiversity of invasive *Typha* vegetation. A pilot study on species and their densities on the Senegal Delta. Veenwouden/Dakar: Altenburg & Wymenga/Wetlands International.

Bryant, D.M. & G. Jones. 1995. Morphological changes in a population of Sand Martins *Riparia riparia* associated with fluctuations in population size. Bird Study 42: 57-65.

van der Burg, G. & J. Poutsma. 2000. Analyse van vogeltellingen langs de Friese IJsselmeerkust, 1975-1999. Velp: Larenstein.

Busche, G. 2004. Zum Durchzug des Wendehalses (*Jynx torquilla*) an der Deutschen Bucht (Helgoland und schleswig-holsteinische Küste) 1965-1998. Vogelwarte 42: 344-351.

Buss, I. O. 1946. Bird detection by radar. Auk 63: 315-318.

Busse, P. 1994a. General patterns of population trends of migrating passerines at the southern Baltic coast based on trapping results (1961-1990). *In*: E.J.M. Hagemeijer & T.J. Verstrael (eds). Bird Numbers 1992. Distribution, monitoring and ecological aspects: 427-434. Voorburg/Heerlen & Beek-Ubbergen: Statistics Netherlands & SOVON.

Busse, P. 1994b. Population trends of some migrants at the southern Baltic coast - autumn catching results 1961-1990. Ring 16: 115-158.

C

Calvert, M. 2005. Reed Warblers at Rostherne Mere. Shrewsbury: English Nature.

Camara, A.M. 1995. Les peuplements d'*Acacia nilotica* de la plaine alluviale du Sénégal: problèmes de conservation. Dakar: ENS-UCAD.

Camberlin, P., N. Martiny, N. Philippon & Y. Richard. 2007. Determinants of the interannual relationships between remote-sensed photosynthetic activity and rainfall in tropical Africa. Remote Sensing of Environment 106: 199-216.

Campos, F. & J.M. Lekuona. 2001. Are rice fields a suitable foraging habitat for Purple Herons during the breeding season? Waterbirds 24: 450-452.

Campredon, P. 1987. La reproduction des oiseaux d'eau sur le Parc National du Banc d'Arguin (Mauritanie) en 1984-1985. Alauda 55: 187-210.

Carmi, N., B. Pinshow, W. P. Porter & J. Jaeger. 1992. Water and energy limitations on flight duration in small migrating birds. Auk 109: 268-276.

Carmouze, J.P. & J. Lemoalle. 1983. The lacustrine environment. *In*: J.P. Carmouze, J.R. Durand & C. Lévêque (eds). Lake Chad: 27-63. The Hague: Junk.

Caro, T. & P. Scholte. 2007. When protection falters. Afr. J. Ecol. 45: 233-235.

Casazza, M.L., J.P. Fleskes, D.A. Haukos, M.R. Miller, D.L. Orthmeyer, W.M.

Perry & J.Y. Takekawa. 2005. Flight speeds of northern pintails during migration determined using satellite telemetry. Wilson Bull. 117: 364-374.

Castelijns, H., E.C.L. Marteijn, B. Krebs & G. Burggraeve. 1988. Overwinterende Kemphanen *Philomachus pugnax* in ZW-Nederland en NW-België. Limosa 61: 119-124.

Castelijns, H. 1994. Grutto en Kemphaan overwinteren in toenemende mate in Zeeuws-Vlaanderen. Limosa 67: 113-115.

Castelijns, H. & W. Castelijns. 2008. Het overwinteren van de Bruine Kiekendief in Zeeland. Limosa 81: 41-49.

Catry, P., M. Lecoq, A. Araujo, G. Conway, M. Felgueiras, J.M.B. King, S. Rumsey, H. Salima & P. Tenreiro. 2005. Differential migration of chiffchaffs *Phylloscopus collybita* and *P. ibericus* in Europe and Africa. J. Avian Biol. 36: 184-190.

Cavé, A. J. 1983. Purple Heron survival and drought in tropical West Africa. Ardea 71: 217-224.

Chamberlain, C.P., S. Bensch, X. Feng, S. Åkesson & T. Andersson. 2000. Stable isotopes examined across a migratory divide in Scandinavian willow warblers (*Phylloscopus trochilus trochilus* and *Phylloscopus trochilus acredula*) reflect their African winter quarters. Proc. R. Soc. London B 267: 43-48.

Chamberlain, D.E., R.J. Fuller, R.G.H. Bunce, J.C. Duckworth & M. Shrubb. 2000. Changes in the abundance of farmland birds in relation to the timing of agricultural intensification in England and Wales. J. appl. Ecol. 37: 771-788.

Chamberlain, D.E. & R.J. Fuller. 2000. Local extinctions and changes in species richness of lowland farmland birds in England and Wales in relation to recent changes in agricultural land-use. Agriculture, Ecosystems & Environment 78: 1-17.

Chang, P., L. Ji & H. Li. 1997. A decadal climate variation in the tropical Atlantic Ocean from thermodynamic air-sea interactions. Nature 385: 516-518.

Chapman, L.J., L. Kaufman & C.A. Chapman. 1994. Why swim upside down?: A comparative study of two mochokid catfishes. Copeia 1994(1): 130-135.

Charney, J.G. 1975. Dynamics of deserts and drought in Sahel. Q. J. Royal Meteor. Soc. 101: 193-202.

Cheke, R.A. & J.F. Walsh. 1996. The birds of Togo. b.o.u. Checklist No. 14. Tring: British Ornithologists' Union.

Cheke, R.A., W.C. Mullié & A. Baoua Ibrahim. 2006. Avian predation of adult Desert Locust *Schistocerca gregaria* affected by *Metarhizium anisopliae* var. *acridum* (Green Muscle®) during a large scale field trial in Aghéliough, northern Niger, in October and November 2005. Chatham Maritime/Dakar/ Niamey: FAO/NRI.

Chernetsov, N. 1996. Preliminary hypotheses on migration of the Sedge Warbler (*Acrocephalus schoenobaenus*) in the Eastern Baltic. Vogelwarte 38: 201-210.

Chernetsov, N. & A. Manukyan. 2000. Foraging strategy of the Sedge Warbler (*Acrocephalus schoenobaenus*) on migration. Vogelwarte 40: 189-197.

Chernetsov, N. & N. Titov. 2001. Movement patterns of European Reed Warblers *Acrocephalus scirpaceus* and Sedge Warblers *A. schoenobaenus* before and during autumn migration. Ardea 89: 509-515.

Chernetsov, N., M. Kaatz, U. Querner & P. Berthold. 2005. Vierjährige Satelliten-Telemetrie eines Weißstorchs *Ciconia ciconia* vom Selbstständigwerden an - Beschreibung einer Odyssee. Vogelwarte 43: 39-42.

Chomitz, K.M. & C.W. Griffiths. 1997. An economic analysis of woodfuel management in the Sahel; the case of Chad. World Bank Policy Research Working Paper 1788: 1-26.

Chu, P.-S., Z.-P. Yu & S. Hastenrath. 1994. Detecting climate change concurrent with deforestation in the Amazon basin: which way has it gone? Bull. Am. Meteor. Soc. 75: 579-583.

Claassen, T.H.L. 2000. Water management in a large Dutch, agricultural used catchment area of shallow lakes, The Netherlands. *In:* P.K. Jha, S.B. Karmacharya, S.R. Baral & P. Lacoul (eds). Environment and Agriculture – At the crossroad of the new millennium: 421-434. Kathmandu: Ecological Society.

Claffey, P. 1999. Dams as new habitat in West African savannah. Bull. African Bird Club 6: 117-120.

Clarke, R., A. Bourgonje & H. Castelijns. 1993. Food niches of sympatric Marsh Harriers *Circus aeruginosus* and Hen Harriers *C. cyaneus* on the Dutch coast in winter. Ibis 135: 424-431.

Clarke, R. 1996a. Preliminary observations on the importance of a large communal roost of wintering harriers in Gujarat (NW India) and comparison with a roost in Senegal (W. Africa). J. Bombay Nat. Hist. Soc. 93: 44-50.

Clarke, R. 1996b. Montagu's Harrier. Chelmsford: Arlequin Press.

Clarke, R., V. Prakash, W.S. Clark, N. Ramesh & D. Scott. 1998. World record count of roosting harriers *Circus* in Blackbuck National Park, Velavadar, Gujarat, north-west India. Forktail 14: 70-71.

Clarke, R. 2002. British Montagu's Harriers - what governs their numbers? Ornithol. Anz. 41: 143-158.

Claro, J.C. 2002. Perspectives for conservation of Montagu's Harrier in southern Portugal. Ornithol. Anz. 41: 211-212.

Cobb, S. 1981. Wild Life in Southern Sudan. Swara 4: 28-31.

Cody, M.L. 1985. Habitat selection in birds. London: Academic Press.

Coe, M.T. & J.A. Foley. 2001. Human and natural impacts on the water resources of the Lake Chad basin. J. Geophysical Research 106: 3349-3356.

Coe, M.T. & C.M. Birkett. 2004. Calculation of river discharge and prediction of lake height from satellite radar altimetry: Example for the Lake Chad basin. Water Resources Research 40: W10205, doi:10.1029/2003WR002543.

Colvin, J. & J. Holt. 1996. A model to investigate the effects of rainfall, predation and egg quiescence on the population dynamics of the

Senegalese grasshopper, *Oedaleus senegalensis*. Sécheresse 7: 145-150.
Conrad, B. 1974. Bestehen Zusammenhänge zwischen dem Bruterfolg der Dorngrasmücke (*Sylvia communis*) und ihrer gegenwärtigen Bestandsverminderung? Vogelwelt 95: 186-198.
Conway, D. 2005. From headwater tributaries to international river: Observing and adapting to climate variability and change in the Nile basin. Global Environmental Change 15: 99-114.
Conway, D., E. Allison, R. Felstead & M. Goulden. 2005. Rainfall variability in East Africa: implications for natural resources management and livelihoods. Phil. Trans. R. Soc. A 363: 49-54.
Corbacho, C., J.M. Sánchez & A. Sánchez. 1999. Effectiveness of conservation measures on Montagu's Harriers in agricultural areas in Spain. J. Raptor Research 33: 117-122.
Cormier, J.-P. & F. Baillon. 1991. Concentrations de Busards cendrés *Circus pygargus* (L.) dans la région de M'Bour (Sénégal) durant l'hiver 1988-1989: Utilisation du milieu et régime alimentaire. Alauda 59: 163-168.
Cotton, P.A. 2003. Avian migration phenology and global climate change. PNAS 100: 12219-12222.
Cowley, E. 1979. Sand Martin population trends in Britain, 1965-1978. Bird Study 26: 113-116.
Cowley, E. & G.M. Siriwardena. 2005. Long-term variation in survival rates of Sand Martins *Riparia riparia*: dependence on breeding and wintering ground weather, age and sex and their population consequences. Bird Study 52: 237-251.
Cox, R.R. & A.D. Afton. 1996. Evening flights of female northern pintails from a major roost site. Condor 98: 810-819.
Cramp, S. & K.E.L. Simmons. 1977. The Birds of the Western Palearctic. Vol. I: Ostrichs to Ducks. Oxford: Oxford University Press.
Cramp, S. & K.E.L. Simmons. 1983. The Birds of the Western Palearctic. Vol. III. Oxford: Oxford University Press.
Cramp, S. & K.E.L. Simmons. 1985. The Birds of the Western Palearctic. Vol. IV. Oxford: Oxford University Press.
Cramp, S. & K.E.L. Simmons. 1988. The Birds of the Western Palearctic. Vol. V. Oxford: Oxford University Press.
Cramp, S. 1992. The Birds of the Western Palearctic. Vol. VI: Warblers. Oxford: Oxford University Press.
Cresswell, W.R.L., J.M. Wilson, J. Vickery, P. Jones & S. Holt. 2007. Changes in densities of Sahelian bird species in response to recent habitat degradation. Ostrich 78: 247-253.
Crétaux, J.F. & C. Birkett. 2006. Lake studies from satellite radar altimetry. C. R. Geoscience 338: 1098-1112.
Cristol D.A., M.B. Baker & C. Carbone. 1999. Differential migration revisited: Latitudinal segregation by age and sex class. New York: Kluwer Academic/Plenum Publishers. 33-88.
Crousse, B. P., P. Mathieu & S. M. Seck. 1991. La vallée du fleuve Sénégal. Evaluations et perspectives d'une décennie d'aménagements (1980-1990). Paris: Karthala.
CSE. 2003. Project LADA au Sénégal (Land Degradation Assessment). Power point presentation. Dakar: Centre du Suivi d'Ecologie.
Culmsee, H. 2002. The habitat functions of vegetation in relation to the behaviour of the desert locust *Schistocerca gregaria* (Forskål) (Acrididae: Orthoptera) – a study in Mauritania (West Africa). Phytocoenologia 32: 645-664.
Curry-Lindahl, K. 1981. Bird migration in Africa, Vol. 1 and 2. London: Academic Press.
Curry, J. 1974. The occurrence and behaviour of turtle doves in the inundation zone of the Niger, Mali. Bristol Ornith. 7: 67-71.
Curry, J. & J.A. Sayer. 1979. The inundation zone of the Niger as an environment for Palaearctic migrants. Ibis 121: 20-40.

D

Dai, A.G., P.J. Lamb, K.E. Trenberth, M. Hulme, P.D. Jones & P.P. Xie. 2004. The recent Sahel drought is real. Intern. J. Clim. 24: 1323-1331.
Dallinga, H. & M. Schoenmakers. 1984. Populatieverandering bij de ooievaar *Ciconia ciconia* in de period 1850-1975. Zeist/Haren: Nederlandse Vereniging tot Bescherming van Vogels/University of Groningen.
Dallinga, J.H. & M. Schoenmakers. 1989. Population changes of the White Stork *Ciconia ciconia* since the 1850s in relation to food resources. *In:* G. Rheinwald, J. Ogden & H. Schulz (eds). Weißstorch – White Stork. Proceedings I International Stork Conservation Symposium, Schriftenreihe DDA 10.
Danfa, A., A.L. Bâ, H. van der Valk, C. Rouland-Lefèvre, W.C. Mullié & J.W. Everts. 2000. Long-term effects of chlorpyrifos and fipronil on epigeal beetles and soil arthropods in the semi-arid savanna of northern Senegal. Dakar: FAO.
Danko, Š., A. Daralová & A. Krištín. 2002. [Birds distribution in Slovakia.] Bratislava: VEDA.
Dansoko, D. & B. Kassibo. 1989. Étude des systèmes de productions halieutiques en 5ème région. Projet de développement; la région de Mopti (ODEM II-III); Études sur les systèmes de productions rurales en 5ème région.
Dean, G. J. W. 1964. Stork and egret as predators of the Red Locust in the Rukwa Valley outbreak area. Ostrich 35: 95-100.
DeCandido, R., R.O. Bierregaard Jr., M.S. Martell & K.L. Bildstein. 2006. Evidence of nocturnal migration by Osprey (*Pandion haliaetus*) in North America and Western Europe. J. Raptor Research 40: 156-158.
DeGeorges, A. & B.K. Reilly. 2006. Dams and large scale irrigation on the Senegal river: impacts on man and the environment. Intern. J. Environ. Studies 63: 633-644.
Dejoux, C. 1983. The exploitation of fish stocks in the Lake Chad region. *In:* J.P. Carmouze, J.R. Durand & C. Lévêque (eds) Lake Chad. Ecology and productivity of a shallow tropical ecosystem: 519-525. The Hague: Junk.
del Hoyo, J., A. Elliott & J. Sargatal. 1992. Handbook of the Birds of the World, Vol. 1. Barcelona: Lynx Edicions.
del Hoyo, J., A. Elliott & J. Sargatal. 1997. Handbook of the Birds of the World, Vol. 3. Barcelona: Lynx Edicions.

Delany, S. & D. Scott. 2006. Waterfowl Population Estimates - Fourth Edition. Wageningen: Wetlands International.

Dement'ev, G.P., N.A. Gladkov & E.P. Spangenberg. 1969. Birds of the Soviet Union, Vol II. Jeruzalem: IPSP.

Dennis, R. 1995. Ospreys *Pandion haliaetus* in Scotland - a study of recolonization. Vogelwelt 116: 193-196.

Dennis, R. 2002. Osprey *Pandion haliaetus*. *In:* C.V. Wernham, M.P. Toms, J.A. Clark, G.M. Siriwardena & S.R. Baillie (eds). The migration atlas: movements of the birds of Britain and Ireland: 243-245. London: Poyser.

Dennis, R. & F. A. McPhie. 2003. Growth of the Scottish Osprey (*Pandion haliaetus*) population. *In:* D.B.A. Thompson, S.M. Redpath, A.H. Fielding, M. Marquiss & C.A. Galbraith (eds). Birds of prey in a changing environment: 163-171. Edinburgh: The Stationery Office.

Dennis, R. 2008. A life of Ospreys. Dunbeath: Whitles Publishing.

Denny, P. 1984. Permanent swamp vegetation of the upper Nile. Hydrobiologica 110: 79-90.

Denny, P. 1993. Wetlands of Africa: introduction. *In:* D. Whigham, D. Dykyjova & S. Hejny (eds). Wetlands of the World: inventory, ecology and management: 1-31. Dordrecht: Kluwer.

Dervieux, A., J.-D. Lebreton & A. Tamisier. 1980. Technique et fiabilité des dénombrements aériens de canards et de foulques hivernant en Camargue. Terre Vie 34: 69-99.

Deutsch, M. 2007. Der Ortolan *Emberiza hortulana* im Wendland (Niedersachsen) - Bestandszunahme durch Grünlandumbruch und Melioration? Vogelwelt 128: 105-115.

Deuzeman, S.B., T.M. van der Have, W.T. de Nobel & B. van den Brink. 2004. European Swallows *Hirundo rustica* and other songbirds of wetlands in Ghana, December 1997. WIWO-report 80. Zeist: WIWO.

Diadicheva, E. & N. Matsievskaya. 2000. Migration routes of waders using stopover sites in the Azov-Black Sea region, Ukraine. Vogelwarte 40: 161-178.

Diagana, C.H., Z.E. ould Sidaty, Y. Diawara & M. ould Daddah. 2006. Oiseaux nicheurs au Parc du Diawling et dans sa zone périphérique (Mauritanie). Unpublished report.

Diawara, Y. & C.H. Diagana. 2006. Impacts of the restoration of the hydrological cycle on bird populations and socio-economic benefits in and around the Parc National du Diawling in Mauritania. *In:* G.C. Boere, C.A. Galbraith & D.A. Stroud (eds). Waterbirds around the world: 725-728. Edinburgh: The Stationery Office.

van Dijk, A.J. 1980. Waarnemingen aan de rui van de Grutto *Limosa limosa*. Limosa 53: 49-57.

van Dijk, A.J. & B.L.J. van Os. 1982. Vogels van Drenthe. Assen: van Gorcum.

van Dijk, A.J., K. van Dijk, L. Dijksen, T. van Spanje & E. Wymenga. 1984. Wintering waders and waterfowl in the Gulf of Gabès, Tunisia, January-March 1984. Zeist: WIWO.

van Dijk, A.J. & B. Dijkstra. 2000. Heeft de Grutto *Limosa limosa* toekomst in Drenthe? Drentse Vogels 13: 10-26.

van Dijk, A.J., L. Dijksen, F. Hustings, R. Oosterhuis, C. van Turnhout, M.J.T. van der Weide, D. Zoetebier & C.L. Plate. 2006. Broedvogels in Nederland in 2004. SOVON-monitoringrapport 2006/01. Beek-Ubbergen: SOVON Vogelonderzoek Nederland.

van Dijk, A.J., A. Boele, L. van den Bremer, F. Hustings, W. van Manen, A. van Kleunen, K. Koffijberg, W. Teunissen, C. van Turnhout, B. Voslamber, F. Willems, D. Zoetebier & C.L. Plate. 2007. Broedvogels in Nederland in 2005. SOVON-monitoringrapport 2007/10. Beek-Ubbergen: SOVON Vogelonderzoek Nederland.

van Dijk, A.J., A. Boele, F. Hustings, K. Koffijberg & C. Plate. 2008. Broedvogels van Nederland in 2006. Beek-Ubbergen: SOVON Vogelonderzoek Nederland.

van Dijk, G. 1983. De populatie-omvang (broedparen) van enkele weidevogelsoorten in Nederland en de omringende landen. Vogeljaar 31: 117-133.

Dijkstra, B., W. Ganzevles, G. Gerritsen & S. de Kort. 2002. Waders and waterfowl in the floodplains of the Logone, Cameroon. January/February 1999. Zeist: WIWO.

Dijkstra, C., N. Beemster, M. Zijlstra & M. van Eerden. 1995. Roofvogels in de Nederlandse wetlands. Lelystad: Rijkswaterstaat Directie IJsselmeergebied.

Diop, I. & P. Triplet. 2000. Un fléau végétal menace le delta du fleuve Sénégal. Bull. Liaison et Information OMPO 22: 63-65.

Diop, M. & P. Triplet. 2006. Parc National des Oiseaux du Djoudj. Plan d'Actions 2006-2008. Direction Parcs Nationaux, Dakar / Centre du Patrimoine Mondial, UNESCO.

Diop, M.D., J. Peeters, B. Faye & R. Diop. 1998. Typologie et problématique environnementale des zones humides de la rive gauche du bassin du fleuve Sénégal. Commission Fleuve Sénégal, Réseau National Zones Humides. Rapport de mission, IUCN.

Diouf, S., M. Diouf, P. Triplet & L. Hiliaire. 2001. Programme de lutte biologique contre *Salvinia molesta* dans le Parc National des Oiseaux du Djoudj et sa périphérie 2001-2003. Rapport Direction des Parcs Nationaux, UNESCO.

Díaz, M., B. Asensio & J.L. Tellería. 1996. Aves Ibéricas. Vol. I: No passeriformes. Madrid: J.M. Reyero Editor.

Dobrynina, I.N. & S.P. Kharitonov. 2006. The Russian waterbird migration atlas: temporal variation in migration routes. *In:* G.C. Boere, C.A. Galbraith & D.A. Stroud (eds). Waterbirds around the world: 582-589. Edinburgh: The Stationery Office.

Dodman, T. & C.H. Diagana. 2003. African Waterbird Census 1999, 2000 & 2001. Wageningen: Wetlands International.

Dodman, T. & C. Diagana. 2007. Movements of waterbirds within Africa and their conservation implications. Ostrich 78: 149-154.

Doligez, B., D.L. Thomson & A.J. van Noordwijk. 2004. Using large-scale data analysis to assess life history and behavioural traits: the case of the reintroduced White stork *Ciconia ciconia* population in the Netherlands. Animal Biodiver. Conserv. 27: 387-402.

Donald, P.F., F.J. Sanderson, I.J. Burfield & P.J. van Bommel. 2006. Further evidence of continent-wide impacts of agricultural intensification on European farmland birds, 1990–2000. Agriculture, Ecosystems & Environment 116: 189-196.

Dornbusch, G., S. Fischer & A. Hochbaum. 2004. Der Langzeit-Vogelschutzversuch der Vogelschutzwarte Steckby - Langfristige Trends und Brutergebnisse 2003. Berichte des Landesamtes für Umweltschutz Sachsen-Anhalt, Sonderheft 2004(4): 65-68.

Dowsett-Lemaire, F. & R.J. Dowsett. 1987. European reed and marsh warblers in Africa: migration patterns, moult and habitat. Ostrich 58: 65-85.

Dowsett, R.J. 1968. Migrants at Malamfatori, Lake Chad, spring 1968. Bull. Nigerian Orn. Soc. 5: 53-56.

Dowsett, R.J. 1969. Migrants at Malamfatori, Lake Chad, autumn 1968. Bull. Nigerian Orn. Soc. 6: 39-45.

Dowsett, R.J. & C.H. Fry. 1971. Weigt losses of trans-Saharan migrants. Ibis 113: 531-535.

Dowsett, R.J. 1980. The migration of coastal waders from the Palaearctic across Africa. Gerfaut 70: 3-35.

Dowsett, R.J., G.C. Backhurst & T.B. Oatley. 1988. Afrotropical ringing recoveries of Palaearctic migrants (I. Passerines). Tauraco 1: 29-63.

Dowsett, R.J. & P.M. Leonard. 1999. Results from bird ringing in Zambia. Zambian bird report 1999: 16-46.

Dowsett, R.J., D.R. Aspinwall & F. Dowsett-Lemaire. 2008. The birds of Zambia. Liège: Tauraco Press and Aves a.s.b.l.

Duhart, F. & M. Descamps. 1963. Notes sur l'avifaune du Delta Central Nigerien et régions avoisinantes. L'Oiseau et RFO 33 - no spécial: 1-107.

Dunning Jr., J.B. 1993. CRC handbook of avian body masses. Boca Raton: CRC Press.

Dupuy, A. & J. Verschuren. 1978. Note sur les oiseaux, principalement aquatiques, de la région du Parc National du Delta du Saloum (Sénégal). Gerfaut 68: 321-345.

Dupuy, A.R. 1975. Nidification de Hérons pourprés (*Ardea purpurea*) au Parc National des Oiseaux du Djoudj, Sénégal. L'Oiseau et RFO 45: 289-290.

Dupuy, A.R. 1982. Reproduction du Marabou (*Leptoptilos crumeniferus*) au Sénégal. L'Oiseau et RFO 52: 52-53.

Durand, J.R. 1983. The exploitation of fish stocks in the Lake Chad region. *In:* J.P. Carmouze, J.R. Durand & C. Lévêque (eds). Lake Chad. Ecology and productivity of a shallow tropical ecosystem: 425-481 The Hague: Junk.

Duvail, S., O. Hamerlynck & M.L. ould Baba. 2002. Une alternative à la gestion des eaux du fleuve Sénégal? *In:* G. Bergkamp, J.-Y. Pirot & S. Hostettler (eds). Integrated Wetlands and Water Resources Management: Proceedings of a workshop held at the Second International Conference on Wetlands and Development, Dakar: 89-97. Gland: Wetlands International, IUCN, WWF.

Duvail, S. & O. Hamerlynck. 2003. Mitigation of negative ecological and socio-economic impacts of the Diama dam on the Senegal River Delta wetland (Mauritania), using a model based decision support system. Hydrol. Earth Syst. Sci. 7: 133-146.

Dybbro, T. 1978. Oversigt over Danmarks fugle. Kobenhavn: Dansk Ornithologisk Forening.

E

Ebbinge, B.S. 1989. A multifactorial explanation for variation in breeding performance of Brent Geese *Branta bernicla*. Ibis 131: 196-204.

Ebbinge, B.S. 1990. Reply by Barwolt S. Ebbinge. Ibis 132: 481-482.

Elbersen, H.W. 2006. Typha for bio-energy. *In:* L. Kuiper (ed.). Quick-scans on upstream biomass. The Biomass Upstream Consortium, Wageningen. http://www.probos.net/biomassa-upstream/pdf/reportBUSB1.pdf.

Elgood, J.H., R.E. Sharland & P. Ward. 1966. Palaearctic migrants in Nigeria. Ibis 108: 84-116.

Elgood, J.H., J.B. Heigham, A.M. Moore, A.M. Nason, R.E. Sharland & N.J. Skinner. 1994. The birds of Nigeria. B.O.U. Checklist No. 4 (Second edition). Tring: British Ornithologists' Union.

Ellegren, H. & T. Fransson. 1992. Fat load and estimated flight-ranges in four *Sylvia* species analysed during autumn migration at Gotland, South-East Sweden. Ring. & Migr. 13: 1-12.

Elliott, H.F.I. 1962. Birds as locust predators. Ibis 104: 444.

Elts, J., A. Kuresoo, E. Leibak, A. Leito, V. Lilleleht, A. Luigujõe, A. Lõhmus, E. Mägi & M. Ots. 2003. [Status and numbers of Estonian birds, 1998-2002.] Hirundo 16: 58-83.

Enemar, A., B. Sjöstrand, G. Andersson & T. von Proschwitz. 2004. The 37-year dynamics of a subalpine passerine bird community, with special emphasis on the influence of environmental temperature and *Epirrita autumnata* cycles. Ornis Svecica 14: 63-106.

Engelhard, P. & B.G. Abdallah. 1986. Enjeux de l'après-barrage. Vallée du Sénégal. ENDA Tiers Monde / Ministère de la Coopération, France.

Ens, B.J., T. Piersma, W.J. Wolff & L. Zwarts (eds). 1990. Homeward bound: problems waders face when migrating from the Banc d'Arguin, Mauritania, to their northern breeding grounds in spring. Ardea 78: 1-364.

Epple, W. 1992. Einführung in das Artenschutzsymposium Wendehals. Beih. Veröff. Naturschutz Landschaftspflege Bad.-Württ. 66: 7-8.

Eraud, C., J.-M. Boutin, M. Rivière, J. Brun, C. Barbraud & H. Lormée. 2009. Survival of Turtle Doves *Streptopelia turtur* in relation to western Africa environmental conditions. Ibis 151: 186-190.

Escandell, V. 2006. Monitoring common breeding birds in Spain. The SACRE programme, report 1996-2005. Madrid: SEO/BirdLife.

Eskildsen, K. & B. Hälterlein. 2002. Lachseeschwalbe - *Gelochelidon nilotica*. *In:* R.K. Berndt, B. Koop & B. Struwe-Juhl (eds). Vogelwelt Schleswig-Holsteins: 206-207. Neumünster: Wachholtz Verlag.

Evans, K.L., J.D. Wilson & R.B. Bradbury. 2003. Swallow *Hirundo rustica* population trends in England: data from repeated historical surveys. Bird Study 50: 178-181.

Evans, K.L. & R.A. Robinson. 2004. Barn Swallows and agriculture. British Birds 97: 218-230.

Eykman, C., P.A. Hens, F.C. van Heurn, C.G.B. ten Kate, J.G. van Marle, M.J. Tekke & T.Gs. de Vries. 1941. De Nederlandsche Vogels, tweede deel. Wageningen: Wageningsche Boek- en Handelsdrukkerij.

Ezealor, A.U. & R.H. Giles. 1997. Wintering Ruff *Philomachus pugnax* are not pests of rice *Oryza spp.* in Nigeria's Sahelian wetlands. Wildfowl 48: 202-209.

F

Fairhead, J. & M. Leach. 2000. Webs of power: forest loss in Guinea. www.india-seminar.com.

Falk, K., F.P. Jensen, K.D. Christensen & B.S. Petersen. 2006. The diet of nestling Abdim's Stork *Ciconia abdimii* in Niger. Waterbirds 29: 215-220.

Fall, O., I. Fall & N. Hori. 2004. Assessment of the abundance and distribution of the aquatic plants and their impacts on the Senegal River Delta: The case of Khouma and Djoudj streams. Weed Technology 18: 1209.

FAO. 1980. Trilingual glossary of terms used in acridology. Rome: FAO.

FAO. 2006. Global Forest Resources Assessment 2005. Rome: FAO.

Farago, S. & P. Zomerdijk. 1997. Garganey. *In:* W.J.M. Hagemeijer & M.J. Blair (eds). The EBCC Atlas of European Breeding Birds: their Distribution and Abundance: 96-97. London: Poyser.

Fasola, M., L. Canova & N. Saino. 1996. Rice fields support a large portion of herons breeding in the Mediterranean region. Colonial Waterbirds 19 (Special Publ.1): 129-134.

Fasola, M., H. Hafner, J. Prosper, H. van der Kooij & I.V. Schogolev. 2000. Population changes in European herons in relation to African climate. Ostrich 71: 52-55.

Felix, K. & L. Felix. 2004. Bestandsentwicklung des Gartenrotschwanzes *Phoenicurus phoenicurus* in der Gemeinde Horgen 1965-2003. Ornithol. Beob. 101: 109-114.

Fenech, N. 1992. Fatal flight. The Maltese obsession with killing birds. London: Quiller Press.

Fernández-Cruz, M., G. Fernández-Alcázar, F. Campos & P.C. Días. 1992. Colonies of ardeids in Spain and Portugal. *In:* M. Finlayson, T. Hollis & T. Davis (eds). Managing Mediterranean wetlands and their birds: 76-78. Slimbridge: IWRB.

Fernández-Cruz, M. & F. Campos. 1993. The breeding of Grey Herons (*Ardea cinerea*) in western Spain: the influence of age. Colonial Waterbirds 16: 53-58.

Fiedler, W. 1998. Trends in den Beringungszahlen von Gartenrotschwanz (*Phoenicurus phoenicurus*) und Wendehals (*Jynx torquilla*) in Süddeutschland. Vogelwarte 39: 233-241.

Finlayson, C. 1992. Birds of the Strait of Gibraltar. London: Poyser.

Fishpool, L.D.C. & M.I. Evans. 2001. Important bird areas in Africa and associated islands. Priority sites for conservation. Newbury & Cambridge: Pisces Publications & BirdLife International.

Fisker, E.N., J. Bak & A. Niassy. 2007. A simulation model to evaluate control strategies for the grasshopper *Oedaleus senegalensis* in West Africa. Crop Protection 26: 592-601.

Flade, M. & K. Steiof. 1990. Bestandstrends häufiger norddeutscher Brutvögeln 1950-1985: Analyse von über 1400 Siedlungsdichte-Untersuchungen. *In:* R. v. d. Elzen, K.-L. Schuchmann & K. Schmidt-Koenig (eds). Current topics in avian biology: 249-260. Bonn: Deutsche Ornithologen-Gesellschaft.

Flade, M. 2007. Searching for wintering sites of the Aquatic Warbler *Acrocephalus paludicola* in Senegal. Eberswalde: Birdlife International Aquatic Warbler Conservation Team.

Fleskes, J.P., R.L. Jarvis & D.S. Gilmer. 2002. September-March survival of female northern pintails radiotagged in San Jaoquin Valley, California. J. Wildl. Manage 66: 901-911.

Fleskes, J.P., D.S. Gilmer & R.L. Jarvis. 2005. Pintail distribution and selection of marsh types at Mendota wildlife area during fall and winter. California Fish and Game 91: 270-285.

Fokin, S., V. Kuzyakin, H. Kalchreuter & J.S. Kirby. 2000. The Garganey in the former USSR: a compilation of the life-history information. Wetlands International Global Series 7: 1-50.

Foley, J.A., M.T. Coe, M. Scheffer & G.L. Wang. 2003. Regime shifts in the Sahara and Sahel: Interactions between ecological and climatic systems in northern Africa. Ecosystems 6: 524-539.

Folland, C.K., T.N. Palmer & D.E. Parker. 1986. Sahel rainfall and worldwide sea temperatures, 1901-85. Nature 320: 602-607.

Fontaine, B. & S. Janicot. 1996. Sea surface temperature fields associated with West African rainfall anomaly types. J. Climate 9: 2935-2940.

Foppen, R., C.J.F. ter Braak, J. Verboom & R. Reijnen. 1999. Dutch sedge warblers *Acrocephalus schoenobaenus* and West-African rainfall: empirical data and simulation modelling show low population resilience in fragmented marshlands. Ardea 87: 113-127.

Fornam, M.P. & J.A. Oguntola. 2004. Lake Chad basin. Kalmar: UNEP/University of Kalmar.

Fournier, O. & E.C. Smith. 1981. Effets des aménagements hydro-agricoles du fleuve Sénégal sur l'écosystème du delta, particulièrement sur le parc des oiseaux du Djoudj. Rapport UNESCO.

Fournier, O. 2001. Impact des braconnages sur les populations de la Tourterelle des Bois *Streptopelia turtur* passant en mai par la Médoc (Gironde, France). Alauda 69: 162.

Fowler, A.C., R.L. Knight, T.L. George & L.C. McEwen. 1991. Effects of avian predation on grasshopper populations in North Dakota grasslands. Ecology 72: 1775-1781.

Franc, A. 2007. Impact des transformations mésologiques sur la dynamique des populations et la grégarisation du criquet nomade dans le bassin de la Sofia (Madagascar). Thèse de doctorat de l'Université Paul Valéry - Montpellier III. Unité de Formation et de Recherche / Sciences Humaines et Sciences de l'Environnement. Discipline / Biologie des Populations et Écologie.

Fransson, T. & J. Pettersson. 2001. Swedish Bird Ringing Atlas, Vol. 1. Stockholm: Naturhistoriska riksmuseet & Sveriges Ornitologiska Förening.

Fransson, T., S. Jakobsson & C. Kullberg. 2005. Non-random distribution of ring recoveries from trans-Saharan migrants indicates species-specific stopover areas. J. Avian Biol. 36: 6-11.

Fransson, T. & B.-O. Stolt. 2005. Migration routes of North European Reed Warblers *Acrocephalus scirpaceus*. Ornis Svecica 15: 153-160.

Freidel, J. W. 1979. Population dynamics of the water hyacinth *Eichornia crassipes* (Mart.) Solms, with special reference to the Sudan. Berichte aus dem Fachgebied Herbologie der Universität Hohenheim 17: 1-132.

Frieze, M., R. Kuhn & D. Pointe. 2001. Biodiversity impact of cattail *(Typha dominguensis)* dominated marsh areas in the seasonal wetlands at Palo Verde National Park. Costa Rica: Coastal Rice Environmental Science Institute.

Fry, C.H., J.S. Ash & I.J. Ferguson-Lees. 1970. Spring weights of some Palaearctic migrants at Lake Chad. Ibis 112: 58-82.

Fry, C.H. 1970. Birds in Waza National Park, Cameroun. Bull. Nigerian Orn. Soc 7: 1-5.

Fry, C.H. 1971. Migration, moult and weights of birds in northern Guinea savanna in Nigeria and Ghana. Ostrich Suppl.8: 239-263.

Fry, C.H. 1982. Destruction of European White Storks in Nigeria by shooting. Malimbus 4: 47.

Fry, C.H., S. Keith & E.K. Urban. 1988. The birds of Africa, Vol. III. London: Academic Press.

Fuller, M.R., W.S. Seegar & L.S. Schueck. 1998. Routes and travel rates of migrating Peregrine Falcons *Falco peregrinus* and Swainson's Hawks *Buteo swainsoni* in the Western Hemisphere. J. Avian Biol. 29: 433-440.

G

Gallais, J. 1967. Le Delta Intérieur du Niger. Etudes de géographie régionale. Paris: Larose.

Ganesh, T. & P. Kanniah. 2000. Roost counts of harriers *Circus* spanning seven winters in Andhra Pradesh, India. Forktail 16: 1-3.

Ganusevich, S.A., T.L. Maechtle, W.S. Seegar, M.A. Yates, M.J. McGrady, M. Fuller, L.S. Schueck, J. Dayton & C.J. Henny. 2004. Autumn migration and wintering areas of Peregrine Falcons *Falco peregrinus* nesting on the Kola Peninsula, northern Russia. Ibis 146: 291-297.

Ganzevles, W. & J. Bredenbeek. 2005. Waders and waterbirds in the floodplains of the Logone, Cameroon and Chad, February 2000. Zeist: WIWO.

Garba Boyi, M. & G. Polet. 1996. Birdlife under water stress. *In:* R.D. Beilfuss, W.R. Tatboton & N.N. Gichuki (eds). Proceedings of the African Crane and wetland training workshop, 8-15 August 1993, Maun, Botswana: 8-15. Baraboo: International Crane Foundation.

García Novo, F. & C. Marín Cabrera. 2006. Doñana, water and biosphere. Madrid: Spanish Ministry of the Environment.

García, J.T. & B.E. Arroyo. 1998. Migratory movements of western European Montagu's Harriers *Circus pygargus*: a review. Bird Study 45: 188-194.

Garfield, B. 2007. The Meinertzhagen mystery: the life and legend of a colossal fraud. Washington: Potomac Books, Inc.

Garrido, J.R. & M. Fernández-Cruz. 2003. Effects of power lines on a White Stork *Ciconia ciconia* population in central Spain. Ardeola 50: 191-200.

Gatter, W. 1987. Zugverhalten und Überwinterung von paläarktischen Vögeln in Liberia (Westafrika). Verh. orn. Ges. Bayern 24: 479-508.

Gatter, W. & H. Mattes. 1987. Anpassungen von Schafstelze *Motacilla flava* und afrikanischen Motacilliden an die Waldzerstörung in Liberia (Westafrika). Verh. orn. Ges. Bayern 24: 467-477.

Gatter, W. 1988. Coastal wetlands of Liberia: their importance for wintering waterbirds. Study Report No. 26. Cambridge: ICBP.

Gatter, W. 1997. Birds of Liberia. Wiesbaden: AULA-Verlag.

Gatter, W. 2000. Vogelzug und Vogelbestände in Mitteleuropa: 30 Jahre Beobachtung des Tagzugs am Randecker Maar. Wiebelsheim: AULA-Verlag.

Gatter, W. 2007. Bestandsentwicklung des Gartenrotschwanzes *Phoenicurus phoenicurus* in Wäldern Baden-Württembergs. Ornithol. Anz. 46: 19-36.

Geesing, D., M. Al-Khawlani & M.L. Abba. 2004. Management of introduced *Prosopis* species: can economic exploitation control an invasive species? Unasylva 55: 36-44.

Geiser, S., R. Arlettaz & M. Schaub. 2008. Impact of weather variation on feeding behaviour, nestling growth and brood survival in Wrynecks *Jynx torquilla*. J. Ornithol. 149: 597-606.

George, K. 2004. Veränderungen der ostdeutschen Agrarlandschaft und ihrer Vogelwelt. Apus 12: 1-138.

Gerritsen, G.J. & J. Lok. 1986. Vogels in de IJsseldelta. Kampen: IJsselakademie.

Gerritsen, G.J. 1990. Slaapplaatsen van Grutto's *Limosa limosa* in Nederland in 1984-85. Limosa 63: 51-63.

Geyr von Schweppenburg, H. 1917. Vogelzug in der westlichen Sahara. J. Ornithol. 65: 48-65.

Geyr von Schweppenburg, H. 1930. Zum Zuge von *Sylvia curruca*. J. Ornithol. 78: 49-52.

GFCC. 1980. Assessment of environmental effects of proposed developments in the Senegal River Basin. Organisation pour la Mise en Valeur du fleuve Sénégal (OMVS). Dakar: Gannett Fleming Corddry & Carpenter Inc.; ORGATEC.

Giannini, A., R. Saravanan & P. Chang. 2003. Oceanic forcing of Sahel rainfall on interannual to interdecadal time scales. Science 302: 1027-1030.

Giannini, A., R. Saravanan & P. Chang. 2005. Dynamics of the boreal summer African monsoon in the NSIPP1 atmospheric model. Climate Dynamics 25: 517-535.

Gibbons, D.W., J.B. Reid & R.A. Chapman. 1993. The new atlas of breeding birds in Britain and Ireland: 1988-1991. London: Poyser.

Gill, J.A., L. Hatton & P. Potts. 2002. Black-tailed Godwit. *In*: C. Wernham, M. Toms, J. Marchant, J. Clark, G. Sriwardena & S. Baillie (eds). The

Migration Atlas: Movements of the Birds of Britain and Ireland: 323-325. London: Poyser.
Gillon, D. & Y. Gillon. 1974. Comparaison du peuplement d'invertébrés de deux milieux herbacés Ouest-Africains: Sahel et savane préforestière. Terre Vie 28: 429-474.
Gillon, Y. & D. Gillon. 1973. Recherches écologiques sur une savane sahélienne du Ferlo septentrional, Sénégal: données quantitatives sur les arthropodes. Terre Vie 27: 297-323.
Gilruth, P.T. & C.F. Hutchinson. 1990. Assessing deforestation in the Guinea highlands of West Africa using remote sensing. Potogramm. Eng. Rem. 56: 1375-1382.
Gilyazov, A.S. 1998. Long-term changes in wader populations at the Lapland Nature Reserve and its surroundings: 1887-1991. International Wader Studies 10: 170-174.
Girard, O., B. Trolliet, M. Fouquet, F. Ibañez, F. Léger, S.I. Sylla & J. Rigoulet. 1991. Dénombrement des anatidés dans le delta du Sénégal, janvier 1991. Bull. mens. O.N.C. 160: 9-13.
Girard, O., P. Triplet, S.I. Sylla & A. Ndiaye. 1992. Dénombrement des anatidés dans le Parc national du Djoudj et ses environs (janvier 1992). Bull. mens. O.N.C. 169: 18-21.
Girard, O. & J. Thal. 1996. Quelques observations ornithologiques dans la région de Garoua, Cameroun. Malimbus 18: 142-148.
Girard, O. & J. Thal. 1999. Mise en place d'un réseau de suivi de populations d'oiseaux d'eau en Afrique subsaharienne. Rapport de mission au Mali 8-29 janvier 1999. Paris: ONC.
Girard, O. & J. Thal. 2000. Mise en place d'un réseau de suivi de populations d'oiseaux d'eau en Afrique subsaharienne. Rapport de mission au Mali 11-31 janvier 2000. Paris: ONC.
Girard, O. & J. Thal. 2001. Mise en place d'un réseau de suivi de populations d'oiseaux d'eau en Afrique subsaharienne. Rapport de mission au Mali 9-23 janvier 2001. Paris: ONCFS.
Girard, O., J. Thal & B. Niagate. 2004. Les Anatidés hivernants dans le delta intérieur du Niger (Mali): une zone humide d'importance internationale. Game Wildl. Sci. 21 (www.ornithomedia.com).
Girard, O., J. Thal & B. Niagaté. 2006. Dénombrements d'oiseaux d'eau dans le delta intérieur du Niger (Mali) en janvier 1999, 2000 et 2001. Malimbus 28: 7-17.
Giraudoux, P., R. Degauquiter, P.J. Jones, J. Weigel & P. Isenmann. 1988. Avifaune du Niger: états de connaissances en 1986. Malimbus 10: 1-140.
Giraudoux, P. & E. Schüz. 1978. Fang von Weißstörchen auch in Niger. Vogelwarte 29: 276-277.
Glue, D.E. 1970. Extent and possible causes of a marked reduction in population of the Common Whitethroat (*Sylvia communis*) in Great Britain in 1969. Den Haag: Abstract XV Congress International Ornithology: 110-112.
Glutz von Blotzheim, U.N., K.M. Bauer & E. Bezzel. 1977. Handbuch der Vögel Mitteleuropas, Band 7/I. Wiesbaden: Akademische Verlagsgesellschaft.
Glutz von Blotzheim, U.N. & K.M. Bauer. 1980. Handbuch der Vögel Mitteleuropas, Band 9. Wiesbaden: Akademische Verlagsgesellschaft.
Glutz von Blotzheim, U.N. & K.M. Bauer. 1982. Handbuch der Vögel Mitteleuropas, Band 8/II. Wiesbaden: Akademische Verlagsgesellschaft.
Glutz von Blotzheim, U.N. & K.M. Bauer. 1988. Handbuch der Vögel Mitteleuropas, Band 11/I. Wiesbaden: AULA-Verlag.
Glutz von Blotzheim, U.N. & K.M. Bauer. 1991. Handbuch der Vögel Mitteleuropas. Band 12/1. Wiesbaden: AULA-Verlag.
Goes, B. 1997. Waterbeheer van een grotendeels door dammen gecontroleerde rivier in noord Nigeria. Stromingen 3: 17-30.
Goes, B.J.M. 1999. Estimate of shallow groundwater recharge in the Hadejia-Nguru Wetlands, semi-arid northeastern Nigeria. Hydrogeology Journal 7: 294-304.
Goes, B.J.M. 2002. Effects of river regulation on aquatic macrophyte growth and floods in the Hadejia-Nguru wetlands and flow in the Yobe River, northern Nigeria; Implications for future water management. River Research and Applications 18: 81-95.
Gonzalez, P. 2001. Desertification and a shift of forest species in the West African Sahel. Climate Research 17: 217-228.
Goodman, S.M. & P.L. Meininger. 1989. The birds of Egypt. Oxford: Oxford University Press.
Goodman, S.M. & C.V. Haynes. 1992. The diet of the Lanner (*Falco biarmicus*) in a hyper-arid region of the eastern Sahara. J. Arid Environ. 22: 93-98.
de Gooijer, J. 1982. Zestig jaar nestkastonderzoek in het National Park "De Hoge Veluwe". Leusden: Privately published.
Goosen, H. & B. Kone. 2005. Livestock in the Inner Niger Delta. *In*: L. Zwarts, P. van Beukering, B. Kone & E. Wymenga (eds). The Niger, a lifeline: 121-135. Lelystad: Rijkswaterstaat/IVM/Wetlands International/A&W.
Gordo, O., L. Brotons, X. Ferrer & P. Comas. 2005. Do changes in climate patterns in wintering areas affect the timing of the spring arrival of trans-Saharan migrant birds? Global Change Biology 11: 12-21.
Gordo, O. & J.J. Sanz. 2006. Climate change and bird phenology: a long-term study in the Iberian Peninsula. Global Change Biology 12: 1993-2004.
Gordo, O. 2007. Why are bird migration dates shifting? A review of weather and climate effects on avian migratory phenology. Climate Research 35: 37-58.
Gordo, O., J.J. Sanz & J.M. Lobo. 2007. Spatial patterns of white stork (*Ciconia ciconia*) migratory phenology in the Iberian Peninsula. J. Ornithol. 148: 293-308.
Gordo, O. & J.J. Sanz. 2008. The relative importance of conditions in wintering and passage areas on spring arrival dates: the case of long-distance Iberian migrants. J. Ornithol. 149: 199-210.
Gore, M.E.J. 1982. Millions of Turtle Doves. Malimbus 2: 78.
Goriup, P.D. & H. Schulz. 1991. Conservation management of the White

Stork; an international need and opportunity. ICBP Technical Publication 12: 97-127.

Gottschalk, T. 1995. Gartenrotschwanz - *Phoenicurus phoenicurus*. *In:* Hessische Gesellschaft für Ornithologie und Naturschutz. Avifauna von Hessen, 2. Lieferung: 1-15. Echzell: Hessische Gesellschaft für Ornithologie und Naturschutz e.V.

Goudie, A.S. & N.J. Middleton. 2001. Saharan dust storms: nature and consequences. Earth-Science reviews 56: 179-204.

Gowthorpe, P.B., B. Lamarche, R. Biaux, A. Gueye, S.M. Lehlou, M.A. Sall & A.C. Sakho. 1996. Les oiseaux nicheurs et les principaux limicoles paléarctiques du Parc National du Banc d'Arguin (Mauritanie). Dynamiques des effectifs et variabilité dans l'utilisation spatio-temporelle du milieu. Alauda 64: 81-126.

Greathead, D.J. 1966. A brief survey of the effects of biotic factors on populations of the Desert Locust. J. appl. Ecol. 3: 239-250.

Green, A.J. 2000. Threatened wetlands and waterbirds in Morocco; a final report. Sevilla: Estación Biológica de Doñana.

Green, R.E. 1976. Adult survival rates for Reed and Sedge Warblers. Wicken Fen Group Report 8: 23-26.

Gregory, R.D., P. Vorisek, A. van Strien, A.W. Gmelig Meyling, F. Jiquet, L. Fornasari, J. Reif, P. Chylarecki & I.J. Burfield. 2007. Population trends of widespread woodland birds in Europe. Ibis 149 (Suppl. 2): 78-97.

Grell, M.B. 1998. Fuglenes Danmark. Copenhagen: Gads Forlag.

Grimes, L.G. 1987. The Birds of Ghana. B.O.U. Checklist No. 9. London: British Ornithologists' Union.

Groen, N.M. & L. Hemerik. 2002. Reproductive success and survival of Black-tailed Godwits *Limosa limosa* in a declining local population in The Netherlands. Ardea 90: 239-248.

Groppali, R. 2006. Djoudj et ses oiseaux. L'avifaune du Parc National et du Sénégal atlantique et Gambie. Publication Parco Adda Sud, Collaborazione Internazionale-A.

Grote, H. 1928. Uebersicht über die Vogelfauna des Tchadsgebiets. J. Ornithol. 76: 739-783.

Grote, H. 1930. Wanderungen und Winterquartiere der paläarktischen Zugvögel in Afrika. Mitteilungen aus dem Zoologischen Museum in Berlin 16: 1-116.

Grote, H. 1937. Neue Beiträge zur Kenntnis der palaearktischen Zugvögel in Afrika. Mitteilungen aus dem Zoologischen Museum in Berlin 22: 45-85.

Grüll, A. & A. Ranner. 1998. Populations of the Great Egret and Purple Heron in relation to ecological factors in the reed belt of the Neusiedler See. Colonial Waterbirds 21: 328-334.

Gschweng, M., E.K.V. Kalko, U. Querner, W. Fiedler & P. Berthold. 2008. All across Africa: highly individual migration routes of Eleonora's falcon. Proc. R. Soc. B 275: 2887-2896.

Guichard, K.M. 1947. Birds of the inundation zone of the River Niger, French Soudan. Ibis 80: 450-489.

Guillemain, M., H. Fritz, M. Klaassen, A.R. Johnson & H. Hafner. 2004. Fuelling rates of garganey (*Anas querquedula*) staging in the Camargue, southern France, during spring migration. J. Ornithol. 145: 152-158.

Gunnarsson, T.G., P.M. Potts, J.A. Gill, R.E. Croger, G. Gélinaud, P.W. Atkinson, A. Gardarsson & W.J. Sutherland. 2005. Estimating population size in Icelandic Black-tailed Godwits *Limosa limosa islandica* by colour-marking. Bird Study 52: 153-158.

Gurtovaya, E.N. 2002. Bird ringing in the USSR and Russia in 1988-1999. *In:* I.N. Dobrynina (ed.). Bird ringing and marking in Russia and adjacent countries 1988-1999: 301-413. Moscow: Russian Academy of Sciences.

Gustafsson, R., C. Hjort, U. Ottosson & P. Hall. 2003. Birds at Lake Chad and in the Sahel of NE Nigeria 1997-2000; The Lake Chad Bird Migration Project. Degerhamm: Special Report from Ottenby Bird Observatory.

Gwinner, E. & W. Wiltschko. 1978. Endogenously controlled changes in migratory direction of Garden Warbler, *Sylvia borin*. J. Comp. Physiol. 125: 273.

Gyurácz, J. & T. Csörgö. 1994. Autumn migration dynamics of the Sedge Warbler (*Acrocephalus schoenobaenus*) in Hungary. Ornis Hungarica 4: 31-37.

H

Haas, W. 1974. Beobachtungen paläarktischer Zugvögel in Sahara und Sahel (Algerien, Mali, Niger). Vogelwarte 27: 194-202.

Haas, W. & P. Beck. 1979. Zum Frühjahrszug paläarktischer Vögel über die westliche Sahara. J. Ornithol. 120: 237-246.

Hafner, H., O. Pineau & Y. Kayser. 1994. Ecological determinants of annual fluctuations in numbers of breeding Little Egrets (*Egretta garzetta* L.) in the Camargue, S France. Terre Vie 49: 53-62.

Hafner, H., Y. Kayser, V. Boy, M. Fasola, A.C. Julliard, R. Pradel & F. Cézilly. 1998. Local survival, natal dispersal & recruitment in Little Egrets *Egretta garzetta*. J. Avian Biol. 29: 216-227.

Hafner, H., R.E. Bennetts & Y. Kayser. 2001. Changes in clutch size, brood size and numbers of nesting Squacco Herons *Ardeola ralloides* over a 32-year period in the Camargue, southern France. Ibis 143: 11-16.

Hagemeijer, W.J.M. & M.J. Blair. (eds). 1997. The EBCC atlas of European breeding birds: their distribution and abundance. London: Poyser.

Hagemeijer, W.J.M., C.J. Smit, P. de Boer, A.J. van Dijk, N. Ravenscroft, M. van Roomen & M. Wright. 2004. Wader and waterbird census at the Banc d'Arguin, Mauritania, January 2000. Beek-Ubbergen: WIWO.

Haines, A.M., M.J. McGrady, M.S. Martell, B.J. Dayton, M.B. Henke & W.S. Seegar. 2003. Migration routes and wintering locations of Broad-winged Hawks tracked by satellite telemetry. Wilson Bull. 115: 166-169.

Hake, M., N. Kjellén & T. Alerstam. 2001. Satellite tracking of Swedish Ospreys *Pandion haliaetus*: Autumn migration routes and orientation. J. Avian Biol. 32: 47-56.

Hall, J.B. 1994. *Acacia seyal* - multipurpose tree of the Sahara desert. NFT Highlights 94-07: 1-6.

Hall, K.S.S. & T. Fransson. 2001. Wing moult in relation to autumn migration in adult Common Whitethroats *Sylvia communis communis*. Ibis 143: 580-586.

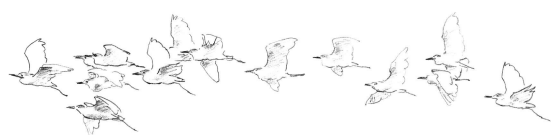

Hall, M.R., E. Gwinner & M. Bloesch. 1987. Annual cycles in moult, body mass, luteinizing hormone, prolactin and gonadal steroids during the development of sexual maturity in the White Stork (*Ciconia ciconia*). J. Zool. Lond. 211: 467-486.

Hall, S. 1996. The timing of post-juvenile moult and fuel deposition in relation to the onset of autumn migration in Reed Warblers *Acrocephalus scirpaceus* and Sedge Warblers *Acrocephalus schoenobaenus*. Ornis Svecica 6: 89-96.

Halupka, L., A. Dyrcz & M. Borowiec. 2008. Climate change affects breeding of reed warblers *Acrocephalus scirpaceus*. J. Avian Biol. 39: 95-100.

Hamerlynck, O. 1997. Plan Directeur d'Aménagement du Parc National du Diawling et de sa zone périphérique,1997-2000. Nouakchott: Ministère du Développement Rural et de l'Environnement.

Hamerlynck, O., M.L. ould Baba & S. Duvail. 1999. The Diawling National Park: joint management for the rehabilitation of a degraded coastal wetland. Vida Sylvestre Neotropical 7: 59-69.

Hamerlynck, O. & B. ould Messaoud. 2000. Suspected breeding of lesser flamingo *Phoenicopterus minor* in Mauritania. Bull. African Bird Club 7: 109-110.

Hamerlynck, O., B. ould Messaoud, R. Braund, C.H. Diagana, Y. Diawara & D. Ngantou. 2002. Crues artificielles et congestion: la réhabilitation des plaines inondables au Sahel. Le Waza Logone (Cameroun) et le bas-delta du fleuve Sénégal (Mauritanie). *In:* D. Orange, R. Arfi, M. Kuper, P. Morand & Y. Poncet (eds). Gestion intégrée des ressources naturelles en zones inonables tropicales: 475-500. Paris: IRD.

Hamerlynck, O. & S. Duvail. 2003. The rehabilitation of the delta of the Senegal River in Mauritania; fielding the ecosystem approach. Gland & Cambridge: IUCN.

Hamerstrom, F. 1994. My double life. Memoirs of a naturalist. Madison: University of Wisconsin Press.

Handrinos, G. & T. Akriotis. 1997. The birds of Greece. London: Christopher Helm.

Harengerd, M. 1982. Beziehungen zwischen Zug und Mauser beim Kampfläufer, *Philomachus pugnax* (Linné 1758, Aves, Charadriiformes, Charadriidae). Dissertation. Bonn: Universität Bonn.

Hario, M., T. Kastepold, M. Kilpi & T. Stjernberg. 1987. Status of Caspian Terns *Sterna caspia* in the Baltic. Ornis Fennica 64: 154-157.

Hario, M. & T. Stjernberg. 1996. [The Caspian Tern in Finland. A monitoring project on the Baltic Caspian Tern in 1984-1996.] Linnut-vuosikirja 1996: 15-24.

Harrison, J.A., D.G. Allan, L.G. Underhill, M. Herremans, A.J. Tree, V. Parker & C.J. Brown. 1997. The atlas of southern African birds, Vol. 1: Non-Passerines. Johannesburg: BirdLife South Africa.

Hartert, E. 1886. Ornithologische Ergebnisse einer Reise in den Niger-Benuë-Gebieten. J. Ornithol. 34: 570-613.

Hastenrath, S. & P.J. Lamb. 1977. Some aspects of circulation and climate over the eastern equatorial Atlantic. Monthly Weather Review 105: 1019-1023.

van der Have, T. & D. A. Jonkers. 1996. Zeven misverstanden over Ooievaars *Ciconia ciconia* in Nederland. Limosa 69: 47-50.

van der Have, T. 1998. The Mediterranean flyway: a network of wetlands for waterbirds. International Wader Studies 10: 81-84.

Hartmann R. 1863. Ornithologische Reiseskizzen aus Nordost-Afrika. J. Ornithol. 11: 299-320

Haverschmidt, F. 1963. The Black-tailed Godwit. Leiden: E.J.Brill.

Heath, M., C. Borggreve & N. Peet. 2000. European bird populations: estimates and trends. Cambridge: BirdLife International.

Hedenström, A., S. Bensch, D. Hasselquist, M. Lockwood & U. Ottosson. 1993. Migration, stopover and moult of the Great Reed Warbler *Acrocephalus arundinaceus* in Ghana, West-Africa. Ibis 135: 177-180.

den Held, J.J. 1981. Population changes in the Purple Heron in relation to drought in the wintering area. Ardea 69: 185-191.

Held, I.M., T.L. Delworth, J. Lu, K.L. Findell & T.R. Knutson. 2005. Simulation of Sahel drought in the 20th and 21st centuries. PNAS 102: 17891-17896.

Heldbjerg, H. 2005. De almindelige fugles bestandsudvikling i Danmark 1975-2004. Dansk Orn. Foren. Tidsskr. 99: 182-195.

Heldbjerg, H. & A.D. Fox. 2008. Long-term population declines in Danish trans-Saharan migrant birds. Bird Study 55: 267-279.

Helldén, U. 1991. Desertification - time for an assessment. Ambio 20: 373-383.

Hellsten, S., C. Dieme, M. Mbengue, G.A. Janauer, N. den Hollander & A.H. Pieterse. 1999. *Typha* control efficiency of a weed-cutting boat in the Lac de Guiers in Senegal: a preliminary study on mowing speed and re-growth capacity. Hydrobiologica 415: 249-255.

Hennig, R.C. 2006. Forests and deforestation in Africa - the wasting of an immense resource. www.afrol.com/features/10278.

Henning, R.K. 2002. Valorisation du Typha comme combustible domestique en Afrique de l'Ouest et en Europe. Workshop on Typha in Saint Louis, Senegal. 23.-25.7.2002.

Henny, C.J. 1973. Drought displaced movement of North American pintails into Siberia. J. Wildl. Manage 37: 23-29.

Hepper, F.N. 1968. Flora of West Tropical Africa. Volume 3 part I. Second edition. London: Whitefriars Press.

Heptner, V. G., A.A. Nasimovich & A.G. Bannikov. 1989. Mammals of the Soviet Union, Vol. I. Leiden: E.J. Brill.

Hering, J. 2008. Weißstörche in der Zentralsahara entdeckt. Falke 55: 390-394.

Herremans, M. 2003. The study of bird migration across the Western Sahara; a contribution with sound luring. www.ifv.terramare.de/ESF/Herremans2003.pdf

Hess, T., W. Stephens & G. Thomas. 1996. Modelling NDVI from decadal rainfall data in the north east Arid Zone of Nigeria. J. Environ. Manag. 48: 249-261.

Hiernaux, P. & L. Diarra. 1983. Pâturages de la zone d'inondation du Niger. *In:* R.T. Wilson, P.N. de Leeuw & C. de Haan (eds). Recherches sur les systèmes des zones arides du Mali: résultats préliminaires. Rapport de recherche 5: 42-48. Addis Abeba: ILCA.

Hiernaux, P.H.Y., M.I. Cissé, L. Diarra & P.N. de Leeuw. 1994. Fluctuations saisonnières de la feuillaison des arbres et des buissons sahéliens. Conséquences pour la quantification des ressources fourragères. Revue d'élevage et de médecine vétérinaire des pays tropicaux 47: 117-125.

Hirschfeld, A. & A. Heyd. 2005. Mortality of migratory birds caused by hunting in Europe: bag statistics and proposals for the conservation of birds and animal welfare. Ber. Vogelschutz 42: 47-74.

Hjort, C., J. Pettersson, Å. Lindström & J.M.B. King. 1996. Fuel deposition and potential flight ranges of blackcaps *Sylvia atricapilla* and whitethroats *Sylvia communis* on spring migration in The Gambia. Ornis Svecica 6: 137-144.

Hobson, K.A., G.J. Bowen, L.I. Wassenaar, Y. Ferrand & H. Lormee. 2004. Using stable hydrogen and oxygen isotope measurements of feathers to infer geographical origins of migrating European birds. Oecologia 141: 477-488.

Hoffmann, L. 1957. Le passage d'automne du chevalier sylvain (*Tringa glareola*) en France Méditerranéenne. Alauda 25: 30-42.

Hogg, P., P.J. Dare & J.V. Rintoul. 1984. Palaearctic migrants in the central Sudan. Ibis 126: 307-331.

Hollis, G.E., S.J. Penson, J.R. Thompson & A.R. Sule. 1993. Hydrology of the river basin. *In:* G.E. Hollis, W.M. Adams & M. Aminu-Kano (eds). The Hadejia-Nguru wetlands: environment, economy and sustainable development of a Sahelian floodplain wetland: 19-67. Gland: IUCN.

Hollis, G.E. & J.R. Thompson. 1993. Water resource developments and their hydrological impacts. *In:* G.E. Hollis, W.M. Adams & M. Aminu-Kano (eds). The Hadejia-Nguru wetlands: environment, economy and sustainable development of a Sahelian floodplain wetland: 149-190. Gland: IUCN.

Holloway, S. 1996. The historical atlas of breeding birds in Britain and Ireland: 1875-1900. London: Poyser.

Holmes, P.R., S.E. Christmas & A.J. Parr. 1987. A study of the return rate and dispersal of Sand Martins *Riparia riparia* at a single colony. Bird Study 34: 12-19.

Holt, J. & J. Colvin. 1997. A differential equation model of the interaction between the migration of the Senegalese grasshopper, *Oedaleus senegalensis*, its predators, and a seasonal habitat. Ecological Modelling 101: 185-193.

le Houérou, H.N. 1980. The rangelands of the Sahel. J. Range Management 33: 41-46.

le Houérou, H.N. 1989. The grazing land ecosystems of the African Sahel. Berlin: Springer-Verlag.

Hovestadt, T. & P. Nowicki. 2008. Process and measurement errors of population size: their mutual effect on precision and bias of estimates for demographic parameters. Biodivers. Conserv. 17: 3417-3429.

Howell, P., M. Lock & S. Cobb. 1988. The Jonglei canal. Impact and opportunity. Cambridge: Cambridge University Press.

Hölker, M. 2002. Beiträge zur Ökologie der Wiesenweihe *Circus pygargus* in der Feldlandschaft der Hellwegbörde/Nordrhein-Westfalen. Ornithol. Anz. 41: 201-206.

Hölzinger, J. 1999. Die Vögel Baden-Württembergs. Band 3.1: Singvögel 1. Stuttgart: Eugen Ulmer.

Hötker, H., H.A. Bruns & S. Dietrich. 1990. Northward migration of waders wintering in Senegal in January. Wader Study Group Bull. 59: 20-24.

Hötker, H., J. Jeromin & J. Melter. 2007. Entwicklung der Brutbestände der Wiesen-Limikolen in Deutschland - Ergebnisse eines neuen Ansatzes im Monitoring mittelhäufiger Brutvogelarten. Vogelwelt 128: 49-65.

Hubalek, Z. 2004. Global weather variability affects avian phenology: a long-term analysis, 1881-2001. Folia Zool. 53: 227-236.

Hudleston, J.A. 1958. Some notes on the effects of bird predators on hopper bands of the Desert Locust (*Schistocerca gregaria* Forsk.). Entomologist's Monthly Magazine 94: 210-214.

Hughes, R.H. & J.S. Hughes. 1992. A directory of African wetlands. Gland: IUCN.

Huin, N. & T.H. Sparks. 1998. Arrival and progression of the Swallow *Hirundo rustica* through Britain. Bird Study 45: 361-370.

van Huis, A., K. Cressman & J. Magor. 2007. Mini Review. Preventing desert locust plagues: optimizing management interventions. Entomologia Experimentalis et Applicata 122: 191-214.

Hulme, M. 1996. Recent climatic change in the world's drylands. Geophysical Research Letters 23: 61-64.

Hulme, M. 2001. Climatic perspectives on Sahelian desiccation: 1973-1998. Global Environmental Change 11: 19-29.

Hulme, M., R. Doherty, T.Ngara, M. New & D. Lister. 2001. African climate change: 1900-2100. Climate Research 17: 145-168.

Hulme, M. 2007. The density and diversity of birds on farmland in West Africa. PhD Thesis. St. Andrews: University of St. Andrews.

Huntley, B., R.E. Green, Y.C. Collingham & S.G. Willis. 2007. A climatic atlas of European breeding birds. Barcelona: Durham University, The RSPB & Lynx Edicions.

Husain, M.A. & H.R. Bhalla. 1931. Some bird enemies of the Desert Locust (*Schistocerca gregaria*, Forsk.) in the Ambala district (Punjab). Indian J. Agric. Sc. 77: 210-219.

Hustings, F. 1992. European monitoring studies on breeding birds: an update. Bird Census News 5(2): 1-56.

Hustings, F. & Vergeer, J.-W. 2002. Atlas van de Nederlandse broedvogels 1998-2000. Leiden: Nationaal Natuurhistorisch Museum Naturalis, KNNV Uitgeverij & European Invertebrate Survey-Nederland.

Hustings, F., J. van der Coelen, B. van Noorden, R. Schols & P. Voskamp. 2006. Avifauna van Limburg. Roermond: Stichting Natuurpublicaties Limburg.

Hüppop, O. & K. Hüppop. 2003. North Atlantic Oscillation and timing of spring migration in birds. Proc. R. Soc. London B 270: 233-240.

Impekoven, M. 1964. Zugwege und Verbreitung der Knäkente (*Anas querquedula*); eine Analyse der europäischen Beringungsresultate. Ornithol. Beob. 61: 1-34.

Insley, H. & R.C. Boswell. 1978. The timing of arrivals of Reed and Sedge Warblers at south coast ringing sites during autumn passage. Ring. & Migr. 2: 1-9.

Isenmann, P. & A. Moali. 2000. Oiseaux d'Algérie/Birds of Algeria. Paris: Société d'Études Ornithologiques de France.

Isenmann, P., T. Gaultier, A. El Hili, H. Azafzaf, H. Dlensi & M. Smart. 2005. Oiseaux de Tunisie/Birds of Tunisia. Paris: Société d'Études Ornithologiques de France.

Isenmann, P. 2006. Les Oiseaux du Banc d'Arguin. Le Sambuc: Fondation Internationale du Banc d'Arguin.

J

Janicot, S., V. Moron & B. Fontaine. 1996. Sahel droughts and ENSO dynamics. Geophysical Research Letters 23: 515-518.

Janicot, S., A. Harzallah, B. Fontaine & V. Moron. 1998. West African monsoon dynamics and eastern equatorial Atlantic and Pacific SST anomalies (1970-88). J. Climate 11: 1874-1882.

Janicot, S., S. Trzaska & I. Poccard. 2001. Summer Sahel-ENSO teleconnection and decadal time scale SST variations. Climate Dynamics 18: 303-320.

Jarry, G., F. Roux & A.M. Czajkowski. 1986. L'importance des zones humides du Sahel occidental pour les oiseaux migrateurs paléarctiques. Paris: CREPO, Muséum National d'Histoire Naturelle.

Jarry, G. & F. Baillon. 1991. Hivernage de la Tourterelle des bois (*Streptopelia turtur*) au Sénégal. Étude d'une population dans la région de Nianing. Paris: Centre de Recherches sur la Biologie des Populations d'Oiseaux (CRBPO).

János, O. 1996. Spring migration of Black-tailed Godwit *Limosa limosa* and Ruff *Philomachus pugnax* in the Tiszasüly area between 1993-1995. Partimadár 5: 63-65.

Jensen, F.P., K. Falk & B.S. Petersen. 2006. Migration routes and staging areas of Abdim's Storks *Ciconia abdimii* identified by satellite telemetry. Ostrich 77: 210-219.

Jiguet, F. 2007. Suivi temporel des oiseaux communs. Bilan du programme STOC pour la France en 2007. http://www2.mnhn.fr/vigie-nature/IMG/pdf/STOC-bilan_2007.pdf (accessed 14 February 2009).

Jobin, W. 1999. Dams and disease: ecological design and health impacts of large dams. canals and irrigation systems. London: E & FN Spon.

Joern, A. 1986. Experimental study of avian predation on coexisting grasshopper populations (Orthoptera: Acrididae) in a sandhill grassland. Oikos 46: 243-249.

Joern, A. 1992. Variable impact of avian predation on grasshopper assemblies in sandhill grassland. Oikos 46: 243-249.

John, J.R.M. & J.D.L. Kabigumila. 2007. Impact of *Eucalyptus* plantations on the avian breeding community in the East Usambaras, Tanzania. Ostrich 78: 265-269.

Jones, G. 1987. Selection against large size in the Sand Martin *Riparia riparia* during a dramatic population crash. Ibis 129: 274-280.

Jones, P. 1995. Migration strategies of Palaearctic passerines in Africa: an overview. Israel J. Zool. 41: 393-406.

Jones, P., J. Vickery, S. Holt & W. Cresswell. 1996. A preliminary assessment of some factors influencing the density and distribution of palearctic passerine migrants wintering in the Sahel zone of West Africa. Bird Study 43: 73-84.

Jones, P. 1998. Community dynamics of arboreal insectivorous birds in African savannas in relation to seasonal rainfall patterns and habitat change. *In:* D.M. Newberry, H.H.T. Prins & N.D. Brown (eds). Dynamics of tropical communities: 421-447. Oxford: Blackwell Scientific Press.

de Jonge, K., J. van der Klei, H. Meilink & R. Storm. 1978. Les migrations en Basse Casamance (Sénégal). Leiden: Afrika Studie Centrum.

Jonkers, D.A. 1987. Foerageergebieden en voedsel van de Ooievaars in Schoonrewoerd. Vianen: Natuur- en Vogelwacht "De Vijfheerenlanden".

Jonkers, D.A. & P. Maréchal. 1990. De achteruitgang van de Gekraagde Roodstaart *Phoenicurus phoenicurus* voer voor nadere discussie. Vogeljaar 38: 49-61.

Jonzén, N., A. Lindén, T. Ergon, E. Knudsen, J.O. Vik, D. Rubolini, D. Piacentini, C. Brinch, F. Spina, L. Karlsson, M. Stervander, A. Andersson, J. Waldenström, A. Lehikoinen, E. Edvardsen, R. Solvang & N.C. Stenseth. 2006. Rapid advance of spring arrival dates in long-distance migratory birds. Science 312: 1959-1961.

Jørgensen, H.E. 1989. Danmarks rovfugle - en statusoversigt. Frederikshus: Øster Ulslev.

Jourdain, E., M. Gauthier-Clerc, Y. Kayser, M. Lafaye & P. Sabatier. 2008. Satellitte-tracking migrating juvenile Purple Herons *Ardea purpurea* from the Camargue area, France. Ardea 96: 121-124.

Jukema, J., T. Piersma, L. Louwsma, C. Monkel, U. Rijpma, K. Visser & D. van der Zee. 1995. Rui en gewichtsveranderingen van doortrekkende Kemphanen in Friesland in 1993 en 1994. Vanellus 48: 55-61.

Jukema, J. & T. Piersma. 2000. Contour feather moult of Ruffs *Philomachus pugnax* during northward migration, with notes on homology of nuptial plumages in scolopacid waders. Ibis 142: 289-298.

Jukema, J., T. Piersma, J.B. Hulscher, A.J. Bunskoeke, A. Koolhaas & A. Veenstra. 2001a. Goudplevieren en wilsterflappers; eeuwenoude fascinatie voor trekvogels. Ljouwert/Utrecht: Fryske Akademy/KNNV.

Jukema, J., E. Wymenga & T. Piersma. 2001b. Opvetten en ruien in de Zuidwesthoek: Kemphanen *Philomachus pugnax* op voorjaarstrek in Friesland. Limosa 74: 17-26.

Jukema, J. & T. Piersma. 2004. Kleine mannelijke Kemphanen met vrouwelijk broedkleed: bestaat er een derde voortplantingsstrategie? Limosa 77: 1-10.

Jukema, J. & T. Piersma. 2006. Permanent female mimics in a lekking shorebird. Biology Letters 2: 161-164.

Junk, W.J., P.B. Bayley & R.E. Sparks. 1989. The flood pulse concept in river-floodplain systems. Can. Spec. Publ. Fish. Aquat. Sci. 106: 110-127.

K

van der Kamp, J. 1989. Herkomst van overwinterende grutto's en kemphanen in Mali. Intern rapport 89/24 Arnhem: Rijksinstituut voor Natuurbeheer.

van der Kamp, J. & M. Diallo. 1999. Suivi écologique du Delta Intérieur du Niger: les oiseaux d'eau comme bio-indicateurs. Recensements crue 1998-1999. Veenwouden: A&W/Wetlands International.

van der Kamp, J., M. Diallo & B. Fofana. 2002a. Dynamique des populations d'oiseaux d'eau. In: E. Wymenga, B. Kone, J. van der Kamp & L. Zwarts (eds). Delta intérieur du fleuve Niger. Ecologie et gestion durable des ressources naturelles: 87-138. Veenwouden: A&W/WetlandsInternational/Rijkswaterstaat.

van der Kamp, J., L. Zwarts & M. Diallo. 2002b. Niveau de crue, oiseaux d'eau et ressources alimentaires disponibles. In: E. Wymenga, B. Kone, J. van der Kamp & L. Zwarts (eds). Delta intérieur du fleuve Niger. Écologie et gestion durable des ressources naturelles: 141-161. Veenwouden: A&W/WetlandsInternational/Rijkswaterstaat.

van der Kamp, J., M. Diallo, B. Fofana & E. Wymenga. 2002c. Colonies nicheuses d'oiseaux d'eau. In: E. Wymenga, B. Kone, J. van der Kamp & L. Zwarts (eds). Delta intérieur du fleuve Niger. Écologie et gestion durable des ressources naturelles: 163-186. Veenwouden: A&W/WetlandsInternational/Rijkswaterstaat.

van der Kamp, J., B. Fofana & E. Wymenga. 2005a. Ecological values of the Inner Niger Delta. In: L. Zwarts, P. van Beukering, B. Kone & E. Wymenga (eds). The Niger, a lifeline: 156-176. Lelystad: Rijkswaterstaat/IVM/Wetlands International/A&W.

van der Kamp, J., B. Fofana & E. Wymenga. 2005b. Sélingué reservoir. In: L. Zwarts, P. van Beukering, B. Kone & E. Wymenga (eds). The Niger, a lifeline: 179-187. Lelystad: Rijkswaterstaat/IVM/Wetlands International/A&W.

van der Kamp, J., M. Diallo & B. Fofana. 2005c. Ecological valuation of major man-made (Sélingué reservoir, irrigation zone of Office du Niger) and floodplain habitats in the Upper Niger Basin. Veenwouden: Altenburg & Wymenga ecological consultants.

van der Kamp, J., I. Ndiaye & B. Fofana. 2006. Post-breeding exploitation of rice habitats in West Africa by migrating Black-tailed Godwit. Veenwouden: Altenburg & Wymenga ecological consultants

van der Kamp, J., D. Kleijn, I. Ndiaye, S. I. Sylla & L. Zwarts. 2008. Rice farming and Black-tailed Godwits in the Casamance, Senegal. Veenwouden: Altenburg & Wymenga ecological consultants.

Kane, A. 2002. Crues et inondations dans la basse vallée du fleuve Sénégal. In: D. Orange, R. Arfi, M. Kuper, P. Morand & Y. Poncet (eds). Gestion intégrée des ressources naturelles en zones inondables tropicales: 197-208. Paris: IRD.

Kanyamibwa, S., A. Schierer, R. Pradel & J.-D. Lebreton. 1990. Changes in adult annual survival rates in a western European population of the White Stork *Ciconia ciconia*. Ornis Scand. 24: 297-302.

Kanyamibwa, S., F. Bairlein & A. Schierer. 1993. Comparison of survival rates between populations of the White Stork *Ciconia ciconia* in Central-Europe. Ornis Scand. 24: 297-302.

Karlionova, N., P. Pinchuk, W. Meissner & Y. Verkuil. 2007. Biometrics of Ruffs *Philomachus pugnax* migrating in spring through southern Belarus with special emphasis on the occurrence of 'faeders'. Ring. & Migr. 23: 134-140.

Karlsson, L., S. Ehnbom, K. Persson & G. Walinder. 2002. Changes in numbers of migrating birds at Falsterbo, South Sweden, during 1980-1999, as reflected by ringing totals. Ornis Svecica 12: 113-137.

Karlsson, L. 2007. Övervakning av beståndsväxlingar hos svenska småfåglar med vinterkvarter i tropikerna via ringmärkningssiffror vid Falsterbo Fågelstation. Falsterbo: Länsstryrelsen i Såne Län.

Kaspari, M.E. & A. Joern. 1993. Prey choice by three insectivorous grassland birds: reevaluating opportunism. Oikos 68: 414-430.

Kassibo, B. & J. Bruner-Jailly. 2003. La pirogue, monture du bozo, hier et aujourd'hui. Djenné Patrimoine Informations 14.

ten Kate, C.G.B. 1936. De vogels van het Zuiderzeegebied. Flora en fauna der Zuiderzee, Supplement: 1-82. Den Helder: de Boer Jr.

Kayser, Y., C. Girard, G. Massez, Y. Chérain, D. Cohez, H. Hafner, A. Johnson, N. Sadoul, A. Tamisier & P. Isenmann. 2003. Compte-rendu ornithologique Camarguais pour les années 1995-2000. Terre Vie 58: 5-76.

Källander, H. 1993. Commensal feeding associations between Yellow Wagtails *Motacilla flava* and cattle. Ibis 135: 97-100.

Keddy, P. 2000. Wetland ecology. Principles and conservation. Cambridge: Cambridge University Press.

Keith, J.O. & W.C. Mullié. 1990. Birds. In: J.W. Everts (ed.). Environmental effects of chemical locust and grasshopper control. A pilot study: 235-270. Rome: FAO.

Keith, J.O. & D.C.H. Plowes. 1997. Considerations of wildlife resources and land use in Chad. S.D. Technical paper No. 45. Washington, D.C.: U.S. Agency for International Development Bureau for Africa.

Keith, S., E.K. Urban & C.H. Fry. 1992. The Birds of Africa, Vol. IV. London: Academic Press.

Keïta, N., J.-F. Bélières & B. Sidibé. 2002. Extension de la zone aménagée de l'Office du Niger: exploitation rationelle et durable des ressources naturelles au service d'un enjeu national de développement. In: D. Orange, R. Arfi, M. Kuper, P. Morand & Y. Poncet (eds). Gestion intégrée des ressources naturelles en zones inondables tropicales: 929-952. Paris: IRD.

Kenward, R.E. & R.M. Sibly. 1977. A Woodpigeon (*Columba palumbus*) feeding preference explained by a digestive bottle-neck. J. appl. Ecol. 14: 815-826.

Kerr, J.T. & M. Ostrovsky. 2003. From space to species: ecological applications for remote sensing. Trends in Ecology & Evolution 18: 299-305.

Kersten, M. & C. Smit. 1984. The Atlantic coast of Morocco. In: P.R. Evans, J.D. Goss-Custard & W.G. Hale (eds). Coastal waders and wildfowl in winter: 276-292. Cambridge: Cambridge University Press.

Kéry, M. 2008. Grundlagen der Bestandserfassung am Beispiel von

Vorkommen und Verbreitung. Ornithol. Beob. 105: 353-386.

Khokhlov, A. N. 1995. [Ornithological observations in West Turkmenia.] Stavropol: State Pedagogical University.

Khoury, F. 2004. Seasonal variation in body fat and weight of migratory *Sylvia* warblers in central Jordan. Vogelwarte 42: 191-202.

Kilian, D., J. Hölzinger, U. Mahler & R. Stegmayer. 1993. Der Graureiher (*Ardea cinerea*) in Baden-Württemberg 1985-1991. Ökologie der Vögel 15: 1-36.

Kilpi, M. & P. Saurola. 1984. Migration and survival areas of Caspian Terns *Sterna caspia* from the Finnish coast. Ornis Fennica 61: 24-29.

King, D.I., J.H. Rappole & J.P. Buonaccorsi. 2006. Long-term population trends of forest-dwelling Nearctic-Neotropical migrant birds: a question of temporal scale. Bird Populations 7: 1-9.

King, J.M. & J.M.C. Hutchinson. 2001. Site fidelity and recurrence of some migrant bird species in The Gambia. Ring. & Migr. 20: 292-302.

Kirk, D.A., M.D. Evenden & P. Mineau. 1996. Past and current attempts to evaluate the role of birds as predators of insect pests in temperate agriculture. Current Ornithology 13: 175-269.

Kirwan, G.M., K. Boyla, P. Castell, B. Demirci, M. Ozen, H. Welch & T. Marlow. 2008. The birds of Turkey. London: Christopher Helm.

Kiwango, Y.A. & E. Wolanski. 2008. Papyrus wetlands, nutrients balance, fisheries collapse, food security & Lake Victoria level decline in 2000-2006. Wetlands Ecol. Manage. 16: 89-96.

Kjeldsen, J.P. 2008. Ynglyfugle i Vejlerne efter inddæmningen, med særlig vægt på feltstationsårene 1978-2003. Dansk Orn. Foren. Tidsskr. 102: 1-238.

Kjellén, N., M. Hake & T. Alerstam. 2001. Timing and speed of migration in male, female and juvenile Ospreys *Pandion haliaetus* between Sweden and Africa as revealed by field observations, radar and satellite tracking. J. Avian Biol. 32: 57-67.

Klaassen, R.H.G., R. Strandberg, M. Hake & T. Alerstam. 2008a. Flexibility in daily travel routines causes regional variation in bird migration speed. Behav. Ecol. Sociobiol. 62: 1427-1432.

Klaassen, R.H.G., R. Strandberg, M. Hake, P. Olofsson, A.P. Tøttrup & T. Alerstam. 2008b. Loop migration in adult Marsh Harriers *Circus aeruginosus* is explained by wind rather than habitat availability. *In*: R. Strandberg (ed.). Migration strategies of raptors - spatio-temporal adaptations and constraints intravelling and foraging: 97-104. Lund: Lund University, Department of Animal Ecology.

Kleefstra, R. 2005. Grutto's jaar na jaar te vroeg, massaal en zonder kroost op Friese slaapplaatsen. Twirre 16: 211-215.

Kleefstra, R. 2007. Slapende Grutto's in de Frieswijkpolder revisited. Twirre 18: 94-97.

Kleijn, D. & W.J. Sutherland. 2003. How effective are European agri-environment schemes in conserving and promoting biodiversity? J. appl. Ecol. 40: 947-969.

Kloff, S. & A.H. Pieterse. 2006. A resource planning review of the Senegal River. www.typha.net. GTZ/KIT.

Kluijver, H.N. 1951. The population ecology of the Great Tit *Parus m. major*. Ardea 39: 1-135.

Knox, A.G. 1993. Richard Meinertzhagen - a case of fraud examined. Ibis 135: 320-325.

Kodio, A., P. Morand, K. Diénépo & R. Laë. 2002. Dynamique de la pêcherie du delta intérieur du Niger revisitée à la lumière des données récentes. *In*: D. Orange, R. Arfi, M. Kuper, P. Morand & Y. Poncet (eds). Gestion intégrée des ressources naturelles en zones inondables tropicales: 431-453. Paris: IRD.

Koks, B.J., C.W.M. van Scharenburg & E.G. Visser. 2001. Grauwe Kiekendieven *Circus pygargus* in Nederland: balanceren tussen hoop en vrees. Limosa 74: 121-136.

Koks, B.J. & E.G. Visser. 2002. Montagu's Harriers *Circus pygargus* in the Netherlands: Does nest protection prevent extinction? Ornithol. Anz. 41: 159-166.

Koks, B.J., C. Trierweiler, H. Hut, A. Harouna, H. Issaka & J. Brouwer. 2006. Grauwe Kiekendief missie naar Niger en Burkina Faso. Groningen: Stichting Werkgroep Grauwe Kiekendief.

Koks, B., C. Trierweiler, H. Hut, H. Harouna, H. Issaka & J. Brouwer. 2007a. Grauwe Kiekendief missie naar Niger en Burkina Faso. Groningen: Stichting Werkgroep Grauwe Kiekendief.

Koks, B.J., C. Trierweiler, E.G. Visser, C. Dijkstra & J. Komdeur. 2007b. Do voles make agricultural habitat attractive to Montagu's Harrier *Circus pygargus*? Ibis 149: 575-586.

Kone, B., M. Diallo & A.M. Maiga. 1999. L'exploitation des oiseaux d'eau dans le Delta Intérieur du Niger. Mali-Pin 99-03. Sévaré: Wetlands International / Altenburg & Wymenga.

Kone, B. & B. Fofana. 2001. Statut de la grue couronnée et son exploitation au Mali. Sévaré: Wetlands International.

Kone, B., M. Diallo & B. Fofana. 2002. Exploitation des oiseaux d'eau. *In*: E. Wymenga, B. Kone, J. van der Kamp & L. Zwarts (eds). Delta intérieur du fleuve Niger. Ecologie et gestion durable des ressources naturelles: 201-207. Veenwouden: A&W/Wetlands International/Rijkswaterstaat.

Kone, B., M. Diallo & D. Bos. 2005. Exploitation des oiseaux d'eau dans le Delta Intérieur du Niger, Mali. Sévaré: Wetlands International.

Kone, B. 2006. Exploitation des oiseaux d'eau et son importance socio-économique au Mali, Niger et Burkina Faso. Sévaré: Wetlands International.

Kone, B., B. Fofana, R. Beilfuss & T. Dodman. 2007. The impact of capture, domestication and trade on Black Crowned Cranes in the Inner Niger Delta, Mali. Ostrich 78: 195-203.

van der Kooij, H. 1991. Nesthabitat van de Purperreiger *Ardea purpurea* in Nederland. Limosa 64: 103-112.

van der Kooij, H. 1991. Het broedseizoen 1990 van de Purperreiger in Nederland: een dieptepunt! Vogeljaar 39: 251-255.

van der Kooij, H. 1992. De Havik *Accipiter gentilis* als broedvogel in purperreigerkolonies *Ardea purpurea*: gaat dat samen? Limosa 65: 53-56.

van der Kooij, H. 1994. Het broedseizoen 1993 van de Purperreiger in Nederland. Vogeljaar 42: 218-220.

van der Kooij, H. 1995a. Werkt de Vos *Vulpes vulpes* de Purperreiger *Ardea*

purpurea in de nesten? Limosa 68: 137-142.
van der Kooij, H. 1995b. Het broedseizoen 1994 van de Purperreiger in Nederland. Vogeljaar 43: 220-222.
van der Kooij, H. 1996. Het broedseizoen 1995 van de Purperreiger in Nederland. Vogeljaar 44: 179-181.
van der Kooij, H. 1997a. Het broedseizoen 1996 van de Purperreiger in Nederland. Vogeljaar 45: 115-117.
van der Kooij, H. 1997b. Wordt het broedresultaat van Purperreigers *Ardea purpurea* beïnvloed door de nesthoogte? Limosa 70: 145-151.
van der Kooij, H. 1998. Het broedseizoen 1997 van de Purperreiger in Nederland. Vogeljaar 46: 53-57.
van der Kooij, H. 1999. Het broedseizoen 1998 van de Purperreiger in Nederland. Vogeljaar 47: 204-207.
van der Kooij, H. 2000. Het broedseizoen 1999 van de Purperreiger in Nederland. Vogeljaar 48: 120-122.
van der Kooij, H. 2001. Het broedseizoen 2000 van de Purperreiger in Nederland. Vogeljaar 49: 67-71.
van der Kooij, H. 2002. Het broedseizoen 2001 van de Purperreiger in Nederland. Vogeljaar 50: 106-110.
van der Kooij, H. 2003. Het broedseizoen 2002 van de Purperreiger in Nederland. Vogeljaar 51: 254-260.
van der Kooij, H. 2005. De broedseizoenen 2003 en 2004 van de Purperreiger in Nederland. Vogeljaar 53: 151-156.
van der Kooij, H. 2007. De broedseizoenen 2005 en 2006 van de Purperreiger in Nederland. Vogeljaar 55: 150-157.
Koopman, K. & P.W. Bouma. 1979. Slaaptrekonderzoek aan steltlopers in Fryslân. Voorlopig verslag. Rapport 6. Leeuwarden: FFF.
Koopman, K. 1986. Primary moult and weight changes of Ruffs in the Netherlands in relation to migration. Ardea 74: 69-77.
Kooyman, C. & I. Godonou. 1997. Infection of *Schistocerca gregaria* (Orthoptera: Acrididae) hoppers by *Metarhizium flavoviride* (Deuteromycotina: Hyphomycetes) conidia in an oil formulation applied under desert conditions. Bull. Entomol. Res 87: 105-107.
Kooyman, C., M. Ammati, K. Moumème, A. Chaouch & A. Zeyd. 2005. Essai de Green Muscle® sur des nymphes du Criquet pèlerin dans la Willaya d'El Oued, Nord-Est Algérie, Avril-mai 2005. Algers/Rome/Cotonou: FAO.
Kooyman, C. 2006. Final technical teport of the Regional Programme for Environmentally Sound Grasshopper Control in the Sahel (Phase I). Cotonou/Niamey/Copenhagen.
Kooyman, C., W.C. Mullié & M. ould Sid'Ahmed. 2007. Essai de Green Muscle® sur des nymphes du Criquet pèlerin dans la zone de Benichab, Ouest Mauritanie, Octobre-novembre 2006. Rome: FAO/Centre de Lutte Anti-Acridienne.
Kosicki, J., T. Sparks & P. Tryjanowski. 2004. Does arrival date influence autumn departure of the White Stork *Ciconia ciconia*? Ornis Fennica 81: 91-95.
Kosicki, J.Z., P. Profus, P.T. Dolata & M. Tobólka. 2006. Food composition and energy demand of the White Stork *Ciconia ciconia* breeding population. Literature survey and preliminary results from Poland. *In:* P. Tryjanowski, T. Sparks & L. Jerzak (eds). The White Stork in Poland: studies in biology, ecology and conservation: 169-183. Poznań: Bogucki Wydawnictwo Naukowe.
Koskimies, P. & P. Saurola. 1985. Autumn migration strategies of the Sedge Warbler *Acrocephalus schoenobaenus* in Finland: a preliminary report. Ornis Fenn.62: 145-152.
Koskimies, P. & R.A. Väisänen. 1991. Monitoring bird populations. Helsinki: Zoological Museum, Finnish Museum of Natural History.
Koskimies, P. 2005. Suomen lintuopas. Helsinki: Werner Söderström Osakeyhtiö.
Kostin, Y. 1983. [Birds of the Crimea.] Moscow: Nauka.
Kostrzewa, A. & G. Speer. 2001. Greifvögel in Deutschland: Bestand, Situation, Schutz. Wiebelsheim: AULA-Verlag.
Krogulec, J. 2002. Distribution and population trend of Montagu's Harrier *Circus pygargus* in Poland. Ornithol. Anz. 41: 212.
Kröpelin, S., D. Verschuren, A.-M. Lézine, H. Eggermont, C. Cocquyt, P. Francus, J.P. Cazet, M. Fagot, B. Rumes, J.M. Russell, F. Darius, D.J. Conley, M. Schuster, H. von Suchodoletz & D.R. Engstrom. 2008. Climate-driven ecosystem succession in the Sahara: the past 6000 years. Science 320: 765-768.
Kruk, M., M.A.W. Noordervliet & W.J. ter Keurs. 1997. Survival of Black-tailed Godwit chicks *Limosa limosa* in intensively exploited grassland areas in the Netherlands. Biol. Conserv. 80: 127-133.
Kube, J., A.I. Korzyukov, D.N. Nankinov, OAG Münster & P. Weber. 1998. The northern and western Black Sea region - the Wadden Sea of the Mediterranean Flyway for wader populations. International Wader Studies 10: 379-393.
Kuijper, D.P.J., E. Wymenga, J. van der Kamp & D. Tanger. 2006. Wintering areas and spring migration of the black-tailed godwit: bottlenecks and protection along the migration route. Veenwouden: Altenburg & Wymenga.
Kumar, L., M. Rietkerk, F. van Langevelde, J. van de Koppel, J. van Andel, J. Hearne, N. de Ridder, L. Stroosnijder, A.K. Skidmore & H.H.T. Prins. 2002. Relationship between vegetation growth rates at the onset of the wet season and soil type in the Sahel of Burkina Faso: implications for resource utilisation at large scales. Ecological Modelling 149: 143-152.
Kuresoo, A. & R. Mänd. 1991. The results of point counts in Estonia. Bird Census News 4(1): 34-39.
Kuschert H. & F. Ziesemer. 1991. Knäkente - *Anas querquedula*. Vogelwelt Schleswig-Holsteins. Band 3: 168-172. Neumünster: Karl Wachholtz Verlag.
Kushlan, J.A. & H. Hafner. 2000. Heron conservation. London: Academic Press.
Kushlan, J.A. & J.A. Hancock. 2005. Herons. Bird families of the world. Oxford: Oxford University Press.

L

Lack, D. & G.C. Varley. 1945. Detection of birds by radar. Nature 156: 446.
Lack, D. 1966. Population studies of birds. Oxford: Oxford University Press.

Lack, P.C. 1983. The movements of palaearctic landbird species in Tsavo East National Park, Kenya. J. Anim. Ecol. 52: 513-524.

Lack, P.C. 1986. Ecological correlates of migrants and residents in a tropical African savanna. Ardea 74: 111-119.

Laë, R., M. Maïga, J. Raffray & J. Troubat. 1994. Évolution de la pêche. *In:* J. Quensière (ed.). La Pêche dans le Delta Central du Niger: 143-163. Paris: Karthala.

Laë, R. 1995. Climatic and anthropogenic effects on fish diversity and fish yields in the Central Delta of the Niger River. Aquat. Living Resour. 8: 43-58.

Laë, R. & C. Levêque. 1999. La pêche. *In:* C. Levêque & D. Paugy (eds). Les poissons des eaux continentales africaines: 385-424. Paris: IRD.

Lamarche, B. 1980. Liste commentée des oiseaux du Mali. 1ère partie: Non-passeraux. Malimbus 2: 121-158.

Lande, R., S. Engen, B.-U. Sæther, F. Filli, E. Matthysen & H. Weimerskirch. 2002. Estimating density dependence from population time series using demographic theory and life-history data. Am. Nat. 159: 321-337.

Lang, M. 2007. Niedergang des süddeutschen Ortolan-Populaton *Emberiza hortulana* – liegen die Ursachen außerhalb des Brutgebiets. Vogelwelt 128: 179-196.

Lange, H., W. Emeis & E. Schüz. 1938. Weitere Angaben über Heimkehr-Verzögerung und Bestand des Weißen Storches 1937. Vogelzug 9: 97-102.

van Langevelde, F., C.A.D.M. van de Vijver, L. Kumar, J. van de Koppel, N. de Ridder, J. van Andel, A.K. Skidmore, J.W. Hearne, L. Stroosnijder, W.J. Bond, H.H.T. Prins & M. Rietkerk. 2003. Effects of fire and herbivory on the stability of savanna ecosystems. Ecology 84: 337-350.

Langslow, D.R. 1977. Weight increases and behaviour of Wrynecks on the Isle of May. Scottish Birds 9: 262-267.

Lartiges, A. & P. Triplet. 1988. L'aménagement du bas-delta mauritanien du fleuve Sénégal et ses conséquences possibles pour l'avifaune. Bull. mens. O.N.C. 123: 40-48.

Latif, M., D. Anderson, T. Barnett, M. Cane, R. Kleeman, A. Leetmaa, J. O'Brien, A. Rosati & E. Schneider. 1998. A review of the predictability and prediction of ENSO. J. Geophysical Research Oceans 103: 14375-14393.

Launois, M. 1978. Modélisation écologique et simulation opérationnelle en acridologie. Application à *Oedaleus senegalensis* (Krauss, 1877). Paris: Ministère de la Coopération, GERDAT.

Lausten, M. & P. Lyngs. 2004. Trækfugle på Christiansø 1976-2001. Gudhjem: Christiansøs Naturvidenskabelige Feltstation.

Le Mao, P. 1980. Les migrations et l'hivernage des limicoles en Maine-et-Loire de 1961 à 1978. Bull. GAEO 19 (no. 30): 180-236.

Lebedeva, E. A. 1998. Waders in agricultural habitats of European Russia. International Wader Studies 10: 315-324.

Lebedeva, M.I., K. Lambert & I.N. Dobrynina. 1985. Wood Sandpiper - *Tringa glareola*. *In:* J.A. Viksne & H.A. Mihelsons (eds). Migrations of birds of eastern Europe and southern Asia: 97-105. Moscow: Nauka..

Leblanc, M.J., C. Leduc, F. Stagnitti, P.J. van Oevelen, C. Jones, L.A. Mofor, M. Razack & G. Favreau. 2006. Evidence for Megalake Chad, north-central Africa, during the late Quaternary from satellite data. Palaeogeography Palaeoclimatology Palaeoecology 230: 230-242.

Lebret, T. 1959. De dagelijkse verplaatsingen tussen dagverblijf en nachtelijk voedselgebied bij smienten *Anas penelope* L. in enige terreinen in het lage midden van Friesland. Ardea 47: 199-210.

Lecoq, M. 1978. Biologie et dynamique d'un peuplement acridien de la zone soudanienne en Afrique de l'Ouest (Orthoptera, Acrididae). Ann. Société Entomol. France 14: 603-681.

Leguijt, R., D. Tanger & P. Zomerdijk. 1995. Report visit Tejo and Sado-estuary, Portugal, 28 January 1995 - 4 February 1995, with special references to census and behaviour of Black-tailed Godwits *Limosa limosa*. Unpubl. report.

Lehikoinen, E., T.H. Sparks & M. Zalakevicius. 2006. Arrival and departure dates. *In:* A.P. Møller, W. Fiedler & P. Berthold (eds). Birds and climate change: 1-31. Burlington: Academic Press.

Leibak, E., V. Lilleleht & H. Veromann. 1994. Birds of Estonia. Status, distribution & numbers. Tallinn: Estonian Academy of Science.

Lemly, A.D., R.T. Kingsford & J.R. Thompson. 2000. Irrigated agriculture and wildlife conservation: Conflict on a global scale. Environm. Manage. 25: 485-512.

Lemoalle, J. 2005. The Lake Chad basin. *In:* L.H. Fraser & P.A. Keddy (eds). The World's Largest Wetland: Ecology and Conservation: 316-346. Cambridge: Cambridge University Press.

Leopold, M.F., C.J.W. Bruin, C.J. Camphuysen, C. Winter & B. Koks. 2003. Waarom is de Visarend in Nederland geen zeearend? Limosa 76: 129-140.

Leuzinger, H. & L. Jenni. 1993. Durchzug des Bruchwasserläufers *Tringa glareola* am Ägelsee bei Frauenfeld. Ornithol. Beob. 90: 169-188.

Lewis, A. & D. Pomeroy. 1989. A bird atlas of Kenya. Rotterdam/Brookfield: Balkema.

Leys, H.N. 1987. Inventarisatie van de Oeverzwaluw (*Riparia riparia*) in 1986 in Nederland. Vogeljaar 35: 119-131.

Leys, H.N., G.M. Sanders & W.C. Knol. 1993. Avifauna van Wageningen en wijde omgeving. Wageningen: KNNV Vogelwerkgroep Wageningen.

Li Zuo Wei, D. & T. Mundkur. 2004. Numbers and distribution of waterfowl and wetlands in the Asian-Pacific region. Results of the Asian Waterbird Census: 1997-2001. Kuala Lumpur: Wetlands International.

Li, J., J. Lewis, J. Rowland, G. Tappan & L.L. Tieszen. 2004. Evaluation of land performance in Senegal using multi-temporal NDVI and rainfall series. J. Arid Environ. 59: 463-480.

Liechti, F., M. Klaassen & B. Bruderer. 2000. Predicting migratory flight altitudes by physiological migratory models. Auk 117: 205-214.

Liechti, F. & H. Schmaljohann. 2007. Wind-governed flight altitudes of nocturnal spring migrants over the Sahara. Ostrich 78: 337-341.

Limiñana, R., A. Soutullo, V. Urios & M. Surroca. 2006. Vegetation height selection in Montagu's Harriers *Circus pygargus* breeding in a natural habitat. Ardea 94: 280-284.

Limiñana, R., A. Soutullo & V. Urios. 2007. Autumn migration of Montagu's Harriers *Circus pygargus* tracked by satellite telemetry. J. Ornithol. 148: 517-523.

Limiñana, R., A. Soutullo, P. López-López & V. Urios. 2008. Pre-migratory movements of adult Montagu's Harriers *Circus pygargus*. Ardea 96: 81-90.

Lind, M., K. Rasmussen, H. Adriansen & A. Ka. 2003. Estimating vegetative productivity gradients around watering points in the rangelands of northern Senegal based on NOAA AVHRR data. Danish J. Geography 103: 1-15.

Lindqvist, S. & A. Tengberg. 1993. New evidence of desertification from case studies in northen Burkina Faso. Geografiska Annaler 18: 127-135.

Lindström, Å. & S. Svensson. 2005. Övervakning av fåglarnas populationsutveckling. Årsrapport för 2004. Lund: Ekologiska institutionen, Lunds universitet.

Lindström, Å., S. Svensson, M. Green & R. Ottvall. 2007. Distribution and population changes of two subspecies of Chiffchaff *Phylloscopus collybita* in Sweden. Ornis Svecica 17: 137-147.

Lindström, Å., M. Green, R. Ottvall & S. Svensson. 2008. Övervakning av fåglarnas populationsutveckling. Årsrapport 2007. Lund: Ekologiska institutionen, Lunds Universitet.

Literák, I., M. Honza & D. Kondelka. 1994. Postbreeding migration of the Sedge Warbler *Acrocephalus schoenobaenus* in the Czech Republic. Ornis Fennica 71: 151-155.

Loske, K.-H. & W. Lederer. 1987. Bestandsentwicklung und Fluktuationstrate von Weitstreckenziehern in Westfalen: Uferschwalbe (*Riparia riparia*), Rauchschwalbe (*Hirundo rustica*), Baumpieper (*Anthus trivialis*) und Grauschnäpper (*Muscicapa striata*). Charadrius 23: 101-127.

Loske, K.-H. 1994. Untersuchungen zu Überlebensstrategien der Rauchschwalbe (*Hirundo rustica*) im Brutgebiet. Göttingen: Cuvillier Verlag.

Loske, K.-H. 1996. Ein wichtiger Schlafplatz europäischer Rauchschwalben *Hirundo rustica* in Nigeria und seine Bedrohung. Limicola 10: 42-48.

Loske, K.-H. & T. Laumeier. 1999. Bestandsentwicklung der Uferschwalbe *Riparia riparia* in Mittelwestfalen. Vogelwelt 120: 133-140.

Loske, K.-H. 2008. Der Niedergang der Rauchschwalbe *Hirundo rustica* in den westphälischen Hellwegbörden 1977-2007. Vogelwelt 129: 57-71.

Loth, P. 2004. The return of the water: restoring the Waza Logone floodplain in Cameroon. Gland: IUCN.

Lounsbury, C.P. 1909. Third annual report of the Committee of Control of the South African Central Locust Bureau. Cape Town: Cape Times Ltd.

Lourenco, P.M. & T. Piersma. 2008. Stopover ecology of Black-tailed Godwits *Limosa limosa limosa* in Portuguese rice fields: a guide on where to feed in winter. Bird Study 55: 194-202.

Lowe-McConnell, R.H. 1987. Ecological studies in tropical fish communities. Cambridge: Cambridge University Press.

LWVT/SOVON. 2002. Vogeltrek over Nederland 1976-1993. Haarlem: Schuyt & Co.

Lynes, H. 1924. On the birds of northern and central Darfur, with notes on the west-central Kordofan and north Nuba Provinces of the British Sudan. Ibis 11: 339-446, 648-719.

Lynes, H. 1925. On the birds of northern and central Darfur, with notes on the west-central Kordofan and north Nuba Provinces of the British Sudan. Ibis 12: 71-131, 346-416, 541-550, 757-797.

M

Macías, M., A.J. Green & M.I. Sánchez. 2004. The diet of the Glossy Ibis during the breeding season in Doñana, southwest Spain. Waterbirds 27: 234-239.

Magnin, G. 1991. Hunting and persecution of migratory birds in the Mediterranean area. *In:* T. Salathé (ed.). Conserving migratory birds: 63-75. Cambridge: International Council for Bird Preservation.

Magor, J.L., P. Ceccato, H.M. Dobson, J. Pender & L. Ritchie. 2007. Preparedness to prevent Desert Locust plagues in the Central Region, a historical review. Part 1. Text & Part 2. Appendices. Rome: FAO.

Mahé, G., J.C. Olivry, R. Dessouassi, D. Orange, F. Bamba & E. Servat. 2000. Surface water and groundwater relationships in a tropical river of Mali. C.R. Acad. Sc. série IIa 330: 689-692.

Mahé, G., Y. L'Hôte, J.C. Olivry & G. Wotling. 2001. Trends and discontinuities in regional rainfall of West and Central Africa: 1951-1989. Hydrol. Sci. 46: 211-226.

Maiga, I.H., M. Lecoq & C. Kooyman. 2008. Ecology and management of the Senegalese grasshopper, *Oedaleus senegalensis* (Krauss, 1877) (Orthoptera: Acrididae) in West Africa. Review and prospects. Annales de la Société Entomologique de France 44: 271-288.

Malzy, P. 1962. La faune avienne du Mali (bassin du Niger). L'Oiseau et RFO 32 no spécial: 1-81.

Mammen, U. & M. Stubbe. 2006. Die Bestandsentwicklung der Greifvögel und Eulen Deutschlands von 1988 bis 2002. Populationsökologie Greifvogel- und Eulenarten 5: 21-40.

van Manen, W. 2001. Wat is er aan de hand met de broedvogels in de polderbossen? Sovon Nieuws 14: 16-17.

van Manen, W. 2008. Broedvogels van Arkemheen in 2007. Beek-Ubbergen: SOVON Vogelonderzoek Nederland.

Mangnall, M.J. & T.M. Crowe. 2003. The effects of agriculture on farmland bird assemblages on the Agulhas Plain, Western Cape, South Africa. Afr. J. Ecol. 41: 266-276.

Marchant, J.H., R. Hudson, S.P. Carter & P. Whittington. 1990. Population trends in British breeding birds. Tring: British Trust for Ornithology.

Marchant, J.H. 1992. Recent trends in breeding populations of some common trans-Saharan migrant birds in northern Europe. Ibis 134 suppl. 1: 113-119.

Marchant, J.H., S.N. Freeman, H.Q.P. Crick & L.P. Beaven. 2004. The BTO Heronries Census of England and Wales 1928-2000: new indices and a comparison of analytical methods. Ibis 146: 323-334.

Marchant, S. 1963. Migration in Iraq. Ibis 105: 369-398.

Marie, J. 2002. Enjeux spatiaux et fonciers dans le delta intérieur du Niger (Mali). In: D. Orange, R. Arfi, M. Kuper, P. Morand & Y. Poncet (eds). Gestion intégrée des ressources naturelles en zones inondables tropicales: 557-586. Paris: IRD.

Marion, L. 1980. Dynamique d'une population de Hérons cendrés Ardea cinerea L. Exemple de la plus grande colonie d'Europe: le lac de Grand-Lieu. L'Oiseau et RFO 50: 219-261.

Marion, L., P. Ulenaers & J. van Vessem. 2000. Herons in Europe. In: J.A. Kushlan & H. Hafner (eds). Heron conservation: 1-31. London: Academic Press.

Markovets, M. & R. Yosef. 2004. Phenology of spring migration of passerines in Eilat (Israel). Rybachy: Biological Station Rybachy.

Marquiss, M., L. Robinson & E. Tindal. 2007. Marine foraging by Ospreys in southwest Scotland: implications for the species' distribution in western Europe. British Birds 100: 456-465.

Martell, M.S., C.J. Henny, P.E. Nye & M.J. Solensky. 2001. Fall migration routes, timing, and wintering sites of North American Ospreys, as determined by satellite telemetry. Condor 103: 715-724.

Massemin-Challet, S., J.P. Gendner, S. Samtmann, L. Pichegru, A. Wulgue & Y. Le Maho. 2006. The effect of migration strategy and food availability on White Stork Ciconia ciconia breeding success. Ibis 148: 503-508.

Mathiasson, S. 1963. Visible migration in the Sudan. Proc. XII Intern. Ornith. Congr.: 430-435.

Mathiasson, S. 1971. Untersuchungen an Klappergrasmücken (Sylvia curruca) im Niltal in Sudan. Vogelwarte 26: 212-221.

Matthiesen, C. 1931. Eine Schwalbenstatistik. Beitr. Fortpfl. Biol. Vögel 7: 47-49.

McCann, J.C. 1999. Climate and causation in African history. Intern. J. Afr. Hist. Studies 32: 261-279.

McClure, H.E. 1974. Migration and survival of the birds of Asia. Bangkok: Army Component SEATO Medical Research Laboratory.

Mebs, T. & D. Schmidt. 2006. Greifvögel Europas, Nordafrikas und Vorderasiens. Stuttgart: Franckh-Kosmos.

Mefit-Babtie, Srl. 1983. Development studies in the Jonglei Canal area. Technical Assistance Contract for Range Ecology Survey, Livestock Investigations and Water Supply. Final Report. Volume 5. Wildlife Studies. Glasgow.

Meinertzhagen, R. 1959. Pirates and predators. Edinburgh/London: Oliver & Boyd.

Meinertzhagen, R. 1930. Nicholl's Birds of Egypt. London: Hugh Rees.

Meinertzhagen, R. 1956. Roosts of wintering harriers. Ibis 98: 535.

Meininger, P.L. & W.C. Mullié. 1981. Egyptian wetlands as threatened wintering areas for waterbirds. Sandgrouse 3: 62-77.

Meininger, P.L. 1989. Palearctic coastal waders wintering in Senegal. Wader Study Group Bull. 55: 18-19.

Meininger, P.L. & G.A.M. Atta. 1994. Ornithological studies in Egyptian wetlands 1989/90. Vlissingen: Foundation for Ornithological Research in Egypt.

Meininger, P.L., G. Nikolaus & E.E. Khounganian. 1994. Ringing recoveries, mainly resulting from the Egyptian wetland project 1989/1990. In: P.L. Meininger & G.A.M. Atta (eds). Ornithological studies in Egyptian wetlands 1989/90: 245-260. Vlissingen: Foundation for Ornithological Research in Egypt.

Meissner, W. 1997. Autumn migration of Wood Sandpipers (Tringa glareola) in the region of the Gulf of Gdansk. Ring 19: 75-91.

Meissner, W. & P. Ziecik. 2005. Biometrics of juvenile Ruffs (Philomachus pugnax) migrating in autumn through Puck Bay region. Ring 27: 91-98.

Melter, J. 2005. Tages- und jahreszeitliche Muster des Verhaltens rastender Kampfläufer Philomachus pugnax in den Rieselfeldern Münster. Vogelwelt 116: 19-33.

Melter, J. & A. Sauvage. 1997. Measurements and moult of Ruffs Philomachus pugnax wintering in West Africa. Malimbus 19: 12-18.

Merikallio, E. 1958. Finnish birds: their distribution and numbers. Fauna Fennica 5: 1-181.

Merle, S. & F. Chapalain. 2005. Recensement hivernal des Cigognes blanche Ciconia ciconia et noire C. nigra en France en 2004. Ornithos 12: 321-327.

Messager, C., H. Gallée & O. Brasseur. 2004. Precipitation sensitivity to regional SST in a regional climate simulation during the West African monsoon for two dry years. Climate Dynamics 22: 249-266.

Meyburg, B.-U., W. Scheller & C. Meyburg. 1995a. Zug und Überwinterung des Schreiadlers Aquila pomarina: Satellitentelemetrische Untersuchungen. J. Ornithol. 136: 401-422.

Meyburg, B.-U., J.M. Mendelsohn, D.H. Ellis, D.G. Smith, C. Meyburg & A.C. Kemp. 1995b. Year-round movements of a Wahlberg's Eagle Aquila wahlbergi tracked by satellite. Ostrich 66: 135-140.

Meyburg, B.-U., C. Meyburg & C. Pacteau. 1996. Migration automnale d'un circaète Jean-Le-Blanc Circaetus gallicus suivi par satellite. Alauda 64: 339-344.

Meyburg, B.-U., C. Meyburg & J.-C. Barbraud. 1998. Migration strategies of an adult short-toed eagle (Circaetus gallicus) tracked by satellite. Alauda 66: 39-48.

Meyburg, B.-U., D.H. Ellis, J.M. Mendelsohn & W. Scheller. 2001. Satellite tracking of two Lesser Spotted Eagles, Aquila pomarina, migrating from Namibia. Ostrich 72: 35-40.

Meyburg, B.-U., M. Gallardo, C. Meyburg & E. Dimitrova. 2004a. Migrations and sojourn in Africa of Egyptian Vultures (Neophron percnopterus) tracked by satellite. J. Ornithol. 145: 273-280.

Meyburg, B.-U., C. Meyburg, T. B lka, O. Šreibr & J. Vrana. 2004b. Migration, wintering and breeding of a lesser spotted eagle (Aquila pomarina) from Slovakia tracked by satellite. J. Ornithol. 145: 1-7.

Meyburg, B.-U., C. Meyburg, J. Matthes & H. Matthes. 2007. Heimzug, verspätete Frühjahrsankunft, vorübergehender Partnerwechsel und Bruterfolg beim Schreiadler Aquila pomarina. Vogelwelt 128: 21-31.

Meyer, S.K., R. Spaar & B. Bruderer. 2003. Sea crossing behaviour of falcons and harriers at the southern Mediterranean coast of Spain. Avian Science 3: 153-162.

Michard-Picamelot, D., T. Zorn, J.P. Gendner, A.J. Mata & Y. Le Maho. 2002. Body protein does not vary despite seasonal changes in fat in the White Stork *Ciconia ciconia*. Ibis 144: E1-E10.

Michard, D., T. Zom, J.-P. Gendner & Y. Le Maho. 1997. La biologie et le comportement de la Cigogne blanche (*Ciconia ciconia*) révélés par le marquage électronique. Alauda 65: 53-58.

Micheloni, P. 2000. The hunting of migratory swallows in Africa. Paris: Pro Natura International.

Mietton, M., D. Dumas, O. Hamerlynck, A. Kane, A. Coly, S. Duvail, F. Pesneaud & M.L. ould Baba. 2007. Water management in the Senegal River Delta: a continuing uncertainty. Hydrol. Earth Syst. Sci. Discuss. 4: 4297-4323.

Mihelsons, H.A. & L. Haraszthy. 1985. *Circus pygargus. In:* V. Il'chev (ed.). Migrations of birds of eastern Europe and northern Asia: 279-284. Moscow: Nauka.

Miller, M.R. & D.C. Duncan. 1999. The northern pintail in North America: status and conservation needs of a struggling population. Wildlife Soc. Bull. 27: 788-800.

Miller, M.R., D.C. Duncan, K. Guyn, P. Plint & J. Austin. 2001. The Northern Pintail in North America: The problem and prescription for recovery. Part 1. Proceeding of the northern pintail workshop 23-25 March 2001, Sacramento, CA..

Miller, M.R., J.Y. Takekawa, J.P. Fleskes, D.L. Orthmeyer, M.L. Casazza & W.M. Perry. 2005. Spring migration of Northern Pintails from California's central valley wintering area tracked with satellite telemetry: routes, timing and destinations. Can. J. Zool. 83: 1314-1332.

Millon, A., J.-L. Bourrioux, V. Riols & V. Bretagnolle. 2002. Comparative breeding biology of the Hen Harrier and Montagu's Harrier: an 8-year study in north-eastern France. Ibis 144: 94-105.

Mohamed, Y.A., W.G.M. Bastiaanssen & H.H.G. Savenije. 2004. Spatial variability of evaporation and moisture storage in the swamps of the upper Nile studied by remote sensing techniques. J. Hydrol. 289: 145-164.

Molina, K.C. & R.M. Erwin. 2006. The distribution and conservation status of the Gull-billed Tern (*Gelochelidon nilotica*) in North America. Waterbirds 29: 271-295.

Monval, J.-Y. & J.-Y. Pirot. 1989. Results of the International Waterfowl Census 1967-1986. Slimbridge: IWRB.

Morand, P., J. Quensière & C. Herry. 1991. Enquête pluridisciplinaire auprès des pêcheurs du delta Central du Niger: plan de sondage et estimateurs associés. Le Transfert d'Echelle, Séminfort 4: 195-211.

Moreau, R.E. & W.L. Sclater. 1938. The avifauna of the mountains along the Rift Valley in North Tanganyika Territory (Mbulu District) – Part II. Ibis (14) 1: 760-786.

Moreau, R.E. 1952. The place of Africa in the Palaearctic migration system. J. Anim. Ecol. 21: 250-271.

Moreau, R.E. 1953. Migration in the Mediterranean area. Ibis 95: 329-364.

Moreau, R.E. 1961. Problems of Mediterranean-Saharan migration. Ibis 103: 373-427.

Moreau, R.E. 1966. The bird faunas of Africa and its islands. London: Academic Press.

Moreau, R.E. 1969. The recurrence in winter quarters (*Ortstreue*) of trans-Saharan migrants. Bird Study 16: 108-110.

Moreau, R.E. & R.M. Dolp. 1970. Fat, water, weights, and winglengths of autumn migrants in transit on the north-west coast of Egypt. Ibis 112: 209-228.

Moreau, R.E., D. Lack, J.F. Monk, A. Landsborough Thompson & W.H. Thorpe. 1970. Autobiographical sketch and obituaries: R.E. Moreau. Ibis 112: 549-564.

Moreau, R.E. 1972. The Palaearctic-African bird migration systems. London: Academic Press.

Morel, G. & M.-Y. Morel. 1961. Une héronnière mixte sur le Bas-Senegal. Alauda 29: 99-117.

Morel, G. & F. Bourlière. 1962. Relations écologiques des avifaunes sédentaire et migratrice dans une savane sahélienne du bas Sénégal. Terre Vie 102: 371-393.

Morel, G. & F. Roux. 1966a. Les migrateurs paléarctiques au Sénégal. I. Non-passereaux. Terre Vie 20: 19-72.

Morel, G. & F. Roux. 1966b. Les migrateurs paléarctiques au Sénégal. II. Passereaux et synthèse générale. Terre Vie 20: 143-176.

Morel, G. 1968. Contribution à la synécologie des oiseaux du Sahel sénégalais. Mém. ORSTOM. 29: 1-179.

Morel, G. & M.-Y. Morel. 1972. Recherches écologiques sur une savane sahélienne du Ferlo septentrional, Sénégal: l'avifaune et son cycle annuel. Terre Vie 26: 410-439.

Morel, G. 1973. The Sahel zone as an environment for Palaearctic migrants. Ibis 115: 413-417.

Morel, G.J. & M.-Y. Morel. 1973. Recherches écologiques sur une savane sahélienne du Ferlo septentrional, Sénégal. Etude d'une communauté avienne. Cah. ORSTOM, sér. Biol. 13: 3-34.

Morel, G. & F. Roux. 1973. Les migrateurs paléarctiques au Sénégal: notes complémentaires. Terre Vie 27: 523-545.

Morel, G. & M.-Y. Morel. 1974. Recherches écologiques sur une savane sahélienne du Ferlo septentrional, Sénégal: influence de la sécheresse de l'année 1972-1973 sur l'avifaune. Terre Vie 28: 95-123.

Morel, G. J. & M.-Y. Morel. 1978. Recherches écologiques sur une savane sahélienne du Ferlo septentrional, Sénégal. Etude d'une communauté avienne. Cah. ORSTOM, sér. Biol. 13: 3-34.

Morel, G. & M.-Y. Morel. 1979. La Tourterelle des bois dans l'extrême Ouest-Africain. Malimbus 1: 66-67.

Morel, G.J. & M.-Y. Morel. 1980. Structure of an arid tropical bird community. Proc. IV Pan-Afr. Orn. Congr.: 125-133.

Morel, G.J. & M.-Y. Morel. 1983. Treize années de comptages d'oiseaux dans un quadrat de steppe arbustive dans le Ferlo (Nord Sénégal). Atelier méthodes d'inventaire et de surveillance continue des écosystèmes pastoraux sahéliens - application au développement. Dakar, 16-18 Novembre 1983.

Morel, G. & M.-Y. Morel. 1988. Nouvelles données sur l'hivernage de la

tourterelle des bois, *Streptopelia turtur*, en Afrique de l'Ouest: Nord de la Guinée. Alauda 56: 85-91.

Morel, G. & M.-Y. Morel. 1990. Les Oiseaux de Sénégambie. Paris: ORSTOM.

Morel, G.J. & M.-Y. Morel. 1992. Habitat use by Palearctic migrant passerine birds in West Africa. Ibis 134: 83-88.

Morel, M.-Y. 1975. Comportement de sept espèces des tourterelles aux points d'eau naturels et artificiels dans une savane sahélienne du Ferlo septentrional Sénégal. L'Oiseau et RFO 45: 97-125.

Morel, M.-Y. 1985. La Tourterelle des bois *Streptopelia turtur* en Sénégambie: évolution de la population au cours de l'année et identification des races. Alauda 53: 100-110.

Morel, M.-Y. 1986. Mue et engraissement de la tourterelle des bois *Streptopelia turtur*, dans une steppe arbustive du Nord Sénégal, région de Richard-Toll. Alauda 54: 121-137.

Morel, M.-Y. 1987. La Tourterelle des bois, *Streptopelia turtur*, dans l'Ouest africain: mouvements migratoires et régime alimentaire. Malimbus 9: 23-42.

Moritz, D. 1993. Long-term monitoring of Palaearctic-African migrants at Helgoland/German Bight, North Sea. Proc. VIII Pan-Afr. Orn. Congr.: 579-586.

Moritz, M., P. Scholte & S. Kari. 2002. The demise of the nomadic contract: arrangements and rangelands under pressure in the Far North of Cameroon. Nomadic People 6: 127-146.

Morony M.G. 2000. Michael the Syrian as a source for Economic history. Hugoye: Journal of Syriac Studies 3.

Mortimore, M. & F. Harris. 2005. Do small farmers' achievements contradict the nutrient depletion scenarios for Africa? Land Use Policy 22: 43-56.

Mortimore, M. & B. Turner. 2005. Does the Sahelian smallholder's management of woodland, farm trees, rangeland support the hypothesis of human-induced desertification? J. Arid Environ. 63: 567-595.

Mouafo, D., T. Fotsing, D. Sighomnou & L. Sigha. 2002. Dam, environment and regional development: Case study of the Logone floodplain in northern Cameroon. Intern. J. Water Resources Devel. 18: 209-219.

Moussa, B.I. 1999. Evolution de l'occupation des sols dans deux terroirs du sud-ouest nigérien: Bogodjotou et Ticko. *In:* C. Floret & R. Pontanier (eds). Jachères et systèmes agraires: 15-24. Dakar: CORAF/IRD ex ORSTOM/Union Européenne.

Møller, A.P. 1975a. Ynglebestanden af Sandterne *Gelochelidon n. nilotica* Gmel. i 1972 i Europa, Afrika og Vestasien, med et tilbageblik over bestandsændringer i dette århundrede. Dansk Orn. Foren. Tidsskr. 69: 1-8.

Møller, A.P. 1975b. Migration of European Gull-billed Terns (*Gelochelidon n. nilotica*) according to recoveries. Danske Fugle 27: 61-77.

Møller, A.P. 1989. Population dynamics of a declining swallow *Hirundo rustica* population. J. Anim. Ecol. 58: 1051-1063.

Møller, A.P. 1992. Nestboxes and the scientific rigour of experimental studies. Oikos 63: 309-311.

Møller, A.P. & F. de Lope. 1999. Senescence in a short-lived migrating bird: age-dependent morphology, migration, reproduction and parasitism. J. Anim. Ecol. 68: 163-171.

Møller, A.P. & T. Szép. 2002. Survival rate of adult barn swallows *Hirundo rustica* in relation to sexual selection and reproduction. Ecology 83: 2220-2228.

Møller, A.P. 2002. North Atlantic Oscillation (NAO) effects of climate on the relative importance of first and second clutches in a migratory passerine bird. J. Anim. Ecol. 71: 201-210.

Møller, A.P., E. Flensted-Jensen & W. Mardal. 2007. Adaptation to climatic change by change in the timing of the annual cycle. J. Anim. Ecol. 76: 515-525.

Mulder, Th. 1972. De Grutto (*Limosa limosa*) in Nederland. Hoogwoud: KNNV.

Mullié, W.C., E.E. Khounganian & M.H. Amer. 1989. A preliminary list of Egyptian bird ringing recoveries 1908-1988. Wageningen: Foundation for Ornithological Research in Egypt.

Mullié, W.C. & J.O. Keith. 1991. Notes on the breeding biology, food and weight of the Singing Bush-Lark *Mirafra javanica* in northern Senegal. Malimbus 13: 24-39.

Mullié, W.C., J. Brouwer & C. Albert. 1992. Gregarious behaviour of African Swallow-tailed Kite *Chelictinia riocourii* in response to high grasshopper densities near Ouallam, western Niger. Malimbus 14: 19-21.

Mullié, W.C. & J.O. Keith. 1993. The effects of aerially applied fenitrothion and chlorpyrifos on birds in the savannah of northern Senegal. J. appl. Ecol. 30: 536-550.

Mullié, W.C., J. Brouwer & P. Scholte. 1995. Numbers, distribution and habitat of wintering white storks in the east-central Sahel in relation to rainfall, food and anthropogenic influences. *In:* O. Biber, P. Enggist, C. Marti & T. Salathé (eds). Proc. Intern. Symp. White Stork (Western Population), Basel 1994: 219-240. Sempach: Schweizerische Vogelwarte.

Mullié, W.C., J. Brouwer, S.F. Codio & R. Decae. 1999. Small isolated wetlands in the Central Sahel: a resource shared between people and waterbirds. *In:* A.J. Beintema & J. van Vessem (eds). Strategies for Conserving Migratory Waterbirds: 30-38. Wageningen: Wetlands International.

Mullié, W.C., C. Rouland-Lefèvre, M. Sarr, A. Danfa & J.W. Everts. 2003. Impact of Adonis® 2UL and 5UL (fipronil) on ant and termite communities in a tropical semi-arid savannah. Report lctx 9902. Dakar: CERES-Locustox.

Mullié, W.C. & P. Mineau. 2004. Comparative avian risk assessment of acridid control in the Sahel based on probability of kill and hazard ratios. Fourth SETEC World Congress and 25th Annual Meeting in North Amercica, Portland, Oregon.

Mullié, W.C. 2007a. Observations sur l'utilisation du Green Muscle® (*Metarhizium anisopliae* var. *acridum*) en lutte antiacridienne au

Sénégal en 2007. Dakar: Fondation Agir pour l'Education et la Santé.

Mullié, W.C. 2007b. Synergy of predation and entomopathogens: an essential element of biocontrol. Poster presentation at International Workshop on the future of Biopesticides in Desert Locust management.

Mullié, W.C. & Y. Gueye. In press. Efficacité du Green Muscle (*Metarhizium anisopliae* var. *acridum*) en dose reduite en lutte antiacridienne au Sénégal en 2008. Dakar, Rapport au Ministère de l'Agriculture.

Mumba, M. & J.R. Thompson. 2005. Hydrological and ecological impacts of dams on the Kafue Flats floodplain system, southern Zambia. Physics and Chemistry of the Earth 30: 422-447.

Muraoka, Y., C.H. Schulze, M. Pavličev & G. Wichmann. 2009. Spring migration dynamics and sex-specific patterns in stopover strategy in the Wood Sandpiper *Tringa glareola*. J. Ornithol. 150: 313-319.

Murton, R.K. 1968. Breeding, migration and survival of Turtle Doves. British Birds 61: 193-212.

Musgrove, A.J. 2002. The non-breeding status of the Little Egret in Britain. British Birds 95: 62-80.

Mustafa, M.A. & M.A.Y. Elsheikh. 2007. Variability, correlation and path co-efficient analysis for yield and its components in rice. African Crop Science Journal 15: 183-189.

Myers, J.P. 1981. A test of three hypotheses for latitudinal segregation of the sexes in wintering birds. Can. J. Zool. 59: 1527-1534.

Myneni, R.B., S.O. Los & C.J. Tucker. 1996. Satellite-based identification of linked vegetation index and sea surface temperature anomaly areas from 1982-1990 for Africa, Australia and South America. Geophysical Research Letters 23: 729-732.

N

N'tchayi, G.M., J. Bertrand, M. Legrand & J. Baudet. 1994. Temporal and spatial variations of the atmospheric dust loading throughout West-Africa over the last 30 years. Annales Geophysicae 12: 265-273.

N'tchayi, G.M., J. Bertrand & S.E. Nicholson. 1997. The diurnal and seasonal cycles of wind-borne dust over Africa north of the equator. J. appl. Meteor. 36: 868-882.

Nagy, K.E. 2005. Review field metabolic rate and body size. J. exp. Biol. 208: 1621-1625.

Nankinov, D.N. 1978. Migrations of Night Herons on the Balkans. *In*: Ilyichev, V.D. (ed). Migrations of birds of Eastern Europe and Northern Asia: 112-114. Moscow: Nauka.

Nankinov, D.N. & A.A. Kistchinski. 1978. Migrations of Little Egrets on the Balkans. *In*: Ilyichev, V.D. (ed). Migrations of birds of Eastern Europe and Northern Asia: 140-142. Moscow: Nauka.

Nankinov, D.N. 1998. Wood Sandpiper *Tringa glareola* and Green Sandpiper *Tringa ochropus* in Bulgaria. International Wader Studies 10: 370-374.

de Naurois, R. 1959. Premiers recherches sur l'avifaune des îles du Banc d'Arguin (Mauritanie). Alauda 27: 241-308.

de Naurois, R. 1965. Une colonie reproductrice du Petit Flamant rose, *Phoeniconaias minor* (Geoffroy) dans l'Aftout es Sahel (Sud-ouest mauritanien). Alauda 33: 166-175.

de Naurois, R. 1965. L'avifaune aquatique du delta du Sénégal et son destin. Bull. I. Fau 27b: 1196-1207.

de Naurois, R. 1969. Peuplement et cycles de reproduction des oiseaux de la côte occidentale d'Afrique. Mémoire du Musée d'Histoire Naturelle série A. Zoologie 57: 1-312.

Nef, L., P. Dutilleul & B. Nef. 1988. Estimation des variations quantitatives de populations de passereaux à partir des bilans de 24 années de baguage au Limbourg, Belge. Gerfaut 78: 173-207.

Negri, A.J., R.F. Adler, L.M. Xu & J. Surratt. 2004. The impact of Amazonian deforestation on dry season rainfall. J. Climate 17: 1306-1319.

Neiland, A.E., C. Béné, T. Jolley, B.M.B. Ladu, S. Ovie, O. Sule, M. Baba, E. Belal, F. Tiotsop, K. Mindjimba, L. Dara & J. Quensière. 2004. Fisheries. *In*: C. Batello, M. Marzot & A. Harouna Touré (eds). The future is an ancient lake; traditional knowledge, biodiversity and genetic resources for food and agriculture in Lake Chad Basin ecosystems: 189-225. Rome: FAO.

Nelson, A. 2004. Population Density for Africa in 2000. Sioux Falls: UNEP/GRID.

Neumann, O. 1917. Über die Avifauna des unteren Senegal-Gebiets. J. Ornithol. 65: 189-214.

Nevo, D. 1996. The Desert Locust, *Schistocerca gregaria*, and its control in the land of Israel and the Near East in antiquity, with some reflections on its appearance in Israel in modern times. Phytoparasitica 24: 7-32.

Newby, J.E. 1979. The birds of the Quadi Rime – Quadi Achim Faunal Reserve, a contribution to the study of the Chadian avifauna. Malimbus 1: 90-109.

Newton, I. 1979. Population ecology of raptors. Berkhamsted: Poyser.

Newton, I. & L. Dale. 1996. Relationship between migration and latitude among west European birds. J. Anim. Ecol. 65: 137-146.

Newton, I. 2004. Population limitation in migrants. Ibis 146: 197-226.

Newton, I. 2006. Can conditions experienced during migration limit the population levels of birds? J. Ornithol. 147: 146-166.

Newton, I. 2007. Weather-related mass mortality events in migrants. Ibis 150: 453-467.

Newton, I. 2008. The migration ecology of birds. London: Academic Press.

Newton, S.F. 1996. Wintering range of Palaearctic-African migrants includes southwest Arabia. Ibis 138: 335-336.

Ngatcha, B.N., J. Mudry, L.S. Nkamdjou, R. Njitchoua & E. Naah. 2005. Climate variability and impacts on an alluvial aquifer in a semiarid climate, the Logone-Chari plain (south of Lake Chad). IHAS 295: 94-102.

Niassy, A. 1990. The grasshopper complex. *In*: J.W. Everts (ed.). Environmental effects of chemical locust and grasshopper control. Rome: FAO.

Nicholson, S.E. 1981a. Rainfall and atmospheric circulation during drought periods and wetter years in West-Africa. Monthly Weather Review 109: 2191-2208.

Nicholson, S.E. 1981b. The historical climatology of Africa. *In:* T.M.L. Wigley, M.J. Ingran & G. Farner (eds). Climate and history: 249-270. Cambridge: Cambridge University Press.

Nicholson, S. 1982. The Sahel, a climatic perspective. Paris: Club du Sahel.

Nicholson, S.E. 1986. The spatial coherence of African rainfall anomalies: interhemispheric teleconnections. J. Climate Appl. Meteor. 25: 1365-1381.

Nicholson, S.E., M.L. Davenport & A.R. Malo. 1990. A comparison of the vegetation response to rainfall in the Sahel and East-Africa, using Normalized Difference Vegetation Index from NOAA AVHRR. Climatic Change 17: 209-241.

Nicholson, S.E. & I.M. Palao. 1993. A reevaluation of rainfall variability in the Sahel.1. Characteristics of rainfall fluctuations. Intern. J. Clim. 13: 371-389.

Nicholson, S.E., C.J. Tucker & M.B. Ba. 1998. Desertification, drought, and surface vegetation: An example from the West African Sahel. Bull. Am. Meteor. Soc. 79: 815-829.

Nicholson, S. 2000. Land surface processes and Sahel climate. Reviews of Geophysics 38: 117-139.

Nicholson, S.E. 2001. Climatic and environmental change in Africa during the last two centuries. Climate Research 17: 123-144.

Nicholson, S.E. & J.P. Grist. 2001. A conceptual model for understanding rainfall variability in the West African Sahel on interannual and interdecadal timescales. Intern. J. Clim. 21: 1733-1757.

Nicholson, S.E. & J. P. Grist. 2003. The seasonal evolution of the atmospheric circulation over West Africa and Equatorial Africa. J. Climate 16: 1013-1030.

Nicholson, S. 2005. On the question of the "recovery" of the rains in the West African Sahel. J. Arid Environ. 63: 615-641.

Nicolai, J. 1976. Beboachtungen an einigen paläarktischen Wintergästen in Ost-Nigeria. Vogelwarte 28: 274-278.

Nikiforov, M. 2003. Distribution trends of breeding bird species in Belarus under conditions of global climate change. Acta Zoologica Lituanica 13: 255-262.

Nikiforov, M.E. & E.A. Mongin. 1998. [Breeding waders in Belarus: number estimates and recent population trends.] *In:* P.S. Tomkovich & E.A. Lebdeveda (eds). Breeding waders in eastern Europe – 2000: 93-96. Moscow: Russian Bird Conservation Union.

Nikolaus, G. 1987. Distribution atlas of Sudan's birds with notes on habitat and status. Bonn: Zoologische Forschungsinstitut und Museum Alexander Koenig.

Nikolaus, G. 1989. Birds of South Sudan. Scopus Special Suppl. 3: 1-124.

Nikolaus, G., J. Ash, P. Hall & J. Barker. 1995. The Boje Ebok Swallow roost. Safring News 24: 85-86.

van Noordwijk, A.J. & D.L. Thomson. 2008. Survival rates of Black-tailed Godwits *Limosa limosa* breeding in The Netherlands estimated from ring recoveries. Ardea 96: 47-57.

Norman, S.C. 1992. Dispersal and site fidelity in Lesser Whitethroats *Sylvia curruca*. Ring. & Migr. 13: 167-174.

O

O'Connor, R.J. & M. Shrubb. 1986. Farming & birds. Cambridge: Cambridge University Press.

OAG Münster. 1989a. Zugphänologie und Rastbestandsentwicklung des Kampfläufers (*Philomachus pugnax*) in den Rieselfeldern Münster anhand von Fangergebnissen und Sichtbeobachtungen. Vogelwarte 35: 132-155.

OAG Münster. 1989b. Beobachtungen zur Heimzugstrategie des Kampfläufers *Philomachus pugnax*. J. Ornithol. 130: 175-182.

OAG Münster. 1991a. Mauser und intraindividuelle Variation des Handschwingenwechsels beim Kampläufer (*Philomachus pugnax*). J. Ornithol. 132: 1-28.

OAG Münster. 1991b. Report of the ornithological expedition to northern Cameroon. January/February 1991. Münster: OAG Münster.

OAG Münster. 1996. Do females really outnumber males in ruff *Philomachus pugnax* wintering in Africa? J. Ornithol. 137: 91-100.

OAG Münster. 1998. Mass of ruffs *Philomachus pugnax* wintering in West Africa. International Wader Studies 10: 435-440.

Oatley, T.B. 2000. Migrant European Swallows *Hirundo rustica* in southern Africa: a southern perspective. Ostrich 71: 205-209.

Odsjö, T. & J. Sondell. 1977. Populationsutveckling och häckningsresultat hos brun kärrhök *Circus aeruginosus* i relation till förekomsten av DDT, PCB och kvicksilver. Vår Fågelvärld 36: 152-160.

Ofori-Danson, P.K., C.J. Vanderpuye & G.J. de Graaf. 2001. Growth and mortality of the catfish, *Hemisynodontis membranaceus* (Geoffrey St. Hilaire), in the northern arm of Lake Volta, Ghana. Fisheries Management and Ecology 8: 37-45.

Olioso, G. 1991. L'Hirondelle de rivage *Riparia riparia* dans le sud-est de la France et plus particulièrement dans la vallée de la Durance. L'Oiseau et RFO 61: 185-202.

Oliver, P. 2005. Roosting behaviour and wintering of Eurasian Marsh Harriers *Circus aeruginosus* in south-east England. Ardea 93: 137-140.

Olivry, J.C. 1987. Les conséquences durables de la sécheresse actuelle sur l'écoulement du fleuve Sénégal et l'hypersalinisation de la Basse-Casamance. IAHS Publ. 168. 501-512.

Olivry, J.C., G. Chouret, G. Vuillaume, J. Lemoalle & J.P. Bricquet. 1996. Hydrologie du Lac Tchad. Paris: ORSTOM..

Olsson, L., L. Eklundh & J. Ardo. 2005. A recent greening of the Sahel - Trends, patterns and potential causes. J. Arid Environ. 63: 556-566.

Oschadleus, H.-D. 2002. The Wood Sandpiper (*Tringa glareola*) in South Africa - data from counting, atlasing and ringing. Ring 24: 71-78.

Ottosson, U., S. Rumsey & C. Hjort. 2001. Migration of four *Sylvia* warblers through northern Senegal. Ring. & Migr. 20: 344-351.

Ottosson, U., D. Bengtson, R. Gustafsson, P. Hall, C. Hjort, A.P. Leventis, R. Neumann, J. Pettersson, P. Rhönsstad, S. Rumsey, J. Waldenström & W. Velmala. 2002. New birds for Nigeria observed during the Lake Chad Bird Migration Project. Bull. African Bird Club 9: 52-55.

Ottosson, U., J. Waldenström, C. Hjort & R. McGregor. 2005. Garden Warbler *Sylvia borin* migration in sub-Saharan West Africa: phenology and body mass changes. Ibis 147: 750-757.

Ouédraogo, S.J., J. Bayala, C. Dembélé, A. Kaboré, B. Kaya, A. Niang & A.N. Somé. 2006. Establishing jujube trees in sub-Saharan Africa: response of introduced and local cultivars to rock phosphate and water supply in Burkina Faso, West Africa. Agroforestry Systems 68: 69-80.

Ouweneel, G.L. 2008. Een halve eeuw overwinterende Bruine Kiekendieven *Circus aeruginosus* in de noordelijke Delta. De Takkeling 16: 124-129.

Owino, A.O. & P.G. Ryan. 2006. Habitat associations of papyrus specialist birds at three papyrus swamps in western Kenya. Afr. J. Ecol. 44: 438-443.

Oyebande, L. 2001. Stream flow regime change and ecological response in the Lake Chad basin in Nigeria. IAHS Publ. 266.

Österlöf, S. 1977. Migration, wintering area & site tenacity of the European Osprey *Pandion h. haliaetus* (L.). Ornis Scand. 8: 61-78.

Österlöf, S. & B.-O. Stolt. 1982. Population trends indicated by birds ringed in Sweden. Ornis Scand. 13: 135-140.

P

Pacteau, C. 2003. Vingt-cinq ans de sauvegarde des Busards en France; le cas du Busard cendré *Circus pygargus*. Alauda 71: 347-356.

Pain, D.J. & M.W. Pienkowski. 1997. Farming and birds in Europe: The common agricultural policy and its implications for bird conservation. San Diego: Academic Press.

Palmer, T.N. 1986. Influence of the Atlantic, Pacific and Indian Oceans on Sahel rainfall. Nature 322: 251-253.

Palmer, T.N., C. Brankovic, P. Viterbo & M.J. Miller. 1992. Modeling interannual variations of summer monsoons. J. Climate 5: 399-417.

Palmgren, P. 1930. Quantitative Untersuchungen über die Vogelfauna in den Wäldern Südfinnlands mit besonderer berücksichtigung Ålands. Acta Zoologica Fennica 7: 1-218.

Panuccio, M., B. D'Amicis, E. Canale & A. Roccella. 2005. Sex and age ratios of marsh harriers *Circus aeruginosus* wintering in central-southern Italy. Avocetta 29: 13-17.

Patrikeev, M. 2004. The birds of Azerbaijan. Sofia: Pensoft.

Pavlova, A., R.M. Zink, S.V. Drovetski, Y. Red'kin & S. Rohwer. 2003. Phylogeographic patterns in *Motacilla flava* and *Motacilla citreola*: species limits and population history. Auk 120: 744-758.

Payevski, V.A. 2006. Mechanisms of population dynamics in trans-Saharan migrant birds: a review. Entomological Review, Supplement 86: 368-381.

Payevsky, V.A. 2000. Demographic studies of migrating bird populations: the aims and possibilities. Ring 22: 57-65.

Peach, W., S. Baillie & L. Underhill. 1991. Survival of British Sedge Warblers *Acrocephalus schoenobaenus* in relation to West African rainfall. Ibis 133: 300-305.

Peal, R.E.F. 1968. The distribution of the Wryneck in the British Isles. Bird Study 15: 111-126.

Pearson, D.J., G.C. Backhurst & D.E.G. Backhurst. 1979. Spring weights and passage of Sedge Warblers *Acrocephalus schoenobaenus* in Central Kenya. Ibis 121: 8-19.

Pearson, D.J. 1981. The wintering and moult of Ruffs *Philomachus pugnax* in the Kenyan Rift Valley. Ibis 123: 158-182.

Pearson, D.J., G. Nikolaus & J.S. Ash. 1988. The southward migration of Palearctic passerines through northeast and east tropical Africa: a review. Proc. Sixth Pan-Afr. Orn. Congr.: 243-261.

Pearson, D.J. & P.C. Lack. 1992. Migration patterns and habitat use by passerine and near-passerine migrant birds in eastern Africa. Ibis 134, suppl.: 89-98.

Pedersen, E.M. 1998. Bestandsindeks for ynglende danske skovfugle 1976-1997. Dansk Orn. Foren. Tidsskr. 92: 275-282.

Peeters, J. 2003. Étude pour la restauration du réseau hydraulique du bassin du fleuve Sénégal. Evaluation environnementale: rapport de première phase. Report OMVS FAD/FAT et SOGED AGRER-SERADE-SETICA.

Penning de Vries, F.W.T. & M.A. Djitèye. 1982. La productivité des pâturages sahéliens, une étude des sols, des végétations et de l'exploitation de cette ressource naturelle. Wageningen: PUDOC.

Persson, B. 1971. Chlorinated hydrocarbons and reproduction of a South Swedish population of whitethroat *Sylvia communis*. Oikos 22: 248-255.

Persson, B. 1974. Degradation and seasonal variation of DDT in whitethroats *Sylvia communis*. Oikos 25: 216-222.

Persson, C. 1998. Weight studies in Wood Sandpipers (*Tringa glareola*) migrating over south-western Scania in late summer and spring, and notes on related species. Ring 20: 95-105.

Petersen, G., J.A. Abeya & N. Fohrer. 2007a. Spatio-temporal water body and vegetation changes in the Nile swamps of southern Sudan. Advances in Geosciences 11: 113-116.

Petersen, B.S., K.D. Christensen & F.P. Jensen. 2007b. Bird population densities along two precipitation gradients in Senegal and Mali. Malimbus 29: 101-121.

Petersen, B.S., K.D. Christensen, K. Falk, F.P. Jensen & Z. Ouambama. 2008. Abdim's Stork *Ciconia abdimii* exploitation of Senegalese Grasshopper *Oedaleus senegalensis* in South-eastern Niger. Waterbirds 31: 159-168.

Pettet, A. 1976. The avifauna of Waza National Park, Cameroun, in December. Bull. Nigerian Orn. Soc. 12: 18-24.

Pettorelli, N., J.O. Vik, A. Mysterud, J.M. Gaillard, C.J. Tucker & N.C. Stenseth. 2005. Using the satellite-derived NDVI to assess ecological responses to environmental change. Trends in Ecology & Evolution 20: 503-510.

Peveling, R., A.N. McWilliam, P. Nagel, H. Rasolomanna, L. Raholijaona, A. Rakotomianina & C. Dewhurst. 2003. Impact of locust control on harvester termites and endemic vertebrate predators in Madagascar. J. appl. Ecol. 40: 729-741.

Pérez-Tris, J., Á. Ramírez & J.L. Tellería. 2003. Are Iberian Chiffchaffs *Phylloscopus (collybita) brehmii* long-distance migrants? An analysis of flight-related morphology. Bird Study 50: 146-152.

Pérez-Trist, J., J. De La Puente, J. Pinilla & A. Bermejo. 2001. Body moult and autumn migration in the barn swallow *Hirundo rustica*: is there a cost of moulting late? Ann. Zool. Fennici 139-148.

Piersma, T. 1983. Gezamenlijk overnachten van grutto's *Limosa limosa* op de Mokkebank. Limosa 56: 1-8.

Piersma, T. 1986. Breeding waders in Europe. Wader Study Group Bull. Suppl. 48.

Piersma, T., L. Zwarts & J.H. Bruggemann. 1990. Behavioural aspects of the departure of waders before long-distance flights: flocking, vocalizations, flight paths and diurnal timing. Ardea 78: 157-184.

Piersma, T. & J. Jukema. 1990. Budgeting the flight of a long-distance migrant: changes in nutrient reserve levels of Bar-tailed Godwits at successive spring staging sites. Ardea 78: 315-338.

Piersma, T. & Y. Ntiamoa-Baidu. 1995. Waterbird ecology and the management of coastal wetlands in Ghana. Accra / Den Burg: Ghana Wildlife Society / Netherlands Institute for Sea Research.

Pieterse, A.H., S. Kettunen, S. Diouf, I. Ndao, K. Sarr, A. Tarvainen, S. Kloff & S. Hellsten. 2003. Effective biological control of *Salvinia molesta* in the Senegal River by means of the weevil *Cyrtobagous salviniae*. Ambio 32: 458-462.

Pilard, P. 2007. Conservation du Faucon crécerellette *Falco naumanni* en hivernage au Sénégal. Paris: LPO.

Pilard, P., G. Jarry & V. Lelong. 2008. Compte-rendu de la mission LPO de janvier 2008 concernant la conservation du dortoir de rapaces insectivores de l'île de Kousmar (Sénégal). Paris: LPO/BirdLife International.

Pinchuk, P., N. Karlionava & D. Zhurauliou. 2005. Wader ringing at the Turov ornithological station, Pripyat Valley (S Belarus) in 1996-2003. Ring 27: 101-105.

Piper, S.E. 2007. Barn Swallow. Draft Final Report. http://eia.dubetradeport.co.za/Documents/Documents/2007Jan26/Draft%20Barn%20Swallow%20Report.pdf.

Platteeuw, M., J. Botond Kiss, N.Y. Zhmud & N. Sadoul. 2004. Colonial waterbirds and their habitat use in the Danube Delta as an example of a large-scale natural wetland. Lelystad: RIZA.

Polet, G. 2000. Waterfowl and flood extent in the Hadejia-Nguru wetlands of north-east Nigeria. Bird Conserv. Intern. 10: 203-209.

Poole, A.F. 1989. Ospreys: A natural and unnatural history. Cambridge: Cambridge University Press.

Poorter, E.P.R., J. van der Kamp & J. Jonker. 1982. Verslag van de Nederlandse lepelaarsexpeditie naar de Senegaldelta in de winter van 1980/81. Lelystad: Rijksdienst voor de IJsselmeerpolders.

Poorter, E.P.R. & L. Zwarts. 1984. Résultats d'une première mission ornitho-écologique de l'IUCN/WWF en Guinée-Bissau. Zeist: Stichting Internationale Vogelbescherming.

Porkert, J. & J. Zajíc. 2005. The breeding biology of the common redstart, *Phoenicurus phoenicurus*, in the Central European pine forest. Folia Zool. 54: 111-122.

Poulin, B., G. Lefebvre & A. Mauchamp. 2002. Habitat requirements of passerines and reedbed management in southern France. Biol. Conserv. 107: 315-325.

Poulin, B., G. Lefebvre & A.J. Crivelli. 2007. The invasive red swamp crayfish as a predictor of Eurasian bittern density in the Camargue, France. J. Zool. Lond. 273: 98-105.

Poupon, H. 1980. Structure et dynamique de la strate ligneuse d'une steppe sahélienne au nord du Sénégal. Travaux et Documents de l'ORSTOM 115: 1-351.

da Prato, S.R.D. & E.S. da Prato. 1983. Movements of Whitethroats *Sylvia communis* ringed in the British Isles. Ring. & Migr. 4: 193-201.

Prévost, Y.A. 1982. Le Balbuzard pêcheur (*Pandion haliaetus*) dans les parcs nationaux. Mémoires L'IFAN 92: 211-219.

PRG (Pesticide Referee Group). 2004. Evaluation of field trials data on the efficacy and selectivity of insecticides on locusts and grasshoppers. Rome: FAO.

Prince, S.D., E. Brown De Colstoun & L.L. Kravitz. 1998. Evidence from rain-use efficiencies does not indicate extensive Sahelian desertification. Global Change Biology 4: 359-374.

Procházka, P. & J. Reif. 2002. Movements and settling patterns of Sedge Warblers (*Acrocephalus schoenobaenus*) in the Czech Republic and Slovakia - an analysis of ringing recoveries. Ring 24: 3-13.

Procházka, P., K.A. Hobson, Z. Karcza & J. Kralj. 2008. Birds of a feather winter together: migratory connectivity in the Reed Warbler *Acrocephalus scirpaceus*. J. Ornithol. 149: 141-150.

Prop, J. & T. Vulink. 1992. Digestion by barnacle geese in the annual cycle: the interplay between retention time and food quality. Funct. Ecol. 6: 180-189.

Prosper, J. & H. Hafner. 1996. Breeding aspects of the colonial Ardeidae in the Albufera de Valencia, Spain: Population changes, phenology, and reproductive success of the three most abundant species. Colonial Waterbirds 19: 98-107.

Prospero, J.M., P. Ginoux, O. Torres, S.E. Nicholson & T.E. Gill. 2002. Environmental characterization of global sources of atmospheric soil dust identified with the NIMBUS 7 total ozone mapping spectrometer (TOMS) absorbing aerosol product. Reviews of Geophysics 40: 1-31.

R

Raes, D., J. Deckers & M. Diallo. 1995. Water requirements for salt control in rice schemes in the Senegal river delta and valley. Irrigation and drainage systems 9: 129-141.

Rahmani, A.R. & M.Y. Shobrak. 1992. Glossy Ibises (*Plegadis falcinellus*) and Black-tailed Godwits (*Limosa limosa*) feeding on sorghum in flooded fields in southwestern Saudi Arabia. Colonial Waterbirds 15: 239-240.

Rappoldt, C., M. Kersten & C.J. Smit. 1985. Errors in large-scale shorebird counts. Ardea 73: 13-24.

Rasmussen, K., B. Fog & J.E. Madsen. 2001. Desertification in reverse? Observations from northern Burkina Faso. Global Environmental Change 11: 271-282.

Rasmussen, L.M. & K. Fischer. 1997. The breeding population of Gull-billed Terns *Gelochelidon nilotica* in Denmark 1976-1996. Dansk Orn. Foren. Tidsskr. 91: 101-108.

Redpath, S.M., R. Clarke, M. Madders & S.J. Thirgood. 2001. Assessing raptor diet: comparing pellets, prey remains, and observational data at Hen Harrier nests. Condor 103: 184-188.

Reichlin, T.S., M. Schaub, M.H.M. Menz, M. Mermod, P. Portner, R. Arlettaz & L.Jenni. 2009. Migration patterns of Hoopoe *Upupa epops* and Wryneck *Jynx torquilla*: an analysis of European ring recoveries. J. Ornithol. 150: 393-400.

Reif, J., P. Voříšek, K. Šťastný & V. Bejček. 2006. [Population trends of birds in the Czech Republic during 1982-2005.] Sylvia 42: 22-37.

Remisiewicz, M. & L. Wennerberg. 2006. Differential migration strategies of the Wood Sandpiper (*Tringa glareola*) - genetic analyses reveal sex differences in morphology and spring migration phenology. Ornis Fennica 83: 1-10.

Remisiewicz, M., W. Meissner, P. Pinchuk & M. Ściborski. 2007. Phenology of spring migration of Wood Sandpiper *Tringa glareola* through Europe. Ornis Svecica 17: 3-14.

Rey, C., G. Tappan & A. Belemvire. 2005. Changing land management practices and vegetation on the Central Plateau of Burkina Faso (1968-2002). J. Arid Environ. 63: 642-659.

van Rhijn, J.G. 1985. The Ruff. Individuality in a gregarious wading bird. London: Poyser.

Ribot, J.C. 1993. Forestry policy and charcoal production in Senegal. Energy Policy May 1993: 559-585.

Ribot, J.C. 1999. A history of fear: imagining deforestation in the West African dryland forests. Global Ecology and Biogeography 8: 291-300.

de Ridder, N., H. Breman, H. van Keulen & T.J. Stomph. 2004. Revisiting a 'cure against land hunger': soil fertility management and farming systems dynamics in the West African Sahel. Agricultural Systems 80: 109-131.

Robinson, R.A., D.E. Balmer & J.H. Marchant. 2003a. Survival rates of hirundines in relation to British and African rainfall. Ring. & Migr. 24: 1-6.

Robinson, R.A., H.Q.P. Crick & W.J. Peach. 2003b. Population trends of Swallows *Hirundo rustica* breeding in Britain. Bird Study 50: 1-7.

Rodwell, S.P., A. Sauvage, S.J.R. Rumsey & A. Bräunlich. 1996. An annotated checklist of birds occurring at the Parc National des Oiseaux du Djoudj in Senegal, 1984-1994. Malimbus 18: 74-111.

Roffey, J. 1979. Locusts and Grasshoppers of Economic Importance in Thaïland. Anti-Locust Memoir 14: 1-200.

Rogacheva, H.V. 1992. Birds of the Central Siberia. Husum: Husum-Druck & Verlagsgesellschaft.

Rosa, G., D. Leitão, C. Mendes, F. Courinha, H. Costa, C. Pacheco & J. Pereira. 2001. [Status of the Marsh Harrier *Circus aeruginosus* in Portugal: survey of the wintering population (1998/99).] Airo 11: 23-27.

Roux, F. 1959a. Captures de migrateurs paléarctiques dans la Basse Vallée du Sénégal. Bull. Muséum Hist. Nat. Paris 31: 458-462.

Roux, F. 1959b. Quelques données sur les Anatidés et les Charadriidés paléarctiques hivernant dans la vallée du Sénégal et sur leur écologie. Terre Vie 13: 315-321.

Roux, F. & G. Morel. 1966. Le Sénégal, région privilégiée pour les migrateurs paléarctiques. Ostrich Suppl. 6: 249-254.

Roux, F. 1973. Censuses of Anatidae in the central delta of the Niger and the Senegal delta - January 1972. Wildfowl 24: 63-80.

Roux, F. 1974. The status of wetlands in the West African Sahel: their value for waterfowl and their future. Proc. Int. Conf. Conservation of Wetlands and Waterfowl: 272-287. Slimbridge: IWRB.

Roux, F., G. Jarry, R. Mahéo & A. Tamisier. 1976. Importance, structure et origine des populations d'Anatidés hivernant dans le Delta du Sénégal. L'Oiseau et RFO 46: 299-336.

Roux, F., G. Jarry, R. Mahéo & A. Tamisier. 1977. Importance, structure et origine des populations d'Anatidés hivernant dans le delta du Sénégal (Fin). L'Oiseau et RFO 47: 1-24.

Roux, F., R. Mahéo & A. Tamisier. 1978. L'exploitation de la basse vallée du Sénégal (quartier d'hiver tropicale) par trois espèces de canards paléarctiques et éthiopien. Terre Vie 32: 387-416.

Roux, F. & G. Jarry. 1984. Numbers, composition and distribution of populations of Anatidae wintering in West Africa. Wildfowl 35: 48-60.

Rowan, M.K. 1968. The origins of European Swallows "wintering" in South Africa. Ostrich 39: 76-84.

Rowell, D.P., C.K. Folland, K. Maskell & M.N. Ward. 1995. Variability of summer rainfall over tropical North-Africa (1906-92) - Observations and modeling. Q.J. Royal Meteor. Soc. 121: 669-704.

Røstad, O.W. 1986. The autumn migration of the Sedge Warbler *Acrocephalus schoenobaenus* in East Finmark. Ring. & Migr. 9: 57-61.

Rubolini, D., A.G. Pastor, A. Pilastro & F. Spina. 2002. Ecological barriers shaping fuel stores in barn swallows *Hirundo rustica* following the central and western Mediterranean flyways. J. Avian Biol. 33: 15-22.

Rudenko, A. 1996. Present status of gulls and terns nesting in the Black Sea Biosphere Reserve. Colonial Waterbirds 19 (Spec. Publ.1): 41-45.

Ruiter, C.J.S. 1941. Waarnemingen omtrent de levenswijze van de Gekraagde Roodstaart, *Phoenicurus ph. phoenicurus* (L.). Ardea 30: 175-214.

Rutschke, E. 1989. Ducks in Europe. Berlin: VEB Deutscher Landwirtschaftverlag.

Ryabitsev, V.K. & N.S. Alekseeva. 1998. Nesting density dynamics and site fidelity of waders on the middle and northern Yamal. International Wader Studies 10: 195-200.

Rydzewski, W. 1956. The nomadic movements and migrations of the European Grey Heron, *Ardea cinerea* L. Ardea 44: 71-188.

Ryttman, H. 2003. Breeding success of Wryneck *Jynx torquilla* during the last 40 years in Sweden. Ornis Svecica 13: 25-28.

SAED. 1997. Recueil des statistiques de la vallée du fleuve Sénégal – Annuaire 1995/1996. Dakar: Ministère de l'Agriculture, SAED.

Sæther, B.E., V. Grøtan, P. Tryjanowski, C. Barbraud & M. Fulin. 2006. Climate and spatio-temporal variation in the population dynamics of a long distance migrant, the white stork. J. Anim. Ecol. 75: 80-90.

Saino, N., T. Szép, M. Romano, D. Rubolini, F. Spina & A.P. Møller. 2004. Ecological conditions during winter predict arrival date at the breeding quarters in a trans-Saharan migratory bird. Ecology Letters 7: 21-25.

Saino, N., T. Szép, R. Ambrosini, M. Romano & A.P. Møller. 2004. Ecological conditions during winter affect sexual selection and breeding in a migratory bird. Proc. R. Soc. London B 271: 681-686.

Saino, N., D. Rubolini, N. Jonzén, T. Ergon, A. Montemaggiori, N.C. Stenseth & F. Spina. 2007. Temperature and rainfall anomalies in Africa predict timing of spring migration in trans-Saharan migratory birds. Climate Research 35: 123-134.

Salamolard, M., A. Butet, A. Leroux & V. Bretagnolle. 2000. Responses of an avian predator to variations in prey density at a temperate latitude. Ecology 81: 2428-2441.

Salem-Murdock, M., M. Niasse, J. Magistro, C. Nutall, O. Kane, K. Grimm & K. Sella. 1994. Les barrages de la controverse. Le cas du fleuve Sénégal. Paris: Harmattan.

Salewski, V., F. Bairlein & B. Leisler. 2000. Recurrence of some palaearctic migrant passerine species in West Africa. Ring. & Migr. 20: 29-30.

Salewski, V., K. H. Falk, F. Bairlein & B. Leisler. 2002a. Numbers, body mass and fat scores of three Palearctic migrants at a constant effort mist netting site in Ivory Coast, West Africa. Ardea 90: 479-487.

Salewski, V., F. Bairlein & B. Leisler. 2002b. Different wintering strategies of two Palearctic migrants in West Africa - a consequence of foraging strategies? Ibis 144: 85-93.

Salewski, V., R. Altwegg, B. Erni, K. H. Falk, F. Bairlein & B. Leisler. 2004. Moult of three Palaearctic migrants in their West African winter quarters. J. Ornithol. 145: 109-116.

Salewski, V., H. Schmaljohann & M. Herremans. 2005. New bird records from Mauritania. Malimbus 27: 19-32.

Salewski, V. & P. Jones. 2006. Palearctic passerines in Afrotropical environments: a review. J. Ornithol. 147: 192-201.

Sanderson, F.J., P.F. Donald, D.J. Pain, I.J. Burfield & P.J. van Bommel. 2006. Long-term population declines in Afro-Palearctic migrant birds. Biol. Conserv. 131: 93-105.

Sapetin, Y.V. 1978a. Migrations of Little Egrets on the Azov Black Sea and Caspian basins, USSR. In: V.D. Ilyichev (ed.). Migrations of birds of Eastern Europe and Northern Asia: 142-150. Moscow: Nauka.

Sapetin, Y.V. 1978b. Migrations of Squacco Herons of the Azov-Black Sea and Caspian Basin. In: V.D. Ilyichev (ed.). Migrations of birds of Eastern Europe and Northern Asia: 127-133. Moscow: Nauka.

Sapetin, Y.V. 1978c. Results of Glossy Ibis' banding in the USSR. In: V.D. Ilyichev (ed.). Migrations of birds of Eastern Europe and Northern Asia: 245-255. Moscow: Nauka.

Saurola, P. 1994. African non-breeding areas of Fennoscandian Ospreys Pandion haliaetus - A ring recovery analysis. Ostrich 65: 127-136.

Saurola, P. 1995. Persecution of raptors in Europe assessed by Finnish and Swedish ring recovery data. In: I. Newton (ed.). Conservation studies on raptors: 439-448. Cambridge: International Council for Bird Preservation.

Saurola, P. 2002. Satelliit sauraavat sääksiämme. Linnut-vuosikirja 2002(4): 11-14.

Saurola, P. 2007. Monitoring and conservation of Finnish Ospreys Pandion haliaetus in 1971-2005. In: P. Koskimies & N.V. Lapshin (eds). Status of raptor populations in Eastern Fennoscandia: 125-132. Petrozavodsk: Karelian Research Centre of the Russian Academy of Sciences / Finnish-Russian Working Group on Nature Conservation.

Sauvage, A., S. Rumsey & S. Rodwell. 1998. Recurrence of Palaearctic birds in the lower Senegal river valley. Malimbus 20: 33-53.

Sauvage, A. & S.P. Rodwell. 1998. Notable observations of birds in Senegal (excluding Parc National des Oiseaux du Djoudj), 1984-1994. Malimbus 20: 75-122.

Sánchez, J.M., A.M. Del Viejo, C. Corbacho, E. Costillo & C. Fuentes. 2004. Status and trends of gull-billed tern Gelochelidon nilotica in Europe and Africa. Bird Conserv. Intern. 14: 335-351.

Sánchez-Guzmán, J.M., R. Morán, J.A. Masero, C. Corbacho, E. Costillo, A. Villegas & F. Santiago-Quesada. 2007. Identifying new buffer areas for conserving waterbirds in the Mediterranean basin: the importance of the rice fields in Extremadura, Spain. Biodivers. Conserv. 16: 3333-3344.

Sánchez-Zapata, J.A. & J.F. Calvo. 1998. Importance of birds and potential bias in food habit studies of Montagu's Harriers (Circus pygargus) in southeastern Spain. J. Raptor Research 32: 254-256.

Scebba, S. & G. Moschetti. 1996. Migration pattern and weight changes of Wood Sandpipers Tringa glareola in a stopover site in southern Italy. Ring. & Migr. 17: 101-104.

Schaub, M. & L. Jenni. 2000a. Body mass of six long-distance migrant passerine species along the autumn migration route. J. Ornithol. 141: 441-460.

Schaub, M. & L. Jenni. 2000b. Fuel deposition of three passerine bird species along the migration route. Oecologia 122: 306-317.

Schaub, M. & R. Pradel. 2004. Assessing the relative importance of different sources of mortality from recoveries of marked animals. Ecology 85: 930-938.

Schaub, M., W. Kania & U. Köppen. 2005. Variation of primary production during winter induces synchrony in survival rates in migratory white storks Ciconia ciconia. J. Anim. Ecol. 74: 656-666.

Schäffer, N., B.A. Walther, K. Gutteridge & C. Rahbek. 2006. The African migration and wintering grounds of the Aquatic Warbler Acrocephalus paludicola. Bird Conserv. Intern. 16: 33-56.

Schekkerman, H. & G. Müskens. 2000. Produceren grutto's Limosa limosa in agrarisch grasland voldoende jongen voor een duurzame populatie? Limosa 73: 121-134.

Schekkerman, H., W.A. Teunissen & E. Oosterveld. 2008. The effect of 'mosaic management' on the demography of black-tailed godwit *Limosa limosa* on farmland. J. appl. Ecol. 45: 1067-1075.

Schekkerman, H., W.A. Teunissen & E. Oosterveld. 2009. Mortality of Black-tailed Godwit *Limosa limosa* and Northern Lapwing *Vanellus vanellus* chicks in wet grasslands: influence of predation and agriculture. J. Ornithol. 150: 133-145.

Schepers, F. & E.C.L. Marteijn. 1993. Coastal waterbirds in Gabon. WIWO-report Nr. 41. Zeist: Foundation Working Group International Wader and Waterfowl Research.

Schimmel, H.J.W. & J.G. de Molenaar. 1982. Hoogstamboomgaarden. Vogeljaar 30: 252-257.

Schimmer, R. 2008. Tracking the genocide in Darfur: Population displacement as recorded by remote sensing; Genocide Studies Working Paper No. 36. Yale University Genocide Studies Program, Remote Sensing Project.

Schipper, W.J.A. 1973. A comparison of prey selection in sympatric harriers, *Circus*, in Western Europe. Gerfaut 63: 17-120.

Schlenker, R. 1988. Zum Zug der Neusiedlersee (Österreich)-Population des Teichrohrsängers (*Acrocephalus scirpaceus*) nach Ringfunden. Vogelwarte 34: 337-343.

Schlenker, R. 1995. Änderungen von Wiederfundquoten beringter Vögel im Arbeitsbereich der Vogelwarte Radolfzell. Vogelwarte 38: 108-109.

Schlesinger, W.H. & N. Gramenopoulos. 1996. Archival photographs show no climate-induced changes in woody vegetation in the Sudan, 1943-1994. Global Change Biology 2: 137-141.

Schmaljohann, H., F. Liechti & B. Bruderer. 2007a. Songbird migration across the Sahara: the non-stop hypothesis rejected! Proc. R. Soc. London B 274: 735-739.

Schmaljohann. H., F. Liechti & B. Bruderer. 2007b. An addendum to 'Songbird migration across the Sahara: the nonstop hypothesis rejected!'. Proc. R. Soc. London B 274: 1919-1920.

Schmaljohann, H., F. Liechti & B. Bruderer. 2008. First records of lesser black-backed gulls *Larus fuscus* crossing the Sahara non-stop. J. Avian Biol. 39: 233-237.

Schmid, H., R. Luder, B. Naef-Daenzer, R. Graf & N. Zbinden. 1998. Schweizer Brutvogelatlas. Verbreitung der Brutvögel in der Schweiz und im Fürstentum Liechtenstein 1993-1996. Sempach: Schweizerische Vogelwarte.

Schmid, H., M. Burkhardt, V. Keller, P. Knaus, B. Volet & N. Zbinden. 2001. Die Entwicklung der Vogelwelt in der Schweiz. Avifauna Report Sempach 1, Annex. Sempach: Schweizerische Vogelwarte.

Schmidt, D. & D. Roepke. 2001. Zugrouten und Überwinterungsgebiete von in Deutschland beringten Fischadlern *Pandion haliaetus*. Vogelwelt 122: 141-146.

Schmidt, E. 1978. Migrations of Hungarian Night Herons. *In*: V.D. Ilyichev (ed.). Migrations of birds of Eastern Europe and Northern Asia: 108-112. Moscow: Nauka.

Schmidt, M.B. & P.J. Whitehouse. 1976. Moult and mensural data of ruff on the Witwatersrand. Ostrich 47: 179-190.

Schogolev, I.V. 1996a. Migration and wintering grounds of Glossy Ibises (*Plegadis falcinellus*) ringed at the colonies of the Dnestr Delta, Ukraine, Black Sea. Colonial Waterbirds 19: 152-158.

Schogolev, I. V. 1996b. Fluctuations and trends in breeding populations of colonial waterbirds in the Dnestr Delta, Ukraine, Black Sea. Colonial Waterbirds (Spec. Publ.1): 91-97.

Scholte, P. 1996. Conservation status of cranes in north Cameroon and western Chad. Proc. 1993 African Crane and Wetland Training Workshop: 153-156.

Scholte, P. 1998. Status of vultures in the Lake Chad Basin, with special reference to Northern Cameroon and Western Chad. Vulture News 39: 3-19.

Scholte, P., S. Kort de & M. van Weerd. 1999. The Birds of the Waza-Logone Area, Far North Province Cameroon. Malimbus 21: 16-50.

Scholte, P. 2003. Immigration: A potential time bomb under the integration of conservation and development. Ambio 32: 58-64.

Scholte, P., W.C. Mullié, C. Batello, M. Marzot, A.H. Touré & D. Williamson. 2004. Wildlife. The Future is an Ancient Lake.Traditional knowledge, biodiversity and genetic resources for food and agriculture in Lake Chad Basin ecosystems: 227-257. Rome: FAO.

Scholte, P. 2006. Waterbird recovery in Waza-Logone (Cameroon), resulting from increased rainfall, floodplain rehabilitation and colony protection. Ardea 94: 109-125.

Scholte, P., S. Adam & B.K. Serge. 2007. Population trends of antelopes in Waza National Park (Cameroon) from 1960 to 2001: the interacting effects of rainfall, flooding and human interventions. Afr. J. Ecol. 45: 431-439.

Scholte, P. 2007. Maximum flood depth characterizes above-ground biomass in African seasonally shallowly flooded grasslands. J. Trop. Ecol. 23: 63-72.

Scholte, P. & J. Brouwer. 2008. Relevance of key resource areas for large-scale movements of livestock. *In:* H.H.T. Prins & F. van Langevelde (eds). Resource Ecology. Spatial and temporal aspects of foraging: 211-232. Dordrecht: Kluwer/Springer.

Scholte, P. & I. Hashim. 2008. *Eudorcas rufifrons - Nanger dama - Gazella dorcas. In:* J.S. Kingdon & M. Hoffmann (eds). The Mammals of Africa. Vol. 6. Amsterdam: Academic Press.

Schoonmaker Freudenberger, K. 1991. Mbegué : L'habile destruction d'une forêt sahélienne. London, IIED, Programme Réseaux des Zones Arides. Dossier No. 29.

Schricke, V., P. Triplet, B. Tréca, S.I. Sylla & M. Perrot. 1990 Dénombrements des anatidés dans le bassin du Sénégal (janvier 1989). Bull. mens. O.N.C. 144: 15-24.

Schricke, V., P. Triplet, B. Tréca, S.I. Sylla & I. Diop. 1991. Dénombrements des anatidés dans le Parc national du Djoudj et ses environs (janvier 1990). Bull. mens. O.N.C. 153: 29-34.

Schricke, V., M. Benmergui, S. Ndiaye, B. ould Messaoud, S. Diouf, C.O. Mbare, S.I. Sylla, B. Amadou, J.Y. Mondain-Monval, J.B. Mouronval, P. Triplet, J.P. Lafond & J. Mehn. 1998. Oiseaux d'eau dans le delta du Sénégal en 1998. Bull. mens. O.N.C. 239: 4-15.

Schricke, V., M. Benmergui, S. Diouf, B. ould Messaoud & P. Triplet. 1999. Oiseaux d'eau dans le delta du Sénégal et ses zones humides environnantes en janvier 1999. Bull. mens. O.N.C. 247: 23-33.

Schricke, V. 2001. Elements for a garganey (Anas querquedula) management plan. Game Wildl. Sci. 18: 9-41.

Schulz, H. 1988. Weißstorchzug - Ökologie, Gefährdung und Schutz des Weißstorchs in Afrika und Nahost. Königslutter-Lelm: Holger Schulz.

Schulz, H. 1998. White Stork. BWP Update 2: 69-105.

Schulz, H. 1999. Weißstorch im Aufwind? - White Storks on the up? Proc. Internat. Symp. on the White Stork: 335-350. Bonn: NABU.

Schulz, J.C. 1981. Adaptive changes in antipredator behavior of a grasshopper during development. Evolution 175-179.

Schulze-Hagen, K. 1993. Habitatansprüche und für den Schutz relevante Aspekte der Biologie des Teichrohrsängers. Beih. Veröff. Naturschutz Landschaftspflege Bad.-Württ. 68: 15-40.

Schüz, E. 1937. Vom Heimzug des Weißen Storches 1937. Vogelzug 8: 175-183.

Schüz, E. 1953. Die Zugscheide des Weißen Storches nach den Beringungs-Ergebnissen. Bonn. zool. Beitr. 4: 31-72.

Schüz, E. 1955. Störche und andere Vögel als Heuschreckenvertilger in Afrika. Vogelwarte 18: 93-95.

Schüz, E. 1971. Grundriß der Vogelzugskunde. Berlin: Verlag Paul Parey.

Schwilch R., R. Mantovani, F. Spina & L. Jenni. 2001. Nectar consumption of warblers after long-distance flights during spring migration. Ibis 143: 24-32.

Seebohm, H. 1901. The birds of Siberia: a record of a naturalist's visits to the valleys of the Petchora and Yenesei. Reprint 1976, Alan Sutton, Dursley.

Seitz, J. 1985. Knäkente Anas querquedula L., 1758. In: F. Goethe, H. Heckenroth & H. Schumann (eds). Die Vögel Niedersachsens und des Landes Bremen. Naturschutz und Landschaftspflege in Niedersachsen, Sonderreihe B Heft 2.2: 81-83.

Sen Gupta, A. & M.M. Chakrabarty. 1964. Composition of the seed fats of the Capparidaceae family. J. Science of Food and Agriculture 15: 69-73.

Serle, W. 1957. A contribution to the ornithology of the eastern region of Nigeria. Ibis 99: 628-685.

Serra, L., A. Magnani & N. Baccetti. 1990. Weights and duration of stays in Ruffs Philomachus pugnax during spring migration: some data from Italy. Wader Study Group Bull. 58: 19-22.

Seto, K.C., E. Fleishman, J.P. Fay & C.J. Betrus. 2004. Linking spatial patterns of bird and butterfly species richness with Landsat TM derived NDVI. Intern. J. Remote Sensing 25: 4309-4324.

Shamoun-Baranes, J., A. Baharad, P. Alpert, P. Berthold, Y. Yom-Tov, Y. Dvir & Y. Leshem. 2003. The effect of wind, season and latitude on the migration speed of white storks Ciconia ciconia, along the eastern migration route. J. Avian Biol. 34: 97-104.

Shebareva, T. P. 1962. [New data on the location of banded Caspian Terns (Hydroprogne tschegrava Lepechin).] Animal Migration (Moscow Acad. Sci. USSR) 3: 92-105.

Shirihai, H. 1996. The birds of Israel. London: Academic Press.

Shrubb, M. 2003. Birds, scythes and combines. Cambridge: Cambridge University Press.

Shurulinkov, P., I. Nikolov, D. Demerdzhiev, K. Bedev, H. Dinkov, G. Daskalova, S. Stoychev, I. Hristov & A. Ralev. 2007. A new census of heron and cormorant colonies in Bulgaria (2006). Bird Census News 20: 70-84.

Simmons, R. 1986. Why is the foraging success of Ospreys wintering in southern Africa so low? Gabar 1: 14-19.

Simmons, R.E., D.M. Avery & G. Avery. 1991. Biases in diets determined from pellets and remains: correction factors for a mammal and bird-eating raptor. J. Raptor Research 25: 63-67.

Siriwardena, G.M., S.R. Baillie, H.Q.P. Crick, J.D. Wilson & S. Gates. 2000. The demography of lowland farmland birds. In: N.J. Aebischer, A.D. Evans, P.V. Grice & J.A. Vickery (eds). Ecology and conservaton of lowland farmland birds: 117-133. Tring: British Ornithologists' Union.

Skibbe, A. 2008. Die Stille kommt von Westen! Die relativen Dichten der Indikatorarten der Agrarlandschaft im deutsch-polnischen Tiefland. Vogelwarte 46: 343.

Skidmore, A.K., B.O. Oindo & M.Y. Said. 2003. Biodiversity assessment by remote sensing. Proc. 30th Intern. symp. remote sensing of the environment: 1-4.

Skinner, J.R. 1987. Complément d'information sur les Tourterelles des bois dans la zone d'inondation du Niger au Mali. Malimbus 9: 133-134.

Skinner, J., P.J. Wallace, W. Altenburg & B. Fofana. 1987a. The status of heron colonies in the Inner Niger Delta, Mali. Malimbus 9: 65-82.

Skinner, J.R., B. Fofana & B. Tréca. 1987b. Dénombrement d'anatidae hivernant dans le bassin du fleuve Niger; février 1986 et janvier 1987. Gland: IUCN.

Skinner, J.R., B. Fofana & B. Niagate. 1989. Dossier relatif à la création de "sites de Ramsar" dans le delta intérieur du Niger, Mali. Gland: IUCN.

Skokova, N.N. 1978. Migrations of Grey Herons breeding on the Rybinsk Reservoir, USSR. In: V.D. Ilyichev (ed.). Migrations of birds of Eastern Europe and Northern Asia: 179-188. Moscow: Nauka.

Skov, H. 1999. Nogle resultater af ringmærkningen af Hvid Stork i Danmark 1901-98. Dansk Orn. Foren. Tidsskr. 93: 230-234.

Smart, M., H. Azafzaf & H. Diensi. 2007. The 'Eurasian' Spoonbill (Platalea leucorodia) in Africa. Ostrich 78: 495-500.

Smith, K.D. & G.B. Popov. 1953. On birds attacking Desert Locust swarms in Eritrea. Entomologist 86: 3-7.

Smith, K.D. 1968. Spring migration through southeast Morocco. Ibis 110: 452-492.

Smith, V.W. & D. Ebbutt. 1965. Notes on Yellow Wagtails Motacilla flava wintering in central Nigeria. Ibis 107: 390-393.

Smith, V.W. 1966. Autumn and spring weights of some Palaearctic migrants in central Nigeria. Ibis 108: 492-512.

Soikkeli, M. 1970. Mortality rates of Finnish Caspian Terns *Hydroprogne caspia*. Ornis Fennica 47: 117-119.

Sokolov, L.V., V.D. Yefremov, M.Y. Markovets, A.P. Shapoval & M.E. Shumakov. 2000. Monitoring of numbers in passage populations of passerines over 42 years (1958-1999) on the Courish Spit of the Baltic Sea. Avian Ecol. Behav. 4: 31-53.

Sokolov, L. V., J. Baumanis, A. Leivits, A.M. Poluda, V.D. Yefremov, M.Y. Markovets, Y.G. Morozov & A.P. Shapoval. 2001. Comparative analysis of long-term monitoring data on numbers of passerines in nine European countries in the second half of the 20th century. Avian Ecol. Behav. 7: 41-74.

Soutullo, A., R. Liminaña, V. Urios, M. Surroca & J.A. Gill. 2006. Density-dependent regulation of population size in colonial breeders: Allee and buffer effects in the migratory Montagu's Harrier. Oecologia 149: 543-552.

SOVON. 2002. Atlas van de Nederlandse broedvogels 1998-2000: verspreiding, aantallen, verandering. Leiden: Nationaal Natuurhistorisch Museum Naturalis / KNNV Uitgeverij / European Invertebrate Survey – Nederland.

Söderström, B., S. Kiema & R.S. Reid. 2003. Intensified agricultural land-use and bird conservation in Burkina Faso. Agriculture, Ecosystems & Environment 99: 113-124.

Spaar, R. & B. Bruderer. 1997. Migration by flapping or soaring: flight strategies of Marsh, Montagu's and Pallid Harriers in southern Israel. Condor 99: 458-469.

Spina, F. 2001. EURING Swallow project. EURING Newsletter 3.

Spinage, C.A. 1968. The natural history of antelopes. Beckenham: Croom Helm.

Staav, R. 1977. Étude du passage de la Sterne caspienne *Hydroprogne caspia* en Méditerranée à partir de reprises d'oiseaux en Suède. Alauda 45: 265-270.

Staav, R. 1979. Dispersal of Caspian Terns *Sterna caspia* in the Baltic. Ornis Fennica 56: 13-17.

Staav, R. 1985. Projekt Skräntärna. Vår Fågelvärld 44: 163-166.

Staav, R. 1988. Projekt Skräntärna 1986 och 1987. Vår Fågelvärld 47: 97-100.

Staav, R. 2001. Svenska skräntärnors flyttning. Presentation av återfyndsmaterial med kartor. Fauna och Flora 95: 159-168.

Stanevičius, V. 1999. Nonbreeding avifauna and water ecosystem succession in the lakes of different biological productivity in south Lithuania. Acta Zoologica Lituanica 9: 90-118.

Šťastný, K., A. Randík & K. Hudec. 1987. [The atlas of breeding birds in Czechoslovakia 1973-77.] Praha: Academia Praha.

Šťastný, K., V. Bejček & K. Hudec. 2005. [Atlas of breeding birds in Czech Republic 2001-2003.] Praha: Aventinum.

Steedman, A. 1990. Locust Handbook, 3rd edition. Chatham: Natural Resources Institute.

Stienen, E.W.M., A. Brenninkmeijer & M. Klaassen. 2008. Why do Gull-billed terns *Gelochelidon nilotica* feed on fiddler crabs *Uca tangeri* in Guinea-Bissau? Ardea 96: 243-250.

Stikvoort, E.C. 1994. Stomach and faeces contents of waterbirds in Egypt. *In*: P.L. Meininger and G.A.M. Atta (eds). Ornithological studies in Egyptian wetlands 1989/90: 193-212. Vlissingen: Foundation for Ornithological Research in Egypt.

Stoate, C. 1995. The Impact of Desert Locust *Schistocerca gregaria* swarms on pre-migratory fattening of Whitethroats *Sylvia communis* in the western Sahel. Ibis 137: 420-422.

Stoate, C. & S.J. Moreby. 1995. Premigratory diet of trans-Saharan migrant passerines in the western Sahel. Bird Study 42: 101-106.

Stoate, C. 1997. Abundance of whitethroats *Sylvia communis* and potential invertebrate prey, in two Sahelian sylvi-agricultural habitats. Malimbus 19: 7-11.

Stoate, C., R.M. Morris & J.D. Wilson. 2001. Cultural ecology of Whitethroat (*Sylvia communis*) habitat management by farmers: winter in farmland trees and shrubs in Senegambia. J. Environ. Management 62: 343-356.

Stolt, B.-O. 1993. Notes on reproduction in a declining population of the Ortolan Bunting *Emberiza hortulana*. J. Ornithol. 134: 59-68.

Stoorvogel, J.J., E.M.A. Smaling & B.H. Janssen. 1993. Calculating soil nutrient balances in Africa at different scales.1. Supra-national scale. Fertilizer Research 35: 227-235.

Stower, W.J. & D.J. Greathead. 1969. Numerical changes in a population of the Desert Locust, with special reference to factors responsible for mortality. J. appl. Ecol. 6: 203-235.

Strandberg, R., T. Alerstam & M. Hake. 2006. Wind-dependent foraging flight in the osprey *Pandion haliaetus*. Ornis Svecica 16: 150-163.

Strandberg, R. & T. Alerstam. 2007. The strategy of fly-and-forage migration, illustrated for the osprey (*Pandion haliaetus*). Behav. Ecol. 61: 1865-1875.

Strandberg, R. & P. Olofsson. 2007. Svenska kärrhökar bland afrikanska juveler. Vår Fågelvärld 66(1): 8-13.

Strandberg, R., R.H.G. Klaassen, M. Hake, P. Olofsson, K. Thorup & T. Alerstam. 2008. Complex timing of Marsh Harrier *Circus aeruginosus* migration due to pre- and post-migratory movements. Ardea 96: 159-171.

Strandberg, R. 2008. Converging migration routes of Eurasian Hobbies *Falco subbuteo* crossing the African equatorial rain forest. *In*: R. Strandberg (ed.). Migration strategies of raptors - spatio-temporal adaptations and constraints in travelling and foraging: 107-116. Lund: Department of Ecology, Lund University.

Stronach, N.R.H. 1991. Wintering harriers in Serengeti National Park, Tanzania. Afr. J. Ecol. 29: 90-92.

Stroud, D.A., N.C. Davidson, R. West, D.A. Scott, L. Haanstra, O. Thorup, B. Ganter & S. Delany. 2004. Status of migratory wader populations in Africa and Western Eurasia in the 1990s. International Wader Studies 15: 1-259.

Summers, R.W., L.G. Underhill, D.J. Pearson & D.A. Scott. 1987. Wader migration systems in southern and eastern Africa and western Asia. Wader Study Group Bull. 49, Suppl.: 15-34.

Sutcliffe, J.V. & Y.P. Parks. 1987. Hydrological Modeling of the Sudd and Jonglei Canal. Hydrol. Sci. 32: 143-159.

Sutcliffe, J.V. & Y.P. Parks. 1989. Comparative water balances of selected African wetlands. Hydrol. Sci. 34: 49-62.

Sutcliffe, J.V. & Y.P. Parks. 1999. The Hydrology of the Nile. Wallingford: IAHS Press, Institute of Hydrology.

Sutcliffe, J.V. 2005. Comment on 'Spatial variability of evaporation and moisture storage in the swamps of the upper Nile studied by remote sensing technique' by Y.A. Mohamed et al., 2004. Journal of Hydrology 289, 145-164. J. Hydrol. 314: 45-47.

Sutherland, W.J. 1992. Evidence for flexibility and constraints in migration systems. J. Avian Biol. 29: 441-446.

Svensson, S.E.. 1985. Effects of changes in tropical environments on the North European avifauna. Ornis Fennica 62: 56-63.

Svensson, S. 1986. Number of pairs, timing of egg-laying and clutch size in a sub-alpine Sand Martin *Riparia riparia* colony, 1968-1985. Ornis Scand. 17: 221-229.

Svensson, S., M. Svensson & M. Tjernberg. 1999. Svensk fågelatlas. Vår Fågelvärld, supplement nr. 31. Stockholm: Sveriges Ornitologiska Förening.

Symmons, P.M. & K. Cressman. 2001. Desert Locust guidelines: Biology and Behaviour (2nd edition). Rome: FAO.

Szép, T. 1993. Changes of the Sand Martin (*Riparia riparia*) population in Eastern Hungary: the role of the adult survival and migration between colonies in 1986-1993. Ornis Hungarica 3: 56-66.

Szép, T. 1995a. Relationship between West African rainfall and the survival of central European Sand Martins *Riparia riparia*. Ibis 137: 162-168.

Szép, T. 1995b. Survival rates of Hungarian Sand Martins and their relationship with Sahel rainfall. J. appl. Stat. 22: 891-904.

Szép, T. 1999. Effects of age- and sex-biased dispersal on the estimation of survival rates of the Sand Martin *Riparia riparia* population in Hungary. Bird Study 46: 169-177.

Szép, T., D. Szabó & J. Vallner. 2003a. Integrated population monitoring of sand martin *Riparia riparia* - an opportunity to monitor the effects of environmental disasters along the river Tisza. Ornis Hungarica 12-13: 169-183.

Szép, T., A.P. Møller, J. Vallner, A. Kovács & D. Norman. 2003b. Use of trace elements in feathers of sand martin *Riparia riparia* for identifying moulting areas. J. Avian Biol. 34: 307-320.

Szép, T., A.P. Møller, S. Piper, R. Nuttall, Z.D. Szabó & P.L. Pap. 2006. Searching for potential wintering and migration areas of a Danish Barn Swallow population in South Africa by correlating NDVI with survival estimates. J. Ornithol. 147: 245-253.

T

Tait, W.C. 1924. The birds of Portugal. London: H.F. & G. Witherby.

Tappan, G.G., M. Sall, E.C. Wood & M. Cushing. 2004. Ecoregions and land cover trends in Senegal. J. Arid Environ. 59: 427-462.

Tarboton, W. & D. Allan. 1984. The status and conservation of birds of prey in the Transvaal. Pretoria: Transvaal Museum.

Taupin, J.D. 1997. Caractérisation de la variabilité spatiale des pluies aux échelles inférieures au kilomètre en région semi-aride (région de Niamey, Niger). C. R. Acad. Sc. série IIa 325: 251-256.

Taupin, J.D. 2003. Accuracy of the precipitation estimate in the Sahel depending on the rain-gauge network density. C. R. Geoscience 335: 215-225.

Tchamba, M.N. 2008. The impact of elephant browsing on the vegetation in Waza National Park, Cameroon. Afr. J. Ecol. 33: 184-193.

Tegen, I. & I. Fung. 1995. Contribution to the atmospheric mineral aerosol load from surface land modification. J. Geophysical Research 100: 18707-18726.

Terborgh, J. 1989. Where have all the birds gone? Princeton: Princeton University Press.

Teunissen, W.A., W. Altenburg & H. Sierdsma. 2005. Toelichting op de Gruttokaart van Nederland. Veenwouden: SOVON/A&W.

Teunissen, W.A. & L.L. Soldaat. 2006. Recente aantalsontwikkelingen van weidevogels in Nederland. De Levende Natuur 107: 70-74.

Thaxter, C.B., C.P.F. Redfern & R.M. Bevan. 2006. Survival rates of adult Reed Warblers *Acrocephalus scirpaceus* at a northern and southern site in England. Ring. & Migr. 23: 65-79.

Théboud, B. & S. Batterbury. 2001. Sahel pastoralists: opportunism, struggle, conflict and negotiation. A case study from eastern Nigeria. Global Environmental Change 11: 69-78.

Thévenot, M., R. Vernon & P. Bergier. 2003. The birds of Morocco. BOU Checklist No. 20. Tring: British Ornithologists' Union.

Thiam, A. K. 2003. The causes and spatial pattern of land degradation risk in southern Mauritania using multitemporal AVHRR-NDVI imagery and field data. Land Degradation & Development 14: 133-142.

Thibault, J.-C. & G. Bonaccorsi. 1999. The birds of Corsica. BOU Checklist No. 17. Tring: British Ornithologists' Union.

Thiollay, J.-M. 1971. L'exploitation des feux de brousse par les oiseaux en Afrique occidentale. Alauda 39: 54-72.

Thiollay, J.-M. 1977. Distribution saisonnière des rapaces diurnes en Afrique occidentale. L'Oiseau et RFO 47: 253-294.

Thiollay, J.-M. 1978a. The birds of Ivory Coast. Malimbus 7: 1-59.

Thiollay, J.-M. 1978b. Les plaines du Nord Cameroun. Centre d'hivernage de rapaces paléarctiques. Alauda 46: 314-326.

Thiollay, J.-M. 1978c. Les migration de rapaces en Afrique occidentale: adaptation écologiques aux fluctuations saisonnières de production des écosystèmes. Terre Vie 32: 89-133.

Thiollay, J.-M. & V. Bretagnolle. 2004. Rapaces nicheurs de France. Distribution, effectifs et conservation. Paris: Delachaux et Niestlé.

Thiollay, J.-M. 2006a. The decline of raptors in West Africa: long-term assessment and the role of protected areas. Ibis 148: 240-254.

Thiollay, J.-M. 2006b. Severe decline of large birds in the Northern Sahel of West Africa: a long-term assessment. Bird Conserv. Intern. 16: 353-365.

Thiollay, J.-M. 2006c. Raptor population decline in West Africa. Ostrich 78: 405-413.

Thiollay, J.-M. 2007. Large bird declines with increasing human pressure in savanna woodland (Burkina Faso). Biodivers. Conserv. 15: 2085-2108.

Thomas, D.H.L., M.A. Jimoh & H. Matthes. 1993. Fishing. In: G.E. Hollis, W.M. Adams & M. Aminu-Kano (eds). The Hadejia-Nguru Wetlands: Environment, Economy and Sustainable development of a Sahelian Floodplain Wetland: 97-115. Gland: IUCN.

Thomas, D.H.L. 1996. Dam construction and ecological change in the riparian forest of the Hadejia-Jama'are floodplain, Nigeria. Land Degradation & Development 7: 279-295.

Thomas, D.H.L. & W.M. Adams. 1999. Adapting to dams: Agrarian change downstream of the Tiga Dam, Northern Nigeria. World Development 27: 919-935.

Thomas, M.B. 1999. Ecological approaches and the development of "truly integrated" pest management. PNAS 96: 5944-5951.

Thompson, J.R. & G. Polet. 2000. Hydrology and land use in a Sahelian floodplain wetland. Wetlands 20: 639-659.

Thonnerieux, Y. 1988. État des connaissances sur la reproduction de l'avifauna du Burkina Faso (ex Haute-Volta). L'Oiseau et RFO 58: 120-146.

Thorup, K., T. Alerstam, M. Hake & N. Kjellén. 2003. Bird orientation: compensation for wind drift in migrating raptors is age dependent. Proc. R. Soc. London B 270: 8-11.

Thorup, K., T.E. Ortvad & J. Rabøl. 2006. Do Nearctic Northern Wheatears (Oenanthe oenanthe leucorhoa) migrate nonstop to Africa? Condor 108: 446-451.

Thorup, O. 2004. Status of populations and management of Dunlin Calidris alpina, Ruff Philomachus pugnax and Black-tailed Godwit Limosa limosa in Denmark. Dansk Orn. Foren. Tidsskr. 98: 7-20.

Thorup, O. 2006. Breeding waders in Europe 2000. International Waders Studies 14. Thetford: Wader Study Group.

Timmerman, A. 1985. Grutto (Limosa limosa). Tweede verslag van de Steltloperringgroep FFF, over 82, met speciale aandacht voor Scholekster Haematopus ostralegus en Grutto Limosa limosa. Leeuwarden: FFF.

Tinarelli, R. 1998. Observations on Palearctic waders wintering in the Inner Niger Delta of Mali. International Wader Studies 10: 441-443.

Tomiałojć, L. & Z. Głowaciński. 2005. [Changes in the Polish avifauna, its past and future, different interpretations.] In: J. J. Nowakowski, P. Tryjanowski & P. Indykiewicz (eds). Ornitologia Polska na progy XXI stulecia - Dokonania i perspektywy: 39-85. Olsztyn: Sekcja Ornitologiczna PTZool., Kat. Ekologii i Ochrony Środowiska UWM.

Tortosa, F. S., J. M. Caballero & J. Reyes-Lopez. 2002. Effect of rubbish dumps on breeding success in the White Stork in southern Spain. Waterbirds 25: 39-43.

Tourenq, C., R.E. Bennetts, H. Kowalski, E. Vialet, J.L. Lucchesi, Y. Kayser & P. Isenmann. 2001. Are ricefields a good alternative to natural marshes for waterbird communities in the Camargue, southern France? Biol. Conserv. 100: 335-343.

Tourenq, C., N. Sadoul, N. Beck, F. Mesleard & J.L. Martin. 2003. Effects of cropping practices on the use of rice fields by waterbirds in the Camargue, France. Agriculture, Ecosystems & Environment 95: 543-549.

Touré, I., A. Ickowicz, C. Sagna & J. Usengumuremyi. 2001. Étude de l'impact du bétail sur les ressources du Parc National des Oiseaux du Djoudj (PNOD, Sénégal). Atelier Regional Niamey du 16-19 Janvier 2001. Faune sauvage et bétail: complémentaire et coexistence, ou compétition. Niamey.

Tøttrup, A. P., K. Thorup & C. Rahbek. 2006. Patterns of change in timing of spring migration in North European songbird populations. J. Avian Biol. 37: 84-92.

Tøttrup, A. P. & K. Thorup. 2008. Sex-differentiated migration patterns, protandry and phenology in North European songbird populations. J. Ornithol. 149: 161-167.

Tree, A. J. 1985. Analysis of ringing recoveries of Ruff involving southern Africa. Safring News 14: 75-79.

Tréca, B. 1975. Les oiseaux d'eau et la riziculture dans le delta du Sénégal. L'Oiseau et RFO 45: 259-265.

Tréca, B. 1977. Le problème des oiseaux d'eau pour la culture du riz au Sénégal. Bull. IFAN 39A: 682-692.

Tréca, B. 1981a. Régime alimentaire de la Sarcelle d'été (Anas querquedula L.) dans le delta du Sénégal. L'Oiseau et RFO 51: 33-58.

Tréca, B. 1981b. Le régime alimentaire du Dendrocygne veuf (Dendrocygna viduata) dans le delta du Sénégal. L'Oiseau et RFO 51: 219-238.

Tréca, B. 1983. L'influence de la sécheresse sur le rythme nycthéméral des Chevaliers combattants Philomachus pugnax au Sénégal. Malimbus 5: 73-77.

Tréca, B. 1984. La Barge à queue noire (Limosa limosa) dans le delta du Sénégal: régime alimentaire, données biométriques, importance économique. L'Oiseau et RFO 54: 247-262.

Tréca, B. 1989. Waterfowl catches by fishermen in Mali. Proc. VI Pan-Afr. Orn. Congr. 47-55.

Tréca, B. 1990. Régimes et préférences alimentaires d'Anatidés et de Scolopacides dans le Delta du Sénégal. Étude de leurs capacités d'adaptions aux modifications du milieu. Paris: ORSTOM.

Tréca, B. 1992. Quelques exemples de possibilités d'adaptation aux aménagements hydro-agricoles chez les oiseaux d'eau, et leurs limites. L'Oiseau et RFO 62: 335-344.

Tréca, B. 1993. Oiseaux d'eau et besoins énergétiques dans le delta du Sénégal. Alauda 61: 73-82.

Tréca, B. 1994. The diet of Ruffs and Black-tailed Godwits in Senegal. Ostrich 65: 256-263.

Tréca, B. 1999. Ricefields and fulvous wistling ducks in Senegal. Terre Vie 54: 43-57.

Triay, R. 2002. Seguimiento por satélite de tres juveniles de Águila Pescadora nacidos en la isla de Menorca. Ardeola 49: 249-257.

Trierweiler, C., B.J. Koks, R.H. Drent, K.-M. Exo, J. Komdeur, C. Dijkstra & F. Bairlein. 2007a. Satellite tracking of two Montagu's Harriers (*Circus pygargus*): dual pathways during autumn migration. J. Ornithol. 148: 513-516.

Trierweiler, C., J. Brouwer, B. Koks, L. Smits, A. Harouna & K. Moussa. 2007b. Montagu's Harrier Expedition to Niger, Benin and Burkina Faso, 9 January - 7 February 2007. Scheemda: Stichting Werkgroep Grauwe Kiekendief.

Trierweiler, C., R.H. Drent, J. Komdeur, K.-M. Exo, F. Bairlein & B.J. Koks. 2008. De jaarcyclus van de Grauwe Kiekendief: een leven gedreven door woelmuizen en sprinkhanen. Limosa 81: 107-115.

Triplet, P., B. Tréca & V. Schricke. 1993. Oiseaux consommateurs de *Schistocerca gregaria*. L'Oiseau et RFO 63: 224-225.

Triplet, P. & P. Yésou. 1994. Oiseaux d'eau dans le delta du Sénégal en janvier 1994. Bull. mens. O.N.C.. 190: 2-11.

Triplet, P., P. Yésou, S.I. Sylla, E.O. Samba, B. Tréca, A. Ndiaye & O. Hamerlynck. 1995. Oiseaux d'eau dans le delta du Sénégal en janvier 1995. Bull. mens. O.N.C. 205: 8-21.

Triplet, P. & P. Yésou. 1995. Concentrations inhabituelles d'oiseaux consommateurs de criquets dans le Delta du Fleuve Sénégal. Alauda 63: 236.

Triplet, P., S.I. Sylla, J.B. Mouronval, M. Benmergui, B. ould Messaoud, A. Ndiaye & S. Diouf. 1997. Oiseaux d'eau dans le delta du Sénégal en janvier 1997. Bull. mens. O.N.C. 224: 37.

Triplet, P. & P. Yésou. 1998. Mid-winter counts of waders in the Senegal delta, West Africa, 1993-1997. Wader Study Group Bull. 85: 66-73.

Triplet, P., A. Tiéga & D. Pritchard. 2000. Rapport de mission au Parc National du Djoudj, Sénégal et au Parc National du Diawling, Mauritanie du 14 au 21 septembre 2000. UNESCO/RAMSAR/Birdlife International.

Triplet, P. & P. Yésou. 2000. Controlling the flood in the Senegal Delta: do waterfowl populations adapt to their new environment? Ostrich 71: 106-111.

Triplet, P., I. Diop & P. Yésou. 2006. Liste des Oiseaux du Parc National des Oiseaux du Djoudj. *In:* I. Diop & P. Triplet (eds). Parc National des Oiseaux du Djoudj. Plans d'Action 2006-2008. Direction Parcs Nationaux, Dakar / Centre du Patrimoine Mondial, UNESCO.

Triplet, P. & V. Schricke. 2008. Les conséquences de l'ouverture d'une brèche dans la Langue de Barbarie sur les stationnements hivernaux de limicoles dans le delta du Sénégal. Alauda 76: 157-159.

Triplet, P., O. Overdijk, M. Smart, S. Nagy, M. Scheider-Jacoby, E.S. Karauz, Cs. Pigniczki, S. Baha El Din, J. Kralj, A. Sandor & J.G. Navedo. 2008. Plan d'actions international pour la conservation de la Spatule d'Europe *Platalea leucorodia*. Bonn: AEWA.

Trocińska, A., A. Leivitis, C. Nitecki & I. Shydlovsky. 2001. Field studies of directional preference of the Reed Warbler (*Acrocephalus scirpaceus*) and the Sedge Warbler (*A. schoenobaenus*) on autumn migration along the eastern and southern coast of the Baltic Sea and in western part of Ukraine. Ring 23: 109-117.

Trolliet, B., O. Girard & M. Fouquet. 2003. Evaluation des populations d'oiseaux d'eau en Afrique de l'Ouest. Rapport Scientifique 2002 ONCFS: 51-55.

Trolliet, B. & O. Girard. 1991. On the Ruff *Philomachus pugnax* wintering in the Senegal delta. Wader Study Group Bull. 62: 10-12.

Trolliet, B., O. Girard, M. Fouquet, F. Ibañez, P. Triplet & F. Léger. 1992. L'effectif de Combattants (*Philomachus pugnax*) hivernants dans le Delta du Sénégal. Alauda 60: 159-163.

Trolliet, B., M. Fouquet, P. Triplet & P. Yésou. 1993. Oiseaux d'eau dans le delta du Sénégal en janvier 1993. Bull. mens. O.N.C. 185: 2-9.

Trolliet, B. & P. Triplet. 1995. A propos de l'hivernage de la Barge à queue noire *Limosa limosa* dans le Delta du Sénégal. Alauda 63: 246-247.

Trolliet, B. & M. Fouquet. 2001. La population ouest-africaine du Flamant nain *Phoeniconaias minor*: Effectifs, répartition et isolement. Malimbus 23: 87-92.

Trolliet, B. & O. Girard. 2001. Numbers of Ruff *Philomachus pugnax* wintering in West Africa. Wader Study Group Bull. 96: 74-78.

Trolliet, B. & M. Fouquet. 2004. Wintering waders in coastal Guinea. Wader Study Group Bull. 103: 56-62.

Trolliet, B., O. Girard, M. Benmergui, V. Schricke & P. Triplet. 2007. Oiseaux d'eau en Afrique subsaharienne. Bilan des dénombrements de janvier 2006. Faune sauvage no. 275/février 2007: 4-11.

Trolliet, B., O. Girard, M. Benmergui, V. Schricke, J.-M. Boutin, M. Fouquet & P. Triplet. 2008. Oiseaux d'eau en Afrique subsaharienne. Bilan des dénombrements de janvier 2007. Faune sauvage no. 279/février 2008: 4-11.

Trotignon, J. 1976. La nidification sur le Banc d'Arguin (Mauritanie) au printemps 1974. Alauda 44: 119-133.

Tryjanowski, P. & R. Bajczyk. 1999. Population decline of the Yellow Wagtail *Motacilla flava* in an intensively used farmland of western Poland. Vogelwelt 120: 205-207.

Tryjanowski, P., S. Kuzniak & T. Sparks. 2002. Earlier arrival of some farmland migrants in western Poland. Ibis 144: 62-68.

Tryjanowski, P., S. Kuzniak & T.H. Sparks. 2005. What affects the magnitude of change in first arrival dates of migrant birds? J. Ornithol. 146: 200-205.

Tucker, C.J. 1979. Red and photographic infrared linear combinations for monitoring vegetation. Remote Sensing of Environment 8: 127-150.

Tucker, C.J., C.L. Vanpraet, M.J. Sharman & G. van Ittersum. 1985. Satellite Remote-Sensing of Total Herbaceous Biomass Production in the Senegalese Sahel - 1980-1984. Remote Sensing of Environment 17: 233-249.

Tucker, C.J., H.E. Dregne & W.W. Newcomb. 1991. Expansion and Contraction of the Sahara Desert from 1980 to 1990. Science 253: 299-301.

Tucker, C.J. & S.E. Nicholson. 1998. Variations in the size of the Sahara Desert from 1980 to 1997. Ambio 28: 587-591.

Tucker, C.J., J.E. Pinzon, M.E. Brown, D.A. Slayback, E.W. Pak, R. Mahoney, E.F. Vermote & N. El Saleous. 2005. An extended AVHRR 8-km NDVI dataset compatible with MODIS and SPOT vegetation NDVI data. Intern. J. Remote Sensing 26: 4485-4498.

Tucker, G.M., M.N. McCulloch & S.R. Baillie. 1990. The conservation of migratory birds in the Western Palaearctic-African flyway: Review of losses incurred to migratory birds during migration. Research Report No. 58. Tring: British Trust for Ornithology.

Tucker, G.M. & M.F. Heath. 1994. Birds in Europe: their conservation status. Cambridge: BirdLife International.

Turner, A. 2006. The Barn Swallow. London: Poyser.

Turner, M.D. 2004. Political ecology and the moral dimensions of "resource conflicts": the case of farmer-herder conflicts in the Sahel. Political Geography 23: 863-889.

Turner, M.D., P. Hiernaux & E. Schlecht. 2005. The distribution of grazing pressure in relation to vegetation resources in semi-arid West Africa: the role of herding. Ecosystems 8: 668-681.

Turner, M.D. & P. Hiernaux. 2008. Changing access to labor, pastures & knowledge: The extensification of grazing management in Sudano-Sahelian West Africa. Human Ecology 36: 59-80.

Turner, W., S. Spector, N. Gardiner, M. Fladeland, E. Sterling & M. Steininger. 2003. Remote sensing for biodiversity science and conservation. Trends in Ecology & Evolution 18: 306-314.

U

Underhill-Day, J.C. 1984. Population and breeding biology of Marsh Harriers in Britain since 1900. J. appl. Ecol. 21: 773-787.

Underhill-Day, J.C. 1993. The foods and feeding rates of Montagu's Harriers *Circus pygargus* breeding in arable farmland. Bird Study 40: 74-80.

Underhill-Day, J. 1998. Breeding Marsh Harriers in the United Kingdom, 1983-95. British Birds 91: 210-218.

Underhill, L.G. & R.W. Summers. 1990. Multivariate analyses of breeding performance in Dark-bellied Brent Geese *Branta b. bernicla*. Ibis 132: 477-480.

Underhill, L.G., A.J. Tree, H.D. Oschadleus & V. Parker. 1999. Review of Ring Recoveries of Waterbirds in Southern Africa. Cape Town: Avian Demography Unit, University of Cape Town.

Urban, E.K., C.H. Fry & S. Keith. 1986. The Birds of Africa, Volume II. London: Academic Press..

Urban, E.K. 1993. Status of Palearctic wildfowl in Northeast and East Africa. Wildfowl 44: 133-148.

Urban, E.K., C.H. Fry & S. Keith. 1997. The Birds of Africa, Volume V. San Diego: Academic Press.

US Congress OTA. 1990. Special Report. A plague of Locusts. Congress of the United States, Office of Technology Assessment.

V

van der Valk, H. 2007. Review of the efficacy of *Metarhizium anisopliae* var. *acridum* against the Desert Locust. Rome: FAO Technical Series, No. AGP/DL/TS/34.

Vandekerkhove, K., A. Vande Walle, M. Cassaert & N. Lievrouw. 2007. Habitatvoorkeur en populatieontwikkeling van Grauwe Kiekendief *Circus pygargus* in de Franse Lorraine: hebben beschermingsacties het gewenste effect? Natuur. oriolus 73: 17-24.

Väisänen, R. 2001. [Steep recent decline in Finnish populations of Wryneck, Wheatear, Chiffchaff and Ortolan Bunting.] Linnut 36: 14-15.

Väisänen, R.A., E. Lammi & P. Koskimies. 1998. Muuttuva pesimälinnusto. Helsinki: Kustannusosakeyhtiö Otava.

Väisänen, R.A. 2005. [Monitoring population changes of 84 land bird species breeding in Finland in 1983-2004.] Linnut-vuosikirja 2004: 105-119.

Väisänen, R.A. 2006. [Monitoring population changes of 84 land bird species breeding in Finland in 1983-2005.] Linnut-vuosikirja 2006: 83-89.

Veen, J., H. Dallmeijer & C.H. Diagana. 2006. Monitoring colonial nesting birds along the West African Seaboard / Final report. Dakar: Wetlands International.

Vepsäläinen, V., T. Pakhala, M. Piha & J. Tiainen. 2005. Population crash of the ortolan bunting *Emberiza hortulana* in agricultural landscapes in southern Finland. Ann. Zool. Fennici 42: 91-107.

Verhoef, H. 1996. Health aspects of Sahelian plaine d'inondation development. *In:* M.C. Acreman & G.E. Hollis (eds). Water management and Wetlands in Sub-Saharan Africa: 35-50. Gland: IUCN.

Verhulst, J., D. Kleijn & F. Berendse. 2007. Direct and indirect effects of the most widely implemented Dutch agri-environment schemes on breeding waders. J. appl. Ecol. 44: 70-80.

Vermeersch, G., A. Anselin, K. Devos, M. Herremans, J. Stevens, J. Gabriëls & B. van der Krieken. 2004. Atlas van de Vlaamse broedvogels 2000-2002. Brussel: Mededelingen van het Instituut voor Natuurbehoud 23.

Vickery, J., M. Rowcliffe, W. Cresswell, P. Jones & S. Holt. 1999. Habitat selection by Whitethroats *Sylvia communis* during spring passage in the Sahel zone of northern Nigeria. Bird Study 46: 348-355.

Vickery, J.A., J.R. Tallowin, R.E. Feber, E.J. Asteraki, P.W. Atkinson, R.J. Fuller & V.K. Brown. 2001. The management of lowland neutral grasslands in Britain: effects of agricultural practices on birds and their food resources. J. appl. Ecol. 38: 647-664.

Vielliard, J. 1972. Recensement et statut des populations d'Anatidés du bassin tchadien. Cah. ORSTOM sér. Hydrobiol. 6: 85-100.

von Vietinghoff-Riesch, A. 1955. Die Rauchschwalbe. Berlin: Duncker & Humblot.

Vīksne, J., A. Mednis, M. Janaus & A. Stipniece. 2005. Changes in the breeding bird fauna, waterbird populations in particular, on Lake Engure (Latvia) over the last 50 years. Acta Zoologica Lituanica 15: 188-194.

Vīksne, J.A. & H.A. Michelson. 1985. [Migration of birds of Eastern Europe and Northern Asia.] Moscow: Nauka.

Vincke, C. 1995. La dégradation des systèmes écologiques Sahéliens. Effets de la sécheresse et des facteurs anthropiques sur l'évolution de la végétation ligneuse du Ferlo (Sénégal). Mémoire de fin d'études, Université Catholique de Louvain, Fac. Sci. Agron. 1-82.

Visser, E., B. Koks, C. Trierweiler, J. Arisz & R.-J. van der Leij. 2008. Grauwe Kiekendieven *Circus pygargus* in Nederland in 2007. De Takkeling 16: 130-145.

Vizy, E.K. and K.H. Cook. 2001. Mechanisms by which Gulf of Guinea and eastern North Atlantic sea surface temperature anomalies can influence African rainfall. J. Climate 14: 795-821.

Voinstvenskiy, M.N. 1986. [Colonial hydrophilous birds of the South of Ukraine.] Kiev: Naukova Dumba.

Voisin, C. 1983. Les Ardéidés du delta du fleuve Sénégal. L'Oiseau et RFO 53: 335-360.

Voisin, C. & J.-F. Voisin. 1984. Observations sur l'avifaune du delta du Sénégal. L'Oiseau et RFO 54: 351-359.

Voisin, C. 1994. Bihoreau gris *Nycticorax nycticorax*. In: D. Yeatman-Berthelot & G. Jarry (eds). Nouvel atlas des oiseaux nicheurs de France 1985-1989: 90-91. Paris: Société Ornithologique de France.

Voisin, C. 1996. The migration routes of Purple Herons *Ardea purpurea* ringed in France. Vogelwarte 38: 155-168.

Voisin, C., J. Godin & A. Fleury. 2005. Status and behaviour of Little Egrets wintering in western France. British Birds 98: 468-475.

Vuillaume, G. 1981. Bilan hydrologique mensuel et modélisation sommaire du régime hydrologique du Lac Chad. Cah. ORSTOM sér. Hydrobiol. 18: 23-72.

W

Wahl, R. & C. Barbraud. 2005. Dynamique de population et conservation du Balbuzard pêcheur *Pandion haliaetus* en région centre. Alauda 73: 365-373.

Waldenström, J. & U. Ottosson. 2002. Moult strategies in the common whitethroat *Sylvia c. communis* in northern Nigeria. Ibis 144: E11-E18.

Walsh, J.F. & L.G. Grimes. 1981. Observations on some Palaearctic land birds in Ghana. Bull. Brit. Orn. Club. 101: 327-334.

Walther, B.A. & C. Rahbek. 2002. Where do Palearctic migratory birds overwinter in Africa? Dansk Orn. Foren. Tidsskr. 96: 4-8.

Wang, G. 2003. Reassessing the impact of North Atlantic Oscillation on the sub-Saharan vegetation productivity. Global Change Biology 9: 493-499.

Wang, G., E.A. B. Eltahir, J.A. Foley, D. Pollard & S. Levis. 2004. Decadal variability of rainfall in the Sahel: results from the coupled GENESIS-IBIS atmosphere-biosphere model. Climate Dynamics 22: 625-637.

Wanink, J.H. 1999. Prospects for the fishery on the small pelagic *Rastrineobola argentea* in Lake Victoria. Hydrobiologica 407: 183-189.

Wanink, J.H., P.C. Goudswaard & M.C. Berger. 1999. *Rastrineobola argentea*, a major resource in the ecosystem of Lake Victoria. In: W.L.T. van Densen & M.J. Morris (eds). Fish and fisheries of lakes and reservoirs in Southeast Asia and Africa: 295-309. Otley: Westbury Publishing.

Ward, P. 1963. Lipid levels in birds preparing to cross the Sahara. Ibis 105: 109-111.

Ward, P. 1964. The fat reserves of Yellow Wagtail *Motacilla flava* wintering in southwest Nigeria. Ibis 106: 370-375.

Wardell, D.A., A. Reenberg & C. Tøttrup. 2003. Historical footprints in contemporary land use systems: forest cover changes in savannah woodlands in the Sudano-Sahelian zone. Global Environmental Change 13: 235-254.

Wassink, A. & G.J. Oreel. 2007. The birds of Kazakhstan. De Cocksdorp, Texel: Arend Wassink.

WDPA Consortium. 2005. 2005 World Database on Protected Areas. Cambridge: IUCN UNEP.

Weesie, P.D.M. 1996. Les oiseaux d'eau du Sahel Burkinabe: peuplement d'hiver, capacité de charge. Alauda 65: 263-278.

Weggler, M. & M. Widmer. 2000. Vergleich der Brutvogelbestände im Kanton Zürich 1986-88 und 1999. I. Was hat der ökologische Ausgleich in der Kulturlandschaft bewirkt? Ornithol. Beob. 97: 123-146.

Weggler, M. 2005. Entwicklung der Brutvogelbestände 1976-2003 in den Reservaten der Ala-Schweizerische Gesellschaft für Vogelkunde und Vogelschütz. Ornithol. Beob. 102: 205-227.

Weis, J.S. 1923. Life of the harrier in Denmark. London: Wheldon & Wesley Ltd.

Weisshaupt, N. 2007. Habitat selection by foraging Wryneck *Jynx torquilla* during the breeding season: identifying optimal species habitat. Bern: Philosophisch-naturwissenschaftlichen Fakultät der Universität Bern.

Wernham, C., M. Toms, J. Marchant, J. Clark, G. Siriwardena & S. Baillie. 2002. The migration atlas: movements of the birds of Britain and Ireland. London: Poyser.

Wesołowski, T. & M. Stańska. 2001. High ectoparasite loads in hole-nesting birds - a nestbox bias? J. Avian Biol. 32: 281-285.

van Wetten, J.C.J., C. ould Mbaré, M. Binsbergen & T. van Spanje. 1990. Zones humides du sud de la Mauritanie. Leersum: R.I.N.

van Wetten, J.C.J. & P. Spierenburg. 1998. Waders and waterfowl in the floodplains of the Logone, Cameroon. January, 1993. Zeist: WIWO.

Wichmann, G., J. Barker, T. Zuna-Kratky, K. Donnerbaum & M. Rössler. 2004. Age-related stopover strategies in the Wood Sandpiper *Tringa glareola*. Ornis Fennica 81: 169-179.

Wijmenga, J. & Y. Komnotougo. 2005. Wintering Ruff (*Philomachus pugnax*) in Lac Débo, in the Inner Niger Delta, Mali. Development of mass and fat reserves and sex ratios in a wetland of international importance. Mission report. Sévaré / Groningen: Wetlands International / RUG.

Wilson, A.M. & J.A. Vickery. 2005. Decline in Yellow Wagtail *Motacilla flava flavissima* breeding on lowland wet grassland in England and Wales between 1982 and 2002. Bird Study 52: 88-92.

Wilson, J.D., A.J. Morris, B.E. Arroyo, C.S. Clark & R.B. Bradbury. 1999. A review of the abundance and diversity of invertebrate and plant foods of granivorous birds in Northern Europe in relation to agricultural

change. Agriculture Ecosystems & Environment 75: 13-30.
Wilson, J.M. 2004. Factors determining the density and distribution of Palearctic migrants wintering in sub-Saharan Africa. PhD Thesis. St. Andrews: University of St. Andrews.
Wilson, J.M. & W. Cresswell. 2006. How robust are Palearctic migrants to habitat loss and degradation in the Sahel? Ibis 148: 789-800.
Wilson, J.M. & W.R.L. Cresswell. 2007. Identification of potentially competing Afrotropical and Palaearctic bird species in the Sahel. Ostrich 78: 363-368.
Wilson, R.T. 1982. Environmental changes in western Darfur, Sudan, over half a century and their effects on selected bird species. Malimbus 4: 15-26.
van der Winden, J. & P.W. van Horssen. 2001. Voedselgebieden van de purperreiger in Nederland. Culemborg: Bureau Waardenburg.
van der Winden, J., K. Krijgsveld, R. van Eekelen & D.M. Soes. 2002. Het succes van de Zouweboezem als foerageergebied voor purperreigers. Culemborg: Bureau Waardenburg.
van der Winden, J. 2002. The odyssey of the Black Tern *Chlidonias niger*: migration ecology in Europe and Africa. Ardea 90: 421-435.
van der Winden, J., A. Siaka, S. Dirksen & M.J.M. Poot. 2007. Coastal wetland bird census Sierra Leone, January-February 2005. Zeist: WIWO.
Winkel, W. 1992. Der Wendehals (*Jynx torquilla*) als Brutvogel in Nisthöhlen-Untersuchungsgebieten bei Braunschweig. Beih. Veröff. Naturschutz Landschaftspflege Bad. -Württ. 66: 31-41.
Winkler, D.W. 2006. Roosts and migrations of swallows. Hornero 21: 85-97.
Winstanley, D., R. Spencer & K. Williamson. 1974. Where have all the Whitethroats gone? Bird Study 21: 1-14.
de Wit, M. & J. Stankiewicz. 2006. Changes in Surface Water Supply Across Africa with Predicted Climate Change. Science 311: 1917-1921.
Witherby, H.F., F.C.R. Jourdain, N.F. Ticehurst & B.W. Tucker. 1940. The Handbook of British Birds, Vol. IV. London: H.F. & G. Witherby Ltd.
Wittemyer, G., P. Elsen, W.T. Bean, A. Coleman, O. Burton & J.S. Brashares. 2008. Accelerated human population growth at protected area edges. Science 321: 123-126.
Włodarczyk, R., P. Minias, K. Kaczmarek, T. Janiszewski & A. Kleszcz. 2007. Different migration strategies used by two inland wader species during autumn migration, case of Wood Sandpiper *Tringa glareola* and Common Snipe *Gallinago gallinago*. Ornis Fennica 84: 119-130.
Wolda, G. 1918. Ornithologische Studies. 's-Gravenhage: J. & H. van Langenhuysen.
Woldhek, S. 1980. Bird killing in the Mediterranean. Zeist: European committee for the prevention of mass destruction of migratory birds.
Wood, B. 1975. The distribution of races of the Yellow Wagtail overwintering in Nigeria. Bull. Nigerian Orn. Soc. 11: 19-26.
Wood, B. 1976. The biology of Yellow Wagtails *Motacilla flava* L. overwintering in Nigeria. Aberdeen: University of Aberdeen.
Wood, B. 1978. Weights of Yellow Wagtails wintering in Nigeria. Ring. & Migr. 2: 20-26.
Wood, B. 1979. Changes in numbers of over-wintering Yellow Wagtails *Motacilla flava* and their food supplies in a west African savanna. Ibis 121: 228-231.
Wood, B. 1982. The trans-Saharan spring migration of Yellow Wagtails (*Motacilla flava*). J. Zool. Lond. 197: 267-283.
Wood, B. 1992. Yellow Wagtail *Motacilla flava* migration from West Africa to Europe: pointers towards a conservation strategy for migrants on passage. Ibis 134 suppl. 1: 66-76.
Wood, L.C., G.G. Tappan & A. Hadj. 2004. Understanding the drivers of agricultural land use change in south-central Senegal. J. Arid Environ. 59: 565-582.
Wulffraat, S. 1993. Beyond the Diama dam. The impact of changing hydrology on the ecology of Djoudj National Park and its surrounding area. Enschedé: Thesis, International Institute for Aerospace survey and Earth sciences (ITC).
Wymenga, E., M. Engelmoer, C.J. Smit & T. van Spanje. 1990. Geographical breeding origin and migration of waders wintering in west Africa. Ardea 78: 83-112.
Wymenga, E. & W. Altenburg. 1992. Short note on the occurrence of terns in Guinea-Bissau in winter. *In:* W. Altenburg, E. Wymenga & L. Zwarts (eds). Ornithological importance of the coastal wetlands of Guinea-Bissau: 69-77. Zeist: WIWO.
Wymenga, E. 1997. Grutto's *Limosa limosa* in de zomer van 1993 vroeg op de slaapplaats: aanwijzing voor een slecht broedseizoen. Limosa 70: 71-75.
Wymenga, E. 1999. Migrating Ruffs *Philomachus pugnax* through Europe, spring 1998. Wader Study Group Bull. 88: 43-48.
Wymenga, E., J. van der Kamp & B. Fofana. 2005. The irrigation zone of Office du Niger. *In:* L. Zwarts, P. van Beukering, B. Kone & E. Wymenga (eds). The Niger, a lifeline: 189-209. Lelystad: Rijkswaterstaat/IVM/Wetlands International/A&W.
Wymenga, E. 2005. Steltlopers op slaapplaatsen in Fryslân 1998-2004. Twirre 16: 200-210.

X

Xiao, J. & A. Moody. 2005. Geographical distribution of global greening trends and their climatic correlates: 1982-1998. Intern. J. Remote Sensing 26: 2371-2390.
Xie, P.P. & P.A. Arkin. 1996. Analyses of global monthly precipitation using gauge observations, satellite estimates and numerical model predictions. J. Climatol. 9: 840-858.

Y

Yeatman-Berthelot, D. & G. Jarry. 1994. Nouvel atlas des oiseaux nicheurs de France 1985-1989. Paris: Société Ornithologique de France.
Yeatman, L. 1976. Atlas des oiseaux nicheurs en France de 1970 à 1975. Paris: Société Ornithologique de France.

Yésou, P., P. Triplet, S.I. Sylla, M. Diarra, A. Ndiaye, O. Hamerlynck, S. Diouf & B. Tréca. 1996. Oiseaux d'eau dans le delta du Sénégal en janvier 1996. Bull. mens. O.N.C. 217: 2-9.

Yésou, P. & P. Triplet. 2003. Taming the delta of the Senegal River, West Africa: effects on Long-tailed and Great Cormorant *Phalacrocorax africanus*, *P. carbo lucidus* and Darter *Anhinga melanogaster rufa*. In: T. Keller, D. Carss, A. Helbig & M. Flade (eds). Cormorants: Ecology and management at the start of the 21th century. Vogelwelt 124, Supplement: 99-103.

Yohannes, E., S. Bensch & R. Lee. 2008. Philopatry of winter moult area in migratory Great Reed Warblers *Acrocephalus arundinaceus* demonstrated by stable isotope profiles. J. Ornithol. 149: 261-265.

Yohannes, E., H. Biebach, G. Nikolaus & D.J. Pearson. 2009. Passerine migration strategies and body mass variation along geographic sectors across East Africa, the Middle East and the Arabian Peninsula. J. Ornithol. 150: 369-381

Yosef, R., P. Tryjanowski & M. Remisiewicz. 2002. Migration characteristics of the Wood Sandpiper (*Tringa glareola*) at Eilat (Israel). Ring 24: 61-69.

Yosef, R. & N. Chernetsov. 2004. Stopover ecology of migratory Sedge Warblers (*Acrocephalus schoenobaenus*) at Eilat, Israel. Ostrich 75: 52-56.

Yosef, R. & N. Chernetsov. 2005. Longer is fatter: body mass changes of migrant Reed Warblers (*Acrocephalus scirpaceus*) staging at Eilat, Israel. Ostrich 76: 142-147.

Z

Zakala, O., I. Shydlovsky & P. Busse. 2004. Variation in body mass and fat reserves of the Sedge Warbler *Acrocephalus schoenobaenus* on autumn migration in the L'Viv province (W Ukraine). Ring 26: 55-69.

Zalakevicius, M., G. Bartkeviciene, L. Raudonikis & J. Janulaitis. 2006. Spring arrival response to climate change in birds: a case study from eastern Europe. J. Ornithol. 147: 326-343.

Zang, H. & H. Heckenroth. 2001. Die Vögel Niedersachsens und des Landes Bremen. Band 2.8. Hannover: Naturschutz und Landschaftspflege in Niedersachsen.

Zang, H., H. Heckenroth & P. Südbeck. 2005. Die Vögel Niedersachsens und des Landes Bremen. Band 2.9. Hannover: Naturschutz und Landschaftspflege in Niedersachsen.

Zbinden, N., V. Keller & H. Schmid. 2005. Bestandsentwicklung von regelmässig brütenden Vogelarten der Schweiz 1990-2004. Ornithol. Beob. 102: 271-282.

Zedler, J.B. & S. Kerber. 2004. Cause and consequences of invasive plants in wetlands: opportunities, opportunists and outcomes. Critical Reviews in Plant Sciences 23: 431-452.

Zedlitz, O. 1910. Meine ornithologische Ausbeute in Nordost-Afrika. J. Ornithol. 58: 731-808.

Zedlitz, O. 1921. Die Avifauna des westlichen Pripjet-Sumpfes im Lichte der Forschung deutscher Ornithologen in den Jahren 1915-1918. J. Ornithol. 69: 269-399.

Zehtindjiev, P., M. Ilieva, A. Ożarowska & P. Busse. 2003. Directional behaviour of the Sedge Warbler (*Acrocephalus schoenobaenus*) studied in two types of oriental cages during autumn migration - a case study. Ring 25: 53-63.

Zheng, X. Y., E.A.B. Eltahir & K.A. Emanuel. 1999. A mechanism relating tropical Atlantic spring sea surface temperature and west African rainfall. Q. J. Royal Meteor. Soc. 125: 1129-1163.

Zijlstra, M. 1987. Bruine Kiekendief *Circus aeruginosus* in Flevoland in de winter. Limosa 60: 57-62.

Zink, G. 1973. Der Zug europäischer Singvögel, 1. Lieferung. Möggingen: Vogelwarte Radolfzell.

Zink, G. 1975. Der Zug europäischer Singvögel, 2. Lieferung. Möggingen: Vogelzug-Verlag.

Zink, G. 1981. Der Zug europäischer Singvögel, 3. Lieferung. Möggingen: Vogelzug-Verlag.

Zöckler, C. 2002. Declining Ruff *Philomachus pugnax* populations: a response to global warming? Wader Study Group Bull. 97: 19-29.

Zubakin, V.A. 2001. [Current distribution and numbers of Black-tailed Godwit *Limosa limosa* in Moscow region.] Ornitologia 29: 229-232.

Zwarts, L. 1972. Bird counts in Merja-Zerga, Morocco. Ardea 60: 120-124.

Zwarts, L. 1988. Numbers and distribution of coastal waders in Guinea-Bissau. Ardea 76: 42-55.

Zwarts, L. 1990. Increased prey availabilty drives premigratory hyperphagia in Whimbrels and allows them to leave the Banc d'Arguin, Mauritania, in time. Ardea 78: 279-300.

Zwarts, L., B.J. Ens, M. Kersten & T. Piersma. 1990. Moult, mass and flight range of waders ready to take off for long-distance migrations. Ardea 78: 339-364.

Zwarts, L. 1993. Het voedsel van de Grutto. Graspieper 13: 53-57.

Zwarts, L., J. van der Kamp, O. Overdijk, T.M. van Spanje, R. Veldkamp, R. West & M. Wright. 1998. Wader counts on the Banc d'Arguin, Mauritania, in January/February 1997. Wader Study Group Bull. 86: 53-69.

Zwarts, L. & M. Diallo. 2002. Éco-hydrologie du Delta. In: E. Wymenga, B. Kone, J. van der Kamp & L. Zwarts (eds). Delta intérieur du fleuve Niger: écologie et gestion durable des ressources naturelles: 45-63. Wageningen: Mali-PIN.

Zwarts, L. & M. Diallo. 2005. Fisheries in the Inner Niger Delta. In: L. Zwarts, P. van Beukering, B. Kone & E. Wymenga (eds). The Niger, a lifeline: 89-107. Lelystad: Rijkswaterstaat/IVM/Wetlands International/A&W.

Zwarts, L. & I. Grigoras. 2005. Flooding of the Inner Niger Delta. In: L. Zwarts, P. van Beukering, B. Kone & E. Wymenga (eds). The Niger, a lifeline: 43-77. Lelystad: Rijkswaterstaat/IVM/Wetlands International/A&W.

Zwarts, L. & B. Kone. 2005a. People in the Inner Niger Delta. In: L. Zwarts, P. van Beukering, B. Kone & E. Wymenga (eds). The Niger, a lifeline: 79-86. Lelystad: Rijkswaterstaat/IVM/Wetlands International/A&W. 79-86.

Zwarts, L. & B. Kone. 2005b. Rice production in the Inner Niger Delta.

In: L. Zwarts, P. van Beukering, B. Kone & E. Wymenga (eds). The Niger, a lifeline: 137-153. Lelystad: Rijkswaterstaat/IVM/Wetlands International/A&W. & & &

Zwarts, L., N. Cissé & M. Diallo. 2005a. Hydrology of the Upper Niger. *In:* L. Zwarts, P. van Beukering, B. Kone & E. Wymenga (eds). The Niger, a lifeline: 15-40. Lelystad: Rijkswaterstaat/IVM/Wetlands International/A&W.

Zwarts, L., I. Grigoras & J. Hanganu. 2005b. Vegetation of the lower inundation zone of the Inner Niger Delta. *In:* L. Zwarts, P. van Beukering, B. Kone & E. Wymenga (eds). The Niger, a lifeline: 109-119. Lelystad: Rijkswaterstaat/IVM/Wetlands International/A&W.

Zwarts, L, P. van Beukering, B. Kone & E. Wymenga 2005c. The Niger, a lifeline: Lelystad: Rijkswaterstaat/IVM/Wetlands International/A&W.

Endnote

1 The list of references is arranged alphabetically, but note that *da Prato* is listed under the *P*, *de Bont* under the *B*, etc. The same rule applies to other names as: van den Bergh, ould Boubouth, van den Brink, van der Burg, van Dijk, van der Have, den Held, le Houérou, van Huis, de Jonge, van der Kamp, ten Kate, van der Kooij, van Langevelde, van Manen, de Naurois, van Noordwijk, van Rhijn, de Ridder, van der Valk, van der Winden, and de Wit.

Index

A

Acacia albida, Inner Niger Delta, 76
   phenology, 428, 481
   roost, 428
Acacia kirkii, bird density, 56
Acacia nilotica, Inner Niger Delta, 76
   phenology, 481
   Senegal Delta, 115, 118, 124
   tree density, 461
Acacia raddiana, Senegal Delta, 118
Acacia senegal, 461
Acacia seyal, bird density, 56-57
   Hadejia-Nguru, 140
   Inner Niger Delta, 66, 76
   phenology, 481
   Sudd, 164
   tree density, 461
   wood production, 76
Acanthacris ruficornis citrina, 219
Accipiter brevipes, see Sparrowhawk, Levant
Acorypha clara, 213, 216, 219
Acrididae, see Grasshopper, see Locust
Acrocephalus arundinaceus, see Warbler, Great Reed
Acrocephalus baeticatus, see Warbler, African Reed
Acrocephalus paludicola, see Warbler, Aquatic
Acrocephalus palustris, see Warbler, Marsh
Acroephalus schoenobaenus, see Warbler, Sedge
Acrocephalus scirpaceus, see Warbler, European Reed
Actitis hypoleucos, see Sandpiper, Common
Actophilornis africana, see Jacana, African
Addax nasomaculatus, see Antelope, Addax
Aeschynomene elaphroxylon, see Ambatch
Aeschynomene nilotica, see Poro
Africa, density livestock, 50-51
   land cover, 51
   population density, 57
   protected areas, 57
Alopochen aegyptiacum, see Goose, Egyptian
Alysicarpus, as food, 385
Amaurornis flavirostris, see Crake, Black
Ambatch, in Kafue, 184
   Lake Chad, colonisation of, 149
   Lake Fitri, 150
Anacardium occidentale, 459

Anacridium melanorhodon, see Locust, Tree
Anas acuta, see Pintail, Northern
Anas clypeata, see Shoveler, Northern
Anas crecca, see Teal, Eurasian
Anas platyrhynchos, see Mallard
Anas querquedula, see Garganey
Anastomus lamelligerus, see Stork, African Openbill
Anhinga rufa, see Darter, African
Antelope, Addax, 59
Antelope bubalis, see Oxen, Wild
Anthropoides virgo, see Crane Demoiselle
Anthus campestris, see Pipit, Tawny
Anthus cervinus, see Pipit, Red-throated
Anthus cinnamomeus, see Pipit, Grassland
Anthus leucophrys, see Pipit, Plain-backed
Anthus trivialis, see Pipit, Tree
Apus apus, see Swift, Common
Apus pallida, see Swift, Pallid
Aquila pomarina, see Eagle, Lesser Spotted
Aquila nipalensis, see Eagle, Steppe
Aquila wahlbergi, see Eagle, Wahlberg's
Ardea cinerea, see Heron, Grey
Ardea goliath, see Heron, Goliath
Ardea melanocephala, see Heron, Black-headed
Ardea purpurea, see Heron, Purple
Ardeola ralloides, see Heron, Squacco
Ardeotis arabs, see Bustard, Arabian
Asio capensis, see Owl, Marsh
Avocet, Pied, numbers, Africa, 186
   numbers, Asia, 186
      Debo/Walado, 85
      Europe, 186
      Hadejia-Nguru, 186-187
      Inner Niger Delta, 186-187
      Senegal Delta, 132, 186-187
Aythya nyroca, see Duck, Ferruginous
Azadirachta indica, see Neem

B

Balearica pavonina, see Crane, Black Crowned
Balaeniceps rex, see Shoebill
Balanites aegyptiaca, density of, 434
   invertebrates in, 459
   phenology, 481
   Sudd, 164
Bani River, discharge 31
Bee-eater, Blue-cheeked,
   habitat in Africa, 192, 486
   Sahel dependence, 192

   trend, 486
Bee-eater, European, habitat, 486
   trend, 486
Bee-eater, Little, 133
Bird, Secretary, 168
bird, killing, 499-450
   catching versus body mass, 198
   in colonies, 156
   drought-related, 286-290, 357-359
   methods in Africa, 197
   shooting versus body mass, 198
Bittern, Great, crayfish, 236
   density per habitat, 96
   estimate Inner Niger Delta, 100
   habitat in Africa, 192, 485
   killing, 499
   Sahel dependence, 192
   Senegal Delta, 132
   trend, 485
Bittern, Little, density per habitat, 96
   habitat in Africa, 192, 485
   Sahel dependence, 192
   trend, 485
Blackcap, 201
   carry-over effects, 477
   habitat, 487
   migratory connectivity, 193-194
   mortality, rainfall, impact of, 470
      seasonal & geographical variation, 465-469
   trend, 487
Bluethroat, density per habitat, 77, 96
   estimate Inner Niger Delta, 100
   habitat in Africa, 192
   Sahel dependence, 192
   Senegal Delta, 134
   trend, 486
Bolanha, see Rice
Bonga, 295, 296
Boscia, 481
Bostrychia hagedash, see Ibis, Hadada
Botaurus stellaris, see Bittern, Great
Bourgou, bird densities in, 182-183
   Inner Niger Delta, 53, 66-68
   Kafue Flats, 184
   Lake Fitri, 150
   Senegal Delta, 116
   Sudd, 164
Bourgoutière, bird densities in, 96
   buoancy and heron weight, 98-99

definition, 97
Brown, Leslie, and African Fish Eagle, 293
   number of African birds, 189
Bubulcus ibis, see Heron, Cattle
Bulbul, Common, in Neem, 57
Bunting, Cretzschmar's, 201, 487
Bunting, Ortolan, decline, 487
   habitat, Africa, 192, 487
      changes, 492
      return rate, 492
   Sahel impact, 192, 487
   trend, 480, 487
Bunting, Reed, 477
Burhinus oedicnemus, see Stone-curlew
Burhinus senegalensis, see Thick-knee, Senegal
Bustard, Arabian, in Inner Niger Delta, 77
   in Sudd, 165
Bustard, Black-bellied, in Sudd, 165
Bustard, Nubian, in Sahel, 77
Buteo buteo vulpinus, see Buzzard, Steppe
Buteo platypterus, see Hawk, Broad-winged
Buteo swainsoni, see Hawk, Swainson's
Butorides striatus, see Heron, Green-backed
Buzzard, Steppe, 201

C

Caelatura aegyptiaca, availability 91
   biomass Debo/Walado, 92-93
Calandrella brachydactyla, see Lark, Greater
   Short-toed
Calidris alba, see Sanderling
Calidris canutus, see Knot, Red
Calidris ferruginea, see Sandpiper, Curlew
Calidris minuta, see Stint, Little
Calidris temminckii, see Stint, Temminck's
Calliptamus italicus, see Locust, Italian
Canis aureus, see Jackal, Golden
Canthium, 460, 468
Capparis aphylla, 384
Caprimulgus europaeus, see Nightjar, European
Caprimulgus ruficollis, see Nightjar, Red-necked
Carry-over effects, 472-479
   climate change, impact of, 476
   phenology, 476
   reproduction, 477
   weather, impact of, 478
Cassia siebeniana, 461
Cataloipus cymbiferus, 219
Catching, see Bird, Killing
Catfish, countershading, reverse, 292

Moustache, 292
Cattail, bird densities, 133, 135
   Hadejia Nguru, 421
   Kafue, 184
   Lake Chad, 149
   Lake Guiers, 117
   Okavango, 184
   in rice fields, 174
   as roost, 411
   Senegal Delta, 120
   species in West Africa, 137
   Sudd, 169
Census methods, 76-101
   aerial, 82-86, 165, 282
   breeding birds, colonial, 78-81
   colonies, 76-77
   density counts, 94-101
   effort, Senegal Delta, 122-123
   habitat scores, 319
   monitoring schemes, 482-483
   pitfalls, 482-483
   plots, distribution Inner Niger Delta, 95
   prey abundance, 322
   reliability, 100-101
   rice fields, 175
   road transects, 320
   roosts, 81-82, 320-322
   timing, 75
Ceratophyllum demersum, 116
Cercopithecus patas, 121
Cercothrichas galactotes, see Robin, Rufous
   Scrub
Chad Basin, see Lake Chad Basin
Chari/Logone, discharge, 33
Charadrius alexandrinus, see Plover, Kentish
Charadrius dubius, see Plover, Little Ringed
Charadrius hiaticula, see Plover, Common
   Ringed
Charadrius marginatus, see Plover, White-
   fronted
Charadrius pecuarius, see Plover, Kittlitz's
Cheilopogon, Osprey, food, 297
Chelictinia riocourii, see Kite, Swallow-tailed
Chiffchaff, Common, in *Acacia kirkii*, 432
   carry-over effects, 477
   Cattail, 133
   density in *Acacia*, 56-57
   habitat in Africa, 192, 487
   migration, *abietienus*, 494

   collybita, 494
   migratory connectivity, 194
   migratory divide, 494
   mortality, rainfall, impact of, 470
      seasonal & geographical variation, 467, 469
   phenology, 476
   Sahel dependence, 192
   trend, 487
      by subspecies, 494-495
   wintering grounds, 438, 494-495
Chiffchaff, Iberian, migration, 494-495
Chlidonias hybridus, see Tern, Whiskered
Chlidonias leucopterus, see Tern, White-winged
Chlidonias niger, see Tern, Black
Ciconia abdimii, see Stork, Abdim's
Ciconia ciconia, see Stork, White
Ciconia episcopus, see Stork, Woolly-necked
Ciconia nigra, see Stork, Black
Circaetus gallicus, see Eagle, Short-toed
Circus aeruginosus, see Harier, Eurasian Marsh
Circus macrourus, see Harrier, Pallid
Circus pygargus, see Harrier, Montagu's
Cisticola galactotes, see Cisticola, Winding
Cisticola juncidis, see Cisticola, Zitting
Cisticola, Winding, in Cattail, 133
Cisticola, Zitting
   density, rice fields, 176
Clamator glandarius, see Cuckoo, Great Spotted
Cleopatra bulinoides, availability, 91-93
   biomass Debo/Walado, 92
Cocculus pendulus, 385
Coracias abyssinicus, see Roller, Abyssinian
Coracias garrulus, see Roller, European
Corbicula fluminalis, availability, 91-93
   biomass Debo/Walado, 91-93
   depletion, 92-93
Cormorant, Great, breeding, Diawling NP, 124
   breeding, Senegal Delta, 126
Cormorant, Long-tailed,
   breeding, Banc d'Arguin, Mauritania, 137
   Diawling NP, 124
   Hadejia-Nguru, 143
   Inner Niger Delta, 78
   Senegal Delta, 126
   Sudd, 165
   density per habitat, 96
   mass mortality, 77, 92
   numbers, coastal rice, 186-187
      Debo/Walado, 85, 88-89

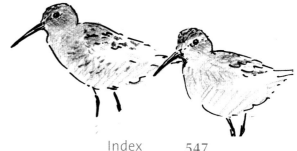

    Hadejia-Nguru, 186-187
    Inner Niger Delta, 186-187
    Lake Fitri, 154, 186-187
    Logone, 154, 186-187
    Senegal Delta, 186-187
    Sudd, 186-187
  roosting, 81-82
  trend, wintering, 88
Corn-crake, 201, 485
Coturnix coturnix, see Quail, Common
Counting, see Census, methods
Courser, Cream-coloured, 215
Crab, Fiddler, 364
  as food, 365
Crake, Baillon's, 201
  Senegal Delta, 133
Crake, Black, in Cattail, 133
Crake, Little, 201
  Senegal Delta, 133
Crake, Spotted, 201
Crane, Black Crowned, breeding, Mali, 80-81
  captivity, Mali, 81
  Debo/Walado, 88-89
  numbers, Inner Niger Delta, 186-187
    Lake Chad Basin, 152-153, 156
    Lake Fitri, 186-187
    Logone, 186-187
    Senegal Delta, 128-130, 186-187
    Sudd, 165-166, 169, 186-187
  past abundance, Inner Niger Delta, 77
    Hadejia Nguru, 143
  trend, wintering Debo/Walado, 88
Crane, Common, 201
Crane, Demoiselle, 201, 485
  trend, 485
Crawfish, Red Swamp, 236, 255
Crex crex, see Corn-crake
Crocodylus niloticus, 121
Cuckoo, Common, 201
  habitat, 486
  trend, 486
Cuckoo, Great Spotted, habitat in Afica, 192, 486
  Sahel dependence, 192
  trend, 486
Cuculus canorus, see Cuckoo, Common
Curlew, Eurasian, bill length, 191
  numbers Debo/Walado, 85, 88-89
  origin, 195
  trend, wintering, 88

Cursorius cursor, see Courser, Cream-coloured
Cyperus, in Senegal Delta, 115-116
Cyperus articulatus, in Inner Niger Delta, 66
Cyperus papyrus, see Papyrus
Cyrtobagous salviniae, 120

D

Dactyloctenium, as food, 385
Dam, impact of, 65
  Challawa Gorge, 139
  Djenne, 65
  Fomi, 65
  Maga, 150
  Markala, 65
  Selingue, 65
  Talo, 65
  Tiga, 140
Damaliscus korrigum, see Topi
Damaliscus korrigum korrigum, see Tiang
Darter, African, breeding, Diawling NP, 124
  breeding, Inner delta, 78
    Senegal Delta, 125-126
  human exploitation, 77
  maximum Debo/Walado, 85, 88-89
  trend, wintering, 88
Deforestation, 496-497
  Burkina Faso, 54
  Ghana, 54
  Senegal, 52-53
Delichon urbica, see Martin, Common House
Dendrocygna bicolor, see Duck, Fulvous Whistling
Dendrocygna viduata, see Duck, White-faced Whistling
Desertification, dust, 41
  trigger of, 16-18, 40
Diabolocatantops axillaris, 220
Didere, bird densities, 182-183
  in Inner Niger Delta, 66-68
  in Lake Chad, 149
  in Lake Fitri, 150
  in Senegal Delta, 114, 116
  in Sudd, 163
Dociostaurus maroccanus, see Locust, Moroccan
Dove, African Collared, food, 385
Dove, African Mourning, food, 385
Dove, European Turtle, 379-389
  breeding, 379
    arrival date, 388
    conditions, 387-388

    departure date, 388
    ecology, 387
  daily rhythm, 384, 389
  drinking, 383, 384
  food & feeding, on breeding grounds, 387
    competition, interspecific, 385
    digestion, 384
    fat load, 379
    retention time, 384
    in Sahel, 383-385
  habitat in Africa, 192, 381-383
    in Inner Niger Delta, 77
  migration, 379-381
    age-specific, 380-381
    migratory connectivity, 194
    nocturnal, 381
  mortality, dehydration, 380
    regional, 469
    predation, 380
    shooting, 388
  Sahara, crossing of, 380-381, 389
  Sahel dependence, 192, 485-487
    food abundance, 385
    impact drought, 382, 384-385
  roosting, 378, 382-384
  survival, 389
  trends, in Europe, 385-388, 486
    in Inner Niger Delta, 383
    in Senegal Delta, 382
Dove, Laughing, food, 385
Dove, Namaqua, food, 385
Dove, Vinaceous, food, 385
Duck, Ferruginous, habitat in Africa, 192, 485
  numbers, Hadejia-Nguru, 141
    Inner Niger Delta, 82-83
    Lake Fitri, 155, 186-187
    Senegal Delta, 186-187
  Sahel dependence, 192
  trend, 485
Duck, Fulvous Whistling,
  numbers, Debo/Walado, 85, 88-89
    Hadejia-Nguru, 141-142, 186-187
    Inner Niger Delta, 82-83, 186-187
    Lake Fitri, 154, 186-187
    Logone, 154, 186-187
    rice fields, abandoned, 497
    Senegal Delta, 128-129, 132, 186-187
    Sudd, 165, 186-187
  trend, wintering Debo/Walado, 88

Duck, Knob-billed,
    numbers, Debo/Walado, 85, 88-89
        Hadejia-Nguru, 141-142, 186-187
        Inner Niger Delta, 82-83, 186-187
        Lake Fitri, 155, 186-187
        Logone, 155, 186-187
        Senegal Delta, 128-129, 186-187
        Sudd, 165, 186-187
    trend, wintering Debo/Walado, 88
Duck, White-faced Whistling,
    numbers, coastal rice, 186-187
        Debo/Walado, 85, 88-89
        Hadejia-Nguru, 141-142, 186-187
        Inner Niger Delta, 82-83, 186-187
        Lake Fitri, 154, 186-187
        Logone, 154, 186-187
        Senegal delta, 128-129, 132, 186-187
        Sudd, 165, 186-187
    trend, wintering, 88

E

Eagle, African Fish, diel pattern, 293
    in Inner Niger Delta, 77
    numbers Debo/Walado, 85
Eagle, Booted, habitat in Africa, 192, 485
    Sahel dependence, 192
    trend, 485
Eagle, Lesser Spotted, 201
    Habitat, 485
    phenology, dealyed, 474
    satellite-tracking, 325
    trend, 485
Eagle, Short-toed, habitat in Africa, 485
    Sahel dependence, 192
    satellite-tracking, 325
    trend, 485
Eagle, Steppe, 201
    habitat, 485
    trend, 485
Eagle, Wahlberg's, satellite-tracking, 325
Echinochloa haploclada, 164
Echinochloa pyramidalis, 164
Echinochloa stagnina, see *Bourgou*
Egret, Cattle, breeding, Diawling NP, 124
    breeding, Hadejia-Nguru, 143
        Inner Niger Delta, 78
        Senegal Delta, 126-127
    density per habitat, 96, 130
        in rice fields, 176
        seasonal, 176

estimate Inner Niger Delta, 100
increase, 80
numbers, coastal rice, 186-187
    Debo/Walado, 83, 88-89
    Hadejia-Nguru, 186-187
    Inner Niger Delta, 186-187
    Lake Fitri, 154, 186-187
    Lake Télé, 95
    Logone, 154, 186-187
    Office du Niger, 186-187
    Senegal Delta, 131
    Sudd, 165, 186-187
predation on grasshoppers, 216
roosting, 84, 428
trend Debo/Walado, 88
Egret, Great, in *bourgou*, 96-97
    breeding Diawling NP, 124
        Hadejia-Nguru, 143
        Inner Niger Delta, 78
        Senegal Delta, 126-127
    density per habitat, 96, 130
        in rice fields, 176
        seasonal, 176
    estimate Inner Niger Delta, 100
    numbers, coastal rice, 186-187
        Debo/Walado, 85, 88-89
        Hadejia-Nguru, 186-187
        Inner Niger Delta, 186-187
        Lake Fitri, 154, 186-187
        Lake Télé, 95
        Logone, 154, 186-187
        Sudd, 165, 186-187
    trend Debo/Walado, 88
Egret, Intermediate, in *bourgou*, 96-97
    breeding, Hadejia-Nguru, 143
        Inner Niger Delta, 78
        Senegal Delta, 126
    density per habitat, 96
        in rice fields, 176
    estimate Inner Niger Delta, 100
    numbers, coastal rice, 186-187
        Debo/Walado, 85, 88-89
        Hadejia-Nguru, 186-187
        Inner Niger Delta, 186-187
        Lake Télé, 95
        Logone, 154, 186-187
        Office du Niger, 186-187
    trend Debo/Walado, 88
Egret, Little, 246-251

bourgou, 96-97
breeding, Africa, 247
    Diawling NP, 124
    Europe, 247
    Hadejia-Nguru, 143
    Inner Niger Delta, 78, 80
    Senegal Delta, 126-127, 131
dark morph, 78
density per habitat, 96, 130
    in rice fields, 176
estimate Inner Niger Delta, 100
feeding, communal, 246
habitat in Africa, 192
migration, 247-248
    dispersal, 250
    residency, 250
migratory connectivity, 194
numbers, coastal rice, 186-187, 250
    Debo/Walado, 85, 88-89
    Hadejia-Nguru, 186-187
    Inner Niger Delta, 186-187
    Lake Fitri, 154, 186-187
    Liberia, 250
    Logone, 154, 186-187
    Office du Niger, 186-187
population change, 250-251
    European conditions, 250
    Sahel conditions, 192, 250
trend, Camargue, 248
    Debo/Walado, 88
    Italy, 248
    Spain, 248
wintering, Africa, 248-250
    Europe, 247, 250
Egret, Western Reef
    maximum Debo/Walado, 85
    morph, white, 248
Egretta alba, see Egret, Great
Egretta ardesiaca, see Heron, Black
Egretta garzetta, se Egret, Little
Egretta gularis, see Egret, Western Reef
Egretta intermedia, see Egret, Intermediate
Eichhornia crassipes, see Hyacinth, Water
Elephant, African, 59, 150, 164
Eleocharis mutata, 116
Emberiza caesia, see Bunting, Cretzschmar's
Emberiza hortulana, see Bunting, Ortolan
Emberiza schoeniclus, see Bunting, Reed
Ephippiorhynchus senegalensis, see Stork,

Saddle-billed
Ethmalosa fimbricata, see Bonga
Eucalyptus, bird density, 56-57
　surface area Senegal, 54
Eupodotis melanogaster, see Bustard, Black-
　bellied

F

Falco amurensis, see Falcon, Amur
Falco biarmicus, see Falcon, Lanner
Falco cherrug, see Falcon, Saker
Falco eleonorae, see Falcon, Eleonora's
Falco peregrinus, see Falcon, Peregrine
Falco subbuteo, see Hobby
Falco naumanni, see Kestrel, Lesser
Falco tinnunculus, see Kestrel, Common
Falco vespertinus, see Falcon, Red-footed
Falcon, Eleonora's, satellite-tracking, 325
Falcon, Lanner, 215, 240
Falcon, Peregrine, satellite-tracking, 325
Falcon, Red-footed, 201, 485
Falcon, Saker, habitat in Africa, 192, 485
　Sahel dependence, 192
　trend, 485
Fat, daily mass gain, 284-285
　fattening, drought, 93, 284-290
Ficedula albicollis, see Flycatcher, Collared
Ficedula hypoleuca, see Flycatcher, European
　Pied
Ficus verruculosa, 184
Fimbristylis, as food, 385
Flamingo, Lesser, Senegal Delta, 127-128
Floodplain, importance for birds, 180-187
　Hadeija-Nguru, 138-157
　　bird density, 182
　　change in size, 182
　　permanent water, 182
　　rainfall, impact on extent, 195
　　vegetation cover, 182
　Inner Niger Delta, 60-103
　　bird density, 182
　　change in size, 182
　　permanent water, 182
　　rainfall, impact on extent, 195
　　vegetation cover, 182
　Logone
　　bird density, 182
　　change in size, 182
　　permanent water, 182
　　vegetation cover, 182

　Senegal Delta, 102-135
　　bird density, 182
　　change in size, 182
　　permanent water, 182
　　rainfall, impact on extent, 195
　　vegetation cover, 182
　Sudd, 156-167
　　bird density, 182
　　change in size, 182
　　permanent water, 182
　　vegetation cover, 182
Flycatcher, Collared, 201, 487
Flycatcher, European Pied, 201, 411
　habitat, 487
　migratory connectivity, 194
　mortality, rainfall, impact of, 470
　　seasonal & geographical variation, 467, 469
　phenology, 476
　trend, 487
　wintering, strategy, 494
Flycatcher, Spotted, 201
　habitat, 487
　migratory connectivity, 194
　mortality, 469
　phenology, 476
　trend, 487
Food, berries, 468
　depletion, 468
　and grazing, 468
　and rainfall, 468

G

Galerida cristata, see Lark, Crested
Gallinago gallinago, see Snipe, Common
Gallinago media, see Snipe, Great
Gallinula angulata, see Moorhen, Lesser
Gallinula chloropus, see Moorhen, Common
Gallinule, Allen's
Garganey, 278-291
　breeding, 279-280
　fattening, 284-285, 286
　food, Inner Niger Delta, 281, 284-285
　　Senegal, 281
　habitat in Africa, 192, 485-487
　management of breeding grounds, 490
　matrimonial market, 285
　numbers, Debo/Walado, 85, 88-89
　　flood level, 282-283
　　Hadejia-Nguru, 141, 186-187
　　Lake Chad, 150, 155

　　Inner Niger Delta, 82-83, 186-187
　　Lake Fitri, 155, 186-187
　　Logone, 155, 186-187
　　Senegal Delta, 128-129, 132, 186-187
　　Uganda, 281
　migration, 279-281
　　daily, 290
　　timing, 284-285
　migratory connectivity, 194
　mortality, catching, 287-289
　　and flood extent, 286-287, 289-290
　　and lunar cycle, 288
　　numbers killed, 285, 289
　　shooting, 285, 289
　moult, 281
　population change,
　　breeding grounds, 283-285, 290
　　flood level, 286, 290
　　management, breeding grounds, 290, 490
　　wintering grounds, 285-290
　Sahel dependence, 192
　trend, 88, 485
　Veljerne, 500
　wintering, Africa, East, 281, 285
　　Africa, West, 281-283, 285
　　Europe, 281
　　flood level, 282-283
　　Sahel, 281-283
Gazella dama, see Gazelle, Dama
Gazella dorcas, see Gazelle, Dorcas
Gazella rufifrons, see Gazelle, Red-fronted
Gazella leptocero, see Gazelle, Slender-horned
Gazella thomsonii albonotata, see Gazelle,
　Mongalla
Gazelle, Dama, 58
　Dorcas, 58, 121
　Mongalla, 59, 164
　Red-fronted, 58, 121
　Slender-horned, 58
Gelochelidon nilotica, see Tern, Gull-billed
Giraffe, 59
Gisekia pharnacoides, as food, 385
Glareola nordmanni, see Pratincole, Black-
　winged
Glareola pratincola, see Pratincole, Collared
Godwit, Bar-tailed, pre-migratory fattening, 93
　Senegal Delta, 132
Godwit, Black-tailed, 328-343
　breeding, 329-330

        failure, 335, 342-371
            timing, forward shift, 334-335
            trends, Europe, 340-343
Corbicula as food, 91-93
density per habitat, 96
fattening, Inner Niger Delta, 93
    post-breeding, 335
food and feeding, 330
    assimilation rate, 93
    digestive bottleneck, 335
    Inner Niger Delta, 91-93
    intake rate, 93, 335
    leatherjackets, 335
    processing rate, 335
    rice, 332
    sorghum, 343
    time budget, 335
habitat, in Africa, 192, 330-334
    changes, in Europe, 342-343
migration, 334-338
    arrival, Africa, 336
    first-years, 338
    islandica, 329, 343
    migratory connectivity, 194
    phenology, 474
    stopovers, 334-338, 474
    summering, 338
    timing, 337-338, 343
mortality,
    causes of, 339
numbers, Africa, West, 330-334
    Asia, 186
    Casamance, 334
    coastal rice, 186-187
    Debo/Walado, 85, 88-89
    Europe, 184, 329-330
    Gambia, 334
    Guinea, 334
    Guinea-Bissau, 332, 334
    Hadejia-Nguru, 141, 186-187
    Inner Niger Delta, 82-83, 186-187
    Lake Fitri, 155, 186-187
    Logone, 155, 186-187
    Senegal Delta, 128-129, 132, 186-187, 332
    Sine Saloum, 334
predation pressure on *Corbicula*, 92
ring density, by region, 338
rice fields, 177
roost, post-breeding, 334-336

    timing, forward shift, 335-336
    Sahel dependence, 192, 341-342
    survival, 339-343
        age-specific, 339
        longevity, 340
        rate, temporal changes, 338
        Sahel rainfall, 339
        shooting, impact of, 340-341
    trend, Europe, 339-343, 485
        Inner Niger Delta, 88
        Senegal Delta, 332
        West Africa, 334
    wintering, 334-336
        numbers, 343
Goose, African Pygmy,
    numbers Debo/Walado, 85, 88-89
    numbers, Hadejia-Nguru, 141
    Inner Niger Delta, 82-83
    Lake Fitri, 155
    Logone, 155
    Senegal Delta, 128-129
Goose, Egyptian, Debo/Walado, 85, 88-89
    numbers, Hadejia-Nguru, 141, 186-187
    Inner Niger Delta, 82-83, 186-187
    Lake Fitri, 155, 186-187
    Logone, 155, 186-187
    Senegal Delta, 128-129, 186-187
    trend, wintering Debo/Walado, 88
Goose, Spur-winged,
    maximum Debo/Walado, 85, 88-89
    numbers, Hadejia-Nguru, 141, 186-187
    Inner Niger Delta, 82-83, 186-187
    Lake Fitri, 155, 186-187
    Logone, 155, 186-187
    Senegal Delta, 128-129, 186-187
    Sudd, 165, 186-187
    trend, wintering, 88
Grande Sécheresse, see Great Drought
Grass, Elephant, 399
Grasshopper, annual cycle, 208
    biomass, 214
    control, dieldrin, 220
        entomopathogen, 221-222
        fipronil, 220-221
        organophosphates, 221
    depredation, size-related, 220
    increase, causes of, 219
    infestations, 210, 213
    Intertropical Convergence Zone,

        impact of, 211-212
    seasonal cycle, 208
        continuous reproduction, 208
        diapausing adults, 208
        diapausing nymphs, 208
        habitat-related, 209
Grasshopper, Senegalese, area infested, 210
    densities, 219
    importance of, 214
    monthly infestation, 210
    outbreaks in 1974-1989, Sahel, 214
    outbreaks, rainfall, 222
Great Drought, 15, 22-23
Greenshank, Common,
    density Senegal Delta, 130
    in rice fields, 176
    habitat in Africa, 192, 486
    migratory connectivity, 194
    numbers, Asia, 186
        coastal rice, 186-187
        Debo/Walado, 85, 88-89
        Europe, 186
        Hadejia-Nguru, 186-187
        Inner Niger Delta, 186-187
        Logone, 155
    Sahel dependence, 192
    trend, 88, 486
Grewia, 481
Grus grus, see Crane, Common
Guiera senegalensis, 459
Guinea, vegetation zone, 37
Gull, Black-headed, maximum Debo/Walado, 85
Gull, Grey-headed Debo/Walado, 85, 88-89
    Logone, 155
    trend, wintering, 88
Gull, Lesser Black-backed, 85, 465

H

Hadejia-Nguru, 138-143
    comparison, with Inner Niger Delta, 142
    fish, 140
    floodplain size, 139
    hydrology, 139-140
    land use, 140
    people, density, 140
    waterbirds, 141-142
Haliaeetus vocifer, see Eagle, African Fish
Hamerkop, in rice fields, 176
Harmattan, 15, 41, 404, 465, 469, 470
Harpezocatantops stylifer, 213

Harrier, Eurasian Marsh, 304-311
  breeding, in Africa, 305
    Europe, numbers, 305
    expansion, 309
    numbers, and flood extent, 310
    range, 305
    reclamations, and numbers, 309
    trends, 308-310
  habitat, in Africa, 192, 307, 485
    Circus-complex, interspecific, 307
    sex- and age-specific, 94, 309
  migration, 305-307
    migratory connectivity, 194, 305-306
    migratory tendency, 305
    satellite-tracking, 325
    site fidelity, 307-308
    stopovers, 307
  mortality, and agro-chemicals, 309
    flood extent, 308
    culling, 309
    persecution, 308
  numbers, in Africa, 307-308
    Debo/Walado, 85, 88-89
    Inner Niger Delta, 307
    Lake Fitri, 155
    Logone, 155
    Senegal Delta, 133, 307
  Sahel dependence, 192
    trend, 88, 92
  trend, 306-308, 485
  wintering, in Africa, 307-308, 485
    in Europe, 305
    and flood extent, 308
    by latitude, age, 305
    return, 316
Harrier, Montagu's, 312-327
  breeding, 313, 324-325
    farmland, 325
    food, impact of, 325
    nesting habitat, 324
    numbers, 325
    range, 313
    protection, 325
    reproduction, 324
    trends, and Sahel rainfall, 324
  deforestation, impact of, 497
  density, Africa, 326
  food, 216, 220
    composition, 321-322
    grasshoppers, 321-322
    intake, Senegal, 218
    locusts, 321-322
    pellets, 321-322
  loafing, 218
  habitat, Africa, 192, 320-321, 485
    and food availability, 320
    degradation, 320
    Europe, 324-325
    and tree cover, 321
  maximum Debo/Walado, 85
  migration, 313-318
    distance covered, daily, 326
    flight paths, satellite-tracked, 316
    loop, 316
    pre-migratory, 314-315
    Mediterranean, crossing of, 315
    migratory connectivity, 315, 316
    return, 318
    satellite-tracking, 325
    speed, 315
    wing loading, 315
  numerical response, 220
  predation on Acrididae, 217
  roost, 321-322
    daytime, 321
    night, 322
    size, 322
    timing, 322
  Sahel dependence, 192, 313-321
    and breeding numbers, 324-325
  Senegal Delta, 133-134
  trend, 485
  water holes, 218
  wintering, 485
    Africa, West 313
    conditions, 320
    and food supply, 320
    India, 313
    Intertropical Convergence Zone, 317-318
    Kenya, 313
    numbers, trend, 324, 485
    Sahel, 324
    site fidelity, 318
Harrier, Pallid, Debo/Walado, 85
  Ethiopia, 327
  habitat in Africa, 192, 307, 485
  Sahel dependence, 192
  trend, 485

Hartert, Ernst, 188
Hawk, Broad-winged, satellite-tracking, 325
Hawk, Swainson's, satellite-tracking, 325
Hemisynodontis membranacea, see Catfish, Moustache
Heranguina, 460, 468
Heron, Black, breeding, Inner Niger Delta, 78, 80
  breeding, Senegal Delta, 126
  numbers, Debo/Walado, 85, 88-89
    Inner Niger Delta, 186-187
    Lake Fitri, 154, 186-187
    Logone, 154, 186-187
  trend Debo/Walado, 88
Heron, Black-crowned Night, 234-239
  breeding, Africa 235
    Diawling NP, 124
    Europe, 235
    Inner Niger Delta, 78
  breeding, Senegal Delta, 126-127, 131
  habitat in Africa, 192, 485
  migration, 235
  migratory connectivity, 194
  mortality, 235
  numbers, Africa, 236
    Debo/Walado, 85
    Inner Niger Delta, 186-187, 236
    Logone, 154, 186-187
    Senegal Delta, 186-187, 236
    Sudd, 186-187
  population change, flood extent, 239
    hydrology, 238
    population size, impact of, 238
    Sahel dependence, 192, 236-239
  roosting, 238
  trend, 236-236, 485
Heron, Black-headed, colonies in Inner Niger Delta, 78
  density per habitat, 96
    in rice fields, 176
  Logone, protection colony Andirni, 156
  numbers, coastal rice, 186-187
    Hadejia-Nguru, 186-187
    Inner Niger Delta, 186-187
    Logone, 186-187
    Lake Fitri, 154
    Logone, 154
    Sudd, 165
Heron, Goliath, in Inner Niger Delta, 77

Sudd, 165
Heron, Green-backed,
  breeding, Senegal Delta, 126
  density, in rice fields, 176
    seasonal, 176
  numbers, coastal rice, 186-187
Heron, Grey, 224-227
  Africa, 226
    age-distribution, 226
    flood extent, impact of, 226
    proportion in West Africa, 226
  bourgou, 96-97
  breeding, Europe, 225
    Hadejia-Nguru, 143
    Inner Niger Delta, 78, 80
    Senegal Delta, 126, 131
    Sudd, 165
  density per habitat, 96
    in rice fields, 176
  estimate Inner Niger Delta, 100
  habitat in Africa, 192, 485
  migration, 225
  migratory connectivity, 194
  mortality, 227
  numbers, coastal rice, 186-187
    Chad Basin, 226
    Debo/Walado, 85, 88-89
    Guinea, 226
    Hadejia-Nguru, 186-187
    Inner Niger Delta, 186-187
    Lake Fitri, 154, 186-187
    Lake Télé, 95
    Logone, 154
    Office du Niger, 186-187
    Senegal Delta, 186-187
    Sudd, 186-187
  population change
    European conditions, 227
      legal protection, 227
      severe winters, 227
    Sahel conditions, 226-227
  reporting rate, Africa, 227
  Sahel dependence, 192
  trend, Debo/Walado, 88
    Europe, 224, 485
  wintering, 224, 485
Heron, Purple, 228-233
  bourgou, 96-97
  breeding, Europe, 229

Inner Niger Delta, 78, 80
Senegal Delta, 126, 131
Sudd, 165
density per habitat, 96, 130
  in rice fields, 182
  seasonal, 176
estimate Inner Niger Delta, 100
  Senegal Delta, 136
habitat in Africa, 192, 232, 485
migration, 229
  itinerancy, 231
  satellite telemetry, 230-231
migratory connectivity, 194
mortality, 229-230
  density-dependence, 233
numbers, coastal rice, 186-187
  Debo/Walado, 85, 88-89
  Hadejia-Nguru, 186-187
  Inner Niger Delta, 186-187, 230
  Lake Fitri, 154, 186-187
  Lake Télé, 95
  Liberia, 230
  Logone, 154, 186-187
  Sudd, 186-187
population change, 230-233
  conservation, 233
  European conditions, 233
  Sahel conditions, 230-233
roosting, 81
Sahel dependence, 192, 233
trend, Debo/Walado, 88
  Europe, 232, 485
Heron, Squacco, 240-244
  bourgou, 96-97
  breeding, Diawling NP, 124
    Europe, 241
    Inner Niger Delta, 78, 241
    Senegal Delta, 126-127
  density per habitat, 96, 130
    rice fields, 176
    seasonal, 176
  estimate Inner Niger Delta, 100
  flood extent, impact of, 243, 244
    on clutch size, 244
  habitat in Africa, 192, 485
  migration, 241
  numbers, Africa, West, 242
    coastal rice, 186-187, 242
    Debo/Walado, 85

Hadejia-Nguru, 186-187
Inner Niger Delta, 186-187
Lake Fitri, 154, 186-187
Lake Télé, 95
Logone, 154, 186-187
Sudd, 165, 186-187
population change, 243-244
  carry-over effects, 244
  European conditions, 244
  Sahel conditions, 243-244
Sahel dependence, 192
trend, 242-243, 485
Hieraaetus pennatus, see Eagle, Booted
Himantopus himantopus, see Stilt, Black-winged
Hippo, 101, 164
Hippolais icterina, see Warbler, Icterine
Hippolais olivetorum, see Warbler, Olive-tree
Hippolais pallida, see Warbler, Olivaceous
Hippolais polyglotta, see Warbler, Melodious
Hirundo daurica, see Swallow, Red-rumped
Hirundo rustica, see Swallow, Barn
Hobby, Eurasian, 201, 485
  satellite-tracking, 325
  trend, 485
Honey Buzzard, European, 201, 485
  habitat, 485
  satellite-tracking, 325
  trend, 485
Hoopoe, Eurasian, decline, 480, 486
  habitat in Africa, 192, 486
  Sahel dependence, 192
  trend, 486
Hyacinth, Water, 120, 174, 182
Hyaena hyaena, 497
Hyparrhenia rufa, 164
Ibis, Glossy, 266-271
  breeding, Europe, 267
    Inner Niger Delta, 78, 80
    Spain, increase, 267
  density per habitat, 96
  habitat in Africa, 192, 485
  hunting, 271
  food, 91-93
    Corbicula, 90
    intake, 89
  migration, 267-268
  migratory connectivity, 194
  numbers, Africa, West, 269
    Chad, Central, 269

Debo/Walado, 85, 88-89
Hadejia-Nguru, 141, 186-187
Inner Niger Delta, 82-83, 186-187
Lake Chad, 268
Lake Fitri, 154, 186-187
Lake Télé, 95
Logone, 154, 186-187
Mauritania, 268
Saudi Arabia, 268
Senegal Delta, 128-129, 131, 186-187, 268
Sudd, 165-166, 186-187
population change, decline, 269-270
  flood extent, impact of, 270
  survival rate, 270
Sahel dependence, 192, 268-269
  summering, 269
trend, 485
  adjustment, 503
  Debo/Walado, 88
  Inner Niger Delta, 269

I

Ibis, Hadada, in Inner Niger Delta, 77
  Logone, 154
  Sudd, 165
Ibis, Sacred, breeding, Diawling NP, 124
  breeding, Inner Niger Delta, 78
    Senegal Delta, 126-127
  numbers, coastal rice, 186-187
    Debo/Walado, 85, 88-89
    Hadejia-Nguru, 186-187
    Inner Niger Delta, 186-187
    Lake Fitri, 154, 186-187
    Logone, 154, 186-187
    Senegal Delta, 186-187
    Sudd, 165, 186-187
  trend, wintering, 88
Inner Niger Delta, 60-103
  bird density/ha, 100
  Debo/Walado, bird numbers, trends, 88, 92
  Debo/Walado, maxima waterbirds, 85
  Debo/Walado, water level, 84
  colonies, 77
  dams, impact of, 65
  farming, rice, 71
  fishing, 73-76
  flood level, Debo/Walado, 84
  flooding, annual variation, 61-66
  flooding, seasonal variation, 62-63
  floodplain, size, 61

Lake Debo, 62
Lake Horo, 61
Lake Korientzé, 61
Lake Télé, 61
  birds in, 95
Lake Walado, 62
Lake Télé, bird numbers, 95
livestock, 71-72
people, 69-76
threats, 100-103
vegetation, zones, 66-69
  bourgou, 66-68
  didere, 66-68
  forest, flood, 68-69
  Guinea rush, 67
  rice, cultivated, 68-69
  rice, wild, 66
  wood, need of, 77
Intertropical Convergence Zone, 15, 17, 25, 209-215, 317
Invasive, plants, Lake Chad, 149
  Senegal Delta, 118-120
Ipomoea aquatica, in Senegal Delta, 116
Isotope, stable, 397, 436, 494-495
ITCZ, see Intertropical Convergence Zone
Ixobrychus minutus, see Bittern, Little

J

Jacana, African, in Cattail, 133
  density per habitat, 96
    in rice fields, 176
    seasonal, 176
  numbers, Inner Niger Delta, 100
    Lake Télé, 95
    Logone, 155
Jacana, Lesser, density per habitat, 96
  estimate Inner Niger Delta, 100
  numbers Lake Télé, 95
Jackal, Golden, 497
Jynx torquilla, see Wryneck, Eurasian

K

Khaya senegalensis, see Mahogany
Kestrel, Common, 201
Kestrel, Lesser, food, 220
  habitat in Africa, 192, 485
  numerical response, 220
  roost Senegal, 207
  Sahel dependence, 192
  trend, 485
Kite, African Swallow-tailed, deforestation, 497

food, 220
numerical response, 220
roost Senegal, 206-207
Kite, Black, habitat in Africa, 192, 485
  Sahel dependence, 192
  trend, 485
Kite, Yellow-billed, maximum, Debo/Walado, 85
Knot, Red, 116, 132
Kob, White-eared, 57, 164
Kobus kob leucotis, see Kob, White-eared
Kobus leche smithemani, see Lechwe, Black
Kobus megaceros, see Lechwe, Nile
Kraussella amabile, 219

L

Lady, Painted, 364
Lake Chad Basin, 144-157
  bird densities, 152
  cattle, 149
  Chari, 150
  cropland, 149
  fishing, 149
  hydrology, 32-33, 145-149
  irrigation, impact of, 147-149
  people, 149
  vegetation, 149
  water level, Lake Chad, 147, 157
  waterbirds versus river discharge, 152
Lake Fitri, 150
Lake Maga, 150
Lake Victoria, 159-160
Lake Volta, 290
Lanius collurio, see Shrike, Red-backed
Lanius minor, see Shrike, Lesser Grey
Lanius nubicus, see Shrike, Masked
Lanius senator, see Shrike, Woodchat
Lapwing, African Wattled, Debo/Walado, 85
  density, in rice fields, 176
    seasonal, 176
Lapwing, Black-headed,
  numbers Inner Niger Deltas, 85
Lapwing, Spur-winged,
  density per habitat, 96, 130
    in rice field, 176
    seasonal, 176
  estimate Inner Niger Delta, 100
  numbers, Africa, 186
    coastal rice, 186-187
    Debo/Walado, 85, 88-89
    Europe, 186

Hadejia-Nguru, 186-187
Inner Niger Delta, 186-187
Lake Fitri, 155, 186-187
Logone, 155, 186-187
Office du Niger, 186-187
trend, wintering, 88
Lark, Crested,
density per habitat, 96
in rice fields, 176
estimate Inner Niger Delta, 100
numbers Lake Télé, 95
Lark, Greater Short-toed,
habitat in Africa, 192, 486
Sahel dependence, 192, 486
trend, 486
Larus cirrocephalus, see Gull, Grey-headed
Larus fuscus, see Gull, Lesser Black-backed
Larus ridibundus, see Gull, Black-headed
Lechwe, Black, 52
Lechwe, Nile, 59, 164
Leptoptilos crumeniferus, see Stork, Marabou
Lettuce, Water, 120, 149, 164
Lily, Water, in Lake Chad, 149
in Lake Fitri, 150
in Office du Niger, 174
in Senegal Delta, 116
Limosa lapponica, see Godwit, Bar-tailed
Limosa limosa, see Godwit, Black-tailed
Lion, 59
Liza falcipennis, see Mullet, Sicklefin
Locust, Brown, 203
Locust, Desert, control, dieldrin, 222
entomopathogen, 215-216, 221-222
organochlorines, 220
organophosphates, 220
timing, 222
densities, 209
distribution northern Africa, swarms, 205
groups, 205
solitary, 205
gregarisation, 207-208
insecticides, 208
Intertropical Convergence Zone, 209
migration, 209, 364
predation rate by birds, 215-216
rainfall versus swarms, 206
recession, 208
seasonal distribution per latitude, 204
upsurge, 209

Locust, Italian, 203
Locust, Migratory, 203-204
Locust, Moroccan, 203
Locust, Red, Madagascar, 220
Locust, Tree, 207
Locusta migratoria, see Locust, Migratory
Locustana pardalina, see Locust, Brown
Locustella fluviatilis, see Warbler, Eurasian River
Locustella luscinioides, see Warbler, Savi's
Locustella naevia, see Warbler, Grasshopper
Ludwigia stolonifera, 96
Luscinia luscinia, see Nightingale, Thrush
Luscinia megarhynchos, see Nightingale, Common
Luscinia svecica, see Bluethroat

M

Maerua, 481
Mahogany, for building pinasse, 76
Mangifera indica, see Mango
Mango, 459
Martin, Common House, 201, 486
Martin, Common Sand, 396-405
breeding, 397
food, 404
laying date, 403-404
rainfall, 404
reproduction, 404
habitat in Africa, 192, 398-399, 486
migration, 397-399
migratory connectivity, 194, 397
mortality, 441
moult, 398
roosting, 82, 134, 398
mixed 399
Sahel, dependence, 192
drought, 400-403
floodplains, 134, 398
survival, 401-402
age, 401
body size, 402
density dependence, 403
return rate, 401-402
sex, 401
trend, 399-404, 486
climate change, 403
flooding, Europe, 403
floodplain size, 400
NDVI, 400
pollution, 403

Sahel rainfall, 399, 402
wintering, 398-399, 486
Meinertzhagen, Richard, reputation, 327
Merops albicollis, see Bee-eater, White-throated
Merops apiaster, see Bee-eater, European
Merops persicus, see Bee-eater, Blue-cheeked
Merops pusillus, see Bee-eater, Little
Mesquite, Lake Chad, 149
Microparra capensis, see Jacana, Lesser
Microtus arvalis, see Vole, Common
Migration, 464-471
altitude, 466, 468
dehydration, 465
migratory connectivity, 194
refuelling, 466
mortality, age-dependent, 469
geographical variation, 466-467
seasonal variation, 466-467
visible, 468
phenology, 470
pre-migratory fattening, 468-469
strategies, 465-466
water, constraint, 466
wind, Sahara, 461
within-Africa, 493-494
Milvus migrans, see Kite, Black
Milvus migrans parasitus, see Kite, Yellow-billed
Mimosa pigra, bird densities, 96
Kafue, 184
Miscanthus junceus, 184
Mochokiella, see Catfish
Monticola saxatilis, see Thrush, Common Rock
Monticola solitarius, see Thrush, Blue Rock
Moorhen, Common, Debo/Walado, 88-89
density per habitat, 96
trend, wintering, 88
Moorhen, Lesser, density per habitat, 96
Moreau, Reginald Ernst, 8
number of Palearctic migrants, 190-191
Mortality, annual variation, 470
geographical variation, 469-470
and Sahel rainfall, 470
Motacilla alba, see Wagtail, White
Motacilla flava, see Wagtail, Yellow
Mouse, Harvest, 309
Mugil cephalus, see Mullet, Flathead
Mugil curema, see Mullet, White
Mullet, Flathead, 297
Mullet, Sicklefin, 297

Mullet, White, 297
Muscicapa striata, see Flycatcher, Spotted
Mycromys minutus, see Mouse, Harvest
Mycteria ibis, see Stork, Yellow-billed

N

NAO, see North Atlantic Oscillation
NDVI, see Normalized Difference Vegetation Index
Neem, 55-57, 460
    bird density, 56-57
    properties, antimicrobial, 57
Neotis nuba, see Bustard, Nubian
Nephron percnopterus, see Vulture, Egyptian
Nettapus auritus, see Goose, African Pygmy
Niger Delta, see Inner Niger Delta
Niger River Basin, Delta mort, 31-32
    dams, impact of, 29-32, 65
    discharge, 35
    discharge, reconstructed, 35
    hydrology, 29-32
    Office du Niger, 31-32
Nightingale, Common, habitat in Africa, 192, 487
    Sahel dependence, 192
    trend, 487
Nightingale, Thrush, 201, 486
Nightjar, European, habitat in Africa, 192, 486
    Sahel dependence, 192
    trend, 487
Nightjar, Red-necked, habitat in Africa, 192, 486
    Sahel dependence, 192
    trend, 486
Nomadacris septemfasciata, see Locust, Red
Normalized Difference Vegetation Index
    annual variation in Sahel, 44
    formula, 45
    Sahel, 38
    seasonal variation, 39
North Atlantic Oscillation, definition, 479
    phenology, spring, 476
Numenius arquata, see Curlew, Eurasian
Numenius phaeopus, see Whimbrel
Nycticorax nycticorax, see Heron, Black-crowned Night
Nymphea lotus, see Lily, Water
Nymphea maculata, see Lily, Water

O

Oena capensis, see Dove, Namaqua
Oenanthe cypriaca, see Wheatear, Cyprus
Oenanthe deserti, see Wheatear, Desert
Oenanthe hispanica, see Wheatear, Black-eared
Oenanthe isabellina, see Wheatear, Isabelline
Oenanthe oenanthe, see Wheatear, Northern
Oenanthe oenanthe leucorhoa, see Wheatear Northern
Oenanthe pleschanka, see Wheatear, Pied
Okavango Delta, 184
    bird density per habitat, 184
    papyrus, 184
Oriole, Eurasian Golden, 201, 487, 497
Oriolus oriolus, see Oriole, Eurasian Golden
Ornithacris cavroisi, biomass, 213
    predictable food source, 213
    prey of, 207
Oryza barthii, see Rice, wild
Oryza breviligulata, see Rice, wild
Oryza glaberrima, see Rice, cultivated
Oryza sativa, see Rice, cultivated
Oryx, 58
Oryx dammah, see Oryx
Osprey, 292-303
    breeding range, 293
        colonisation, 302
        reproduction, improved, 300
        numbers, 300
        trends, 302
    Debo/Walado, 85
    food, 292
        energy requirements, 295
        Inner Niger Delta, fish size, 302
    habitat in Africa, 192, 295-302, 485
    hunting, 300
    Inner Niger Delta, 297
    Lake Chad, 156
    Lake Volta, 292
    migration, 293-295
        stopovers, 293, 465
        tracking, satellite, 293, 325
        travelling speed, 293, 325, 465
    migratory connectivity, 194, 293
    mortality, 300-301
        first-year, habitat, 300
        Inner Niger Delta, hook lines, 301
        Niger River, shooting, 301-302
        recovery rate, long-term, 298
        shooting, by latitude, 301
    numbers, West Africa, 300
    ringing effort, Finland, 197
    Sahel dependence, 192
    sink, 301-302
    summering, in West Africa, 300
    trend, 301-302, 485
    wintering, 295-301
        coastal, 297
        floodplain size, 297
        habitat-related, 297
        lake size, 275
        reservoirs, 297, 302
        rivers, 296, 302
        spacing, 295
        water, fresh vs. salt, 295
Ostrich, in Sudd, 165
Otus scops, see Owl, European Scops
Owl, European Scops, habitat in Africa, 192, 486
    Sahel dependence, 192
    trend, 486
Owl, Marsh, maximum Debo/Walado, 85
Oxen, Wild, 58

P

Palearctic-African migration system,
    distribution in Africa, 190
    numbers involved, 189, 192
    species involved, 192, 201
Pandion haliaetus, see Osprey
Panicum laetum, caloric value, 385
    as food, 385
Papyrus, Lake Chad, 149
    Okavango, 184
    Sudd, 162, 164
Park, National, 57
    Diawling, 121
    Djoudj, 121
    Waza, 156
Parkia biglobosa, nectar, 460
Passer luteus, see Sparrow, Sudan Golden
Passer montanus, see Sparrow, Tree
Passer simplex, see Sparrow, Desert
Pelican, Great White, 201
    breeding, Dogon, 80
    Diawling NP, 124
    Senegal Delta, 126-127
    numbers, Debo/Walado, 85
    Lake Fitri, 154
    Logone, 154
    Office du Niger, 186-187
    Senegal Delta, 128-129
    Sudd, 165
Pelican, Pink-backed,
    breeding, Diawling NP, 124
    breeding, Inner Niger Delta, 80

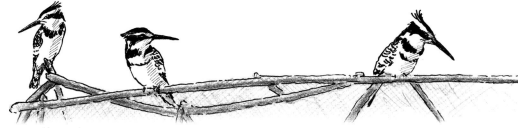

Senegal Delta, 126-127
numbers, Lake Fitri, 154
  Logone, 154
  Office du Niger, 186-187
  Senegal Delta, 128-129
  Sudd, 165
Pelecanus onocrotalus, see Pelican, Great White
Pelecanus rufescens, see Pelican, Pink-backed
Pennisetum purpureum, see Grass, Elephant
Pernis apivorus, see Honey Buzzard, European
Phacochoerus aethiopicus, 121
Phalacrocorax africanus, see Cormorant, Long-tailed
Phalacrocorax carbo lucidus, see Cormorant, Great
Philomachus pugnax, see Ruff
Phoeniconaias minor, see Flamingo, Lesser
Phoenicopterus ruber, see Flamingo, Greater
Phoenicurus phoenicurus, see Redstart, Common
Phragmites australis, see Reed
Phylloscopus bonelli, see Warbler, Western Bonelli's
Phylloscopus collybita, see Chiffchaff, Common
Phylloscopus collybita abietinus, see Chiffchaff, Common
Phylloscopus collybita collybita, see Chiffchaff, Common
Phylloscopus ibericus, see Chiffchaff, Iberian
Phylloscopus orientalis, see Warbler, Eastern Bonelli's
Phylloscopus sibilatrix, see Warbler, Wood
Phylloscopus trochilus, see Warbler Willow
Phylloscopus trochilus acredula, see Warbler Willow
Phylloscopus trochilus trochilus, see Warbler, Willow
Piliostigma reticulatum, bird density Nigeria, 57
  loss of, 434
  phenology, 481
  Whitethroat habitat, 459
Pintail, Northern, 272-277
  breeding, Europe, 273
  catching, Inner Niger Delta, 289
  habitat in Africa, 192, 275, 485
  migration, 273-274
  migratory connectivity, 194
  numbers, Africa, West, 275
    Debo/Walado, 85, 88-89
    Hadejia-Nguru, 141, 186-187
    Inner Niger Delta, 186-187
    Lake Fitri, 155, 186-187

Logone, 155, 186-187
Senegal Delta, 130, 186-187
population change
  breeding conditions, 276
  wintering, 275-276
Sahel dependence, 192, 276
trend, 88, 277, 485
Pipit, Grassland, 96
Pipit, Plain-backed, 96, 419
Pipit, Red-throated, density per habitat, 96
  habitat in Africa, 192, 487
  Sahel dependence, 192
  trend, 487
Pipit, Tawny, decline, 487
  habitat in Africa, 192, 487
  in Inner Niger Delta, 77
  Sahel dependence, 192
  trend, 487
Pipit, Tree, habitat in Africa, 192, 487
  Sahel dependence, 192
  trend, 487
Pistia stratiotes, see Lettuce, Water
Platalea alba, see Spoonbill, African
Platalea leucorodia, see Spoonbill, Eurasian
Platalea leucorodia balsaci, see Spoonbill, Eurasian
Plectropterus gambensis, see Goose, Spur-winged
Plegadis falcinellus, see Ibis, Glossy
Ploceus cucullatus, see Weaver, Village
Ploceus vitellinus, see Weaver, Vitelline Masked
Plover, Common Ringed,
  density per habitat, 96, 130
  numbers, Asia, 186
    coastal rice, 186-187
    Debo/Walado, 85, 88-89
    Europe, 186
    Hadejia-Nguru, 186-187
    Inner Niger delta, 186-187
  trend, wintering, 88
Plover, Egyptian, Debo/Walado, 88-89
  Logone, 155
  trend, wintering, 88
Plover, Grey, 201
  maximum Debo/Walado, 85
  Senegal Delta, 132
Plover, Kentish, maximum Debo/Walado, 85
  Senegal Delta, 132
Plover, Kittlitz's, density per habitat, 96
  numbers, Africa, 186
    Debo/Walado, 85

Hadejia-Nguru, 186-187
Inner Niger Delta, 186-187
Plover, Little Ringed, 201
  habitat, 486
  maximum Debo/Walado, 85
  numbers Lake Télé, 95
  trend, 486
Plover, Egyptian, numbers, Africa, 186
  numbers, Debo/Walado, 85
  Inner Niger Delta, 186-187
  Logone, 186-187
Plover, White-fronted, numbers, Africa, 186
  Inner Niger Delta, 186-187
Pluvialis aegyptius, see Plover, Egyptian
Pluvialis squatarola, see Plover, Grey
Polygonum, bird densities, 95
Polygonum senegalense, 96, 174
Poro, 97, 148
  Habitat for Sedge Warbler, 438
Porphyrio alleni, see Gallinule, Allen's
Porphyrio porphyrio, see Swamphen, Purple
Porzana parva, see Crake Little
Porzana porzana, see Crake, Spotted
Porzana pusilla, see Crake, Baillon's
Potamogeton, in Lake Fitri, 150
  in Senegal Delta, 116
Pratincole, Collared, density per habitat, 96
  habitat in Africa, 192, 485
  numbers, Africa, 186
    Asia, 186
    coastal rice, 186-187
    Debo/Walado, 85, 88-89
    Europe, 186
    Hadejia-Nguru, 186-187
    Inner Niger Delta, 186-187
    Lake Télé, 95
    Logone, 155, 186-187
    Office du Niger, 186-187
  Sahel dependence, 192
  trend, 88, 485
Pratincole, Black-winged, 201, 485
Prinia fluviatilis, see Prinia, River
Prinia subflava, see Prinia, Tawny-flanked
Prinia, River, in Cattail, 133
Prinia, Tawny-flanked, 96, 133
Procambarus clarkii, see Crawfish, Red Swamp
Prososis juliflora, see Mesquite
Pycnonotus barbatus, see Bulbul, Common
Python sebae, 121

Index 557

Q

Quail, Common, habitat in Africa, 192, 485
  in Inner Niger Delta, 77
  Lake Chad, 156
  Sahel dependence, 192
  trend, 485

R

Rainfall
  anomaly, calculation of, 25
    North Africa, 23
    Sahel, 22-23
    South Africa, 23
    synchronisation within Africa, 23
  annual variation, Senegal 1848-2005, 16
    Sahel countries 1960-1993, 21
  average, Africa, 16, 18
  and flood extent, Hadejia-Nguru, 195
    Inner Niger Delta, 195
    Senegal Delta, 195
  impact on vegetation, 21
  latitudinal, 15, 18-19
  monthly variation, western Sahel, 18
  seawater, impact on, 25
Recurvirostra avosetta, see Avocet, Pied
Redshank, Common, density per habitat, 96
  maximum, Debo/Walado, 85
Redshank, Spotted,
  density per habitat, 96
    in rice fields, 176
  habitat in Africa, 192, 486
  feeding, communal, 246
  numbers, Asia, 186
    coastal rice, 186-187
    Debo/Walado, 85, 88-89
    Europe, 186
    Hadejia-Nguru, 186-187
    Inner Niger Delta, 186-187
    Lake Fitri, 155, 186-187
    Logone, 155, 186-187
    Senegal Delta, 128-129, 186-187
  Sahel dependence, 192
  trend, 88, 486
Redstart, Common, 426-435
  abundance, past, 77
  breeding, 427
    laying date, shift in, 477
    nestboxes, 426, 429, 431
    orchard, 431, 434
    reproduction, 435

carry-over effects, 477
fattening, 427
  pre-migratory, 429
food, 429
  Salvadora, 429
habitat in Africa, 192, 433-434, 486
  density in *Acacia*, 56-57
  loss of, 434
migration, 427
  migratory connectivity, 194, 427
mortality, rainfall, impact of, 470
  seasonal & geographical variation, 467, 469
phenology, 476-477, 479
recurrence, 427
Sahel, deforestation, impact of, 433-434
  dependence, 192
  rainfall, 429, 431
site fidelity, 427
trend, adjustment, 503
  in Europe, 429-432, 486
    habitat loss, 434
Redunca redunca, see Reedbuck
Reed, Lake Chad, 149
  Okavango, 184
  Senegal Delta, 106, 115-117
  The Netherlands, 309
Reedbuck, 59, 164
Rice, 170-179
  abandoned, 497
  African *Oryza glaberrina*, 171
  Asian *Oryza sativa*, 171
  birds, densities, 174, 176
    numbers, coastal, 179
    numbers, Office du Niger, 179
  bolanha, 171
  coastal, 171-173
  cultivated, caloric value, 385
    food for doves, 381-382
    in Inner Niger Delta, 68-69
  expansion, 173-174
  fallow, 173
  in floodplains, 171
  import, Mali, 48
  import, Senegal, 48
  irrigated, 173-174
  in mangrove zone, 172
  production, coastal, 171-173
    in floodplains, 171
    in irrigation schemes, 173-174

West Africa, 48-49
substitute for wetlands, 177-178
wild, caloric value, 385
  food for doves, 381-382
  in Inner Niger Delta, 66
  in Sudd, 164
Ringing data, distribution recoveries, Sahel, 196
  mortality, derived from, 197
  pitfalls, 196-196
  reporting rate, 468
  resightings, 197
  use of, 193-198
Riparia riparia, see Martin, Common Sand
Robin, Rufous Scrub,
  habitat in Africa, 192, 487
  Sahel dependence, 192
  trend, 487
Roller, European, 201, 486
Ruff, 350-363
  breeding, 360
  density per habitat, 96, 130
    in rice fields, 176
  distribution, in Africa, 351-352
    sex-related, 353
  fattening, 359-360
    body mass, 360, 363
    and flood extent, 360
    sex-specific, 360
    weight gain, daily, 361
  food, 91-93, 351
    Corbicula, 91-92
  habitat in Africa, 192, 351-352
    in Europe, 351
  migration, 353-356
    pathways, 353-355
    migratory connectivity, 194, 353-355
    pre-migratory fattening, 360
    seasonal, 354
    sex-specific, 356, 363
    timing, 356
  mortality, age-specific, 360
    annual variation, 357
    catching, 358-359
    and flooding, 357-9
    and lunar cycle, 358
    seasonal variation, 357
    shooting, 357
  moult, 363
  numbers, Africa, 352

Asia, 186
coastal rice, 186-187
Debo/Walado, 85, 88-89
Europe, 351
Hadejia-Nguru, 141, 186-187
Inner Niger Delta, 82-83, 100, 186-187, 352
Lake Chad, 352
Lake Fitri, 155, 186-187
Lake Télé, 95
Logone, 155, 186-187
Senegal Delta, 128-129, 186-187, 352
phenology, 354-356, 363
protandry, 363
population, 351
climate change, 351-352, 360
habitat change, 361
trend, 88, 362, 502
predation risk,
Sahel dependence, 192
carry-over effects, 473-474
sex ratio, operational, 356
stomach content, 91-92
wintering, in Europe, 351, 356
Rynchops flavirostris, see Skimmer, African
Sagittarius serpentarius, see Bird, Secretary

S

Sahel, 480-503
agriculture, 497
habitat degradation, 497
bird exploitation, 499-500
climate change, impact of, 496
regime shift, 496
cropland, 49-50
deforestation, 496-497
firewood, 47, 54
floodplains, wildlife, 52
food production, 47-48,
forest, 53-55
grazing, 50-51, 497-498
land degradation, favouring grasshoppers, 219
gradual, 218
impact on birds, 218
Senegal, 219
land use, 46-59
latitude, impact on birds, 493-495
livestock, 50-53, 54
longitude, impact on birds, 491-492
nutrient mining, 49

poisoning, use of strychnine, 497
population growth, 47
protection, importance of, 121, 156
rainfall, 14-34
sahelisation, 497
transhumance, 50, 52
trees, Palearctic birds in, 56-57
exotic, 55-57
indigenous, 55-57
phenology, 481
urbanisation, 47
vegetation, 36-45
watering points, 52
wetlands, loss of, 498-499
Niger River, 64, 499
protection of, 499
Senegal Delta, 112, 498
Waza Logone, 498
wildlife, 52, 58-59
wintering, strategies, 481, 484
zone, 37
Salsola baryosma, 117
Salvadora persica, berries, abundance, 452
as food, 385, 429, 459-460, 468
lipid, 469
phenology, 481
Senegal Delta, 116, 134, 359
Salvinia molesta, see Weed, Kariba
Sandpiper, Common, 201
density Senegal Delta, 130
in rice fields, 176
habitat, 486
maximum Debo/Walado, 85
migratory connectivity, 194
trend, 486
Sandpiper, Curlew, density per habitat, 96, 130
maximum Debo/Walado, 85, 88-89
Senegal Delta, 132
trend, wintering, 88
Sandpiper, Green, 201
density per habitat, 96
in rice fields, 176
estimate Inner Niger Delta, 100
habitat, 486
maximum Debo/Walado, 85
trend, 486
Sandpiper, Marsh, density per habitat, 96, 130
estimate Inner Niger Delta, 100
habitat in Africa, 192, 486

migratory connectivity, 194
numbers, Asia, 186
coastal rice, 186-187
Debo/Walado, 85
Europe, 186
Hadejia-Nguru, 186-187
Inner Niger Delta, 186-187
Logone, 186-187
Office du Niger, 186-187
Sahel dependence, 192
trend, 486
Sandpiper, Temminck's, 201, 486
Sandpiper, Wood, 344-349
breeding, 345
density per habitat, 96, 130
in rice fields, 176
habitat in Africa, 192, 486
density per habitat, 346
migration, 345-346
fattening, 345
stopovers, 345, 346
numbers, Asia, 186
coastal rice, 186-187
Debo/Walado, 85, 88-89
Europe, 186
Hadejia-Nguru, 186-187
Inner Niger Delta, 100, 186-187
Lake Télé, 95
Logone, 155, 186-187
Office du Niger, 186-187
Senegal Delta, 132, 136
Sahel dependence, 192
survival, flood extent, 347
monthly variation, 347
trend, 88, 486
Sanderling, 201
Sardinella, Osprey, food, 297
Sarkidiornis melanotos, see Duck, Knob-billed
Saxicola rubetra, see Whinchat
Schistocerca migratoria, see Locust, Desert
Scirpus littoralis, in Senegal Delta, 116
Scirpus maritimus, in Senegal Delta, 116
Scopus umbretta, see Hamerkop
Senegal Delta, 104-137
bird density/ha, 100
Cattail, 120
colonies, 122, 124-129
dams, construction, 107
Diawling NP, 121

Djoudj NP, 121
flood control, 106-108
flooding, artificial, 112, 116
    natural, 105-106
    reconstructed, 112-114
floodplain, size, 137
forests, protected, 118
invasive plants, 118-120
land use, 115
people, 109-110
rice cultivation, off season, 109
rainy season, 109
waterbirds, 128-129
Senegal River Basin, discharge 35
  hydrology, 27-29
  impact of dams, 28-29
  reconstructed discharge, 31
Shoebill, Bangweulu, Zambia, 166
  Sudd, 165-166, 186-187
Shoveler, Northern, habitat in Africa, 192, 485
  migratory connectivity, 194
  numbers, Inner Niger Delta, 82-83
    Hadejia-Nguru, 141
    Lake Fitri, 155, 186-187
    Logone, 155, 186-187
    Senegal Delta, 128-129, 132, 186-187
  Sahel dependence, 192
  trend, 485
Shrike, Masked, 201, 487
Shrike, Red-backed, 201
  Kalahari, 198
  trend, 487
Shrike, Lesser Grey, 201, 487
Shrike, Woodchat, decline, 487
  density in *Acacia*, 57
  habitat in Africa, 192, 487
  Sahel dependence, 192
Skimmer, African, in Inner Niger Delta, 77
  Lake Chad, 156
Snipe, Common, 201, 486
  density per habitat, 96
Snipe, Great, abundance in past, 77
  density per habitat, 96
  estimate Inner Niger Delta, 100
  habitat in Africa, 192, 485
  maximum Debo/Walado, 85
  Sahel dependence, 192
  trend, 485
Snipe, Greater Painted, density per habitat, 96

estimate Inner Niger Delta, 100
maximum Debo/Walado, 85
numbers, Lake Télé, 95
Sparrow, Desert,
  predation on hopper bands, 215
Sparrow, Sudan Golden
  predation on hopper bands, 215-216
Sparrow, Tree, 426
Sparrowhawk, Levant, 201, 485
Spoonbill, African, breeding, Diawling NP, 124
  breeding, Inner Niger Delta, 78
    Senegal Delta, 126-127
  numbers, Debo/Walado, 85, 88-89
    Hadejia-Nguru, 186-187
    Inner Niger Delta, 186-187
    Lake Fitri, 154, 186-187
    Logone, 154, 186-187
    Senegal Delta, 186-187
  trend, wintering, 88
Spoonbill, *balsaci*, 490-491
  Eurasian, 490-491
    Debo/Walado, 85, 88-89
    Mauritania, 490-491
  habitat in Africa, 192, 485
  Hadejia-Nguru, 186-187
  Inner Niger Delta, 186-187
  Sahel dependence, 192
  Senegal Delta, 186-187
  trend, 88, 485, 90-491
  adjustement, 503
Sporobolus robustus, 114-116, 121, 183
Sporobolus pyramidalis, 164, 183
Sterna albifrons, see Tern, Little
Sterna caspia, see Tern, Caspian
Stilt, Black-winged, *Corbicula* as food, 91
  density per habitat, 96, 130
  in rice fields, 176
  food, 91
  habitat in Africa, 192, 485
  Mauritanian Senegal Delta, 137
  numbers, Africa, 186
    coastal rice, 186-187
    Debo/Walado, 85, 87, 88-89
    Europe, 186
    flood level, 88
    Hadejia-Nguru, 186-187
    Inner Niger Delta, 186-187
    Lake Fitri, 155, 186-187
    Lake Télé, 95

Logone, 155, 186-187
Office du Niger, 186-187
Senegal Delta, 132, 186-187
Sahel dependence, 192
stomach content, 91
trend, 88, 485
water level, 87
Stint, Little,
  density per habitat, 96, 130
  in rice fields, 176
  habitat in Africa, 192, 486
  numbers, Asia, 186
    coastal rice, 186-187
    Debo/Walado, 85, 88-89
    Europe, 186
    Hadejia-Ngruru, 186-187
    Inner Niger Delta, 100, 186-187
    Lake Télé, 95
    Logone, 155, 186-187
    Senegal Delta, 132
  Sahel dependence, 192
  trend, 486
Stone-curlew, decline, 480
  habitat in Africa, 192, 486
  Sahel dependence, 192
  trend, 485
Stork, Abdim's, in Inner Niger Delta, 77
  breeding Inner Niger Delta, 80
  food intake, Niger, 218
  numbers, Sudd, 165
  Tanzania, 256
Stork, African Openbill, colonies in Inner Niger Delta, 78
  foraging method, 165
  numbers, Inner Niger Delta, 77, 80
    Lake Fitri, 154, 186-187
    Logone, 154, 186-187
    Sudd, 165-166, 186-187
Stork, Black, habitat in Africa, 192, 485
  migration routes, 192
  Sahel dependence, 192
  Senegal Delta, 131
  trend, 485
Stork, Marabou, in Inner Niger Delta, 77, 80
  Lake Fitri, 154, 186-187
  Logone, 154, 186-187
  Senegal Delta, 125
  Sudd, 165, 186-187
Stork, Saddle-billed, Inner Niger Delta, 77, 80

Lake Fitri, 154
Logone, 154, 186-187
Sudd, 165, 186-187
Stork, White, 252-265
  breeding, Europe, 253, 257
  carry-over effects, delayed breeding, 261
    geographic variation, 262, 265
    mortality NW Africa, 261
    reproduction, 261-262
  fattening, 256-257, 265
  food, armyworm, 256
    crawfish, 254
    demand, 263
    earthworms, 263
    grasshoppers, 256
    locusts, 207, 215, 257
    voles, 263
  habitat in Africa, 192, 485
  Lake Fitri, 154
  Logone, 154
  migration, 253-255
    daily distance covered, 253
    dispersal, 264
    glide ratio, 265
    migratory connectivity, 194
    migratory divide, 253
    residency, increase in, 264
    satellite tracking, 253-254
    soaring, energy expenditure, 265
    wing-loading, 265
  mortality, causes of, 260
    impact Sahel rainfall, 259-260, 265
    methods of killing, 259
    powerlines, 260
    recovery rate, annual, 259
    Sahara, 259, 265
    Sahel, regional differences, 259
    seasonal, 259
    survival rates, 264-265
    trend, long-term, 259
  population change, dynamics, 264
    habitat deterioration, 264
    impact locust outbreaks, 257
    Sahel rainfall, impact of, 257-258, 264-265
  reintroductions, 252-253, 264
  reproduction, geographic variation, 265
    impact on numbers, 264
    long-term changes, 265
    Sahel rainfall, impact of, 262, 265

  Sahel, dependence, 192
    impact rainfall on survival, 265
  Störungsjahre, 262
  Sudd, 165
  trend, 485
    Alsace, 258
    Baden-Württemberg, 258
    The Netherlands, 258
  wintering, Africa, 255-257
    Europe, 254
    Libya, 491
    Middle East, 257
    Sudd, 265
Stork, Woolly-necked, Lake Fitri, 154
  Logone, 154, 186-187
  Sudd, 186-187
Stork, Yellow-billed,
  breeding, Senegal Delta, 126-127
  past breeding, Inner Niger Delta, 77, 80
  Hadejia-Nguru, 186-187
  Lake Fitri, 154, 186-187
  Logone, 154, 186-187
  Sudd, 165, 186-187
Streptopelia decipiens, see Dove, African Mourning
Streptopelia roseogrisea, see Dove, African Collared
Streptopelia senegalensis, see Dove, Laughing
Streptopelia turtur, see Dove, European Turtle
Streptopelia vinacea, see Dove, Vinaceous
Struthio camelus, see Ostrich
Sudan, vegetation zone, 37
  sudanisation, 497
Sudd, 158-169
  cattle, 164
  dams, 159-160
  farming, 164
  fishing, 164
  hydrology, 161-162
  mammals, wild, 163
  Nile Basin, 159-160
  people, 164
  vegetation, 162-164
Swallow, Barn, 406-417
  Africa, distribution, 408-411
    habitat, 192, 410-411, 486
    reporting rate, 411
    wintering, geographic variation, 411
  breeding, carry-over effects, 473

    conditions, 415-416
    numbers, 407
    trends, 414-416
  migration, 407-382
    migratory connectivity, 194
    ringing, 408
    Sahel and Sahara, 412-413
  mortality, age-dependent, 469, 471
    and body mass, 411
    human predation, 412-413
    mass, 406, 412, 473
    moult, 411
    and rainfall, 470
    reporting rate, 413
    seasonal, 413, 469, 471
  moult, 473
  phenology, 475
  reproduction, and carry-over effects, 473
    and weather, 416
  ringing effort, 470
  roosting, dangers, 412
    habitat, 410
    mixed, 399
    numbers, 382-411
  Sahel dependence, 192, 413-414
  survival, carry-over effects, 413, 416,
    cold spell, 406
    rainfall, Africa, 411, 416
    Sahara, 413
  trend, 414-416, 486
  wintering, 408-411
Swallow, Red-rumped, habitat in Africa, 192, 486
  Sahel dependence, 192
  trend, 487
Swamphen, Purple, density per habitat, 96
  numbers, Africa, 186
    Debo/Walado, 88-89
    Inner Niger Delta, 186-187
    Lake Télé, 95
  trend, 88
Swift, Alpine, 201, 486
Swift, Common, 201, 486
Swift, Pallid, habitat in Africa, 192, 486
  Sahel dependence, 192
  trend, 486
Sylvia atricapilla, see Blackcap
Sylvia borin, see Warbler, Garden
Sylvia cantillans, see Warbler, Subalpine

Sylvia communis, see Whitethroat, Common
Sylvia conspicillata, see Warbler, Spectacled
Sylvia curruca, see Whitethroat, Lesser
Sylvia hortensis, see Warbler, Orphean
Sylvia melanocephala, see Warbler, Sardinian
Sylvia mystacea, see warbler, Ménétries's
Sylvia nana, see Warbler, Desert
Sylvia nisoria, see Warbler, Barred
Sylvia rueppelli, see Warbler, Rüppell's
Synodontis, see Catfish

T

Tachymarptis melba, see Swift, Alpine
Tamarindus indicus, 452
Tamarix senegalensis, 116, 452
Teal, Eurasian, 201
Temperature, global warming, 24
　northern Africa, 24
Tern, Black, 201
Tern, Caspian, 370-377
　breeding, Africa, 371
　　conditions, 376-377
　　Europe, 377
　food and feeding, 370, 376
　habitat in Africa, 192, 486
　migration, 371-373
　　timing, 372
　migratory connectivity, 194
　mortality, causes of death, 374
　　seasonal, 375
　numbers, Africa, West, 186, 373
　　Debo/Walado, 85-86, 88-89, 375
　　Europe, 186, 371
　　Hadejia-Nguru, 186-187
　　Inner Niger Delta, 186-187
　Sahel dependence, 192, 372
　　flood extent, 86, 374
　　summering, 375-376
　survival, age-specific, 374
　　and flood level, 88, 374-375
　trend, 373-377, 486
　wintering, distribution, 372
　　monthly variation, 86, 375
　　trend, long-term, 88
　water level, impact on numbers, 86
Tern, Gull-billed, 364-369
　breeding, 365
　　trends, 368-369, 375-377, 486
　food, 365-366
　habitat in Africa, 192, 486
　　floodplains, 366

　　lakes, 366-367
　　tidal flats, 366
　migration, 365
　numbers, Africa, 186, 367
　　Asia, 186
　　Debo/Walado, 85, 88-89
　　Europe, 186, 365
　　Hadejia-Nguru, 186-187
　　Inner Niger Delta, 186-187
　　Lake Victoria, 367
　　Logone, 186-187
　　Mauritania, 368-369
　　Sahel dependence, 192
　　and flood extent, 368
　　summering, non-breeding, 365
　trend, breeding, 368, 486
　　wintering, 88, 366
Tern, Little, breeding, Lake Debo, 81
　Debo/Walado, 85, 88-89
　Logone, 155
　trend, wintering, 88
Tern, Whiskered, breeding, Lake Debo, 81
　habitat in Africa, 192, 486
　numbers, Debo/Walado, 85, 88-89
　　Europe, 186
　　Hadejia-Nguru, 186-187
　　Inner Niger Delta, 186-187
　　Logone, 155, 186-187
　　Sahel dependence, 192
　　trend, 88, 486
Tern, White-winged, Debo/Walado, 85, 88-89
　habitat in Africa, 192, 486
　Logone, 155
　Sahel dependence, 192
　trend, 88, 486
Thick-knee, Senegal,
　maximum Debo/Walado, 85
Threskiornis aethiopica, see Ibis, Sacred
Thrush, Blue Rock
Thrush, Common Rock,
　habitat in Africa, 192, 486
　Sahel dependence, 192
　trend, 486
Tiang, 57, 163-164
Topi, 56
Trees, Sahelian, phenology, 481
Tribulus terrestris, see Vine, Puncture
Tringa erythropus, see Sandpiper, Spotted
Tringa glareola, see Sandpiper, Wood
Tringa nebularia, see Greenshank, Common

Tringa ochropus, see Sandpiper, Green
Tringa stagnatilis, see Sandpiper, Marsh
Troglodytes troglodytes, see Wren, Winter
Tsetse fly, 50, 52
Turdoides fulvus, see Babbler, Fulvous
Typha australis, see Cattail

U

Uca tangeri, see Crab, Fiddler
Upupa epops, see Hoopoe, Eurasian
Utricularia stellaris, 117

V

Vanellus senegallus, see Lapwing, African
　Wattled
Vanellus spinosus, see Lapwing, Spur-winged
Vanellus tectus, see Lapwing, Black-headed
Vanessa cardui, see Lady, Painted
Varanus niloticus, 117, 121
Vegetation, anomaly, 45
　floodplains, 45
　NDVI, 38-39
　West Africa, zones in, 37
Vetiveria nigritana, in Inner Niger Delta, 66
　in Senegal Delta, 114
Vine, Puncture, caloric value, 385
　food for doves, 384-385
Vole, Common, 264
Vossia cuspidata, see Didere
Vulture, Egyptian, abundance, past, 77
　habitat in Africa, 192, 485
　in Inner Niger Delta, 77
　Sahel dependence, 192
　satellite-tracking, 325
　trend, 485

W

Wagtail, White, 201
　habitat in Africa, 192, 486
　Sahel dependence, 192
　trend, 486
Wagtail, Yellow, 418-425
　breeding, 419
　　range contraction, 424
　　trend, 423
　density in *Acacia*, 57
　　Cattail, 133
　　Inner Niger Delta, 420
　　mangroves, 424, 448
　　rice fields, 176, 419
　　Senegal Delta, 130, 420
　estimate Inner Niger Delta, 100
　fattening, pre-migratory, 421

timing, by sex, 421
food, 420
  biomass, seasonal, 420
  cattle, associated with, 420-421
  food intake, 216, 223
  foraging, 420
habitat, in Africa, 192, 486
  loss, 421
migration, 419-420,
  latitudinal variation, by sex, 420
  leap-frog, 419, 421
  migratory connectivity, 194, 419
  recovery rate, Sahara, 422
  subspecies, 419, 421, 423, 495
mortality, 421
  density-dependent, 421
  rainfall, impact of, 470
  seasonal & geographical variation, 467, 469
numbers Lake Télé, 95
predation on locust nymphs, 216
roosting, 82, 399
Sahel dependence, 192
  floodplain size, 423-424
  rainfall, 422
  recovery rate, 422
trend, 422-424, 486
wintering, floodplains, 423
  rainfall, 422
  strategy, 420-421, 495
Warbler, African Reed, in Cattail, 133
  competition with European Reed Warbler, 446
Warbler, Aquatic, habitat in Africa, 192, 487
  Inner Niger Delta, 135
  Sahel dependence, 192
  Senegal Delta, 135
  trend, 487
Warbler, Barred, 201, 487
Warbler, Eurasian River, 201, 487
Warbler, European Reed, 444-421
  breeding, 445
    habitat quality, 449
    laying date, 449
    reproduction, 449
    return rate, 449
  carry-over effects, 477
  competition, with African Reed Warbler, 446
  habitat in Africa, 192, 448, 487
    Cattail, 133

mangrove, importance of, 446, 448
migration, 446-448
  migratory connectivity, 194, 446
  migratory divide, 446
  strategy, age-specific, 446
  by subspecies, 448
mortality, rainfall, impact of, 470
  seasonal & geographical variation, 467, 469
phenology, 476
Sahel dependence, 192
  rainfall, 447
Senegal Delta, 134
trend, 447-449, 487
Warbler, Garden, 201
  carry-over effects, 477
  habitat, 487
  migration, migratory connectivity, 194
    strategy, 487
  mortality, 469
  phenology, 476
  trend, 487
Warbler, Grasshopper, density in *Acacia*, 57
  habitat in Africa, 192, 487
  Sahel dependence, 192
  Senegal Delta, 134
  trend, 487
Warbler, Great Reed, habitat in Africa, 192, 487
  migration, migratory connectivity, 194
    strategy, 484
  Sahel dependence, 192
  trend, 487
Warbler, Icterine, 201, 476, 487
Warbler, Marsh, 201, 477, 487, 502
Warbler, Melodious, habitat in Africa, 192, 487
  mangroves, 446, 448
  Sahel dependence, 192
  trend, 487
Warbler, Ménétries's, 201, 487
Warbler, Olivaceous, in *Acacia kirkii*, 432
  density, *Acacia*, 56-57
    Neem, 57
  habitat in Africa, 192, 487
  Sahel dependence, 192, 487
  trend, 487
Warbler, Olive-tree, 201, 487
Warbler, Orphean, habitat in Africa, 192, 487
  Sahel dependence, 192
  trend, 487
Warbler, Rüppell's, 201, 487

Warbler, Savi's, habitat in Africa, 192, 487
  Sahel dependence, 192
  Senegal Delta, 134
  trend, 487
Warbler, Sedge, 436-443
  breeding, 437
  carry-over effects, 477
  density in *Acacia kirkii*, 57
    bourgou, 439
    Cattail, 133, 439
    floodplains, 439
    Mimosa, 439
    rice fields, 176, 439
    Senegal Delta, 130, 134
  fattening, 438
    rate of, 438
  food, 437
  habitat in Africa, 192, 487
  migration, age-specific, 437
    direction, 437-438
    migratory connectivity, 194, 438
    site fidelity, 438
    stopover, 437-438
  mortality, 469
  numbers, Inner Niger Delta, 100, 439
    Lake Télé, 95
    rice fields, 439
  phenology, 476
  Sahel, dependence on, 192
    floodplains, 440-441
    rainfall, 440-441
  survival, age-specific, 440-441
  trend, density-dependent, 441-442
    in Europe, 412-414, 487
    habitat change, 442
Warbler, Spectacled
Warbler, Subalpine, in *Acacia kirkii*, 432
  Cattail, 133
  density in *Acacia*, 56
    mangroves, 446, 448
    in N Nigeria, 57
  habitat in Africa, 192, 452, 463, 487
  Sahel dependence, 192
  trend, 487
Warbler, Western Bonelli's, density in *Acacia seyal*, 56-57
  density in *A. kirkii*, 56
  habitat in Africa, 192, 487
  Sahel dependence, 192, 487

trend, 487
Warbler, Willow, carry-over effects, 477
　climbing rate, 468
　density, in *Acacia*, 57
　　mangroves, 446, 448
　habitat, 487
　migration, *acredula*, 495
　　itinerancy, 493
　　migratory connectivity, 193, 194
　　migratory divide, 495
　　strategy, 493
　　trochilus, 495
　mortality, rainfall, impact of, 470
　　seasonal & geographical variation, 467, 469
　phenology, 476
　Sahel, dependence on, 201, 493
　trend, *acredula*, 494
　　trochilus, 494
Warbler, Wood, 201, 487
Weaver, Village
　predator of locusts, 216
　rice consumption, 177
Weaver, Vitelline Masked, 133
Weed, Kariba, invasive, 115, 118, 172
Wetlands, importance for birds, 182-187
　Lake Chad, 151-157
　　bird density, 182
　　change in size, 182
　　permanent water, 182
　　vegetation cover, 182
　Lake Fitri, 154-155
　　bird density, 182
　　change in size, 182
　　permanent water, 182
　　vegetation cover, 182
　resilience, 187
　rice fields as substitute, 178
　Sahel, 180-201
Wheatear, Black-eared,
　habitat in Africa, 192, 487
　Sahel dependence, 192
　trend, 487
Wheatear, Cyprus, 201, 487
Wheatear, Isabelline,
　habitat in Africa, 192, 487
　Sahel dependence, 192
　trend, 487
Wheatear, Northern, carry-over effects, 476
　density in *Acacia seyal*, 57

　in Senegal, 220
　habitat in Africa, 192, 487
　migration from Greenland/Canada, 198
　Sahel dependence, 192
　trend, 487
Wheatear, Pied, 201, 487
Weaver, Village, in Cattail, 133
Whimbrel, 132, 201
　fattening, 473
Whinchat, carry-over effects, 476
　habitat in Africa, 192, 487
　migratory connectivity, 194
　mortality, 469
　Sahel dependence, 192
　trend, 487
Whitethroat, Common, 456-463
　breeding, 457
　　contamination, 462
　　habitat changes, 461
　carry-over effects, on laying date, 476
　　on productivity, 477
　density in *Acacia*, 57
　　N Nigeria, 54
　fattening, 457
　　pre-migratory, 457
　habitat in Africa, 192, 461-433, 487
　migration, 457-458
　　migratory connectivity, 194, 461
　　migratory divide, 457
　　phenology, 458, 476
　　recurrence rate, 495
　　stopover, 457
　　strategy, 457-458
　mortality, 468-469
　moult, 457
　phenology, 476
　Sahel, density, 459-461
　　dependence, 192
　　NDVI, 461
　　rainfall, impact of, 454, 461-462
　　territoriality, 495
　survival, age-specific, 462
　　sex-specific, 462
　trend, adjustement, 503
　　in Europe, 454, 462, 487
　　recovery, habitat-related, 462
Whitethroat, Lesser, 450-455
　breeding, 451, 453
　carry-over effects, on productivity, 477

　fattening, 451-452
　food and feeding, 452
　habitat, in Africa, 192, 452-453, 487
　　Darfur, 451
　　degraded, 450, 452-553
　　densities in, 452
　　loss, resilience to, 453
　migration, 451-452
　　dispersal, post-juvenile, 451
　　migratory connectivity, 194
　　phenology, spring, 452, 453
　　stopovers, 452
　　West Africa, 451
　moult, 451
　phenology, 476
　Sahel dependence, 192
　　Ethiopia, 453
　　rainfall, 453
　　Sudan, 452, 427
　survival, and rainfall, 453
　　return rate by sex, 453, 455
　trend, in Europe, 450, 453-454, 487
Wren, Winter, 478
Wryneck, Eurasian, 390-395
　breeding, 391
　　food abundance, 392
　　habitat loss, 392
　　weather, 395
　foraging, 391
　habitat in Africa, 192, 391, 487
　migration, 391
　　fattening, 391, 395
　　recurrence, 391
　　stopovers, 395
　Sahel dependence, 192, 393-394
　　deforestation, 392
　　rainfall, 392
　　survival, 392
　trends, Europe, 391-393, 487

Z

Zambia, Bengweulu Basin, 52
　Kafue, 184
Ziziphus, 56, 459, 461, 463, 481
　mucronata, bird density, 57
Zornia glochidiata, as food, 385

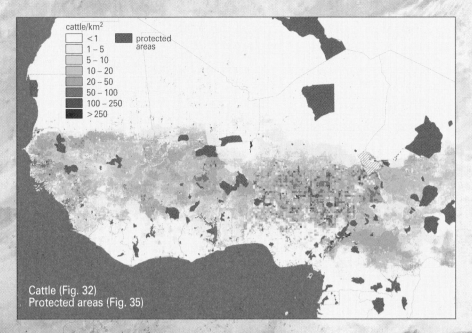
Cattle (Fig. 32)
Protected areas (Fig. 35)

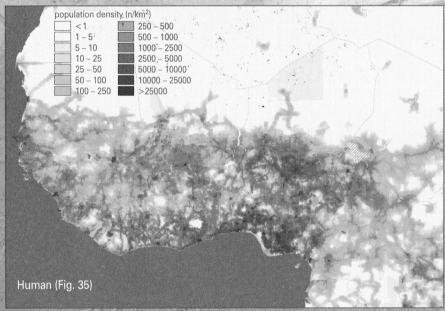
Human (Fig. 35)

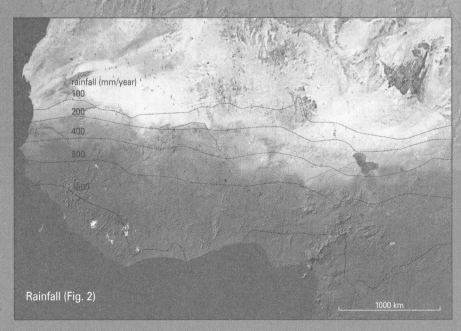
Rainfall (Fig. 2)

1000 km